CAMPYLOBACTER

MOLECULAR AND CELLULAR BIOLOGY

Copyright © 2005
Horizon Bioscience
32 Hewitts Lane
Wymondham
Norfolk NR18 0JA
U.K.

www.horizonbioscience.com

British Library Cataloguing-in-Publication Data

A catalogue record for this book is available from the British Library

ISBN: 1-904933-05-X

Printed and bound in Great Britain

Contents

Contributors

Masato Akiba
Department of Veterinary Microbiology
and Preventive Medicine
College of Veterinary Medicine
Iowa State University
Ames, IA
USA

Brenda Allan
Vaccine and Infectious Disease
Organization
Saskatoon, SK
Canada

Olivia L. Champion
Department of Infectious and Tropical
Diseases
London School of Hygiene and Tropical
Medicine
London
UK

Kate E. Dingle PhD
Nuffield Department of Clinical Sciences
University of Oxford
John Radcliffe Hospital
Oxford
UK

Nick Dorrell PhD
Department of Infectious and Tropical
Diseases
London School of Hygiene and Tropical
Medicine
London
UK

Hubert Ph. Endtz
Department of Medical Microbiology and
Infectious Diseases
Erasmus University Medical Center
Rotterdam
Rotterdam
The Netherlands

Paul Everest MRCPath PhD
Institute of Comparative Medicine
Department of Veterinary Pathology
University of Glasgow Veterinary School
Glasgow
UK

Patricia I. Fields PhD
Foodborne and Diarrheal Diseases Branch
Centers for Disease Control and
Prevention
Atlanta, GA
USA

Collette Fitzgerald PhD
Foodborne and Diarrheal Diseases Branch
Centers for Disease Control and
Prevention
Atlanta, GA
US

Michel Gilbert
Institute for Biological Sciences
National Research Council Canada
Ottawa, ON
Canada

Peggy C. R. Godschalk
Department of Medical Microbiology and
Infectious Diseases
Erasmus University Medical Center
Rotterdam
Rotterdam
The Netherlands

Scarlett Goon
Enteric Diseases Department
Naval Medical Research Center
Silver Spring, MD
USA

Patricia Guerry
Enteric Diseases Department
Naval Medical Research Center
Silver Spring, MD
USA

Pawel Herzyk
The Sir Henry Wellcome Functional
Genomics Facility
Institute of Biomedical and Life Sciences
University of Glasgow
Glasgow
UK

Lan Hu MD PhD
Enteric Diseases Department
Naval Medical Research Center
Silver Spring, MD
USA

Aparna Jagannathan PhD
School of Biosciences
University of Birmingham
UK

Lynn A. Joens
Department of Veterinary Science and
Microbiology
University of Arizona
Tucson, AZ
USA

George W. P. Joshua
London School of Hygiene and Tropical
Medicine
London
UK

Andrey V. Karlyshev PhD
London School of Hygiene and Tropical
Medicine
London
UK

David J. Kelly PhD
Department of Molecular Biology and
Biotechnology
University of Sheffield
Sheffield
UK

Julian Ketley PhD
Department of Genetics
University of Leicester
Leicester
UK

John D. Klena
School of Molecular Biosciences
Washington State University
Pullman, WA
USA

Michael E. Konkel
School of Molecular Biosciences
Washington State University
Pullman, WA
USA

Dennis J. Kopecko PhD
Laboratory of Enteric and Sexually
Transmitted Diseases
FDA-CBER
Bethesda, MD
USA

Victoria Korolik PhD
Institute for Glycomics
Microbial Glycobiology
Griffith University
Gold Coast, QLD
Australia

Joseph C. Larsen
Department of Microbiology and
Immunology
F. Edward Hebert School of Medicine
Uniformed Services University of the
Health Sciences
Bethesda, MD
USA

Robert B. Lee PhD
Department of Microbiology, Immunology
and Molecular Genetics
University of Kentucky
Chandler Medical Center
Lexington, KY
USA

Jun Lin
Department of Veterinary Microbiology
and Preventive Medicine
College of Veterinary Medicine
Iowa State University
Ames, IA
USA

Amanda MacCallum PhD
Institute of Comparative Medicine
Department of Veterinary Pathology
University of Glasgow Veterinary School
Glasgow
UK

Martin C. J. Maiden PhD
The Peter Medawar Building for Pathogen
Research
University of Oxford
Oxford
UK

Robert E. Mandrell
Produce Safety and Microbiology
Research Unit
Western Regional Research Center
Agricultural Research Service
US Department of Agriculture
Albany, CA
USA

Linda S. Mansfield
Department of Microbiology and
Molecular Genetics
Colleges of Human Medicine and
Veterinary Medicine
Michigan State University
East Lansing, MI
USA

William G. Miller
Produce Safety and Microbiology
Research Unit
Western Regional Research Center
Agricultural Research Service
US Department of Agriculture
Albany, CA
USA

Marshall R. Monteville
School of Molecular Biosciences
Washington State University
Pullman, WA
USA

Stephen L. W. On PhD
Danish Institute for Food and Veterinary
Research
Copenhagen
Denmark

Simon F. Park PhD
School of Biomedical and Life Sciences
University of Surrey
Guildford
UK

Craig T. Parker
United States Department of Agriculture
Agricultural Research Service
Produce Safety and Microbiology
Research
Albany, CA
USA

Charles Penn BSc PhD
School of Biosciences
University of Birmingham
UK

Carol L. Pickett PhD
Department of Microbiology, Immunology
and Molecular Genetics
University of Kentucky
Chandler Medical Center
Lexington, KY
USA

Contributors

Brian H. Raphael
School of Molecular Biosciences
Washington State University
Pullman, WA
USA

Andrew D. Sails PhD
Health Protection Agency
Newcastle Laboratory
Institute of Pathology
Newcastle General Hospital
Newcastle upon Tyne
UK

Christine M. Szymanski
National Research Council Canada
Ottawa, ON
Canada

Diane E. Taylor
Departments of Medical Microbiology and
Immunology and Biological Sciences
University of Alberta
Edmonton, AL
Canada

Dobryan M. Tracz
Department of Medical Microbiology and
Immunology
University of Alberta
Edmonton, AL
Canada

Arnoud H. M. van Vliet PhD
Department of Gastroenterology and
Hepatology
Erasmus MC – University Medical Center
Rotterdam
The Netherlands

Warren W. Wakarchuk
Institute for Biological Sciences
National Research Council Canada
Ottawa, ON
Canada

Karl G. Wooldridge PhD
Molecular Bacteriology and Immunology
Group
Division of Microbiology and Infectious
Diseases
University of Nottingham
Queens Medical Centre
Nottingham
UK

Brendan W. Wren PhD
Department of Infectious and Tropical
Diseases
London School of Hygiene and Tropical
Medicine
London
UK

Catriona Young
The Sir Henry Wellcome Functional
Genomics Facility
Institute of Biomedical and Life Sciences
Joseph Black Building, University of
Glasgow
Glasgow
UK

Vincent B. Young MD PhD
Departments of Internal Medicine,
Microbiology and Molecular Genetics
National Food Safety and Toxicology
Center
Colleges of Human Medicine and
Veterinary Medicine
Michigan State University College of
Human Medicine
East Lansing, MI
USA

Qijing Zhang
Department of Veterinary Microbiology
and Preventive Medicine
College of Veterinary Medicine
Iowa State University
Ames, IA
USA

Preface

Research on *Campylobacter* has come along way since it became clear in the late 1970s that the thermophilic campylobacters are a major cause of diarrhoeal disease. Expanding on the initial basic microbiological and epidemiological characterization of *Campylobacter* spp., recent research efforts have utilized a more molecular approach. Indeed, the new millennium heralded the entry of *C. jejuni* into the genome age with the publication of the complete genome sequence of NCTC 11168 (Parkhill *et al.*, 2000). Over the next few years this first sequence will be joined by complete genome sequences from at least one other *C. jejuni* strain and other *Campylobacter* species. These new sequences will expand comparative genomics beyond that presently available using other members of the epsilon proteobacteria (http://campy.bham.ac.uk/), notably *Helicobacter hepaticus* (Suerbaum *et al.*, 2003) and *Wolinella succinogenes* (Baar *et al.*, 2003).

The analysis of the *C. jejuni* genome sequence produced many interesting observations, including an unappreciated capacity for polysaccharide biosynthesis and the presence of hypervariable homopolymeric tract sequences. Although initially disappointing, the apparent lack of genes encoding virulence determinants similar to those expressed in other more extensively characterized enteric pathogens such as *Escherichia coli* and serovars of *Salmonella enterica* presents us with an exciting challenge. The new insight into the biology of campylobacters provided by the genome sequence emphasized that campylobacters offer an alternative and novel model for intestinal colonization and pathogenicity to that provided by the current enterobacterial paradigm.

Now that *C. jejuni* has entered the often cited 'post-genome age', we believed that the time is right for a new volume to review our current progress in *Campylobacter* research. Where applicable, we asked contributors to review progress in their area from a molecular and genomic perspective. Moreover, we encouraged writers to synthesize speculative models upon which to base future research efforts. The book begins by setting the clinical context for the investigation of *Campylobacter* biology. The next few chapters explore the taxonomy and genetic diversity of the campylobacters and provide an important framework for detailed studies on particular strains. The book continues with an extensive review of the incidence of campylobacters in various habitats and a discussion of new methods of epidemiological analysis. Next, the long overlooked topic of *Campylobacter* plasmid biology is addressed with a review of pVir, and this is followed by chapters on the increasingly important area of antimicrobial resistance. One of the most extensively studied areas of *Campylobacter* biology since the publication of the genome sequence has been polysaccharide biosynthesis. Consequently, we have included chapters that cover lipooligosaccharide and capsule biosynthesis as well as the exciting area of protein glycosylation. The genome provides the blueprint for a review of the metabolism and bioenergetics of *C. jejuni*, and this is given from the perspective of the organism's intestinal lifestyle. The next chapters address the organism's response to stress with a description of iron transport and regulation and a review of the response of campylobacters to stresses associated with food and intestinal colonization. One of the best-characterized areas of *Campylobacter* biology 'pre genome' was the flagellum, and this topic remains an important area of research in the 'post-genomic' era, as indicated by the inclusion of a review on motility. This chapter is complemented by another making genome-based predictions of the mechanism of signal transduction in chemotaxis. The final chapters focus more directly on interactions between campylobacters and the host that are likely to contribute to disease pathology. The characteristics of the one verified toxin produced by *C. jejuni* are reviewed, as are the mechanisms of host cell invasion, interactions with phagocytes and the inflammatory response elicited during infection of the human intestine. Finally, we return to the clinical expression of infection. The last chapter discusses potential mechanisms for the production of diarrhoea arising from a microarray-based analysis of host cell gene expression in response to infection by *C. jejuni*.

Each chapter is independent and can be read in isolation, but as a whole the book provides an important resource summariszing our current knowledge of *Campylobacter* molecular and cellular biology. Our contributors include many of the internationally recognized experts in the field of *Campylobacter* research, and we are extremely grateful to them for producing a timely and state-of-the-art collection of authoritative reviews. Finally, we would like to thank Annette Griffin of Horizon Scientific Press for approaching us to edit this book. We have appreciated her patience, support and understanding of the pressures that have affected both contributors and editors alike!

Julian Ketley and Mike Konkel
April 2005

References

Baar, C. *et al*. 2003. Complete genome sequence and analysis of *Wolinella succinogenes*. Proc. Natl. Acad. Sci. USA 100: 11690–11695.

Parkhill, J. *et al*. 2000. The genome sequence of the food-borne pathogen *Campylobacter jejuni* reveals hypervariable sequences. Nature 403: 665–668.

Suerbaum, S. *et al*. 2003. The complete genome sequence of the carcinogenic bacterium *Helicobacter hepaticus*. Proc. Natl. Acad. Sci. USA 100: 7901–7906.

Chapter 1

Campylobacter Infection – Clinical Context

Vincent B. Young and Linda S. Mansfield

Abstract

Campylobacter jejuni can cause a spectrum of diseases including gastroenteritis, proctitis, septicaemia, meningitis, abortion, and autoimmune diseases such as Reiter's arthritis and Guillain–Barré syndrome (GBS). The most common clinical syndrome seen in humans infected with *C. jejuni* is gastroenteritis. In most individuals, *C. jejuni*-associated gastroenteritis is self-limiting, resolving within a week unless underlying illness is present such as human immunodeficiency virus (HIV) infection. Such individuals are also more prone to develop invasive/systemic disease. *Campylobacter* spp. are rarely isolated from healthy individuals in developed countries, but commonly isolated from nonsymptomatic individuals in developing countries. *Campylobacter jejuni* has been linked to GBS, a debilitating inflammatory polyneuritis characterized by fever, pain and weakness that progresses to paralysis, and may result in long-term disability. Also, reactive arthritis can follow diarrhoeic episodes of *Campylobacter* or asymptomatic exposure to the organism and is attributed to autoimmune response in joints.

    Campylobacter jejuni is a known cause of disease in animals. Natural infections with *C. jejuni* resulting in enteritis have been reported in juvenile macaques, weaning-age ferrets, dogs, cats and swine. Additionally, chickens, rodents, ferrets, dogs, primates, rabbits and pigs have been inoculated experimentally by various routes with *C. jejuni* to mimic the various syndromes of infection in humans. Recently, concern has been focused on the role that both normal and diseased animals play in zoonotic transmission of campylobacters to humans.

INTRODUCTION

C. jejuni belongs to the 16S rRNA superfamily VI of spiral, microaerobic Gram-negative bacteria as proposed by Vandamme *et al.* (1991). Particular members of this group are causative agents of gastroenteritis in humans and animals. Infection with *C. jejuni* is one of the most common causes of gastroenteritis worldwide. In the United States, data from the FoodNet surveillance network indicate that *C. jejuni* is one of the major causes of acute gastroenteritis (Anonymous, 2003). Additionally, *C. jejuni* is the most common antecedent infection among patients who develop Guillain–Barré syndrome (GBS), an acute, inflammatory peripheral polyneuropathy (Blaser, 1997).

    The recognition of the prevalence of *C. jejuni* in cases of human gastroenteritis has occurred only in the past 20 years. In major part, this had to await the development of selective and transport media for the isolation of the organism from clinical specimens

(Bolton and Coates, 1983; Bolton *et al.*, 1988). However, it should be noted that *Campylobacter* species have been recognized as important veterinary pathogens for at least a century. In this chapter we will discuss the clinical features of *C. jejuni* infection in both humans and animals. This will provide not only the historical context, but also a discussion of animal models that can lead to important insights into the pathogenesis of disease in humans.

DISEASE IN HUMANS

Gastroenteritis
The most common clinical syndrome seen in humans infected with *C. jejuni* is gastroenteritis (Blaser, 1997). The specific signs and symptoms can vary from individual to individual presumably due to variation in the host and the infecting organisms. Following an incubation period typically in the order of 24–72 hours, an acute diarrhoeal illness develops. A nonspecific syndrome of fever, chills, myalgia and headache may follow. Abdominal cramping and fever often accompany the diarrhoea. Patients typically have 8–10 bowel movements a day at the peak of illness. The diarrhoea can range from loose and watery to frankly bloody. Whether or not blood is obvious in the stools, faecal leucocytes and erythrocytes are present in the majority of cases. The inflammatory nature of *Campylobacter* enteritis, along with endoscopic and/or histopathological findings, produce a clinical picture that resembles inflammatory bowel disease (Farmer, 1990; Perkins and Newstead, 1994).

In the majority of individuals, *C. jejuni*-associated gastroenteritis is self-limiting, with the resolution of symptoms within a week. Extended illness can occur, particularly in individuals with underlying illness including human immunodeficiency virus (HIV) infection (see below). Occasionally individuals infected with *C. jejuni* may present with a pseudoappendicitis syndrome similar to that seen with *Yersinia pseudotuberculosis*, *Yersinia enterocolitica* and *Salmonella enteritidis*. Patients with pseudoappendicitis syndrome present with acute abdominal pain that is localized to the right lower quadrant and do not have diarrhoea.

Disease in individuals in developed countries
In primary infections with *C. jejuni* in developed countries, infection with mucosal disease predominates with symptoms of diarrhoea, abdominal pain, and blood in the stool (Black *et al.*, 1988b; Karmali, 1979; Ketley *et al.*, 1996). Young adults are the most common group with disease. However infrequently, infection with systemic spread, infection without disease with short-term bacterial persistence, and infection with resistance and no bacterial persistence occur (Black *et al.*, 1988b; Karmali, 1979; Ketley *et al.*, 1996).

Disease in individuals in developing countries
The clinical manifestations of *C. jejuni* infection differ in a number of important ways in developing areas of the world. The disease spectrum includes severe inflammatory illness, mild secretory diarrhoea or an asymptomatic carrier state. *C. jejuni* enteritis usually occurs in infants and adults rarely have symptomatic disease in developing countries (Blaser, 1980). *C. jejuni* infection is almost ubiquitous in children under 2 years of age. In this population there is significant morbidity and mortality associated with *C. jejuni* infection, mainly characterized by acute gastroenteritis. However, the duration of disease is shorter

in children in hyperendemic areas than in those with little previous exposure. In older children and adults, *C. jejuni*-associated gastroenteritis is much less common. In fact, asymptomatic shedding of *C. jejuni* can be detected in this group (Pazzaglia *et al.*, 1991).

There are likely multiple explanations for the differences in disease seen in developing versus developed countries. *Campylobacter* species possess significant genotypic and phenotypic diversity and, recently, have been shown to undergo phase variation in genes that encode surface structures (Guerry *et al.*, 2002). However, *C. jejuni* strain variation does not account for all of these differences; the same isolates associated with watery diarrhoea in children in developing countries have been recovered from visitors with acute inflammatory disease (Skirrow and Blaser, 1992). In another study, there was no difference in serotypes isolated from symptomatic and asymptomatic children (Sjogren *et al.*, 1989). One significant explanation for the different disease expression in developing versus developed countries has been the early acquisition of immunity by children exposed to hyperendemic and/or polymicrobial infection (Calva *et al.*, 1988; Richardson, 1981; Taylor, 1988). This is supported by documentation of a concurrent parasitic infection, such as are common in the developing world, precipitating *C. jejuni* enteritis in an animal model (Mansfield *et al.*, 2003; Mansfield *et al.*, 1996). From these insights it is clear that the epidemiological situations are complex involving development of a protective immune response to primary *C. jejuni* infection coupled with continuous re-exposure to various *C. jejuni* isolates, and other pathogens as well.

Invasive disease
Bacteraemia is a less common manifestation of *C. jejuni* infection, generally appearing in individuals at the extremes of age or in those with underlying immune compromise such as concurrent HIV infection. In HIV-infected patients, *C. jejuni* bacteraemia is generally a severe illness with a significant mortality rate, whereas non-HIV-infected individuals who present with bacteraemia commonly do so in the setting of a self-limiting gastroenteritis. In the latter cases, bacteraemia is generally transient and may be discovered only incidentally (Tee and Mijch, 1998). A case of neonatal *C. jejuni* sepsis was recently reported in a 3-week-old infant who acquired the infection through transmission from a recently acquired household puppy. The *Campylobacter* isolates recovered from the puppy and child were genetically indistinguishable from one another when genotyped by amplified fragment-length polymorphism (AFLP) analysis and grouped on the basis of their genetic relationship, thus providing evidence for *C. jejuni* dog–human transmission (Wolfs *et al.*, 2001)

Guillain–Barré syndrome
Guillain–Barré syndrome (GBS) is the most common form of acute neuromuscular paralysis with a yearly incidence of 1–2/100 000 in many parts of the world. Symptoms generally begin with motor and sensory deficits of the lower extremities, which can subsequently spread to the upper extremities and trunk. This can lead to the need for ventilatory support. The majority of patients will recover completely, but others can be left with severe neurological impairment.

Until recently, little was known about the aetiology of GBS. However, an observation was made that in the majority of cases of GBS, patients had an acute respiratory or intestinal illness several days or weeks before the onset of symptoms (Mishu and Blaser, 1993). It is now known that in the majority of cases where a specific pathogen can be identified, *C. jejuni* is the aetiological agent mostly frequently identified (Mishu-Allos *et al*., 1998). It has been hypothesized that patients who suffer GBS following *C. jejuni* infection develop antibodies against the bacterial lipooligosaccharides that cross-react with peripheral nerve cell surface gangliosides. *Campylobacter jejuni* with Penner heat-stable (HS) serotypes HS:O19 and HS:O41 are over-represented among GBS-associated isolates. However, other currently undefined bacterial and host variables also are considered to be important in the pathogenesis of GBS. This is based on evidence that *C. jejuni* belonging to other serotypes can be found in GBS patients, and *C. jejuni* of HS:O19 and HS:O41 can be found in patients with gastroenteritis alone and not GBS.

Disease in HIV-infected patients

Immunocompromised patients, including those infected with HIV, have a higher incidence of infection with *Campylobacter* than the general population. In these patients, disease manifestations are generally more severe and may result in profuse diarrhoea and persistence of the organism (Snijders *et al*., 1997; Tee and Mijch, 1998). In addition, some have suggested that HIV-infected individuals are often infected with atypical *Campylobacter* species, i.e. species other than *C. jejuni* or the closely related *Campylobacter coli* that have been linked to the majority of cases. *Campylobacter upsaliensis* is an atypical *Campylobacter* species that was originally identified in dogs (Olson and Sandstedt, 1987). Recently, this species has been noted to be associated with prolonged diarrhoea in HIV-infected patients (Jenkin and Tee, 1998). In a separate cross-sectional study, a variety of atypical *Campylobacter* species from HIV-infected patients with diarrhoea were found including *Campylobacter sputorum*, *Campylobacter mucosalis* and *Campylobacter hyointestinalis* (Snijders *et al*., 1997).

Role of immunity in clinical presentation

From the above discussion, it is clear that the host immune system plays a major in role in the pathogenesis of *C. jejuni* infection. As described above, the role of host immunity in limiting *C. jejuni* infection is illustrated by comparing the clinical manifestations of *C. jejuni* disease in developing versus developed countries. Another obvious example is the association between *C. jejuni* and subsequent development of GBS, which is characterized by the development of autoantibodies. The importance of immune surveillance in the clearance of *C. jejuni* infection becomes apparent given the predisposition of HIV-infected patients to prolonged and/or invasive *C. jejuni* disease. The development of protective immunity has been demonstrated in volunteer infection studies where individuals challenged with a particular *C. jejuni* strain were protected from disease (but not re-infection) when re-challenged with the given strain (Black *et al*., 1988a). Finally, additional evidence has been presented by studies of poultry abattoir workers who develop antibody responses during their employment. Long-term workers may asymptomatically shed *C. jejuni* unlike new workers who develop clinical signs of *C. jejuni*-associated gastroenteritis (Cawthraw *et al*., 2000). These observations support the notion that this organism is capable of producing a spectrum of disease scenarios depending on the immune status of the host, virulence determinants of the bacterium, complexity of concurrent infections, and the host's propensity for manifesting autoimmune disease.

DISEASE IN DOMESTICATED ANIMALS

Campylobacter jejuni is a known cause of disease in animals (FSIS, 1996). Natural infections with *C. jejuni* resulting in enteritis have been reported in juvenile macaques (Russell, 1993), weaning age ferrets (Bell, 1990), dogs (Fox *et al.*, 1988; Davies *et al.*, 1984), cats (Fox *et al.*, 1986) and swine (Babakhani *et al.*, 1993; Boosinger and Powe, 1988; Mansfield and Urban, 1996; Vítovec, 1989). Chickens (Meinersmann *et al.*, 1991), rodents (Baquar, 1996; Baquar, 1993; Humphrey, 1985), ferrets (Bell, 1990), dogs (Prescott, 1981), primates (Fitzgeorge, 1981; Russell *et al.*, 1989), rabbits (McSweegan, 1987; Walker, 1988), and pigs (Babakhani *et al.*, 1993; Boosinger and Powe, 1988; Mansfield *et al.*, 2003; Vítovec, 1989) have been inoculated experimentally by various routes with *C. jejuni* to mimic the course of infection in humans.

Disease in dogs and cats

Campylobacter jejuni causes diarrhoea in dogs and cats and they are considered a significant source of the bacterium for the human population (Bruce and Ferguson, 1980; Bruce *et al.*, 1980; Deming *et al.*, 1987). 0.8% of client-owned dogs that presented to a veterinary teaching hospital in North America for evaluation of acute small-bowel, large-bowel, or mixed-bowel diarrhoea ($n=71$) and control age-matched, healthy dogs ($n=59$) had *C. jejuni* in their faeces (Hackett and Lappin, 2003). Positive test results occurred in dogs with or without gastrointestinal signs of disease. In another case–control study of dogs with acute hemorrhagic diarrhoeal syndrome, *Campylobacter* spp. were isolated from 13 of 260 (5%) of dogs with diarrhoea and 21 of 74 (28.4%) of control dogs (Cave *et al.*, 2002). In a study at a European veterinary college, 12.7% of dog diarrhoea cases were considered caused by *C. jejuni* (Varga *et al.*, 1990). In experimental studies, exposure of dogs to *C. jejuni* causes mild diarrhoea followed by bacteraemia (Davies *et al.*, 1984; Fox, 1992; Fox *et al.*, 1983a; Fox *et al.*, 1983b). Infection is most common in puppies and kittens, but *Campylobacter* spp. can also be isolated from clinically normal adult dogs and cats (up to 30%), as well. Additionally, *C. jejuni* has been isolated in pure culture from the vaginal discharge from three German Shepherd bitches after late-pregnancy abortions (Odendaal, 1994). The main clinical sign occurring in the bitches was a profuse and odourless haemorrhagic vaginal discharge. *Campylobacter jejuni* was cultured from the faeces of 3 of 206 (1.0%) cats in a Colorado study designed to identify enteric zoonotic organisms. The bacterium was detected in samples from cats with and without diarrhoea, although the sample size was small (Hill, 2000). Rectal swabs from healthy cats and dogs, and from cats and dogs with clinical diarrhoea were collected from small-animal veterinary practices throughout Norway (Sandberg *et al.*, 2002). Of the 529 healthy dogs, 124 (23%) were positive for *Campylobacter*, compared to 18 of 66 (27%) dogs with diarrhoea. *Campylobacter jejuni* was isolated from 20 (3%) and *C. upsaliensis* from 117 (20%) of the dogs sampled. Of the 301 healthy cats sampled, 54 (18%) were positive for *Campylobacter*, compared to 5 out of 31 (16%) cats with diarrhoea. *Campylobacter jejuni* was isolated from 11 (3%), *C. upsaliensis* from 42 (13%) and *C. coli* from 2 (0.6%) of the cats sampled. It seems clear from these data that dogs and cats are commonly infected with campylobacters and sometimes suffer disease, although careful clinical workups are needed to form an association of this bacterium with disease signs.

Veterinarians report that both dogs and cats commonly have clinical signs after *C. jejuni* infection. Diarrhoea lasting 5–15 days is the most common clinical sign in dogs less than 6 months old. Diarrhoea may be watery to bloody with mucus, and is sometimes bile-stained. Occasionally chronic diarrhoea lasting for months can result, which may be accompanied by an increased body temperature and an increased white cell (leucocyte) count. Cats less than 6 months of age commonly have bloody (or non-bloody) diarrhoea. Most cat cases reported in the literature had other infectious agents present such as *Toxoplasma* and *Giardia*. Additionally, some infected cats show no clinical signs.

Disease in cattle and sheep

Before 1972, when methods were developed for its isolation from faeces, *C. jejuni* was believed to be primarily an animal pathogen causing abortion and enteritis in sheep and cattle. To date, three studies have compared prevalence between healthy and sick adult cattle (Manser, 1985; Munroe, 1983; Prescott, 1981). By definition, cattle were considered 'sick' because of diarrhoea. In these studies examining 198 to 314 samples, investigators concluded that there was not a significant difference in the frequency of *Campylobacter* isolation between these groups. However, in studies of abortion rates, 3.2% of cattle and 21.7% of sheep had abortions attributed to *C. jejuni*. Both beef or dairy cattle have significant levels of *Campylobacter* organisms with prevalence rates of 2.5 to 60% (Wesley, 2000). In a recent USDA-sponsored, collaborative study headed by the Population Medicine Centre at Michigan State University, *Campylobacter* preliminary prevalence in cows on conventional Midwest dairies was determined to be 12% ($n=25\,155$) (Green *et al.*, 2002). In this study set, antibiotic resistance was highly prevalent (i.e. 55% of isolates had tetracycline resistance). In another study, cattle checked at slaughter harboured the organism in gallbladders, large and small intestines, and liver (Garcia and Ruckerbauer, 1985). Faecal shed in cattle leads to contamination of milk (Jayarao, 2001) and beef (Dilworth *et al.*, 1988).

Campylobacter jejuni has been associated with abortion in sheep (Delong *et al.*, 1996; Varga, 1990). In a study designed to examine *Campylobacter* spp. isolated from cases of sheep abortion for biochemical, antigenic, and genetic differences investigators found 15 isolates of *Campylobacter* spp. during a single lambing season (Delong *et al.*, 1996). Eight strain variants were detected among the 15 isolates. Fourteen of the isolates were *C. jejuni*, 13 of which were biotype I and one of which was biotype II, and the remaining isolate was identified as *C. fetus* subsp. *fetus* (Delong *et al.*, 1996).

Disease in swine

Campylobacters can contribute to colitis in weaned pigs. Swine commonly carry *C. coli* and *C. jejuni* as intestinal commensals (Stern, 1992) and studies in the US, Netherlands, Great Britain and Germany show that more than half of commercially raised pigs excrete the organisms (Blaser *et al.*, 1983; Moore *et al.*, 2002a; Moore *et al.*, 2002b), suggesting that exposure occurs at an early age. Gnotobiotic or colostrum-deprived piglets inoculated orally with pathogenic strains of *C. jejuni* develop clinical signs and pathology observed in acute infections in humans with diarrhoea (Babakhani *et al.*, 1993; Boosinger and Powe, 1988; Vítovec, 1989). In this case, pigs have anorexia, fever, and diarrhoea for 1–5 days followed by remission of clinical signs, but continue to shed *C. jejuni* in the faeces. Likewise, weaned, outbred pigs that have never been exposed to campylobacters develop disease and pathology in response to 10^9 cfu *C. jejuni* (81-176) given as an oral challenge (Mansfield L.S, 2003) (Figure 1.1). However, most immunocompetent pigs with a full complement of enteric bacteria and immune system components, and with colonized campylobacters are refractory to infection with *C. jejuni*, presumably due to

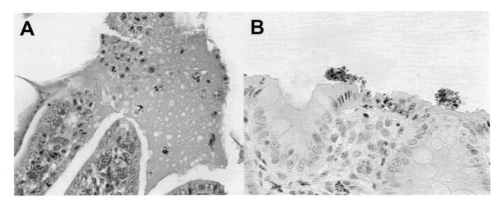

Figure 1.1 Images of pathological changes elicited in the small intestines of a 21-day-old pig 48 hours after being given 10^9 colony-forming units of low passage *Campylobacter jejuni* 81-176 parent strain expressing CDT toxin. (A) The pathological changes in the jejunum at the villus tip with haematoxylin and eosin staining. Note the damaged and absent epithelial cells, the proteinaceous exudate, the haemorrhage within the lamina propria and the mainly neutrophilic cellular response. (B) The location of *Campylobacter jejuni* 81-176 parent strain at the villus tip in the jejunum after staining with a monoclonal antibody directed against outer membrane proteins. Notice the presence of the bacteria beneath the epithelial tip cells and the lifting off of the epithelial layer.

acquired immunity (Babakhani *et al.*, 1993). When these pigs were challenged with a TH-2-inducing *Trichuris suis* infection, they developed a self-limiting colitis with watery, bloody diarrhoea that correlated with the presence of fourth stage larvae and adult worms in the proximal colon and resolved at 50–60 days of infection when worms were expelled (Mansfield *et al.*, 1996). Gnotobiotic swine infected with *C. jejuni* from a human with enteritis (ATTC 33292) had organisms colonizing the colon with no attachment or invasion (Mansfield *et al.*, 2003). *Trichuris suis* exposure converted these pigs with commensal *C. jejuni* to a severe disease status with signs and pathology comparable to that of conventionally reared pigs and humans. In this experiment, germ-free pigs inoculated with *T. suis* alone or *C. jejuni* alone had no clinical signs or pathology. Others have shown that concurrent infections with viruses (Konkel and Joens, 1990) and bacteria (Bukholm, 1987) increase the disease and pathology caused by *C. jejuni*. Taken together these studies suggest that host immunity is significant in determining disease outcome and that certain superimposed pathogens appear to ablate this resistance.

Disease in birds

Campylobacter jejuni rarely causes disease in birds. However, in pet birds especially psittaciforms (parrots), and passeriforms (finches and canaries), the organism has been isolated from the small intestines of clinically ill birds with hepatitis, lethargy, loss of appetite, weight loss and yellow diarrhoea (Pesek, 1998). Mortality may be high.

Birds appear to have a higher infection rate and carriage of *C. jejuni* than other animals (Montrose, 1985; Pearson, 1993). Domesticated poultry are the most significant reservoir for *C. jejuni* for humans causing 50–70% of cases (Allos, 2001), and chicken is the number one source (Adak *et al.*, 1995; Deming *et al.*, 1987; Fukushima *et al.*, 1987; Harris *et al.*, 1986a; Harris *et al.*, 1986b). Transmission in broilers is extremely rapid, which may be due to palatine colonization leading to transmission through communal water troughs and standard faecal–oral spread (Montrose, 1985; Pearson *et al.*, 1993). Additionally,

Campylobacter spp. have been isolated from free-living birds including migratory birds and waterfowl, crows, gulls and domestic pigeons (Hill and Grimes, 1984; Maruyama *et al.*, 1990). Despite this high rate of colonization, disease in naturally infected birds due to *C. jejuni* is rare.

Disease in exotic pets (ferrets, mink, primates, hamsters, guinea pigs and rats)

Campylobacter gastrointestinal disease has been reported in ferrets, mink, primates, hamsters, guinea pigs, and rats (Dyer, 2001). Although there are variations in clinical disease and signs observed in these species, clinical signs generally include mucoid, watery, bile-streaked diarrhoea sometimes with blood, anorexia, vomiting, and fever. Prolonged infections are possible but uncommon; most infections are self-limiting with mild clinical signs. In particular, ferrets are one of the few animals that develop a *Campylobacter*-induced diarrhoeal disease similar to that seen in humans (Rollins, 2000). Ferrets orally fed *C. jejuni* developed diarrhoeal disease in a dose-dependent fashion and developed protection against reinfection and disease when rechallenged with either the same strain or a different serotype.

SUMMARY

Campylobacters including *C. jejuni* can cause gastroenteritis in humans and animals. Infection is initiated in the gastrointestinal tract, but can become extraintestinal in severe cases particularly in immunocompromised hosts. In clinical reports describing primary infections with *C. jejuni* in developed countries, infection with mucosal disease predominates with symptoms of diarrhoea, abdominal pain, and blood in the stool. However infrequently, infection with systemic spread, infection without disease with short-term bacterial persistence, and infection with resistance and no bacterial persistence occur. In developing countries, *C. jejuni* enteritis usually occurs in infants; adults rarely have symptomatic disease. Here, disease spectrum includes severe inflammatory illness, mild secretory diarrhoea or an asymptomatic carrier state. Also, antecedent *C. jejuni* has been linked to autoimmune diseases such as GBS. It is clear that the epidemiological situations are complex involving development of a protective immune response to primary *C. jejuni* infection coupled with continuous re-exposure to various *C. jejuni* isolates, and other pathogens as well. These observations support the notion that this organism is capable of producing a spectrum of disease scenarios depending on the immune status and regulation within the host, virulence determinants of the bacterium and complexity of concurrent infections.

References

Adak, G.K., Cowden, J.M., Nicholas, S., and Evans, H.S. 1995. The Public Health Laboratory Service national case–control study of primary indigenous sporadic cases of *Campylobacter* infection. Epidemiol. Infect. 115: 15–22.

Allos, B.M. 2001. *Campylobacter jejuni* infections: update on emerging issues and trends. Clin. Infect. Dis. 32: 1201–1206.

Anonymous 2003. Preliminary FoodNet data on the incidence of food-borne illnesses – selected sites, United States, 2002. MMWR Morb. Mortal Wkly Rep. 52: 340–343.

Babakhani, F.K., Bradley, G.A., and Joens, L.A. 1993. Newborn piglet model for campylobacteriosis. Infect. Immun. 61: 3466–3475.

Baqar, S., Bourgeois A.L., Applebee, L.A., Mourad, A.S. Kleinosky, M.T., Mohran, Z., and Murphy, J.R. 1996. Murine intranasal challenge model for the study of *Campylobacter* pathogenesis and immunity. Infect. Immun. 64: 4933–4939.

Baquar, S., Pacheco, N.D., and Rollwagen, F.M. 1993. Modulation of mucosal immunity against *C. jejuni* by orally administered cytokines. Antimicr. Agents Chemother. 37: 2688–2692.

Bell, J.A., and Manning, D.D. 1990. A domestic ferret model of immunity to *C. jejuni*-induced enteric disease. Infect. Immun. 58: 1848–1852.

Black, R.E., Levine, M.M., Clements, M.L., Hughes, T.P., and Blaser, M.J. 1988a. Experimental *Campylobacter jejuni* infection in humans. J. Infect. Dis. 157: 472–479.

Black, R.E., Levine, M.M., Clements, M.L., Hughes, T.P., and Blaser, M.J. 1988b. Experimental *Campylobacter jejuni* infection in humans. J. Infect. Dis. 157: 472–479.

Blaser, M.J. 1997. Epidemiologic and clinical features of *Campylobacter jejuni* infections. J. Infect. Dis. 176: S103–105.

Blaser, M.J., Glass, R.I., Hug, M.I., Stoll, B., Kibrya, G.M., Alim, A.R.M.A. 1980. Isolation of *Campylobacter fetus* subsp. *jejuni* from Bangladeshi children. J. Clin. Microbiol. 12: 744–747.

Blaser, M.J., Taylor, D.N., and Feldman, R.A. 1983. Epidemiology of *C. jejuni* infections. Epidemiol. Rev. 5: 157–176.

Bolton, F.J., and Coates, D. 1983. A comparison of microaerobic systems for the culture of *Campylobacter jejuni* and *Campylobacter coli*. Eur. J. Clin. Microbiol. 2: 105–110.

Bolton, F.J., Hutchinson, D.N., and Parker, G. 1988. Reassessment of selective agars and filtration techniques for isolation of *Campylobacter* species from faeces. Eur. J. Clin. Microbiol. Infect. Dis. 7: 155–160.

Boosinger, T.R., and Powe, T.A. 1988. *C. jejuni* infections in gnotobiotic pigs. Am. J. Vet. Res. 49: 456–458.

Bruce, D., and Ferguson, I.R. 1980. *Campylobacter jejuni* in cats. Lancet 2: 595–596.

Bruce, D., Zochowski, W., and Fleming, G. 1980. *Campylobacter* infections in cats and dogs. Vet. Rec. 107: 200–201.

Bukholm, G., and Kapperud, G. 1987. Expression of *C. jejuni* invasiveness in cell cultures coinfected with other bacteria. Infect. Immun. 55: 2816–2821.

Calva, J.J., Ruiz-Palacios, G.M., Lopez-Vidal, A.B., Ramos, A., and Bojalil, R. 1988. Cohort study of intestinal infection with *Campylobacter* in Mexican children. Lancet 1: 503–506.

Cave, N.J., Marks, S.L., Kass, P.H., Melli, A.C., and Brophy, M.A. 2002. Evaluation of a routine diagnostic fecal panel for dogs with diarrhea. J. Am. Vet. Med. Assoc. 221: 52–59.

Cawthraw, S.A., Lind, L., Kaijser, B., and Newell, D.G. 2000. Antibodies, directed towards *Campylobacter jejuni* antigens, in sera from poultry abattoir workers. Clin. Exp. Immunol. 122: 55–60.

Davies, A.P., Gebhart, C.J., and Meric, S.A. 1984. *Campylobacter*-associated chronic diarrhoea in a dog. J. Am. Vet. Med. Assoc. 184: 469–471.

Delong, W.J., Jaworski, M.D., and Ward, A.C. 1996. Antigenic and restriction enzyme analysis of *Campylobacter* spp. associated with abortion in sheep. Am. J. Vet. Res. 57, 163–167.

Deming, M.S., Tauxe, R.V., Blake, P.A., Dixon, S.E., Fowler, B.S., Jones, T. S., Lockamy, E.A., Patton, C.M., and Sikes, R.O. 1987. *Campylobacter* enteritis at a university: transmission from eating chicken and from cats. Am. J. Epidemiol. 126: 526–534.

Dilworth, C.R., Lior, H., and Belliveau, M.A. 1988. *Campylobacter* enteritis acquired from cattle. Can. J. Publ. Health 79: 60–62.

Dyer, N.W., and Stoltenow, C.L. 2001. Campylobacteriosis caused by the bacterium *Campylobacter jejuni*. In: eds NDSU Extension Service, North Dakota State University of Agriculture and Applied Science, and US Department of Agriculture, NDSU Diagnostic Laboratory publication V-1211.

Farmer, R.G. 1990. Infectious causes of diarrhoea in the differential diagnosis of inflammatory bowel disease. Med. Clin. N. Am. 74: 29–38.

Fitzgeorge, R.B., Baskerville, A., and Lander, K.P. 1981. Experimental infection of Rhesus monkeys with a human strain of *Campylobacter jejuni*. J. Hyg. Lond. 86: 343–351.

Fox, J.G., Ackerman, J.I., Taylor, N., Claps, M., Murphy, J.C. *Campylobacter jejuni* infection in the ferret: an animal model of human campylobacteriosis. Am. J. Vet. Res. 48 (1): 85–90.

Fox, J.G. 1992. In vivo models of enteric campylobacteriosis: natural and experimental infections. In *Campylobacter jejuni*: Current Status and Future Trends, I. Nachamkin, M.J. Blaser, and L.S. Tompkins, eds. (Washington, DC, ASM Press), pp. 131–138.

Fox, J.G., Ackerman, J.I., Maxwell, K., and Essigman, E. 1983a. *Campylobacter jejuni* associated diarrhoea in juvenile beagle dogs. Lab. Anim. Sci. 33: 483–484.

Fox, J.G., Moore, R., and Ackerman, J. I. 1983b. *Campylobacter jejuni*-associated diarrhoea in dogs. J. Am. Vet. Med. Assoc. 183: 1430–1433.

Fox, J.G., Claps, M., and Beaucage, C. M. 1986. Chronic diarrhea associated with *Campylobacter jejuni* infection in a cat. J. Am. Vet. Med .Assoc. 189: 455–456.

Fox, J.G., Claps, M.C., Taylor, N.S., Maxwell, K.O., Ackerman, J.I., and Hoffman, S.B. 1988. *Campylobacter jejuni–coli* in commercially reared beagles – prevalence and serotypes. Lab. Anim. Sci. 38: 262–265.

FSIS 1996. Pathogen reduction, Hazard Analysis and Critical Control point (HACCP) Systems; Final Rule. Department of Agriculture 61 *FR*, 38805.

Fukushima, H., Hoshina, K., Nakamura, R., and Ito, Y. 1987. Raw beef, pork and chicken in Japan contaminated with *Salmonella* sp., *Campylobacter* sp., *Yersinia enterocolitica*, and *Clostridium perfringens* – a comparative study. Zentralbl. Bakteriol. Mikrobiol. Hyg. [B] 184: 60–70.

Garcia, M.M., Lior, H., Stewart, R.B., Ruckerbauer, G.M., Trudel, J.R., and Skljarevski, A. 1985. Isolation, characterization, and serotyping of *Campylobacter jejuni* and *Campylobacter coli* from slaughter cattle. Appl. Environ. Microbiol. 49: 667–672.

Green, A.M., Kaneene, J.B., Ruegg, P.L., Warnick, L.D., Wells, S.J., Halbert, L.W., Miller, R.A., Fossler, C. and Mansfield, L.S. 2002. Patterns of occurrence of *Campylobacter jejuni* in dairy farms in Midwestern and Northeastern United States: A herd level analysis. Proceedings of the American Association of Veterinary Medicine Meeting 2001.

Guerry, P., Szymanski, C.M., Prendergast, M.M., Hickey, T.E., Ewing, C.P., Pattarini, D.L., and Moran, A.P. 2002. Phase variation of *Campylobacter jejuni* 81-176 lipooligosaccharide affects ganglioside mimicry and invasiveness *in vitro*. Infect. Immun. 70: 787–793.

Hackett, T., and Lappin, M.R. 2003. Prevalence of enteric pathogens in dogs of north-central Colorado. J. Am. Anim. Hosp. Assoc. 39: 52–56.

Harris, N.V., Thompson, D., Martin, D.C., and Nolan, C.M. 1986a. A survey of *Campylobacter* and other bacterial contaminants of pre-market chicken and retail poultry and meats, King County, Washington. Am. J. Publ. Health 76: 401–406.

Harris, N.V., Weiss, N.S., and Nolan, C.M. 1986b. The role of poultry and meats in the aetiology of *Campylobacter jejuni/coli* enteritis. Am. J. Publ. Health 76: 407–411.

Hill, G.A., and Grimes, D.J. 1984. Seasonal study of a freshwater lake and migratory waterfowl for *Campylobacter jejuni*. Can. J. Microbiol. 30: 845–849.

Hill, S.L., Cheney, J.M., Taton-Allen, G.F., Reif, J.S., Bruns, C., and Lappin, M.R. 2000. Prevalence of enteric zoonotic organisms in cats. J. Am. Vet. Med. Assoc. 216: 687–692.

Humphrey, C.D., Montag, D.M., and Pittman, F.E. 1985. Experimental infection of hamsters with *C. jejuni*. J. Infect. Dis. 151: 485–493.

Jayarao, B.M., Henning, D.R. 2001. Prevalence of food-borne pathogens in bulk tank milk. J. Dairy Sci. 84: 2157–2162.

Jenkin, G.A., and Tee, W. 1998. *Campylobacter upsaliensis*-associated diarrhoea in human immunodeficiency virus-infected patients. Clin. Infect. Dis. 27: 816–821..

Karmali, M.A., and Fleming, P.C. 1979. *Campylobacter* enteritis in children. .J. Paediatr. 94: 527–533.

Ketley, J., Guerry, P., and Panigrahi, P. 1996. Pathogenic mechanisms. In Campylobacters, Helicobacters, and Related Organisms, D.G. Newell, J.M. Ketley, and R.A. Feldman, eds. (New York, Plenum Press), pp. 537–544.

Konkel, M.E., and Joens, L.A. 1990. Effect of enteroviruses on adherence to and invasion of Hep-2 cells by *Campylobacter* isolates. Infect. Immun. 58: 1101–1105.

McSweegan, E., Burr D.H., and Walker, R.I. 1987. Intestinal mucus gel and secretory antibody are barriers to *C. jejuni* adherence to INT 407 cells. Infect. Immun. 55: 1431–1435.

Manser, P.A., and Dalziel, R.W. 1985. A survey of *Campylobacter* in animals. J. Hyg. (Lond.) 95: 15–21.

Mansfield, L.S., and Urban, J.F. 1996. The pathogenesis of necrotic proliferative colitis in swine is linked to whipworm induced suppression of mucosal immunity to resident bacteria. Vet. Immunol. Immunopathol. 50: 1–17.

Mansfield, L.S., Gauthier, D.T., Abner, S.R., Jones, K.M., Wilder, S.R., and Urban, J.F. 2003. Enhancement of disease and pathology by synergy of *Trichuris suis* and *Campylobacter jejuni* in the colon of immunologically naive swine. Am. J. Trop. Med. Hyg. 68: 70–80.

Mansfield, L.S., Hill, D.E., and Urban, J.F. 1996. Worm induced suppression of immunity to intracellular and extracellular colonic bacteria in swine. J. Cell. Biochem., 259–259.

Maruyama, S., Tanaka, T., Katsube, Y., Nakanishi, H., and Nukina, M. 1990. Prevalence of thermophilic campylobacters in crows (*Corvus levaillantii, Corvus corone*) and serogroups of the isolates. Nippon Juigaku Zasshi 52: 1237–1244.

Meinersmann, R.J., Rigsby, W. E., Stern, N.J., Kelley, L.C., Hill, J.E., and Doyle, M.P. 1991. Comparative study of colonizing and noncolonizing *Campylobacter jejuni*. Am. J. Vet. Res. 52: 1518–1522.

Mishu, B., and Blaser, M.J. 1993. Role of infection due to *Campylobacter jejuni* in the initiation of Guillain–Barré syndrome. Clin. Infect. Dis. 17: 104–108.

Mishu-Allos, B., Lippy, F.T., Carlsen, A., Washburn, R.G., and Blaser, M.J. 1998. *Campylobacter jejuni* strains from patients with Guillain–Barré syndrome. Emerging Infect. Dis. 4: 263–268.

Montrose, M.S., Shane, S.M., and Harrington, K.S. 1985. Role of litter in the transmission of *Campylobacter jejuni*. Avian Dis. 29: 392–399.

Moore, J. E., Garcia, M.M., and Madden, R.H. 2002a. Subspecies characterization of porcine *Campylobacter coli* and *Campylobacter jejuni* by multilocus enzyme electrophoresis typing. Vet. Res. Commun. 26: 1–9.

Moore, J.E., Lanser, J., Heuzenroeder, M., Ratcliff, R.M., Millar, B.C., and Madden, R.H. 2002b. Molecular diversity of *Campylobacter coli* and *C. jejuni* isolated from pigs at slaughter by flaA-RFLP analysis and ribotyping. J. Vet. Med. B Infect. Dis. Vet. Publ. Health 49: 388–393.

Munroe, D.L., Prescott, J.F., and Penner, J.L. 1983. *Campylobacter jejuni* and *Campylobacter coli* serotypes isolated from chickens, cattle, and pigs. J. Clin. Microbiol. 18: 877–881.

Odendaal, M.W., de Cramer, K.G., van der Walt, M.L., Botha, A.D., and Pieterson, P.M. 1994. First isolation of *Campylobacter jejuni* from the vaginal discharge of three bitches after abortion in South Africa. Onderstepoort J. Vet. Res. 61: 193–195.

Olson, P., and Sandstedt, K. 1987. *Campylobacter* in the dog: a clinical and experimental study. Vet. Rec. 121: 99–101.

Pazzaglia, G., Bourgeois, A.L., el Diwany, K., Nour, N., Badran, N., and Hablas, R. 1991. *Campylobacter* diarrhoea and an association of recent disease with asymptomatic shedding in Egyptian children. Epidemiol. Infect. 106: 77–82.

Pearson, A.D., Greenwood, M., Healing, T.D., Rollins, D., Shahamat, M., Donaldson, J., and Colwell, R.R. 1993. Colonization of broiler chickens by waterborne *Campylobacter jejuni*. Appl. Environ. Microbiol. 59: 987–996.

Pearson, A.D., Greenwood, M., Healing, T.D., Rollins, D., Shahamat, M., Donaldson, J., and Colwell, R.R. 1993. Colonization of broiler chickens by waterborne *Campylobacter jejuni*. Appl Environ Microbiol 59: 987–996.

Perkins, D.J., and Newstead, G.L. 1994. *Campylobacter jejuni* enterocolitis causing peritonitis, ileitis and intestinal obstruction. Austr. NZ J. Surg. 64: 55–58.

Pesek, L. 1998. Allergic alveolitis, campylobacteriosis, Newcastle's disease. In: Zoonotic Diseases, Part II, Bird to Human Transmission, Pet Bird Magazine June, 1–2.

Prescott, J.F., Barker, I.K. Manninen, K.I., and Miniats, O.P. 1981. *C. jejuni* colitis in gnotobiotic dogs. Can. J. Comp. Med. 45: 37–383.

Richardson, N.J., Koornhoff, H.J., Bockenheuser, V.D. 1981. Long-term infections with *Campylobacter fetus* subsp. *jejuni*. J. Clin. Microbiol. 13: 846–849.

Rollins, D.M., and Burr, D.H. 2000. Protective immunity in animals following infection with *Campylobacter jejuni*. The 100th General Meeting of the American Society for Microbiology, May 21–25: 2000, Los Angeles, California, Session 214/D, Paper D-229.

Russell, R.G., Blaser, M.J., Sarmiento, J.I., and Fox, J. 1989. Experimental *Campylobacter jejuni* infection in *Macacca nemestrina*. Infect. Immun. 57: 1438–1444.

Russell, R.G, O'Donnoghue, M., Blake, D.C. Jr., Zulty, J., and DeTolla, L.J. 1993. Early colonic damage and invasion of *Campylobacter jejuni* in experimentally challenged infant *Macaca mulatta*. J. Infect. Dis. 168: 210–215.

Sandberg, M., Bergsjo, B., Hofshagen, M., Skjerve, E., and Kruse, H. 2002. Risk factors for *Campylobacter* infection in Norwegian cats and dogs. Prev. Vet. Med. 55: 241–253.

Sjogren, E., Ruiz-Palacios, G., and Kaijser, B. 1989. *Campylobacter jejuni* isolations from Mexican and Swedish patients, with repeated symptomatic and/or asymptomatic diarrhoea episodes. Epidemiol. Infect. 102: 47–57.

Skirrow, M.B., and Blaser, M.J. 1992. Clinical and epidemiological considerations. In *Campylobacter jejuni*: Current Status and Future Trends, I. Nachamkin, M. J. Blaser, and L.S. Tompkins, eds. (Washington, D.C., ASM Press), pp. 3–8.

Snijders, F., Kuijper, E.J., de Wever, B., van der Hoek, L., Danner, S.A., and Dankert, J. 1997. Prevalence of *Campylobacter*-associated diarrhoea among patients infected with human immunodeficiency virus. Clin. Infect. Dis. 24: 1107–1113.

Stern, N.J. 1992. Reservoirs for *Campylobacter jejuni* and approaches for intervention in poultry. In: *Campylobacter jejuni*: Current Status and Future Trends, Nachamkin, I, Blaser, MJ, Tompkins, LS, eds. Washington, DC: American Society for Microbiology, pp. 49–60.

Taylor, D.N., Echeverria, P., Pitarangsi, C., Seriwatana, J., Bodhidatta, L., and Blaser, M. 1988. Influence of strain characteristics and immunity on the epidemiology of *Campylobacter* infections in Thailand. J. Clin. Microbiol. 26: 863–868.

Tee, W., and Mijch, A. 1998. *Campylobacter jejuni* bacteremia in human immunodeficiency virus (HIV)-infected and non-HIV-infected patients: comparison of clinical features and review. Clin. Infect. Dis. 26: 91–96.

11

Vandamme, P., Falsen, E., Rossau, R., Hoste, B., Segers, P., Tytgat, R., and De Ley, J. 1991. Revision of *Campylobacter*, *Helicobacter*, and *Wolinella* taxonomy: emendation of generic descriptions and proposal of *Arcobacter* gen. nov. Int. J. Syst. Bacteriol. 41: 88–103.

Varga, J., Mezes, B., and Fodor, L. 1990. Serogroups of *Campylobacter jejuni* from man and animals. Zentralbl Veterinarmed [B] 37: 407–411.

Varga, J., Mezes, B., Fodor, L., Hajtos, I. 1990. Serogroups of *Campylobacter fetus* and *Campylobacter jejuni* isolated in cases of ovine abortion. Zentralbl. Veterinarmed. [B] 37: 148–152.

Walker, R.I., Schmauder-Chock, E.A., Parker, J.L., and Burr, D. 1988. Selective association and transport of *C. jejuni* through M cells of rabbit Peyer's patches. Can. J. Microbiol. 34: 1142–1147.

Wesley, I.V., Wells, S.J., Harmon, K.M., Green, A., Schroeder-Tucker, L., Glover, M., Siddique, I. 2000. faecal shedding of *Campylobacter* and *Arcobacter* spp. in dairy cattle. Appl. Environ. Microbiol. 66: 1994–2000.

Wolfs, T.F., Duim, B., Geelen, S.P., Rigter, A., Thomson-Carter, F., Fleer, A., and Wagenaar, J. A. 2001. Neonatal sepsis by *Campylobacter jejuni*: genetically proven transmission from a household puppy. Clin. Infect. Dis. 32: E97–99.

Chapter 2

Taxonomy, Phylogeny, and Methods for the Identification of *Campylobacter* Species

Stephen L. W. On

Abstract

The taxonomy of the genus *Campylobacter* has changed dramatically since its inception in 1963 by Sebald and Véron, at which time the genus contained just two taxa, *C. fetus* and '*C. bubulus*'. At present, *Campylobacter* contains 16 species, six subspecies and several validly named biovars, most of which are of substantial clinical and economic importance. Moreover, a variety of taxa previously described as *Campylobacter* species have been reclassified into other genera. This chapter reviews the historical development of *Campylobacter* taxonomy, outlines the phylogeny within and between *Campylobacter* and related bacteria, summarizes the characteristics and taxonomic problems of each species and provides an overview of current methods applied to identify members of this important group of bacteria.

INTRODUCTION

Taxonomy is a discipline comprising three areas: *classification*, where organisms are ordered into groups on the basis of one or more properties; *identification*, whereby a strain of unknown taxonomic affiliation is assigned to a defined taxon by comparison of their characteristics; and *nomenclature*, the naming of distinct taxonomic groups. Taxonomy may be considered to be a part of the wider field of *systematics*, a subject area that encompasses biodiversity in a broad scale, and is closely allied to *phylogeny*, which is the study of evolution. In brief, taxonomy is the science of relationships between all organisms capable of replication.

Taxonomy – and microbial taxonomy in particular – is a dynamic and rapidly changing field. Hundreds of new species (many representing new genera) are described in scientific journals each year. In addition, established species are (not infrequently) reassigned to different genera and new or altered higher-order groupings (families, orders, divisions and many others) may also be proposed. Such changes are often to the chagrin of clinicians, veterinarians, epidemiologists, ecologists and researchers, i.e. those who use the taxonomic systems in practice. Nonetheless, it is important to realize that the principal aim of good taxonomy is to provide a stable and sensible classification system that accurately reflects organismal relationships for the benefit of all. Changes in taxonomy often reflect developments in methods and/or hypotheses. On occasion, changes are required to amend errors of judgement.

The taxonomy of *Campylobacter* has changed dramatically since its inception in 1963. The changes have reflected the sweeping developments in bacterial taxonomy as a whole, and the intense interest in the bacteria since their recognition as clinically, economically and ecologically important organisms. This chapter aims to give an overview of the major developments in the field and of its current status.

GENESIS OF THE GENUS

Bacterial classifications prior to 1963 were predominantly based on cell morphology, growth requirements, and biochemical and immunological tests. Despite these limited criteria, it was recognized that the genus *Vibrio* (essentially defined as curved- or spiral-shaped bacteria that resembled the causative agent of cholera) was highly diverse. In addition to the 'true' *Vibrio* species (e.g. *V. cholerae, V. metchnikovii*), the genus pre-1963 contained various microaerobic and anaerobic taxa from such diverse habitats as bovine and ovine abortions ('*V. fetus*', '*V. bubulus*': Smith and Taylor, 1919; Florent, 1953), the human oral cavity ('*V. sputorum*': Prévot, 1940), pig faeces ('*V. coli*': Doyle, 1948) and bovine intestinal contents ('*V. jejuni*': Jones *et al.*, 1931). The confused taxonomy of *Vibrio* was addressed by Sebald and Véron (1963), who applied Hugh and Leifson's test for fermentative metabolism and the G+C ratio in genomic DNA to show that '*V. fetus*' and '*V. bubulus*' were notably different from other *Vibrio* species. Thus, a new genus – *Campylobacter* – was proposed to encompass these taxa (Sebald and Véron, 1963). The other microaerobic and/or anaerobic '*Vibrio*' taxa (in addition to '*V. faecalis*': Firehammer, 1965) were reclassified as *Campylobacter* spp. in a later and more comprehensive taxonomic study that used various biochemical and serological tests, and the G+C ratio in DNA (Véron and Chatelain, 1973). The latter study found a greater level of acceptance in the wider scientific community, and few workers referred to the microaerobic species as '*Vibrio*' spp. post-1973.

FROM LITTLE ACORNS, MIGHTY OAKS DO GROW

The most dramatic changes in *Campylobacter* taxonomy took place over the next two decades following the seminal study of Véron and Chatelain (1973). There were two fundamental reasons for this.

Up until 1973, *Campylobacter* (or 'related vibrios' as they were often referred to pre-1973) was principally considered a problem for veterinarians, since they were mainly associated with reproductive problems in cattle and sheep. The report of a 'related vibrio' in human diarrhoea in 1957 seemed to be anomalous (King, 1957). The studies by Butzler *et al.* (1973) and Skirrow (1977) showing *Campylobacter* spp. were frequently associated with diarrhoea generated tremendous interest in the wider scientific community. The concurrent description of an inexpensive isolation protocol allowed researchers in many areas of human and animal health to investigate the possible involvement of these bacteria in their disease of interest. Since spiral bacteria had been observed, but never before cultured from diarrhoeal and gastric samples (to name but two sources) for over a century (e.g. Escherich, 1886; Salomon, 1898), this enthusiasm was hardly misplaced. As a consequence of this interest, various species were described during the next 20 years from a wide range of sources. These included *C. mucosalis* (first designated '*C. sputorum*' subsp. *mucosalis*, from pig intestinal contents; Lawson *et al.*, 1975), *C. concisus* (oral cavity, Tanner *et al.*, 1981), *C. hyointestinalis* (pig intestine, Gebhart *et al.*, 1983), '*C. nitrofigilis*' (plant roots, McClung *et al.*, 1983), *C. lari* (gulls, first named '*C. laridis*': Benjamin *et al.*,

1983), 'C. pylori' (human gastric mucosa, first named 'C. pyloridis' by Skirrow, 1983), 'C. cryaerophilus' (pig and cattle abortions: Neill et al., 1985), 'C. cinaedi' and 'C. fennelliae' (human intestine: Totten et al., 1985), 'C. mustelae' (ferret gastric mucosa: first named 'C. pylori subsp. mustelae' by Fox et al., 1988), C. jejuni subsp. doylei (human enteritis and gastritis: Steele and Owen, 1988), 'C. intracellularae' (porcine proliferative enteritis: McOrist et al., 1990), C. upsaliensis (dog faeces: Sandstedt and Ursing, 1991), and 'C. butzleri' (human diarrhoea: Kiehlbauch et al., 1991).

The second reason for the dramatic changes in the taxonomy of Campylobacter during this time was related to the ever-increasing sophistication in the methods used to classify bacteria. Numerical taxonomy had become relatively commonplace, largely due to the groundbreaking work of Sneath and Sokal (1973). Electron microscopy could reveal key ultrastructural features of bacterial strains. Chemotaxonomic methods such as cellular fatty acid and isoprenoid quinone analysis, protein profiling, and, critically, DNA–DNA hybridization (which remains a salient defining criterion for a bacterial species: Wayne et al., 1987) proved invaluable for identifying and describing new species. However, a major aim of taxonomy was to classify organisms according to phylogenetic principles, i.e. to group organisms in such a way that reflected the natural or evolutionary pathway from which they developed. Methods for reconstructing bacterial phylogeny proved elusive until the concept of the 'molecular chronometer' was elucidated: widely distributed, but functionally conserved genes that changed slowly over time (see Woese, 1987, for a review). By tracking changes in such genes, it was proposed that phylogenetic relationships among all living organisms could be reconstructed. The model gene for this purpose was rRNA and the 16S subunit in particular. This concept gained substantial credibility over time and technical developments in sequencing (and later PCR) enabled studies to be performed in laboratories worldwide. As a consequence, several studies of 16S rRNA gene sequence divergence in Campylobacter demonstrated the diversity of the genus as then defined (Romaniuk et al., 1987; Lau et al., 1987; Paster and Dewhirst, 1988), the most taxonomically complete investigation clearly delineating three major groups (Thompson et al., 1988). In all studies, 'C. pylori' and 'C. mustelae' were especially distinct. These taxa possessed sheathed flagella (unlike other campylobacters: Han et al., 1989), and with other notable phenotypic differences (notably cellular fatty acid composition) also in evidence (Goodwin et al., 1989a), 'C. pylori' and 'C. mustelae' were reassigned to a new genus, Helicobacter (Goodwin et al., 1989b). These findings were corroborated in an authoritative study by Vandamme et al (1991a), who, using rRNA–DNA hybridization as the means to determine divergence in the 16S rRNA gene, and with protein profile, phenotypic and immunotyping data to examine closer genetic relationships, proposed a substantial restructuring of the group. The so-called aerotolerant campylobacters were assigned to a new genus, Arcobacter, as A. nitrofigilis and A. cryaerophilus respectively, with A. butzleri reassigned later (Vandamme et al., 1992b). Two enteric species were added to Helicobacter as H. cinaedi and H. fennelliae, broadening the ecological niches inhabited by such bacteria from the gastric environment favoured by H. pylori and H. mustelae. One free-living strain (Laanbroek et al., 1977) was not immediately placed, but later assigned to a novel genus, Sulfurospirillum, thus far containing only species of environmental origin (Schumacher et al., 1992). The remaining Campylobacter species were seen to belong to a single phylogenetic group, to which the generically misnamed Bacteroides ureolyticus is clearly but distinctly affiliated. Two oral anaerobic species, 'Wolinella curva' and 'W. recta', were also shown to belong to Campylobacter. While distinct, Campylobacter,

Arcobacter and *Sulfurospirillum* share enough characteristics to be considered in the same family, *Campylobacteraceae* (Vandamme and de Ley, 1991; Vandamme *et al.,* in press). All of the above taxa form a distinct, but diverse phylogenetic group, variously referred to as rRNA superfamily VI (Vandamme *et al.,* 1991a), the epsilon division of the *Proteobacteria* (Stackebrandt, 1992) or the Epsilobacteria (Cavalier-Smith, 2002). In contrast, the unculturable '*C. intracellularae*' was later renamed *Lawsonia intracellularis* and represents an organism phylogenetically distant from the Epsilobacteria (McOrist *et al.,* 1995). Similarly, a bile-tolerant bacterium first isolated from intestinal tract infections first described as '*C. gracilis*-like' was subsequently confirmed as more closely related to *Alcaligenes* and *Bordetella,* and was assigned to a new genus as *Sutterella wadsworthensis* (Wexler *et al.,* 1996).

SPECIES OF THE GENUS *CAMPYLOBACTER*

It is beyond the scope of this chapter to discuss the genera *Arcobacter, Helicobacter, Sulfurospirillum* or the other Epsilobacterial taxa. The interested reader is directed elsewhere for further reading (Mansfield and Forsythe, 2000; On, 2001; Solnick and

Table 2.1 Phenotypic test results useful for differentiating *Campylobacter* spp. and the generically misnamed *Bacteroides ureolyticus*. The figures in parentheses after the species name denote the number of isolates examined. All other values describe the percentage of isolates that gave a positive result in the test concerned.

	Oxidase	Catalase	Urease	Indoxyl acetate hydrolysis	Hippurate hydrolysis
C. coli (18)	100	100	0	100	0
C. concisus (21)	57	0	0	0	0
C. curvus (5)	100	0	0	60	20
C. fetus subsp. *fetus* (17)	100	100	0	0	0
C. fetus subsp. *venerealis* (29)	100	93	0	0	0
C. gracilis (7)	0	29	0	71	0
C. helveticus (9)	100	0	0	100	0
C. hominis (5)	100	0	0	0	0
C. hyointestinalis subsp. *hyointestinalis* (17)	100	100	0	0	0
C. hyointestinalis subsp. *lawsonii* (9)	100	100	0	0	0
C. jejuni subsp. *doylei* (10)	100	70	0	100	100
C. jejuni subsp. *jejuni* (22)	100	100	0	100	91
C. lanienae (5)	100	100	0	0	0
C. lari (14)	100	100	64§	7	0
C. mucosalis (10)	100	0	0	0	0
C. rectus (5)	100	20	0	100	0
C. showae (2)	50	100	0	50	0
C. sputorum (40)	100	Biovar dependent*	Biovar dependent*	0	0
C. upsaliensis (9)	100	0	0	100	0
B. ureolyticus (10)	100	20	100	10	0

*Catalase- and urease-negative strains, bv. sputorum; catalase-positive strains, bv. fecalis; urease-positive strains, bv. paraureolyticus. §Urease-positive strains are referred to as the urease-positive thermophilic *Campylobacter* (UPTC) group. †Weak reaction. ‡Strong reaction in most (87–100%) strains that give a

Schauer, 2001). However, it is important to note that the close resemblance of these genera to *Campylobacter* spp. generally, and some species in particular, present particular problems to diagnosticians wishing to identify a strain of clinical relevance. This is discussed in more detail below and elsewhere (On, 1996a, 2001).

At this time (July 2003), *Campylobacter* contains 16 species, of which three may be divided into defined subspecies. Distinguishing characteristics are given in Table 2.1 and phylogenetic relationships as inferred by 16S rRNA gene sequence comparisons shown in Figure 2.1. It must be emphasized that reconstructing phylogenetic relationships is not simply a matter of examining a cluster analysis based on aligned sequence data. Indeed, this approach (especially when based on the 16S rRNA gene) has been strongly criticized in recent times (Gupta, 1998; Cavalier-Smith, 2002). It is important to consider all types of data to justify a given evolutionary hypothesis. Here, the grouping of *Campylobacter* spp. in the 16S rRNA gene sequence tree is used as a basis, with clades of probable phylogenetic significance given substance by identifying common traits.

Campylobacter coli was first isolated from pigs afflicted with infectious dysentery (Doyle, 1948). It remains a frequently encountered species in pigs, although it is generally

Data were derived from various published (On *et al.* 1996, 1998a; On and Vandamme, 1997; Lawson *et al.*, 2001; On and Harrington, 2001) and unpublished studies. All tests were performed using recommended media and/or methods (On and Holmes, 1991a,b, 1992, 1995). H_2S, hydrogen sulfide; TSI, triple sugar iron agar.

Nitrate reduction	Selenite reduction	H_2S/TSI production	Growth at 42°C	Growth on minimal medium	Nalidixic acid resistance	Growth on 1% glycine medium
100	100	50†	100	100	0	94
14	24	5	76	0	71	24
100	0	20	60	80	100	100
100	94	0	70	71	94	100
100	21	0	0	90	69	7
86	0	0	71	29	57	100
100	0	0	100	0	0	44
100	0	0	0	0	20	100
100	100	71†	100	65	100	100
100	100	89‡	100	66	100	22
0	0	0	0	0	0	20
100	73	0	100	10	0	91
100	100	0	100	0	80	80
100	50	0	100	0	29†	100
10	10	100ʃ	100	0	80	50
100	0	0	20	0	80	100
100	0	50†	50	50	0	50
97	45	100‡	97	NR	82	100
100	100	0	89	0	0	100
100	0	0	60	30	0	100

positive test result. ʃStrong reaction in a few (20%) of strains that give a positive test result. †'Classical' strains are resistant: this value was derived from a taxonomically biased sampling. NR, test not reproducible.

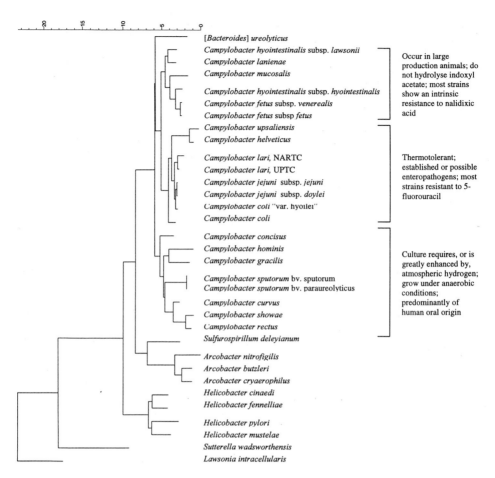

Figure 2.1 Phylogeny of *Campylobacter* species and related bacteria as inferred by comparison of 16S rRNA gene sequences. Important unifying traits of the major clades in *Campylobacter* are annotated on the right-hand side of the dendrogram. All of the species shown are, or have previously been described as, a *Campylobacter* (see text). The scale bar indicates 25% sequence dissimilarity. Sequences were aligned using the program BioNumerics (version 2.5, Applied Maths, Kortrijk, Belgium) and gaps and unknown bases were not considered in the analysis. A 250 bp region corresponding to nucleotide positions 180–430 in the *C. sputorum* bv. sputorum sequence was excluded in the comparison to obviate the influence of the intervening sequence present in the 16S rRNA gene of this species (see text and Van Camp *et al.*, 1993). Data were corrected for multiple base changes by the method of Jukes and Cantor (1969) and relationships depicted in a dendrogram by use of the neighbour-joining method. The tree is rooted to *L. intracellularis*.

not thought to be a cause of disease in these animals. Some strains may be pig pathogens (see below). *C. coli* may also be found in cattle, poultry, ostriches and dogs. It is occasionally associated with hepatitis in birds (Stephens *et al.*, 1998). In humans, infection causes diarrhoea and *C. coli* is often regarded as the second most common *Campylobacter* species reported in human gastrointestinal illness: in some regions, the species accounts for up to 50% of human campylobacteriosis (Skirrow, 1994). The closest genetic relative to *C. coli* is *C. jejuni*, with which DNA–DNA hybridization values of 21–63% have been described (Harvey and Greenwood, 1983; Morris *et al.*, 1985; Vandamme *et al.*, 1997). These two species are phenotypically homogeneous and differentiation between the two is problematic. The most common test for this purpose is hippurate hydrolysis, in which *C. coli* gives a negative result. Most PCR tests described for *C. coli* identification are effective (On and Jordan, 2003) and there is some evidence to suggest that *C. coli* is a less diverse and more clonal species than *C. jejuni* (On and Harrington, 2000; Duim *et al.*, 2001).

'*Campylobacter hyoilei*' (Alderton *et al.*, 1995) was isolated from pigs afflicted with porcine proliferative enteritis (PPE). Its close phenotypic and genotypic resemblance to both *C. coli* and *C. jejuni* was noted, but some phenotypic test differences, divergence in the 16S rRNA gene and relatively low binding ratios in dot-blot genomic DNA hybridization assays to type strains of the other species suggested '*C. hyoilei*' to be novel. A later study using a range of established, standardized methods and some novel assays found the phenotypic differences were not reliable and all molecular methods (including a classical DNA–DNA hybridization technique) classified '*C. hyoilei*' as *C. coli* (Vandamme *et al.*, 1997). Later studies based on PCR and amplified fragment length polymorphism (AFLP) analyses also identified '*C. hyoilei*' strains as *C. coli* (On and Harrington, 2000; Duim *et al.*, 2001; On and Jordan, 2003). A few quantitative differences in enzyme activities, and a genetic marker detectable by PCR has been noted in so-called '*C. hyoilei*' isolates (Dep *et al.*, 2001), but such results do not alter the taxonomic position of these strains, given the existing criteria used for species delineation (Wayne *et al.*, 1987). However, certain *C. coli* strains associated with the PPE syndrome may represent a pathogenic variety (pathovar) of the species and could be referred to as '*C. coli* pv. *hyoilei*'. Further work is needed to clarify this issue.

Campylobacter concisus was first isolated from the oral cavity of humans affected with periodontal disease (Tanner *et al.*, 1981). No animal reservoir has yet been identified. The role of *C. concisus* in periodontal disorders is as yet unclear, since the organism can be found in both healthy (Macuch and Tanner, 2000) and diseased (Kamma *et al.*, 1994; 2000) sites, and most frequently in association with other oral bacteria (Socransky *et al.*, 1998). Similarly, the potential of *C. concisus* as a gastrointestinal pathogen is uncertain. Although these bacteria can be isolated from cases of diarrhoea in humans in frequencies equivalent to those of *C. jejuni* (Lastovica and Skirrow 2000; Engberg *et al.*, 2000; Aabenhus *et al.*, 2002), they may also be isolated from healthy control patients at a similar rate (van Etterick *et al.*, 1996). Cytotoxic activity has not been observed in *C. concisus* but a cytotonic effect on CHO cells has been seen (Musmanno *et al.*, 1998). Although the pathogenicity of this species is controversial, its genetic diversity is without question and has been well established by use of PCR- and PFGE-based typing methods (van Etterick *et al.*, 1996; On and Harrington, 2000; Matsheka *et al.*, 2002). Crucially, some diarrhoeal strains exhibit ca. 50% DNA–DNA homology to the type strain (of oral origin), clearly indicating them to represent a genetically distinct species (Vandamme *et al.*, 1989). The absence of distinguishing characteristics at the biochemical or even protein electrophoretypic level for these genomospecies prevents their formal nomenclature or easy identification, making it extremely difficult to determine whether certain genomospecies of *C. concisus* act as primary pathogens in periodontal or gastrointestinal disease. At present, the species may be considered a taxonomic complex, comprising at least two genomospecies and probably more (Matsheka *et al.*, 2002). The wider application of rapid genetic profiling methods may help resolve some of these issues (On and Harrington, 2000).

On

Campylobacter curvus was first described as '*Wolinella curva*' (Tanner *et al.*, 1981) and included anaerobic, hydrogen-dependent strains from the oral cavity of humans that were once referred to as '*Vibrio succinogenes*'-like (Smibert and Holdeman, 1976). By use of comprehensive phenotyping and DNA–DNA hybridization studies, Tanner *et al.* (1981) proposed the genus *Wolinella* to encompass anaerobic, curved rod-like strains from the bovine rumen (*W. succinogenes*) and phenotypically similar, but genetically distinct oral isolates ('*W. curva*'). The latter was assigned to *Campylobacter* when its phylogenetic position had been determined: *W. succinogenes* is more closely related to *Helicobacter* (Vandamme *et al.*, 1991a; On, 2001).

C. curvus is seldom reported in bacteriological studies of periodontal disease (e.g. Kamma *et al.*, 1994, 2000; Socransky *et al.*, 1998; Macuch and Tanner, 2000) and does not invade human gingival epithelial cells (Han *et al.*, 2000). It is not associated with pathogenesis of periodontitis (Macuch and Tanner, 2000). The species, or strains closely resembling it, is occasionally isolated from human diarrhoea (Koga *et al.*, 1999; Engberg *et al.*, 2000), but its clinical significance is unknown. Certain strains demonstrate an intervening sequence (IVS) in their 16S rRNA gene (Etoh *et al.*, 1998) of 140–202 bp. This could have implications for their identification by PCR-based methods (On, 2001 and below).

Campylobacter fetus comprises two subspecies. *Campylobacter fetus* subsp. *fetus* may be found in the intestine of cattle and sheep. It is a recognized cause of sporadic abortion in both these animals (Skirrow, 1994). In humans, *C. fetus* subsp. *fetus* may cause diarrhoea, abortion, bacteraemia, endocarditis, and meningitis (Farrugia *et al.*, 1994; Skirrow, 1994; On, 1996a). In contrast, *C. fetus* subsp. *venerealis* appears considerably more restricted in ecological terms. It does not survive the bovine intestinal environment (Bryner *et al.*, 1964), and is only occasionally isolated from humans in cases of bacteraemia (Salama *et al.*, 1992a; On and Harrington, 2001). However, *C. fetus* subsp. *venerealis* seems to be highly adapted for the genital tract of cattle and sheep, where it causes a form of infectious infertility (Skirrow, 1994). This disease can have a substantial economic impact on conventional breeding programmes (Hum, 1996) and may cause considerable losses if contaminated semen is used in artificial insemination-based breeding.

Despite the divergent ecological and pathological behaviour of the two subspecies, they are essentially inseparable in DNA–DNA hybridization experiments, protein profile and certain AFLP analyses and 16S rRNA gene sequences (Harvey and Greenwood, 1983; Vandamme *et al.*, 1990; On and Harrington, 2000, 2001). Phenotypically, the subspecies are also highly similar but may be distinguished by a few biochemical tests (On *et al.*, 1996; On and Harrington 2001). A differential tolerance to 1.0% glycine is most often used to distinguish the two subspecies, with *C. fetus* subsp. *fetus* considered to be capable of growth on media containing the agent. Glycine-resistant strains of *C. fetus* subsp. *venerealis* have been described and are referred to as *C. fetus* subsp. *venerealis* bv. intermedius (Véron and Chatelain, 1973), although some evidence suggests that these strain characteristics are artefacts due to inappropriate standardization of biochemical tests (On and Harrington, 2001). It has however been noted that strains classified as *C. fetus* subsp. *fetus*, *C. fetus* subsp. *venerealis* and *C. fetus* subsp. *venerealis* bv. intermedius differ in their genome size, with values averaging 1.1 Mb, 1.4 Mb and 1.5 Mb respectively for each taxon (Salama *et al.*, 1992a). This trait may only be a general rule, since a well-characterized laboratory strain (23D) of *C. fetus* subsp. *fetus* has an estimated genome size of 1.4 Mb (S.L.W. On, unpublished results). The subtle genetic differences between the two subspecies have

20

been demonstrated in a multiplex PCR assay (Hum *et al.*, 1997) and in an AFLP profile comparison (Duim *et al.*, 2001), but differentiation of certain strains remains equivocal (On and Harrington, 2001; Wagenaar *et al.*, 2001). A full understanding of the genetic basis for the ecological, pathological and thus taxonomic differences between the two subspecies can probably only be made by use of whole-genome sequence comparisons.

Campylobacter gracilis was originally described as a *Bacteroides* species associated with periodontal disease (Tanner *et al.*, 1981). Its initial classification as '*B. gracilis*' is understandable when consideration of its anaerobic growth requirements, straight rod-shaped cell body and absence of flagella are considered, but phylogenetic analysis clearly assigns it to *Campylobacter* (Vandamme *et al.*, 1991a; Figure 2.1). A closer inspection of the fatty acid- and menaquinone composition of the organism in addition to consideration of extant data led Vandamme and colleagues (1995) to formally transfer it to *Campylobacter* as *C. gracilis*. The unusual morphological and metabolic characteristics of this species are no longer unique to the genus (see *C. hominis* below). Like *C. concisus*, the role of *C. gracilis* in periodontal disease is unclear (Kamma *et al.*, 2000; Macuch and Tanner, 2000). The organism has also been reported from various types of deep tissue and pleuropnomony infections (Johnson *et al.*, 1985; Lee *et al.*, 1993).

Campylobacter helveticus was described by Stanley *et al.* (1992) and represented a group of strains that had been isolated from faecal samples of healthy and diarrhoetic cats and dogs (Burnens and Nicolet, 1992). The species closely resembles *C. upsaliensis*, also a common *Campylobacter* in the enteric tract of domestic pets; in semi-quantitative DNA–DNA hybridization experiments, these two species demonstrate approximately 35% DNA homology (Stanley *et al.*, 1992) and share more than 97% homology in their 16S rRNA gene sequences (Figure 2.1). Numerical analysis of whole-cell protein profiles (Stanley *et al.*, 1992) also show *C. helveticus* and *C. upsaliensis* to be distinct, but closely related species.

C. helveticus appears to be far more common in cats than dogs (Stanley *et al.*, 1992) and has not at present been described in association with human disease, although its closest taxonomic neighbour *C. upsaliensis* is established as a human pathogen (Bourke *et al.*, 1998). It cannot be discounted that the species is being underreported. The inabilities of *C. helveticus* to grow on potato starch agar or reduce selenite are the only phenotypic traits known to distinguish it from *C. upsaliensis* (Stanley *et al.*, 1992). Since these tests are infrequently used in routine laboratories, it is possible that *C. helveticus* is frequently misidentified as *C. upsaliensis* and thus underreported in animal and/or human disease. The wider use of genetic methods, in particular species-specific PCR tests (Linton *et al.*, 1996) is suggested to rectify this oversight.

Campylobacter hominis was initially detected in human faecal samples on the basis of its 16S rRNA gene sequence alone (Lawson *et al.*, 1998), since in this study the organism could not be cultured. Consequently, Lawson and colleagues (1998) referred to the organism as *Candidatus Campylobacter* hominis. The term *Candidatus* is a taxonomic category initiated to record the properties of 'incompletely described prokaryotes': bacterial species that appear to be distinct, but cannot be subjected to closer scrutiny since they resist cultivation (Murray and Schliefer, 1994). A close phylogenetic association of *Candidatus* C. hominis with predominantly anaerobic campylobacters was noted and used to formulate an amended isolation protocol that involved screening of *C. hominis*-PCR positive stool samples for growth under anaerobic growth conditions. The modified approach was successful and a subsequent polyphasic study determined *C. hominis* to represent a distinct species (Lawson *et al.,* 2001). A commensal role for *C. hominis* in the human gut has been proposed since it is found in both healthy and diarrhoetic stools (Lawson *et al.,* 1998, 2001).

Campylobacter hyointestinalis is divided into two subspecies. Strains referred to as *C. hyointestinalis* subsp. *hyointestinalis* are predominantly of enteric origin, typified by strains originally isolated from the pig intestine (Gebhart *et al.,* 1983). *C. hyointestinalis* subsp. *hyointestinalis* has since been found in healthy (Atabay and Corry, 1998) and diarrhoetic (Diker *et al.,* 1990) cattle, hamsters (Gebhart *et al.,* 1985), monkeys (Misawa *et al.,* 2000), healthy (Hänninen *et al.,* 2002) and diarrhoetic (Hill *et al.,* 1987) reindeer, and elephants (Misawa *et al.,* 2000). In humans, it has been associated with sporadic- (Edmonds *et al.,* 1987; Lawson *et al.,* 1999) and outbreak (Salama *et al.,* 1992b) cases of diarrhoea.

C. hyointestinalis subsp. *lawsonii* was first proposed to encompass a group of distinct *C. hyointestinalis* isolates from the pig stomach (On *et al.,* 1995). A few faecal isolates exist (On and Vandamme, 1997). Unlike similar infections with gastric *Helicobacter* species, no clear association of gastric ulceration with *C. hyointestinalis* subsp. *lawsonii* has yet been made and the role of the organism in the pig stomach remains unknown. Since no urease activity is exhibited, even the means by which *C. hyointestinalis* subsp. *lawsonii* survives in the porcine gastric tract is unclear.

In contrast to the taxonomic structure of the *C. fetus* subspecies, there is a clear distinction between the two *C. hyointestinalis* subspecies. Numerical analysis of phenotypic tests (On *et al.,* 1995; On and Vandamme, 1997), whole-cell protein profiles (On and Vandamme, 1997), PFGE (On and Vandamme, 1997; On, 2001)- and AFLP profiles (On and Harrington 2000; Duim *et al.,* 2001) all show the subspecies represent distinct yet closely related taxa. This distinction is also strikingly observed in comparisons of 16S rRNA gene sequences (Harrington and On, 1999), where the species as a whole demonstrates substantial diversity (over 6% divergence), more than double the value generally expected for strains of a single bacterial species (Stackebrandt and Goebel, 1994).

Campylobacter jejuni comprises two subspecies that differ substantially in their ubiquity and to some extent ecology. *C. jejuni* subsp. *jejuni* is often simply referred to as *C. jejuni* and represents the taxon first described by Jones *et al.* (1931) as 'Vibrio jejuni' from bovine intestinal contents. Since the 1970s, this organism has been widely recognized as the most commonly isolated bacterial cause of human gastroenteritis the world over (Skirrow, 1994). A wide range of other sequelae including septicaemia and neuropathic disorders may also occur (Skirrow and Blaser, 2000). *C. jejuni* subsp. *jejuni* usually occurs as a commensal in a wide range of animal hosts, including chickens, cattle, pigs, sheep, dogs, and ostriches (Skirrow, 1994); its predominance in food animals make it one of the most important food-borne pathogens today. It is a genetically diverse species with a range of phenomena shaping its genomic plasticity (Parkhill *et al.,* 2000; Wassenaar *et al.,* 2000; Dingle *et al.,* 2001). Whole-genome sequences for two strains are available (Parkhill *et al.,* 2000; http://www.tigr.org/tdb/mdb/mdbinprogress.html; http://campy.bham.ac.uk/).

C. jejuni subsp. *doylei* (Steele and Owen, 1988) differs considerably from *C. jejuni* subsp. *jejuni* in its distribution, ecology and characteristics. To date, no animal host has been found. It has been found in cases of enteritis, gastritis and septicaemia (Lastovica and Skirrow, 2000). Moreover, it fails to reduce nitrate or grow at 42°C, properties that were instrumental in its initial recognition as a distinct taxon (Steele *et al.,* 1985). Whole-cell protein (Owen *et al.,* 1988) and AFLP (On and Harrington 2000) profile analyses suggest *C.*

jejuni subsp. *doylei* to be more homogeneous than *C. jejuni* subsp. *jejuni*. These analyses, in addition to DNA–DNA hybridization data (Steele and Owen, 1988) demonstrate the close but distinct relationship between these taxa. The genetic distinctiveness of *C. jejuni* subsp. *doylei* may help explain the relatively poor performance of many PCR tests intended to identify *C. jejuni* at the species level, since most assays were designed with only *C. jejuni* subsp. *jejuni* in mind (On and Jordan, 2003).

C. *jejuni* shares considerable DNA–DNA relatedness with *C. coli* (see above), a species with a similar host range. Differentiation between the latter and *C. jejuni* subsp. *jejuni* is problematic, with atypical hippurate-negative strains of *C. jejuni* essentially indistinguishable from those of *C. coli* in routine laboratories (Morris *et al.,* 1985). Additional tests such as growth on a minimal medium and alpha-haemolytic activity may be of some use but require stringent standardization and, like hippurate hydrolysis, do not provide unequivocal discrimination (On *et al.,* 1996). The 16S rRNA gene sequences of these taxa are highly similar (Alderton *et al.,* 1995, Vandamme *et al.,* 1997) or identical (Harrington *et al.,* 1999). The genetic diversity of *C. jejuni* was used to help explain the performance of PCR tests developed to identify this species, since no assay proved entirely sensitive or specific (On and Jordan, 2003). Determination and comparison of a genome sequence of *C. coli* with those of *C. jejuni* could lead to the development of more effective rapid methods for differentiating these two important species.

Campylobacter lanienae was described by Logan and colleagues (2000) to define isolates from asymptomatic abbatoir workers. Strains have subsequently been found in pigs and cattle (Inglis and Kalischuk, 2003; Sasaki *et al.,* 2003). The description is somewhat controversial: although various taxonomic methods showed *C. lanienae* to be distinct from the other taxa studied, no strains of *C. hyointestinalis* subsp. *lawsonii* (also of porcine origin) were examined (Logan *et al.,* 2000). The results presented indicated that these taxa could not be distinguished phenotypically (cf. On *et al.,* 1995 and Logan *et al.,* 2000; see Table 2.1). 16S rRNA gene sequence comparisons clustered *C. lanienae* in between each of the *C. hyointestinalis* subspecies (On, 2001) and a PCR assay described as *C. lanienae*-specific (Logan *et al.,* 2000) produced amplicons with *C. hyointestinalis* subsp. *lawsonii* strains (Inglis and Kalischuk, 2003; S.L.W. On, unpublished results). However, both AFLP (On and Harrington, 2000; C. Harrington, unpublished data) and protein profiling (P. Vandamme, personal communication) affirm that *C. lanienae* is a genetically distinct species, albeit one closely related to *C. hyointestinalis*. This resemblance even extends to the distribution of intervening sequences in the 16S rRNA gene, found in some strains of both species (Harrington and On, 1999; Sasaki *et al.,* 2003). It must be noted that the absence of useful differential tests to discriminate *C. lanienae* from other *Campylobacter* spp. is contrary to normal recommended procedures to describe new species (Wayne *et al.,* 1987; Ursing *et al.,* 1994).

Campylobacter lari was first isolated from gulls and distinguished from the morphologically similar species *C. coli* and *C. jejuni* by virtue of its resistance to nalidixic acid (Skirrow and Benjamin, 1980), then a rare trait among campylobacters. An integrated study of phenotypic and genotypic characteristics of the organism led to the proposal of a new species, '*C. laridis*' (Benjamin *et al.,* 1983), a name later corrected to *C. lari* to conform to accepted rules of nomenclature (von Graevenitz, 1990). The species is ecologically, phenotypically and genetically diverse. Strains have been isolated from the intestinal tract of wild birds, poultry, cattle, shellfish and untreated water (Skirrow 1994; Aarestrup *et al.,* 1997; Lastovica and Skirrow; 2000). Phenotypically, three broadly distinguishable groups have been described; the nalidixic acid-resistant 'classical' biotype, a urease-positive thermophilic campylobacter (UPTC) biotype and a nalidixic-acid sensitive variant. Numerical analysis of protein profiles from 37 isolates from Dutch shellfish delineated several groups and RAPD-based DNA fingerprinting indicated substantial genetic diversity among these strains (Endtz *et al.*, 1997). Recent studies using AFLP profile analyses have shown that 'classical' and UPTC *C. lari* strains are closely related but distinct (On and Harrington, 2000; Duim *et al.*, 2001). The degree of relatedness between these groups in these analyses are concordant with those observed between *C. hyointestinalis* and *C. jejuni* subspecies in AFLP analyses (On and Harrington, 2000; Duim *et al.*, 2001). A formal proposal to describe a subspecies structure for *C. lari* is expected, broadly following the suggestions made earlier (On *et al.*, 2001).

Campylobacter mucosalis was first described as a subspecies of *C. sputorum*, due to their close resemblance in classical biochemical tests (Lawson *et al.*, 1975, 1981). Its classification as '*C. sputorum* subsp. *mucosalis*' was revised when a subsequent analysis of the DNA–DNA relatedness between these taxa revealed them to be poorly (3–5%) related (Roop *et al.*, 1985a). Consequently, '*C. sputorum* subsp. *mucosalis*' was elevated to species status (Roop *et al.*, 1985b).

C. mucosalis is closely related to *C. concisus* at the genetic level, with these species sharing 9–11% DNA–DNA relatedness in hybridization experiments (Roop *et al.*, 1985a) and clustering together in AFLP profile analyses (On and Harrington, 2000). Conversely, these species invariably fall into different clades in phylogenetic trees based upon 16S rRNA gene sequence comparisons (On, 2001; Figure 2.1), showing that divergence at the single gene level does not always concur with divergence at the whole genome level. However, the inferred phylogenetic association of *C. mucosalis* with *C. hyointestinalis* in such comparisons is interesting given that both species are frequent colonizers of the porcine intestine. *C. mucosalis* was originally suspected to be responsible for porcine proliferative enteritis (PPE), a serious disease in pigs, but the principal causal agent in PPE is now believed to be the unrelated organism *Lawsonia intracellularis* (Lawson and Gebhart, 2000). No other animal host has yet been confirmed: human isolates originally believed to have been *C. mucosalis* (Figura *et al.*, 1993) were later identified as *C. concisus* (On, 1994).

Campylobacter rectus was originally described as '*Wolinella recta*' (Tanner *et al.*, 1981) prior to the extensive taxonomic revision of the *Campylobacter* phylogenetic group (Vandamme *et al.*, 1991a). As with the closely related species *C. curvus* and *C. sputorum*, the 16S rRNA gene of certain *C. rectus* strains contains an intervening sequence (Etoh *et al.*, 1998), which may complicate phylogenetic and/or identification analyses (On, 2001; see below). *C. rectus* is generally considered to be a periodontal pathogen in humans (Ebersole *et al.*, 1995; Macuch and Tanner, 2000) and shows a unique cell surface layer (S-layer) structure comprising a distinctive array of hexagonal shaped, closely packed, macromolecular subunits (Dokland *et al.*, 1990). Although the sequence of the gene coding for the *C. rectus* S-layer shows some homology to the S-layer locus of *C. fetus*, only the *C. rectus* locus possesses a domain exhibiting some homology to Gram-negative bacterial RTX toxins (Miyamoto *et al.*, 1998). The evolutionary origin of this feature is difficult to fathom, given the inferred phylogenetic distance between *C. rectus* and *C. fetus* and their host adaptations. It is interesting to note that some *C. rectus*-like strains in domestic pets have demonstrated the same cell surface ultrastructure as *C. rectus* (Love *et al.*, 1984), perhaps representing a 'missing link' between the two species.

Campylobacter showae was first considered a subspecies of '*Wolinella curva*', following the discovery that two isolates from the human gingival crevice shared 53–60% DNA–DNA relatedness with the '*W. curva*' type strain and were phenotypically similar, but distinct from this species (Etoh *et al.*, 1988). A subsequent study involved more strains of the so-called '*W. curva* subsp. *intermedius*' and the use of 16S rRNA gene sequence comparisons to make a more thorough study of its taxonomic position (Etoh *et al.*, 1993). As a consequence, the organism was elevated to species status as *C. showae*. There is some evidence to suggest that *C. showae* is a primary pathogen in periodontal disease, since it has been found more frequently, and in higher levels, in diseased sites compared with healthy areas (Macuch and Tanner, 2000). However, many oral *Campylobacter* spp. occur in association with each other (Socransky *et al.*, 1998) and the complex nature of periodontal disease does require this hypothesis to be further investigated.

Campylobacter sputorum has a long and chequered taxonomic history. The morphological and biochemical similarities between catalase-negative human oral ('*V. sputorum*', Prévot, 1940) and bovine genital ('*V. bubulus*', Florent, 1953) strains had been noted by Véron and Chatelain (1973) who proposed these taxa as subspecies, '*C. sputorum* subsp. *sputorum*' and '*C. sputorum* subsp. *bubulus*' respectively. In a separate study, Firehammer (1965) had described '*V. faecalis*' from ovine faeces, noted to be biochemically similar to '*V. bubulus*' apart from the production of catalase. Although '*V. faecalis*' was first regarded by Véron and Chatelain (1973) as a misclassified *C. coli*, DNA–DNA hybridization experiments demonstrated each of the above taxa shared over 70% DNA homology by this method, thus representing biochemical variants – biovars – of the same species, *C. sputorum* (Roop *et al.*, 1985a). The biovar structure was based upon reactions in two tests (catalase production and tolerance to 3.5% NaCl) and initially thought to denote source-specificity appertaining to the environments from which these taxa were first isolated. This supposition was questioned when a study of over 60 phenotypic characteristics and whole-cell protein profiles failed to differentiate strains from humans, pigs, cattle and sheep (On *et al.*, 1994). In addition, this study showed that the salient test for designating a strain as *C. sputorum* bv. bubulus (growth on 3.5% NaCl medium) was not reproducible. A further study of the species confirmed that tolerance to 3.5% NaCl was an unreliable taxonomic marker for biovar delineation but confirmed that urease-producing campylobacters from cattle and humans were also *C. sputorum* (On *et al.*, 1998a). These data were used to propose the current taxonomic structure of the species, comprising *C. sputorum* bv. sputorum, *C. sputorum* bv. faecalis, and *C. sputorum* bv. paraureolyticus. There is some evidence to suggest that at least two of the biovars may each represent a distinct clonal lineage (On *et al.*, 1999). *C. sputorum* is generally considered a commensal in its hosts, but cases of diarrhoea and abscess formation have been described (reviewed by On *et al.*, 1998a).

Campylobacter upsaliensis is the name given to a group of campylobacters first isolated from canine faeces and designated the catalase negative or weak (CNW) group, in recognition of their most prominent characteristic (Sandstedt *et al.*, 1983). Although catalase-negative *Campylobacter* species were known at this time (e.g. *C. concisus*, *C. mucosalis*, *C. sputorum*), none of these were thermotolerant in the way the CNW-group were. In that respect, the CNW group resembled the more frequently encountered *C. jejuni* subsp. *jejuni*, *C. coli* and *C. lari*, to which a close relationship was seen in comparisons of 16S rRNA gene sequences (Thompson *et al.*, 1988; Figure 2.1). DNA base composition, cell morphology and isoprenoid quinone content were all concordant with the CNW group representing a novel *Campylobacter* spp. (Sandstedt *et al.*, 1983; Moss *et al.*, 1990) and the name *C. upsaliensis* was proposed for this species by Sandstedt and Ursing (1991). Evidence has accumulated for a pathogenic role for *C. upsaliensis* and this species has been isolated from sporadic and outbreak cases of gastroenteritis in humans as well as episodes of septicaemia (reviewed by Bourke *et al.*, 1998). Its natural host appears to be domestic pet dogs and cats, where it may on occasion cause diarrhoea. The closest taxonomic neighbour to *C. upsaliensis* is *C. helveticus*: problems with distinguishing these species are discussed above.

The species ***Bacteroides ureolyticus*** is worthy of discussion. This name was proposed to refer to rod-shaped bacteria that formed 'corroding' colonies on solid agar media, and that were readily distinguished from *Eikenella corrodens* in producing urease and by their obligate anaerobic growth characteristics: both species had originally been considered a single taxon, '*Bacteroides corrodens*' (Jackson and Goodman, 1978). Assignation to the genus *Bacteroides* was principally made on the basis of the cell morphology and anaerobic nature of the organism, but a later study showed that *B. ureolyticus* bore little resemblance to other *Bacteroides* spp. in protein profile- and DNA–DNA hybridization studies (Taylor *et al.*, 1986). Subsequent phylogenetic analyses using the 16S rRNA gene clearly demonstrated *B. ureolyticus* was more closely related to *Campylobacter* than *Bacteroides* spp. (Paster and Dewhirst, 1988; Vandamme *et al.*, 1991, 1995). Nonetheless, reclassification of *B. ureolyticus* as a *Campylobacter* species is not straightforward. The fatty acid content, disease associations, proteolytic metabolism and urease production of *B. ureolyticus* are atypical of *Campylobacter* spp. (Vandamme *et al.*, 1995). However, numerical analysis of 67 phenotypic characters cluster *B. ureolyticus* with other anaerobic *Campylobacter* spp. (On and Holmes, 1995), it can be found in the intestine (Kamiya *et al.*, 1993), a common habitat of campylobacters, and both *C. lari* (UPTC group) and *C. sputorum* bv. paraureolyticus also produce urease. Nevertheless, since the disease associations of *B. ureolyticus* include both deep wound infections and non-gonococcal urethritis, with some strains from each source potentially representing different taxa (Taylor *et al.*, 1987), it is prudent to follow the recommendations of Vandamme and colleagues (1995) and wait until further data are gathered, before reclassifying *B. ureolyticus* as either a *Campylobacter* spp. or into a new genus.

UNCLASSIFIED CAMPYLOBACTERS

Apart from those bacteria that have now been reclassified into the genera *Arcobacter, Lawsonia, Helicobacter, Sulfurospirillum,* and *Sutterella* (see above), several *Campylobacter*-like organisms have been described that may represent distinct, novel species. There are three notable examples of potentially novel species, however, it must be emphasized that their taxonomic position requires clarification through further study. First, isolates from human diarrhoea have been described that demonstrate between 30% and 60% DNA–DNA homology to *C. coli* and *C. jejuni* in hybridization experiments but that cannot be distinguished from the former species phenotypically (Morris *et al.*, 1985). According to current criteria for species delineation (Wayne *et al.*, 1987), such strains should be considered as distinct genetic species. Second, by protein profile analysis, some isolates from cases of periodontal disease appear distinct from other oral *Campylobacter* spp. and have been referred to as '*Campylobacter* X' (Macuch and Tanner, 2000). Third, a total of 64 anaerobic, agar-corroding strains from cats and dogs that resembled *B. ureolyticus, C. curvus, C. gracilis, C. rectus* and *W. succinogenes* phenotypically and in their DNA base composition, but distinct by protein profile, were described by Love *et al.* (1984). A total of five groups were delineated on the basis of protein profile and cell wall ultrastructure.

IDENTIFICATION OF *CAMPYLOBACTER* SPECIES

In almost every area of microbiology – clinical studies, epidemiology, population genetics, comparative genomics and beyond – the identification of a strain to species, and sometimes subspecies, or even variant-level is required. Without accurate identification methods, we cannot estimate the prevalence and significance of different species in a given environment, nor rely upon estimates of population genetic structure of a species if the study set includes misclassified isolates. Clinical management is often facilitated if the identity of a strain is known. The value of identification in these, and many other examples, is self-evident, and its practice forms the microbiologist's most frequent encounter with taxonomy.

Identification involves matching data sets derived from the unknown isolate to those of defined taxa. An isolate is identified when phenotypic (e.g. biochemical tests, fatty acid- or protein profiles) or genotypic (e.g. DNA fingerprints, sequences, PCR primers) data matches that determined for a defined taxon to an acceptable level. Accurate identification is dependent upon (i) a sound classification and (ii) effective methods. Where strains have been misclassified and data made publicly available, then there is potential for further misidentifications by others using the data. In that respect, the emphasis is on the practising taxonomist to perform high quality, polyphasic studies when describing new species to avoid forming a source for future error (discussed in On, 2001).

The efficacy of an identification method is also reliant upon the stability of the taxon-specific marker (s). Excessive variation in such a marker may make identification of atypical strains difficult, particularly in the differentiation of species that are highly related (e.g. *C. jejuni* and *C. coli*). In such circumstances it is important to carefully screen a newly developed identification tool with a representative set of isolates that properly circumscribe the diversity of the target species. Inappropriate assessment of marker diversity invariably leads to false-positive or -negative results, as discussed previously (On, 1996a, 2001; On and Jordan, 2003).

The clinical and economic importance of *Campylobacter* spp. coupled to their taxonomic complexity has led to a wide range of phenotypic and genetic methods being developed to identify them. The most commonly used methods, and recent developments, are briefly reviewed below. The interested reader is referred elsewhere for more detailed discussions (On, 1996a, 2001).

Phenotypic tests remain a commonly used approach to identify campylobacters. Tables or dichotomous-key type schema are frequently used in routine laboratories, but the use of tables rapidly becomes complex when manually comparing test results described for 16 taxa. Often, specific tests are given more credence than others when determining whether an unknown belongs to one species or another. To that extent, the latter approach resembles dichotomous trees, where reactions in a single test lead the user to different branches in the scheme, ultimately resulting in a named taxon (Priest and Austin, 1993). For campylobacters, and indeed most taxonomically complex groups, manual assessment of biochemical test data is substantially flawed, particularly where particular test results are 'weighted' to favour particular species. Atypical strains are extensively described in the literature and results from such strains will invariably lead to misidentifications. The problem is particularly acute in *Campylobacter* since a number of phylogenetically related species such as *Arcobacter* and *Helicobacter* spp. will often not be included in such phenotypic schema but can give results that closely resemble, or even be identical with, those of *Campylobacter* species. Of special note are the enteric species *H. canadensis* and *H. pullorum* that, in limited biochemical schema, cannot be distinguished from *C. coli* or *C. lari* respectively (Steinbrueckner *et al.,* 1997; S.L.W. On, unpublished observations).

A more effective approach to the use of phenotypic tests involves computer-aided assessment of the data. Tables in which the percentage of strains giving a positive result can be used as probability matrices, with test results from an unknown strain numerically compared with those taxa in the table and the 'goodness of fit' of the results for each given. The resultant identification score should reach a predetermined threshold limit before the unknown isolate is considered identified. This approach has been shown to be very effective for many species (On *et al.,* 1996, 1998b) but is most efficacious when a large number of tests are used (On, 1996b). For particular groups of strains, 'special purpose' matrices may be of value (On, 1996b; On and Harrington, 2001). The approach, however, is not foolproof, and identification of highly atypical strains remains problematic (On *et al.,* 1998b; On and Harrington, 2001). There is also a risk of misidentification if the unknown isolate belongs to a taxon not present in the database, particularly if a species closely resembling that taxon is already present.

Whether or not manual or computerized methods of data evaluation are used, it is important that the tests used are of a good quality and standardized for optimal results. Different methods for performance of phenotypic tests may result in a different result and that has obvious implications for identification schema. Recommendations for optimal performance of most tests used to identify campylobacters are available (On and Holmes 1991a,b, 1992).

Cellular fatty acid (CFA) profiling, whole-cell protein profiling and mass spectrometry (MS) methods all examine for differences in particular moieties present in (first two methods), or derived from (MS-based methods), the entire bacterial cell. The phenotypic data so derived are of a complex nature, and data analysis is invariably undertaken by computerized methods. Results are commonly given in dendrogramatic form, where the position of the unknown strain relative to known taxa is shown and identifications made on the basis of a cluster analysis. Certain commercial systems for CFA (Sherlock Microbial Identification System) and matrix-assisted laser desorption ionization time-of-flight (MALDI-TOF) MS (Micromass MicrobeLynx) analyses use software that simply give the name of the species to which the profile of the unknown isolate most closely resembles, similar to results from phenotypic probability matrices.

The taxonomic value of CFA- and, particularly, protein profiling is well established (Priest and Austin, 1993). The fact that the number of data extracted from such whole-cell analyses tends to greatly exceed that from conventional biochemical test analysis should, in principle, result in a more stable and secure identification result. Nonetheless, such approaches are not free from difficulties. One problem common to all methods lies in the use of dendrograms as a basis for assigning an unknown isolate to a defined taxon. In general, typical strains are readily identified, but a strain with atypical characteristics will often be placed as a single outlier to defined clusters, making it difficult to determine if it represents a novel species or an atypical strain of an existing one. Where systems employ a more probabilistic approach for data analysis, similar problems to those discussed above for biochemical test matrices apply. Thus the quality of the database to which strains are compared plays a vital role in the security of identifications and ideally, several strains for each taxon that adequately represent its diversity should be present.

Advantages and problems specific to each of the whole-cell methods listed above are evident also. CFA profiles of *Campylobacter* spp. and *Bacteroides ureolyticus* allow good differentiation from the related *Arcobacter* and *Helicobacter* genera, a useful trait that has been exploited to distinguish the phenotypically similar species *C. lari* and *H. pullorum* (Steinbrueckner *et al.*, 1997, 1998). However, a number of species are not adequately discriminated, with several studies showing that species such as *C. coli*, *C. jejuni* and *C. lari*; *C. concisus* and *C. hyointestinalis;* and *C. fetus* and *C. hyointestinalis* cannot always be differentiated (On, 1996a; Rosseel *et al.*, 1998). Numerical analysis of whole-cell protein profiles has been shown to have an excellent correlation with results from DNA–DNA hybridizations (Vandamme *et al.*, 1990) and its efficacy for discriminating most *Campylobacter* species extensively documented (e.g. Vandamme *et al.*, 1990, 1991b, 1992a). Nevertheless, the approach shows problems in differentiating *C. coli* and *C. jejuni* (Vandamme *et al.*, 1992a) and best results require stringent standardization of the method as well as expertise in computer analysis of the patterns, particularly since major protein bands can significantly influence the cluster analysis (Costas *et al.*, 1987).

Two types of MS analysis have, to my knowledge, been applied to campylobacters. Pyrolysis mass spectrometry (Py-MS) generates whole-cell spectra by thermal degradation in an inert atmosphere or a vacuum. Pyrolysed products are separated by their mass/charge ratio in a mass spectrometer. The entire process is completed in a few minutes and this rapid response time is one of the most attractive features for clinical microbiology laboratories. In the only study of Py-MS analysis of campylobacters to date, the species *C. gracilis* and *B. ureolyticus* were differentiated, but many strains from superficial skin ulcers and the oral cavity formed groups distinct from either of these species (Duerden *et al.*, 1989). Reference strains of species such as *C. showae*, *C. sputorum*, *C. rectus*, *C. curvus* and *C. concisus* were not included in the analysis, which is unfortunate since these species are of oral origin and could have shed some light as to the identity of the dental strains. Moreover, a number of the unclassified ulcer-associated isolates closely resembled *B. ureolyticus* in phenotypic tests (Duerden *et al.*, 1989), probably reflecting the diversity of this species as described previously (Taylor *et al.*, 1987) and potentially illustrative of problems with identification of complex taxa, as discussed above. In addition, exceptional standardization is required to produce comparable results generated over a period of time, although methods of data handling such as artificial neural networks have been shown to improve the efficacy of matching data sets for certain bacteria (Chun *et al.*, 1997; Wilkes *et al.*, 2002).

MALDI-TOF MS may also be applied to the analysis of intact bacterial cells, but differs from Py-MS in that the spectral fingerprint is derived from the cell surface moieties only. The method involves laser bombardment of bacterial cells in a matrix that assists the ionization process and subsequent separation of gas-phase ions by virtue of their mass/charge ratio. As with Py-MS, the process is rapid and, in principle, MALDI-TOF analysis holds great potential for bacterial identification (Bright *et al.*, 2002a). In practice, rigid standardization is required for optimal results (Bright *et al.*, 2002b), results that are arguably indicative of the same problems that are evident with Py-MS. Thus far, studies have shown that *C. jejuni*, *C. coli* and *C. fetus* (and the related helicobacters *H. pylori* and *H. mustelae*) display species-specific markers when analysed by MALDI-TOF MS

(Mandrell *et al.*, 1999; Winkler *et al.*, 1999). Arguably the true test of the efficacy of this method will come when applied to the full taxonomic spectrum of *Campylobacter*.

Polymerase chain reaction (PCR)-based assays involve the amplification of specific DNA segments by annealing complementary primer sequences on either side of the target DNA region and subsequently synthesizing the DNA sequence in between by use of a DNA polymerase. In principle, the amplification process allows for a high level of sensitivity since very low levels of target DNA can be replicated at an exponential rate, and the PCR assay can be designed with almost any level of specificity in mind. Sequences specific to genera, species, subspecies or even strain level can be targeted and amplified by PCR, making it an invaluable research and diagnostic tool for microbiologists.

The design of PCR assays requires that sequence data are available and the most frequently used target for the design of species-specific tests is the 16S rRNA gene: the 23S rRNA gene is also widely used. Species-specific PCR assays based on such rRNA data have been described for many *Campylobacter* species, including *C. coli*, *C. jejuni*, *C. fetus*, *C. hyointestinalis*, *C. lari*, *C. helveticus*, *C. upsaliensis*, *C. concisus*, *C. sputorum*, *C. mucosalis*, *C. hominis* and *C. lanienae* (Eyers *et al.*, 1993; Bastyns *et al.*, 1994, 1995; Linton *et al.*, 1996; Lawson *et al.*, 1998; Logan *et al.*, 2000). Non-rRNA target sequences have also been used. A multiplex PCR test for identification of the two *C. fetus* subspecies has been described (Hum *et al.*, 1997) and appears to be effective (Wagenaar *et al.*, 2001; On and Harrington, 2001). A PCR-detectable marker for the so-called 'C. coli var. hyoilei' taxon is available (Dep *et al.*, 2001). In addition, at least nine non-23S rRNA-derived assays have been described for *C. coli* and/or *C. jejuni*; a total of 11 PCR tests derived from genes including *mapA*, *ceuE*, *hipO* and several random sequences have been compared (On and Jordan, 2003).

It is important to note that the accuracy of a PCR test is as good as the taxonomic resolution of the sequence it is based upon. The fact that rRNA sequence divergence is often insufficient to discriminate between closely related taxa readily explains the ability of a species-specific 16S rRNA gene-based PCR for *C. lanienae* (Logan *et al.*, 2000) to react with isolates of *C. hyointestinalis* subsp. *lawsonii* (Inglis and Kalischuk, 2003; S.L.W. On, unpublished results). Inadequate sequence divergence also explains the failure of PCR tests based on the 23S rRNA gene to effectively differentiate between *C. coli*, *C. jejuni* and occasionally *C. lari* (Linton *et al.*, 1996; On and Jordan, 2003). The 16S-23S rRNA internal spacer region (ISR) differs in length and sequence between *C. coli*, *C. jejuni* and *C. lari*, but the size of the spacer is not consistent in all strains of a single species (Christensen *et al.*, 1999). Sequence similarities of up to 98% between *C. lari* and *C. jejuni* ISRs have been reported (Miyajima *et al.*, 2002) and for these reasons the ISR is unlikely to be a useful target for species-specific PCR tests for these species.

The observation that none of the PCRs for identification of *C. jejuni* currently available are entirely specific or sensitive (On and Jordan, 2003) emphasizes the importance of careful and thorough screening of PCR assays for species known to be genetically diverse and with particularly close taxonomic relatives. Interestingly, with the exception of the aforementioned 23S rRNA-based test, PCR assays for *C. coli* proved both accurate and sensitive, a result attributed by the authors to its relatively clonal nature (On and Jordan, 2003).

In a variation of the taxon-specific PCR approach, broad-range schema involving restriction fragment analysis of PCR-amplified regions (0.5–3.0 kbp in size) of 16S (Cardarelli-Leite *et al.*, 1996) or 23S (Hurtado and Owen, 1997; Fermer and Engvall,

1999) rRNA genes have been described. Unknown strains are identified by comparison of the fragment patterns with those of defined taxa. The schemes currently available recognize *Campylobacter* and *Arcobacter* (Hurtado and Owen, 1997), together with *Helicobacter* (Marshall *et al.*, 1999), and also *Wolinella succinogenes* (Cardarelli-Leite *et al.*, 1996). A more limited scheme aimed at the thermotolerant species *C. coli, C. jejuni, C. lari* and *C. upsaliensis* is also available (Fermer and Engvall, 1999). These schemes work well where there is sufficient divergence in the sequences to allow discrimination. However, in each of the assays described, discrimination between all or some strains of certain closely related taxa is not possible (Cardarelli-Leite *et al.*, 1996; Hurtado and Owen, 1997; Fermer and Engvall, 1999; Marshall *et al.*, 1999; Engvall *et al.*, 2002). Patterns may also be obtained that do not conform to those present in the database (e.g. Engvall *et al.*, 2002) and these could represent strains with atypical rRNA sequences (see below), taxa not included in the schema, or potentially novel species. In principle the approach is a rapid and convenient means for routine laboratories to identify a relatively broad range of campylobacteria, but their use requires careful consideration by laboratories according to their needs. Additionally, each assay requires further validation to fully determine their specificity and sensitivity.

16S rRNA gene sequence comparisons have become an increasingly popular means of identifying a wide range of bacteria, especially species that are encountered rather infrequently in routine laboratories. It is easy to understand why. PCR methods for sequencing the 16S rRNA gene are universally applicable to all bacteria and comparison of the DNA sequence to those in public databases (constantly updated to include sequences from new species) readily attained by use of rapid and simple search algorithms such as BLAST. The approach circumvents many of the disadvantages associated with bacterial identification, particularly the need to have a wide range of specialised reagents for a particular group of organisms. Strains of *C. sputorum, C. upsaliensis, C. fetus* and other spiral bacteria have been successfully identified by this approach (Tee *et al.*, 1998; Hayden *et al.*, 2001; Misawa *et al.*, 2000). Despite the general advantages of PCR-16S rDNA comparisons, several problems are evident for identification of *Campylobacter* species and other bacteria also. Firstly, it is increasingly evident that some genetically distinct species share identical or near-identical 16S rDNA sequences. At or above 97% identity, some groups of species such as *C. jejuni, C. coli* and *C. lari*; *C. upsaliensis* and *C. helveticus*; *C. fetus, C. hyointestinalis* and *C. lanienae*; and *C. showae, C. curvus* and *C. rectus* cannot be confidently distinguished from each other (On, 2001; Figure 2.1). Conversely, substantial infraspecific diversity in the 16S rRNA gene of some species is also known, whereby strains of each of the two *C. hyointestinalis* subspecies share only 95.7% homology (Harrington and On, 1999). Intervening sequences (IVSs) may also occur in the 16S rRNA gene, resulting in an enlarged PCR-16S rDNA product and making sequence analysis more taxing. Thus far, IVSs (ranging in length from 140 to 244 bp) have been found in all *C. sputorum* strains, and some *C. curvus, C. rectus, C. helveticus, C. hyointestinalis* subsp. *lawsonii* and *C. lanienae* strains (Van Camp *et al.*, 1993; Linton *et al.*, 1994; Etoh *et al.*, 1998; Harrington and On 1999; Sasaki *et al.*, 2003). IVSs have also been found in the 23S rRNA gene of *C. fetus, C. upsaliensis, C. sputorum, C. coli* and *C. jejuni* strains (Van Camp *et al.*, 1993; Trust *et al.*, 1994; On and Jordan, 2003). Interoperon differences in both 16S and 23S rRNA genes due to differential IVS insertion are also evident (Linton *et al.*, 1994; On and Harrington, 1999; On, 2001). In other bacteria, there is evidence for horizontal gene transmission of the entire rRNA operon from related species (Yap *et al* 1999) and a nonlinear relationship between 16S rDNA sequence and DNA–DNA hybridization level has been seen in approximately one-third of species belonging to 20 bacterial genera (Keswani and Whitman, 2001). The influence of mislabelled sequences in public databases should also not be discounted (On, 2001). In conclusion, the accuracy of 16S and 23S rRNA gene sequence comparisons as an identification tool for *Campylobacter* and other bacteria is strongly dependent upon individual species diversity and taxonomic resolution between closely related species. Results must therefore be carefully considered alongside such data.

Whole-genome profiling is defined here as the determination of polymorphisms in the genetic content of the whole bacterial cell. The techniques employed include the use of restriction endonucleases, PCR and/or hybridization techniques to whole-cell DNA and are most frequently used for epidemiological typing. However, certain methods have also been used for speciation.

Restriction endonuclease analysis (REA) involves the electrophoretic separation (in agarose gels) of DNA fragments derived from digestion of DNA with high-frequency cutting endonucleases such as *Hin*dIII and *Hae*III. Bruce *et al.* (1988) and Owen *et al.* (1989) indicated that several *Campylobacter* species could be differentiated by numerical analysis of the complex patterns, although relatively few strains were included and not all strains of *C. jejuni* clustered together, indicating possible problems for interpretation of data. REA is unlikely to find wider usage in bacterial taxonomy. Since the patterns are highly complex and the method difficult to standardize, REA has now been superseded by other techniques.

One means by which complex REA patterns can be simplified for analysis involves the hybridization of an oligonucleotide probe to the restriction pattern in order to map the position of a specific gene. When the probe used is the rRNA operon then the technique is usually referred to as ribotyping. Species-specific markers for *C. jejuni, C. coli, C. lari, C. fetus* and *C. upsaliensis* were seen in *Xho*I–*Bgl*II-derived ribotype patterns when a 23-mer oligonucleotide probe based on a conserved region in the 16S rRNA gene was used (Moreau *et al.*, 1989). In general, the laborious nature of the method and relatively low discriminatory power for typing of *Campylobacter* spp. has limited the interest of workers in this area. However, an automated ribotyping system (the Riboprinter) is now available and a numerical analysis of 24 *C. coli* and 26 *C. jejuni* strain 'riboprints' clearly separated these species (de Boer *et al.*, 2000).

Rare-cutting restriction endonucleases have become a powerful means of determining DNA polymorphisms in whole-cell DNA when fragments are separated by pulsed-field gel electrophoresis (PFGE). PFGE typing is widely used in epidemiological studies of *Campylobacter* (Newell *et al.*, 2000) and many other bacteria, and has been shown to discriminate between the closely related taxa *C. fetus* subsp. *fetus*, *C. fetus* subsp. *venerealis*, *C. hyointestinalis* subsp. *hyointestinalis*, *C. hyointestinalis* subsp. *lawsonii* (On and Vandamme, 1997; On, 2001; On and Harrington, 2001), *C. coli* and *C. jejuni* (de Boer *et al.*, 2000). In these studies, profiles were usually compared by numerical analysis. This approach may be useful for reference laboratories where databases of PFGE profiles are established for epidemiological studies, but the taxonomic range of the approach will be limited to that of the available restriction enzymes. The above studies used *Sma*I for PFGE profiling but this enzyme does not cut DNA from *C. concisus* (Matsheka *et al.*, 2002) or *C. helveticus* (Moser *et al.*, 2001). Furthermore, some (but not all) strains of *C. hyointestinalis, C. upsaliensis* and *C. jejuni* resist restriction analysis with *Sma*I and, for *C. hyointestinalis, Sal*I (On and Vandamme, 1997; Bourke *et al.*, 1996; S.L.W. On, unpublished data). In addition, some *C. coli* strains yield PFGE profiles that resemble those of *C. jejuni* in terms of fragment size distribution and are not separated from the latter in cluster analyses (S.L.W. On, unpublished data).

Amplified fragment length polymorphism (AFLP) analysis is an alternative approach to resolving complex DNA polymorphisms. Here, whole-cell DNA is digested with each of two restriction enzymes and a subset of the resulting fragments detected by a PCR technique (Savelkoul *et al.*, 1999). The resultant AFLP profile represents a semi-random sampling of the whole DNA content as a simplification of thousands of restriction fragments to 40–400 bp in the size range of the detection apparatus, commonly an automated sequencer (Savelkoul *et al.*, 1999). The high genotypic resolution of AFLP has resulted in its application to both taxonomic and molecular epidemiological issues, often simultaneously. Two studies have addressed *Campylobacter* using slightly different methods. One study examined *Hind*III and *Hha*I polymorphisms in 69 strains of nine species (Duim *et al.*, 2001), whereas another determined *Bgl*II and *Csp*6I-based polymorphisms in 138 strains representing the known taxonomic spectrum of 16 species and three subspecies (On and Harrington, 2000). Both studies demonstrated that most species examined were readily distinguished when AFLP profiles were compared by numerical analysis. Moreover, the subspecies of both *C. jejuni* and *C. hyointestinalis*, and two distinct biotypes of *C. lari*, were also differentiated. Differentiation of *C. fetus* was attained in one method (Duim *et al.*, 2001), although a later study indicated that differentiation was less clear-cut when a larger number of strains were considered (Wagenaar *et al.*, 2001). Conversely, biovars of *C. sputorum* were differentiated in the analysis presented by On and Harrington (2000), concurring with predictions of clonality among the biovars made by PFGE and biochemical analyses (On *et al.*, 1999). Interestingly, both methods experienced problems with characterization of *C. helveticus* and *C. upsaliensis*, suggesting that further method development may be required to optimize the use of AFLP profiling for classification and identification of all *Campylobacter* species.

FINAL REMARKS

The taxonomy of *Campylobacter* has seen substantial change since its inception in 1963. Given the continued interest in this group of bacteria as human and animal pathogens involved in diseases as diverse as abortion, gastroenteritis, abscesses and periodontitis, there is every indication that this trend will continue and that more species will be described in due course. A number of taxonomic problems and questions remain concerning some extant species, and the application of modern chemotaxonomic methods is likely to help resolve some of these at least. The use of well-documented, established methods will also play an important role in such investigations: despite their faults, their strengths are equally notable. In that respect, the wider practice of polyphasic taxonomy in studies of *Campylobacter* must be encouraged.

The whole genomic sequence of an organism represents the ultimate taxonomic resource. The availability of such data is rapidly expanding, itself the result of technical advances that allow a whole bacterial genome to be sequenced in one day (Ussery, 2001). Additional genome sequences of other *Campylobacter* and related species will provide invaluable data for novel method development, and for a critical reappraisal of taxonomic and phylogenetic relationships within and outside of the group. Moreover, the advent of microarray technology, where oligonucleotide probes representing all genes annotated in a genome sequence are spotted onto glass- or polycarbonate slides, also enable taxonomic analysis relative to the genome in question to be performed (see Dorrell *et al.,* Chapter 5). Hybridization of DNA from different *Shewanella* species to whole-genome microarrays containing ORFs from *S. oneidensis* has been shown to give similar results to that of classical DNA–DNA hybridization data (Murray *et al.,* 2001). It was reported recently that *C. coli* shares around 40% genes with NCTC 11168, the sequenced strain of *C. jejuni,* in comparative microarray experiments (B. Wren, personal communication; Dorrell *et al.,* Chapter 5) and this value correlates well with DNA–DNA hybridization values for these species (On, 2002). Random genomic fragments have also been spotted onto microarrays and used to identify *Pseudomonas* species (Cho and Tiedje, 2001) and whole-chromosomal DNA from various *Streptococcus, Mycobacterium* and *Enterobactereaceae* has also been used in a microarray format (Ezaki *et al.,* 2002) as a tool for speciation. The latter resembles a development of the 'chequerboard' DNA–DNA hybridization array that has been applied to identify oral *Campylobacter* species (Socransky *et al.,* 1998). The wider availability of microarrays and whole genome sequences will probably not find immediate application in routine laboratories but nonetheless illustrate interesting developments in the field.

Major technical advances in microbiological methods have been made in recent years. It is possible in just a few days to sequence entire bacterial genomes, assess the transcription of thousands of genes, and characterize the microbial flora in complex communities such as soil, freshwater or the gastrointestinal tract. These, and other advances have helped to shape our knowledge of biodiversity and evolution of *Campylobacter* and many other bacteria, and further developments will continue to elucidate major issues in their biology. In the light of such technical progress, the importance of well-planned studies involving an appropriately diverse choice of strains becomes even more vital, such that the full benefit of the method can be gained. In that respect, taxonomy will continue to benefit from, and contribute to, major scientific advances well into the next century as it has during the last. In *Campylobacter* research, change remains the only constant.

References

Aabenhus, R., Permin, H., On, S.L.W., and Andersen, L.P. 2002. Prevalence of *Campylobacter concisus* in diarrhoea of immunocompromised patients. Scand. J. Infect. Dis. 34: 248–252.

Aarestrup F.M., Nielsen, E.M., Madsen, M., and Engberg, J. 1997. Antimicrobial susceptibility patterns of thermophilic *Campylobacter* spp. from humans, pigs, cattle, and broilers in Denmark. Antimicrob. Agents. Chemother. 41: 2244–2250.

Alderton, M.R., Korolik, V., Coloe, P.J., Dewhirst, F.E., and Paster, B.J. 1995. *Campylobacter hyoilei* sp. nov., associated with porcine proliferative enteritis. Int. J. Syst. Bacteriol. 45: 61–66.

Atabay, H.I., and Corry, J.E. 1998. The isolation and prevalence of campylobacters from dairy cattle using a variety of methods. J. Appl. Microbiol. 84: 733–740.

Bastyns, K., Chapelle, S., Vandamme, P., Goossens, H., and De Wachter, R. 1994. Species-specific detection of campylobacters important in veterinary medicine by PCR amplification of 23S rDNA areas. System. Appl. Microbiol. 17: 563–568.

Bastyns, K., Chapelle, S., Vandamme, P., Goossens, H., and De Wachter, R. 1995. Specific detection of *Campylobacter concisus* by PCR amplification of 23S rDNA areas. Mol. Cell. Probes 9: 247–250.

Benjamin, J., Leaper, S., Owen, R.J. and Skirrow, M.B. 1983. Description of *Campylobacter laridis*, a new species comprising the Nalidixic Acid Resistant Thermophilic *Campylobacter* (NARTC) group. Curr. Microbiol. 8: 231–238.

Bourke, B., Chan, V.L., and Sherman, P. 1998. *Campylobacter upsaliensis*: waiting in the wings. Clin. Microbiol. Rev. 11: 440–449.

Bourke, B., Sherman, P.M., Woodward, D., Lior, H., and Chan, V.L. 1996. Pulsed-field gel electrophoresis indicates genotypic heterogeneity among *Campylobacter upsaliensis* strains. FEMS Microbiol. Lett. 143: 57–61.

Bright, J.J., Claydon, M.A., Soufian, M., and Gordon, D.B. 2002a. Rapid typing of bacteria using matrix-assisted laser desorption ionization time-of-flight mass spectrometry and pattern recognition software. J. Microbiol. Methods. 48: 127–138.

Bright, J.J., Sutton, H.E., Edwards-Jones, V., Dare, D.J., Keys, C.J., Shah, H., McKenna, T., Wells, G., Lunt, M., and Devine, D.A. 2002b. Rapid identification of anaerobic bacteria using matrix-assisted laser desorption ionization time-of-flight mass spectrometry. Poster presented at the 10th International Union of Microbiological Societies Congress, downloaded from http: //www.micromass.co.uk/litpdf/MMP239.pdf

Bruce, D., Hookey, J.V., and Waitkins, S.A. 1988. Numerical classification of campylobacters by DNA-restriction endonuclease analysis. Zentralbl. Bakteriol. Mikrobiol. Hyg. Ser. A. 269: 284–297.

Bryner, J.H.,. O'Berry, P.A., and Frank, A. H. 1964. Vibrio infection of the digestive organs of cattle. Am. J. Vet. Res. 25: 1048–1050.

Burnens, A.P., and Nicolet, J. 1992. Detection of *Campylobacter upsaliensis* in diarrheic dogs and cats, using a selective medium with cefoperazone. Am. J. Vet. Res. 53: 48–51.

Butzler, J.P., Dekeyser, P., Detrain, M., and Dehaen, F. 1973. Related vibrio in stools. J. Paediatr. 82: 493–495.

Cardarelli-Leite, P., Blom, K., Patton, C.M., Nicholson, M.A., Steigerwalt, A.G., Hunter, S.B., Brenner, D.J., Barrett, T.J., and Swaminathan, B. 1996. Rapid identification of *Campylobacter* species by restriction fragment length polymorphism analysis of a PCR-amplified fragment of the gene coding for 16S rRNA. J. Clin. Microbiol. 34: 62–67.

Cavalier-Smith, T. 2002. The neomuran origin of archaebacteria, the negibacterial root of the universal tree and bacterial megaclassification. Int. J. Syst. Evol. Microbiol. 52: 7–76.

Cho, J.C., and Tiedje, J.M. 2001. Bacterial species determination from DNA–DNA hybridization by using genome fragments and DNA microarrays. Appl. Environ. Microbiol. 67: 3677–3682.

Christensen, H., Jorgensen, K., and Olsen, J.E. 1999. Differentiation of *Campylobacter coli* and *C. jejuni* by length and DNA sequence of the 16S-23S rRNA internal spacer region. Microbiology 145: : 99–105.

Chun, J., Ward, A.C., Kang, S.O., Hah, Y.C., and Goodfellow, M. 1997. Long-term identification of streptomycetes using pyrolysis mass spectrometry and artificial neural networks. Zentralbl. Bakteriol. 285: 258–266.

Costas, M., Owen, R.J., and Jackman, P.J.H. 1987. Classification of *Campylobacter sputorum* and allied campylobacters based on numerical analysis of electrophoretic protein patterns. System. Appl. Microbiol. 9: 125–131.

de Boer, P., Duim, B., Rigter, A., van Der Plas, J., Jacobs-Reitsma, W.F., and Wagenaar, J.A. 2000. Computer-assisted analysis and epidemiological value of genotyping methods for *Campylobacter jejuni* and *Campylobacter coli*. J. Clin. Microbiol. 38: 1940–1946.

Dep, M.S., Mendz, G.L., Trend, M.A., Coloe, P.J., Fry, B.N., and Korolik, V. 2001. Differentiation between *Campylobacter hyoilei* and *Campylobater coli* using genotypic and phenotypic analyses. Int. J. Syst. Evol. Microbiol. 51: 819–826.

Diker, K.S., Diker, S., and Ozlem, M.B. 1990. Bovine diarrhoea associated with *Campylobacter hyointestinalis*. Zentralbl. Veterinarmed. B. 37: 158–160.

Dingle, K.E., Colles, F.M., Wareing, D.R., Ure, R., Fox, A.J., Bolton, F.E., Bootsma, H.J., Willems, R.J., Urwin, R., and Maiden, M.C. 2001. Multilocus sequence typing system for *Campylobacter jejuni*. J. Clin. Microbiol. 39: 14–23.

Dokland, T., Olsen, I., Farrants, G., and Johansen, B.V. 1990. Three-dimensional structure of the surface layer of *Wolinella recta*. Oral Microbiol. Immunol. 5: 162–165.

Doyle, L.P. 1948. The aetiology of swine dysentery. Am. J. Vet. Res. 9: 50–51.

Duerden, B.I., Eley, A., Goodwin, L., Magee, J.T., Hindmarch, J.M., and Bennett, K.W. 1989. A comparison of *Bacteroides ureolyticus* isolates from different clinical sources. J. Med. Microbiol. 29: 63–73.

Duim, B., Vandamme, P.A., Rigter, A., Laevens, S., Dijkstra, J.R., and Wagenaar, J.A. 2001 Differentiation of *Campylobacter* species by AFLP fingerprinting. Microbiology 147: 2729–2737.

Ebersole, J.L., Kesavalu, L., Schneider, S.L., Machen, R.L., and Holt, S.C. 1995. Comparative virulence of periodontopathogens in a mouse abscess model. Oral Dis. 1: 115–128.

Edmonds, P., Patton, C.M., Griffin, P.M., Barrett, T.J., Schmid, G.P., Baker, C.N., Lambert, M.A., and Brenner, D.J. 1987. *Campylobacter hyointestinalis* associated with human gastrointestinal disease in the United States. J. Clin. Microbiol. 25: 685–691.

Endtz, H.P., Vliegenthart, J. S., Vandamme, P., Weverink, H. W., van den Braak, N.P., Verbrugh, H.A., and van Belkum, A. 1997. Genotypic diversity of *Campylobacter lari* isolated from mussels and oysters in The Netherlands. Int. J. Food Microbiol. 34: 79–88.

Engberg, J., On, S.L.W., Harrington, C.S., and Gerner-Smidt, P. 2000. Prevalence of *Campylobacter, Arcobacter, Helicobacter* and *Sutterella* spp. in human fecal samples estimated by a reevaluation of isolation methods for campylobacters. J. Clin. Microbiol. 38: 286–291.

Engvall, E.O., Brandstrom, B., Gunnarsson, A., Morner, T., Wahlstrom, H., and Fermer, C. 2002. Validation of a polymerase chain reaction/restriction enzyme analysis method for species identification of thermophilic campylobacters isolated from domestic and wild animals. J. Appl. Microbiol. 92: 47–54.

Escherich, T. 1886. Beitraege zur kenntniss der Darmbacterien. III. Ueber das vorkommen von Vibrionen im darmcanal und den stuhlgaengen der saeuglinge. Müenchener medicinische Wochenschrift 33: 815–817 and 833–835.

Etoh, Y., Takahashi, M. and Yamamoto, A. 1988. *Wolinella curva* subsp. *intermedius* subsp. nov.: isolated from human gingival crevice. J. Showa Univ. Dent. Soc. 8: 349–354.

Etoh, Y., Dewhirst, F.E., Paster, B.J., Yamamoto, A. and Goto, N. 1993. *Campylobacter showae* sp. nov., isolated from the human oral cavity. Int. J. Syst. Bacteriol. 43: 631–639.

Etoh, Y., Yamamoto, A., and Goto, N. 1998. Intervening sequences in 16S rRNA genes of *Campylobacter* sp.: diversity of nucleotide sequences and uniformity of location. Microbiol. Immunol. 42: 241–243.

Eyers, M., Chapelle, S., Van Camp, G., Goossens, H., and De Wachter, R. 1993. Discrimination among thermophilic *Campylobacter* species by polymerase chain reaction amplification of 23S rRNA gene fragments. J. Clin. Microbiol. 31: 3340–3343.

Ezaki, T., Kawamura, Y., and Hata, H. 2002. Chromosomal DNA microarray to measure genetic relatedness among bacterial strains. In: Abstracts of the 10th International Congress of Bacteriology and Applied Microbiology, July 27-August 1 2002, Paris, France. EDK, Paris, p. 291 no. B-865.

Farrugia, D.C., Eykyn, S.J., and Smyth, E.G. 1994. *Campylobacter fetus* endocarditis: two case reports and review. Clin. Infect. Dis. 18: 443–446.

Fermer, C., and Engvall, E.O. 1999. Specific PCR identification and differentiation of the thermophilic campylobacters, *Campylobacter jejuni, C. coli, C. lari*, and *C. upsaliensis*. J. Clin. Microbiol. 37: 3370–3373.

Figura, N., Guglielmetti, P., Zanchi, A., Partini, N., Armellini, D., Bayeli, P. F., Bugnoli, M. and Verdiani, S. 1993. Two cases of *Campylobacter mucosalis* enteritis in children. J. Clin. Microbiol. 31: 727–728.

Firehammer, B. D. 1965. The isolation of vibrios from ovine feces. Maizeell Vet. 55: 482–494.

Florent, A. 1953. Isolement d'un vibrion saprophyte du sperme du Taureau et du vagin de la vache (*Vibrio bubulus*). C. R. Soc. Biol. 147: 2066–2069.

Fox, J.G., Taylor, N. S., Edmonds, P., and Brenner, D.J. 1988. *Campylobacter pylori* subsp. *mustelae*, subsp. nov. isolated from the gastric mucosa of ferrets (*Mustela putorius furo*), and an emended description of *Campylobacter pylori*. Int. J. Syst. Bacteriol. 38: 367–370.

Gebhart, C.J., Edmonds, P., Ward, G.E., Kurtz, H.J., and Brenner, D.J. 1985. '*Campylobacter hyointestinalis*' sp. nov.: a new species of *Campylobacter* found in the intestines of pigs and other animals. J. Clin. Microbiol. 21: 715–720.

Gebhart, C.J., Ward, G.E., Chang, K., and Kurtz, H.J. 1983. *Campylobacter hyointestinalis* (new species) isolated from swine with lesions of proliferative enteritis. Am. J. Vet. Res. 44: 361–367.

Goodwin, C.S., Armstrong, J.A., Chilvers, T., Peters, M., Collins, M.D., Sly, L., McConnell, W., and Harper, W.E.S. 1989a. Transfer of *Campylobacter pylori* and *Campylobacter mustelae* to *Helicobacter* gen. nov. as *Helicobacter pylori* comb. nov. and *Helicobacter mustelae* comb. nov. respectively. Int. J. Syst. Bacteriol. 39: 397–405.

Goodwin, C.S., McConnell, W., McCulloch, R.K., McCullough, C., Hill, R., Bronsdon, M.A. and Kasper, G. 1989b. Cellular fatty acid composition of *Campylobacter pylori* from primates and ferrets compared with of other campylobacters. J. Clin. Microbiol. 27: 938–943.

Gupta, R.S. 1998. Protein phylogenies and signature sequences: a reappraisal of evolutionary relationships among archaebacteria, eubacteria, and eukaryotes. Microbiol. Mol. Biol. Rev. 62: 1435–1491.

Han, Y.-H., Smibert, R.M., and Krieg, N.R. 1989. Occurrence of sheathed flagella in *Campylobacter cinaedi* and *Campylobacter fennelliae*. Int. J. Syst. Bacteriol. 39: 488–490.

Han, Y.W., Shi, W., Huang, G.T., Kinder Haake, S., Park, N.H., Kuramitsu, H., and Genco, R.J. 2000. Interactions between periodontal bacteria and human oral epithelial cells: *Fusobacterium nucleatum* adheres to and invades epithelial cells. Infect. Immun. 68: 3140–3146.

Hänninen, M.-L., Sarelli, L., Sukura, A., On, S.L.W., Harrington, C. S., Matero, P., and Hirvelä-Koski, V. 2002. *Campylobacter hyointestinalis* subsp. *hyointestinalis*, a common *Campylobacter* species in reindeer. J. Appl. Microbiol. 92: 717–723.

Harrington, C.S. and On, S.L.W. 1999. Extensive 16S ribosomal RNA gene sequence diversity in *Campylobacter hyointestinalis* strains: taxonomic, and applied implications. Int. J. Syst. Bacteriol. 49: 1171–1175.

Harrington, C.S., On, S.L.W., Carter, P.E., and Thomson-Carter, F.M. 1999. Identification and discrimination of *Campylobacter* species by means of 16S rRNA gene PCR and RFLP analysis: evidence for 16S rRNA gene identity in *C. jejuni* and *C. coli*. In: abstracts, 10th International Workshop on *Campylobacter, Helicobacter*, and related organisms, 12–16 September 1999, Baltimore, USA. p. 26 no. CD15.

Harvey, S.M. and Greenwood, J.R. 1983. Relationships among catalase-positive campylobacters determined by deoxyribonucleic acid-deoxyribonucleic acid hybridization. Int. J. Syst. Bacteriol. 33: 275–284.

Hayden, R.T., Kolbert, C.P., Hopkins, M.K., and Persing, D.H. 2001. Characterization of culture-derived spiral bacteria by 16S ribosomal RNA gene sequence analysis. Diagn. Microbiol. Infect. Dis. 39: 55–59.

Hill, B.D., Thomas, R.J., and Mackenzie, A.R. 1987. *Campylobacter hyointestinalis*-associated enteritis in Moluccan rusa deer (*Cervus timorensis* subsp. *moluccensis*). J. Comp. Pathol. 97: 687–694.

Hum, S. 1996. Bovine venereal campylobacteriosis: a diagnostic and economic perspective. In: Campylobacters, Helicobacters, and Related Organisms. D. G. Newell, J. M. Ketley, R. A. Feldman eds. Plenum Press, Delaware, pp. 355–358.

Hum, S., Quinn, K., Brunner, J., and On, S.L.W. 1997. Evaluation of a PCR assay for identification and differentiation of *Campylobacter fetus* subspecies. Aust. Vet. J. 75: 827–831.

Hurtado, A., and Owen, R.J. 1997. A molecular scheme based on 23S rRNA gene polymorphisms for rapid identification of *Campylobacter* and *Arcobacter* species. J. Clin. Microbiol. 35: 2401–2404.

Inglis, G. D. and Kalischuk, L. D. 2003. Use of PCR for direct detection of *Campylobacter* species in bovine faeces. Appl. Environ. Microbiol. 69: 3345–3447.

Jackson, F.L., and Goodman, Y.E. 1978. *Bacteroides ureolyticus*, a new species to accommodate strains previously identified as '*Bacteroides corrodens*, anaerobic'. Int. J. Syst. Bacteriol. 28: 197–200.

Johnson, C.C., Reinhardt, J.F., Edelstein, M.A., Mulligan, M.E., George, W.L., and Finegold, S.M. 1985. *Bacteroides gracilis*, an important anaerobic bacterial pathogen. J. Clin. Microbiol. 22: 799–802.

Jones, F.S., Orcutt, M., and Little, R.B. 1931. Vibrios (*Vibrio jejuni*, N. Sp.) associated with intestinal disorders of cows and calves. J. Exp. Med. 53: 853–867.

Jukes, T.H., and Cantor, C.R. 1969. Evolution of protein molecules. In: H.N. Munro, ed. Mammalian protein metabolism, vol. 3. Academic Press, New York, pp. 21–132.

Kamiya, S., Taniguchi, I., Yamamoto, T., Sawamura, S., Kai, M., Ohnishi, N., Tsuda, M., Yamamura, M., Nakasaki, H., Yokoyama, S., *et al.*, 1993. Analysis of intestinal flora of a patient with congenital absence of the portal vein. FEMS Immunol. Med. Microbiol. 7: 73–80.

Kamma, J.J., Nakou, M., and Manti, F.A. 1994. Microbiota of rapidly progressive periodontitis lesions in association with clinical parameters. J. Periodontol. 65: 1073–1978.

Kamma, J.J., Diamanti-Kipioti, A., Nakou, M., and Mitsis, F.J. 2000. Profile of subgingival microbiota in children with mixed dentition. Oral Microbiol. Immunol. 15: 103–111.

Keswani, J., and Whitman, W.B. 2001. Relationship of 16S rRNA sequence similarity to DNA hybridization in prokaryotes. Int. J. Syst. Evol. Microbiol. 51: 667–678.

Kiehlbauch, J.A., Brenner, D.J., Nicholson, M.A., Baker, C.N., Patton, C.M., Steigerwalt, A.G., and Wachsmuth, I. K. 1991. *Campylobacter butzleri* sp. nov. isolated from humans and animals with diarrhoeal illness. J. Clin. Microbiol. 29: 376–385.

King, E.O. 1957. Human infections with *Vibrio fetus* and a closely related vibrio. J. Infect. Dis. 101: 119–128.

Koga, M., Yuki, N., Takahashi, M., Saito, K., and Hirata, K. 1999. Are *Campylobacter curvus* and *Campylobacter upsaliensis* antecedent infectious agents in Guillain–Barré and Fisher's syndromes? J. Neurol. Sci. 1: 53–57.

Laanbroek, H.J., Kingma, W., and Veldkamp, H. 1977. Isolation of an aspartate-fermenting, free-living *Campylobacter* species. FEMS Microbiol. Lett. 1: 99–102.

Lastovica, A.J. and Skirrow, M.B. 2000. Clinical significance of *Campylobacter* and related species other than *Campylobacter jejuni* and *C. coli*. In: *Campylobacter,* 2nd edition. I. Nachamkin and M.J. Blaser, eds. ASM Press, Washington DC, pp. 89–120.

Lau, P.P., DeBrunner-Vossbrink, B., Dunn, B., Miotto, K., Macdonell, M.T., Rollins, D.M., Pillidge, C.J., Hespell, R.B., Colwell, R.R., Sogin, M.L. and Fox, G.E. 1987. Phylogenetic diversity and position of the genus *Campylobacter*. System. Appl. Microbiol. 9: 231–238.

Lawson, A.J., Logan, J.M., O'Neill, G.L., Desai, M., and Stanley, J. 1999. Large-scale survey of *Campylobacter* species in human gastroenteritis by PCR and PCR-enzyme-linked immunosorbent assay. J. Clin. Microbiol. 37: 3860–3864.

Lawson, A.J., On, S.L.W., Logan, J.M.J., and Stanley, J. 2001. *Campylobacter hominis* sp. nov. from the human gastrointestinal tract. Int. J. Syst. Evol. Microbiol. 51: 651- 660.

Lawson, A.J., Shafi, M.S., Pathak, K., and Stanley, J. 1998. Detection of *Campylobacter* in gastroenteritis: comparison of direct PCR assay of faecal samples with selective culture. Epidemiol. Infect. 121: 547–553.

Lawson, G.H.K. and Gebhart, C.J. 2000. Proliferative enteropathy. J. Comp. Pathol. 122: 77–100.

Lawson, G.H.K., Leaver, J.L., Pettigrew, G.W. and Rowland, A.C. 1981. Some features of *Campylobacter sputorum* subsp. *mucosalis* subsp. nov., nom. rev. and their taxonomic significance. Int. J. Syst. Bacteriol. 31: 385–391.

Lawson, G.H.K., Rowland, A.C. and Wooding, A.C. 1975. The characterization of *Campylobacter sputorum* subspecies *mucosalis* isolated from pigs. Res. Vet. Sci. 18: 121–126.

Lee, D., Goldstein, E.J., Citron, D.M., and Ross, S. 1993. Empyema due to *Bacteroides gracilis*: case report and *in vitro* susceptibilities to eight antimicrobial agents. Clin. Infect. Dis. 16 Suppl 4: S263–265.

Linton, D., Dewhirst, F.E., Clewley, J.P., Owen, R.J., Burnens, A.P., and Stanley, J. 1994. Two types of 16S rRNA gene are found in *Campylobacter helveticus*: analysis, applications and characterization of the intervening sequence found in some strains. Microbiology 140: 847–855.

Linton, D., Owen, R.J., and Stanley, J. 1996. Rapid identification by PCR of the genus *Campylobacter* and of five *Campylobacter* species enteropathogenic for man and animals. Res. Microbiol. 147: 707–718.

Logan, J.M., Burnens, A., Linton, D., Lawson, A.J., and Stanley, J. 2000. *Campylobacter lanienae* sp. nov., a new species isolated from workers in an abattoir. Int. J. Syst. Evol. Microbiol. 50: 865–872.

Love, D.N., Jones, R.F., Bailey, M., and Calverley, A. 1984. Comparison of strains of gram-negative, anaerobic, agar-corroding rods isolated from soft tissue infections in cats and dogs with type strains of *Bacteroides gracilis, Wolinella recta, Wolinella succinogenes*, and *Campylobacter concisus*. J. Clin. Microbiol. 20: 747–750.

McClung, C. R., Patriquin, D. G., and Davis, R. E. 1983. *Campylobacter nitrofigilis* sp. nov., a nitrogen-fixing bacterium associated with roots of *Spartina alterniflora* Loisel. Int. J. Syst. Bact. 33: 605–612.

McOrist, S., Lawson, G.H.K., Roy, D.J. and Boid, R. 1990. DNA analysis of intracellular *Campylobacter*-like organisms associated with the porcine proliferative enteropathies: novel organism proposed. FEMS Microbiol. Lett. 69: 189–194.

McOrist, S., Gebhart, C.J., Boid, R., and Barns, S.M. 1995. Characterization of *Lawsonia intracellularis* gen. nov., sp. nov., the obligately intracellular bacterium of porcine proliferative enteropathy. Int. J. Syst. Bacteriol. 45: 820–825.

Macuch, P.J., and Tanner, A.C. 2000. *Campylobacter* species in health, gingivitis, and periodontitis. J. Dent. Res. 79: 785–792.

Mandrell, R.E., and Wachtel, M.R. 1999. Novel detection techniques for human pathogens that contaminate poultry. Curr. Opin. Biotechnol. 10: 273–278.

Mansfield, L.P., and Forsythe, S.J. 2000. *Arcobacter butzleri* and *A. cryaerophilus* – newly emerging human pathogens. Rev. Med. Microbiol. 11: 161–170.

Marshall, S.M., Melito, P.L., Woodward, D.L., Johnson, W.M., Rodgers, F.G., and Mulvey, M.R. 1999. Rapid identification of *Campylobacter, Arcobacter*, and *Helicobacter* isolates by PCR-restriction fragment length polymorphism analysis of the 16S rRNA gene. J. Clin. Microbiol. 37: 4158–4160.

Matsheka, M.I., Elisha, B.G., Lastovica, A.L., and On, S.L.W. 2002. Genetic heterogeneity of *Campylobacter concisus* determined by PFGE-based macrorestriction profiling. FEMS Microbiol. Lett. 211: 17–22.

Misawa, N., Shinohara, S., Satoh, H., Itoh, H., Shinohara, K., Shimomura, K., Kondo, F., and Itoh, K. 2000. Isolation of *Campylobacter* species from zoo animals and polymerase chain reaction-based random amplified polymorphism DNA analysis. Vet. Microbiol. 71: 59–68.

Miyajima, M., Matsuda, M., Haga, S., Kagawa, S., Millar, B.C., and Moore, J.E. 2002. Cloning and sequencing of 16S rDNA and 16S-23S rDNA internal spacer region (ISR) from urease-positive thermophilic *Campylobacter* (UPTC). Lett. Appl. Microbiol. 34: 287–289.

Miyamoto, M., Maeda, H., Kitanaka, M., Kokeguchi, S., Takashiba, S., and Murayama, Y. 1998. The S-layer protein from *Campylobacter rectus*: sequence determination and function of the recombinant protein. FEMS Microbiol. Lett.166: 275–281.

Morris, G.K., el Sherbeeny, M.R, Patton, C.M., Kodaka, H., Lombard, G.L., Edmonds, P., Hollis, D.G., and Brenner, D.J. 1985. Comparison of four hippurate hydrolysis methods for identification of thermophilic *Campylobacter* spp. J. Clin. Microbiol. 22: 714–718.

Moser, I., Rieksneuwohner, B., Lentzsch, P., Schwerk, P., and Wieler, L.H. 2001. Genomic heterogeneity and O-antigenic diversity of *Campylobacter upsaliensis* and *Campylobacter helveticus* strains isolated from dogs and cats in Germany. J. Clin. Microbiol. 39: 2548–2557.

Moss, C.W., Lambert-Fair, M.A., Nicholson, M.A. and Guerrant, G.O. 1990. Isoprenoid quinones of *Campylobacter cryaerophila, C. cinaedi, C. fennelliae, C. hyointestinalis, C. pylori* and '*C. upsaliensis*'. J. Clin. Microbiol. 28: 395–397.

Moureau, P., Derclaye, I., Gregoire, D., Janssen, M., and Maizeelis, G.R. 1989. *Campylobacter* species identification based on polymorphism of DNA encoding rRNA. J. Clin. Microbiol. 27: 1514–1517.

Murray, R.G.E., and Schliefer, K.H. 1994. Taxonomic notes: a proposal for recording the properties of putative taxa of procaryotes. Int. J. Syst. Bacteriol. 44: 174–176.

Murray, A.E., Lies, D., Li, G., Nealson, K., Zhou, J., and Tiedje, J.M. 2001. DNA/DNA hybridization to microarrays reveals gene-specific differences between closely related microbial genomes. Proc. Natl. Acad. Sci. USA 98: 9853–9858.

Musmanno, R. A., Russi, M., Figura, N., Guglielmetti, P., Zanchi, A., Signori, R., and Rossolini, A. 1998. Unusual species of campylobacters isolated in the Siena, Tuscany area, Italy. New Microbiol. 21: 15–22.

Neill, S.D., Campbell, J.N., O'Brien, J.J., Weatherup, S.T. and Ellis, W.A. 1985. Taxonomic position of *Campylobacter cryaerophilus* sp. nov. Int. J. Syst. Bacteriol. 35: 342–356.

Newell, D.G., Frost, J.A., Duim, B., Wagenaar, J.A., Madden, R.H., Van der Plas, J. and On, S.L.W. 2000. New developments in the subtyping of campylobacters. In: I. Nachamkin and M.J. Blaser, eds. *Campylobacter*, 2nd edition. ASM Press, Washington, pp. 27–44.

On, S.L.W. 1994. Confirmation of human *Campylobacter concisus* isolates misidentified as *Campylobacter mucosalis* and suggestions for improved differentiation between the two species. J. Clin. Microbiol. 32: 2305–2306.

On, S. L. W. 1996a. Identification methods for campylobacters, helicobacters, and related organisms. Clin. Microbiol. Rev. 9: 405–422.

On, S.L.W. 1996b. Computer-assisted strategies for identifying campylobacteria. In: D.G. Newell, J. M. Ketley, and R.A. Feldman, eds. *Campylobacter, Helicobacter* and related organisms. Plenum Press, Delaware, pp. 221–226.

On, S.L.W. 2001. Taxonomy of *Campylobacter, Arcobacter, Helicobacter,* and related bacteria: current status, future prospects, and immediate concerns. Symp. Suppl. J. Appl. Microbiol. 90: 1S–15S.

On, S.L.W. 2002. International Committee on Systematic Bacteriology Subcommittee on the Taxonomy of *Campylobacter* and Related Bacteria: Minutes of the Meetings 2nd and 4th September 2001, Freiburg, Germany. Int. J. Syst. Evol. Microbiol. 52: 2339–2341.

On, S.L.W., and Holmes, B. 1991a. Effect of inoculum size on the phenotypic characterization of *Campylobacter* spp. J. Clin. Microbiol. 29: 923–926.

On, S.L.W., and Holmes, B. 1991b. Reproducibility of tolerance tests that are useful in the identification of campylobacteria. J. Clin. Microbiol. 29: 1785–1788.

On, S.L.W., and Holmes, B. 1992. Assessment of enzyme detection tests useful in identification of campylobacteria. J. Clin. Microbiol. 30: 746–749.

On, S.L.W., and Holmes, B. 1995. Classification and identification of campylobacters, helicobacters and allied taxa by numerical analysis of phenotypic characters. System. Appl. Microbiol. 18: 374–390.

On, S.L.W., and Harrington, C.S. 2000. Identification of taxonomic and epidemiological relationships among *Campylobacter* species by numerical analysis of AFLP profiles. FEMS Microbiol. Lett. 193: 161–169.

On, S.L.W. and Harrington, C.S. 2001. Evaluation of numerical analysis of PFGE-DNA profiles for differentiating *Campylobacter fetus* subspecies by comparison with phenotypic, PCR, and 16S rDNA sequencing methods. J. Appl. Microbiol. 90: 285–293.

On, S.L.W. and Jordan, P.J. 2003. Evaluation of 11 PCR assays for species-level identification of *Campylobacter jejuni* and *Campylobacter coli*. J. Clin. Microbiol. 41: 330–336.

On, S.L.W., and Vandamme, P. 1997. Identification and epidemiological typing of *Campylobacter hyointestinalis* subspecies by phenotypic and genotypic methods and description of novel subgroups. Syst. Appl. Microbiol. 20: 238–247.

On, S.L.W., Costas, M., and Holmes, B. 1994. Classification and identification of *Campylobacter sputorum* by numerical analyses of phenotypic tests and of one-dimensional electrophoretic protein profiles. System. Appl. Microbiol. 17: 543–553.

On, S.L.W., Bloch, B., Holmes, B., Hoste, B., and Vandamme, P. 1995. *Campylobacter hyointestinalis* subsp. *lawsonii* subsp. nov., isolated from the porcine stomach, and an emended description of *Campylobacter hyointestinalis*. Int. J. Syst. Bacteriol. 45: 767–774.

On, S.L.W., Holmes, B., and Sackin, M.J. 1996. A probability matrix for the identification of campylobacters, helicobacters, and allied taxa. J. Appl. Bacteriol. 81: 425–432.

On, S.L.W., Atabay, H.I., Corry, J.E.L., Harrington, C.S., and Vandamme, P. 1998a. Emended description of *Campylobacter sputorum* and revision of its infrasubspecific (biovar) divisions, including *C. sputorum* bv. *paraureolyticus*, a urease-producing variant from cattle and humans. Int. J. Syst. Bacteriol. 48: 195–206.

On, S.L.W., Atabay, H.I., and Harrington, C.S. 1998b. Evaluation of a probability matrix for identifying campylobacteria. In: A. Lastovica, D.G. Newell, and E.E. Lastovica, eds. *Campylobacter, Helicobacter* and Related Organisms. University of Cape Town, Cape Town, pp. 232–235.

On, S.L.W., Atabay, H.I., and Corry, J.E.L. 1999. Clonality of *Campylobacter sputorum* bv. paraureolyticus determined by macrorestriction profiling and biotyping, and evidence for long-term persistence in cattle. Epidemiol. Infect. 122: 175–182.

On, S.L.W., Fields, P.I., Broman, T., Helsel, L.O., Fitzgerald, C., Harrington, C.S., Laevens, S., Steigerwalt, A.G., Olsen, B, and Vandamme, P.A.R. 2001. Polyphasic taxonomic analysis of *Campylobacter lari*: delineation of three subspecies. Abstract P-21, 11th International Workshop on *Campylobacter*, Helicobacter and Related Organisms, Freiburg, Germany, 1–5 September 2001. Int. J. Med. Microbiol. 291; 144.

Owen, R.J., Costas, M., and Dawson, C. 1989. Application of different chromosomal DNA restriction digest fingerprints to specific and subspecific identification of *Campylobacter* isolates. J. Clin. Microbiol. 27: 2338–2343.

Owen, R.J., Costas, M., and Sloss, L. L. 1988. Electrophoretic protein typing of *Campylobacter jejuni* subspecies *doylei* (nitrate-negative campylobacter-like organisms) from human faeces and gastric mucosa. Eur. J. Epidemiol. 4: 277–283.

Parkhill, J., Wren, B.W., Mungall, K., Ketley, J.M., Churcher, C., Basham, D., Chillingworth, T., Davies, R.M., Feltwell, T., Holroyd, S., Jagels, K., Karlyshev, A.V., Moule, S., Pallen, M.J., Penn, C.W., Quail, M.A., Rajandream, M.A., Rutherford, K.M., van Vliet, A.H., Whitehead, S., and Barrell, B.G. 2000. The genome sequence of the food-borne pathogen *Campylobacter jejuni* reveals hypervariable sequences. Nature 403: 665–668.

Paster, B.J., and Dewhirst, F.E. 1988. Phylogeny of campylobacters, wolinellas, *Bacteroides gracilis*, and *Bacteroides ureolyticus* by 16S ribosomal ribonucleic acid sequencing. Int. J. Syst. Bacteriol. 38: 56–62.

Prévot, A. R. 1940. Études de systématique bactérienne. V. Essai de classification des vibrions anaérobies. Ann. Inst. Pasteur 64: 117–125.

Priest, F. and Austin, B. 1993. Modern Bacterial Taxonomy, 2nd edition. Chapman & Hall, London, U.K.

Romaniuk, P.J., Zoltowska, B., Trust, T.J., Lane, D.J., Olsen, G.J., Pace, N.R., and Stahl, D.A. 1987. *Campylobacter pylori*, the spiral bacterium associated with human gastritis, is not a true *Campylobacter* sp. J. Bacteriol. 169: 2137–2141.

Roop II, R.M., Smibert, R.M., Johnson, J.L., and Krieg, N.R. 1985a. DNA homology studies of the catalase-negative campylobacters and 'Campylobacter fecalis,' an emended description of *Campylobacter sputorum*, and proposal of the neotype strain of *Campylobacter sputorum*. Can. J. Microbiol. 31: 823–831.

Roop II, R.M., Smibert, R.M., Johnson, J.L., and Krieg, N.R. 1985b. *Campylobacter mucosalis* (Lawson, Leaver, Pettigrew, and Rowland 1981) comb. nov.: emended description. Int. J. Syst. Bacteriol. 35: 189–192.

Rosseel, P., Muyldermans, G., Breynaert, J., Vandamme, P., and Lauwers, S. 1998. Identification of *Campylobacter* species isolated from human faeces by cellular fatty acid analysis. In: *Campylobacter, Helicobacter* and Related Organisms. A.J. Lastovica, D.G. Newell, and E.E. Lastovica, eds. University of Cape Town, S. Africa, pp. 236–241.

Salama, S.M., Garcia, M.M., and Taylor, D.E. 1992a. Differentiation of the subspecies of *Campylobacter fetus* by genomic sizing. Int. J. Syst. Bacteriol. 42: 446–450.

Salama, S.M., Tabor, H., Richter, M., and Taylor, D.E. 1992b. Pulsed-field gel electrophoresis for epidemiologic studies of *Campylobacter hyointestinalis* isolates. J. Clin. Microbiol. 30: 1982–1984.

Salomon, H. 1898. Ueber das Spirillum des Säugetiermagens und sein Verhalten zu den Belegzellen. Zentralbl. Bakteriol. Parasitenkd. Infektionskr. Hyg. Abt. 1 Orig. 19: 422–441.

Sandstedt, K. and Ursing, J. 1991. Description of *Campylobacter upsaliensis* sp. nov. previously known as the CNW group. System. Appl. Microbiol. 14: 39–45.

Sandstedt, K., Ursing, J. and Walder, M. 1983. Thermotolerant *Campylobacter* with no or weak catalase activity isolated from dogs. Curr. Microbiol. 8: 209–213.

Sasaki, Y., Fujisawa, T., Ogikubo, K., Ohzono, T., Ishihara, K. and Takahashi, T. 2003. Characterization of *Campylobacter lanienae* from pig feces. J. Vet. Med. Sci. 65: 129–131.

Sasser, M. 2001. Identification of bacteria by gas chromatography of cellular fatty acids. Technical note 101. http: //www.midi-inc.com/media/pdfs/TechNote_101.pdf

Savelkoul, P.H., Aarts, H.J., de Haas, J., Dijkshoorn, L., Duim, B., Otsen, M., Rademaker, J.L., Schouls, L., and Lenstra, J.A. 1999. Amplified-fragment length polymorphism analysis: the state of an art. J. Clin. Microbiol. 37: 3083–3091.

Schumacher, W., Kroneck, P.M.H., and Pfennig, N. 1992. Comparative systematic study on 'Spirillum' 5175, *Campylobacter* and *Wolinella* species. Description of 'Spirillum' 5175 as *Sulfurospirillum deleyianum* gen. nov., spec. nov. Arch. Microbiol. 158: 287–293.

Sébald, M. and Véron, M. 1963. Teneur en bases da l'ADN et classification des vibrions. Ann. Inst. Pasteur 105: 897–910.

Skirrow, M.B. 1977. *Campylobacter* enteritis: a 'new' disease. Br. Med. J. 2: 9–11.

Skirrow, M.B. 1983. Taxonomy and biotyping. Molecular aspects. In: *Campylobacter* II. Proceedings of the Second International Workshop on *Campylobacter* Infections. A.D Pearson, M.B. Skirrow, B. Rowe, J.R. Davies, and D.M. Jones, eds. Public Health Laboratory Service, London, pp. 33–38.

Skirrow, M.B. 1994. Diseases due to *Campylobacter, Helicobacter* and related bacteria. J. Comp. Pathol. 111: 113–149.

Skirrow, M.B., and Benjamin, J. 1980. '1001' campylobacters: cultural characteristics of intestinal campylobacters from man and animals. J. Hyg. (Camb.) 85: 427–442.

Skirrow, M.B., and Blaser, M.J. 2000. Clinical aspects of *Campylobacter* infection. In: *Campylobacter*, 2nd edition. I. Nachamkin and M.J. Blaser, eds. ASM Press, Washington DC, pp. 69–88.

Smibert, R.M. and Holdeman, L. V. 1976. Clinical isolates of anaerobic gram-negative rods with a formate-fumarate energy metabolism: *Bacteroides corrodens, Vibrio succinogenes*, and unidentified strains. J. Clin. Microbiol. 3: 432–437.

Smith, T. and Taylor, M.S. 1919. Some morphological and biological characters of the spirilla (*Vibrio fetus*, N. Sp.) associated with disease of the fetal membranes in cattle. J. Exp. Med. 30: 299–312.

Sneath, P.H.A., and Sokal, R.R., 1973. Numerical Taxonomy: the Principles and Practice of Numerical Classification. W.H. Freeman and Co., San Francisco.

Socransky, S.S., Haffajee, A.D., Cugini, M.A., Smith, C., and Kent, R.L., Jr. 1998. Microbial complexes in subgingival plaque. J. Clin. Periodontol. 25: 134–144.

Solnick, J.V., and Schauer, D.B. 2001. Emergence of diverse *Helicobacter* species in the pathogenesis of gastric and enterohepatic diseases. Clin. Microbiol. Rev. 14: 59–97.

Stackebrandt, E. 1992. Unifying phylogeny and phenotypic diversity. In: The Prokaryotes A. Balows, H.G. Trüper, M. Dworkin, W. Harder, and K.-H. Schleifer, eds. Springer-Verlag, New York, pp. 19–47.

Stackebrandt, E. and Goebel, B.M. 1994. Taxonomic note: a place for DNA–DNA reassociation and 16S rRNA sequence analysis in the present species definition in bacteriology. Int. J. Syst. Bacteriol. 44: 846–849.

Stanley, J., Burnens, A.P., Linton, D., On, S.L.W., Costas, M., and Owen, R. J. 1992. '*Campylobacter helveticus*' sp. nov., a new thermophilic group from domestic animals: characterization of the species and cloning of a species-specific DNA probe. J. Gen. Microbiol. 138: 2293–2303.

Steele, T.W., Lanser, J.A., and Sangster, N. 1985. Nitrate-negative campylobacter-like organisms. Lancet i: 394.

Steele, T.W. and Owen, R.J. 1988. *Campylobacter jejuni* subsp. *doylei* subsp. nov., a subspecies of nitrate-negative campylobacters isolated from human clinical specimens. Int. J. Syst. Bacteriol. 38: 316–318.

Steinbrueckner, B., Haerter, G., Pelz, K., Burnens, A., and Kist, M. 1998. Discrimination of *Helicobacter pullorum* and *Campylobacter lari* by analysis of whole cell fatty acid extracts. FEMS Microbiol. Lett. 168: 209–212.

Steinbrueckner, B., Haerter, G., Pelz, K., Weiner, S., Rump, J.A., Deissler, W., Bereswill, S., and Kist, M. 1997. Isolation of *Helicobacter pullorum* from patients with enteritis. Scand. J. Infect. Dis. 29: 315–318.

Stephens, C.P., On, S.L.W., and Gibson, J.A. 1998. An outbreak of infectious hepatitis in commercially reared ostriches associated with *Campylobacter coli* and *Campylobacter jejuni*. Vet. Microbiol. 61: 183–190.

Tanner, A.C.R., Badger, S., Lai, C.-H, Listgarten, M.A., Visconti, R.A. and Socransky, S.S. 1981. *Wolinella* gen. nov., *Wolinella succinogenes* (*Vibrio succinogenes* Wolin *et al.,*) comb. nov., and description of *Bacteroides gracilis* sp. nov., *Wolinella recta* sp. nov., *Campylobacter concisus* sp. nov., and *Eikenella corrodens* all from humans with periodontal disease. Int. J. Syst. Bacteriol. 31: 432–445.

Taylor, A.J., Costas, M., and Owen, R.J. 1987. Numerical analysis of electrophoretic protein patterns of *Bacteroides ureolyticus* clinical isolates. J. Clin. Microbiol. 25: 660–666.

Taylor, A.J., Dawson, C.A., and Owen, R.J. 1986. The identification of *Bacteroides ureolyticus* from patients with non-gonococcal urethritis by conventional biochemical tests and by DNA and protein analyses. J. Med. Microbiol. 21: 109–116.

Tee, W., Luppino, M., and Rambaldo, S. 1998. Bacteremia due to *Campylobacter sputorum* biovar *sputorum*. Clin. Infect. Dis. 27: 1544–1545.

Thompson III, L.M., Smibert, R.M., Johnson, J.L., and Krieg, N.R. 1988. Phylogenetic study of the genus *Campylobacter*. Int. J. Syst. Bacteriol. 38: 190–200.

Totten, P.A., Fennell, C.L., Tenover, F.C., Wezenberg, J.M., Perine, P.L., Stamm, W.E., and Holmes, K.K. 1985. *Campylobacter cinaedi* (sp. nov.) and *Campylobacter fennelliae* (sp. nov.): two new *Campylobacter* species associated with enteric disease in homosexual men. J. Infect. Dis. 151: 131–139.

Trust, T.J., Logan, S.M., Gustafson, C.E., Romaniuk, P.J., Kim, N.W., Chan, V.L., Ragan, M.A., Guerry, P., and Gutell, R.R. 1994. Phylogenetic and molecular characterization of a 23S rRNA gene positions the genus *Campylobacter* in the epsilon subdivision of the Proteobacteria and shows that the presence of transcribed spacers is common in *Campylobacter* spp. J. Bacteriol. 176: 4597–4609.

Ursing, J.B., Lior, H., and Owen, R.J. 1994. Proposal of minimal standards for describing new species of the family *Campylobacteraceae*. Int. J. Syst. Bacteriol. 44: 842–845.

Ussery, D. 2001. Databases, sequenced genomes. In: S. Brenner and J.H. Miller, eds. Encyclopaedia of Genetics. Academic Press, London, pp. 517–521.

Van Camp, G., Van De Peer, Y., Nicolai, S., Neefs, J.-M., Vandamme, P. and De Wachter, R. 1993. Structure of 16S and 23S ribosomal RNA genes in *Campylobacter* species: phylogenetic analysis of the genus *Campylobacter* and presence of internal transcribed spacers. System. Appl. Microbiol. 16: 361–368.

On

Vandamme, P. and De Ley, J. 1991. Proposal for a new family, *Campylobacteraceae*. Int. J. Syst. Bacteriol. 41: 451–455.

Vandamme, P., Falsen, E., Pot, B., Hoste, B., Kersters, K. and De Ley, J. 1989. Identification of EF Group 22 campylobacters from gastroenteritis cases as *Campylobacter concisus*. J. Clin. Microbiol. 27: 1775–1781.

Vandamme, P., Pot, B., Falsen, E., Kersters, K. and De Ley, J. 1990. Intra- and interspecific relationships of veterinary campylobacters revealed by numerical analysis of electrophoretic protein profiles and DNA:DNA hybridizations. System. Appl. Microbiol. 13: 295–303.

Vandamme, P., Falsen, E., Rossau, R., Hoste, B., Segers, P., Tytgat, R. and De Ley, J. 1991a. Revision of *Campylobacter, Helicobacter,* and *Wolinella* taxonomy: emendation of generic descriptions and proposal of *Arcobacter* gen. nov. Int. J. Syst. Bacteriol. 41: 88–103.

Vandamme, P., Pot, B. and Kersters, K. 1991b. Differentiation of campylobacters and *Campylobacter*-like organisms by numerical analysis of one-dimensional electrophoretic protein patterns. System. Appl. Microbiol. 14: 57–66.

Vandamme, P., Dewettnick, D. and Kersters, K. 1992a. Application of numerical analysis of electrophoretic protein profiles for the identification of thermophilic campylobacters. System. Appl. Microbiol. 15: 402–408.

Vandamme, P., Vancanneyt, M., Pot, B., Mels, L., Hoste, B., Dewettinck, D., Vlaes, L., Van Den Borre, C., Higgins, R., Kersters, K., Butzler, J.-P. and Goossens, H. 1992b. Polyphasic taxonomic study of the emended genus *Arcobacter* with *Arcobacter butzlerii* comb. nov. and *Arcobacter skirrowii* sp. nov., an aerotolerant bacterium isolated from veterinary specimens. Int. J. Syst. Bacteriol. 42: 344–356.

Vandamme, P., Daneshvar, M.I., Dewhirst, F.E., Paster, B.J., Kersters, K., Goossens, H., and Moss, C.W. 1995. Chemotaxonomic analyses of *Bacteroides gracilis* and *Bacteroides ureolyticus* and reclassification of *B. gracilis* as *Campylobacter gracilis* comb. nov. Int. J. Syst. Bacteriol. 45: 145–152

Vandamme, P., Van Doorn, L.J., al Rashid, S.T., Quint, W.G., van der Plas, J., Chan, V.L., and On, S.L.W. 1997. *Campylobacter hyoilei* Alderton *et al.,* 1995 and *Campylobacter coli* Véron and Chatelain 1973 are subjective synonyms. Int. J. Syst. Bacteriol. 47: 1055–1060.

Vandamme, P., Dewhirst, F.E., Paster, B.J. and On, S.L.W. Family *Campylobacteraceae*. Bergey's Manual of Systematic Bacteriology, 2nd Edition. Springer-Verlag, New York. In press.

Véron, M. and Chatelain, R. 1973. Taxonomic study of the genus *Campylobacter* Sebald and Veron and designation of the neotype strain for the type species, *Campylobacter fetus* (Smith and Taylor) Sebald and Veron. Int. J. Syst. Bacteriol. 23: 122–134.

Van Etterick, R., Breynaert, J., Revets, H., Devreker, T., Vandenplas, Y., Vandamme, P. and Lauwers, S. 1996. Isolation of *Campylobacter concisus* from feces of children with and without diarrhoea. J. Clin. Microbiol. 34: 2304–2306.

von Graevenitz, A. 1990. Revised nomenclature of *Campylobacter laridis*, *Enterobacter intermedium*, and '*Flavobacterium branchiophila*'. Int. J. Syst. Bacteriol. 40: 211.

Wagenaar, J.A., van Bergen, M.A., Newell, D.G., Grogono-Thomas, R., and Duim, B. 2001. Comparative study using amplified fragment length polymorphism fingerprinting, PCR genotyping, and phenotyping to differentiate *Campylobacter fetus* strains isolated from animals. J. Clin. Microbiol. 39: 2283–2286.

Wassenaar, T.M., On, S.L.W., and Meinersmann, R. 2000. Genotyping and the consequences of genetic instability. In: I. Nachamkin and M.J. Blaser, eds. *Campylobacter*, 2nd edition. ASM Press, Washington DC, pp. 369–380.

Wayne, L.G., Brenner, D.J., Colwell, R.R., Grimont, P.A.D., Kandler, O., Krichevsky, M.I., Moore, L.H., Moore, W.E.C., Murray, R.G.E., Stackebrandt, E., Starr, M.R., and Trüper, H.G. 1987. Report of the Ad Hoc committee on reconciliation of approaches to bacterial systematics. Int. J. Syst. Bacteriol. 37: 463–464.

Wexler, H.M., Reeves, D., Summanen, P.H., Molitoris, E., McTeague, M., Duncan. J., Wilson, K.H., and Finegold, S.M. 1996. *Sutterella wadsworthensis* gen. nov., sp. nov., bile-resistant microaerophilic *Campylobacter gracilis*-like clinical isolates. Int. J. Syst. Bacteriol. 46: 252–258.

Wilkes, J.G., Glover, K.L., Holcomb, M., Rafii, F., Cao, X., Sutherland, J.B., McCarthy, S.A., Letarte, S., and Bertrand, M.J. 2002. Defining and using microbial spectral databases. J. Am. Soc. Mass Spectrom. 13: 875–887.

Winkler, M.A., Uher, J., and Cepa, S. 1999. Direct analysis and identification of *Helicobacter* and *Campylobacter* species by MALDI-TOF mass spectrometry. Anal. Chem. 71: 3416–3419.

Woese, C.R. 1987. Bacterial evolution. Microbiol. Rev. 51: 221–271.

Yap, W.H., Zhang, Z., and Wang, Y. 1999. Distinct types of rRNA operons exist in the genome of the actinomycete *Thermomonospora chromogena* and evidence for horizontal transfer of an entire rRNA operon. J. Bacteriol. 181: 5201–5209.

Chapter 3

Population Genetics of *Campylobacter jejuni*

Kate E. Dingle and Martin C.J. Maiden

Abstract

Population genetic studies have an important role to play in the investigation of bacterial pathogens and have proved particularly informative in exploring the epidemiology of *Campylobacter jejuni*. Improvements in the technology available for high-throughput nucleotide sequence determination, together with substantial reductions in its cost and the development of novel analysis techniques, have facilitated multilocus investigations of large isolate collections. The data generated by techniques such as multilocus sequence typing (MLST) are highly accurate and readily comparable among laboratories. Several MLST studies of *C. jejuni* have confirmed the genetic diversity of this bacterium and shown that whilst there is extensive evidence for recombination in populations of this organism, certain groups of related genotypes, called clonal complexes, persist over time and during geographical spread. Most human disease is caused by a relatively limited subset of these clonal complexes and certain genotypes appear to be associated with particular animal host species. Single phenotypic or genotypic characteristics, such as serotype or the *flaA* gene type, are inconsistent indicators of clonal complex, explaining why it can be difficult to interpret these data epidemiologically. Together with MLST data, the clonal complex model provides the prospect of a single unified typing procedure for *C. jejuni*.

INTRODUCTION

Studies of the epidemiology of bacterial pathogens are enhanced by an understanding of their population structures and in recent years, improved models of the bacterial population structures have become available, revealing new insights into bacterial epidemiology. The development of these models has been promoted by the combination of improvements in high-throughput nucleotide sequence determination with novel analytical techniques and theoretical frameworks. Such studies enable the elucidation of the relationships of bacterial genotypes with phenotypes such as association with particular host species or disease syndromes. This general approach is particularly useful in studies of bacterial transmission and pathogenesis, especially for zoonotic bacteria such as *Campylobacter jejuni* that have a wide host range. It does, however, rely on access to representative isolate collections and appropriate sampling of the bacterial genome. The insights gained are potentially invaluable in the design and implementation of novel public health interventions (reviewed by Musser, 1996).

Defining the transmission routes of *C. jejuni* from birds and mammals to man has been a particular challenge due to the many possible sources of infection (Frost, 2001; Gillespie *et al.*, 2002) and the genetic and antigenic diversity of the organism, which renders the use of a single genotypic or phenotypic characteristic of limited value for epidemiological surveys (Dingle *et al.*, 2002). The application of multilocus sequence typing (MLST) to the characterization of *C. jejuni* isolates has overcome many of the problems experienced with other methods and has provided data that have been useful in the development of models of the population structure of this organism. This chapter reviews the progress made in the understanding of *C. jejuni* population genetics and epidemiology by the application of MLST to a number of isolate collections. This approach to *Campylobacter* molecular epidemiology has the potential to clarify the relationships among *C. jejuni* genotypes, mammalian and avian hosts, and human disease.

THEORETICAL MODELS OF BACTERIAL POPULATION STRUCTURE

Bacteria are haploid organisms which reproduce asexually: in the absence of mutation each cell division event results in two daughter cells which are genetically identical to their mother cell. If there is no genetic exchange (a strictly asexual organism), genetic variation can only spread in the population by the accumulation of point mutations passed down vertically from the cells in which they arose to their descendants (Figure 3.1a). The variation is therefore distributed non-randomly among the population and chromosomal polymorphisms are in linkage disequilibrium. In combination with diversity reduction events, which can be either selective (e.g. periodic selection) or stochastic (bottle-necking), this process results in a clonal population structure. In strictly clonal populations, genetic relationships among individuals is accurately modelled by phylogenetic trees, with the branches generated by mutational events.

This rather simple paradigm of evolution is disrupted by the movement of genetic material among individuals which do not share a mother cell. In bacteria this can happen by the process of horizontal genetic exchange (Figure 3.1b) which involves only parts of the chromosome (so-called localized sex). Such genetic exchange is mediated by the parasexual processes of conjugation, transduction or transformation and, if it occurs at a sufficiently high rate relative to mutation, will disrupt or even erase the lineage structure characteristic of a clonal population and result in a non-clonal population with genetic variation randomized, and therefore at linkage equilibrium, in the population (Spratt and Maiden 1999; Ochman *et al.*, 2000). Homologous recombination can occur when the donor and recipient bacteria have as little as 70% nucleotide sequence identity (Lorenz and Wackernagel, 1994) potentially allowing the movement of genetic material among bacterial cells which are evolutionarily distant, although the efficiency is greater the higher the level of sequence identity. Repeated recombination events involving diverse sequences creates a genome which is a mosaic of genetic variation with widely different evolutionary histories. The phylogenetic relationships among alleles in a population at linkage equilibrium, are better represented by a network than a tree, since each individual has multiple ancestors.

A highly clonal population structure with limited genetic variation is exemplified by *Mycobacterium tuberculosis* (Kapur *et al.*, 1994), which may represent a recently emerged single clone which is genetically isolated from its ancestral population. In contrast, *Helicobacter pylori* is the most frequently recombining species yet identified (Suerbaum *et al.*, 1998) and any two independent isolates are very unlikely to be the same and cannot be linked by a tree-like phylogeny. The contrasting population structures of *M. tuberculosis* and *H. pylori* illustrate the effects of a complete lack of horizontal genetic exchange (change occurring only by point mutation), and very frequent recombination such that any two isolates examined are likely to be genetically distinct. The populations of many bacterial species appear to contain elements of both of these structures, consisting of groups or clusters of closely related isolates, referred to as clonal complexes. The clonal complex model has now become a major paradigm for the study of bacterial population structures (Gupta and Maiden, 2001).

EVOLUTIONARY EVENTS GIVING RISE TO CLONAL COMPLEXES

It has been proposed that the relative rates at which genetic variation is generated by point mutation and recombination greatly influence bacterial population structure (Feil and Spratt, 2001). However the rate at which these mutations become integrated into a proportion of the population, is also likely to be important and this will be related to the selection pressures exerted by the environment, which may change over time, and not simply the relative rates of occurrence of particular genetic events. Potentially, therefore, four inter-related evolutionary processes impact on the population structure: point mutation; recombination; selection (positive and negative); and demographic processes such as population growth or reduction (Figure 3.1).

The selection pressures exerted by a given set of conditions affect the ability of different genotypes of the same species to survive and reproduce. Deleterious mutations cause a reduction in fitness and are removed from the population by negative or purifying selection. A change in environmental conditions can change the selection pressures, conferring an advantage on different genotypes. Such selective sweeps or bottlenecks can eliminate part of the population, lowering its overall diversity (Figure 3.1c). Negative selection therefore has a stabilizing effect on population diversity. Positive selection may increase population diversity (Figure 3.1d) by allowing advantageous mutations to expand within the population by becoming established in a large proportion of individuals. However, in other circumstances, if a given allele becomes fixed (i.e. present in 100% of individuals), it may reduce the diversity of the population through its success. Neutral variation occurs in loci which are not subject to selection, and confers on the host cell neither selective advantage nor disadvantage: this type of variation is best understood from a theoretical perspective, but in practice most bacterial genes are subject to either positive or negative selection. In many population analyses variation in housekeeping genes which are under negative (stabilizing) selection is assumed to be neutral.

MULTILOCUS SEQUENCE TYPING

The first technique employed for studies of bacterial population genetics was multilocus enzyme electrophoresis (MLEE) (Milkman, 1973; Selander *et al.*, 1986; Musser, 1996). In MLEE, variation at multiple chromosomal locations is assessed indirectly by measuring the mobility of variants of core metabolic enzymes during starch gel electrophoresis. As any enzyme which can be stained by a colour reaction can be visualized from crude cell suspensions, MLEE can relatively quickly and easily generate data for many loci. Several bacteria have been investigated with this method and a MLEE study of *Campylobacter* species established their high level of genetic diversity and identified clonal groupings among 59 *C. jejuni* isolates (Meinersmann *et al.*, 2002). The same principle is used during MLST, a bacterial typing technique providing data for molecular epidemiology, which can also be used to obtain population genetic inferences.

(a)

(b)

(c)

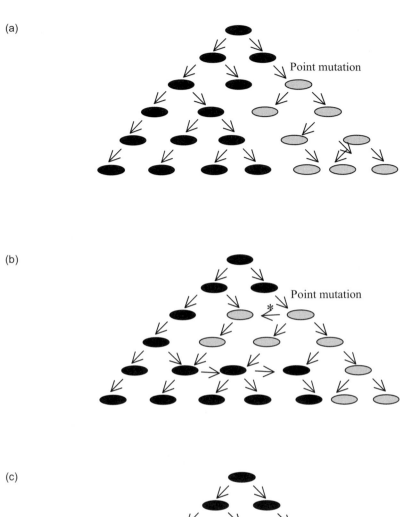
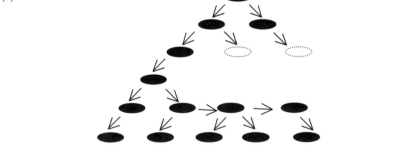

Figure 3.1 Factors shaping bacterial population structures. (a) **Vertical transmission of point mutations**. A point mutation conferring a variant genotype occurs and is transmitted vertically to daughter cells. (b) **Horizontal genetic exchange**. The gene conferring the variant genotype may be transmitted vertically and also by horizontal genetic exchange*. However, the original genotype may be restored by reversion (back mutation) or re-acquisition of the original allele. (c) **Negative or stabilizing selection**. Negative selection (for example, a change in conditions) results in the elimination of the variant genotype if it is poorly adapted to these conditions. Overall diversity is reduced. (d) **Positive or diversifying selection**. Positive selection, for example imposed by rapidly changing conditions leads to the emergence of additional genotypes, more quickly than they would otherwise. (e) **Stochastic events**. Chance events; for example the transfer to a new niche of a genotype which it is better adapted than the resident colonizer, causes the new colonizer to out compete the pre-existing colonizer.

(d)

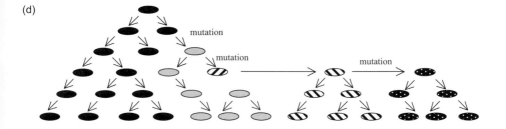

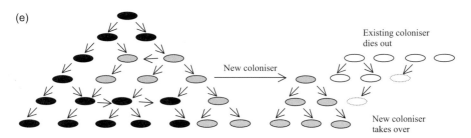

(e)

Figure 3.1 Continued.

There were a number of problems with MLEE, including: difficulties in comparing the results obtained in different laboratories; the fact that only differences that change the mobility of the enzyme were detected; and the indirect nature of genotype determination. To overcome these difficulties MLST was developed. This method takes the same approach as MLEE, in that housekeeping metabolic gene loci are targeted, but in this case the nucleotide sequences of short (~500 bp) DNA fragments of the genes are determined directly. This approach has numerous advantages in its accuracy and portability of the data, and seven loci, distributed around the chromosome, provide a level of discrimination roughly equivalent to that obtained with 15 to 20 MLEE loci (Maiden *et al.*, 1998). To ensure that the level of nucleotide sequence diversity identified among each of the seven loci is similar, they should not be adjacent to a gene under diversifying selection. The fact that nucleotide sequences are determined means that both the level of diversity and the selection pressure experienced by each locus can be assessed. In addition to providing isolate characterization information, MLST data allow the population structure of a bacterial species to be investigated.

The MLST approach has been applied successfully to studies of several bacterial pathogens, including *C. jejuni* (Enright and Spratt, 1998; Suerbaum *et al.*, 1998; Achtman *et al.*, 1999; Maiden *et al.*, 1998 Dingle *et al.*, 2001; Suerbaum *et al.*, 2001). The alleles and their fragment lengths chosen for the *C. jejuni* MLST scheme (Dingle *et al.*, 2001) are summarized in Table 3.1. In common with other MLST schemes, each unique allele is assigned a number in the order of discovery, and the seven numbers form an allelic profile or sequence type (ST) for each isolate, which is indicative of its genotype. At the time of writing (September 2003) the number of STs identified was 813, with data from a total of over 2000 isolates. The numbers of alleles described ranged from 63 for the *uncA* locus to 141 for the *pgm* locus (Table 3.1). A study employing a diverse collection of 43 *C. jejuni* isolates from five countries has confirmed that MLEE (using 11 loci), and

Table 3.1 Genetic diversity of the seven housekeeping gene loci used in MLST.

Locus	Fragment size (bp)	No. of alleles	No. of variable sites (%)
aspA	477	80	90 (18.9)
glnA	477	102	124 (26.0)
gltA	402	77	91 (22.6)
glyA	507	111	174 (34.3)
pgm	498	141	186 (37.3)
tkt	459	109	141 (30.7)
uncA	489	63	122 (24.9)

MLST with seven loci provide an equivalent level of discrimination (Sails *et al.*, 2001). Another MLST system for *C. jejuni* using six different loci demonstrated very similar results (Manning *et al.*, 2001).

MLST has a number of advantages for the study of *C. jejuni* epidemiology. For example, the technique is unaffected by the changes in gene order along the chromosome which occur as a result of intragenomic recombination (Wassenaar *et al.*, 1998; Hanninen *et al.*, 1999). Furthermore, MLST is easily reproduced in different laboratories and the data are electronically portable and amenable to storage in internet-accessible databases. A *Campylobacter* MLST database for the collation and exchange of MLST data has been established at http://pubmlst.org/campylobacter/. MLST data are highly discriminatory for isolate characterization, but have also enabled investigation of the population structure of the organism.

C. JEJUNI CHARACTERISTICS WHICH LED TO EPIDEMIOLOGICAL CONFUSION

The bacterial typing method of choice for much of the last century was serotyping, the Kauffman–White scheme (Popoff *et al.*, 2001) for *Salmonella enterica* providing a paradigm for other species. Serotyping was particularly useful for studies of *S. enterica* epidemiology because given genotypes are associated with particular phenotypic characters, perhaps as a consequence of infrequent recombination. Each *S. enterica* genotype could therefore be identified from any one of several characteristics, including the immunological reactivity of the surface polysaccharides, the property used in serotyping (Selander *et al.*, 1990). This approach proved invaluable in studies of *S. enterica* epidemiology because serotype was an extremely reliable indicator of the strain responsible in a particular disease outbreak.

A similar serotyping scheme, recognizing lipooligosaccharide (LOS) and capsule, was developed for *C. jejuni* (and also *C. coli*) (Penner *et al.*, 1983). However, in *C. jejuni* there is little evidence for a tree-like phylogeny, probably as a consequence of frequent horizontal genetic exchange (Dingle *et al.*, 2001; Suerbaum *et al.*, 2001). As a result, a given serotype may be associated with more than one genotype (i.e. ST or clonal complex) and, conversely, more than one serotype may occur in combination with a given genotype (Figure 3.2) (Dingle *et al.*, 2002). Therefore, the surface polysaccharides of *C. jejuni* cannot be relied upon to provide reliable epidemiological data since they are a very inconsistent indicator of genotype. This explains some of the difficulties which have been encountered in understanding *C. jejuni* epidemiology using serological methods.

C. jejuni isolates undergo intrachromosomal rearrangements which affect the whole chromosome as well as individual loci such as the flagellin genes (Parkhill *et al.*, 2000; Harrington *et al.*, 1997; Wassenaar, 1998; Hanninen *et al.*, 1999; Boer *et al.*, 2002).

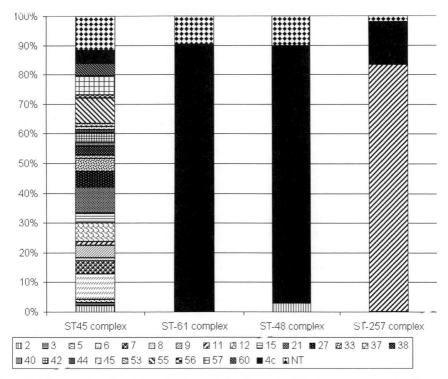

Figure 3.2 Serotype is an inconsistent indicator of clonal complex. The Penner serotypes found among four clonal complexes (ST-45 complex, ST-61 complex, ST-48 complex and ST-257 complex) are shown. The specific serotypes are indicated in the legend. 4c indicates the Penner serotype 4 complex containing serotypes 4, 13, 16, 50, and 64. NT indicates non-typeable isolates. ST-61 complex and ST-48 complex both contained predominantly isolates of Penner serotype 4 complex, ST-257 complex contained predominantly serotype 11, but also Penner 4 complex and ST-45 complex contained 25 different serotypes in addition to non-typeable isolates.

This complicates the interpretation of another bacterial typing technique, pulse-field gel electrophoresis (PFGE) fingerprinting, for epidemiological studies. Meinersmann (2000) observed that independently collected *C. jejuni* isolates rarely share the same PFGE profile and PFGE data for a collection of *C. jejuni* isolates from Curaçao did not always agree with MLST data (Duim *et al.*, 2003). A third characteristic of *C. jejuni*, which is likely to have contributed to the confusion over its epidemiology, is the presence of homopolymeric tracts within the genome. These result in phase variability of various genes involved in flagella, LOS and capsule biosynthesis (Parkhill *et al.*, 2000). These antigens are recognized in the Lior and Penner serotyping schemes (Lior *et al.*, 1982; Penner *et al.*, 1983) and their phase variation could confuse the interpretation of serotyping data. This may also account for the high numbers of untypeable isolates.

C. JEJUNI POPULATION STRUCTURE
The majority of *C. jejuni* isolates are naturally competent during growth, taking up DNA (particularly *Campylobacter* DNA), without special treatment (Wang and Taylor, 1990; Richardson and Park, 1997). This, and its extremely high level of genetic diversity,

suggested that the *C. jejuni* chromosome could be subject to frequent horizontal genetic exchange. Analysis of the nucleotide sequences of the fragments of the housekeeping genes of *C. jejuni* employed for MLST revealed mosaic genes with a net-like phylogeny (Figure 3.3), which is consistent with frequent recombination. High levels of recombination in *C. jejuni* are also suggested by several calculations of the index of association, a measure of the amount of free recombination within a bacterial population (Maynard Smith *et al.*, 1993). An I_A close to zero suggests the nucleotide sequences are at linkage equilibrium, whereas an I_A much greater than zero suggests a clonal population. An I_A of 0.256 was reported for a collection of 32 *C. jejuni* isolates (Suerbaum *et al.*, 2001), and a value of 1.9 for a set of 69 isolates (Manning *et al.*, 2001). Suerbaum *et al.* (2001) reported that the I_A was consistent with an amount of recombination which did not completely destroy linkage among loci. The importance of recombination in *C. jejuni* is also indicated by the observation that the number of allele combinations (referred to as sequence types or STs) exceeds the numbers of alleles. In one study, a total of 155 STs were identified among 194 isolates, the number of alleles ranging from 27 to 46 at different loci (Dingle *et al.*, 2001). In another, 31 STs occurred among 32 isolates, with the number of alleles identified between 9 and 15 per locus (Suerbaum *et al.*, 2001).

Further evidence for extensive recombination shaping *C. jejuni* populations is provided by the comparison of phylogenetic trees generated for different parts of the genome. An important prediction of clonality generated by an asexual process of reproduction is congruence, whereby different parts of the genome exhibit identical, or at least very similar,

(a)   (b)

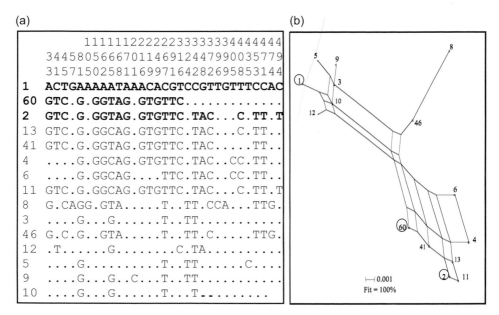

Figure 3.3 Analysis of 15 *C. jejuni pgm* alleles to indicate variation by point mutation and recombination (a) Variable sites identified among the 15 alleles. The allele numbers within the *C. jejuni* MLST scheme are listed to the left. The position of the variable sites within the allele sequence is indicated along the top (numbers to be read vertically). Allele 60 appears to be a mosaic of alleles 1 and 2, shown in bold. (b) Analysis of the same 15 sequences by split decomposition which created a network showing the relationship among the *pgm* alleles. This was produced using the program splitstree (Huson). Alleles 1, 2 and 60 which form the mosaic in (a) are ringed.

phylogenies. The construction of maximum likelihood trees using each of the MLST loci demonstrated a lack of congruence in that each tree was different. In addition, comparison of the trees generated for each MLST locus with each other and to 100 randomly generated trees (Figure 3.4) indicated that the data from a given locus were equally poorly modelled by a tree generated from any of the other loci examined and random tree topologies. These data therefore do not conform to the clonal model for population structure and are more consistent with frequent horizontal genetic exchange.

Despite high levels of recombination and a lack of congruence, the MLST data derived for *C. jejuni* can be described in the context of a clonal complex model and at least 17 *C. jejuni* clonal complexes have been identified to date (Dingle *et al.*, 2002). The practical steps in the identification of a clonal complex using MLST data are as follows. Firstly, the central genotype (sometimes referred to as the founder genotype) is identified. After the central genotype has been identified, additional STs differing at one, two or three loci are then assigned to the clonal complex. Each complex is thought to be descended from the central genotype and the complex is named after the ST of this genotype (Dingle *et al.*, 2002). The hypothesis that the central genotype is also the fittest in evolutionary terms, and therefore the most abundant, is supported by the observation that the majority *C. jejuni* outbreaks examined to date were caused by isolates with central genotypes (unpublished data).

Heuristic methods such as the BURST algorithm within the program START (Jolley *et al.*, 2001), and split decomposition within the program Splitstree (Huson, 1998) have

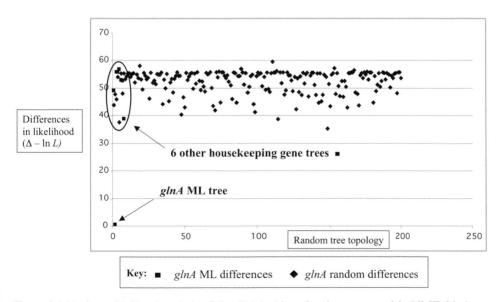

Figure 3.4 Maximum likelihood analysis of the *C. jejuni* housekeeping genes used in MLST. Maximum likelihood trees were constructed for each of the seven housekeeping genes used in MLST. In the example shown, the tree for *glnA* was compared to the trees for the six other housekeeping genes and to 100 trees generated at random. The tree for *glnA* was no more similar to the six other housekeeping gene trees than it was to the 100 randomly generated trees. This demonstrates that frequent recombination has disrupted any phylogenetic signal among the housekeeping genes of *C. jejuni*. If no recombination had occurred, the *glnA* tree would have been similar to the trees for the other six housekeeping genes.

been used to identify the central genotypes. Software incorporated into the MLST database (http://pubmlst.org/) now enables an ST to be assigned automatically to its clonal complex. STs which are members of newly identified complexes can be recognized by the fact that they are not assigned by this software to a previously known complex. Collections of these 'unique' STs can then be examined to identify the central genotypes of additional clonal complexes. Examination of the collection of 'unique' STs facilitates the identification of central genotypes of novel clonal complexes that arise.

The 17 *C. jejuni* clonal complexes described to date were first identified using a diverse collection of 814 UK and Dutch *C. jejuni* isolates from cases of human disease, food animals or products, and the environment (Dingle *et al.*, 2002). These clonal complexes vary in their frequency among human disease isolates and in their diversity. The occurrence of single 'unique' isolates which have yet to be assigned a clonal complex, suggests that there may be many more complexes to be found. The combined impact of the selection processes described earlier is likely to have resulted in the weakly clonal population structure observed in *C. jejuni*. However, although clonal complexes can be defined in *C. jejuni*, the pattern of descent is extremely difficult to discern due to the frequency of horizontal genetic exchange.

Of the clonal complexes identified to date, 14 contain a central genotype distinct from that of any other complex. The three exceptions are the ST-21, ST-48 and ST-206 complexes, which share identical allele combinations at four of the seven loci. These central genotypes and their relatives may form a 'supercomplex' of closely related genotypes which undergo very frequent horizontal genetic exchange. They have been described as separate entities because each has an abundant central genotype (Dingle *et al.*, 2002), and each central genotype had many single, double and triple locus variants. Together, these three complexes contained 385 (47%) of the collection of 814 isolates. The abundance of the STs within the ST-21 supercomplex is shown in Figure 3.5. The supercomplex exhibits a high level of antigenic diversity in terms of their Penner serotype and variants of *flaA* short variable region (SVR) (Dingle *et al.*, 2002). This diversity may be a consequence of rapid clonal expansion and a high level of competence conferring greater potential for horizontal genetic exchange and adaptation to new environments.

Among the other clonal complexes, some are associated with a single Penner serotype and are uniform in terms of the variability within their *flaA* SVR (Dingle *et al.*, 2002). The *C. jejuni* clonal complex ST-362 (containing isolates of Penner serotype 41 only) is extremely homogeneous with only isolates with the central genotype have been identified to date. The ST-22 complex (mainly Penner serotype 19) is slightly more diverse, with 14 further STs in addition to ST-22. Other examples include the ST-49 complex (Penner serotype 18) and the ST-42 complex (Penner serotypes 23 and 36). These clonal complexes are genotypically homogeneous, containing a more limited number of STs than the ST-21 supercomplex, and the similarly large and diverse ST-45 complex. They may also have a lower level of competence for DNA uptake, so that the uptake of foreign DNA by transformation is limited. Evidence of a limited level of competence among isolates of Penner serotypes 19 and 41 (found in ST-22 and ST-362 complexes) and a high level of natural competence among Penner serotypes 1 and 2 (most numerous in the ST-21 supercomplex) was provided by de Boer *et al.* (2001).

Islands of genetic stability may also occur within the large heterogeneous clonal complexes. A genetically stable clone associated with Penner serotype 6 has been described

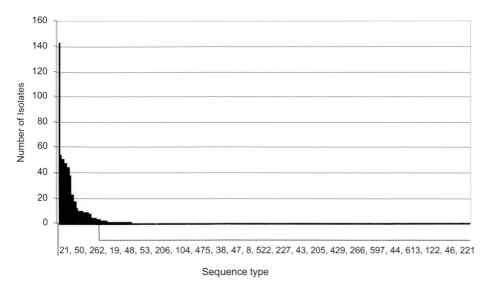

Figure 3.5 The relative abundance of the isolates having STs assigned to the ST-21 supercomplex. This comprises the ST-21 complex, the ST-206 complex and the ST-48 complex. STs numbers are shown only for isolates which occur four times or more.

(Manning *et al.*, 2001), a serotype which falls within the ST-45 complex. Penner serotype 55 isolates have also been reported to lack natural competence (de Boer *et al.*, 2001), and these occur within the ST-45 complex. ST-45 complex may represent a collection of very closely related clones, which appear to be one clonal complex due to their genetic similarity. The complex may contain more than one predominant genotype.

FREQUENCY AND GEOGRAPHICAL DISTRIBUTION OF CLONAL COMPLEXES

Isolates of the most frequent clonal complexes in the largest *C. jejuni* MLST study to date (814 isolates, Dingle *et al.*, 2002) were identified with broadly similar frequency in two additional studies (Sails *et al.*, 2001, Manning *et al.*, 2001) including isolates predominantly from the UK, and North America. An Australian study of 157 isolates (Mickan *et al.*, 2002), also found that isolates of the ST-21 supercomplex were commonly isolated from cases of human disease, but ST-48 complex was by far the most frequent of the three clonal complexes within the supercomplex. Although ST-21 did occur, it was very rare. This study found ST-257 complex second in abundance and, as expected, it was dominated by ST-257. However, isolates of ST-45 complex (second most abundant in the study by Dingle *et al.*, 2002) were rare. ST-354 complex was the third most abundant in the Australian data set, but ST-528 predominated within this complex. So, although the same clonal complexes were identified in cases of human disease occurring on a different continent, their frequency, and in some cases the predominant ST differed to those observed in Europe. This may be a reflection of a similar diet, but large geographical separation and climatic differences between the two continents may have allowed different genotypes to dominate.

CLONAL COMPLEX, HOST SPECIES AND HUMAN DISEASE SYNDROMES

The clonal complex is proving to be an epidemiological unit which can be used successfully to examine the associations between *C. jejuni* clones and their mammalian and avian hosts, the potential sources of human infection. Associations among distinct clones and particular clinical syndromes such as invasive disease, demyelinating neuropathies and reactive arthritis may also be identified. The data described by Dingle *et al.* (2001) support the idea that clonal complexes may be maintained by niche adaptation (and therefore geographical separation) as certain clonal complexes were over-represented among particular isolation sources. ST-45 and ST-257 complexes contained isolates mainly from human disease and poultry, and the isolates of ST-61 and ST-48 complexes were mainly from human disease, cattle and sheep. ST-177 and ST-179 complexes contained isolates mainly from the sand of bathing beaches. Three isolates from ST-177 complex have now been identified in three cases of human disease (G. Manning, personal communication), however, this clonal complex is under represented in human disease compared to the clonal complexes commonly identified in humans. Isolates of ST-403 complex are frequently carried by pigs in the UK (G. Manning, personal communication). This clonal complex is rare among human disease in the UK, but it is the most frequently isolated clone from cases of *C. jejuni* gastroenteritis in Curaçao (Duim *et al.*, 2003). Data such as these may allow the importance of different transmission routes in different parts of the world to be determined.

Guillain–Barré syndrome (GBS) and Miller Fisher syndrome (MFS) are autoimmune diseases of the peripheral nervous system which are characterized by acute flaccid paralysis (Nachamkin *et al.*, 1998). As many as 40% of patients who develop GBS or MFS have a *C. jejuni* infection 1 to 3 weeks prior to developing neurological symptoms (Molnar *et al.*, 1982; Yuki *et al.*, 1990). MLST data have been obtained for a set of 25 *C. jejuni* isolates associated with cases of GBS and MFS (Dingle *et al.*, 2001). A total of 13 clonal complexes were identified among these isolates which contained 20 distinct STs and 20 different *flaA* short variable region sequences. The most common clonal complex was ST-21 which is also the most common among the isolates associated with human gastroenteritis (Dingle *et al.*, 2002). ST-22 complex was the next most common among GBS and MFS patients and apparently over-represented among the neuropathy-associated isolates (although the data set was small). This clonal complex is also associated with Penner serotype 19, a clone of which has been proposed to be associated with GBS (Fujimoto *et al.*, 1997). ST-45 complex, the second most common clonal complex in the human isolates (Dingle *et al.*, 2002) was not detected at all among the GBS and MFS isolates, suggesting that this complex is under-represented among the isolates which cause the demyelinating neuropathies. A diverse range of Penner Serotypes are associated with ST-45 complex, but they may be less likely to induce antibodies which are cross-reactive with the gangliosides (Nachamkin *et al.*, 1998) of the host.

CONCLUSIONS

The genetic diversity of *C. jejuni* is consistent with a large effective population size, which would be expected given its wide host range. The isolates which have not yet been assigned to clonal complexes (66 in a study by Dingle *et al.*, 2002, six in a smaller study by Mickan *et al.*, 2002) frequently occur in small groups of 2–4 STs and may represent additional clonal complexes, which occur infrequently among the isolation sources examined to date. Models of population structure will be refined as greater volumes of data are acquired and *C. jejuni* genetic variation is exploited fully to determine and understand transmission routes to humans.

By providing robust, comparable data that can be analysed with a variety of theoretical frameworks, population genetics provides a means by which bacterial population structures can be defined and the adaptation of clones to particular hosts and their relationships with disease frequency and severity can be examined. This may ultimately lead to effective public health interventions which break the cycle of transmission or allow vaccines to be developed for humans or the food animals which carry *Campylobacter*. To halt the persistent increase in *C. jejuni* infections of humans, a more complete understanding of its population structure in relation to its mammalian and avian hosts is required. Studies of *Campylobacter* population genetics in which the complete sequences of carefully chosen genes are determined (in addition to the seven loci used in MLST) from representative isolate collections, are required to further advance the field.

Acknowledgements

This publication made use of the Multilocus Sequence Typing Website located at http://pubmlst.org/, initial development by Man-Suen Chan, University of Oxford, UK, with further work by Dr Keith Jolley, University of Oxford. The web site is funded by the Wellcome Trust and supported by the United Kingdom Department of the Environment, Food and Rural Affairs contract number OZ0604.

The authors thank Dr Eddie Holmes, Department of Zoology, Oxford University, for performing the maximum likelihood analysis.

The serotyping data shown in Figure 3.2 were supplied by Dr David Wareing (Microbiology Research and Development Department, Dynal Biotech Ltd, UK), Professor Eric Bolton and Professor Andrew Fox (Health Protection Agency North West Laboratory, Manchester Medical Microbiology Partnership, Manchester Royal Infirmary, Manchester, UK).

References

Achtman, M., Zurth, K., Morelli, G., Torrea, G., Guiyoule, A., and Carniel, E. 1999. *Yersinia pestis*, the cause of plague, is a recently emerged clone of *Yersinia pseudotuberculosis*. Proc. Natl. Acad. Sci. USA 96: 14043–14048.

de Boer, P., Duim, B., van Putten, J.P.M., and Wagenaar, J.A. 2001. Clonal complexes of *Campylobacter jejuni* are genetically preserved by lack of natural transformation. Abstract H41 *Campylobacter*, *Helicobacter* and Related Organisms Conference 2001. p. 74.

Boer, P., Wagenaar, J.A., Achterberg, R.P., Putten, J.P., Schouls, L.M., and Duim, B. 2002. Generation of *Campylobacter jejuni* genetic diversity *in vivo*. Mol. Microbiol. 44: 351–359.

Dingle, K.E., Colles, F.M., Wareing, D.R., Ure, R., Fox, A.J., Bolton, F.E., Bootsma, H.J., Willems, R.J., Urwin, R., and Maiden, M.C. 2001. Multilocus sequence typing system for *Campylobacter jejuni*. J. Clin. Microbiol. 39: 14–23.

Dingle, K.E., Van Den Braak, N., Colles, F.M., Price, L.J., Woodward, D.L., Rodgers, F.G., Endtz, H.P., Van Belkum, A., and Maiden, M.C. 2001. Sequence typing confirms that *Campylobacter jejuni* strains associated with Guillain–Barré and Miller Fisher syndromes are of diverse genetic lineage, serotype, and flagella type. J. Clin. Microbiol. 39: 3346–3349.

Dingle, K.E., Colles, F.M., Ure, R., Wagenaar, J.A., Duim, B., Bolton, F.J., Fox, A.J., Wareing, D.R., and Maiden, M.C. 2002. Molecular characterization of *Campylobacter jejuni* clones: a basis for epidemiologic investigation. Emerging Infect. Dis. 8: 949–955.

Duim, B., Godschalk, P.C., van den Braak, N., Dingle, K.E., Dijkstra, J.R., Leyde, E., van der Plas, J., Colles, F.M., Endtz, H.P., Wagenaar, J.A., Maiden, M.C., and van Belkum, A. Molecular evidence for dissemination of unique *Campylobacter jejuni* clones in Curaçao, Netherlands Antilles. J. Clin. Microbiol. 41: 5593–5597.

Enright, M.C. and Spratt, B.G. 1998. A multilocus sequence typing scheme for *Streptococcus pneumoniae*: identification of clones associated with serious invasive disease. Microbiology 144: 3049–3060.

Feil, E.J., and Spratt, B.G. 2001. Recombination and the population structures of bacterial pathogens. Annu. Rev. Microbiol. 55: 561–590.

Frost, J.A. 2001; Current epidemiological issues in human campylobacteriosis. J. Appl. Microbiol. 90: 85S–95S.

Fujimoto, S., Allos, B.M., Misawa, N., Patton, C.M., and Blaser, M.J. 1997. Restriction fragment length polymorphism analysis and random amplified polymorphic DNA analysis of *Campylobacter jejuni* strains isolated from patients with Guillain–Barré syndrome. J. Infect. Dis. 176: 1105–1108.

Gillespie, I.A, O'Brien, S.J, Frost, J.A., Adak, G.K., Horby, P., Swan, A.V., Painter, M.J., and Neal, K.R., *Campylobacter* Sentinel Surveillance Scheme Collaborators. 2002. A case–case comparison of *Campylobacter coli* and *Campylobacter jejuni* infection: a tool for generating hypotheses. Emerging Infect. Dis. 8: 937–942.

Gupta, S., and Maiden, M.C. 2001. Exploring the evolution of diversity in pathogen populations. Trends Microbiol. 9: 181–185.

Hanninen, M.L., Hakkinen, and M., Rautelin, H. 1999. Stability of related human and chicken *Campylobacter jejuni* genotypes after passage through chick intestine studied by pulsed-field gel electrophoresis. Appl. Environ. Microbiol. 65: 2272–2275.

Harrington, C.S., Thomson-Carter, F.M., and Carter, P.E. 1997. Evidence for recombination in the flagellin locus of *Campylobacter jejuni*: implications for the flagellin gene typing scheme. J. Clin. Microbiol. 35: 2386–2392.

Huson, D.H. 1998. SplitsTree: analyzing and visualizing evolutionary data. Bioinformatics 14: 68–73.

Jolley, K.A., Feil, E.J., Chan, M.S., and Maiden, M.C. 2001. Sequence type analysis and recombinational tests (START). Bioinformatics 17: 1230–1231.

Kapur, V., Whittam, T.S., Musser, J.M. 1994. Is *Mycobacterium tuberculosis* 15,000 years old? J. Infect. Dis. 170: 1348–1349.

Lior, H., Woodward, D.L., Edgar, J.A., Laroche, L.J., and Gill, P. 1982. Serotyping of *Campylobacter jejuni* by slide agglutination based on heat-labile antigenic factors. J. Clin. Microbiol. 15: 761–768.

Lorenz, M.G., and Wackernagel, W. 1994; Bacterial gene transfer by natural genetic transformation in the environment. Microbiol. Rev. 58: 563–602.

Manning, G., Dowson, C.G., Ahmed, I.H., Bagnall, M., West, M., and Newell, D. 2001. A study of the population structure of *Campylobacter jejuni* using multilocus sequence typing (MLST). Abstract H02 *Campylobacter*, Helicobacter and Related Organisms Conference 2001. Int. J. Med. Microbiol. 291: supplement 31, 63.

Maiden, M.C., Bygraves, J.A., Feil, E., Morelli, G., Russell, J.E., Urwin, R., Zhang, Q., Zhou, J., Zurth, K., Caugant, D.A., Feavers, I.M., and Achtman, M., Spratt, B.G. 1998; Multilocus sequence typing: a portable approach to the identification of clones within populations of pathogenic microorganisms. Proc. Natl. Acad. Sci. USA 95: 3140–3145.

Maynard-Smith, J., Dowson, C.G., and Spratt, B.G. 1993; How clonal are bacteria? Proc. Natl. Acad. Sci USA. 90: 4384–4388.

Meinersmann, R.J. 2000; Population genetics and genealogy of *Campylobacter jejuni*. In: *Campylobacter*. I. Nachamkin and M. Blaser eds. ASM Press, Washington, DC. pp. 351–368.

Meinersmann, R.J., Patton, C.M., Evins, G.M., Wachsmuth, I.K., and Fields, P.I. 2002. Genetic diversity and relationships of *Campylobacter* species and subspecies. Int. J. Syst. Evol. Microbiol. 52: 1789–1797.

Mickan, L., Doyle, R., Dingle, K.E., Unicomb, L., and Ferguson, J. 2002. Multilocus sequence typing of *Campylobacter jejuni*. ASM Conference Poster.

Milkman, R. 1973. Electrophoretic variation in *Escherichia coli* from natural sources. Science 182: 1024–1026.

Molnar, G.K., Mertsola, J., and Erkko, M. 1982. Guillain–Barré syndrome associated with *Campylobacter* infection. Br. Med. J. (Clin. Res. Ed.). 285: 652.

Musser, J.M. 1996. Molecular population genetic analysis of emerged bacterial pathogens: selected insights. Emerging Infect. Dis. 2: 1–17.

Nachamkin, I., Allos, B.M., and Ho, T. 1998. *Campylobacter* species and Guillain–Barré syndrome. Clin. Microbiol. Rev. 11: 555–567.

Ochman, H., Lawrence, J.G., and Groisman, E.A. 2000; Lateral gene transfer and the nature of bacterial innovation. Nature 405: 299–304.

Parkhill, J., Wren, B.W., Mungall, K., Ketley, J.M., Churcher, C., Basham, D., Chillingworth, T., Davies, R.M., Feltwell, T., Holroyd, S., Jagels, K., Karlyshev, A.V., Moule, S., Pallen, M.J., Penn, C.W., Quail, M.A.,

Rajandream, M.A., Rutherford, K.M., van Vliet, A.H., Whitehead, S., and Barrell, B.G. 2000. The genome sequence of the food-borne pathogen *Campylobacter jejuni* reveals hypervariable sequences. Nature 403: 665–668.

Penner, J.L., Hennessy, J.N., and Congi, R.V. 1983. Serotyping of *Campylobacter jejuni* and *Campylobacter coli* on the basis of thermostable antigens. Eur. J. Clin. Microbiol. 4: 378–383.

Popoff, M.Y., Bockemuhl, J., Brenner, F.W., and Gheesling, L.L. 2001. Supplement 2000 (no. 44) to the Kauffmann–White scheme. Res. Microbiol. 152: 907–909.

Richardson, P.T., and Park, S.F. 1997. Integration of heterologous plasmid DNA into multiple sites on the genome of *Campylobacter coli* following natural transformation. J. Bacteriol. 179: 1809–1812.

Sails, A.D., Swaminathan, B., and Fields, P.I. 2001. DNA sequencing-based subtyping of *Camplyobacter jejuni*. ASM Conference Poster C479.

Selander, R.K., Korhonen, T.K., Vaisanen-Rhen, V., Williams, P.H., Pattison, P.E., and Caugant, D.A. 1986. Genetic relationships and clonal structure of strains of *Escherichia coli* causing neonatal septicemia and meningitis. Infect. Immun. 52: 213–222.

Selander, R.K., Beltran, P., Smith, N.H., Helmuth, R., Rubin, F.A., Kopecko, D.J., Ferris, K., Tall, B.D., Cravioto, A., and Musser, J.M. 1990. Evolutionary genetic relationships of clones of *Salmonella* serovars that cause human typhoid and other enteric fevers. Infect. Immun. 58: 2262–2275.

Spratt, B.G., and Maiden, M.C. 1999. Bacterial population genetics, evolution and epidemiology. Phil. Trans. R. Soc. Lond. B Biol. Sci. 354: 701–710.

Suerbaum, S., Smith, J.M., Bapumia, K., Morelli, G., Smith, N.H., Kunstmann, E., Dyrek I., and Achtman, M. 1998. Free recombination within *Helicobacter pylori*. Proc. Natl. Acad. Sci. USA 95: 12619–12624.

Suerbaum, S., Lohrengel, M., Sonnevend, A., Ruberg, F., and Kist, M. 2001. Allelic diversity and recombination in *Campylobacter jejuni*. J. Bacteriol. 183: 2553–2559.

Wassenaar, T.M., Geilhausen, B., and Newell, D.G. 1998; Evidence of genomic instability in *Campylobacter jejuni* isolated from poultry. Appl. Environ. Microbiol. 64: 1816–1821.

Wang, Y., and Taylor, D.E. 1990; Natural transformation in *Campylobacter* species. J. Bacteriol. 172: 949–955.

Yuki, N., Yoshino, H., Sato, S., Miyatake, T. 1990; Acute axonal polyneuropathy associated with anti-GM1 antibodies following *Campylobacter* enteritis. Neurology 40: 1900–1902.

Chapter 4

Campylobacter jejuni Strain Variation

Collette Fitzgerald, Andrew D. Sails and Patricia I. Fields

Abstract
The diversity within *Campylobacter jejuni* is well established and has been detected at both the phenotypic and genotypic level. The application of higher resolution molecular subtyping methods has resulted in increasing questions regarding the (in)stability of the *Campylobacter* genome and the implications of this for molecular epidemiological investigations. Current evidence suggests that genomic rearrangements play a role in strain diversity. But stable clones of *C. jejuni* have also been reported. The aim of this chapter is to review current experimental data on *C. jejuni* strain variation based on phenotypic and genotypic methods.

INTRODUCTION
Campylobacter has been a leading cause of gastroenteritis in the US and other parts of the world (Mead *et al.*, 1999), though recent data indicate that the incidence of laboratory-diagnosed cases of infection with *Campylobacter* may be declining in the US (CDC, 2003). Despite this decline, the burden of disease caused by *C. jejuni* is still substantial, as is the socioeconomic impact (Roberts *et al.*, 2003).

Campylobacters have been isolated from a diverse range of domestic and wild animals and birds (Healing *et al.*, 1992; Butzler, 1984). Poultry is now established as a common source of human infection with *C. jejuni*, though numerous other vehicles have been noted (Sopwith *et al.*, 2003). *Campylobacter* infection is typically sporadic; outbreaks are relatively rare, making it difficult to characterize the epidemiology of the disease.

A wide variety of subtyping methods have been applied to *Campylobacter* isolates and have shown that this organism is very diverse (Bolton *et al.*, 1984; Wareing *et al.*, 2002; Dingle *et al.*, 2001a; Dorrell *et al.*, 2001). The large diversity among human *Campylobacter* isolates may be indicative of the sporadic nature of campylobacterosis and the variety of different sources for *Campylobacter* infection. Recent evidence suggests that genomic rearrangements may play a role in strain diversity (de Boer *et al.*, 2002). However, other reports suggest that not all campylobacters exhibit diversity; stable clones of *C. jejuni* have been reported (Manning *et al.*, 2001; Nielsen *et al.*, 2001). The aim of this chapter is to review current experimental data on *C. jejuni* strain variation based on phenotypic and genotypic methods.

STRAIN VARIATION BASED ON SEROTYPING

Penner and Lior serotyping
Two serotyping schemes were developed in the 1980s for the epidemiological characterization of *Campylobacter* isolates. The heat-stable (HS) serotyping scheme of

Penner (Penner and Hennessy, 1980) and the heat-labile (HL) serotyping scheme of Lior *et al.* (1982) provided the first methods to subtype *C. jejuni*. The application of these techniques to *C. jejuni* isolates from human, farm animal and environmental specimens has proven useful in the investigation of outbreaks and identification of potential reservoirs for human infection (Patton *et al.*, 1985; Woodward and Rodgers, 2002). However, the requirement for a large panel of antisera that is labour intensive to produce and maintain and is not commercially available in order to perform serotyping has restricted it use to larger reference laboratories and prevented its routine application to strain characterization.

Early studies indicated that common serotypes existed among isolates from human infection. The Penner serotypes HS1, HS2 and HS4 complex were identified as common *C. jejuni* serotypes in many countries (Penner, 1983; Mills *et al.*, 1991; Karmali *et al.*, 1983). However, global differences in prevalence of other serotypes were noted (Lastovica *et al.*, 1986; Albert *et al.*, 1992). For example in Sweden, Penner serotypes HS2, HS4, HS6 and HS8 were the four most common serotypes and accounted for more than half of domestically acquired infections during a 5-year period (Sjogren *et al.*, 1989). Similarly, these four serotypes were among the 10 most common identified in the US (Patton *et al.*, 1993). In contrast, HS8, HS17, HS22, HS1,44 and HS19 were the predominant serotypes in an Australian study (Albert *et al.*, 1992). Common serotypes that were identified among human isolates were also found in environmental samples, food, and food animals, indicating that these may be reservoirs for human disease (Munroe *et al.*, 1983; Jones *et al.*, 1984; Fricker and Park, 1989). More recent serotyping data indicate that the same serotypes still predominate in human infections (Table 4.1).

Waring *et al.* (2002) used a combination of phenotypic methods including Penner serotyping, phage typing and biotyping to investigate the diversity of *Campylobacter* isolates from sporadic cases of human enteritis in the UK. The 754 *C. jejuni* isolates were divided into 33 serogroups; Penner HS4 complex was the most common serotype and accounted for 24% of isolates. HS2 and HS1 were the second and third most common types, respectively. The 10 most common serotypes accounted for 66% of the isolates tested. Non-typability is a well-documented shortcoming associated with serotyping and today still remains a major disadvantage of serotyping. In the UK study mentioned above, 14.6% of the 754 isolates were non-typeable using the panel of 42 antisera. The Simpson's diversity index for all three methods combined was 0.997, indicating that isolates from sporadic infections are very diverse. However, significant associations were seen between some serotypes and phage groups (PG) including serotypes HS1, HS2 and HS4 and phage types PG52, 55 and 121. Despite the phenotypic diversity observed, some strains indistinguishable for several phenotypic markers were isolated from sporadic human disease over a period of at least 10 years. The persistence of these strains in the population indicates that they may be derived from a common reservoir which may serve as a persistent source of human infection.

Patton *et al.* (1993) demonstrated that a combination of the most prevalent Penner and Lior serotypes provided a greater level of discrimination for outbreak investigations. Other studies have also indicated that the combination of HS and HL serotyping is a useful approach for investigating the epidemiology of *C. jejuni* infections (Jackson *et al.*, 1998; Woodword and Rodgers, 2002). Jackson *et al.* (1998) described strong, non-random associations between HS and HL serotypes among 9024 sporadic human isolates from the UK. Ten serotypes accounted for 83.1% of the typable isolates; three serotypes were further characterized by biotyping, phagetyping, 16S ribotyping and RFLP analysis.

Table 4.1 Distribution of the most common *Campylobacter* Penner serotypes.

Penner serotype	Canada[1]		United States[2]		UK[3]		Scotland[4]	
	Rank	No. of isolates (% of total)	Rank	No. of isolates (% of total)	Rank	No. of isolates (% of total)	Rank	No. of isolates (% of total)
HS2	1	292 (14.4)	5	24 (8.3)	2	91 (12.1)	2	658 (17.4)
HS4	2	291 (14.3)	1	60 (20.9)	1	178 (23.6)	1	696 (18.4)
HS1	3	209 (10.3)	2	34 (11.8)	3	78 (10.3)	4	301 (7.9)
HS3	4	141 (7.0)	6	17 (5.9)				
HS5	5	81 (4.0)	4	27 (9.4)				
HS8	6	58 (2.9)	3	29 (10.1)			6	145 (3.8)
HS11	7	49 (2.4)			5	22 (2.9)	5	214 (5.7)
HS18	8	42 (2.1)			10	13 (1.7)		
HS21	9	38 (1.9)			8	18 (2.4)	8	52 (1.4)
HS37	10	37 (1.8)					9	52 (1.4)
HS23	11	35 (1.7)			4	26 (3.5)	10	51 (1.8)
HS31	12	34 (1.7)						
HS19	13	32 (1.6)	8	10 (3.5)	7	19 (2.5)		
HS10	14	27 (1.3)	10	9 (3.1)	9	17 (2.2)		
HS42	15	27 (1.3)						
HS58							3	504 (13.3)
HS6			7	14 (14.9)	6	22 (2.9)	7	144 (3.8)
HS15, 38			8	10 (3.5)				
NT (non-typeable)						109 (14.4)		311 (8.2)

[1]Penner (1983), Canada.
[2]Patton *et al.* (1993), USA, July 1989 and June 1990.
[3]Wareing (2001), UK: December 1990 to June 1991.
[4]McKay *et al.* (2001), Scotland, November 1998 to October 1999.

Substantial conservation of both genotypic and phenotypic characteristics within a serotype was evident despite the fact the strains within each serotype were geographically and temporally diverse, indicating that serotypes represented lineages of common descent in *C. jejuni*. Further evidence for clonal relationships within HS/HL serovars were observed by Nachamkin *et al.* (1996) using flagellin gene typing.

LEP serotyping

In an attempt to alleviate the problems associated with non-specific agglutination and cross-reactivity in the Penner scheme and to simplify the method, a modified Penner serotyping scheme (LEP) was developed (Frost *et al.*, 1998). Antigens are detected by direct bacterial cell agglutination rather than a passive haemagglutination and absorbed rather than unabsorbed antisera is used. When this scheme was used to serotype 2407 *C. jejuni* strains, 47 *C. jejuni* serotypes were identified. The 10 most prevalent serotypes accounted for about 50% of the total number of isolates tested in Scotland and the UK (Frost *et al.*, 1998; McKay *et al.*, 2001). Phage typing was used to subdivide the predominant LEP serotypes and together they provided a more discriminatory typing strategy than serotyping alone. Despite the phenotypic diversity observed among the strains, the prevalent subtypes were the same as those described by Wareing *et al.* (2002), who isolated them over 5 years earlier. The combination of LEP serotyping and phage typing has since enabled identification of eight specific subtypes through a 2-year period in England and Wales.

In 2001, the Scottish Reference laboratory performed Penner and LEP serotyping on 3633 *C. jejuni* isolates from human gastrointestinal infections received over a 1-year period to determine if the modified scheme provided improved discrimination (McKay *et al.*, 2001). Certain Penner serotypes appeared more predominant in Scotland during the sampling period; the 10 most common types occurring in 73.2% of isolates (see Table 4.1). Regional differences were recognized in Scotland; the third most common Penner serotype, HS58, had a very low incidence (0.4%) in the Penner study (Penner *et al.*, 1983). No significant advantages for the LEP scheme over the Penner scheme were reported. A significant percentage of isolates were nontypable (36.6%) using the LEP method, compared to 8.2% using the Penner method; greater cross-reactivity was observed using the LEP method, and the results from the two schemes were not directly comparable. The differences in agglutination specificity between the two schemes may reflect the differences in antigens detected by the two methods (Oza *et al.*, 2002).

INFERENCES ON POPULATION STRUCTURE FROM MULTILOCUS ENZYME ELECTROPHORETIC ANALYSIS

Multilocus enzyme electrophoresis (MEE) is a technique that indexes the neutral genetic variation of several housekeeping genes by measuring electrophoretic mobility of enzymes on starch gels. Neutral genetic variation is used as a measure of evolutionary relationships between strains. Early MEE studies predicted a moderately high frequency of intra-species recombination for *C. jejuni* and *C. coli* (Aeschbacher and Piffaretti, 1989), implying that *Campylobacter* species are a largely diverse population with no clonal structure. A more recent MEE study (Meinersmann *et al.*, 2002) also concluded that *C. jejuni* as a population is extremely diverse, although the index of association reported in this study indicated possible clonal development in *C. jejuni*.

Although phenotypic methods such as serotyping have been effectively used for the subtyping of *C. jejuni*, they do have limitations. Many are not universally applicable and are only useful for the species they were developed for. In addition, production and quality control of antiserum reagents for serotyping schemes are costly and not widely available and nontypability is a concern. The development of molecular techniques which target genetic variation within the DNA sequence of the organism have since become widely applied to subtype *C. jejuni*. These molecular methods are often more universally applicable, reduce problems encountered with untypability and provide a much higher level of discriminatory power.

GENETIC STABILITY OF THE *C. JEJUNI* GENOME

Genome instability in *C. jejuni*

The ability to track the sources and routes of transmission in both outbreaks and sporadic infection relies on the stability of the characteristic used as the epidemiological marker. In recent years, there has been increasing evidence of instability in the *C. jejuni* genome that has been shown to affect both PFGE patterns (Wassenaar *et al.*, 1998) and *fla* typing (Harrington *et al.*, 1997; Wassenaar *et al.*, 1995). Further, de Boer *et al.* (2002) demonstrated that interstrain genetic exchange as well as intragenomic alterations occur *in vivo* during experimental *C. jejuni* infection of chickens, even in the absence of selective pressure.

Wassenaar *et al.* (1998) observed that *C. jejuni* isolates from chicken meat had identical serotype and *fla* types, but had minor variations in their *Sma*I PFGE profiles. The PFGE genotypes were then shown to be stable upon *in vitro* subculturing and *in vivo* passage in chickens. Genetic rearrangements were noted as the most likely mechanism for this genetic variation. Finnish investigators observed a more frequent rate of genomic rearrangements than Wassanaar *et al.* (1998) during experimental colonization of chickens as demonstrated by changes in PFGE and ribotype patterns (Hanninen *et al.*, 1999). Two of 12 *C. jejuni* strains changed PFGE profile and ribotype after colonizing chickens for six days. One of the two strains also had a change in Penner serotype from HS57 to HS27; the new PFGE profile, ribotype and serotype pattern for this strain matched that of another isolate used in the study that had a 'stable' genotype. Interestingly, this genotype appears to have long-term stability, as it had also been isolated 10 years earlier from cattle.

Another study investigated six *C. coli* strains subcultured up to 50 times and showed that four of the six isolates underwent changes in their PFGE profile (On *et al.*, 1998). However, Nielsen *et al.* (2001) subsequently noted that the changes in PFGE pattern for three of the four isolates was due to mixed cultures rather than genomic instability. Indirect evidence for recombination within and between flagellin loci was documented by Harrington *et al.* (1997). Similarly, rearrangements at flagellin loci were demonstrated in a natural host during colonization of chicken intestine (Nuijten *et al.*, 2000). Thus, these studies suggest that the propensity of the flagellin locus to undergo genetic change is a limiting factor in its usage in long-term epidemiological studies (Harrington *et al.*, 1997; Petersen and On, 2000). A recent study investigating the utility of MLST, *fla* short variable region (SVR) sequencing, and PFGE for the investigation of outbreaks of *C. jejuni* disease showed that one of the outbreak isolates had a different *fla* SVR sequence compared to the other outbreak isolates (Sails *et al.*, 2003a). The divergent *fla*A sequence contained five nucleotide substitutions within a 150 nucleotide region of the SVR, suggesting the change was due to recombination rather than spontaneous mutation. Thus, the instability in this locus may make *fla*A SVR typing unsuitable either for epidemic or for longitudinal epidemiological studies.

Genetically stable clones of *C. jejuni*

In contrast to the genomic instability just described, some strains of *C. jejuni* have been isolated, apparently unchanged over long periods of time. The laboratory strain of *C. jejuni* 81116 was shown to be genetically and serologically stable over more than 20 years (Manning *et al.*, 2001), according to a spectrum of typing methods. These authors also noted that selected epidemiologically linked *C. jejuni* chicken strains were stable during the short time courses of chicken flock outbreaks.

Nielsen *et al.* (2001) investigated the serological and genotypic stability of three *C. jejuni* strains after *in vivo* and *in vitro* passage. The strains were subcultured 50 times *in vitro* or inoculated into mice and colonized for up to 26 days, and then characterized by serotyping, ribotyping, PFGE and RAPD. No changes in any of the DNA profiles or in serotype were identified. An additional 11 *C. jejuni* strains were tested after 10 *in vitro* passages and shown to have stable PFGE patterns and serotypes.

C. jejuni subtypes, as defined by Penner serotype and genotypic analysis, have been shown to persist for several years in diverse locations throughout Scotland. Harrington *et al.* (1999) reported a molecular epidemiological investigation of an outbreak of *C. jejuni* serotype HS55. They found limited genetic diversity among HS55 strains; outbreak-associated HS55 isolates were indistinguishable from nine sporadic HS55 isolates by three molecular methods (PFGE, ribotyping and flagellin gene typing). Five other HS55 strains isolated during 1996 and the HS55 reference strain, which was isolated from human faeces in North America over 20 years ago, had a second closely related genotype. The existence of indistinguishable genotypes among strains isolated over long periods and from different geographic regions (within Scotland) implies stability of the recurrent genotypes to environmental pressures. Stable clonal lineages appear then to exist in the *C. jejuni* population associated with Penner serotype O55, as noted for the other Penner serotypes, HS19, HS41 and HS 11. The identification of evolutionary lineages within other serotypes that are known to be more genotypically heterogeneous has also been described previously (Owen *et al.*, 1995; Santesteban *et al.*, 1996, Slater and Owen, 1998). Santesteban *et al.* (1996) provided evidence of several distinct subgroups within Penner HS1 and HS4 complex strains by PFGE, ribotying and fla-typing. Two of these accounted for 16/80 HS1 and 29/87 HS4 *C. jejuni* strains. The strains were of diverse origin and host.

Moore *et al.* (2003) performed ERIC-PCR on 272 *Campylobacter* isolates from acute bacterial enteritis in Northern Ireland isolated over an 8-year period. They identified 13 different genotypes; and three genotypes accounted for 56% of the isolates. In contrast, other studies using ERIC-PCR demonstrated greater diversity over a shorter time period (Shi *et al.*, 1996; Iriarte and Owen, 1996; Giesendorf *et al.*, 1994). Moore *et al.* (2003) suggested that the local human *Campylobacter* population in Northern Ireland consisted of a limited number of stable genotypes. Similarly, limited heterogeneity was reported for *C. jejuni* isolates from Northern Ireland poultry flocks when automated ribotyping was used to compare *Campylobacter* populations in Northern Ireland and in US poultry flocks (Englen *et al.*, 2001). At present the mechanisms that preserve clonal lineages of *C. jejuni* are largely unknown and remain to be determined.

STANDARDIZATION OF METHODS THAT DETECT STRAIN VARIATION

Many of the advantages offered by the highly sensitive and discriminating molecular methods used for subtyping *C. jejuni* are negated by the lack of standardization and the inability to compare or exchange data, and so make it impossible to determine if the same strains are circulating in different places (Fitzgerald *et al.*, 2003b). Attempts have been made to standardize the subtyping of *C. jejuni* and *C. coli*. Campynet, a 3-year network project funded by the European Union, was established in 1998 with the aim of standardizing and harmonizing subtyping methods for *C. jejuni* and *C. coli*. Three subtyping methods were targeted for standardization in the Campynet network: (1) *fla*-PCR RFLP, (2) PFGE, and (3) AFLP (www.svs.dk/campynet).

PulseNet, the national molecular subtyping network for food-borne disease surveillance established in the US in 1996, is at the forefront of routine application of standardized subtyping for food-borne pathogenic bacteria (Swaminathan *et al.*, 2001). At this time, PulseNet laboratories use PFGE as the molecular subtyping method of choice. Public health laboratories in all 50 states participate in PulseNet, as do the laboratories of the US Department of Agriculture's Food Safety and Inspection Service and the laboratories of the US Food and Drug Administration. PulseNet addresses the standardization needs to facilitate molecular subtype data comparisons between laboratories, provides a platform for rapid exchange of DNA 'fingerprints' of bacteria between laboratories and a central repository at the Centers for Disease Control and Prevention in Atlanta, GA, and provides a uniform nomenclature for the unique patterns of each pathogen. Currently, PulseNet has participant-accessible databases of PFGE patterns for *E. coli* O157:H7, *Salmonella* serotypes, *L. monocytogenes*, and *Shigella* spp. A rapid and standardized PulseNet *C. jejuni* protocol was developed in 2001 (Ribot *et al.*, 2001) and a national database for *C. jejuni* is under development (Fitzgerald *et al.*, 2003a). Although routine subtyping of human clinical *C. jejuni* isolates has not yet been demonstrated to be useful for detecting outbreaks that otherwise would have been missed, the availability of real-time PFGE data helps epidemiologists separate outbreak-associated cases from sporadic cases occurring in the same geographic locations during the same time period. Also, comparison of the patterns of the outbreak isolates against patterns of environmental and food isolates in the databases may assist epidemiologists to refine their hypotheses about probable sources of infection in the outbreak. The limitations of PFGE for all organisms in the PulseNet network including *C. jejuni* are well noted. The development of DNA-based sequencing techniques as 'next generation' subtyping methods for implementation in the PulseNet network are well under way and will be implemented within the network in the future. Despite these attempts made to standardize subtyping of *C. jejuni*, progress has been slow in applying them in practice. A standardized approach is critical if we are to compare data between different laboratories, so that a more accurate estimate of the subtypes of *C. jejuni* circulating locally, nationally and at the global level can be determined.

STRAIN VARIATION BASED ON DNA SEQUENCE ANALYSIS

DNA sequencing-based subtyping recently has emerged as the next generation subtyping technique. This approach is becoming a viable alternative for genotyping bacterial isolates since the introduction of automated sequencers has made sequencing more rapid and the cost continue to decrease in recent years. DNA sequence analysis is a highly reproducible method, which does not rely on the interpretation of gel patterns. DNA sequencing also provides more precise information on strain relatedness, which is only suggested by other subtyping methods such as PFGE.

MLST

Multilocus sequence typing (MLST) is variation on MEE where changes in the loci under analysis are determined by DNA sequence analysis rather than starch gel electrophoresis (Maiden *et al.*, 1998). MLST is a powerful technique to unambiguously define strain types and has the advantage of easy, inter-laboratory comparison of sequence data when compared to fragment based methods such as PFGE and amplified fragment length polymorphism analysis (AFLP). Application of the technique to *C. jejuni* has confirmed the genetic diversity of this pathogen and has demonstrated its population structure to be weakly clonal (Dingle *et al.*, 2001a; Suerbaum *et al.*, 2001). Isolates can be grouped into clonal complexes defined as groups of two or more individual isolates with a sequence type (ST) sharing identical alleles at four or more loci (Dingle *et al.*, 2001a). Members of clonal complexes are thought to be derived from a common ancestor, and are given the same name as the ST identified as the putative founder of the complex (central genotype), e.g. ST-45 complex (Dingle *et al.*, 2001a). In a recent study, 379 distinct STs were identified in a collection of 814 isolates from humans and animals, demonstrating the diversity of genotypes in these populations (Dingle *et al.*, 2002). Interestingly, 63% (318 of 501) of human isolates belonged to one of six clonal complexes. The clonal complexes, ST-21, ST-206 and ST-48 are thought to be related to each other and form a 'complex group' of related genotypes which are widely distributed and may correspond to the diverse group of strains previously described as the Penner serotype HS1, HS2, and HS4 complex (Dingle *et al.*, 2002).

A recent study demonstrated that the clonal complexes of *C. jejuni* identified by MLST correlate with strain associations previously identified by MEE (Sails *et al.*, 2003b). The ability of MLST to group isolates from diverse sources demonstrates that, amidst all of the genetic diversity observed, the *C. jejuni* population still has a definite structure which can be elucidated using this powerful method. The ability of MLST to identify related genotypes from diverse sources will have a profound effect on our understanding of both the genetic diversity and epidemiology of *C. jejuni* infection. The application of the technique to additional isolates from animal and environmental sources will help to resolve the role of animals as reservoirs in human infection and identify whether lineages are adapted to certain niches or hosts. Preliminary data from MLST studies indicate that certain lineages may be over represented in samples from some sources (Dingle *et al.*, 2002). For example, the ST-177 and ST-179 complexes were found exclusively in sand from bathing beaches and may reflect adaptation of these lineages to the colonization of wild birds or other sources of contamination of the sand.

MLST has also contributed to our understanding of the role of recombination in shaping the population structure of *C. jejuni*. The data from MLST studies can be analysed to assess the relative contributions of recombination and mutation to the diversification of clonal complexes (Feil *et al.*, 1999). Differences between alleles can arise by point mutations or by recombinational replacements. Multiple nucleotide changes between two alleles suggest that the difference between the alleles occurred through recombination, whereas single nucleotide differences can arise by mutation or recombination. Distinguishing between these events is not possible. Nevertheless, a single point mutation will most likely result in a variant allele that is unique and which only occurs once within the database (Spratt *et al.*, 2001). The accuracy of this assumption depends on the database containing most of the alleles that are present at a significant frequency in the population. Rare alleles occur; but they will have a low frequency and will rarely be exchanged by recombination.

MLST of *C. jejuni* was recently used to estimate the relative contributions of mutation and recombination on the overall genetic diversity of the species. The results indicate that the population structure of *C. jejuni* is shaped primarily by recombination rather than mutation (Dingle *et al.*, 2001b). These findings are consistent with split decomposition analysis of the allele sequences, which indicate that interspecies recombination frequently occurs within the population (Dingle *et al.*, 2001a; Suerbaum *et al.*, 2001).

Microarray analysis

The publication of the complete genome sequence of the NCTC 11168 strain of *C. jejuni* was a significant step in understanding of the molecular biology of this pathogen (Parkhill *et al.*, 2000). A new range of possibilities for analysis of gene regulation and genetic diversity have arisen in this post genomic era. Genomotyping is a new technique, which uses DNA microarrays to compare interstrain variations in bacteria at the genome level. PCR fragments corresponding to all of the genes in the sequenced genome are spotted onto a solid support in an ordered array. Fluorescently labelled probes are then generated by PCR using DNA from a test organism. The probes are then hybridized to the array and the hybridization signals quantified using high-resolution fluorescent scanning.

Dorrell and colleagues (2001) constructed a whole genome DNA microarray containing all 1654 annotated genes from the NCTC 11168 genome sequence. Whole genome comparisons with 11 *C. jejuni* strains of various heat-stable serotypes identified a set of 1300 'core' genes common to all strains. These core genes appeared to encode proteins with housekeeping functions such as biosynthetic, metabolic, cellular and regulatory processes. Conversely, 354 (21%) of the genes identified in NCTC 11168 were absent or highly divergent in at least one of the 11 strains tested, suggesting that those genes may not be essential. Many of the absent or highly divergent genes encoded surface-located structures such as the flagellin synthesis and modification system, LOS, and capsule. Since the genes represented on the microarray correspond to only one of the 12 genomes analysed, the genome diversity identified among the 12 strains may actually be an underestimate of the actual diversity. Leonard *et al.* (2003) used an open-reading frame-specific *C. jejuni* DNA microarray to investigate genetic diversity among five clusters of epidemiologically related isolates. They noted similar variability in genes involved in surface modification such as the LOS, capsule and flagellin loci. In addition, divergence was identified within the *C. jejuni* genome that contained genes involved in sugar transport and metabolism (the *uxa*A locus).

DNA microarrays have the potential to identify variable genes encoding pathogenic factors responsible for the varying clinical presentations, post-infection sequelae and adaptation to particular hosts. Application of whole genome analysis to isolates from human disease and those associated with particular clinical syndromes may identify genetic differences, which may correlate with pathogenicity and prove useful as epidemiological markers.

IMPLICATIONS OF STRAIN VARIATION FOR EPIDEMIOLOGY

Strain variation and seasonality of *C. jejuni* infection

While the seasonality of *Campylobacter* infection is well documented (Friedman, 2000), the epidemiological causes of it are not well understood. Little is known about the prevalence and distribution of specific subtypes of *C. jejuni* and their contribution to the seasonal spring peak. Owen and colleagues (Owen *et al.*, 1997) combined Penner serotyping and pulsed-field gel electrophoresis (PFGE) to study *C. jejuni* isolates from sporadic cases of diarrhoea in England from three different locations over a 12-month period. They concluded that the subtypes causing human infections were highly diverse with the HS1and HS4 complex being the predominant types (34% of all strains). The late spring peak appeared to be associated with a rise in many serotypes rather than an emergence of a predominant single serotype. PFGE provided evidence of a high degree of genomic diversity within the HS1 and HS4 complex isolates, however some subtypes were common to more than one location. A study of the serotype distribution in New Zealand (Hudson *et al.*, 1999), noted differences in serotype distribution between strains isolated from environmental samples, food samples, and human infection in winter and summer. Strains of serotype HS33 were not isolated in the summer, but were the most prevalent serotype isolated in the winter accounting for 19.3% of the strains examined. Significant differences were seen in the types isolated in the two seasons when serotype and PFGE type were combined. Thirty one per cent of the total number of isolates clustered into 10 groups based on the PFGE and serotype results. Isolates from human cases were indistinguishable from those isolated from water and raw chicken, suggesting they may serve as possible reservoirs for the isolates infecting humans.

A study in Finland demonstrated that there was a high degree of diversity in *Sma*I PFGE profiles of human sporadic *C. jejuni* isolates from two geographic areas; however, five PFGE types predominated in 42% and 44% of the isolates in the two areas respectively (Hanninen *et al.*, 1998). A further report from Finland described a 3-year study of *C. jejuni* genotypes in humans with domestically acquired infections and chickens from the Helsinki area (Hanninen *et al.*, 2000). Certain genotypes persisted throughout the 3-year period and the predominant genotypes represented 28– 52% of the strains during a restricted period of time. Most of these were also found among the chicken isolates. Some genotypes seen in the chicken isolates were not seen in the human isolates and vice versa, indicating that there may be other important reservoirs for the strains infecting humans. Five of these common genotypes were further characterized by AFLP analysis, ribotyping and serotyping (Hanninen *et al.*, 2001). The data suggested that common Finnish *C. jejuni* genotypes form genetic lineages which colonize both humans and chickens and that these genotypes persist from one year to another.

A Danish survey (Nielsen *et al.*, 1997) reported the serotype distribution in *C. jejuni* strains isolated from food production animals and isolates from human patients with campylobacteriosis during the same time period. The three most common serotypes in humans were HS2, HS4 complex and HS1, 44 and they accounted for 62% of the isolates, with other serotypes representing 4% or less. In broilers these same three serotypes were demonstrated in 50% of the isolates. In contrast, HS2 was the dominant serotype in cattle, with seven other serotypes being relatively common, representing 7–9% each (HS1; HS4; HS6,7; HS11; HS19; HS21; HS23,36). Further work investigating strain diversity of dairy herds, rather than isolates from cattle at slaughter showed similar results with HS2, HS4 complex and HS11 the most commonly occurring serotypes (Nielsen, 2002). Clonal strains of *C. jejuni* defined by indistinguishable types using a combination of multiple typing methods have also been found in cattle and humans in both Denmark and the UK (Nielsen, 2000; Fitzgerald *et al.*, 2001; On *et al.*, 1998) supporting the hypothesis that other animals including cattle may also be reservoirs for the isolates causing disease in humans.

Strain variation in poultry

While sources of sporadic infection are still debatable, the contribution of poultry consumption as a major risk factor for human infection with *C. jejuni* in developed countries is well documented (Friedman, 2000). For many years efforts have focused on identifying and controlling sources of *C. jejuni* contamination in broiler chickens; however, the pathways involved in *C. jejuni* contamination of poultry flocks remain unresolved. Several suspected sources or vectors of contamination have been studied including transmission from parent to progeny, exposure of birds to contaminated water, litter, feed, personnel, flies, rodents, small mammals, contaminated transport crates and environmental exposure. While the introduction of improved hygiene barriers has contributed to the reduction of *Campylobacter*-positive broiler flocks mainly at the farm level, these strategies are not 100% effective (Newell, 2000).

Poultry are colonized with many of the same serotypes that cause disease in humans (Neilsen *et al.*, 1997). Molecular epidemiological studies have been helpful in addressing questions regarding infection dynamics of *C. jejuni* at the flock level (Newell, 2000). Molecular subtype analyses of campylobacters from poultry production and processing environments suggests that in countries such as the US and Australia multiple genotypes are commonly isolated from broiler flocks (Thomas *et al.*, 1997; Hiett *et al.*, 2002). In contrast, broiler flocks in the UK, the Netherlands and Sweden are predominantly colonized with a more limited number of subtypes of *C. jejuni* (Newell and Fearnlcy, 2003).

While thc debate on the role of vertical transmission as a source of flock colonization continues, horizontal transmission is still considered the most significant cause of *C. jejuni* infection in broiler flocks (for a review of sources of poultry colonization see Newell and Fearnley, 2003). In a Danish study, Petersen and Wedderkopp (2001) provided evidence that local contamination reservoirs are of importance for the colonization of broiler farms with *C. jejuni*. Using serotyping, *fla* typing and PFGE, certain clones of *C. jejuni* were shown to persist during successive broiler flock rotations. Highly stable clones were found in 63% of all flocks (47/75). In a study of 10 farms, 7 of the 12 broiler houses carried these farm specific clones that covered an interval of at least four broiler flock rotations, or at least 6 months. This suggests that these particular *C. jejuni* strains are well adapted for the colonization of broiler flocks and for survival around broiler houses.

In the US, molecular analysis of *Campylobacter* spp. from poultry production and processing environments using *fla* typing showed that many of the clones found within a flock are present on the final product, although the diversity of *Campylobacter* on the final product was reported to be reduced relative to that observed in the flock (Hiett *et al.*, 2002). Similar findings were noted by Newell *et al.* (2001), who followed both *Campylobacter*-positive and -negative flocks through commercial abattoirs. They suggested that the reduction of diversity of *Campylobacter* population contaminating poultry carcasses at retail may be related to the number of subtypes that are better able to respond and survive the environmental stresses encountered during meat processing, as some subtypes survived all the processing stages, whereas others only survived to the chilling stage and some were never recovered from the carcasses despite being present in the caeca. While choice of isolation method was noted to possibly result in preferential selection of certain subtypes, it was suggested that there is a considerable differential in the ability of *Campylobacter* strains to survive environmental stresses. Results from the study also demonstrated that *Campylobacter*-negative flocks can quickly become contaminated during processing and that contaminated transport crates constitute a source of contamination following processing. *fla* typing and PFGE were also used in Denmark to show that vertical transmission or horizontal transmission via the hatchery are not significant transmission routes of *C. jejuni* to broiler chickens in Denmark (Petersen *et al.*, 2001a) and that genetic diversity among quinolone-resistant *C. jejuni* isolates was lower than among susceptible isolates (Pedersen and Wedderkopp, 2003). In a comparison of genotypes and serotypes of *C jejuni* isolated from Danish wild mammals, birds, broiler flocks and humans, diversity was seen among the isolates in this study. However, 17 groups of indistinguishable strains (using serotype, *fla* type, PFGE with up to three enzymes) were identified that represented 36% of the total number of examined isolates and 68% were assigned to similarity groups. The dominant serotypes in wildlife were reported to be HS4, HS12 and HS2 (Petersen *et al.*, 2001b). Molecular analysis revealed that the majority of wildlife isolates did not seem to share any relationship with humans or broilers; however, two clones belonging to the HS4 complex were identified among human and hedgehog isolates. HS:12 was also reported in a New Zealand study to be the most common serotype among surface water samples (Hudson *et al.*, 1999). Similarly the PFGE profiles most prevalent in gull isolates did not occur among the isolates from humans and broiler chickens in a recent study in Sweden, where it was suggested that subpopulations of *C. jejuni* adapted to species-specific colonization or environmental survival exist (Broman *et al.*, 2002).

Differences in the colonization potential between *C. jejuni* strains have also been documented (Korolik *et al.*, 1998; Cawthraw *et al.*, 1996). More recently, the colonization ability of *C. jejuni* strains was shown to fall into five distinct phenotypes ranging from immediate sustained colonization to completely non-colonizing (Ringoir and Korolik, 2003). Genetic differences potentially involved in colonization of the chicken intestinal tract were identified using subtractive hybridization by Ahmed and colleagues (2002) when comparing two *C. jejuni* isolates that exhibited different colonization potential.

It is not surprising that contamination of raw meat products with multiple species as well as with multiple subtypes of *C. jejuni* has been documented (Dickens *et al.*, 2002; Kramer *et al.*, 2000; Zhao *et al.*, 2001). To determine if infection with multiple strains of *Campylobacter* species occurs in human cases of *Campylobacter* enteritis and if genetic rearrangements can be observed during an infection, Steinbrueckner *et al.* (2001) investigated *C. jejuni* isolates using PFGE and enterobacterial repetitive intergenic consensus sequence PCR (ERIC-PCR). The authors reported that it is rare for an individual to be infected with more than one *Campylobacter* isolate at any one time. However, for 10% of the patients the banding patterns of PFGE and ERIC-PCR changed over the course of the infection in the *C. jejuni* isolates recovered from the individuals. This finding needs to be considered during epidemiological investigations to ensure that misinterpretation of the data by underestimating isolate relatedness does not happen. Richardson *et al.* (2001) also observed that a majority of 537 individuals were infected with a single isolate of *Campylobacter*; however, four individuals from this group were infected with two strains of *C. jejuni*. Using molecular methods, Lawson *et al.* (1999) detected isolates of different *Campylobacter* species in the stools of 3.6% of 543 PCR-positive individuals. However, they also reported that only a single isolate was identified by culture from these specimens. In the cases of coinfection reported by Richardson *et al.* (2001), one of the isolates was dominant rather than both isolates being equally represented.

Guillain–Barré syndrome and Miller Fisher syndrome

Some Penner serotypes, including HS19 and HS41 have been reported to be overrepresented among *C. jejuni* isolates recovered from patients with Guillian–Barré (GBS) or Miller Fisher (MFS) syndromes compared to isolates from individuals with enteritis in the US, Japan, Mexico and South Africa (Nachamkin *et al.*, 1998). The risk of developing GBS after infection with *C. jejuni* strains of serotype HS19 is also thought to be higher when compared to infections with other serotypes (Nachamkin *et al.*, 1998). Flagellin gene typing and random amplified polymorphic DNA analyses further support the hypothesis that HS19 strains form a clonal population (Fujimoto *et al.*, 1997; Nachamkin *et al.*, 1996). Characterization of a worldwide collection of strains from patients with GBS and gastroenteritis using multilocus enzyme electrophoresis (MEE) demonstrated that HS19 represents a distinct cluster separate from other HS serotypes. (Nachamkin *et al.*, 2001). Serotype HS19 strains do not however form a monomorphic clone as suggested by previous studies (Fujimoto *et al.*, 1997; Misawa *et al.*, 1998), as 17 closely related electrophoretic types (ET) were demonstrated within the HS19 strains tested. One electrophoretic type (ET4) did occur in many of the HS19 strains but was not uniquely associated with GBS when compared to isolates from individuals with uncomplicated enteritis and from animal sources also sharing this serotype. The data also suggested that HS19 was closely related to HS:41 (Goddard *et al.*, 1997), which also has a clonal structure (Wassenaar *et al.*, 2000). Data from the UK and the Netherlands suggest that overrepresentation of specific serotypes is not seen in all parts of the world, since more than 14 other different Penner serotypes have been associated with GBS and MFS (Nachamkin *et al.*, 1998; Endtz *et al.*, 2000). In addition GBS and MFS-related strains of serotypes other than HS19 demonstrate significant genetic diversity (Enberg *et al.*, 2001; Endtz *et al.*, 2000) and do not represent a unique population across serotypes. More recently, MLST confirmed that *C. jejuni* strains associated with GBS and MFS are of diverse genetic lineage, serotype and *fla*A short variable region sequence type (Dingle *et al.*, 2001c). No epidemiological marker for GBS-associated strains has yet been identified (Dingle *et al.*, 2001c; Endtz *et al.*, 2000; Endberg, 2001).

CONCLUSIONS AND FUTURE DEVELOPMENTS

It is apparent that genetic exchange contributes to the generation of diversity in *C. jejuni*. While the mechanism (s) by which this occurs are still not fully understood, it appears that certain isolates are more susceptible to genetic change than others. Since *C. jejuni* must be able to adapt to different environments, it may undergo genetic variation to increase its potential to adapt. Genetic variation could result in phenotypic changes that allow variants to survive during such environmental transitions. The increasing awareness of genomic instability in *C. jejuni* and the consequent effect on molecular fingerprints do not invalidate the use of these methods for outbreak investigations or other epidemiological studies such as sporadic campylobacterosis. However, the limitations of the typing methods should be kept in mind and a combination of typing methods may be needed for reliable determination of strain relatedness

The detection of strain variation within *C. jejuni* populations has come a long way since the introduction of serotyping methods in the early 1980s. Developments in molecular biology have led to DNA-based subtyping methods such as AFLP and PFGE that have allowed us to investigate strain variation throughout the genome of *C. jejuni*. Advances in DNA sequencing technology have provided a means to investigate strain variation at the nucleotide level and has been exploited by methods such as MLST. As sequencing becomes further automated in the future and the costs further diminish the determination of more of the nucleotide sequence of strains will become feasible. Perhaps we may even reach a stage in the future where comparative analysis of the complete genome sequence of individual strains on a routine basis to aid epidemiological investigations may be possible.

The sequencing of the NCTC 11168 strain of *C. jejuni* by The Sanger Centre has laid the groundwork for more extensive studies of genomic variation in clinical and environmental isolates of this bacterium. Knowledge of the complete genome sequence facilitates targeted re-sequencing of genomic DNA from bacterial pathogens. This technique is an important tool to determine variation within the species and between similar species at the nucleotide sequence level. Targeted re-sequencing allows high-resolution comparisons between sequence data from the sampled strains and the reference genome to be carried out. In the future targeted re-sequencing of *C. jejuni* strains will allow more precise comparisons of variation between strains.

A second *C. jejuni* isolate from a chicken is currently being sequenced in the USA, which will enable the first comparative genomic study of *C. jejuni* to be performed between two completed genome sequences. The direct comparison of completed genome sequences provides a comprehensive assessment of species diversity and avoids the bias associated with methods such as MLST which sample very limited parts of the genome (Read *et al.*, 2002). Comparative sequencing can also identify novel lineage-specific markers which may provide additional resolution in molecular subtyping schemes. Comparison of complete genome sequences may also elucidate genetic variation responsible for phenotypic differences observed between strains.

The generation of complete sequence data for bacterial pathogens has also fuelled the development of new analytical tools which take advantage of the ever growing amount of genome sequence data. DNA microarray technology is a promising technology for investigating and defining genetic variation. This technique can be used as an adjunct to genome sequencing providing detailed comparisons of genetic diversity between strains without having to sequence the entire genome of the strains being compared to the reference sequence. The analysis of genetic diversity using microarrays is very much in its infancy and therefore accepted standards for microarray design and standardized experimental conditions have yet to be established. In addition standards for the selection, analysis and interpretation of microarray data, also have to be established.

In this post-genomic era the application of these powerful new tools to *C. jejuni* in a standardized approach will improve our understanding of the population biology of this important pathogen. A clearer understanding of the strain diversity exhibited by *C. jejuni* will lead to improvements in our understanding of the epidemiology of infection. Previous studies of strain variation either retrospective or prospective have proven useful in the past; however, if we are to gain most from these new technologies we need to devise hypotheses which ask more specific questions about the epidemiology of the organism and collect specific data to answer these questions. Understanding the epidemiology of *C. jejuni* disease more definitively will direct the implementation of more targeted intervention strategies and lead to a reduction in the public health burden of campylobacteriosis.

References

Aeschbacher, M., and Piffaretti, J.C. 1989. Population genetics of human and animal enteric *Campylobacter* strains. Infect. Immun. 57: 1432–1437.

Ahmed, I.H., Manning, G., Wassenaar, T.M., Cawthraw, S., and Newell, D.G. 2002. Identification of genetic differences between two *Campylobacter jejuni* strains with different colonization potentials. Microbiol. 148: 1203–1212.

Albert, M.J., Leach, A., Asche, V., Hennessy, J., and Penner, J.L. 1992. Serotype distribution of *Campylobacter jejuni* and *Campylobacter coli* isolated from hospitalized patients with diarrhoea in central Australia. J. Clin. Microbiol. 30: 207–210.

Bolton, F.J., Holt, A.V., and Hutchinson, D.N. 1984. *Campylobacter* biotyping scheme of epidemiological value. J. Clin. Pathol. 37: 677–681.

Broman, T., Palmgren, H., Bergstrom, S., Sellin, M., Waldenstrom, J., Danielsson-Tham, M.L., and Olsen, B. 2002. *Campylobacter jejuni* in black-headed gulls (*Larus ridibundus*): Prevalence, genotypes, and influence on *C. jejuni* epidemiology. J. Clin. Microbiol. 40: 4594–4602.

Butzler, J.P. 1984. *Campylobacter* Infections in Man and Animals. CRC Press, Florida.

Cawthraw, S.A., Wassenaar, T.M., Ayling, R., and Newell, D.G. 1996. Increased colonization potential of *Campylobacter jejuni* strain 81116 after passage through chickens and its implication on the rate of transmission within flocks. Epidemiol. Infect. 117: 213–215.

CDC. 2003. Preliminary FoodNet Data on the Incidence of Food-borne Illnesses-Selected Sites, United States, 2002. MMWR. 52: 340–343.

de Boer, P., Wagenaar, J.A., Achterberg, R. P., van Putten, J.P.M., Schouls, L.M., and Duim, B. 2002. Generation of *Campylobacter jejuni* genetic diversity *in vivo*. Mol. Microbiol. 44: 351–359.

Dickins, M.A., Franklin, S., Stefanova, R., Schutze, G.E., Eisenach, K.D., Wesley, I., and Cave, M.D. 2002. Diversity of *Campylobacter* isolates from retail poultry carcasses and from humans as demonstrated by pulsed-field gel electrophoresis. J. Food Prot. 65: 957–962.

Dingle, K.E., Colles, F.M., Wareing, D.R., Ure, R., Fox, A.J., Bolton, F.E., Bootsma, H.J., Willems, R.J., Urwin, R., and Maiden, M.C. 2001a. Multilocus sequence typing system for *Campylobacter jejuni*. J. Clin. Microbiol. 39: 14–23.

Dingle, K.E., Colles, F.M., Ure, R., Wareing, D.R.A., Fox, A.J., Bolton, F.J., Holmes, H.J., Feil, E.J., Maiden, M.C. 2001b. The population structure of *Campylobacter jejuni* is shaped by genetic recombination in combination with selection. 11th International Workshop on *Campylobacter*, Helicobacter and related Organisms. Int. J. Med. Microbiol. (Suppl. 31): 291: 63.

Dingle, K.E., Van Den Braak, N., Colles, F.M., Price, L.J., Woodward, D.L., Rodgers, F.G., Endtz, H.P., Van Belkum, A. and Maiden, M.C. 2001c. Sequence typing confirms that *Campylobacter jejuni* strains associated with Guillain–Barré and Miller-Fisher syndromes are of diverse genetic lineage, serotype, and flagella type. J. Clin. Microbiol. 39: 3346–3349.

Dingle, K.E., van den Braak, N., Colles, F.M., Price, L.J., Woodward, D.L., Rodgers, F.G., Endtz, H.P., van Belkum, A. and Maiden, M.C. 2001cDingle, K.E., Colles, F.M., Ure, R., Wagenaar, J.A., Duim, B., Bolton, F.J., Fox, A.J., Wareing, D.R., Maiden, M.C. 2002. Molecular characterization of *Campylobacter jejuni* clones: a basis for epidemiologic investigation. Emerging Infect. Dis. 8: 949–955.

Dorrell, N., Mangan, J.A., Laing, K.G., Hinds, J., Linton, D., Al-Ghusein, H., Barrell, B.G., Parkhill, J., Stoker, N.G., Karlyshev, A.V., Butcher, P.D., and Wren, B.W. 2001. Whole genome comparison of *Campylobacter jejuni* human isolates using a low-cost microarray reveals extensive genetic diversity. Genome Res. 11: 1706–1715.

Endtz, H.P., Ang, C.W., van Den Braak, N., Duim, B., Rigter, A., Price, L.J., Woodward, D.L., Rodgers, F.G., Johnson, W.M., Wagenaar, J.A., Jacobs, B.C., Verbrugh, H.A., and van Belkum, A. 2000. Molecular characterization of *Campylobacter jejuni* from patients with Guillain–Barré and Miller Fisher syndromes. J. Clin. Microbiol. 38: 2297–2301.

Engberg, J., Nachamkin, I., Fussing, V., McKhann, G.M., Griffin, J.W., Piffaretti, J.C., Nielsen, E.M., and Gerner-Smidt, P. 2001 Absence of clonality of *Campylobacter jejuni* in serotypes other than HS: 19 associated with Guillain–Barré syndrome and gastroenteritis. J. Infect Dis. 184 (2): 215–220.

Englen, M.D., Harrington, C.S., Fedorka-Cray, P.J. *et al.* 2001. Comparison of *Campylobacter* populations in Northern Ireland and USA poultry flocks by means of automated *Pst*I ribotyping. Int. J. Med. Microbiol. Suppl. 31: 69–70.

Feil, E.J., Maiden, M.C., Achtman, M., and Spratt, B.G. 1999. The relative contributions of recombination and mutation to the divergence of clones of *Neisseria meningitidis*. Mol. Biol. Evol. 16: 1496–1502.

Fitzgerald, C., Stanley, K., Andrew, S., and Jones, K. 2001. Use of pulsed-field gel electrophoresis and flagellin gene typing in identifying clonal groups of *Campylobacter jejuni* and *Campylobacter coli* in farm and clinical environments. Appl. Environ. Microbiol. 67: 1429–1436.

Fitzgerald, C., Sails, A., Hunter, S., Ribot, E., Fields, P.I., and the PulseNet participating laboratories. 2003a. Pulse-field gel electrophoresis (PFGE) Subtyping of *Campylobacter jejuni* – What's the point? Experiences from the PulseNet *C. jejuni* database. Int. J. Med. Microbiol. 293: 105.

Fitzgerald, C., Sails, A., and Swaminathan, B. 2003b. Chapter 17, Genetic techniques: molecular subtyping methods. In: Detecting Pathogens in Food. Woodhouse Publishing, Cambridge, UK.

Fricker, C.R., and Park, R.W. 1989. A two-year study of the distribution of thermophilic campylobacters in human, environmental and food samples from the Reading area with particular reference to toxin production and heat-stable serotype. J. Appl. Bacteriol. 66: 477–490.

Friedman, C.R., Neimann, J., Wegener, H.C., and Tauxe, R.V. 2000. Epidemiology of *Campylobacter jejuni* infections in the United States and other industralized nations. In *Campylobacter jejuni*. 2nd edition, pp. 121–138. Edited by I. Nachamkin and M. J. Blaser, Washington, DC: American Society for Microbiology.

Frost, J.A., Oza, A.N., Thwaites, R.T., and Rowe, B. 1998. Serotyping scheme for *Campylobacter jejuni* and *Campylobacter coli* based on direct agglutination of heat-stable antigens. J. Clin. Microbiol. 36: 335–339.

Fujimoto, S., Allos, B.M., Misawa, N., Patton, C.M., and Blaser, M.J. 1997. Restriction fragment length polymorphism analysis and random amplified polymorphic DNA analysis of *Campylobacter jejuni* strains isolated from patients with Guillain–Barré syndrome. J. Infect. Dis. 176: 1105–1108.

Giesendorf, B.A., Goossens, H., Niesters, H.G., Van Belkum, A., Koeken, A., Endtz, H.P., Stegeman, H., and Quint, W.G. 1994. Polymerase chain reaction-mediated DNA fingerprinting for epidemiological studies on *Campylobacter* spp. J. Med. Microbiol. 40: 141–147.

Goddard, E.A., Lastovica, A.J., and Argent, A.C. 1997. *Campylobacter* 0: 41 isolation in Guillain–Barré syndrome. Arch. Dis. Child. 76: 526–528.

Hanninen, M.L., Pajarre, S., Klossner, M.L., Rautelin, H. 1998. Typing of human *Campylobacter jejuni* isolates in Finland by pulsed-field gel electrophoresis. J. Clin. Microbiol. 36: 1787–1789.

Hanninen, M.L., Hakkinen, M., and Rautelin, H. 1999. Stability of related human and chicken *Campylobacter jejuni* genotypes after passage through chick intestine studied by pulsed-field gel electrophoresis. Appl. Environ. Microbiol. 65: 2272–2275.

Hanninen, M.L., Perko-Makela, P., Pitkala, A., and Rautelin, H. 2000. A three-year study of *Campylobacter jejuni* genotypes in humans with domestically acquired infections and in chicken samples from the Helsinki area. J. Clin. Microbiol. 38: 1998–2000.

Hanninen, M.L., Perko-Makela, P., Rautelin, H., Duim, B., Wagenaar, J.A. 2001. Genomic relatedness within five common Finnish *Campylobacter jejuni* pulsed-field gel electrophoresis genotypes studied by amplified fragment length polymorphism analysis, ribotyping, and serotyping. Appl. Environ. Microbiol. 67: 1581–1586.

Harrington, C.S., Thomson-Carter, F.M., and Carter, P.E. 1997. Evidence for recombination in the flagellin locus of *Campylobacter jejuni*: implications for the flagellin gene typing scheme. J. Clin. Microbiol. 35: 2386–2392.

Harrington, C.S., Thomson-Carter, F.M., and Carter, P.E. 1999. Molecular epidemiological investigation of an outbreak of *Campylobacter jejuni* identifies a dominant clonal line within Scottish serotype HS55 populations. Epidemiol. Infect. 122: 367–375.

Healing, T.D., Greenwood, M.H. and Pearson, A.D. 1992. Campylobacters and enteritis. Rev. Med. Micro. 3: 159–167.

Hiett, K.L., Stern, N.J., Fedorka-Cray, P., Cox, N.A., Musgrove, M.T., and Ladely, S. 2002. Molecular subtype analyses of *Campylobacter* spp. from Arkansas and California poultry operations. Appl. Environ. Microbiol. 68: 6220–6236.

Hudson, J.A., Nicol, C., Wright, J., Whyte, R., and Hasell, S.K. 1999. Seasonal variation of *Campylobacter* types from human cases, veterinary cases, raw chicken, milk and water. J. Appl. Microbiol. 87: 115–124.

Iriarte, M.P., and Owen, R.J. 1996. Repetitive and arbitrary primer DNA sequences in PCR-mediated fingerprinting of outbreak and sporadic isolates of *Campylobacter jejuni*. FEMS Immunol. Med. Microbiol. 15: 17–22.

Jackson, C.J., Fox, A.J., Jones, D.M., Wareing, D.R., and Hutchinson, D.N. 1998. Associations between heat-stable (O) and heat-labile (HL) serogroup antigens of *Campylobacter jejuni*: evidence for interstrain relationships within three O/HL serovars. J. Clin. Microbiol. 36: 2223–2228.

Jones, D.M., Abbott, J.D., Painter, M.J., and Sutcliffe, E.M. 1984. A comparison of biotypes and serotypes of *Campylobacter* spp. isolated from patients with enteritis and from animal and environmental sources. J. Infect. 9: 51–58.

Kramer, J.M., Frost, J.A., Bolton, F.J., and Wareing, D.R. 2000. *Campylobacter* contamination of raw meat and poultry at retail sale: identification of multiple types and comparison with isolates from human infection. J. Food Prot. 63: 1654–1659.

Karmali, M.A., Penner, J.L., Fleming, P.C., Williams, A., and Hennessy, J.N. 1983. The serotype and biotype distribution of clinical isolates of *Campylobacter jejuni* and *Campylobacter coli* over a three-year period. J. Infect. Dis. 147: 243–246.

Korolik, V., Alderton, M.R., Smith, S.C., Chang, J., and Coloe, P.J. 1998. Isolation and molecular analysis of colonizing and non-colonizing strains of *Campylobacter jejuni* and *Campylobacter coli* following experimental infection of young chickens. Vet. Microbiol. 60: 239–249.

Lastovica, A.J., Le Roux, E., Congi, R.V., and Penner, J.L. 1986. Distribution of sero-biotypes of *Campylobacter jejuni* and *C. coli* isolated from paediatric patients. J. Med. Microbiol. 21: 1–5.

Leonard, E.E., 2nd., Takata, T., Blaser, M.J., Falkow, S., Tompkins, L.S., and Gaynor, E.C. 2003. Use of an open-reading frame-specific *Campylobacter jejuni* DNA microarray as a new genotyping tool for studying epidemiologically related isolates. J. Infect. Dis. 187: 691–694.Lawson, A.J., Logan, J.M., O'Neill, G.L., Desai, M., and Stanley J. 1999. Large-scale survey of *Campylobacter* species in human gastroenteritis by PCR and PCR-enzyme-linked immunosorbent assay. J. Clin. Microbiol. 37: 3860–3864.

Lior, H., Woodward, D.L., Edgar, J.A., Laroche, L.J., and Gill, P. 1982. Serotyping of *Campylobacter jejuni* by slide agglutination based on heat-labile antigenic factors. J. Clin. Microbiol.15: 761–768.

McKay, D., Fletcher, J., Cooper, P., and Thomson-Carter, F.M. 2001. Comparison of two methods for serotyping *Campylobacter* spp. J. Clin. Microbiol. 39: 1917–1921.

Maiden, M.C., Bygraves, J.A., Feil, E., Morelli, G., Russell, J.E., Urwin, R., Zhang, Q., Zhou, J., Zurth, K., Caugant, D.A, Feavers, I.M., Achtman, M., and Spratt, B.G. 1998. Multilocus sequence typing: a portable approach to the identification of clones within populations of pathogenic microorganisms. Proc. Natl. Acad. Sci. USA 95: 3140–3145.

Manning, G., Duim, B., Wassenaar, T., Wagenaar, J.A., Ridley, A., and Newell, D.G. Evidence for a genetically stable strain of *Campylobacter jejuni*. 2001. Appl. Environ. Microbiol. 67: 1185–1189.

Matsui, T., Moore, J.E., Patterson, C., and Millar, B.C. 2002. Matsuda M. Molecular characterization of human campylobacteriosis in Northern Ireland: evidence of clonal stability. Irish J. Med. Sci. 171: 33–36.

Mead, P.S., Slutsker, L., Dietz, V., McCraig, L.F., Bresee, J.S., Shapiro, C., Griffin, P.M. and Tauxe, R.V. Food-related illness and death in the United States. Emerg. Infect. Dis. 5: 607–625.

Mills, S.D., Congi, R.V., Hennessy, J.N., and Penner, J.L. 1991. Evaluation of a simplified procedure for serotyping *Campylobacter jejuni* and *Campylobacter coli* which is based on the O antigen. J. Clin. Microbiol. 29: 2093–2098.

Misawa, N., Allos, B.M., and Blaser, M.J. 1998. Differentiation of *Campylobacter jejuni* serotype O19 strains from non-O19 strains by PCR. J. Clin. Microbiol. 36: 3567–3573. [Erratum appears in J. Clin. Microbiol. 2000, 38 (1): 474.]

Meinersmann, R.J., Patton, C.M., Evins, G.M., Wachsmuth, I.K., and Fields, P.I. 2002. Genetic diversity and relationships of *Campylobacter* species and subspecies. Int. J. Sys. Evol. Microbiol. 52: 1789–1797.

Moore, J.E., and Wareing, D.R. 2003. Phenotypic diversity of *Campylobacter* isolates from sporadic cases of human enteritis in Northern Ireland. British J. Biomed. Science 60 (1): 28–30.

Munroe, D.L., Prescott, J.F., and Penner, J.L. 1983. *Campylobacter jejuni* and *Campylobacter coli* serotypes isolated from chickens, cattle and pigs. J. Clin. Microbiol. 18: 877–881.

Nachamkin, I., Ung, H., and Patton, C.M. 1996. Analysis of HL and O serotypes of *Campylobacter* strains by the flagellin gene typing system. J. Clin. Microbiol. 34: 277–281.

Nachamkin, I., Allos, B.M., and Ho, T. 1998. *Campylobacter* species and Guillain–Barré syndrome. Clin. Microbiol. Rev. 11: 555–567.

Nachamkin, I., Engberg, J., Gutacker, M., Meinersman, R.J., Li, C.Y., Arzate, P., Teeple, E., Fussing, V., Ho, T.W., Asbury, A.K., Griffin, J.W., McKhann, G.M., and Piffaretti, J.C. 2001. Molecular population genetic analysis of *Campylobacter jejuni* HS:19 associated withGuillain–Barré syndrome and gastroenteritis. J. Infect. Dis. 184: 221–226.

Newell, D., Frost, J., Duim, B., Wagenaar, J., Madden, R., van der Plas, J., and On, S. 2000. New developments in the subtyping of *Campylobacter* species. In: Nachamkin, I., Blaser, M., eds. *Campylobacter*, Washington DC, American Society of Microbiology.

Newell, D.G., Shreeve, J.E., Toszeghy, M., Domingue, G., Bull, S., Humphrey, T., and Mead, G. 2001. Changes in the carriage of *Campylobacter* strains by poultry carcasses during processing in abattoirs. Appl. Environ. Microbiol. 67: 2636–2640.

Newell, D.G., and Fearnley, C. 2003. Sources of *Campylobacter* colonization in broiler chickens. Appl. Environ. Microbiol. 69: 4343–4351.

Nielsen, E.M., Engberg, J., and Madsen, M. 1997. Distribution of serotypes of *Campylobacter jejuni* and *C. coli* from Danish patients, poultry, cattle and swine. FEMS Immunol. Med. Microbiol. 19: 47–56.

Nielsen, E.M., Engberg, J., Fussing, V., Petersen, L., Brogren, C.H., and On, S.L. 2000. Evaluation of phenotypic and genotypic methods for subtyping *Campylobacter jejuni* isolates from humans, poultry, and cattle. J. Clin. Microbiol. 38: 3800–3810.

Nielsen, E.M., Engberg, J., and Fussing, V. 2001. Genotypic and serotypic stability of *Campylobacter jejuni* strains during *in vitro* and *in vivo* passage. International J. Med. Microbiol. 291: 379–385.

Nielsen, E.M. 2002. Occurrence and strain diversity of thermophilic campylobacters in cattle of different age groups in dairy herds. Lett. Appl. Microbiol. 35: 85–89.

Nuijten, P.J., van den Berg, A.J., Formentini, I., van der Zeijst, B.A. and Jacobs, A.A. 2000. DNA rearrangements in the flagellin locus of an *fla*A mutant of *Campylobacter jejuni* during colonization of chicken ceca. Infect. Immun. 68: 7137–7140.

On, S.L., Nielsen, E.M., Engberg, J., and Madsen, M. 1998. Validity of SmaI-defined genotypes of *Campylobacter jejuni* examined by *Sal*I, *Kpn*I, and *Bam*HI polymorphisms: evidence of identical clones infecting humans, poultry, and cattle. Epidemiol. Infect. 120: 231–237.

Owen, R.J., Sutherland, K., Fitzgerald, C., Gibson, J., Borman, P., and Stanley, J. 1995. Molecular subtyping scheme for serotypes HS1 and HS4 of *Campylobacter jejuni*. J. Clin. Microbiol. 33: 872–877.

Owen, R.J., Slater, E., Telford, D., Donovan, T., and Barnham, M. 1997. Subtypes of *Campylobacter jejuni* from sporadic cases of diarrhoeal disease at different locations in England are highly diverse. Eur. J. Epidemiol. 13: 837–840.

Oza, A.N., Thwaites, R.T., Wareing, D.R., Bolton, F.J., and Frost, J.A. 2002. Detection of heat-stable antigens of *Campylobacter jejuni* and *C. coli* by direct agglutination and passive hemagglutination. J. Clin. Microbiol. 40: 996–1000.

Parkhill, J., Wren, B.W., Mungall, K., Ketley, J.M., Churcher, C., Basham, D., Chillingworth, T., Davies, R.M., Feltwell, T., Holroyd, S., Jagels, K., Karlyshev, A.V., Moule, S., Pallen, M.J., Penn, C.W., Quail, M.A., Rajandream, M.A., Rutherford, K.M., van Vliet, A.H., Whitehead, S. and Barrel, B.G. 2000. The genome sequence of the food-borne pathogen *Campylobacter jejuni* reveals hypervariable sequences. Nature 403: 665–668.

Patton, C.M., Barrett, T.J., and Morris, G.K. 1985. Comparison of the Penner and Lior methods for serotyping *Campylobacter* spp. J. Clin. Microbiol. 22: 558–565.

Patton, C.M., Nicholson, M.A., Ostroff, S.M., Ries, A.A., Wachsmuth, I.K., and Tauxe, R.V. 1993. Common somatic O and heat-labile serotypes among *Campylobacter* strains from sporadic infections in the United States. J. Clin. Microbiol. 31: 1525–1530.

Pedersen, K., and Wedderkopp A. 2003. Resistance to quinolones in *Campylobacter jejuni* and *Campylobacter coli* from Danish broilers at farm level. J. Appl. Microbiol. 94: 111–119.

Penner, J.L., and Hennessy, J.N. 1980. Passive haemagglutination technique for serotyping *Campylobacter fetus* subsp. *jejuni* on the basis of soluble heat-stable antigens. J. Clin. Microbiol. 12: 732–737.

Penner, J.L., Hennessy, J.N. and Congi, R.V. 1983. Serotyping of *Campylobacter jejuni* and *Campylobacter coli* on the basis of thermostable antigens. Eur J. Clin. Microbiol. 2: 378–383.

Petersen, L., and On, S.L. 2000. Efficacy of flagellin gene typing for epidemiological studies of *Campylobacter jejuni* in poultry estimated by comparison with macrorestriction profiling. Lett. Appl. Microbiol. 31: 14–19.

Petersen, L., and Wedderkopp, A. 2001. Evidence that certain clones of *Campylobacter jejuni* persist during successive broiler flock rotations. Appl. Environ. Microbiol. 67: 2739–2745.

Petersen, L., Nielsen, E.M., and On, S.L. 2001a. Serotype and genotype diversity and hatchery transmission of *Campylobacter jejuni* in commercial poultry flocks. Vet. Microbiol. 82: 141–154.

Petersen, L., Nielsen, E.M., Engberg, J., On, S.L., and Dietz, H.H. 2001b. Comparison of genotypes and serotypes of *Campylobacter jejuni* isolated from Danish wild mammals and birds and from broiler flocks and humans. Appl. Environ. Microbiol. 67: 3115–3121.

Read, T.D., Salzberg, S.L., Pop, M., Shumway, M., Umayam, L., Jiang, L., Holtzapple, E., Busch, J.D., Smith, K.L., Schupp, J.M., Solomon, D., Keim, P., and Fraser, C.M. 2002. Comparative genome sequencing for discovery of novel polymorphisms in *Bacillus anthracis*. Science 296 (5575): 2028–2033.

Richardson, J.F., Frost, J.A., Kramer, J.M., Thwaites, R.T., Bolton, F.J., Wareing, D.R., and Gordon, J.A. 2001. Coinfection with *Campylobacter* species: an epidemiological problem? J. Appl. Microbiol. 91: 206–211.

Ringoir, D.D., and Korolik, V. 2003. Colonization phenotype and colonization potential differences in *Campylobacter jejuni* strains in chickens before and after passage *in vivo*. 2003 Vet. Microbiol. 92 (3): 225–235.

Ribot, E. M., Fitzgerald, C., Kubota, K., Swaminathan, B., and Barrett, T. J. 2001. Rapid pulsed-field gel electrophoresis protocol for subtyping of *Campylobacter jejuni*. J. Clin. Microbiol. 39: 1889–1894.

Roberts, J.A., Cumberland, P., Sockett, P.N., Wheeler, J., Rodrigues, L.C., Sethi, D., and Roderick, P.J. 2003. The study of infectious intestinal disease in England: socioeconomic impact. Epidemiol. Infect. 130: 1–11.

Sails, A.D., Swaminathan, B., and Fields, P.I. 2003a. Utility of multilocus sequence typing as an epidemiological tool for investigation of outbreaks of gastroenteritis caused by *Campylobacter jejuni*. J. Clin. Microbiol. 41: 4733–4739.

Sails, A.D., Swaminathan, B. and Fields, P.I. 2003b. Clonal complexes of *Campylobacter jejuni* identified by multilocus sequence typing correlate with strain associations identified by multilocus enzyme electrophoresis. J. Clin. Microbiol. 41: 4058–4067.

Santesteban, E., Gibson, J., and Owen, R.J. 1996. Flagellin gene profiling of *Campylobacter jejuni* heat-stable serotype 1 and 4 complex. Res. Microbiol. 147: 641–649.

Shi, Z.Y., Liu, P.Y., Lau, Y.J., Lin, Y.H., Hu, B.S., and Tsai, H.N. 1996. Comparison of Polymerase chain reaction and pulsed-field gel electrophoresis for the epidemiological typing of *Campylobacter jejuni*. Diagn. Microbiol. Infect. Dis. 26: 103–108.

Sjogren, E., Alestig, K., and Kaijser, B. 1989. *Campylobacter* strains from Swedish patients with diarrhoea. Distribution of serotypes over a five year period. APMIS 97 (3): 221–226.

Slater, E., and Owen, R.J. 1998. Subtyping of *Campylobacter jejuni* Penner heat-stable (HS) serotype 11 isolates from human infections. J. Med. Microbiol. 47: 353–357.

Sopwith, W., Ashton, M., Frost, J.A., Tocque, K., O'Brien, S., Regan, M., and Syed, Q. 2003. Enhanced surveillance of *Campylobacter* infection in the North West of England 1997–1999. J. Infect. 46: 35–45.

Spratt, B.G., Hanage, W.P. and Feil, E J. 2001. The relative contributions of recombination and point mutation to the diversification of bacterial clones. Curr. Opin. Microbiol. 4: 602–606.

Steinbrueckner, B., Ruberg, F., and Kist, M. 2001. Bacterial genetic fingerprint: a reliable factor in the study of the epidemiology of human *Campylobacter* enteritis? J. Clin. Microbiol. 39: 4155–4159.

Suerbaum, S., Lohrengel, M., Sonnevend, A., Ruberg, F., and Kist, M. 2001. Allelic diversity and recombination in *Campylobacter jejuni*. J. Bacteriol. 183: 2553–2559.

Swaminathan, B., Barrett, T., Hunter, S., Tauxe, R., and The PulseNet Force, 'The molecular subtyping network for food-borne bacterial disease surveillance, United States', Emerg. Infect. Dis. 2001. **7**: 382–389.

Thomas, L.M., Long, K.A., Good, R.T., Panaccio, M., and Widders, P.R. 1997. Genotypic diversity among *Campylobacter jejuni* isolates in a commercial broiler flock. Appl. Envir. Microbiol. 63: 1874–1877.

Wareing, D.R., Bolton, F.J., Fox, A.J., Wright, P.A., and Greenway, D.L. 2002. Phenotypic diversity of *Campylobacter* isolates from sporadic cases of human enteritis in the UK. J. Appl. Microbiol. 92: 502–509.

Wassenaar, T.M., Fry, B.N., and van der Zeijst, B.A. 1995. Variation of the flagellin gene locus of *Campylobacter jejuni* by recombination and horizontal gene transfer. Microbiol. 141: 95–101.

Wassenaar, T.M., Geilhausen, B., and Newell, D.G. 1998. Evidence of genomic instability in *Campylobacter jejuni* isolated from poultry. Appl. Environ. Microbiol. 64: 1816–1821.

Wassenaar, T.M., Fry, B.N., Lastovica, A.J., Wagenaar, J.A., Coloe, P.J., and Duim, B. 2000. Genetic characterization of *Campylobacter jejuni* O: 41 isolates in relation with Guillain–Barré syndrome. J. Clin. Microbiol. 38: 874–876.

Woodward, D.L., and Rodgers, F.G. 2002. Identification of *Campylobacter* heat-stable and heat-labile antigens by combining the Penner and Lior serotyping schemes. J. Clin. Microbiol. 40: 741–745.

Zhao, C., Ge, B., De Villena, J., Sudler, R., Yeh, E., Zhao, S., White, D.G., Wagner, D., and Meng, J. 2001. Prevalence of *Campylobacter* spp., *Escherichia coli*, and *Salmonella* serovars in retail chicken, turkey, pork, and beef from the Greater Washington, D.C., area. Appl. Environ. Microbiol. 67: 5431–5436.

Chapter 5

Advances in *Campylobacter jejuni* Comparative Genomics through Whole Genome DNA Microarrays

Nick Dorrell, Olivia L. Champion and Brendan W. Wren

Abstract

DNA microarray technology is revolutionizing the study of bacteria at both the genomic and transcriptional levels. Whole genome comparisons from microarray analysis between bacterial strains and species have expanded our understanding of the genetic diversity and evolution within bacterial populations. *Campylobacter jejuni* strain diversity combined with variable host responses results in a complex spectrum of disease outcomes, ranging from asymptomatic colonization to severe inflammatory diarrhoea. A major reason for continued comparative genomics studies is the absence of an animal model that reflects *C. jejuni*-associated disease. Precise strain comparisons from well-characterized strains of diverse origins will allow correlates of pathogenesis to be determined and the subsequent identification of potential virulence determinants. Also an understanding of genetic differences between *C. jejuni* strains from different ecological niches should allow the identification of host-specific epidemiological markers and the development of rational approaches to reduce this problematic pathogen from the food chain. Here an overview of the different microarray technologies is presented followed by the different types of *C. jejuni* microarrays in use today. Published *C. jejuni* microarray data are discussed along with examples of the different comparative genomic studies for other bacteria that have utilized microarray technology and future developments are proposed.

INTRODUCTION

DNA microarray technology allows whole genome comparisons to be made between bacterial strains and species and has revolutionized our understanding of the genetic diversity and evolution within bacterial populations. This chapter will describe the role of this new discipline in the study of *Campylobacter jejuni*.

The ability to easily distinguish between *C. jejuni* strains is vital for surveillance and epidemiological studies, enabling the routes and sources of *C. jejuni* infection to be traced. Traditionally, *C. jejuni* strains have been classified using Penner (heat-stable) and Lior (heat-labile) serotyping. The identification of serotypes associated with post-infection neuropathies, for example, has been of particular importance (Fujimoto *et al.*, 1997; Kuroki *et al.*, 1993). More recently, molecular DNA-based techniques have been developed, including randomly amplified polymorphic DNA (RAPD) and pulsed-field gel electrophoresis (PFGE). Multilocus sequence typing (MLST) studies have shown that *C. jejuni* strains are genetically very diverse, exhibit a high degree of plasticity, and have a weakly clonal population structure (Dingle *et al.*, 2001; Suerbaum *et al.*, 2001).

However, current typing methods have generally failed to identify strains with phenotypic characteristics associated with pathogenicity, virulence and different ecological habitats.

C. jejuni causes a wide spectrum of disease in humans, ranging from mild diarrhoea to severe bloody diarrhoea and occasionally septicaemia. Some of the differences in the severity of disease that is caused may be due to host susceptibility, but it is thought that differences in the pathogenic potential of *C. jejuni* strains may provide an explanation for the observed spectrum of *C. jejuni*-associated disease. We do know that remarkable differences exist between *C. jejuni* strains with respect to phenotypic properties such as cell invasiveness, rates of translocation across cell monolayers, toxin production and colonization of chickens (Hu and Kopecko, 1999), but little is known how these observations might reflect differences in genetic factors. Similarly, highlighting genetic differences between strains from various sources may lead to the identification of host-specific markers.

Whole genome microarrays have the potential to finally demonstrate the link between genetic characteristics and the ability of a strain to cause disease or inhabit a preferred ecological niche. The availability of the genome sequence for the *C. jejuni* strain NCTC 11168 (Parkhill *et al.*, 2000) has enabled the construction of DNA microarrays that contain reporter elements representing each of the 1654 predicted coding sequences for this strain. Such arrays can be used to scan the genome content of other *C. jejuni* strains, and compare them to the sequenced strain. As more *Campylobacter* strains are sequenced, it will be possible to design arrays that are representative of not just a single strain, but of the species and maybe even the *Campylobacter* genus.

GENETIC DIVERSITY OF *C. JEJUNI* STRAINS

The ability to distinguish between strains is important for monitoring human infections. It contributes to the identification of sources of infection and routes of transmission, aids in the development of targeted control strategies, and allows questions regarding host specificity of strains or association with specific clinical outcomes to be addressed. There are a number of different typing methods available, and the message from studies carried out so far is clear – the phenotypic and genotypic diversity of *C. jejuni* strains is extensive.

Phenotypic methods such as serotyping and phage typing are relatively easy to perform, but they do not allow all strains to be typed (Patton *et al.*, 1991). In the 1980s, two serotyping schemes were developed for campylobacters. The Penner scheme is based on the variation in heat-stable antigens (Penner and Hennessy, 1980), whilst the Lior scheme is based on the variation of heat-labile antigens (Lior *et al.*, 1982). For both methods the precise nature of the antigen detected was unknown at the time, except that it must be a surface structure, with the Penner scheme linked to lipo-polysaccharide (LPS) and the Lior scheme linked to flagellin. The genome sequence of NCTC 11168 has since revealed a previously unidentified capsule biosynthesis locus (Parkhill *et al.*, 2000). Mutation of genes associated with the transport of the capsular polysaccharide to the cell surface resulted in changes to the Penner serotype, suggesting that the capsule is the major determinant for this serotype (Karlyshev *et al.*, 2000). Therefore, the large number of Penner serotypes highlights diversity in the capsule biosynthesis locus amongst *C. jejuni* strains (see Chapter 12).

Serotyping classifies strains into broad groups, but there is further variation within each serotype, which is undetectable by the Penner or Lior methods. Phage typing schemes using bacteriophages to classify *C. jejuni* isolates have been described in the USA, UK and Canada (Grajewski *et al.*, 1985; Khakhria and Lior, 1992; Salama *et al.*, 1990). These allow strains in the same serotype group to be subdivided further.

Genotypic methods discriminate between strains better, and can also be applied to epidemiological studies (Hanninen *et al.*, 1998; Lorenz *et al.*, 1998). One commonly used genotyping technique for campylobacters is PFGE. It involves digesting of chromosomal DNA with rare cutting restriction enzymes, then separating the fragments by coordinated application of pulsed electric fields from different positions in the electrophoresis cell. PFGE is much more sensitive than the phenotypic typing methods (Steele *et al.*, 1998). Another promising technique is amplified fragment length polymorphism (AFLP). With this method, the chromosomal DNA is digested using two different restriction enzymes, after which specific oligonucleotide adapters are ligated to the restriction sites and the fragments amplified by PCR. Other genotypic methods include RAPD, *fla* typing and ribotyping (see Chapter 7 for more details).

Recently, a multilocus sequence typing (MLST) scheme (Maiden *et al.*, 1998) has been developed and validated for *C. jejuni* (Dingle *et al.*, 2001; Suerbaum *et al.*, 2001) (Chapter 3). MLST assesses variation in housekeeping genes, which are usually highly conserved and evolve slowly. Initial analysis of 194 isolates suggests that *C. jejuni* is genetically diverse with a weakly clonal population and that horizontal gene transfer occurs at both between strains and between species (Dingle *et al.*, 2001a). MLST analysis of *C. jejuni* strains associated with Guillain–Barré and Miller Fisher syndromes suggests that these strains are indeed from a diverse genetic lineage (Dingle *et al.*, 2001b). The *Campylobacter* MLST database (http://campylobacter.mlst.net/index.html) currently contains data from nearly 2000 isolates.

But while both phenotypic and genotypic typing methods suggest that *C. jejuni* strains are very diverse, they do not provide any real insight into the nature of this diversity, particularly with regard to pathogenicity. DNA microarrays now allow a whole genome approach to studying the genetic differences between *C. jejuni* strains.

DNA MICROARRAYS EXPLAINED

The term 'microarray' has been applied to high-density arrangements of DNA elements on nitrocellulose or nylon membranes, glass slides and silica wafers. To further confuse the issue, the term 'DNA chip' has often been used interchangeably with the term 'microarray'. The terms used to describe microarray experiments have also been inconsistent, both in the literature and in scientific meetings, over the short time this new technology has been available. The terminology associated with microarray experiments is clarified below, along with details of the common set-ups that have been used to date.

Terminology

The labelled sample in solution that hybridizes to a bound target has traditionally been referred to in molecular biology terms as a 'probe'. However with respect to microarray experiments, the term has often been used to describe not the labelled sample hybridized to the array, but the element attached to the array. This has resulted in considerable confusion. A standard terminology has now been developed, called MAGE-ML (MicroArray Gene Expression Markup Language; http://www.mged.org). MAGE-ML proposes that the term 'reporter' should refer to the elements bound to the microarray surface, and that 'sample' should refer to the labelled nucleic acid in the hybridization solution. The use of the term 'probe' is avoided altogether.

Membrane arrays

In membrane arrays, the reporter elements are printed onto nylon membranes. They are usually referred to as 'macroarrays' because they are larger than microarrays, and the density of the reporter elements on the array is lower. They have included both unknown sequences derived from cDNA libraries and PCR products of known sequence. Labelled samples are hybridized to the reporter elements on the membranes using protocols similar to Southern blots. To compare different samples, the hybridization signals from one membrane must be compared to signals from another membrane. This can lead to problems with normalizing the signal intensities from membranes hybridized at different times.

Glass slide arrays

These are usually referred to as 'microarrays', because the reporter elements are placed at a much higher spot density (>20-fold) than can be achieved using macroarrays. Complete genome sequences enable the design of whole genome arrays, which contain at least one reporter element for every gene in the genome, but may also include intergenic regions, additional genomic features or genes from other organisms. The reporter elements can be double-stranded, generated from gene-specific PCR or by amplifying clone inserts of DNA/cDNA libraries, or single-stranded oligonucleotides synthesized in the size range of 40–70 nucleotides. These reporter elements are then spotted onto specially coated glass microscope slides (Hinds *et al.*, 2002a).

One major advantage of the glass slide arrays is that they can incorporate a 'two-colour' technology. With this method, two samples (test and reference conditions) are each labelled with a different fluorescent dye, so they can be compared directly on a single array. This competitive hybridization is one of the major advantages of microarray-based experiments. The differences in signal intensity from the two hybridized labels can be detected using a dual laser confocal scanner. Another benefit of the microarray approach is the flexibility of design and construction that allows the arrays to be easily customised to suit particular applications.

Affymetrix GeneChips™

Oligonucleotides of around 25 bases in length are synthesized *in situ* on silica wafers through a process of photolithography and solid-phase chemistry. This creates arrays that contain hundreds of thousands of reporter elements packed at extremely high density. Full details of the manufacturing process can be found at http://www.affymetrix.com/technology/manufacturing/index.affx. These arrays are often referred to as 'DNA chips', as the production processes was adapted from the semiconductor industry. A DNA chip can include multiple oligonucleotides for each gene in a genome, as well as oligonucleotides from intergenic regions. The reporter elements are designed to be as sensitive, specific, and reproducible as possible so that specific signals can consistently be distinguished from the background or from closely related target sequences. As with macroarrays, this technology is 'one-colour', meaning that only one channel of information is gained from each array. Comparisons are made between individual chips that have been hybridized with different samples.

CAMPYLOBACTER DNA MICROARRAYS

Of the three types of array described above, it is the microarray that has proved most popular in the study of *C. jejuni*. In the short time since the development of microarray technology, a wide range of different types has been employed. These are described in detail below.

Clone arrays

To our knowledge, the first whole-genome DNA microarray for *C. jejuni* was produced by the Bacterial Microarray Group at St George's Hospital Medical School in London (http://bugs.sghms.ac.uk/index.php) in January 2000. This early *C. jejuni* array contained amplified clone inserts. Most of the cost of producing DNA microarrays based on PCR products is due to synthesizing the oligonucleotide primer pairs needed to amplify the gene fragments that make up the reporter elements present on the array. This cost can be dramatically reduced by amplifying the fragments from clone libraries instead of directly from the genome, so that the same pair of primers can be used for all reporter elements on the array. For example, ordered pUC clone libraries are a by-product of most bacterial genome sequencing projects and it is relatively straightforward to select from the library an optimum clone to represent every gene in the genome. In the case of *C. jejuni* NCTC 11168, the ordered plasmid library generated by the Sanger Institute was used as the template and a single pair of primers based on the pUC19 vector sequence was used to amplify gene fragments representative of the 1731 initially predicted coding sequences, to produce a whole genome microarray. The predicted coding sequences were based on the preliminary annotation of the NCTC 11168 genome sequence (Julian Parkhill, unpublished data), however following completion of this annotation, the number of predicted genes was reduced to 1654 (Parkhill *et al.*, 2000), so the resulting 77 'non-genes' were flagged in the gene lists used by the analysis software as 'not annotated genes' (Dobrindt *et al.*, 2003).

A drawback of this type of microarray is that many of the clones available contain adjacent genes or gene fragments as well as the desired gene. Only 34.5% of the clones selected from the NCTC 11168 ordered plasmid library contained a single gene fragment, with 35.4% containing one adjacent gene fragment and the remaining 30.1% containing more than one adjacent fragment. The presence of these adjacent fragments can result in cross-hybridization if a gene sequence present in the hybridization sample is contained within more than one reporter element on the array. This distorts the hybridization signal intensity for both the desired gene, and the element to which cross-hybridization is occurring. Despite this limitation however, useful biological information and microarray expertise has been gained from this cost-effective approach.

Gene-specific arrays

A second-generation *C. jejuni* microarray was produced by the Bacterial Microarray Group at St George's Hospital in June 2002. This was a gene-specific array, where a specific pair of oligonucleotide primers was designed for each gene. Many bacterial microarrays use PCR to amplify the whole gene. The resulting microarrays are gene-specific, however for some microbial species cross-hybridization between gene fragments representing different but very similar genes can still be a problem. This makes genetic rearrangements, insertions, inversions and duplications difficult to detect. For example, it would be difficult to distinguish separate genes in the Cj0617 or Cj1318 paralogous gene families of *C. jejuni* NCTC 11168, using whole gene reporter elements. So it would be hard to tell whether a particular gene in a paralogous gene family was present or absent in a test strain from the microarray alone.

For this second-generation *C. jejuni* array, the PCR primers were designed more selectively (Hinds *et al.*, 2002b), to minimize cross-hybridization to other genes in the genome. The primers ranged from 22 to 25 bases in size, with products ranging from 70 to 800 bases long. The primers were also designed to have closely matched melting temperature to aid high-throughput PCR amplification. As well as the NCTC 11168 genes, the microarray contains reporter elements representing 73 genes from other strains, to make it more representative of the *C. jejuni* species, rather than just the sequenced strain (http://bugs.sghms.ac.uk/organisms/cj.php?tab=array). This type of gene-specific array is the most commonly used at present. At the time of writing, many groups have developed their own *Campylobacter* gene-specific PCR product microarrays. To our knowledge, there are *C. jejuni* arrays in use in one other group in the UK, at least two groups in the USA and one in Canada (Table 5.1).

Full details of the design and construction of the *C. jejuni* array in use at the NRC Institute for Biological Sciences in Ottawa are available on their website (http://ibs-isb.nrc-cnrc.gc.ca/ibs/immunochemistry/microarray_chips_e.htm).

Oligo arrays

Even when gene-specific arrays are designed to minimize cross-hybridization as described above, cross-hybridization between homologous DNA sequences can still be a problem when using double-stranded DNA reporter elements such as PCR products, particularly for gene expression studies. Microarrays printed with such elements may not allow differential expression patterns for highly homologous or overlapping genes to be detected. In addition, PCR amplification of individual reporter elements is labour intensive and expensive, and can have a high failure rate (typically 5–10%). Oligonucleotides have started to show many advantages over more traditional PCR products. They offer increased specificity and better discrimination between gene families and overlapping genes, while maintaining the sensitivity of PCR products.

The two most popular formats are the Affymetrix GeneChip™ and oligonucleotide systems. The first microarray that Affymetrix produced for a prokaryote was the GeneChip® *E. coli* Genome Array. A custom-made *Campylobacter* GeneChip® should

Table 5.1 Research groups currently producing 'in-house' gene-specific *C. jejuni* glass slide microarrays.

Institute	Website	Reference
Bacterial Microarray Group, St George's Hospital Medical School, London, UK	http://bugs.sghms.ac.uk/index.php	Dorrell *et al.* (2001) Genome Res. 11: 1706–1715
Institute of Food Research, Norwich, UK	http://www.ifr.bbsrc.ac.uk/Safety/ Microarrays	Wells and Bennik (2003) Nutr. Res. Rev. 16: 21–35
Stanford University, USA		Leonard *et al.* (2003) J. Infect. Dis. 187: 691–694
Oklahoma State University, USA		Stintzi (2003) J. Bacteriol. 185: 2009–2016
NRC Institute for Biological Sciences, Ottawa, Canada	http://ibs-isb.nrc-cnrc.gc.ca/ibs/ immunochemistry/microarray_ chips_e.htm	Carillo *et al.* (2004) J. Biol Chem. 279: 20327–20333

be available in 2003 (David Ussery, personal communication). Meanwhile the longer oligonucleotide systems use an optimized oligonucleotide (50-mer or 70-mer in size) to represent each gene in a genome. Each is designed to be as specific as possible for its target gene and all the oligonucleotides are designed to have similar melting temperatures. Two *Campylobacter* oligo arrays are now commercially available, produced by MWG Biotech (www.mwg-biotech.com) and by Qiagen (www.qiagen.com).

Comparative genomic studies in other pathogenic bacteria

Bacterial DNA microarrays have been used in a range of pathogenic species, to compare the genomes of related strains and identify deletions or sequence divergence in one strain by comparison with another reference strain. This technique, allied with the increasing availability of genome sequence data, is set to revolutionize our ability to distinguish bacteria. DNA microarray analyses have revealed a vast genetic diversity both between genera and within species. Deciphering the mechanisms behind this variability will help us to understand of the phylogeny, physiology, ecology and evolution of bacteria. Comparing the genomes of pathogens and non-pathogens within a species can be particularly useful for identifying determinants that are important in virulence, transmission and host specificity. Selected examples of how microarray-based studies have been used to further our understanding of virulence in other bacterial species are highlighted below.

Genome diversity

Microarrays constructed from the genome sequence of one bacterial strain may be used to analyse different strains of the same or closely related bacterial species to give an indication of the extent of genetic diversity. *Photorhabdus* species are found in the guts of nematodes that invade insects. Some strains are orally toxic and kill the insect host, whilst other strains are non-toxic. A microarray was constructed consisting of 96 putative virulence factor genes from an orally toxic *Photorhabdus luminescens* strain W14 (Marokhazi *et al.*, 2003). This study aimed to identify the minimal subset of toxin complex (tc) genes required for oral toxicity by hybridizing genomic DNA from both orally toxic and non-toxic strains to this array. A striking split was found in the distribution of tc genes between orally toxic and non-toxic strains. Orally toxic strains were found to carry all three genes in the *tca* operon (*tcaABC*), whereas those lacking oral toxicity lack *tcaA* and *tcaB*. This study clearly demonstrates that even a very limited microarray can be an extremely powerful tool for correlating phenotype and genotype, with particular reference to virulence.

Genome plasticity

Virulence-associated genes are often located on mobile genetic elements such as plasmids, bacteriophages and pathogenicity islands (PAIs), which may be transferred laterally between bacteria. Microarray analysis can indicate the presence of such horizontally acquired genes and highlight gene deletions which contribute to genome plasticity and consequently to the evolution of new species, subspecies and pathotypes. *Escherichia coli* is an excellent model for such studies owing to its ecological and pathogenic diversity. Genome plasticity in pathogenic and commensal *E. coli* isolates was analysed using two different arrays, one containing all the genes from the non-pathogenic *E. coli* K-12 strain MG1655 and another so-called '*E. coli* pathoarray', consisting of 456 reporter elements specific for typical virulence-associated genes of extraintestinal pathogenic *E.*

coli (ExPEC), intestinal pathogenic *E. coli* (IPEC) and *Shigella* (Dobrindt *et al.*, 2003). Twenty-three isolates from varying sources (including eight commensal strains) were hybridized with these DNA microarrays. ExPEC and IPEC strains, plus one commensal, all carried genomic islands whose products contribute to the individual traits of these strains. The insertion of foreign genetic elements such as PAIs and plasmids associated with tRNA genes is well documented in several bacteria (Hacker and Kaper, 2000; Ochman *et al.*, 2000). This study revealed that tRNA loci represent variable regions of the *E. coli* chromosome that may be due to the integration or reduction of genetic elements and that both acquisition and deletion of DNA elements are important processes involved in the evolution of prokaryotes.

Microarray analysis of *Staphylococcus aureus* has revealed extensive genomic variation including 18 large chromosomal regions differing between 36 strains from various sources (Fitzgerald *et al.*, 2001). RD13 was one such region that was present in all 36 strains but varied in its nucleotide and gene content. RD13 encodes at least seven staphylococcal exotoxin-like (SET) proteins depending on the strain. RD13 is present in all strains representing the major clonal lineages of *S. aureus* and the G+C% content is equivalent to that of the *S. aureus* genome, suggesting that RD13 is an ancient feature of the *S. aureus* chromosome. This suggests that an ancestral strain with a full complement of *set* genes underwent multiple independent losses of *set* genes in parallel in separate lineages. DNA sequence analysis revealed that horizontal gene transfer and recombination have contributed to the diversification of RD13. The SET variants may act as a mechanism of immune evasion employed by *S. aureus* which encounter a host already exposed to SET proteins from a previous staphylococcal infection (Fitzgerald *et al.*, 2003).

Pathological outcome of infections

DNA microarrays were used to address the issue of whether strains responsible for the current seventh cholera pandemic have genes encoding gain-of-function traits that may have displaced pre-existing classical *Vibrio cholerae* strains (Dziejman *et al.*, 2002). The first six recorded cholera pandemics occurred between 1817 and 1923 and were caused by the classical biotype of *V. cholerae*. However in 1961 the El Tor biotype emerged to cause the seventh cholera pandemic and resulted in the eventual global elimination of the classical biotype strains as a cause of disease. A *V. cholerae* microarray based on the El Tor O1 strain N16961, was used to analyse a collection of nine strains of diverse global origin isolated between 1910 and 1992. It was possible to differentiate classical biotype strains from El Tor biotype strains and two putative chromosomal islands (VSP-I and VSP-II) with a deviant G+C content were identified in El Tor biotypes. Genes associated with the VSP-1 and VSP-II islands may encode key properties that led to the global success of the El Tor biotype and studies are in progress to determine their potential role both in human infection and in promoting the fitness of *V. cholerae* in environmental ecosystems.

Similarly, the pathological outcome of *Neisseria* species infections has been investigated using microarrays (Perrin *et al.*, 2002). *Neisseria meningitis* (the causative agent of meningitis) is very closely related to *N. gonorrhoeae* (the causative agent of gonorrhoea) and *N. lactamica* (a harmless commensal of the nasopharynx), yet their disease profiles are very different. This study did not reveal a pathogenicity island responsible for differences in clinical outcome, but a series of relatively small sequences scattered throughout the genome which were either specific to *N. meningitis* or shared with *N. gonorrhoeae*, but absent from *N. lactamica*. This study confirmed that the capsule biosynthesis locus and the RTX toxin family were meningococcal specific.

Studying genomes of related species

Provided that a microarray has been constructed for a closely related species, microarray analysis can allow the rapid identification of genes in the genome of an unsequenced bacteria. *Wigglesworthia glossinidia*, a member of the Enterobacteriaceae, is an obligate endosymbiont of the tsetse fly, which relies on the bacterium for fertility and nutrition. Symbiotic associations with microorganisms are pivotal in many insects, but the functional role of obligate symbionts can be difficult to study due to the problem of growing these organisms *in vitro*. The *W. glossinidia* genome is less than 770 kb, about one-sixth that of the related free-living bacteria *E. coli* (4.6 Mb). In order to gain an insight into the composition of the genome, *W. glossinidia* genomic DNA was hybridized to an *E. coli* DNA microarray, revealing 650 orthologous genes, corresponding to approximately 85% of the predicted size of the *W. glossinidia* genome (Akman and Aksoy, 2001). Many of the genes retained in the *W. glossinidia* genome are involved in cell processes, DNA replication, transcription and translation. However, genes encoding transport proteins, chaperones, biosynthesis of co-factors, and some amino acids were also identified in significant numbers, suggesting an important role for these proteins in the bacterium's symbiotic lifestyle. This is a good example of how a bacterial microarray can be used to obtain broad genome information for a closely related organism in the absence of complete genome sequence data. This may be a useful approach for studying the comparative genome content of different *Campylobacter* species.

The above are only a small number of examples of many recent microarray-based comparative genomic studies of different bacterial species that were selected to highlight the range of comparative genomic studies that can be undertaken using microarrays.

C. JEJUNI COMPARATIVE GENOMIC STUDIES

One of the major advantages of glass slide microarray technology is that the hybridization of the test and control samples occurs simultaneously under identical experimental conditions. The ability to selectively incorporate nucleotides tagged with fluorescent dyes that have different excitation and emission wavelengths means that it is possible to differentially label the samples to be compared. Therefore, samples that have been labelled with different dyes can be co-hybridized to the array in a direct competitive hybridization, the two individually labelled DNA or cDNA samples competing with each other for hybridization to the specific reporter elements immobilised on the array. A DNA-based microarray hybridization involves labelling the test strain genomic DNA with one fluorescent dye (Cy5) and the genomic DNA of the reference strain with a second fluorescent dye (Cy3). This is usually achieved using whole genomic DNA as the template for a randomly primed polymerization reaction, resulting in the direct incorporation of the fluorescent analogues, which are attached to the dCTP nucleotides. These two labels are then combined and hybridized to the reporter elements on the array. The amount of hybridization to each element of DNA within each of the samples reflects their relative abundance in the two samples. The hybridization is quantitated by measuring the light emitted from each of the elements on the array when excited by lasers of different wavelengths. Fluorescent spot intensities are then quantified using a suitable software package, such as ImaGene (BioDiscovery Inc., Marina del Rey, USA). Data analysis is usually performed using a second software package, such as GeneSpring (Silicon Genetics, Redwood City, CA, USA). Generally, for each spot, background fluorescence is subtracted from the median spot fluorescence to produce a channel-specific value, then the geometric mean of the normalized Cy5 signal/Cy3 signal ratio is calculated from three array experiments. In early studies, genes were identified as absent or highly divergent if the normalized Cy5 signal/Cy3 signal ratio was < 0.5 (Dorrell *et al.*, 2001). However more recent studies have begun to use more advanced statistical models for determining whether genes are indeed absent or highly divergent. The GACK method determines whether genes are present or divergent depending on the shape of the signal-ratio distribution and does not require empirical determination of a cut-off. In fact the cut-off is determined on an array-to-array basis, accounting for variation in strain composition and hybridization quality (Kim *et al.*, 2002). This means that many genes previously classified as present using static methods are in fact classified as divergent using this type of analysis. However separating genes which are identified as absent or divergent into those that are absent and those that are just divergent but with a functional gene product is much more complicated. Whole genome comparisons may be used to model phylogeny to determine the relationship between strains associated with a particular clinical outcome or source of infection. A number of heuristic tree building algorithms are available through software packages such as GeneSpring. However, statistical support for phylogenetic relationships may be calculated when using methods including Parsimony, Maximum Likelihood and Bayesian based algorithms.

Evidence for genetic diversity – the link between capsule and Penner serotype

The first whole genome comparison of a group of *C. jejuni* strains was performed using the clone array described above (Dorrell *et al.*, 2001). The aim of this study was to investigate the extent of genetic diversity among *C. jejuni* isolates. Following the discovery of the role of the capsular polysaccharide as the serodeterminant of the Penner serotyping system (Karlyshev *et al.*, 2000), of particular interest was the level of diversity in the capsule locus and whether this could be related to Penner serotype in any way. The organization of the *C. jejuni* capsule locus bears similarity to the *Escherichia coli* paradigm, where the genes (*kpsCDEFMST*) associated with the transport of the polysaccharide are conserved and flank a central region of variable genes involved in the generation of the polysaccharide structures. In *E. coli*, variation of the genes in this central region results in production of diverse polysaccharide structures (Roberts, 1996). In *C. jejuni*, as the capsular polysaccharide is the major serodeterminant of the Penner serotyping system (Karlyshev *et al.*, 2000), then this central biosynthetic region was predicted to show variation between strains with different Penner serotypes. The *C. jejuni* strains analysed in this study are shown in Table 5.2.

Laboratory strains 81116 and 81-176 were analysed initially along with a recent human isolate which caused gastroenteritis (X) and a human isolate associated with Guillain–Barré syndrome (G1). All four strains showed variation in the central capsule biosynthetic region and conservation of the flanking *kps* genes, which could be predicted as all four strains were a different Penner serotype to the sequenced strain NCTC 11168, which is O:2. Four of the Penner reference strains (P1, P2, P3 and P4) were then analysed and whilst strains P1, P3 and P4 also showed variation in the central biosynthetic region and conservation of the flanking *kps* genes, the P2 strain showed complete conservation of the capsule biosynthetic locus. Complete conservation of the capsule locus in O:2 strains

Table 5.2 Initial *Campylobacter jejuni* strains analysed by whole genome DNA microarrays.

Strain	Relevant characteristics and history	Penner serotype	Source/reference
11168	Sequenced strain, human isolate (Worcester 1977)	O:2	NCTC 11168
81116	Human isolate, source: water (Chelmsford 1981)	O:6	NCTC 81116
81-176	Human isolate, source: raw milk (Minnesota 1981)	O:23/36	G. Perez-Perez, Vanderbilt University, Nashville, USA
X	Human isolate, gastroenteritis, source: poultry (London 1997)	UT[d]	A. Karlyshev, LSHTM, London, UK
G1	Human isolate from a patient who later developed Guillain–Barré syndrome	O:1	N. Gregson, Guy's Medical School, London, UK
P1	Penner serotype reference strain	O:1	NCTC 12500
P2	Penner serotype reference strain	O:2	NCTC 12501
P3	Penner serotype reference strain	O:3	NCTC 12502
P4	Penner serotype reference strain	O:4	NCTC 12561
PHLS01	Human isolate, phage type 35	O:2	PHLS (C0118480), Public Health Laboratory Service, Colindale, London, UK
PHLS02	Human isolate, phage type 1	O:2	PHLS (C0178110)
PHLS03	Human isolate, phage type 44	O:2	PHLS (C0125990)

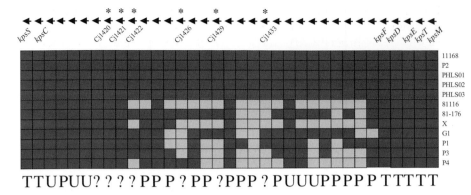

TTUPUU? ? ? ? PP P ? PP ?PPP ? PUUUPPPP P TTTTT

Figure 5.1 Representation of the 42.6 kb *C. jejuni* capsule biosynthetic locus (Cj1413 to Cj1448) showing the genes absent in the 11 strains compared against the sequenced strain NCTC 11168. Genes identified as absent or highly divergent by microarray hybridization analysis are indicated by the lighter grey boxes. The flanking polysaccharide transport-related genes are indicated as T, and polysaccharide biosynthesis-related genes from the central region are indicated as P. Genes with significant similarity to the genes unrelated to polysaccharide biosynthesis are indicated as U, whilst genes with unknown function are indicated by a question mark. The * indicates putative phase variable genes. The direction of transcription is indicated by the arrowhead representing each gene.

was tested further through analysis of three human O:2 isolates, each with a different phage type. Again all three of these strains showed complete conservation of the capsule biosynthetic locus. These data (Figure 5.1) indicate that the central biosynthetic region is responsible for the variation in the composition of the capsular polysaccharide which is detected by the Penner serotyping system.

Conserved and variable genes

Whole genome analysis of genes that were absent or highly divergent in these 11 *C. jejuni* strains compared to the sequenced strain showed that of the 1654 annotated NCTC 11168 genes, 1300 were common to all strains (referred to as species-specific genes), whilst 354 were absent or highly divergent in one or more strains (referred to as strain-specific genes). Species-specific or conserved genes included those involved in metabolic, biosynthetic, cellular and regulatory processes, but also several putative virulence factors, including the genes encoding the cytolethal distending toxin (Pickett *et al.*, 1996), flagellar structural proteins (Wassenaar *et al.*, 1991), phospholipase A (Grant *et al.*, 1997), the PEB antigenic surface proteins (Pei *et al.*, 1991), and proteins potentially involved in host–pathogen interactions, such as CiaB (Konkel *et al.*, 1999), CadF (Konkel *et al.*, 1997) and CheY (Yao *et al.*, 1997). The recently described protein glycosylation locus (Szymanski *et al.*, 1999), which was originally thought to be involved in LPS biosynthesis, was also conserved in all these strains. This supports earlier reports that this locus, which is involved in the N-linked glycosylation of over 30 *C. jejuni* proteins (Young *et al.*, 2002), is highly conserved among strains of both *C. jejuni* and *C. coli* (Szymanski *et al.*, 1999).

Many of the strain-specific or variable genes identified in this study are involved in the modification of surface structures. Three major regions of variation are observed in the capsule, lipo-oligosaccharide (LOS) and flagellar glycosylation loci. The variation observed in the capsule locus has been discussed above. Previous studies suggest that *C. jejuni* LOS consists of a relatively conserved inner core with a terminal outer core region that often varies in structure between strains and can mimic human gangliosides (Moran and Penner, 1999). The microarray data supported this, showing conservation of biosynthesis genes (*waaCVF* and *waaDE*), but also showing that many of the genes involved in the biosynthesis of the LOS outer core to be absent or highly divergent compared to the sequenced strain. Cj1139 (*wlaN*) is a β-1,3-galactosyltransferase responsible for the addition of the terminal galactose residue of the NCTC 11168 ganglioside GM_1-like LOS (Linton *et al.*, 2000a). This gene is absent in four of the strains tested (81116, 81-176, P3 and P4) as are three other putative outer-core sugar transferases, the glycosyltransferases Cj1135 and Cj1136 and the galactosyltransferase Cj1138, suggesting variation of the LOS in these four strains compared to that of NCTC 11168 (Dorrell *et al.*, 2001). The outer core of strains with serotypes O:23/36 (strain 81-176), O:3 (strain P3) and O:4 (strain P4) are known to differ markedly from O:2 strains, providing evidence to support these microarray-based predictions. The genes *neuA1*, *neuB1* and *neuC1* are involved in the sialylation of LOS (Linton *et al.*, 2000b) and are also absent or highly divergent in three of the above four strains (Dorrell *et al.*, 2001). Taken as a whole, the microarray-based comparison of the LOS biosynthesis locus is indicative of extensive genetic diversity between the strains tested, which will result in wide ranging phenotypic variation.

The region involved in the biosynthesis and post-translational modification of the flagellin subunits FlaA and FlaB also showed variation between strains, most strikingly in strains 81116 and 81-176, where much of the flagellin post-translational modification locus in NCTC 11168 is missing. This large deletion in these two widely studied laboratory strains may be the result of long term passage on solid media since isolation from human patients over 20 years ago. The locus is flanked by two identical genes Cj1318 and Cj1336, so a possible mechanism for deletion of this large section of genomic DNA by recombination between these two genes exists. If a stock of either of the original clinical isolates exists, it would be interesting to perform microarray analysis to compare genomic DNA from the laboratory-passaged strains to that of the original clinical isolates to see if the flagellin post-translational locus was present when these strains were isolated from humans.

Variation in iron acquisition systems and also DNA restriction and modification systems was also observed. This study has confirmed that extensive genetic diversity exists between *C. jejuni* strains, but started to reveal the exact genetic location of conserved and variable genes. As further microarray data from other strains are acquired, a spectacular kaleidoscope of the genetic diversity amongst *C. jejuni* isolates will appear.

A new genotyping tool?

One of the major advantages of DNA microarrays is the ability to differentiate between strains in a cluster of *C. jejuni* isolates which can not be separated using more traditional methods. Indeed the use of DNA microarrays as a new genotyping tool for the study of epidemiologically related *C. jejuni* clinical isolates has recently been proposed (Leonard *et al.*, 2003). The isolates tested had been reported in previous epidemiological analyses (Blaser *et al.*, 1981; Blaser *et al.*, 1982; Perlman *et al.*, 1988) and are listed in Table 5.3.

The isolates were analysed using both RAPD and DNA microarrays. RAPD analysis was performed as an initial determination of relatedness and identified five discrete groups on the basis of similar banding patterns. One strain in group B and one in group E showed a single band difference from other members of those groups, but otherwise the RAPD analysis suggested that the groups of strains were highly related, using current RAPD

Table 5.3 Epidemiologically linked groups of *Campylobacter jejuni* isolates differentiated by microarray analysis.

Group	Source
A	Minnesota milk outbreak
B	Raw milk outbreak with two human isolates and a bovine isolate
C	Household cluster
D	Multiple isolates from a patient with hypogammaglobulinaemia
E	Multiple isolates from a patient with human immunodeficiency virus (HIV) infection

interpretation guidelines. Microarray analysis was performed on the same isolates to test whether this analysis would cluster the strains into groups in the same way that the RAPD analysis had done. Overall this was the case; however two strains did not cluster in the manner predicted by RAPD analysis. One of the human isolates from the raw milk outbreak (group B) did not cluster with the other human isolate and the bovine isolate from this group, clearly suggesting that one of the supposed outbreak isolates was acquired from a different source. One of the three isolates from the patient with HIV (group E) was shown to be substantially different from the other two isolates, suggesting that the infection was caused by two separate *C. jejuni* strains. Both the anomalous group B and group E strains were the ones that showed a single band difference from other strains in their respective group. However the microarray analysis clearly differentiated these isolates as separate strains because of the whole genome comparison that this method allowed. This study shows that DNA microarrays can provide a highly detailed whole genome fingerprint of individual strains and, more importantly, can differentiate between strains that appear to be the same using a currently established genotyping tool. Data from this study also supported the findings of the first comparative genomics *C. jejuni* study in terms of identifying areas of significant divergence in the genome which contain genes involved in surface modification. An additional area of divergence was highlighted which contains genes involved in sugar transport and metabolism, suggesting that *C. jejuni* survival during infection may be dependent on both surface modification and sugar use.

Microarrays and MLST: a search for non-pathogenic *C. jejuni* strains

One significant advantage of microarray-based comparative genomics is the ability to identify genes that are absent in a test strain but present in the reference strain. The sequenced strain NCTC 11168 was originally isolated from a human patient and so can be used as a reference for a pathogenic strain. One of the recurring problems of *C. jejuni* research into pathogenesis has been the identification of genuine non-pathogenic strains because of the lack of an appropriate animal model of disease. Initial MLST analysis confirmed the highly diverse genetic nature of the *C. jejuni* species and indicated that the population structure was weakly clonal (Dingle *et al.*, 2001; Suerbaum *et al.*, 2001). Weakly clonal bacterial populations comprise a number of clonal complexes, which correspond to 'lineages'. Lineages are groups of organisms that are presumed to be derived from the same progenitor, however the phylogenetic relationships between distinct clonal complexes are difficult to reconstruct as they have been disrupted by horizontal gene transfer (Holmes *et al.*, 1999). A number of *C. jejuni* clonal complexes can be associated with human disease (Dingle *et al.*, 2001ab). Further MLST analysis of 814 strains identified two clonal complexes containing isolates exclusively from beach sand and separate from clonal complexes associated with human disease and farm animals (Dingle *et al.*, 2002). These isolates are assumed to originate from wild birds and these two clonal complexes are separate from other clonal complexes which include isolates from wild birds. It was therefore hypothesised that isolates in these two clonal complexes (ST-177 and ST-179) were *C. jejuni* strains that were unable to cause disease in humans. These three isolates from ST-177 and three from ST-179 were analysed by microarray hybridization and genes absent or highly divergent in these strains compared with NCTC 11168 were identified. This comparison identified the pantothenate biosynthesis genes *panBCD* as the only genes that were absent or highly divergent in all the environmental isolates, but present in NCTC 11168 and all the other human isolates tested (Nick Dorrell, unpublished data). Pantothenic acid is a precursor of coenzyme A that is thought to be involved in fatty acid synthesis, fatty acid oxidation and pyruvate oxidation. The exact role of pantothenic acid in bacteria is not clear, however a recent study has shown a *Mycobacterium tuberculosis* pantothenate auxotroph is highly attenuated in mice (Sambandamurthy *et al.*, 2002). It is possible that possession of the *panBCD* operon provides *C. jejuni* strains with a selective metabolic advantage under certain conditions and could act as a marker to distinguish strains that are solely avian commensals from those that are capable of causing disease in humans. Further studies will be required to test this hypothesis.

Microarray analysis of the second sequenced *C. jejuni* strain RM1221

The Agricultural Research Service, the in-house research arm of the US Department of Agriculture, is currently in the process of sequencing and annotating the genome of a poultry isolate of *C. jejuni*, in collaboration with The Institute for Genome Research (TIGR). This project (http://www.ars.usda.gov/research/projects/projects.htm?ACCN_ NO=404578) is due for completion by the end of September 2003. The strain RM1221 was isolated from a chicken sold in a supermarket (Miller *et al.*, 2000). The availability of a second fully annotated genome sequence will be invaluable for many aspects of *C. jejuni* research and particularly with regard to microarray analysis in terms of the relationship between divergence in the nucleotide sequence of a gene and the retention of function of the gene product. Currently we can predict that a gene is absent or highly divergent from a microarray-based genomic comparison, but deciding whether a gene is absent or just divergent at the nucleotide level is much more difficult. With this in mind, genomic DNA from strain RM1221 was analysed in a comparison with genomic DNA from strain NCTC 11168, using gene-specific PCR product glass slide arrays.

Microarray analysis of strain RM1221 suggested that 128 NCTC 11168 genes are absent or highly divergent in this strain (Nick Dorrell, unpublished data), with many of these genes associated with modification of surface structures, as had been seen with other *C. jejuni* strains (Dorrell *et al.*, 2001; Leonard *et al.*, 2003). Three major regions of variation are observed in the capsule, lipo-oligosaccharide (LOS) and flagellar loci, whilst the glycosylation locus is conserved, except for the *wlaJ* gene (Figure 5.2).

A static analysis method where spots were excluded for low signal (<100 units in the Cy3 channel) then a cut-off for the normalized Cy5/Cy3 signal intensity ratio of <0.5 was used to identify genes as absent or highly divergent. To investigate whether genes identified by microarray analysis were really absent or just highly divergent, these data were analysed by selecting a number of genes and comparing the NCTC 11168 gene

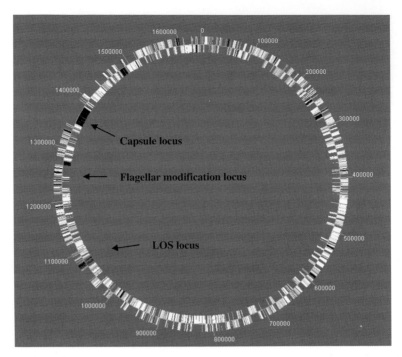

Figure 5.2 GeneSpring 6 generated impression of the RM1221 genome imposed on the NCTC 11168 genome backbone. The genes on the outside represent those transcribed in the same direction as replication whilst those on the inside represent the genes transcribed in the opposite direction. Genes highlighted in white are present in RM1221, whilst those highlighted in black are absent or highly divergent. Regions in the NCTC 11168 genome responsible for synthesizing major cell surface structures are highlighted.

sequences to the incomplete RM1221 sequence data on the TIGR website (http://tigrblast. tigr.org/ufmg/index.cgi?database=c_jejuni|seq) using *blastn* (to compare nucleotide identity) and *tblastx* (to compare predicted amino acid sequences) searches. The results are summarized in Table 5.4.

This analysis shows that some genes which fall below the arbitrary 0.5 cut-off appear to be present in the RM1221 genome. In some cases, this is very clear, such as with *kpsT* and *kpsM*. Both genes fall just below the 0.5 cut-off, but comparison of the NCTC 11168 gene sequences shows very high nucleotide identity in the RM1221 genome by *blastn* analysis whilst *tblastx* analysis predicts a highly conserved protein in both cases. In other cases, such as with the *neuA1*, *neuB1* and *neuC1* genes, which give signal intensity rations of < 0.1, the nucleotide identity in the RM1221 genome sequence is much lower but *tblastx* analysis predicts a conserved protein. Intriguingly, genes such as Cj0008 and *wlaJ* which have a signal intensity ratio much higher than *neuA1*, *neuB1* and *neuC1*, have a similar nucleotide identity in the RM1221 genome sequence but no prediction of a conserved protein. The analysis of the three putative type I restriction enzymes Cj1549, Cj1551 and Cj1553 is harder to interpret. Such variation between microarray signal intensity ratios and nucleotide identity suggests that the dynamics of hybridization between the labelled sample and the PCR product reporter elements on the glass slide are not fully understood and that this relationship is clearly not linear. It is possible that the use of oligo arrays would result in a more linear hybridization relationship and allow easier and more accurate data

Table 5.4 Comparison of selected genes from the *C. jejuni* NCTC 11168 genome sequence with the *C. jejuni* RM1221 genome sequence by microarray and bioinformatic analysis.

Gene	Microarray data	*blastn*	*tblastx*	Absent or divergent?
Cj0008	0.184	54%	0.991	Absent
Cj1122 (*wlaJ*)	0.276	57%	0.90	Absent
Cj1126 (*pglB*)	0.968	98%	0.0	Present
Cj1141 (*neuB1*)	0.085	55%	3.1×10^{-46}	Present, but divergent
Cj1142 (*neuC1*)	0.028	54%	1.0×10^{-45}	Present, but divergent
Ci1143 (*neuA1*)	0.036	54%	4.2×10^{-14}	Present, but divergent
Cj1444 (*kpsD*)	0.954	98%	6.8×10^{-274}	Present
Cj1447 (*kpsT*)	0.484	93%	8.3×10^{-114}	Present, some divergence
Cj1448 (*kpsM*)	0.374	91%	1.1×10^{-120}	Present, some divergence
Cj1549	0.027	53%	2.0×10^{-7}	Unsure
Cj1551	0.021	54%	9.0×10^{-7}	Unsure
Cj1553	0.042	54%	1.2×10^{-22}	Unsure

analysis. These results also suggest that the current methods used to analyse microarray data cannot accurately determine whether a gene is absent or just highly divergent for *C. jejuni* strains and that further work is required to confirm that a gene is really absent.

This study was performed using the second generation *C. jejuni* microarray from the Bacterial Microarray Group at St George's Hospital, which contains reporter elements for an additional 73 non-NCTC 11168 genes (http://bugs.sghms.ac.uk/organisms/cj.php ?tab=array). Data analysis suggested that seven of these genes are present in RM1221 and a *tblastx* comparison of the predicted amino acid sequences from these genes with the incomplete RM1221 genome sequence found significant matches for each one. Once the RM1221 genome is completely annotated it will be possible to confirm whether these genes are indeed present, but these preliminary data do suggest that the expansion of *C. jejuni* microarrays from representing a single strain to a more species-specific array is worthwhile.

FUTURE DEVELOPMENTS

Comparison of the genomes of different *C. jejuni* strains using DNA microarrays reveals detailed information about which genes are present or absent in the genomes of individual isolates. For example, microarray analysis has identified regions within the *C. jejuni* genome that show a high degree of variation between strains, and study of these regions has started to highlight genetic factors that may be linked to phenotypic variation and adaptation to different ecological niches. Specific genes that vary between species have also been identified that may contribute to different disease outcomes and host specificity. The studies reported so far are relatively small scale, but in the future it should be possible to compare larger collections of strains with known clinical histories. The hope is that this will enable the identification of genetic traits that are characteristic of subpopulations of strains that cause different types of disease. As well as being valuable epidemiological markers, such genetic factors would be likely to encode virulence determinants, which should in turn lead the way towards new strategies to reduce the burden of *C. jejuni*-associated disease. Identification of source-specific markers would improve *C. jejuni* epidemiology allowing the factors contributing to human *C. jejuni* infection to be established. This would aid the design of targeted control strategies to eliminate *C. jejuni* from the food chain.

Konkel, M.E., S.G. Garvis, S.L. Tipton, D.E. Anderson, Jr., and W. Cieplak, Jr. 1997. Identification and molecular cloning of a gene encoding a fibronectin- binding protein (CadF) from *Campylobacter jejuni*. Mol. Microbiol. 24: 953–963.

Konkel, M.E., B.J. Kim, V. Rivera-Amill, and S.G. Garvis. 1999. Bacterial secreted proteins are required for the internalization of *Campylobacter jejuni* into cultured mammalian cells. Mol. Microbiol. 32: 691–701.

Kuroki, S., T. Saida, M. Nukina, T. Haruta, M. Yoshioka, Y. Kobayashi, and H. Nakanishi. 1993. *Campylobacter jejuni* strains from patients with Guillain–Barré syndrome belong mostly to Penner serogroup 19 and contain beta-N-acetylglucosamine residues. Ann. Neurol. 33: 243–247.

Leonard II, E.E., T. Takata, M.J. Blaser, S. Falkow, L.S. Tompkins, and E.C. Gaynor. 2003. Use of an open-reading frame-specific *Campylobacter jejuni* DNA microarray as a new genotyping tool for studying epidemiologically related isolates. J. Infect. Dis. 187: 691–694.

Linton, D., M. Gilbert, P.G. Hitchen, A. Dell, H.R. Morris, W.W. Wakarchuk, N.A. Gregson, and B.W. Wren. 2000a. Phase variation of a beta-1,3 galactosyltransferase involved in generation of the ganglioside GM1-like lipo-oligosaccharide of *Campylobacter jejuni*. Mol. Microbiol. 37: 501–514.

Linton, D., A.V. Karlyshev, P.G. Hitchen, H.R. Morris, A. Dell, N.A. Gregson, and B.W. Wren. 2000b. Multiple N-acetyl neuraminic acid synthetase (*neuB*) genes in *Campylobacter jejuni*: identification and characterization of the gene involved in sialylation of lipo-oligosaccharide. Mol. Microbiol. 35: 1120–1134.

Lior, H., D.L. Woodward, J.A. Edgar, L.J. Laroche, and P. Gill. 1982. Serotyping of *Campylobacter jejuni* by slide agglutination based on heat-labile antigenic factors. J. Clin. Microbiol. 15: 761–768.

Lorenz, E., A. Lastovica, and R.J. Owen. 1998. Subtyping of *Campylobacter jejuni* Penner serotypes 9, 38 and 63 from human infections, animals and water by pulsed field gel electrophoresis and flagellin gene analysis. Lett. Appl. Microbiol. 26: 179–182.

Maiden, M.C., J.A. Bygraves, E. Feil, G. Morelli, J.E. Russell, R. Urwin, Q. Zhang, J. Zhou, K. Zurth, D.A. Caugant, I.M. Feavers, M. Achtman, and B.G. Spratt. 1998. Multilocus sequence typing: a portable approach to the identification of clones within populations of pathogenic microorganisms. Proc. Natl. Acad. Sci. USA 95: 3140–3145.

Marokhazi, J., N. Waterfield, G. LeGoff, E. Feil, R. Stabler, J. Hinds, A. Fodor, and R.H. Ffrench-Constant. 2003. Using a DNA microarray to investigate the distribution of insect virulence factors in strains of *Photorhabdus* bacteria. J. Bacteriol. 185: 4648–4656.

Miller, W.G., A.H. Bates, S.T. Horn, M.T. Brandl, M.R. Wachtel, and R.E. Mandrell. 2000. Detection on surfaces and in Caco-2 cells of *Campylobacter jejuni* cells transformed with new *gfp*, *yfp*, and *cfp* marker plasmids. Appl. Environ. Microbiol. 66: 5426–5436.

Moran, A.P. and J.L. Penner. 1999. Serotyping of *Campylobacter jejuni* based on heat-stable antigens: relevance, molecular basis and implications in pathogenesis. J. Appl. Microbiol. 86: 361–377.

Ochman, H., J.G. Lawrence, and E.A. Groisman. 2000. Lateral gene transfer and the nature of bacterial innovation. Nature 405: 299–304.

Parkhill, J., B.W. Wren, K. Mungall, J.M. Ketley, C. Churcher, D. Basham, T. Chillingworth, R.M. Davies, T. Feltwell, S. Holroyd, K. Jagels, A.V. Karlyshev, S. Moule, M.J. Pallen, C.W. Penn, M.A. Quail, M.A. Rajandream, K.M. Rutherford, A.H. van Vliet, S. Whitehead, and B.G. Barrell. 2000. The genome sequence of the food-borne pathogen *Campylobacter jejuni* reveals hypervariable sequences. Nature 403: 665–668.

Patton, C.M., I.K. Wachsmuth, G.M. Evins, J.A. Kiehlbauch, B.D. Plikaytis, N. Troup, L. Tompkins, and H. Lior. 1991. Evaluation of 10 methods to distinguish epidemic-associated *Campylobacter* strains. J. Clin. Microbiol. 29: 680–688.

Pei, Z.H., R.T.D. Ellison, and M.J. Blaser. 1991. Identification, purification, and characterization of major antigenic proteins of *Campylobacter jejuni*. J. Biol. Chem. 266: 16363–16369.

Penner, J.L. and J.N. Hennessy. 1980. Passive hemagglutination technique for serotyping *Campylobacter fetus* subsp. *jejuni* on the basis of soluble heat-stable antigens. J. Clin. Microbiol. 12: 732–737.

Perlman, D.M., N.M. Ampel, R.B. Schifman, D.L. Cohn, C.M. Patton, M.L. Aguirre, W.L. Wang, and M.J. Blaser. 1988. Persistent *Campylobacter jejuni* infections in patients infected with the human immunodeficiency virus (HIV). Ann. Intern. Med. 108: 540–546.

Perrin, A., S. Bonacorsi, E. Carbonnelle, D. Talibi, P. Dessen, X. Nassif, and C. Tinsley. 2002. Comparative genomics identifies the genetic islands that distinguish *Neisseria meningitidis*, the agent of cerebrospinal meningitis, from other *Neisseria* species. Infect. Immun. 70: 7063–7072.

Pickett, C.L., E.C. Pesci, D.L. Cottle, G. Russell, A.N. Erdem, and H. Zeytin. 1996. Prevalence of cytolethal distending toxin production in *Campylobacter jejuni* and relatedness of *Campylobacter* sp. *cdtB* gene. Infect. Immun. 64: 2070–2078.

Roberts, I.S. 1996. The biochemistry and genetics of capsular polysaccharide production in bacteria. Annu. Rev. Microbiol. 50: 285–315.

Salama, S.M., F.J. Bolton, and D.N. Hutchinson. 1990. Application of a new phagetyping scheme to campylobacters isolated during outbreaks. Epidemiol. Infect. 104: 405–411.

Sambandamurthy, V.K., X. Wang, B. Chen, R.G. Russell, S. Derrick, F.M. Collins, S.L. Morris, and W.R. Jacobs, Jr. 2002. A pantothenate auxotroph of *Mycobacterium tuberculosis* is highly attenuated and protects mice against tuberculosis. Nature Med. 8: 1171–1174.

Steele, M., B. McNab, L. Fruhner, S. DeGrandis, D. Woodward, and J.A. Odumeru. 1998. Epidemiological typing of *Campylobacter* isolates from meat processing plants by pulsed-field gel electrophoresis, fatty acid profile typing, serotyping, and biotyping. Appl. Environ. Microbiol. 64: 2346–2349.

Suerbaum, S., M. Lohrengel, A. Sonnevend, F. Ruberg, and M. Kist. 2001. Allelic diversity and recombination in *Campylobacter jejuni*. J. Bacteriol. 183: 2553–2559.

Szymanski, C.M., R. Yao, C.P. Ewing, T.J. Trust, and P. Guerry. 1999. Evidence for a system of general protein glycosylation in *Campylobacter jejuni*. Mol. Microbiol. 32: 1022–1030.

Wassenaar, T.M., N.M. Bleumink-Pluym, and B.A. van der Zeijst. 1991. Inactivation of *Campylobacter jejuni* flagellin genes by homologous recombination demonstrates that *flaA* but not *flaB* is required for invasion. EMBO J. 10: 2055–2061.

Yao, R.J., D.H. Burr, and P. Guerry. 1997. CheY-mediated modulation of *Campylobacter jejuni* virulence. Mol. Microbiol. 23: 1021–1031.

Young, N.M., J.R. Brisson, J. Kelly, D.C. Watson, L. Tessier, P.H. Lanthier, H.C. Jarrell, N. Cadotte, F. St Michael, E. Aberg, and C.M. Szymanski. 2002. Structure of the N-linked glycan present on multiple glycoproteins in the Gram-negative bacterium, *Campylobacter jejuni*. J. Biol. Chem. 277: 42530–4253.

Chapter 6

Prevalence of *Campylobacter* in the Food and Water Supply: Incidence, Outbreaks, Isolation and Detection

William G. Miller and Robert E. Mandrell

Abstract

Campylobacter is the primary cause of bacterial diarrhoeal illness in the developed world, with an estimated 2–3 million *Campylobacter*-related illnesses occurring in the United States per year. Although thermophilic *Campylobacter* species are considered food-borne pathogens, most illnesses caused by *Campylobacter* occur sporadically. *Campylobacter* species rarely cause food- or water-borne outbreaks, but in rare cases serious neuropathy and paralytic illness has occurred in association with a sporadic *Campylobacter* infection.

In this chapter we will review reports of the incidence of *Campylobacter* species in poultry and livestock, and in the food and water supply. We will also review reports of *Campylobacter*-related outbreaks and the links to food and water sources. Additionally, we will review studies of molecular methods that have been developed to rapidly detect, identify and discriminate *Campylobacter* species and strains present in food and water, and that could be used also to potentially source track *Campylobacter* related to outbreaks.

INTRODUCTION

Campylobacter species can be associated with large outbreaks affecting thousands of individuals. On 19 May 2000, several paediatric cases of bloody diarrhoea and gastroenteritis from Walkerton, Ontario, were reported to the Bruce-Grey-Owen Sound Health Unit in Ontario (Health Canada, 2000). Inquiries revealed an increase in gastroenteritis in the Walkerton area and a possible outbreak related to the Walkerton water supply. *Escherichia coli* O157:H7 was isolated from stool cultures (Hrudey *et al.*, 2003) and confirmed to be present in the water supply. The final estimated number of cases was more than 2300 with 65 patients hospitalized, 27 cases of haemolytic uraemic syndrome, and seven deaths (Health Canada, 2000; Hrudey *et al.*, 2003). Eventually, it was determined that the cause of the outbreak was due to heavy rainfall that flushed cattle manure from an adjacent farm into a well that supplied municipal water (Clark *et al.*, 2003).

Even though 41% of the stool samples tested were positive for *Campylobacter*, the outbreak has been described widely as the 'Walkerton *E. coli* outbreak'. This is due to valid concerns focused by the seven deaths, attributed to a single strain of *E. coli* O157:H7, during the outbreak. In contrast, a diverse group of *C. jejuni* strains of different serotypes and genotypes were isolated from humans, cattle and the environment associated with the outbreak, even though only two strains were linked with human illnesses (Clark *et al.*, 2003). However, the presence of *Campylobacter* organisms and their role in the Walkerton outbreak generally have been ignored, even though this was the fourth largest water-borne outbreak of *Campylobacter* in North America (see Table 6.2 below).

In 2001, 4740 laboratory-confirmed cases of campylobacteriosis were identified in the United States by the CDC's Emerging Infections Program Food-borne Diseases Active Surveillance Network (FoodNet), second only to the number of illnesses caused by *Salmonella* (5198 cases) (CDC, 2002b). *Campylobacter* has surpassed *Salmonella enterica* as the primary cause of sporadic bacterial gastroenteritis in much of the developed world, even though it is rarely identified as an infective agent in food- or water-borne outbreaks, causing only 1% of reported outbreaks between 1980 and 1996 (Friedman *et al.*, 2000).

The prevalence of *Campylobacter* species in the environment, including wild and domesticated animals, animal waste, soil and water is related to the ultimate contamination of food with *Campylobacter*. In this chapter, we will review studies reporting the carriage rates of *Campylobacter* species in animals, including poultry and livestock, and the prevalence of *Campylobacter* in the food and water supply. We have documented *Campylobacter*-related outbreaks reported worldwide, and molecular methods that can be used to both detect rapidly and type *Campylobacter* species present in food and water, and to potentially source track *Campylobacter* outbreaks.

RESERVOIRS OF *CAMPYLOBACTER* AND PREVALENCE IN DOMESTIC ANIMALS, FOOD AND WATER

Campylobacter reservoirs

The association of *Campylobacter* with poultry (e.g. chickens, turkeys, ducks, and geese), has been known for at least 30 years (Winkenwerder, 1967; Smith and Muldoon, 1974). Among wild birds, four types are considered reservoirs of *Campylobacter* species: waterfowl, shorebirds, gulls, and corvids (crows, ravens, jays, magpies and jackdaws). In a recent survey, differences in the prevalence of *Campylobacter* infection between different ecological guilds of birds in Sweden was reported (Waldenstrom *et al.*, 2002). The highest incidence (>50%) was found in shoreline-foraging birds and opportunistic feeders (e.g. gulls). Many other ground-foraging and plant-eating guilds also were colonized by moderate concentrations of *Campylobacter* (11–20% incidence), whereas in most insectivores and granivores, the incidence of *Campylobacter* was relatively low (<4%). Other studies have also reported the presence of *Campylobacter* in migratory waterfowl (Luechtefeld *et al.*, 1980; Pacha *et al.*, 1988; Yogasundram *et al.*, 1989; Petersen *et al.*, 2001) (R. Mandrell, unpublished data). *Campylobacter* carriage in waterfowl may reflect diet and feeding habits, with lower *Campylobacter* incidence in birds that feed primarily on vegetable matter compared to birds that feed on aquatic animal life (Luechtefeld *et al.*, 1980). *Campylobacter* is detected often in fresh water, estuarine, and marine environments (Hill and Grimes, 1984; Endtz *et al.*, 1997; Obiri-Danso and Jones, 1999). Thus, birds that feed in shallow water or along shorelines in these environments would have a higher risk of exposure to *Campylobacter*. Additionally, high isolation rates of *Campylobacter* in crows and gulls (Ito *et al.*, 1988; Maruyama *et al.*, 1990) suggests that refuse-feeders may also be a potential reservoir.

Although domesticated animals and avian species are considered major reservoirs, *Campylobacter* is widespread also in non-avian wildlife (Fernie and Park, 1977; Luechtefeld *et al.*, 1981; Manser and Dalziel, 1985; Pacha *et al.*, 1985; Pacha *et al.*, 1987; Fox *et al.*, 1989; Taylor *et al.*, 1989; Cabrita *et al.*, 1992; Petersen *et al.*, 2001). *Campylobacter* has been isolated from several wild mammals, including primates (e.g. monkeys, gorillas, chimpanzees, and lemurs), ungulates (bighorn sheep, deer, antelope, llamas, gazelle, and tapirs), wild cats (bobcats and cheetahs), canines (foxes and coyotes), bears, pandas, ferrets, hedgehogs, and badgers. Additionally, *Campylobacter* has been found in several types of rodents, such as hamsters, hares, rats, voles, muskrats, and squirrels. *Campylobacter* carriage in non-avian wildlife often is less than that reported for avian species.

Transmission of *Campylobacter* from the avian (non-food related) and non-avian taxa is more likely to be a zoonotic or water-borne event rather than food-borne. Since *Campylobacter* carriage in some species of birds and rodents can be very high, the close proximity of these species to farms may result in a cycle of horizontal transmission between wildlife, and domestic livestock and poultry. Wild birds, such as sparrows and pigeons, are often present near broiler houses (Cabrita *et al.*, 1992; Chuma *et al.*, 2000). In addition, black rats and houseflies positive for *C. jejuni* have been identified on poultry and swine farms (Rosef and Kapperud, 1983; Cabrita *et al.*, 1992). However, in at least one instance, the majority of *Campylobacter* types isolated from wild birds and mammals were not the same as those isolated from broilers, suggesting that horizontal transmission between wildlife and broilers is limited (Petersen *et al.*, 2001). In contrast, the isolation of quinolone-resistant *C. jejuni* from sparrows was reported, suggesting that the sparrows either directly acquired the resistant *C. jejuni* from broilers, that are commonly fed quinolone-amended feed, or that existing intestinal microflora became resistant as a result of consumption of the amended feed (Chuma *et al.*, 2000). In either situation, it is possible that broilers become infected with resistant *Campylobacter* through ingestion of sparrow droppings.

Poultry

Campylobacter carriage (mainly *C. jejuni* and *C. coli*) in poultry is very high, often markedly higher than those reported for other game birds and other avian species. *Campylobacter* primarily colonizes the caecum and can be maintained asymptomatically at levels of at least 10^{10} cfu per gram of caecal contents (Shreeve *et al.*, 2000). Although the incidence of *Campylobacter* in poultry is low immediately following hatching, *Campylobacter* species can be detected in birds within 2 to 4 weeks (Berndtson *et al.*, 1996; Evans and Sayers, 2000; Stern *et al.*, 2001). However, this apparent delay in infection in broiler flocks is not due to an age dependency of infection, per se, since newly hatched chicks are colonized efficiently following oral gavage (Stern *et al.*, 1988). A recent study indicates that maternal antibodies have a protective role in young chickens, preventing colonization by *Campylobacter* under natural production conditions (Sahin *et al.*, 2003).

Horizontal transmission, due to contamination of food and water with faecal material and to the behaviour of chickens to ingest faeces (coprophagy), rapidly disseminates *Campylobacter* within poultry flocks. It has been reported that one contaminated bird present in an otherwise *Campylobacter*-free flock can result in contamination of all of the birds in the flock within 3 days (Shanker *et al.*, 1990). Also, a rapid colonization of most of the chicks in a flock within 7 days, when exposed to *C. jejuni*-contaminated water, was reported (Shanker *et al.*, 1990). Horizontal transmission is facilitated by the low infectious dose for *Campylobacter* in chicks, reported to be as low as 35 bacterial cells (Stern *et al.*, 1988). *Campylobacter* carriage within flocks increases with flock size, approaching 100% of birds in larger flocks (Berndtson *et al.*, 1996). Since dissemination within a flock only requires one *Campylobacter*-positive bird, this is most likely due to an increased probability of a positive bird being present in larger flocks. Nevertheless, many flocks have been shown to remain free of *Campylobacter* until slaughter (Berndtson *et al.*, 1996; Pearson *et al.*, 1996; van de Giessen *et al.*, 1998; Stern *et al.*, 2003). However, *Campylobacter*-free flocks can be contaminated during transport, probably from contaminated crates (Newell *et al.*, 2001; Stern *et al.*, 2001; Slader *et al.*, 2002).

Whereas multiple studies report horizontal transmission of *Campylobacter* through a flock, few studies have reported vertical transmission of *C. jejuni* through infected eggs. Although an intriguing possibility that could involve interesting biology, many investigators have discounted the role of vertical transmission based on low or non-detectable levels of *Campylobacter* on eggs (Shanker *et al.*, 1990; Evans, 1992; van de Giessen *et al.*, 1992; Jacobs-Reitsma, 1997; Shane, 2000). However, Pearson *et al.* (1996) suggested that low-level vertical transmission with subsequent horizontal amplification within a shed can occur, based on the spread of a predominant serotype introduced into multiple poultry sheds.

Campylobacter jejuni and *Campylobacter coli* are common contaminants of poultry meat and meat products (Table 6.1, pp. 123–130), but the reported incidence of *Campylobacter* on non-frozen chicken carcasses, as an example, can be highly variable. Turkey and duck meat also are highly contaminated by *Campylobacter*, but the incidence is lower on meat from other game birds (e.g. pheasant and guinea fowl). The incidences of *Campylobacter* on frozen chicken meat and offal usually are lower substantially compared to non-frozen samples. *C. jejuni* strain 81116 was reported to survive for several weeks on chicken skin at $-20°C$, but a significant decrease of 10^4–10^5 cfu was observed (Lee *et al.*, 1998). Strain 81116 failed to survive storage at $-20°C$ in the absence of chicken skin indicating the importance of this interaction in 'preserving' the pathogen.

Livestock, beef, pork, lamb, and livestock offal

Campylobacter carriage in cattle is associated significantly with the age of the animal since *C. jejuni* incidence in calves often is much higher than in adult (>1 year) cattle (Grau, 1988; Giacoboni *et al.*, 1993; Busato *et al.*, 1999; Nielsen, 2002). Maximum carriage in calves occurs 61–120 d after birth (Nielsen, 2002) and subsequently declines as the animals age. Feedlot cattle have been shown to maintain a higher *Campylobacter* carriage rate than adult, pasture-fed animals (Grau, 1988; Beach *et al.*, 2002), which may be due to differences in either animal density or diet (Grau, 1988). Also, the amount of *Campylobacter* in cattle is variable along the length of the gastrointestinal tract, with *C. jejuni* being detected at markedly different concentrations in both rumen (< 100 cells g^{-1}) and faecal samples (10^4–10^6 cells g^{-1}) (Grau, 1988). *C. jejuni* has been isolated from the small and large intestine and the caecum of cows, but not from the true stomach (Stanley *et al.*, 1998b). Similarly, large numbers of *C. jejuni* were isolated from the small intestine of lambs at slaughter (Stanley *et al.*, 1998a).

Although the incidence of *Campylobacter* in livestock can approach 100%, the presence of the organism on beef, pork, and lamb is remarkably low (Jacobs-Reitsma, 2000; Table 6.1), probably due to dehydration during carcass chilling (Grau, 1988; Jacobs-Reitsma, 2000). Spray chilling of carcasses, which reduces the level of dehydration, may not provide the same level of protection against *Campylobacter* (Vanderlinde *et al.*, 1998). The incidence rates for edible livestock offal (i.e. liver, heart, and kidney) are much higher (up to 73% for lamb liver; Table 6.1) compared to carcass contamination. Offal contamination is probably not due to internal colonization (Moore and Madden, 1998), but to intestinal leakage during slaughtering (Bolton *et al.*, 1985; Kramer *et al.*, 2000). Ground meat, minced meat, and sausage also have very low levels of *Campylobacter* contamination (Table 6.1).

Although *C. jejuni* is often isolated from cattle and sheep, *C. coli* is the primary *Campylobacter* isolated from swine and pork (Table 6.1). The predominance of this species on pork and the low incidence of *C. coli*-related sporadic illnesses and outbreaks has led some to suggest that pork and pork products are not a major source of *Campylobacter* illnesses in people (Korsak *et al.*, 1998; Kramer *et al.*, 2000). However, Young *et al.* (2000) reported a much higher ratio of *C. jejuni* to *C. coli* (approximately 5:1) in swine compared to other studies. The authors speculated that *C. jejuni* prevalence differs geographically and that the incidence of *C. coli* may be due, in part, to inaccuracies in the hippuricase test used to distinguish the two species.

Milk

Incidence rates of up to 12% for *Campylobacter* in raw milk have been reported (Jacobs-Reitsma, 2000). Faecal contamination of milk is the most likely explanation, although direct excretion of *C. jejuni* in milk by asymptomatic dairy cows with mastitis has been reported (Orr *et al.*, 1995). In milk-associated outbreaks, *Campylobacter* often cannot be detected in milk samples, even though rectal swabs of dairy cows on the farm test positive for *Campylobacter* and strain types isolated from these cows are indistinguishable from those isolated from patients (Robinson *et al.*, 1979; Robinson and Jones, 1981). Detection of *Campylobacter* in bulk milk samples may be confounded, in some instances, by transient contamination of the milk (Robinson and Jones, 1981). Additionally, only a small proportion of cows may be shedding *Campylobacter* at any given time, thereby diluting the level of contamination. Although *C. jejuni* has been reported to survive for 24h in milk at room temperature and for several weeks in milk at 4°C (Blaser *et al.*, 1980), it cannot grow in milk (Kalman *et al.*, 2000). However, a dose in milk as low as 500 *Campylobacter* cells has been reported to be sufficient to produce gastroenteritis (Robinson, 1981), suggesting that bulk raw milk samples negative for *Campylobacter* at the level of detection sensitivity may remain a source of *Campylobacter* infection. Conventional pasteurization will eliminate *Campylobacter* from milk (Robinson and Jones, 1981). However, instances of sporadic illness and outbreaks in the UK have been attributed to ineffective pasteurization and re-inoculation of pasteurized bottled milk by jackdaws and magpies pecking at the tops of the milk bottles (Hudson *et al.*, 1991).

It is clear that consumption of raw milk has been associated with many *Campylobacter* outbreaks (Table 6.2, pp. 131–145). However, regular consumption of raw milk may provide in some individuals immunity against *Campylobacter* infection (Friedman *et al.*, 2000). In a raw milk-related outbreak, 76% of college students who consumed raw milk for the first time became infected, compared with no infections in those who regularly consumed raw milk (Blaser *et al.*, 1987). The most likely reason for this immunity is the presence of circulating IgG antibodies. However, components of milk, such as oligosaccharides, have been implicated also in protection against *Campylobacter* infection, especially in nursing infants. For example, the binding of *C. jejuni* to epithelial cells has been reported to be inhibited *in vitro* by human milk (Newburg, 1997; Ruiz-Palacios *et al.*, 2003), specifically by α1,2-fucosylated oligosaccharides (Ruiz-Palacios *et al.*, 2003), which are present also on the surface of human epithelial cells as components of the H blood group antigen. It is relevant that binding of *Campylobacter* cells to the H-2 epitope has been reported (Ruiz-

Palacios *et al.*, 2003). Therefore, it can be speculated that fucosylated oligosaccharides present in human milk disrupt *Campylobacter* adherence and colonization, and might provide protection for nursing infants from *Campylobacter* infection. The fact that *Campylobacter* is present in raw bovine milk implicated frequently in both sporadic and outbreak *Campylobacter* illnesses suggests that α1,2-fucosylated oligosaccharides are absent or limited in bovine milk. Indeed, fewer types of oligosaccharides have been identified in bovine milk compared to human milk (Newburg and Newbauer, 1995), and only α1,3-fucosylated oligosaccharides have been reported as present in the neutral oligosaccharides (Gopal and Gill, 2000). Bovine milk has been shown to have protective effects for newborn calves (Gopal and Gill, 2000), but calves can become infected with *Campylobacter* within a few days of birth (Stanley *et al.*, 1998a). Therefore, the species-specific oligosaccharide composition of milk (Newburg, 1997) may be associated with homologous protection of infants (i.e. human milk), but not heterogeneous protection (e.g. bovine milk and human infant). Synthesis of α1,2-fucosylated structures in humans is due to a specific fucosyltransferase, synthesized by individuals with the dominant form of the secretor gene, FUT2. Since non-secretor mothers lack detectable levels of α1,2-fucosylated oligosaccharides in their milk, it will be interesting to determine if infants nursed by non-secretor mothers are at a higher risk for campylobacteriosis, or if other protective oligosaccharides are present in human milk (Ruiz-Palacios *et al.*, 2003).

C. *jejuni* has been isolated occasionally from other dairy products. *C. jejuni* has been shown to survive for up to 13 days in butter held at 5°C (Zhao *et al.*, 2000), but bacterial cells were killed rapidly by the addition of garlic to the butter. Also, *C. jejuni* strains inoculated into yoghurt were killed within 25 minutes (Cuk *et al.*, 1987), presumably by organic acids (i.e. propionic and lactic acid) in the yoghurt, and *C. jejuni* survived poorly in milk after the pH was adjusted with these organic acids. *C. jejuni* and *C. coli* were not detectable in butter, curds, yoghurt, or sour cream made with raw milk, whereas they were detected in one sample of raw milk 'sheep cheese' (Allmann *et al.*, 1995). *Campylobacter* was isolated from 'sheep cheese' in a second study, and also a soft cheese sample, but not from yoghurt (Wegmuller *et al.*, 1993). A 1992 outbreak of *C. fetus* subsp. fetus in Ohio (Table 6.2) was linked to consumption of cottage cheese, suggesting that soft cheeses may be a source of *Campylobacter*. In contrast, *C. jejuni* has been reported to have a low survival rate in hard and semi-hard cheeses (Bachmann and Spahr, 1995).

Water

Surface water may be the most important reservoir of *Campylobacter* species, not only as a potential source of human infection, but also as a source of infection for both poultry and livestock (Pearson *et al.*, 1993; Refregier-Petton *et al.*, 2001). As described above, thermophilic *Campylobacter* species have been found in fresh water, estuarine and marine environments. Entry of *Campylobacter* into surface water is due to: (1) excretion of the organism into water by waterfowl and gulls (Bolton *et al.*, 1987; Brennhovd *et al.*, 1992; Obiri-Danso and Jones, 1999); (2) runoff from agricultural and residential land, especially during periods of heavy rainfall and flooding (Bolton *et al.*, 1987; Daczkowska-Kozon and Brzostek-Nowakowska, 2001); (3) waste water from poultry houses and processing facilities (Koenraad *et al.*, 1995); and (4) releases of municipal sewage (Bolton *et al.*, 1987). The prevalence of *Campylobacter* in surface water is highly variable with reported isolation rates between 0% (Hill and Grimes, 1984) and 82% (Stelzer *et al.*, 1989). However, the concentration of *Campylobacter* in water samples is quite low, often less than 10 bacterial cells per 100 mL. The presence of viable but non-culturable (VBNC) *Campylobacter* cells in water (see below) and the potentially low infectious dose (Robinson, 1981) suggests that even these low concentrations of *Campylobacter* cells could pose a health risk. *Campylobacter* in water is eliminated rapidly by chlorination (Jacobs-Reitsma, 2000). Consistent with this observation is the reported isolation of *Campylobacter* from well, lake and river samples but not from chlorinated water (Moore *et al.*, 2001b).

Seasonal effects on the isolation of *Campylobacter* from surface waters have been reported (Bolton *et al.*, 1987; Brennhovd *et al.*, 1992; Obiri-Danso and Jones, 1999; Daczkowska-Kozon and Brzostek-Nowakowska, 2001). The prevalence of *Campylobacter* in surface water was reported to be highest in late autumn and winter and lowest in spring and summer. Since the incidence of *Campylobacter* in poultry and human campylobacteriosis is highest during the summer months, it is unlikely that this seasonal effect in water is due to differences in the concentrations of *Campylobacter* in wastewater, but is probably due to environmental factors affecting survival (Koenraad *et al.*, 1997). *Campylobacter* seasonality may be due to its ability to survive for longer periods in relatively cold water. Buswell *et al.* (1998) reported mean survival times for 17 different *Campylobacter* strains in bore hole tap water of 202 h at 4°C, but only 22 h at 37°C. However, Thomas *et al.* (1999a) reported much longer survival times in sterile river water (>60 d at 5°C), suggesting that survival might be influenced by the chemical or biological composition of the water. Survival in artificial seawater was also reported to be much longer at 4°C (Obiri-Danso *et al.*, 2001). All three of the above studies reported a significant decrease in survival rates between 10 and 20°C. However, isolation of *Campylobacter* from water was not influenced by temperature, suggesting that seasonal effects may not be due solely to water temperature. (Daczkowska-Kozon and Brzostek-Nowakowska, 2001). Additional factors may include competition from aquatic microflora (Brennhovd *et al.*, 1992), increased runoffs due to heavy rainfall during winter months (Bolton *et al.*, 1987), increases in ultraviolet (UV-B) radiation during the summer months resulting in a lower *Campylobacter* survival rate in shallow waters (Bolton *et al.*, 1987; Obiri-Danso *et al.*, 2001), and the establishment of biofilms in water microcosms (Buswell *et al.*, 1998).

Fruits, vegetables and produce-related products

Studies investigating the incidence and biology of *Campylobacter* species in fresh fruits and vegetables (produce) have been limited compared to studies for other food pathogens such as *Salmonella* species and *E. coli* O157:H7 (Mandrell and Brandl, 2004). A summary of incidence studies of *Campylobacter* in produce, published between 1982 and 2003, is presented in Table 6.1.

Matsusaki and coworkers may have provided the first published report of an attempt to determine the incidence of *Campylobacter* in produce (Matsusaki *et al.*, 1982). Three hundred and forty-five different foods obtained from markets, including 24 samples of vegetables, were tested for contamination with *C. coli* and *C. jejuni*. *Campylobacter* strains were isolated from 14 of 34 chicken meat samples (41%), but not from any vegetable samples or other foods (e.g. beef, pork, frozen food, shellfish, milk). The first published report of the isolation of *Campylobacter* from a 'produce-related' item may be of the isolation of *C. jejuni* from 3 of 200 retail fresh mushroom samples (Table 6.1 and Doyle and Schoeni, 1986). Even though mushrooms are not plants, but fungi, there are similarities between mushroom and vegetable production and processing practices.

Only five of the 17 incidence studies listed in Table 6.1 reported any *Campylobacter* species isolated from produce samples. In one of the most extensive studies from Ottawa, Canada, Park and Sanders (1992) tested 1564 fresh produce samples obtained from farmers' outdoor markets (533 samples) and supermarkets (1031 samples). Of 10 different vegetables tested, *Campylobacter* species were isolated from 2/60, 2/67, 2/74, 1/40, 1/42 and 1/63 samples of spinach, lettuce, radish (with leaves), green onion, parsley and potatoes, respectively. All of the positive samples originated from farmers' markets during the summer months resulting in an incidence of 0.6%, 1.7% and 2.6%, respectively, for the total number of produce samples (1564), farmers' market samples (533), and positive vegetable group samples (346) tested. In addition, multiple strains were isolated from some samples and all *Campylobacter* strains isolated were obtained only after enrichment culture methods; none were isolated from any sample by direct plating. The presence of multiple species and strains of *Campylobacter* with different fitness levels in selective isolation media emphasizes the difficulty in obtaining accurate information for epidemiological investigations of outbreaks, but also the need for developing methods for identifying more of the strains of the suspect species that may be present in a sample at a broad range of concentrations.

Campylobacter have been isolated from modified atmosphere packaged mixed salad vegetables (Phillips, 1998). *Campylobacter* were isolated from 12/36 (33.3%) and 8/54 (14.8%) samples tested on the 'date of purchase' or stored at 4°C and tested on the 'use by date', respectively. The average rate of isolation of *Campylobacter* from these packaged salad vegetables (22.2%), is much higher than that reported for any other study of fresh produce. The number of colony-forming units isolated from the packaged salad samples was low (80–170 cfu/g). These results suggest that the modified atmosphere used for packaging the salad vegetables (reported to be 5–8% CO_2:2–5% O_2:87–93% N_2) might enhance the survival of *Campylobacter* in other produce or food samples, even though the viability of *Campylobacter* may decrease over time in this atmosphere (Phillips, 1998).

In a study of 400 samples of ready-to-eat vegetable samples collected over a 2-year period in the Nantes region of France, 2/150 and 0/250 grated ready-to-eat vegetables and salad samples, respectively, were positive for *C. jejuni* (Federighi *et al.*, 1999). Differences were noted in the efficiency of isolation of *Campylobacter* on two different selection and three different plating media indicating the importance of the media and culture methods for accurate estimates of *Campylobacter* incidence.

Finally, a recent study on the incidence of *Campylobacter* in produce reported that 2/56 vegetable samples collected in a local market in Northern India were positive (Kumar *et al.*, 2001). One each of nine spinach and nine fenugreek samples were positive, whereas no *Campylobacter* were isolated from cauliflower, cabbage, coriander, radish or carrot samples in this study.

It is difficult to draw any definitive conclusions regarding these incidence studies and risks of illness related to fresh produce. It is noteworthy that a recent study of sporadic *Campylobacter* diarrhoeal illnesses in the Cardiff, UK, population during 2001 identified the consumption of uncooked salad vegetables as one of the risk factors (Evans *et al.*, 2003). The significant association with eating tomatoes and cucumbers, rather than lettuce, suggested to the authors that contamination was a result of cross-contamination due to extensive handling during preparation (e.g. cutting board). Nevertheless, the prevalence of *Campylobacter* species in wild animals and animals raised for food (Table 6.1) indicate that soil and manure for soil fertilization could be probable sources for exposure of crop plants to *Campylobacter* and other human pathogens that colonize animal GI tracts. The survival of *Campylobacter* in water, as evidenced by major outbreaks of *Campylobacter* due to contaminated water supplies (Table 6.2), also suggests a potential source of contamination of fruits and vegetables, and the need to determine whether pre-harvest contamination by *Campylobacter* species occurs and is a potential risk factor. A recent study of *C. jejuni* survival on and in the radish, spinach and lettuce plant phyllospheres and rhizospheres indicates that *C. jejuni* survives much better in the presence of soil, both around roots, and even on leaf surfaces when diluted soil is present (Brandl *et al.*, 2004; Mandrell and Brandl, 2004). The fitness of *C. jejuni* in these plant environments was not as good as *Salmonella enterica* or *E. coli* strains tested in similar models (Brandl and Mandrell, 2002; Charkowski *et al.*, 2002; Wachtel *et al.*, 2002a, 2002b); however, the enhanced survival in soil is relevant to understanding the cycle of *Campylobacter* species through the environment and the food chain.

The incidence of *Campylobacter* species in foods relates ultimately to factors that enhance survival and growth at a range of temperatures, pHs, osmotic conditions and humidities. Specific information regarding *Campylobacter* in foods under these types of processing conditions, and interventions to decrease the level of *Campylobacter* species in food, have been reviewed previously (Jacobs-Reitsma, 2000; Stern and Line, 2000).

OUTBREAKS

The first observation of *Campylobacter* organisms may have been as early as 1886 by Theodor Escherich, who described spiral organisms in the large intestines of children who had died of diarrhoeal disease (Escherich, 1886). Subsequently, spiral ('vibrio') bacteria were identified as a probable cause of abortion in sheep (subsequently named *Campylobacter fetus*) (McFadyean and Stockman, 1913) and later as a possible cause of human illness (King, 1957; Cooper and Slee, 1971; Dekeyser *et al.*, 1972). Sebald and Véron (1963) were the first to propose the genus *Campylobacter* in 1963, but a subsequent and more detailed study resulted in acceptance of the genus as distinct from others (Véron and Chatelain, 1973; On, 2001). Improved methods for isolating the 'vibrio' organisms (i.e. *C. coli* and *C. jejuni*) from human stools were developed (Skirrow, 1977) and verified the connection between *Campylobacter* and human illness, and the possible association of food as a vehicle of transmission (King, 1957; Skirrow, 1977).

Campylobacter and illness

In recent years, *Campylobacter* species have become the major bacterial cause of gastrointestinal (GI) illnesses in the developed world (Blaser *et al.*, 1984; Altekruse *et al.*, 1999; Wheeler *et al.*, 1999; Friedman *et al.*, 2000; CDC, 2002b; Frost *et al.*, 2002). Approximately 90% of the illnesses associated with *Campylobacter* species are caused by *C. jejuni*, with *C. coli* causing 5–10% of the additional identified cases (Tam *et al.*, 2003). In England and Wales, for example, *Campylobacter* species have been the most frequent bacterial cause of GI illness since 1981 (Frost *et al.*, 2002), even though they accounted for only 2.1% of the total number of outbreaks between 1995 and 1999 with an aetiological agent identified (50 of 2374 total). In New Zealand, campylobacteriosis identified by a national notification system exceeds 230 cases per 100000 and 300 cases per 100000 in certain regions of the country (Skelly and Weinstein, 2003). Most of this illness is sporadic, unlinked to any known source. The difference in rates of sporadic and outbreak *Campylobacter* illness is consistent with observations in the US and other countries indicating that *Campylobacter* species cause more sporadic illness compared to other food pathogens (Pebody *et al.*, 1997; Friedman *et al.*, 2000). Since *Campylobacter* causes mostly sporadic illness, traceback investigations that might identify possible point sources of contamination do not occur for most *Campylobacter* cases.

Although *Campylobacter* illnesses are sporadic and uncomplicated generally, in rare cases serious manifestations occur following the diarrhoeal illness. Approximately 0.1% of the *C. jejuni*-caused GI illnesses are followed by Guillain–Barré syndrome (GBS), a sometimes life-threatening polyneuropathy that can result in paralysis and other serious complications (Hahn, 1998; Nachamkin *et al.*, 1998). *C. jejuni* has been associated also with other neuropathies such as Miller Fisher syndrome (Willison and O'Hanlon, 1999), an acute motor axonal neuropathy, and acute inflammatory demyelinating polyneuropathy (Yuki *et al.*, 1999), and with inflammatory diseases such as reactive arthritis (Ebright and Ryan, 1984) and myocarditis (Florkowski *et al.*, 1984; Westling and Evengard, 2001).

Outbreaks, 1978–present

Although *Campylobacter* species are considered to mostly cause sporadic illness, outbreaks associated with *Campylobacter* species have occurred, sometimes involving milk or water sources with thousands of individuals sickened. It is of historical interest that the first reported outbreak of human illness caused by a *Campylobacter* species may have been in May, 1938 involving 357 prison inmates sickened by consumption of contaminated milk; selected faecal samples contained mucoid material and 'vibrio-like micro-organisms almost in pure culture' (Figure 6.1) (Levy, 1946).

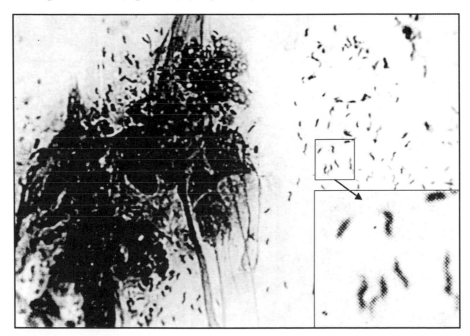

Figure 6.1 *Campylobacter* species associated with an outbreak in 1938? A 'direct smear of a whitish flake' obtained from faeces of a patient who was part of an outbreak of diarrhoeal illness due to consumption of raw milk in 1938. The sample was stained with carbolfuchsin (400×). A region of the image has been enlarged (see inset) for observing the 'vibrio-like' spiral cells. This image from the original article published by A. Levy in 1946 has been used with permission from V. Richardson, Yale *Journal of Biology and Medicine*, New Haven, CT.

Table 6.2 shows a comprehensive list of *Campylobacter* outbreaks reported to be caused by exposure to a variety of contaminated foods, or water sources, or caused by an unknown or unreported source. The information provided in Table 6.2 has been documented through a combination of the peer-reviewed literature; local, state, national, and international public health agency reports; WHO reports; and newspapers. It should be noted, however, that this list represents only a small number of the actual *Campylobacter* cases and outbreaks, since additional outbreaks go unidentified, unreported, or the information is not easily accessible (CDC, 2002b). The outbreaks have been organized into eight different categories based on the identified or suspected source of the outbreak: (a) meat and meat-related, (b) milk and dairy, (c) miscellaneous or mixed food sources not well characterized, (d) poultry and poultry products, (e) produce and produce-related products, (f) seafood/shellfish, (g) water, and (h) outbreaks of unknown source. For some of the outbreaks listed in Table 6.2, a specific source was difficult to identify, thus, the source reported could be inaccurate. Nevertheless, these data provide a context for assessing the risk associated with various sources and *Campylobacter* illness and the relationship of illness to incidence and biology discussed in other sections of this chapter.

The cumulative numbers of *Campylobacter* outbreaks and cases reported from 1978 to 2003 for the eight categories cited in Table 6.2 have been summarized according to source and the total numbers are shown in Figure 6.2. *C. jejuni* has been reported consistently as the predominant cause of outbreaks associated with *Campylobacter* (Table 6.2).

To the best of our knowledge, the first major, well-documented, human outbreaks known to be associated with a *Campylobacter* species occurred in 1978 in England (March, 1978) and subsequently in the United States (June, 1978); these outbreaks were associated

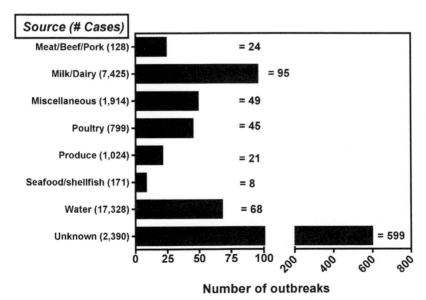

Number of outbreaks

Figure 6.2 The number of outbreaks and cases of campylobacteriosis associated with different food and water sources, 1978–2003. The data for producing this figure were compiled from Table 6.2. The total number of outbreaks for a given source was determined by adding each of the entries plus the total number outbreaks reported in the references cited for those entries designated with 'Multi' (also see footnotes). When the number of cases in an outbreak was uncertain (e.g. 'approximately' or 'greater than'), the least number of cases noted was used to calculate the total.

with the consumption of raw milk (Robinson and Jones, 1981) and contaminated water (Vogt *et al.*, 1982), respectively.

Water and milk have continued to be the major sources responsible for outbreaks since 1978 (Table 6.2). An additional 12 outbreaks of campylobacteriosis linked to milk occurred in the United Kingdom between October 1978 and April 1980 (Robinson and Jones, 1981) and at least eight outbreaks occurred in the US in 1981 (Finch and Blake, 1985). At least 20 outbreaks occurring in 11 US states between 1981 and 1990 were associated with raw milk consumption during youth activities, with children in kindergarten to third grade at highest risk of illness (Wood *et al.*, 1992). Whether greater risk was due to this age group's increased susceptibility, or exposure as a result of increased field trip activity, could not be determined. The high number of outbreaks during this time probably reflected both a corresponding increased awareness of *C. jejuni* as a cause of diarrhoeal illness and the development of improved methods for culturing *C. jejuni* from stool samples (Skirrow, 1977).

Recent reviews of the epidemiology of outbreaks involving *Campylobacter* in water-related environments identified 12 major outbreaks (Koenraad *et al.*, 1997; Lee *et al.*, 2002). The relatively large number of total cases resulting from a water source (Figure 6.2) reflects the fact that 7 of 8 *Campylobacter*-related outbreaks with >1000 cases due to any source have been associated with water (Table 6.2). It is worth noting that a few water-borne outbreaks of GBS have been reported to be associated with *C. jejuni* (Roman, 1995, 1998).

Water-related outbreaks indicate the capability of *Campylobacter* species to survive in aqueous environments at infectious concentrations, possibly even in a VBNC state (Rollins and Colwell, 1986; Tholozan *et al.*, 1999). The ability of organisms to survive in a VBNC form presumably is important in cycling them through different environments relevant to direct contamination of food animals, wild animals (e.g. birds, rodents and insects), and crops. Although the VBNC concept is somewhat controversial, it is probable that *Campylobacter* cells are stressed at different levels in different water environments due to conditions of osmolarity, temperature, nutrients/chemicals and other microorganisms. Some of these conditions may be important in understanding the infectious potential of *Campylobacter* species in water.

Although *Campylobacter* species are frequently present in livestock animals and meat produced from them (Table 6.1), there have been few outbreaks and even fewer cases associated with non-poultry meat sources (Table 6.2). A meat-related outbreak occurring in 1978 due to consumption of raw hamburger (Oosterom *et al.*, 1980) was the only outbreak reported due to a non-poultry meat source for approximately 12 years. This gap for non-poultry meat sources reflects the limited number of outbreaks reported for any source prior to 1990, with the exceptions of the milk and water source outbreaks (Table 6.2).

The 49 outbreaks associated with a 'miscellaneous' source are difficult to assess since multiple items and processes would have been involved in food preparation. The most likely reasons for how these outbreaks occurred are cross-contamination with a food item not part of the final food prepared, or contamination with an added item, and improper cooking or storage (see 'Misc.' in Table 6.2). Avian, meat, seafood and dairy products are listed as ingredients in many of the miscellaneous foods, and in a few of the 'salads' listed under the produce category.

The number of outbreaks and cases reported for poultry products (45 outbreaks/799 cases) over the last 25 years (Pebody *et al.*, 1997) seems low considering the prevalence and of pathogenic *Campylobacter* species in chickens and the concentrations that exist in the GI tract of some of the colonized birds during production and processing (Table 6.1). This may be due to a general awareness of food preparers about the risks associated with uncooked poultry, thus minimizing outbreaks with a large number of cases that would be noticed. Cross-contamination of other foods prepared near raw poultry and a small volume of juices, might result in the contamination of a raw food by a limited inoculum (i.e. low infectious potential), and thus resulting in sporadic illness. Studies of multiple sites in domestic kitchens after normal preparation of chickens with incurred *Campylobacter* contamination revealed, not surprisingly, that multiple species of *Campylobacter* were present on hands, oven handles, countertops and draining boards (Gorman, 2002).

The first outbreak reported of *Campylobacter* associated with fresh produce occurred in 1993 and was associated with melons and strawberries (Table 6.2). The *Campylobacter* outbreaks associated with fresh produce or produce-related products are noteworthy because these foods are often consumed with minimal preparation or uncooked. The results of a recent retrospective study in Wales (UK) have implicated the consumption of uncooked salad vegetables as a risk factor for sporadic *Campylobacter* infection, and second only to consumption of poultry products (Evans *et al.*, 2003). Twenty-one produce-related outbreaks due to *Campylobacter* species are documented in Table 6.2, but it should be noted that produce probably was not the original source of *Campylobacter* contamination for most of these outbreaks; the food source listed for 12 of the 21 outbreaks reported in Table 6.2 suggests that multiple ingredients may have been in the source, including meat or seafood for two outbreaks. Although it cannot be determined conclusively from the outbreak data, it is suspected that the outbreaks and cases associated with produce often are from cross-contamination by another food during preparation (e.g. Table 6.1) (CDC, 1998). Such a scenario could result in sporadic illnesses caused by *Campylobacter*-contaminated produce without a food source being identified, and in contrast to the identification of the food source and aetiology occurring during intensive outbreak traceback investigation. Therefore, an important area for consideration and study in determining the comparative risks associated with *Campylobacter* species is their overall fitness for survival and growth in pre-harvest environments, and environmental conditions related to both pre- and post-harvest produce (Mandrell and Brandl, 2004).

For many outbreaks associated with *Campylobacter* species, few epidemiological details are available. Information is often difficult to obtain because of the perishable nature of commodities (e.g. fruits and vegetables), the rapid decrease in viable *Campylobacter* cells in some foods, and multiple points where contamination could occur during food preparation. Even when the aetiology of a produce outbreak is clear, the predominant hypothesis for *Campylobacter* contamination of produce is through cross-contamination by *Campylobacter* on poultry or meat (Acuff *et al.*, 1986; CDC, 1998) or from food-preparers. It should be emphasized again that none of the produce outbreaks noted in Table 6.2 are attributable definitely to pre-harvest contamination. Additionally, multistate outbreaks of campylobacteriosis, indicative of contamination occurring prior to arrival at a retail market, are few or non-existent regardless of the food source (Friedman *et al.*, 2000).

Nearly 600 outbreaks and more than 2300 cases attributed to *Campylobacter* species from an unknown source have been reported (Table 6.2; note that many entries in the 'Unknown' source category represent multiple outbreaks). Approximately half of the outbreaks involved fewer than 10 cases, suggesting that many of them may have occurred within families and/or by cross-contamination of food consumed by a limited number of people. It is probable that outbreaks with only a few cases identified are less likely to be investigated as thoroughly as those with a larger number of cases and/or clear leads on possible food sources. These unknown source outbreaks emphasize the importance of further studies to identify risk factors and the actual sources of the outbreaks in order to obtain a clearer picture of the epidemiology and public health risks related to *Campylobacter* species.

Finally, it is probable that the number of cases noted for some outbreaks represents only illnesses where a *Campylobacter* was isolated; the actual number of cases associated with an outbreak probably was much higher than reported.

DETECTION OF *CAMPYLOBACTER* IN WATER, MILK AND POULTRY

More sensitive and specific detection methods are needed because of the potentially low concentrations of *Campylobacter* species in foods and water that are sufficient to cause illness. The first evidence in humans of a low infectious dose for *C. jejuni* was illustrated in a study involving a single volunteer receiving 500 organisms; this dose was sufficient to cause diarrhoea, cramps and mucus in the stool (Robinson, 1981). In subsequent volunteer studies, it was reported that 10–60% of 111 adults challenged with 8×10^2 to 2×10^9 *C. jejuni* cells had diarrhoea or fever, with differences noted between strains, inconsistencies in dose and amount of disease, and the method of presenting the inoculum (Black *et al.*, 1988). In a study of food samples causing an outbreak at a restaurant, the potential number of *C. jejuni* in the chicken meal was reported to range from 53 to 750 cfu per cm^2 of raw chicken surface area (e.g. 25 cm^2 sample consumed = 625 to 18 750 cfu) (Rosenfield *et al.*, 1985). These studies illustrate the variability in the virulence of *Campylobacter* (i.e. *C. jejuni*) and the potential role of several factors: host and strain differences, environment, concentration, method of exposure, etc.

Standard culturing methods for detecting *Campylobacter* species in environmental samples require 1–2 days of enrichment, followed by 1–2 days of incubation on selective media and possibly 1–2 days for strain purification. At present, *Campylobacter* species can be identified by biochemical or phenotypic tests that can take 2–3 days longer than current PCR-based speciation methods. Besides the overly lengthy time frame for detection with such methods, conventional culturing methods will not detect viable, non-culturable cells (see below) and often will detect only *C. jejuni* and *C. coli*. Many rapid detection methods for environmental *Campylobacter* have been described and most of these are PCR based. Although PCR-based detection methods generally are quicker than culturing methods (1–2 days), care must be taken to ensure that they are at least as sensitive as existing methods. Environmental samples often contain compounds that could inhibit *Taq* polymerase, resulting in a decrease in sensitivity: humic acids and metal ions in water (Abbaszadegan *et al.*, 1993; Tsai *et al.*, 1993), and blood, charcoal, iron, proteins, lipids, and Ca^{2+} in meat, milk, and culture media (Ng *et al.*, 1997; Wilson, 1997; Thunberg *et al.*, 2000). These compounds have not been reported to be a significant influence in culture methods, but they must be considered a potential variable in molecular methods. Some *Taq* inhibitors can be removed through the use of a Percoll buoyant density centrifugation procedure (Wang *et al.*, 1999) or by using a Chelex-100 resin (Purdy *et al.*, 1996). Rapid molecular detection methods for *Campylobacter* in food and water and comparisons of these methods with standard culturing techniques will be described in the following sections.

PCR-based detection of *Campylobacter* in water

Detection of *Campylobacter* in water presents several challenges since the number of cells in naturally contaminated water often is quite low. Quantification of this contamination by conventional culturing methods usually involves filtration to concentrate the cells and subsequent enrichment and plating (Sails *et al.*, 2002). Standard culturing methods only measure the number of viable, colony-forming cells and not the VBNC cells. The transition from a viable to VBNC state in *Campylobacter* has been reported to correlate with the morphological transition from vibrioid to coccoid (Rhoades, 1954; Rollins and Colwell, 1986; Moran and Upton, 1987; Thomas *et al.*, 1999b). This transition takes place as the cultures age and the cells are exposed to oxidative stresses (Moran and Upton, 1987) or starvation conditions (Rollins and Colwell, 1986). VBNC cells are presumed to be damaged and, although they can be resuscitated under certain conditions, are unable to grow on selective media (Rollins and Colwell, 1986). The epidemiological significance of VBNC cells is controversial; both infectivity (Jones *et al.*, 1991a; Saha *et al.*, 1991; Stern *et al.*, 1994; Cappelier *et al.*, 1999) and non-infectivity (Beumer *et al.*, 1992; Moedema *et al.*, 1992; van de Giessen *et al.*, 1996; Hald *et al.*, 2001) of VBNC cells have been reported. As *Campylobacter* cells enter an aquatic environment, viability decreases over time and the proportion of dead cells increases (Thomas *et al.*, 2002). Nevertheless, water has been reported to maintain a high population of VBNC *Campylobacter* (Rollins and Colwell, 1986; Korhonen and Martikainen, 1991).

Conventional culture-based detection of *Campylobacter* in water involves the most probable number (MPN) (Bolton *et al.*, 1982; Bolton *et al.*, 1987), enrichment (Daczkowska-Kozon and Brzostek-Nowakowska, 2001) or filtration/enrichment methods (Bolton *et al.*, 1982; Bolton *et al.*, 1987; Brennhovd *et al.*, 1992; Moore *et al.*, 2001b). The MPN and filtration/enrichment methods have a reported sensitivity of approximately 10 *Campylobacter* cells/100 mL (Bolton *et al.*, 1982; Bolton *et al.*, 1987). However, culture based methods will not detect the presence of VBNC cells in the water sample (Sails *et al.*, 2002). Therefore, detection methods that quantify all water-borne *Campylobacter* are generally PCR-based. As stated above, total *Campylobacter* levels are low in water systems, so many PCR-based methods employ enrichment and/or sample filtration (Hernandez *et al.*, 1995; Jackson *et al.*, 1996; Purdy *et al.*, 1996; Waage *et al.*, 1999; Moore *et al.*, 2001a; Sails *et al.*, 2002; Moreno *et al.*, 2003). Enrichment minimizes the chance of detecting dead cells (Hernandez *et al.*, 1995) or free DNA (Josephson *et al.*, 1993), dilutes out potential inhibitors of *Taq* polymerase present in water samples, and probably produces a suspension of cells ideal for efficient detection of DNA. However, damaged or VBNC organisms might not be detected by these methods unless the enrichment is designed to resuscitate these populations (Waage *et al.*, 1999).

Methods with designed semi-nested primer sets have been developed to increase sensitivity (Purdy *et al.*, 1996; Waage *et al.*, 1999). Purdy *et al.* (1996) demonstrated that amplification was detectable at 1 fg of template DNA, an approximately 1000-fold improvement over traditional primer sets. While 1 fg roughly corresponds to the DNA in one *Campylobacter* cell (Purdy *et al.*, 1996), detection of *Campylobacter* in spiked water samples was less sensitive [20 cfu/mL (Purdy *et al.*, 1996) or 3–15 cfu/100 mL (Waage *et al.*, 1999)] than cells in pure culture. Similarly, *Campylobacter* cells have been quantified directly with a filtration/PCR method with a sensitivity of approximately 10–20 cfu/mL (Moore *et al.*, 2001a) or 10–100 cfu/mL (Oyofo and Rollins, 1993). Although these studies demonstrate the ability to detect low levels of *Campylobacter* in water samples, it should be noted that in three instances (Oyofo and Rollins, 1993; Purdy *et al.*, 1996; Waage *et al.*, 1999), spiked water samples were tested rather than naturally contaminated water. In one filtration/PCR method (Moore *et al.*, 2001a) using non-spiked water, it was unclear how many of the detected *Campylobacter* cells were viable or in the VBNC state. Sails *et al.* (2002) used a PCR-ELISA assay with *C. jejuni-* and *C. coli*-specific capture probes to detect *Campylobacter* in naturally contaminated water samples. This assay could detect concentrations as low as 20 cfu/mL, but it could not distinguish between viable and VBNC cells. Thus, existing methods are able to detect very low levels of *Campylobacter* contamination in water, but further work is required to distinguish the different potential *Campylobacter* populations (non-viable, VBNC, viable).

PCR-based detection of *Campylobacter* in food samples

Methods exist for releasing and removing *Campylobacter* from food samples. Meat samples are washed usually with phosphate-buffered saline or peptone water, or homogenized in *Campylobacter* blood-free broth by stomaching. Milk and dairy samples are homogenized in an enrichment broth. Wash solutions are filtered through cheesecloth and/or centrifuged at low speed to remove debris. Finally, wash solutions or homogenates are inoculated into enrichment media or are diluted and plated directly onto selective media. Enrichment steps often are required since the concentration of *Campylobacter* cells on food samples is too low for direct plating. Pre-enrichment has been shown also to increase the sensitivity of some PCR methods by increasing the numbers of organisms in general and diluting out potential PCR inhibitors. Also, the duration of enrichment is a determinant of sensitivity: increasing the enrichment time from 24 h to 48 h lowered the detection limit (Thunberg *et al.*, 2000). Similarly, other methods have used a 48 h incubation in Hunt broth (Hunt and Tran, 1997) or Campy-Cefex agar (Winters and Slavik, 1995).

As noted in the previous section, many methods use semi-nested PCR to increase sensitivity and specificity (Winters *et al.*, 1998; Magistrado *et al.*, 2001; Bang *et al.*, 2002). A semi-nested PCR method in which the chicken washes were used directly for PCR following differential centrifugation was reported (Winters *et al.*, 1997). This method was used also to detect *C. jejuni* on poultry carcasses, with a reported sensitivity of 10^2 cfu/mL. The percentage of positive samples detected by this method was less than that reported by culturing methods; however, results were obtained in less than 24 h. Generally, PCR-based methods detect *Campylobacter* cells at sensitivities similar to standard culturing techniques; however, the percentage of positive samples obtained with molecular methods often is less than that obtained with culturing methods (Manzano *et al.*, 1995b; Winters and Slavik, 1995; Winters *et al.*, 1997). Considering that culture methods would not detect sublethally injured cells that might be detected by PCR, this discrepancy in detection sensitivity could be larger than reported.

The sensitivity of PCR-based assays can be enhanced also by the use of a reverse transcriptase PCR (RT-PCR) assay and rRNA primers (Giesendorf and Quint, 1995). rRNA is a good choice for RT-PCR since it is in high copy (10^3–10^4 molecules) in the cell (Giesendorf and Quint, 1995). When used to detect *Campylobacter* on naturally contaminated poultry with an rRNA primer set, RT-PCR provided an approximately 100-fold increase in sensitivity over standard PCR (Giesendorf and Quint, 1995). RT-PCR has the added advantage of being able to detect viable cells specifically. Since most prokaryotic mRNA species have an extremely short half-life and intracellular mRNAs degrade rapidly following cellular death, only viable cells will contain mRNA. A RT-PCR method using a *ccoN* primer set (Jackson *et al.*, 1996) has been described (Sails *et al.*, 1998). DNA isolated from freshly heat-killed cells amplified with this primer set; however, mRNA could no longer be detected after 4 h post-heat treatment. A variation of rRNA amplification, nucleic acid sequence-based amplification (NASBA), was reported by Uyttendaele *et al.* Unlike conventional PCR, NASBA does not use a thermocycler and does not require a separate reverse transcription step (Uyttendaele *et al.*, 1995). NASBA was successful in detection of *Campylobacter* on a number of meat products and in milk, but large numbers of indigenous organisms resulted in a weak or non-existent signal. At least 10^3 CFU were necessary for detection among high numbers of indigenous microflora. A real-time PCR assay for *C. jejuni* was described recently (Sails *et al.*, 2003a). Naturally contaminated food samples were enriched for 48 h and then analysed by quantitative PCR. The assay showed a strong specificity for *C. jejuni*, a good correlation between culture on agar and results by PCR, and a detection limit of 12 genome equivalents.

Some PCR-based methods use hybridization and a colorimetric detection endpoint instead of gel electrophoresis (PCR-ELISA). In these methods, the amplicons are labelled, either by incorporation of digoxygenin (DIG) nucleotides during amplification (Bolton *et al.*, 2002) or with biotinylated primers (O'Sullivan *et al.*, 2000; Grennan *et al.*, 2001). The labelled amplicons are then bound to *C. jejuni*- or *C. coli*-specific capture probes. In the method of Grennan *et al.* (2001), the probes have been aminated and immobilized onto a carboxylated nylon membrane and in the method of Bolton *et al.* (2002), the probes are biotinylated and the probe-amplicon hybrids are bound to a streptavidin-coated microtitre plate. In both methods, a peroxidase conjugate and chromogenic substrate was used to detect the hybrids. The reported limit of detection with purified *Campylobacter* DNA was 100 fg, a 10-fold improvement over standard PCR (Grennan *et al.*, 2001), and detection using PCR-ELISA correlated strongly with detection by conventional culturing methods.

Hydrophobic grid membrane filter (HGMF) detection assays, a variation of PCR-ELISA assays, have been described (Ng *et al.*, 1997; Wang *et al.*, 2000). Food sample washes or enrichments are filtered through HGMFs, which are then placed on selective media for 48 h. In HGMF-EIA (Wang *et al.*, 2000), the loosely bound bacterial cells are removed and the HGMFs are probed using polyclonal antibodies and a membrane peroxidase substrate. Alternatively, bacterial cells are lysed, the DNA is hybridized to DIG-labelled probes, and hybrids are detected with anti-digoxygenin antisera (Ng *et al.*, 1997). A primary concern with immunocapture methods is the specificity of the antisera. Some commercially available *Campylobacter* polyclonal antibodies detect, albeit weakly, non-*Campylobacter* strains (Wang *et al.*, 2000).

Immunomagnetic selection methods have been described for *Campylobacter* (Fierens *et al.*, 1993; Docherty *et al.*, 1996; Lamoureux *et al.*, 1997; Yu *et al.*, 2001). In these methods, monoclonal or polyclonal anti-*Campylobacter* antibodies are bound to magnetic beads. The immunomagnetic beads (IMBs) then are added to food or milk samples to bind target organisms. The method of Docherty *et al.* (1996) coupled IMB capture with enrichment and PCR. Initial direct detection limits were approximately 10^6 cfu/g for chicken and 10^5 cfu/mL for milk. Sensitivity of the assay was improved for poultry samples when a pre-enrichment step was added (420 cfu/g after 18h pre-enrichment, 4.2 cfu/g after 36 h) and similar results were obtained with inoculated milk samples. Lamoureux *et al.* (1997) described a method involving IMB capture and subsequent hybridization with a 23S rDNA probe or a specific RNA probe to detect *Campylobacter* on artificially contaminated poultry samples. This method detected thermophilic *Campylobacter* specifically. The detection limit was 10^4 organisms in pure culture, but insensitive in detecting *Campylobacter* in naturally contaminated samples. The efficiencies of two different IMBs (tosylated and biotin/streptavidin) in the detection of *Campylobacter* in spiked poultry samples were tested (Yu *et al.*, 2001). The limit of detection was 10^4 cfu/g but, again, the method was not tested on naturally contaminated samples.

SOURCE TRACKING RELATED TO OUTBREAKS
Epidemiological studies of *Campylobacter*-related outbreaks are facilitated greatly by the ability to rapidly trace the source of the infection. Source tracking entails typing outbreak isolates in order to establish their clonal nature, then locating food or environmental *Campylobacter* isolates of the same biotype or serotype. This strategy has proven difficult for *Campylobacter*, and this is due partly to the extreme diversity within the taxa and the high incidence of *Campylobacter* species in food and environmental sources. In order to establish unequivocally a source-outbreak link, each *Campylobacter* strain must contain multiple markers and the typing scheme must be able to show discrimination (preferably, if not practicably) at the strain level to provide, essentially, a strain fingerprint. Therefore, rapid and robust typing methods are an essential component of outbreak source tracking.

Serotyping
Multiple *Campylobacter* typing methods have been reported. These methods generally fall into two groups: phenotypic methods, involving outer surface structures as determinants (e.g. Penner heat-stable antigen serotyping and phage typing), and genotypic methods, involving variations in the genomic sequence (e.g. pulsed field gel electrophoresis (PFGE) and amplified fragment length polymorphism (AFLP)). Two of the most commonly used phenotypic typing methods for *Campylobacter* (primarily *C. jejuni* and *C. coli*) are the Penner scheme based on heat-stable (HS) antigens (Penner and Hennessy, 1980) and the Lior scheme based on heat-labile (HL) antigens (Lior *et al.*, 1981). More than 65 Penner serotypes have been reported for *C. jejuni* and *C. coli* (at least 48 HS serotypes for *C. jejuni* and 17 HS serotypes for *C. coli*) (Newell *et al.*, 2000; McKay *et al.*, 2001). Although lipopolysaccharides were assumed to be the primary HS antigen for many years, it has been reported, relatively recently, that the Penner serotypes correspond to variations in the capsular polysaccharide (Karlyshev *et al.*, 2000); the Lior serotype determinant has not been identified. Penner serotyping has been an important primary typing scheme, but only a few laboratories in the world have the necessary antiserum and expertise to serotype clinical and environmental isolates. Also, some *Campylobacter* strains have been shown to be HS untypeable (Newell *et al.*, 2000) and, because of antigenic cross-reactivity, some isolates are assigned multiple serotypes (e.g. HS4 complex is O:4,13,16,43,50,64). The most common Penner HS serotypes identified in screens of predominately clinical specimens are O:1, O:2, O:3, and O:4 (and O:4 complex) (McKay *et al.*, 2001; Woodward and Rodgers, 2002). Recently, a serotyping scheme based on direct agglutination rather than the passive hemagglutination scheme of Penner was described (Frost *et al.*, 1998; Oza *et al.*, 2002). This scheme was able to expand the 65 *C. jejuni/C. coli* HS types to 118. Nevertheless, isolates that cannot be typed with the expanded typing system continue to be identified.

Phage typing

The other phenotypic typing scheme commonly used for *Campylobacter* is phage typing (reviewed in (Frost *et al.*, 1999)). Phage typing relies on the lytic reaction patterns of an isolate to 16 virulent typing bacteriophages. A phage type is defined as two or more unrelated isolates that have the same bacteriophage reaction pattern (Frost *et al.*, 1999). Phage typing is often used in conjunction with HS serotyping; however, as with HS serotyping, many isolates are nontypeable or possess unique phage reaction patterns.

Genotypic typing

Genotypic typing schemes (reviewed in (Newell *et al.*, 2000; Wassenaar and Newell, 2000)) rely mainly on nucleotide differences between strains that alter the genomic restriction pattern. PFGE, a commonly used genotypic typing method, detects polymorphisms in the restriction patterns of enzymes, such as *Sma*I and *Sal*I, that cut the *Campylobacter* chromosome infrequently. Ribotyping detects polymorphisms in the rDNA loci. Genomic DNA is digested and then hybridized with rRNA-specific probes. PCR-RFLP methods detect variation at a particular locus (e.g. flagellin or *fla* typing) through amplification and subsequent digestion with one or more restriction enzymes. An RFLP-related method is AFLP. AFLP protocols involve digesting genomic DNA with two enzymes with different recognition specificities. Labelled primers are designed to amplify only digestion products that are flanked by both restriction sites. Amplicons are often numerous (approximately 40–60) and very small (35–500 bp) and are resolved usually on a sequencing gel.

Restriction-based typing methods have several drawbacks: (1) for many protocols, there are no standard sets of restriction enzymes making comparisons between laboratories difficult, (2) differences in DNA quality, digestion conditions, and electrophoretic conditions can lead to profile differences, (3) some strains cannot be digested by certain restriction enzymes because of a particular methylation pattern, resulting in variations in fragment sizes, and (4) some methods, such as AFLP, are lengthy and require specialized equipment. Since restriction-based methods reveal only a subset of the bp differences between two strains, the greater the number of restriction enzymes used and/or target sites present, the greater the discriminatory potential. Thus, AFLP has a higher level of discrimination than ribotyping. Although different profiles usually indicate different strains, identical profiles do not always indicate identical strains.

The ultimate genotypic marker, however, is the genomic sequence. Genomic sequencing has become progressively feasible, even though full-scale sequencing of outbreak isolates is not practical or logistically feasible. Nevertheless, regions of the genome of these isolates can be sequenced and compared. This is the basis for the recent genotypic method, multilocus sequence typing (MLST) (Dingle *et al.*, 2001), which relies on sequencing short (approximately 500 bp) regions from multiple loci. Over 1018 MLST-based profiles have been identified and are deposited in an Internet database (http://mlst. zoo.ox.ac.uk). Even though genotypic typing methods have drawbacks, the discriminatory power of some methods (e.g. AFLP, MLST) makes them valuable for epidemiological studies. Additionally, since genotypic methods are a function only of the genomic sequence, these same methods can be adapted for studies of non-*jejuni*/*coli Campylobacteraceae*, for many species of which minimal genetic data are available.

Source tracking in outbreaks

Campylobacter typing schemes are not applied to the majority of reported outbreaks, much less sporadic cases. In fact, the strain or strains involved in 119 of the 295 (40%) outbreaks listed in Table 6.2 were not characterized further than the genus level, although it is probable that most of these outbreaks were caused by *C. jejuni*. The long incubation period of campylobacteriosis, the presence of multiple *Campylobacter* types in food and in the environment, and the relatively high incidence of sporadic illness, compared to other food-borne pathogens, makes source tracking difficult. Often, by the time an outbreak is recognized, the source of the infection no longer exists and tracking becomes impossible. An outbreak is recognized by the high degree of clonality among the isolates (Fayos *et al.*, 1993; Evans *et al.*, 1996; Iriarte and Owen, 1996; Evans *et al.*, 1998b; Harrington *et al.*, 1999), but often *Campylobacter* strains with identical serotypes or molecular types cannot be obtained from a food, water, or animal source (Evans *et al.*, 1996; Evans *et al.*, 1998b; Roels *et al.*, 1998; Harrington *et al.*, 1999; Bopp *et al.*, 2003). In some instances, outbreak isolates can have multiple molecular types. For example, in a waterborne *Campylobacter* outbreak, 17% of the isolates had one of six PFGE patterns different from the outbreak pattern (Bopp *et al.*, 2003). Unfortunately, no *C. jejuni* could be cultured from the water samples associated with this outbreak. Additionally, characterization of a waterborne outbreak in Sweden (Lind *et al.*, 1996) revealed five *Campylobacter* types among the patients and three types from water, with only one of the water types corresponding to an outbreak type. However, there are some reports of disease being traced definitively to a single source (Lind *et al.*, 1996; Engberg *et al.*, 1998; Lehner *et al.*, 2000; Nielsen *et al.*, 2000; Olsen *et al.*, 2001).

CONCLUSIONS, EMERGING *CAMPYLOBACTER* SPECIES AND FUTURE RESEARCH NEEDS

After nearly three decades of research involving thousands of studies, with many of the food-related studies descriptive rather than fundamental, many questions remain regarding the biology and epidemiology of *Campylobacter* species and their relationship to food safety risks and public health. It is probable that the risks associated with *Campylobacter*-caused illness to an individual are variable based on factors like geographic location (e.g. intensive animal production, farming practices, climate, water, food processing, travel), host-susceptibility (e.g. immunity, host-genetics), ecology (e.g. strain differences) and habits (e.g. eating out, cooking techniques, awareness of risks, pets). Surveillance studies and studies to identify risk factors related to sporadic *Campylobacter* illness indicate some factors are common in different regions, and some are not. Examples of common factors include consumption of poultry in a restaurant, daily contact with chickens, contact with pets (cats and dogs), travel abroad, drinking unpasteurized milk, involvement in water sports, and barbecuing meat and consumption of salad vegetables (Kapperud *et al.*, 1992; Altekruse *et al.*, 1999; Studahl and Andersson, 2000; Rodrigues *et al.*, 2001; Evans *et al.*, 2003; Sopwith *et al.*, 2003). However, the sources of outbreaks and the number of cases confirm the increased risk of consuming raw milk or untreated water (Table 6.2).

    C. coli and *C. jejuni* (Table 6.1), and possibly other *Campylobacter* species, appear to be ubiquitous in the environment, suggesting that humans are exposed frequently to low concentrations of multiple *Campylobacter* species. The relatively small genome size of *Campylobacter* species (~1.6–2.0 Mb) does not appear to limit their ability to survive and adapt in many environments, including humans. The small genome size may be compensated by the diversity and hypervariability that occurs through genetic exchange and rearrangements (Wassenaar *et al.*, 1998; Meinersmann, 2000; de Boer *et al.*, 2002), plasmids (Lind *et al.*, 1989; Bacon *et al.*, 2000; Luo and Zhang, 2001), and polynucleotide repeats (Parkhill *et al.*, 2000). These mechanisms probably are important for enhancing the survival of *Campylobacter* in a variety of hostile environments and the infectious potential of a *Campylobacter* strain for a given human host. An analysis of the population genetics and genealogy of *C. jejuni*, however, suggests that if selective evolution is occurring in the human host, it is not reflected in the general *C. jejuni* population; rather human strains simply reflect those that are transmitted to humans 'accidentally' thus making humans, at least hypothetically, a dead-end host for *C. jejuni* and possibly all *Campylobacter* species (Meinersmann, 2000). It is also possible, however, that sources other than poultry, the most identifiable and logical source, are responsible for human illnesses and that strains evolved in this unknown niche are more virulent also for the human host (see Table 6.2, 'Unknown'). These provocative hypotheses reflect why fundamental information is needed to help understand the ecology and biology of *Campylobacter* in food.

A recent study of a multidrug-resistant *C. jejuni* strain presumably passed between men through sexual activity and associated with an outbreak of enterocolitis represents a somewhat unexpected route of transmission resulting in GI illness (Gaudreau and Michaud, 2003) and, perhaps, adaptation to a new mode of entry to the human host. Hypervariable gene expression and a high rate of spontaneous mutations could be beneficial especially for survival of certain strains during the stressful conditions encountered in an animal host or food processing (e.g. hot and cold water, oxidative stress, chemical sanitizers). New genotypes and phenotypes of *Campylobacter* could emerge from an initial population of cells and strains in animal or food environments due to increased fitness characteristics that enhance survival in stress conditions (Kelly *et al.*, 2003). It should be noted that the multiple mechanisms of genetic exchange and hypervariability in *Campylobacter* create tremendous challenges for analysis of populations of *Campylobacter* cells present in environments where often multiple strains of *Campylobacter* exist due to intensive animal or food production (e.g. poultry, cattle, pig production). However, the emergence of strains with increased fitness characteristics for surviving and growing under environmental stresses suggests that the strains isolated also contain signatures that may be exploited for developing source tracking methods. These signatures may be specific genes and/or mutations crucial for survival. Understanding the differences between strains obtained from different sources, including food production and animal host environments, will require fundamental molecular, chemical and epidemiological data. These data may provide clues for developing novel strategies for decreasing the prevalence and virulence (i.e. transmission and/or or pathogenic traits) of *Campylobacter* species in food production environments.

Another potentially important area of biology related to *Campylobacter* species in food and water is the formation of biofilms involving *Campylobacter* and other microorganisms. Communities of microorganisms (prokaryotic and eukaryotic) (Brown and Barker, 1999) can interact/aggregate on either biotic or abiotic surfaces, and induce the formation of an exopolysaccharide matrix that protects the biofilm of microorganisms from stressful conditions (e.g. temperature, pH, UV, chemicals) (Potera, 1996; Costerton *et al.*, 1999). It has been reported that *C. jejuni* isolated from chicken house water systems form 'biofilms' on polyvinyl chloride (PVC) surfaces with Gram-positive isolates from the same environment (Trachoo *et al.*, 2002). The *C. jejuni* cells attached approximately

0.5 $\log_{10}$ better to the pre-existing biofilms with Gram-positive bacteria than to PVC alone, they remained viable longer (Trachoo *et al.*, 2002) and they were more difficult to remove or kill with chemical sanitizers (Trachoo and Frank, 2002). Although this work relates to the persistence of *Campylobacter* in water and poultry processing environments, it is likely that *Campylobacter* species form biofilms in many additional food-related environments and that this relates to the ability for them to survive under the same conditions that kill them in less complex, *in vitro* systems. The potential ingestion of *Campylobacter* by protozoa commonly in water, soil, or food relates also to biofilms and, potentially, to the evolution of mechanisms of survival in intracellular vacuoles (King *et al.*, 1998; Brown and Barker, 1999). The availability of more genomic sequence data for *Campylobacter* species will enhance studies to better understand these fundamental aspects of *Campylobacter* biology.

Improved protocols for isolation, detection and identification of emerging *Campylobacteraceae* in the food supply are needed. Filtration plate methods, such as the 'Cape Town Protocol' have been shown to isolate the majority of known *Campylobacter* species, both from blood and faecal samples (Lastovica and Skirrow, 2000). However, faecal samples, especially from sick individuals, often contain many more organisms than would be present on carcasses and meat products, or in milk and water. Because generic *Campylobacter* isolation methods may not be possible and because food-borne *Campylobacter* is present in low levels and often in a VBNC state, molecular methods must be developed to increase detection sensitivity. These methods could include improved immunocapture and PCR detection methods for emerging *Campylobacteraceae*. Extensive genomic data exist for the four thermotolerant *Campylobacter* species, but PCR-based methods for non-*jejuni*/*coli Campylobacteraceae* are hindered by a lack of genomic data for many of the species. For the majority of the remaining species, the only genomic data available are 16S and/or 23S rDNA sequences; minimal sequence data are available for *C. rectus*, *C. concisus*, *C. fetus*, and *Arcobacter* species. Sequencing of the emerging *Campylobacter* species genomes would be very beneficial for comparative genomics with multiple *C. jejuni* genome sequences (TIGR, 2003) and provide genomic data that would be invaluable for development of new and/or improved molecular detection methods.

To address some of the issues described above, a Consortium Grant (CAMPYCHECK) was established through the European Union Fifth Framework Programme (FP5: 1998–2002) (EC, 2002). The goals of the CAMPYCHECK project are: (1) the development of generic isolation procedures and improved atmospheres for emerging *Campylobacteraceae*, (2) the development of novel antibody-based and PCR-based detection methods and the improvement of existing molecular methods, (3) the development of rapid molecular identification methods, (4) demonstration of recovery and isolation of emerging *Campylobacter* from the food supply, (5) the determination of the prevalence of these organisms in the food chain, and (6) formulation of new standard protocols for the isolation and detection of emerging *Campylobacter* (CAMPYCHECK, 2003). Research performed by members of the CAMPYCHECK consortium and others, hopefully, will begin to answer some of the questions regarding emerging *Campylobacteraceae* in the food supply, such as the true prevalence of these organisms in the food chain and whether new zoonotic reservoirs for these organisms exist. Most importantly, improved isolation and detection techniques will help also to delineate more clearly the clinical importance of emerging *Campylobacteraceae* and their impact on public health.

Table 6.1 Prevalence of *Campylobacter* in live animals and the food supply.

Source	Incidence (%)			Country	Reference
	Campylobacter spp. (#)	*C. jejuni*[a]	*C. coli*[a]		
Poultry, birds					
Chicken, live	83 (158)	44	56	Italy	Pezzotti et al. (2003)
Chicken, live	86 (80)			USA	Stern et al. (2002)
Chicken, live	3 (1132)	94	6	Finland	Perko-Makela et al. (2002)
Chicken, live	41 (2325)	95	5	Canada	Nadeau et al. (2002)
Chicken, live		67[b]	20[b]	Brazil	Fernandez (2000)
Chicken, live		45[b]	15[b]	Chile	Fernandez (2000)
Chicken, live		61[b]		Peru	Fernandez (2000)
Chicken, live	82 (98)			UK	Evans and Sayers (2000)
Chicken, live	23 (150)	29	47	Lebanon	Talhouk et al. (1998)
Chicken, live	36 (929)	83	15	Denmark	Nielsen et al. (1997)
Chicken, live		27[b]		UK	Pearson et al. (1996)
Chicken, live		97	3	Sweden	Berndtson et al. (1996)
Chicken, live	57 (30)	13	57	Brazil	Machado et al. (1994)
Chicken, live	64 (175)			UK	Pearson et al. (1993)
Chicken, live		81[b]		UK	Pearson et al. (1993)
Chicken, live		70	28	Tanzania	Kazwala et al. (1993)
Chicken, live	60 (59)	80	20	Portugal	Cabrita et al. (1992)
Chicken, live	59 (56)			India	Kulkarni et al. (1992)
Chicken, live	24 (490)			Finland	Aho and Hirn (1988)
Chicken, live	41 (130)	76	24	Thailand	Rasrinaul et al. (1988)
Average[c]	*33*	*64*	*25*		
Chicken, meat	62 (2099)	26[b]	1[b]	New Zealand	Savill et al. (2003)
Chicken, meat	81 (155)			Iceland	Stern et al. (2003)
Chicken, meat	82 (72)		22	Italy	Pezzotti et al. (2003)
Chicken, meat	83 (241)	82	2	USA	Dickins et al. (2002)
Chicken, meat		98		UK	Jorgensen et al. (2002)
Chicken, meat (frozen/fresh)	77/94 (44/63)			UK	Moore et al. (2002)
Chicken, meat	57 (1127)	69	30	UK	Wilson (2002)

Table 6.1 Continued.

Source	Incidence (%) Campylobacter spp. (#)	C. jejuni[a]	C. coli[a]	Country	Reference
Chicken, meat	50 (198)			Spain	Dominguez et al. (2002)
Chicken, meat	60 (62)	50	50	Brazil	Aquino et al. (2002)
Chicken, meat	21-49 (NT)[g]			USA	Stern et al. (2001)
Chicken, meat	36 (978)			Denmark	Anonymous (2002a)
Chicken, meat	68 (300)			UK	Harrison et al. (2001)
Chicken, meat	71 (719)	55	36	USA	Zhao et al. (2001)
Chicken, carcass (broiler)	76 (141)	70	16	Belgium	Vellinga and Butzler (2000)
Chicken, carcass (layer)	90 (122)	70	25	Belgium	Vellinga and Butzler (2000)
Chicken, breasts	58 (139)	78	14	Belgium	Vellinga and Butzler (2000)
Chicken, meat	24 (859)			Netherlands	Havelaar et al. (2000)
Chicken, meat	79 (NT)[g]			Brazil	Fernandez (2000)
Chicken, meat	83 (198)	95	8	UK	Kramer et al. (2000)
Chicken, meat	29 (772)			Belgium	Uyttendaele et al. (1999)
Chicken, meat	46 (111)			Germany	Atanassova and Ring (1999)
Chicken, meat	46 (72)			Japan	Ono and Yamamoto (1999)
Chicken, meat	38 (120)	50	50	UK	Madden et al. (1998)
Chicken, meat	10 (31)		67	Lebanon	Talhouk et al. (1998)
Chicken, meat	69 (330)			USA	Willis and Murray (1997)
Chicken, meat	33 (1853)			Germany	Geilhausen et al. (1996)
Chicken, meat	88 (1297)			USA	FSIS (1996b)
Chicken, meat	46 (54)	67	33	Italy	Manzano et al. (1995a)
Chicken, meat	50 (30)	6	78	Brazil	Machado et al. (1994)
Chicken, meat (frozen/fresh)	0/45 (58/59)			India	Kulkarni et al. (1992)
Chicken, meat	32 (98)			USA	Jones et al. (1991b)
Chicken, meat	68 (156)			Japan	Tokumaru et al. (1991)
Chicken, meat (retail)	61 (279)			Netherlands	de Boer and Hahne (1990)
Chicken, meat	56 (758)	87	13	UK	Fricker and Park (1989)
Chicken, meat	38 (410)	94	6	Canada	Lammerding et al. (1988)
Chicken, meat (frozen/fresh)	4/48 (24/46)			USA	Hood et al. (1988)
Chicken, meat	98 (45)			France	Marinescu et al. (1987)
Chicken, meat (retail)	23 (862)			USA	Harris et al. (1986)

Chicken, meat (frozen/fresh)	0/13 (40/40)			USA	Stern et al. (1984)
Chicken, meat (retail)	50 (50)			USA	Wesley et al. (1983)
Chicken, wings	65 (153)		26	UK	Flynn et al. (1994)
Chicken, wings (fresh)	83 (94)	45		USA	Kinde et al. (1983)
Chicken, processed	6 (31)			Belgium	Uyttendaele et al. (1999)
Chicken, ground	60 (283)			USA	FSIS (1996d)
Chicken, liver		35[b]	2[b]	Argentina	Fernandez (2000)
Chicken, liver		22[b]	70[b]	Chile	Fernandez (2000)
Chicken, liver	93 (126)	21	79	Chile	Fernandez and Pison (1996)
Chicken, liver	20 (25)	82	18	Italy	Manzano et al. (1995a)
Chicken, liver (frozen/fresh)	15/30 (40/40)			USA	Stern et al. (1984)
Average	*53*	*60*	*31*		
Turkey, meat	14 (172)	15		USA	Zhao et al. (2001)
Turkey, meat	30 (705)			Denmark	DZC (2001)
Turkey, meat	1 (145)			Netherlands	Havelaar et al. (2000)
Turkey, meat	90 (1221)			USA	FSIS (1998)
Turkey, meat	74 (205)		73	Canada	Lammerding et al. (1988)
Turkey, meat (before chilling)	94 (33)	97		USA	Luechtefeld and Wang (1981)
Turkey, meat (after chilling)	34 (83)			USA	Luechtefeld and Wang (1981)
Turkey, processed	13 (39)			Belgium	Uyttendaele et al. (1999)
Turkey, ground	25 (295)			USA	FSIS (1996e)
Average	*56*	*56*	*37*		
Poultry, raw sausages	10 (115)			Canada	Federighi et al. (1999)
Geese, domestic	100 (40)	100	5	Turkey	Aydin et al. (2001)
Geese, live		28[b]	18[b]	Chile	Fernandez (2000)
Ducks, live		65[b]	2[b]	New Zealand	Savill et al. (2003)
Ducks, live		18[b]		Peru	Fernandez (2000)
Ducks, live		66[b]	7[b]	Chile	Fernandez (2000)
Ducks, live	41 (21)	52	48	Portugal	Cabrita et al. (1992)
Duck, meat (wild)	6 (52)			Netherlands	Havelaar et al. (2000)
Duck, meat	90 (10)			France	Marinescu et al. (1987)
Pheasant, meat	4 (27)			Netherlands	Havelaar et al. (2000)

Table 6.1 Continued.

Source	Incidence (%) Campylobacter spp. (#)	C. jejuni[a]	C. coli[a]	Country	Reference
Pheasant, meat	26 (52)			Germany	Atanassova and Ring (1999)
Fowl, domestic	44 (39)			Ghana	Abrahams et al. (1990)
Game hen, retail	17 (29)			USA	Harris et al. (1986)
Guinea fowl, meat	3 (35)			Netherlands	Havelaar et al. (2000)
Guinea fowl, carcass	34 (32)			Belgium	Uyttendaele et al. (1999)
Guinea fowl, meat	25 (20)			France	Marinescu et al. (1987)
Average	*32*	*55*	*16*		
Cow, pig, sheep, goat					
Cattle (dairy cows)	6 (NT)[g]			Australia	Bailey et al. (2003)
Cattle (feedlot beef)	58(NT)[g]			Australia	Bailey et al. (2003)
Cattle (pasture beef)	2 (NT)[g]			Australia	Bailey et al. (2003)
Cattle (dairy cows)		96[b]	10[b]	New Zealand	Savill et al. (2003)
Cattle (beef cattle)		85[b]	12[b]	New Zealand	Savill et al. (2003)
Cattle	54 (89)	22	26	Italy	Pezzotti et al. (2003)
Cattle	23 (360)			Denmark	Nielsen (2002)
Cattle	5 (401)	75	15	USA	Hoar et al. (2001)
Cattle	2 (NT)[g]			Argentina	Fernandez (2000)
Cattle		53[b]	8[b]	Brazil	Fernandez (2000)
Cattle (Japanese Black)	63 (648)	79	9	Japan	Ono and Yamamoto (1999)
Cattle (Holstein)	44 (378)	71	24	Japan	Ono and Yamamoto (1999)
Cattle		32[b]	3[b]	Switzerland	Busato et al. (1999)
Cattle	62 (136)	7		UK	Atabay and Corry (1998)
Cattle	89 (360)			UK	Stanley et al. (1998b)
Cattle	47 (94)	91	7	Denmark	Nielsen et al. (1997)
Cattle, calves	97 (34)	62	3	Japan	Giacoboni et al. (1993)
Cattle, adult	47 (60)	29	0	Japan	Giacoboni et al. (1993)
Cattle	20 (32)	69	31	Portugal	Cabrita et al. (1992)
Cattle	1 (254)	100		Norway	Rosef et al. (1983)
Average	*45*	*62*	*12*		

Product				Country	Reference
Beef	1 (151)			Italy	Pezzotti et al. (2003)
Beef	1 (172)	100		USA	Zhao et al. (2001)
Beef	0.4 (738)			Netherlands	Havelaar et al. (2000)
Beef	3 (NT)[g]			Argentina	Fernandez (2000)
Beef	0 (55)			Japan	Ono and Yamamoto (1999)
Beef	0.3 (657)			Australia	Vanderlinde et al. (1998)
Beef	10 (62)			Belgium	Korsak et al. (1998)
Beef	0 (100)			UK	Madden et al. (1998)
Beef (cows and bulls)	1 (2109)			USA	FSIS (1996a)
Beef (steers and heifers)	4 (2064)			USA	FSIS (1994)
Beef	0 (52)			Japan	Tokumaru et al. (1991)
Beef	24 (127)	100		UK	Fricker and Park (1989)
Beef	23 (598)	86	14	Canada	Lammerding et al. (1988)
Beef	1 (130)	100	0	Thailand	Rasrinaul et al. (1988)
Beef, veal	43 (267)	91	7	Canada	Lammerding et al. (1988)
Beef, ground	0.002 (562)			USA	FSIS (1996c)
Beef, liver (frozen/fresh)	6/15 (36/40)			USA	Stern et al. (1984)
Ox, liver	54 (96)			UK	Kramer et al. (2000)
Average	6	95	7		
Swine	64 (104)	1	64	Italy	Pezzotti et al. (2003)
Swine	6 (NT)[g]			Argentina	Fernandez (2000)
Swine		15[b]	55[b]	Chile	Fernandez (2000)
Swine		5	87	Japan	Ono and Yamamoto (1999)
Swine, gilts	76 (50)	76	21	USA	Young et al. (2000)
Swine, pregnant sows	100 (9)	100	13	USA	Young et al. (2000)
Swine	46 (316)	87	95	Denmark	Nielsen et al. (1997)
Swine	59 (65)	4	97	Portugal	Cabrita et al. (1992)
Swine	100 (114)	3	100	Norway	Rosef et al. (1983)
Pig, liver	72 (99)			UK	Kramer et al. (2000)
Pig, liver	6 (400)	67	30	UK	Moore and Madden (1998)
Pig, liver (frozen/fresh)	0/26 (30/35)			USA	Stern et al. (1984)
Pork, meat	10 (175)			Italy	Pezzotti et al. (2003)
Pork, meat	2 (172)	25	75	USA	Zhao et al. (2001)
Pork, meat	35 (NT)[g]			Brazil	Fernandez (2000)

Table 6.1 Continued.

Source	Incidence (%)			Country	Reference
	Campylobacter spp. (#)	C. jejuni[a]	C. coli[a]		
Pork, meat	21 (154)		79	Belgium	Vellinga and Butzler (2000)
Pork, meat	0 (524)			Netherlands	Havelaar et al. (2000)
Pork, meat	0 (55)			Japan	Ono and Yamamoto (1999)
Pork, meat	0 (50)			UK	Madden et al. (1998)
Pork, meat	2 (49)			Belgium	Korsak et al. (1998)
Pork, meat	32 (2112)			USA	FSIS (1996f)
Pork, meat	2 (94)			Japan	Tokumaru et al. (1991)
Pork, meat	18 (158)	3	97	UK	Fricker and Park (1989)
Pork, meat	62 (400)	2	97	Canada	Mafu et al. (1989)
Pork, meat	17 (463)	41	59	Canada	Lammerding et al. (1988)
Pork, meat	5 (130)	71	29	Thailand	Rasrinaul et al. (1988)
Pork, retail cuts	13 (152)	11	44	Belgium	Vellinga and Butzler (2000)
Pork, ground	2 (149)			Belgium	Vellinga and Butzler (2000)
Pork, sausage (frozen/fresh)	0/3 (45/40)			USA	Stern et al. (1984)
Average	*27*	*29*	*65*		
Sheep	0 (NT)[g]			Australia	Bailey et al. (2003)
Sheep, lambs	92 (360)			UK	Stanley et al. (1998a)
Sheep, adult	15 (27)	56[b]	47[b]	New Zealand	Savill et al. (2003)
Sheep, adult	23 (13)	74	26	Portugal	Cabrita et al. (1992)
Sheep, adult	8 (197)	100		Ghana	Abrahams et al. (1990)
Sheep, adult				Norway	Rosef et al. (1983)
Sheep, carcass	1 (470)			Australia	Vanderlinde et al. (1999)
Lamb	8 (NT)[g]			Australia	Bailey et al. (2003)
Lamb, meat	0 (100)			UK	Madden et al. (1998)
Lamb, meat	0 (30)			Spain	Sierra et al. (1995)
Lamb, meat	16 (103)	94	6	UK	Fricker and Park (1989)
Lamb, meat (frozen/fresh)	2/6 (50/36)			USA	Stern et al. (1984)
Lamb, liver	73 (96)			UK	Kramer et al. (2000)
Lamb, liver (frozen/fresh)	0/32 (35/44)			USA	Stern et al. (1984)
Average	*31*	*81*	*26*		

Food	% (n)			Country	Reference
Meat, sausage	1 (143)			UK	Bolton et al. (1985)
Meat, minced	2 (135)			UK	Bolton et al. (1985)
Goat, live	0 (110)			Norway	Rosef et al. (1983)
Goat, live	33 (72)			Ghana	Abrahams et al. (1990)
Offal[d] (pig, cow, and sheep)	47 (689)	90	9	UK	Fricker and Park (1989)
Offal[d], cattle	11 (153)			UK	Bolton et al. (1985)
Offal[d], sheep	31 (232)			UK	Bolton et al. (1985)
Offal[d], pig	6 (67)			UK	Bolton et al. (1985)
Average	*33*	*90*	*9*		
Seafood					
Shellfish	0 (97)			Netherlands	Havelaar et al. (2000)
Shellfish	52 (100)			Netherlands	Havelaar et al. (2000)
Shellfish (oysters)	1 (660)			Canada	Federighi et al. (1999)
Shellfish (oysters)	6 (49)			UK	Wilson and Moore (1996)
Shellfish (cockles, mussels, scallops)	47 (331)			UK	Wilson and Moore (1996)
Shellfish (oysters)	22 (79)			Australia	Arumugaswamy and Prouford (1987)
Seafood	15 (89)	100		UK	Fricker and Park (1989)
Fish	2 (131)	33	33	Belgium	Vellinga and Butzler (2000)
Average	*16*	*67*	*33*		
Produce and produce-related					
Cereals, nuts, dried fruits, herbs, spices	0 (NT)[g]			Scotland	Candlish et al. (2001)
Cut tomato and onion	0 (NT)[g]	100		South Africa	Mosupye and von Holy (2000)
Fresh produce, retail	0 (127)			US	Thunberg et al. (2002)
Mixed salad vegetables, MAP[e]	22 (90)			UK	Phillips (1998)
Mushrooms, retail	2 (200)	100		US	Doyle and Schoeni (1986)
Preserved fruits, chutney, punch drinks	0 (NT)[g]			Trinidad	Adesiyun (1995)
Salads	0 (12)			South Africa	Mosupye and von Holy (1999)
Vegetables	0 (130)			Thailand	Rasrinaul et al. (1988)
Vegetables	0 (361)			Canada	Odumeru et al. (1997)
Vegetables	4 (56)	100		India	Kumar et al. (2001)
Vegetables, organic	0 (NT)[g]			N. Ireland	McMahon and Wilson (2001)

Table 6.1 Continued.

Source	Incidence (%) Campylobacter spp. (#)	C. jejun[a]	C. coli[a]	Country	Reference
Vegetables, organic, multiple countries, RTE	0 (2883)			UK	Sagoo et al. (2001)
Vegetables, outdoor markets	2 (1564)			Canada	Park and Sanders (1992)
Vegetables, retail, 24 types	0 (NT)[g]			Japan	Matsusaki et al. (1982)
Vegetables, self-serve salad bars	0 (NT)[h]			Singapore	ENB (2002)
Vegetables, RTE	0 (NT)[g]			N. Ireland	Moore et al. (2002)
Vegetables, RTE[f]	1 (150)	100		Canada	Federighi et al. (1999)
Average	*1*	*100*			

[a]Represents percentage of positive samples of the designated species.

[b]Represents percentage of total samples tested for *Campylobacter*.

[c]Multiple number results (e.g. 'frozen/fresh') were averaged for purposes of calculating an average for grouped studies.

[d]Liver, kidney and heart.

[e]MAP, modified atmosphere packaging.

[f]RTE, ready-to-eat.

[g]NT, no total.

[h]Each sample included five different types of vegetables selected from 18 vegetable items.

#Total number tested.

Table 6.2 Outbreaks of campylobacteriosis related to food or water.

Source	Species	Cases[a]	Location	Date	Reference
Meat					
Meat, raw hamburger	*C. jejuni*	13	Netherlands	Nov, 1978	Oosterom *et al.* (1980)
Meat, cold meat tray	*C. jejuni*	20	USA (NY)	Apr, 1991	CDC (2003)
Meat, ham/beef/veal	*C. jejuni* (susp.)	11	USA (IL;)	May, 1992	CDC (2003)
Beef, chopped beef liver	*Campylobacter* sp.	3	USA (NY)	Jun, 1992	CDC (2003)
Beef, veal	*C. jejuni* (susp.)	8	USA (NY)	Jul, 1992	CDC (2003)
Meat, meat products	*Campylobacter* sp.	8	UK	Dec, 1994	Pebody *et al.* (1997)
Meat, meat products	*Campylobacter* sp.	Multi	Sweden[b]	1995–1997	Lindqvist *et al.* (2000)
Meat, meat products	*Campylobacter* sp.	NR	Belgium	1995–1998	WHO (1998)
Beef, BBQ beef sandwich	*Campylobacter* sp.	11	USA (WI)	Jul, 1995	CDC (2003)
Beef, gravy	*Campylobacter* sp.	11	USA (IL;)	Jul, 1995	CDC (2003)
Pork	*C. jejuni*	9	USA (NY)	Sep, 1995	CDC (2003)
Beef	*C. jejuni*	34	USA (MD)	Sep, 1999	CDC (2003)
Total[c]		*128*			
Milk and dairy					
Milk, raw	*C. jejuni*	100	UK	Mar, 1978	Robinson and Jones (1981)
Milk, raw	*C. jejuni*	63	UK	Oct, 1978	Robinson *et al.* (1979); Robinson and Jones (1981)
Milk, raw	*C. jejuni*	16	UK	Nov, 1978	Robinson *et al.* (1979); Robinson and Jones (1981)
Milk, raw	*C. jejuni*	148	Scotland	Jan, 1979	Porter and Reid (1980); Robinson and Jones (1981)
Milk, raw	*C. jejuni*	13	UK	Mar, 1979	Robinson and Jones (1981)
Milk, inadequate pasteurization	*C. jejuni*	3500	UK	Mar, 1979	Jones *et al.* (1981); Robinson and Jones (1981)
Milk, churned, unpasteurized	*C. jejuni*	>75	UK	Apr, 1979	Robinson and Jones (1981)
Milk, raw	*C. jejuni*	4	UK	Jun, 1979	Robinson and Jones (1981)
Milk, faecally contaminated	*C. jejuni*	14	UK	Aug, 1979	Robinson and Jones (1981)
Milk, raw	*C. jejuni*	75	UK[d]	Jan, 1980	Robinson and Jones (1981)
Milk, raw	*C. jejuni*	30	UK	Mar, 1980	Robinson and Jones (1981)
Milk, raw	*C. jejuni*	40	UK	Mar, 1980	Robinson and Jones (1981)
Milk, raw	*C. jejuni*	2	UK	Apr, 1980	Robinson and Jones (1981)

Table 6.2 Continued.

Source	Species	Cases[a]	Location	Date	Reference
Milk, raw	*Campylobacter* sp.	106	USA (OR)	Jan, 1981	Finch and Blake (1985)
Milk, raw	*C. jejuni*	103	USA (KS)	Mar, 1981	Finch and Blake (1985); Kornblatt *et al.* (1985)
Milk, raw	*Campylobacter* sp.	190	USA (AZ)	Apr, 1981	Finch and Blake (1985)
Milk, raw	*Campylobacter* sp.	25	USA (MN)	May, 1981	Finch and Blake (1985); Korlath *et al.* (1985)
Milk, raw	*C. jejuni*	21	USA (GA)	May, 1981	Potter *et al.* (1983); Finch and Blake (1985)
Milk, raw	*Campylobacter* sp.	3	USA (ME)	Oct, 1981	Finch and Blake (1985)
Milk, raw	*Campylobacter* sp.	14	USA (AZ)	Oct, 1981	Finch and Blake (1985)
Milk, raw	*Campylobacter* sp.	14	USA (ME)	Oct, 1981	Finch and Blake (1985)
Milk, raw	*C. jejuni*	>500	Switzerland	Nov, 1981	Stalder *et al.* (1983)
Milk, raw	*C. jejuni*	50	UK	Nov, 1981	Wright *et al.* (1983)
Milk, raw	*C. jejuni*	22	UK	Jan, 1982	Wright *et al.* (1983)
Milk, raw	*Campylobacter* sp.	46	USA (MD)	Apr, 1982	Finch and Blake (1985)
Milk, raw	*Campylobacter* sp.	32	USA (MI)	May, 1982	Finch and Blake (1985)
Milk, raw	*Campylobacter* sp.	32	USA (ME)	Jun, 1982	Finch and Blake (1985)
Milk, raw	*C. jejuni/C. fetus*	16	USA (WI)[e]	Jun, 1982	Finch and Blake (1985); Klein *et al.* (1986)
Milk, raw	*Campylobacter* sp.	15	USA (VT)	Sep, 1982	Finch and Blake (1985)
Milk, raw	*Campylobacter* sp.	4	USA (VT)	Oct, 1982	Finch and Blake (1985)
Milk, raw	*Campylobacter* sp.	57	USA (PA)	May, 1983	CDC (1983)
Milk, raw (goat)	*C. jejuni*	6	USA (WA)	Jul, 1983	Harris *et al.* (1987)
Milk, raw	*C. jejuni*	9	USA (CA)	May, 1984	CDC (1984)
Milk, raw	*C. jejuni*	3	Canada	Jun, 1984	CDC (1986)
Milk, raw	*C. jejuni*	2	USA (WA)	1984	Tauxe *et al.* (1987)
Milk, raw	*C. jejuni*	27	USA (CA)	1984	Tauxe *et al.* (1987)
Milk, inadequate pasteurization	*C. jejuni*	35	USA (VT)	Apr, 1986	Birkhead *et al.* (1988)
Milk, raw	*Campylobacter* sp.	~110	USA[f]	1978–1986	Riley and Finch (1985); Tauxe *et al.* (1988)
Milk, unpasteurized	*Campylobacter* sp.	332	UK (multi)[g]	1987	Sockett (1991)
Milk, heat-treated	*Campylobacter* sp.	526	UK (multi)[h]	1987	Sockett (1991)

Vehicle	Organism	No.	Location	Date	Reference
Milk, heat-treated	Campylobacter sp.	14	UK (multi)	1989	Sockett (1991)
Milk, fecal contamination	C. jejuni	104	USA (OK)	1988	Sails et al. (2003b)
Milk, raw	C. jejuni	87	USA (KS)	1988	Sails et al. (2003b)
Milk, raw	C. jejuni	13	USA (WA)	Mar, 1990	CDC (2003)
Milk, raw	C. jejuni	2	USA (WA)	Apr, 1990	CDC (2003)
Milk, raw	C. jejuni	42	USA (TX)	May, 1990	CDC (2003)
Milk, raw (goat)	C. jejuni	3	USA (WA)	Jul, 1991	CDC (2003)
Milk, raw (susp.)	Campylobacter sp.	11	USA (ME)	Feb, 1992	CDC (2003)
Milk	C. jejuni	23	USA (NY)	Apr, 1992	CDC (2003)
Milk, pecked milk bottles	Campylobacter sp.	23	UK	May, 1992	Pebody et al. (1997)
Milk, raw	C. jejuni	50	USA (MN)	May, 1992	CDC (2003)
Milk, raw	Campylobacter sp.	72	UK	Jun, 1992	Morgan et al. (1994)
Milk, pasteurized	Campylobacter sp.	110	UK	Aug, 1992	Fahey et al. (1995)
Milk, raw	C. jejuni	3	USA (WA)	Oct, 1992	CDC (2003)
Dairy, cottage cheese	C. fetus	13	USA (OH)	Dec, 1992	CDC (2003)
Milk, raw	Campylobacter sp.	22	UK	Jun, 1993	Djuretic et al. (1997); Pebody et al. (1997)
Milk, raw	Campylobacter sp.	23	Wales	Mar, 1994	Evans et al. (1996)
Milk, pecked milk bottles	Campylobacter sp.	12	UK	May, 1995	Stuart et al. (1997)
Milk, raw	Campylobacter sp.	35	UK	May, 1995	Djuretic et al. (1997); Pebody et al. (1997)
Milk	C. jejuni	29	USA (NY)	Jun, 1996	CDC (2003)
Milk	C. jejuni	Multi	Sweden[i]	1995–1997	Lind et al. (1996); Lindqvist et al. (2000)
Milk, raw	C. jejuni/C. coli	34	Hungary	Apr, 1998	Kalman et al. (2000)
Milk	C. jejuni	6	USA (SD)	Jul, 1998	CDC (2003)
Milk	Campylobacter sp.	3	USA (NY)	Jul, 1998	CDC (2003)
Milk, raw	C. jejuni	38	Austria	Sep, 1998	Lehner et al. (2000)
Milk, raw	Campylobacter sp.	3	Sweden	1998	SMI (1999)
Milk, raw	Campylobacter sp.	5	Sweden	1998	SMI (1999)
Milk	C. jejuni	11	Germany	Apr, 1999	Reintjes et al. (1999)
Milk, raw	C. jejuni	2	USA (WA)	Jun, 1999	CDC (2003)
Milk, raw	Campylobacter sp.	2	USA (TX)	Feb, 2000	CDC (2003)
Milk, raw	C. jejuni	4	USA (ID)	Feb, 2000	CDC (2003)

Table 6.2 Continued.

Source	Species	Cases[a]	Location	Date	Reference
Milk, raw	C. jejuni	21	USA (OK)	Feb, 2000	CDC (2003)
Milk, raw	C. jejuni	19	USA (WI)	May, 2000	CDC (2003)
Dairy, malts/unpasteurized milk	Campylobacter sp.	8	USA (MN)	Jun, 2000	MDH (2000)
Milk, raw	Campylobacter sp.	8	USA (MN)	Jun, 2000	CDC (2003)
Milk, raw	C. jejuni	42	USA (ID)	Jun, 2000	CDC (2003)
Milk, raw	C. jejuni	11	USA (OK)	Jun, 2000	CDC (2003)
Milk, raw	C. jejuni	3	USA (PA)	Aug, 2000	CDC (2003)
Milk, raw	C. jejuni	39	USA (NY)	Sep, 2000	CDC (2003)
Milk, raw	C. jejuni	4	USA (MN)	Mar, 2000	MDH (2000)
Milk, raw	C. jejuni	75	USA (WI)	Nov, 2000	CDC (2002a)
Milk, raw	C. jejuni	13	USA (UT)	Jan, 2000	Peterson (2003)
Milk, raw	Campylobacter sp.	3	UK	Oct–Dec, 2002	CDR (2003)
Total		7425			
Miscellaneous items					
Misc., school lunch	Campylobacter sp.	800	Japan[j]	May, 1980	Yanagisawa (1980)
Misc., cake icing	Campylobacter sp.	41	USA (CT)	Jun, 1980	Blaser et al. (1982)
Misc., restaurant	Campylobacter sp.	3	USA (NY)	May, 1981	CDC (2003)
Misc., beef or egg	Campylobacter sp.	19	USA (NY)	Sep, 1981	Finch and Blake (1985)
Misc., fraternity food	Campylobacter sp.	10	USA (NY)	Nov, 1981	Finch and Blake (1985)
Misc., foodhandler	C. jejuni	6	Israel	Jun, 1982	Cohen et al. (1984)
Misc., Chinese food	Campylobacter sp.	2	USA (WA)	Oct, 1982	Finch and Blake (1985)
Misc., travel-associated	Campylobacter sp.	150	USA[k]	1984	Tauxe et al. (1987)
Misc., cabbage stew/meat	C. jejuni	105	Germany	Nov, 1985	Steffen et al. (1986)
Misc., tuna salad	C. jejuni	4	USA (NY)	1986	Sails et al. (2003b)
Misc., chicken/fruit	C. jejuni	10	USA (PA)	May, 1991	CDC (2003)
Misc., roast beef/gravy/dressing	C. jejuni (susp.)	52	USA (ND)	Mar, 1992	CDC (2003)
Misc., pasta salad/au gratin potatoes	C. jejuni	34	USA (IN)	Sep, 1992	CDC (2003)
Misc., chopped liver salad	Campylobacter sp.	>25	USA (NY)	Jun, 1993	Layton et al. (1997)
Misc., meat omelets	C. jejuni	4	USA (WA)	Jun, 1993	CDC (2003)
Misc., shellfish/poultry	Campylobacter sp.	54	UK	Oct, 1993	Pebody et al. (1997)

Food	Organism	No.	Location	Date	Reference
Misc., pate/'vol au vent'	*Campylobacter* sp.	12	UK	Dec, 1993	Pebody *et al.* (1997)
Misc., school lunch	*C. coli/C. jejuni*	Multi	Japan[l]	1993–1996	IASR (1999)
Misc., RTE meals	*C. jejuni*	NR	Croatia	1993–1998	WHO (1998)
Misc., ham/feta cheese	*C. coli*	24	Belgium	May, 1995	Ronveaux *et al.* (2000)
Misc., seafood alfredo/turkey	*Campylobacter* sp.	22	USA (NY)	Jun, 1995	CDC (2003)
Misc., refried beans (susp.)	*Campylobacter* sp.	76	USA (FL)	Sep, 1995	CDC (2003)
Misc., chicken/cole slaw/potato salad	*C. jejuni*	17	USA (FL)	Oct, 1995	CDC (2003)
Misc., dairy/garlic butter/bread	*C. jejuni*	30	USA (LA)	1995	Zhao *et al.* (2000); Sails *et al.* (2003b)
Misc., taco/nacho salad	*C. jejuni*	16	USA (WI)	May, 1998	CDC (2003)
Misc., buffet meal	*Campylobacter* sp.	9	UK	Apr–Jun, 1998	CDR (1998)
Misc., gravy/potato/pineapple	*C. jejuni*	128	USA (KS)	Aug, 1998	Olsen *et al.* (2001)
Misc., mayonnaise	*Campylobacter* sp.	4	UK	Oct–Dec, 1998	CDR (1999b)
Misc., restaurant buffet	*Campylobacter* sp.	11	UK	Oct–Dec, 1998	CDR (1999b)
Misc., hotel meal	*Campylobacter* sp.	21	UK	Oct–Dec, 1998	CDR (1999b)
Misc., tuna/egg/salami/pepperoni/pizza bites	*Campylobacter* sp.	24	UK	Oct–Dec, 1998	CDR (1999b)
Misc., baked macaroni	*Campylobacter* sp.	4	Malta	1998	WHO (1998)
Misc., pizza	*C. jejuni*	81	Slovenia	1998	WHO (1998)
Misc., paella	*Campylobacter* sp.	4	Sweden	1998	SMI (1999)
Misc., unspecif. ethnic	*Campylobacter* sp.	2	USA (FL)	Feb, 1999	CDC (2003)
Misc., club buffet	*Campylobacter* sp.	24	UK	Jul–Sep, 1999	CDR (2000a)
Misc., homemade	*C. jejuni*	18	USA (ML)	Apr, 2000	CDC (2003)
Misc., buffet lunch	*Campylobacter* sp.	19	UK	Jun, 2000	CDR (2001)
Misc., lunch	*C. jejuni*	>30	UK	Mar, 2001	UK-PHLS (2001)
Misc., duck liver	*Campylobacter* sp.	2	Australia	2001	Kirk (2003)
Misc., banquet dish	*Campylobacter* sp.	10	Australia	2001	Kirk (2003)
Misc., aged care setting	*Campylobacter* sp.	7	Australia[m]	2001	Kirk (2003)
Misc., fondue/meat	*C. jejuni*	NR	Canada	Mar, 2002	Anonymous (2002b)
Total		*1,914*			
Poultry					
Chicken	*C. jejuni*	27	Netherlands[n]	1979	Brouwer *et al.* (1979)
Chicken	*Campylobacter* sp.	9	USA (MN)	Sep, 1980	Finch and Blake (1985)

Table 6.2 Continued.

Source	Species	Cases[a]	Location	Date	Reference
Turkey, processed	C. jejuni	11	USA (CA)	Sep, 1980	Finch and Blake (1985); Shandera et al. (1992)
Chicken, barbecued	C. jejuni	11	USA (CO)	Apr, 1982	Istre et al. (1984)
Chicken	Campylobacter sp.	6	USA (GA)	May, 1982	Finch and Blake (1985)
Eggs	Campylobacter sp.	26	USA (MN)	Jul, 1982	Finch and Blake (1985)
Chicken casserole	C. jejuni	7	Australia	Jun, 1983	Rosenfield et al. (1985)
Poultry	Campylobacter sp.	21	UK	Jun, 1992	Pebody et al. (1997)
Chicken, stuffed	C. jejuni	2	USA (WA)	Jul, 1993	CDC (2003)
Chicken	Campylobacter sp.	4	USA (NY)	Sep, 1993	CDC (2003)
Chicken, roasted	C. jejuni	350	Malta	1993	WHO (1998)
Chicken	C. coli/C. jejuni	?	Japan[o]	1993–1998	IASR (1999)
Chicken	C. jejuni	2	USA (WA)	Mar, 1994	CDC (2003)
Turkey, farm	C. jejuni	7	Canada	Jul, 1994	Health Canada (1995)
Poultry	Campylobacter sp.	4	UK	Dec, 1994	Pebody et al. (1997)
Chicken	C. jejuni	19	UK	1994–1995	Pearson et al. (2000)
Chicken	Campylobacter sp.	6	USA (IL)	Apr, 1995	CDC (2003)
Chicken, barbecued (susp.)	C. jejuni	3	Netherlands[p]	Jul, 1995	Ang et al. (2000)
Turkey	Campylobacter sp.	23	USA (IL)	Nov, 1995	CDC (2003)
Chicken	Campylobacter sp.	Multi	Sweden[q]	1995–1997	Lindqvist et al. (2000)
Poultry	Campylobacter sp.	Multi	Belgium[r]	1995–1998	WHO (1998)
Chicken, stir-fried	C. jejuni	12	UK	Feb, 1997	Evans et al. (1998b)
Chicken sandwich	Campylobacter sp.	3	USA (FL)	Jan, 1998	CDC (2003)
Chicken	Campylobacter sp.	16	USA (FL)	Aug, 1998	CDC (2003)
Chicken	Campylobacter sp.	3	UK	Apr–Jun, 1998	CDR (1998)
Poultry	Campylobacter sp.	4	Sweden	1998	SMI (1999)
Chicken	C. jejuni	13	USA (CA)	Jan, 1999	CDC (2003)
Chicken, liver pâté	Campylobacter sp.	9	UK	Apr–Jun, 1999	CDR (1999a)
Chicken	Campylobacter sp.	20	UK	Apr–Jun, 1999	CDR (1999a)
Chicken	Campylobacter sp.	12	UK	Jul–Sep, 1999	CDR (2000a)
Chicken wings	C. jejuni	8	USA (FL)	Feb, 2002	Suarez et al. (2002)

Food	Organism	No.	Location	Date	Reference
Chicken kebabs	*C. jejuni*	3	Australia	Mar, 2001	Australia (CDN)
Poultry, barbecued	*C. jejuni*	5	Germany	Apr, 2001	Allerberger et al. (2003)
Poultry, duck liver	*Campylobacter* sp.	2	Australia	Jul, 2001	Australia (CDN)
Chicken, barbecued (susp.)	*Campylobacter* sp.	3	N. Ireland	Jul, 2001	CDMR-NI (2001)
Chicken, barbecued	*Campylobacter* sp.	93	Singapore[s]	Sep, 2001	ENB (2002)
Chicken, liver parfait	*Campylobacter* sp.	11	UK	Apr–Jun, 2002	CDR (2002a)
Chicken	*C. jejuni*	>24	Australia	Aug, 2002	OZNet (2002)
Chicken, casserole	*C. jejuni*	20	UK	Oct–Dec, 2002	CDR (2003)
Total		799			

Produce

Food	Organism	No.	Location	Date	Reference
Produce, melon/strawberries	*C. jejuni*	48	USA (MN)	Sep, 1993	CDC (2003)
Produce, salad veg.[t]/meat fajitas	*C. jejuni*	8	UK	Dec, 1993	Frost et al. (2002); Long et al. (2002); HPA-UK (2003)
Produce, salad	*C. jejuni*	10	USA (WA)	Jun, 1994	CDC (2003)
Produce, salad	*C. jejuni*	5	USA (WA)	Jul, 1994	CDC (2003)
Produce, fruit salad	*C. jejuni*	62	USA (MN)	Aug, 1994	CDC (2003)
Produce, lettuce	*C. jejuni*	5	USA (WA)	Jun, 1995	CDC (2003)
Produce, cucumber	*Campylobacter* sp.	78	Australia	Oct, 1995	Kirk et al. (1997)
Produce, lettuce	*C. jejuni*	37	France	Nov, 1995	Wight et al. (1997)
Produce, salad/vegetables	*C. jejuni*	NR	Wales	1995, 1996	Evans et al. (1998a)
Produce, lettuce/tomato	*C. jejuni*	16	UK	Apr, 1996	Frost et al. (2002); Long et al. (2002); HPA-UK (2003)
Produce, salad (susp.)	*C. jejuni*	70	USA (NY)	Jun, 1996	CDC (2003)
Produce, lettuce	*C. jejuni*	14	USA (OK)	Aug, 1996	CDC (1998)
Produce, sweet potatoes	*Campylobacter* sp.	17	USA (CT)	Aug, 1997	Winquist et al. (2001)
Produce, salad veg.[u]/curried meat/prawns	*C. jejuni*	61	UK	Apr, 1998	Frost et al. (2002); Long et al. (2002); HPA-UK (2003)
Produce, lettuce/mayonnaise/garlic	*C. jejuni*	30	UK	May, 1998	Frost et al. (2002); Long et al. (2002); HPA-UK (2003)
Produce, lettuce	*C. jejuni*	300	USA (MN)	Jun, 1998	CDC (2003)
Produce, salad	*C. jejuni*	152	USA (MN)	Jun, 1998	CDC (2003)
Produce, lettuce-based salad	*C. jejuni*	13	USA (CT)	May, 2000	CDC (2003)
Produce, lettuce	*C. jejuni*	18	UK	Jun, 2000	Frost et al. (2002); Long et al. (2002); HPA-UK (2003)

Table 6.2 Continued.

Source	Species	Cases[a]	Location	Date	Reference
Produce, orange juice/pasta salad	*C. jejuni*	30	UK	Mar, 2001	Frost *et al.* (2002); Long *et al.* (2002); HPA-UK (2003)
Produce, tomato/cucumber salad	*Campylobacter* sp.	50	Australia	Oct, 2001	Australia CDN)
Total		*1024*			
Seafood					
Seafood, shrimp/water	*Campylobacter* sp.	32	USA (MN)	Nov, 1993	CDC (2003)
Seafood, tuna salad	*C. jejuni*	82	USA (WI)	Jul, 1995	Roels *et al.* (1998)
Seafood, fish/shellfish	*Campylobacter* sp.	Multi	Belgium[v]	1995–1998	WHO (1998)
Seafood, shellfish	*Campylobacter* sp.	4	USA (FL)	Nov, 1997	CDC (2003)
Seafood, raw salmon/tuna	*Campylobacter* sp.	3	USA (FL)	Jan, 1998	CDC (2003)
Seafood, prawn and salmon vol-au-vent	*Campylobacter* sp.	48	UK	Apr–Jun, 1998	CDR (1998)
Seafood, raw oysters	*C. jejuni*	2	USA (WA)	Nov, 1998	CDC (2003)
Total		*171*			
Water					
Water, city	*C. jejuni*	~3000	USA (VT)	Jun, 1978	Vogt *et al.* (1982)
Water, community	*C. jejuni*	~2000	Sweden	Oct, 1980	Mentzing (1981)
Water, untreated surface	*C. jejuni*	NR	USA (WY)	1980–1981	Taylor *et al.* (1983)
Water, community	*C. jejuni*	8	Canada	1981	McMyne *et al.* (1982)
Water, reservoir	*C. jejuni*	150	Israel	Jul, 1982	Rogol *et al.* (1983)
Water, storage tank	*C. jejuni*	257	UK	May, 1981	Palmer *et al.* (1983)
Water, city	*C. jejuni*	865	USA (FL)	May, 1983	Sacks *et al.* (1986)
Water, park	*C. jejuni*	4	USA (NY)	1984	Tauxe *et al.* (1987)
Water, rural/residential	*C. jejuni*	9	USA (MN)	1984	Tauxe *et al.* (1987)
Water, community well/creek	*C. jejuni*	2	USA (OR)	1984	Tauxe *et al.* (1987)
Water, community spring	*C. jejuni*	6	USA (ID)	1984	Tauxe *et al.* (1987)
Water, lake water	*C. lari*	162	Canada	Mar, 1985	Broczyk *et al.* (1987)
Water, community/other	*Campylobacter* sp.	>1000	USA (Multi)[w]	1978–1986	Riley and Finch (1985); Tauxe *et al.* (1988)
Water, hospital	*C. jejuni*	94	Finland	Jun, 1986	Rautelin *et al.* (1990)

Source	Species	No.	Country	Date	Reference
Water, rural, untreated	*C. jejuni*	13	Canada	Jun, 1987	Alary and Nadeau (1990)
Water, town	*Campylobacter* sp.	19	New Zealand	Jun, 1987	Brieseman (1987)
Water, subarctic community	*Campylobacter* sp.	>9	Norway	Aug, 1988	Melby et al. (2000)
Water, untreated	*C. jejuni*	75	Finland	1989	Aho et al. (1989)
Water	*Campylobacter* sp.	44	New Zealand	Aug, 1990	CDC (1991); Stehr-Green et al. (1991)
Water, (meltwater)	*C. jejuni*	>241	Canada	Jun, 1991	Millson et al. (1991)
Water, city	*C. jejuni*	680	Norway	Jun, 1991	Melby et al. (1991)
Water, main supply to farm	*Campylobacter* sp.	42	UK	Jun–Jul, 1992	Gutteridge and Haworth (1994)
Water	*Campylobacter* sp.	8	UK	Apr, 1993	Pebody et al. (1997); Furtado et al. (1998)
Water	*Campylobacter* sp.	43	UK[x]	May, 1993	Duke et al. (1996); Pebody et al. (1997)
Water	*C. coli*	36	UK	Sep, 1993	Pebody et al. (1997); Furtado et al. (1998)
Water	*Campylobacter* sp.	2	Sweden	1993	WHO (1998)
Water	*C. jejuni*	52	UK	Apr, 1994	Pebody et al. (1997); Furtado et al. (1998)
Water	*Campylobacter* sp.	8	UK	Sept, 1994	Pebody et al. (1997); Furtado et al. (1998)
Water	*C. jejuni.*	22	Wales	Sep, 1994	Pebody et al. (1997); Furtado et al. (1998)
Water	*Campylobacter* sp.	NR	Cuba[y]	Jun, 1994	Roman (1998)
Water	*C. coli/C. jejuni*	Multi	Japan[z]	1994–1995	IASR (1999)
Water	*Campylobacter* sp.	1000	Norway[aa]	1994–1995	Varslot et al. (1996)
Water, town	*C. jejuni/ C. coli*	8	UK	Mar, 1995	Jones and Roworth (1996)
Water	*C. jejuni*	16	Switzerland	1995	Maurer and Sturchler (2000)
Water	*C. jejuni*	100	Switzerland	1998	Maurer and Sturchler (2000)
Water, town	*C. jejuni*	>110	Denmark	1995–1996	Engberg et al. (1998)
Water	*Campylobacter* sp.	2800	Denmark[bb]	1996	WHO (1998)
Water	*Campylobacter* sp.	2	Sweden	1996	WHO (1998)
Water	*C. jejuni*	23	Australia	Jun, 1997	Merritt et al. (1999)
Water, bottled (susp.)	*C. jejuni*	106	USA (MN)[cc]	Oct, 1997	DOD-GEIS (1998)
Water	*C. jejuni*	15	Finland	Aug, 1998	Kuusi et al. (1998)
Water, town	*C. jejuni*	28	Switzerland	Aug, 1998	Maurer and Sturchler (2000)

Table 6.2 Continued.

Source	Species	Cases[a]	Location	Date	Reference
Water	*Campylobacter* sp.	12	Finland	1998	WHO (1998)
Water	*Campylobacter* sp.	7	Sweden	1998	WHO (1998)
Water	*Campylobacter* sp.	NR	Iceland	1998	WHO (1998)
Water, private water supply	*Campylobacter* sp.	>2200	Finland	1998, 1999	Miettinen *et al.* (2001)
Water, fairgrounds	*C. jejuni*	44	USA (NY)	Aug, 1999	CDC (1999; Bopp *et al.* (2003)
Water	*Campylobacter* sp.	116	Canada	May, 2000	Health Canada (2000)
Water	*C. jejuni*	15	USA (ID)	Jun, 2000	CDC (2002c)
Water, groundwater	*C. jejuni*	400	Finland	Aug, 2000	Hänninen *et al.* (2003)
Water	*Campylobacter* sp.	281	Wales	Sep, 2000	UK-PHLS (2001)
Water, groundwater well	*C. jejuni*	5	Finland	Aug, 2001	Hänninen *et al.* (2003)
Water, groundwater well	*C. jejuni*	1000	Finland	Oct, 2001	Hänninen *et al.* (2003)
Water, suspected	*Campylobacter* sp.	12	Australia	2001	Kirk (2003)
Water	*C. jejuni*	43	Spain	2001	Godoy *et al.* (2002)
Water, spring	*Campylobacter* sp.	50	UK[dd]	Mar, 2002	CDR (2002b)
Water, stream	*Campylobacter* sp.	4	Wales	Jun, 2002	UK-PHLS (2002)
Water, farm water supply	*Campylobacter* sp.	4	Wales	Jun, 2002	UK-PHLS (2002)
Water, stream	*Campylobacter* sp.	30	UK	Jun, 2002	CDR (2002b)
Water, untreated, farm	*Campylobacter* sp.	30	UK	Jun, 2002	CDR (2002b)
Water, spring-fed supply	*Campylobacter* sp.	16	Wales[ee]	Aug, 2002	UK-PHLS (2002)
Total		*17328*			
Unknown					
Unknown	*C. jejuni*	Multi	Belgium[ff]	1978	Lauwers *et al.* (1978)
Unknown	*C. jejuni*	35	Japan[gg]	Jan, 1979	Itoh *et al.* (1980)
Unknown	*C. jejuni*	606	Israel (multi)[hh]	Jul, 1979, to 1982	Shmilovitz *et al.* (1982)
Unknown	*Campylobacter* sp.	339	USA (multi)[ii]	1978–1986	Riley and Finch (1985); Tauxe *et al.* (1988)
Unknown	*C. jejuni*	14	France	Feb, 1981	Hervé *et al.* (1984)
Unknown	*C. jejuni*	15	Canada (multi)[jj]	1982	McMyne *et al.* (1982)
Unknown	*C. jejuni*	7	Israel[kk]	Aug, 1983	Hershkowici *et al.* (1987)
Unknown	*Campylobacter* sp.	3	England[ll]	Mar/Apr, 1984	Casemore *et al.* (1986)

Unknown	*C. jejuni*	4	France[mm]	1983–1985	Mégraud *et al.* (1987)
Unknown	*C. jejuni*	11	Belgium[nn]	1985	Goossens *et al.* (1986)
Unknown	*C. jejuni*	3	USA (ME)	1989	Sails *et al.* (2003b)
Unknown	*C. jejuni*	17	USA (VA)	Sep, 1990	CDC (2003)
Unknown	*C. upsaliensis*	19	Belgium	1990, 1991	Lauwers *et al.* (1991)
Unknown	*C. jejuni*	50	USA (PA)	Jul, 1991	CDC (2003)
Unknown	*C. jejuni*	7	USA (IL)	Oct, 1991	CDC (2003)
Unknown	*C. jejuni*	3	USA (IL)	Dec, 1991	CDC (2003)
Unknown	*C. jejuni*	7	USA (NY)	May, 1992	CDC (2003)
Unknown	*Campylobacter* sp.	18	UK	Oct, 1992	Pebody *et al.* (1997)
Unknown	*C. fetus*	18	Canada[oo]	Oct, 1992	Rennie *et al.* (1994)
Unknown	*C. jejuni*	21	USA (NY)	Aug, 1993	CDC (2003)
Unknown	*Campylobacter* sp.	3	USA (IL)	Oct, 1993	CDC (2003)
Unknown	*Campylobacter* sp.	2	USA (WA)	Nov, 1993	CDC (2003)
Unknown	*C. coli/jejuni*	Multi	Japan[pp]	1993–1998	IASR (1999)
Unknown	*C. jejuni*	Multi	Hungary[qq]	1993–1998	WHO (1998)
Unknown	*Campylobacter* sp.	Multi	Netherlands[rr]	1993–1998	WHO (1998)
Unknown	*Campylobacter* sp.	Multi	Norway[ss]	1993–1998	WHO (1998)
Unknown	*C. jejuni*	Multi	Spain[tt]	1993–1998	WHO (1998)
Unknown	*Campylobacter* sp.	Multi	Sweden[uu]	1993–1998	Gunnarsson and Svedhem (1998); WHO (1998)
Unknown	*C. jejuni*	Multi	Switzerland[vv]	1993–1998	WHO (1998)
Unknown	*Campylobacter* sp.	Multi	UK[ww]	1993–1998	WHO (1998)
Unknown	*Campylobacter* sp.	113	Wales	May, 1994	Pebody *et al.* (1997)
Unknown	*C. jejuni*	2	USA (MN)	Jun, 1994	CDC (2003)
Unknown	*Campylobacter* sp.	34	UK	Jun, 1994	Pebody *et al.* (1997)
Unknown	*C. jejuni*	13	USA (NY)	Jun, 1994	CDC (2003)
Unknown	*Campylobacter* sp.	6	USA (WA)	Jul, 1994	CDC (2003)
Unknown	*Campylobacter* sp.	12	UK	Jul, 1994	Pebody *et al.* (1997)
Unknown	*Campylobacter* sp.	10	UK	Aug, 1994	Pebody *et al.* (1997)
Unknown	*C. jejuni*	4	USA (MI)	Nov, 1994	CDC (2003)
Unknown	*C. jejuni*	5	USA (WA)	Jun, 1995	CDC (2003)
Unknown	*Campylobacter* sp.	3	USA (NY)	Oct, 1995	CDC (2003)
Unknown	*Campylobacter* sp.	140	Canada (multi)[xx]	1995	CCDR (1998)
Unknown	*Campylobacter* sp.	NR	Belgium[yy]	1995–1998	WHO (1998)

Table 6.2 Continued.

Source	Species	Cases[a]	Location	Date	Reference
Unknown	C. jejuni	40	USA (CA)	Mar, 1996	CDC (2003)
Unknown	Campylobacter sp.	2	USA (FL)	Jun, 1996	CDC (2003)
Unknown	C. jejuni	9	Scotland	Jun, 1996	Harrington et al. (1999)
Unknown	C. jejuni	16	USA (CA)	Nov, 1996	CDC (2003)
Unknown	Campylobacter sp.	33	Canada (multi)[zz]	1996	CCDR (2003)
Unknown	C. jejuni	2	Malta	1996	WHO (1998)
Unknown	Campylobacter sp.	?	Scotland[aaa]	1996–1998	WHO (1998)
Unknown	C. jejuni/C. coli	52	UK (multi)[bbb]	1996–1999	Ribeiro and Frost (2000)
Unknown	C. jejuni	87	USA (CA)	Mar, 1997	CDC (2003)
Unknown	C. jejuni	100	Japan	Nov, 1997	Saito et al. (2002)
Unknown	C. jejuni	32	Japan	Nov, 1997	Saito et al. (2002)
Unknown	C. jejuni	12	UK	Nov, 1997	Gent et al. (1999)
Unknown	Campylobacter sp.	9	Canada (multi)[ccc]	1997	CCDR (2003)
Unknown	C. jejuni	2	Malta	1997	WHO (1998)
Unknown	Campylobacter sp.	3	Malta	1997	WHO (1998)
Unknown	C. jejuni	8	Malta	1997	WHO (1998)
Unknown	Campylobacter sp.	9	USA (NC)	Feb, 1998	CDC (2003)
Unknown	Campylobacter sp.	4	USA (FL)	Mar, 1998	CDC (2003)
Unknown	C. jejuni	6	USA (KS)	Jun, 1998	CDC (2003)
Unknown	Campylobacter sp.	22	UK	Apr–Jun, 1998	CDR (1998)
Unknown	C. jejuni	4	USA (MN)	Aug, 1998	CDC (2003)
Unknown	C. jejuni	2	USA (MN)	Nov, 1998	CDC (2003)
Unknown	Campylobacter sp.	12	UK	Oct–Dec, 1998	CDR (1999b)
Unknown	Campylobacter sp.	18	Canada (multi)[ddd]	1998	CCDR (2003)
Unknown	Campylobacter sp.	6	Sweden	1998	SMI (1999)
Unknown	Campylobacter sp.	5	Sweden	1998	SMI (1999)
Unknown	Campylobacter sp.	3	USA (NY)	Apr, 1999	CDC (2003)
Unknown	Campylobacter sp.	16	UK	Jul–Sep, 1999	CDR (2000a)
Unknown	Campylobacter sp.	25	UK	Jul–Sep, 1999	CDR (2000a)
Unknown	C. jejuni	11	USA (HI)	Nov, 1999	CDC (2003)
Unknown	Campylobacter sp.	12	Canada (multi)[eee]	1999	CCDR (2003)

Unknown	*C. jejuni*	2	USA (CA)	May, 2000	CDC (2003)
Unknown	*Campylobacter* sp.	13	USA (OK)	Jun, 2000	CDC (2003)
Unknown	*Campylobacter* sp.	6	USA (PA)	Jun, 2000	CDC (2003)
Unknown	*Campylobacter* sp.	19	UK	Jun, 2000	CDR (2000b)
Unknown	*C. jejuni*	4	USA (HI)	Aug, 2000	CDC (2003)
Unknown	*Campylobacter* sp.	5	UK	Jun–Sep, 2000	CDR (2000b)
Unknown	*C. jejuni*	2	Spain[fff]	Dec, 2000	Llovo *et al.* (2003)
Unknown	*C. jejuni*	10	Australia[ggg]	May, 2001	Australia CDN (2002); Raupach and Hundy (2003)
Unknown	*Campylobacter* sp.	2	Australia	Nov, 2001	Australia CDN)
Unknown	*Campylobacter* sp.	49	Australia	Nov, 2001	Australia CDN)
Unknown	*C. jejuni*	14	Norway	2001	Stamnes and Grude (2002)
Unknown	*C. jejuni*	65	USA (WA)	Mar, 2002	Associated Press (2002)
Unknown	*C. jejuni*	38	USA (WA)	Jul, 2002	Associated Press (2002)
Unknown	*Campylobacter* sp.	NR	Australia	Nov, 2002	Anonymous (2002c)
Unknown	*C. jejuni*	3	UK	Oct–Dec, 2002	CDR (2003)
Unknown	*C. jejuni*	NR	Israel	2002	Srour *et al.* (2002)
Unknown	*Campylobacter* sp.	12	USA (MN)[hhh]	2002	MDH (2003)
Unknown	*Campylobacter* sp.	>40	USA (multi)[iii]	Jun, 2003	BCPH (2003)
Total		2390			

a   In some outbreaks, more than one microorganism was isolated from patient stool samples. The number of cases listed for some outbreaks represents the number of culture-confirmed cases; however, not all cases were confirmed by culture in some of the large outbreaks. NR, not reported; multi, multiple outbreaks; ML, multistate. Entries designated 'multi' in the 'Location' field represent multiple outbreaks with other details noted in the footnote (see below); these outbreaks are represented in the total outbreaks shown in Figure 6.2.

b   Represents 13 outbreaks.

c   The total number of cases was calculated from only those outbreaks with case numbers reported. The number of cases were not reported for some outbreaks (e.g. 'NR' and 'multi'); thus, the total number of cases listed probably is an underestimate of the total number of cases that occurred.

d   Strains of two different serotypes identified.

e   Four stool samples yielded *C. jejuni* and three *C. fetus*.

f   Represents approximately seven outbreaks between 1978 and 1986 and not reported in other references.

g   Represents six outbreaks.

h   Represents two outbreaks.

i   Represents two outbreaks.

j   School lunch consisted of bread, pasteurized cow's milk, oranges, and vinegared pork with vegetables; the contaminated item was not identified.

Notes to Table 6.2 continued.

k   Travellers returning from Jamaica; the source of the outbreak was unclear.

l   Represents seven total outbreaks: two in 1993, two in 1994, one in 1995 and two in 1996.

m   Suspected person-to-person spread.

n   Eighty-nine military cadets became ill during this outbreak; *C. jejuni* was isolated from 27 stool samples. The exact date of the outbreak (1979 or before) was not reported.

o   Represents 39 total outbreaks: three in 1993, two in 1994, three in 1995, 10 in 1996, eight in 1997 and 13 in 1998. The number of cases associated with these outbreaks could not be determined from the information available.

p   A 10-year-old boy developed Guillain–Bárre syndrome.

q   Represents 11 outbreaks.

r   Represents two outbreaks.

s   *C. coli* isolated from same batch of chicken.

t   Salad composed of vegetables plus meat; the original contaminated food item was not determined.

u   Salad composed of vegetables plus meat or seafood; the original contaminated food item was not determined.

v   Represents two outbreaks.

w   Represents five outbreaks and >1000 cases between 1978 and 1986 and not reported in other references.

x   *Cryptosporidium* species also associated with outbreak.

y   Reported to have caused an epidemic of Guillain–Bárre syndrome.

z   Represents three total outbreaks: two in 1994 and one in 1995; number of cases not reported.

aa Represents two outbreaks; suspected contamination by pink-footed geese.

bb This outbreak is one of 18 general outbreaks due to *Campylobacter* species between 1996 and 1998: water (2), unpasteurized milk (1), poultry (2), cake (1) and unknown sources (12).

cc Minnesota Army National Guard troops returning from Greece; bottled water suspected.

dd *Cryptosporidium* also present.

ee Mixed infections with *E. coli* O157:H7.

ff Represents multiple outbreaks in infants in '5-day-nurseries'; 20–50% of the infants were reported to be infected and had diarrhoea. The exact date of the outbreaks (1978 or before) was not reported.

gg Suspected cause was a communal lunch or snack.

hh Represents 91 institutional and 36 family outbreaks; no source was identified except for two cases associated with a co-infected dog.

ii Represents nine outbreaks.

jj Represents seven family outbreaks and 15 total cases occurring in 1982 or before.

kk Outbreak occurred in a neonatal intensive care unit; person-to-person spread suspected.

ll *Cryptosporidium* was identified initially, then *Campylobacter* sp. also isolated from some samples.

mm Represents two separate family outbreaks with two affected family members for each.

nn Seven female and four male infants had meningitis.

oo Members of this Hutterite colony consumed regularly raw milk and dairy products produced from raw milk.

pp	Represents 130 total outbreaks: 14 in 1993, 25 in 1994, 22 in 1995, 32 in 1996, 39 in 1997 and 37 in 1998.
qq	Represents 39 outbreaks and 295 total cases.
rr	Represents eight outbreaks and 29 total cases.
ss	Represents eight outbreaks.
tt	Represents nine outbreaks.
uu	Represents 34 outbreaks and at least 360 total cases.
vv	Represents 9 outbreaks.
ww	Represents 36 outbreaks in England and Wales; some of these outbreaks are noted separately in this table with known sources.
xx	Represents 39 outbreaks, 27 from Ont
yy	Represents 2 outbreaks.
zz	Represents 16 outbreaks; cases noted represent laboratory confirmed cases.
aaa	Represents 10 outbreaks.
bbb	Represents 23 family clusters of infection; 10 family samples had multiple strains of *C. jejuni*. It was not possible to determine the actual sources of infection.
ccc	Represents four outbreaks; cases noted represent laboratory confirmed cases.
ddd	Represents four outbreaks; cases noted represent laboratory confirmed cases.
eee	Represents six outbreaks; cases noted represent laboratory confirmed cases.
fff	Occurred in neonatal ward.
ggg	Chicken or duck possible food sources.
hhh	Represents four outbreaks: one associated with a restaurant and three with farm animals.
iii	Cases occurred in participants from WV, VA, NY, KY and MD as part of a bike tour.

Acknowledgements

The authors would like to thank Martyn Kirk, Sally Meakins and David Woodward for providing information regarding *Campylobacter* outbreaks, Guilin Wang for data compilation, and Subbarao Ravva and Craig Parker for critical reading of the manuscript. This work was supported by the United States Department of Agriculture, Agricultural Research Service CRIS project 5325-42000-041, and supports a US collaboration in the European Commission Fifth Framework Project QLK1-CT-2002-0220, 'CAMPYCHECK'.

References

Abbaszadegan, M., Huber, M.S., Gerba, C.P. and Pepper, I.L. 1993. Detection of enteroviruses in groundwater with the polymerase chain reaction. Appl. Environ. Microbiol. 59: 1318–1324.

Abrahams, C.A., Agbodaze, D., Nakano, T., Afari, E.A. and Longmatey, H.E. 1990. Prevalence and antibiogram of *Campylobacter jejuni* in domestic animals in rural Ghana. Arch. Environ. Health 45: 59–62.

Acuff, G.R., Vanderzant, C., Hanna, M.O., Ehlers, J.G. and Gardner, F.A. 1986. Effects of handling and preparation of turkey products on the survival of *Campylobacter jejuni*. J. Food Protect. 49: 627–631.

Adesiyun, A.A. 1995. Bacteriologic quality of some Trinidadian ready-to-consume foods and drinks and possible health risks to consumers. J Food Protect. 58: 651–655.

Aho, M. and Hirn, J. 1988. Prevalence of campylobacteria in the Finnish broiler chicken chain from the producer to the consumer. Acta Vet. Scand. 29: 451–462.

Aho, M., Kurki, M., Rautelin, H. and Kosunen, T.U. 1989. Waterborne outbreak of *Campylobacter* enteritis after outdoors infantry drill in Utti, Finland. Epidemiol. Infect. 103: 133–141.

Alary, M. and Nadeau, D. 1990. An outbreak of *Campylobacter* enteritis associated with a community water supply. Can. J. Publ. Health 81: 268–271.

Allerberger, F., Al-Jazrawi, N., Kreidl, P., Dierich, M.P., Feierl, G., Hein, I. and Wagner, M. 2003. Barbecued chicken causing a multi-state outbreak of *Campylobacter jejuni* enteritis. Infection 31: 19–23.

Allmann, M., Hofelein, C., Koppel, E., Luthy, J., Meyer, R., Niederhauser, C., Wegmuller, B. and Candrian, U. 1995. Polymerase chain reaction (PCR) for detection of pathogenic microorganisms in bacteriological monitoring of dairy products. Res. Microbiol. 146: 85–97.

Altekruse, S.F., Stern, N.J., Fields, P.I. and Swerdlow, D.L. 1999. *Campylobacter jejuni* – an emerging food-borne pathogen. Emerging Infect. Dis. 5: 28–35.

Ang, C.W., van Doorn, P.A., Endtz, H.P., Merkies, I.S., Jacobs, B.C., de Klerk, M.A., van Koningsveld, R. and van der Meche, F.G. 2000. A case of Guillain–Barré syndrome following a family outbreak of *Campylobacter jejuni* enteritis. J. Neuroimmunol. 111: 229–233.

Anonymous. 2002a. Annual Report on Zoonoses in Denmark 2001, Ministry of Food, Agriculture and Fisheries. Danish Zoonosis Centre, Copenhagen, Denmark. http: //www.vetinst.dk/high.asp?page_id = 138.

Anonymous. 2002b. Communicable Disease and Epidemiology News: Campylobacteriosis Outbreak Associated with a Fondue Dinner. Public Health-Seattle and King County, Seattle, WA. http: //www.metrokc.gov/health/phnr/prot_res/epilog/vol4204.htm.

Anonymous. 2002c. Investigation into a rise in *Campylobacter* notifications. Queensland Health, South Coast Public Health Unit, Queensland, Australia. http: //www.health.qld.gov.au/phs/Documents/sphun/17258.pdf.

Aquino, M.H.C., Pacheco, A.P.G., Ferreira, M.C.S. and Tibana, A. 2002. Frequency of isolation and identification of thermophilic campylobacters from animals in Brazil. Vet. J. 164: 159–161.

Arumugaswamy, R.K. and Prouford, R.W. 1987. The occurrence of *Campylobacter jejuni* and *Campylobacter coli* in Sydney Rock Oyster. Int. J. Food Microbiol. 4: 101–104.

Associated Press. 2002. Bacteria again sicken scores at state prison; food suspected, Post-Intelligencer, Seattle. http: //seattlepi.nwsource.com/local/79180_prison19.shtml.

Atabay, H.I. and Corry, J.E. 1998. The isolation and prevalence of campylobacters from dairy cattle using a variety of methods. J Appl Microbiol. 84: 733–740.

Atanassova, V. and Ring, C. 1999. Prevalence of *Campylobacter* spp. in poultry and poultry meat in Germany. Int J Food Microbiol. 51: 187–190.

Australia CDN. 2002. Enhancing food-borne disease surveillance across Australia in 2001: The OzFoodNet Working Group. Australia Communicable Diseases Network. http: //www.health.gov.au/pubhlth/cdi/cdi2603/html/cdi2603d3.htm.

Aydin, F., Atabay, H.I. and Akan, M. 2001. The isolation and characterization of *Campylobacter jejuni* subsp. *jejuni* from domestic geese (*Anser anser*). J. Appl. Microbiol. 90: 637–642.

Bacon, D.J., Alm, R.A., Burr, D.H., Hu, L., Kopecko, D.J., Ewing, C.P., Trust, T.J. and Guerry, P. 2000. Involvement of a plasmid in virulence of *Campylobacter jejuni* 81-176. Infect. Immun. 68: 4384–4390.

Bailey, G.D., Vanselow, B.A., Hornitzky, M.A., Hum, S.I., Eamens, G.J., Gill, P.A., Walker, K.H. and Cronin, J.P. 2003. A study of the food-borne pathogens: *Campylobacter*, *Listeria* and *Yersinia*, in faeces from slaughter-age cattle and sheep in Australia. Commun. Dis. Intell. 27: 249–257.

Bang, D.D., Wedderkopp, A., Pedersen, K. and Madsen, M. 2002. Rapid PCR using nested primers of the 16S rRNA and the hippuricase (*hipO*) genes to detect *Campylobacter jejuni* and *Campylobacter coli* in environmental samples. Mol. Cell. Probes. 16: 359–369.

BCPH. 2003. Boulder County Public Health Communicable Disease Control Program. Multistate *Campylobacter* outbreak. http: //www.co.boulder.co.us/health/hpe/pdfs/july 2003.pdf.

Beach, J.C., Murano, E.A. and Acuff, G.R. 2002. Prevalence of *Salmonella* and *Campylobacter* in beef cattle from transport to slaughter. J Food Prot. 65: 1687–1693.

Berndtson, E., Emanuelson, U., Engvall, A. and Danielsson-Tham, M.-L. 1996. A 1-year epidemiological study of campylobacters in 18 Swedish chicken farms. Prevent. Vet. Med. 26: 167–185.

Beumer, R.R., de Vries, J. and Rombouts, F.M. 1992. *Campylobacter jejuni* non-culturable coccoid cells. Int J Food Microbiol. 15: 153–163.

Birkhead, G., Vogt, R.L., Heun, E., Evelti, C.M. and Patton, C.M. 1988. A multiple-strain outbreak of *Campylobacter* enteritis due to consumption of inadequately pasteurized milk. J. Infect. Dis. 157: 1095–1097.

Black, R.E., Levine, M.M., Clements, M.L., Hughes, T.P. and Blaser, M.J. 1988. Experimental *Campylobacter jejuni* infection in humans. J. Infect. Dis. 157: 472–479.

Blaser, M.J., Checko, P., Bopp, C., Bruce, A. and Hughes, J.M. 1982. *Campylobacter* enteritis associated with food-borne transmission. Am. J. Epidemiol. 116: 886–894.

Blaser, M.J., Hardesty, H.L., Powers, B. and Wang, W.L. 1980. Survival of *Campylobacter fetus* subsp. *jejuni* in biological milieus. J. Clin. Microbiol. 11: 309–313.

Blaser, M.J., Hopkins, J.A. and Vasil, M.L. 1984. *Campylobacter* enteritis. N. Engl. J. Med. 305: 1444–1452.

Blaser, M.J., Sazie, E. and Williams, L.P., Jr. 1987. The influence of immunity on raw milk – associated *Campylobacter* infection. JAMA 257: 43–46.

Bolton, F.J., Coates, D., Hutchinson, D.N. and Godfree, A.F. 1987. A study of thermophilic campylobacters in a river system. J. Appl. Bacteriol. 62: 167–176.

Bolton, F.J., Dawkins, H.C. and Hutchinson, D.N. 1985. Biotypes and serotypes of thermophilic campylobacters isolated from cattle, sheep and pig offal and other red meats. J. Hyg. 95: 1–6.

Bolton, F.J., Hinchliffe, P.M., Coates, D. and Robertson, L. 1982. A most probable number method for estimating small numbers of campylobacters in water. J. Hyg. 89: 185–190.

Bolton, F.J., Sails, A.D., Fox, A.J., Wareing, D.R. and Greenway, D.L. 2002. Detection of *Campylobacter jejuni* and *Campylobacter coli* in foods by enrichment culture and polymerase chain reaction enzyme-linked immunosorbent assay. J. Food Prot. 65: 760–767.

Bopp, D.J., Sauders, B.D., Waring, A.L., Ackelsberg, J., Dumas, N., Braun-Howland, E., Dziewulski, D., Wallace, B.J., Kelly, M., Halse, T., Musser, K.A., Smith, P.F., Morse, D.L. and Limberger, R.J. 2003. Detection, isolation, and molecular subtyping of *Escherichia coli* O157: H7 and *Campylobacter jejuni* associated with a large waterborne outbreak. J. Clin. Microbiol. 41: 174–180.

Brandl, M.T., Haxo, A.F., Bates, A.H. and Mandrell, R.E. 2004. A comparison of the survival of *Campylobacter jejuni* in the phyllosphere and rhizosphere of spinach and radish plants. Appl. Environ. Microbiol. 70: 1182–1189.

Brandl, M.T. and Mandrell, R.E. 2002. Fitness of *Salmonella enterica* serovar Thompson in the cilantro phyllosphere. Appl. Environ. Microbiol. 68: 3614–3621.

Brennhovd, O., Kapperud, G. and Langeland, G. 1992. Survey of thermotolerant *Campylobacter* spp. and *Yersinia* spp. in three surface water sources in Norway. Int J Food Microbiol. 15: 327–338.

Brieseman, M.A. 1987. Town water supply as the cause of an outbreak of *Campylobacter* infection. NZ Med. J. 100: 212–213.

Broczyk, A., Thompson, S., Smith, D. and Lior, H. 1987. Water-borne outbreak of *Campylobacter laridis*-associated gastroenteritis. Lancet 1: 164–165.

Brouwer, R., Mertens, M.J., Siem, T.H. and Katchaki, J. 1979. An explosive outbreak of *Campylobacter* enteritis in soldiers. Antonie Van Leeuwenhoek 45: 517–519.

Brown, M.R. and Barker, J. 1999. Unexplored reservoirs of pathogenic bacteria: protozoa and biofilms. Trends Microbiol. 7: 46–50.

Busato, A., Hofer, D., Lentze, T., Gaillard, C. and Burnens, A. 1999. Prevalence and infection risks of zoonotic enteropathogenic bacteria in Swiss cow-calf farms. Vet. Microbiol. 69: 251–263.

Buswell, C.M., Herlihy, Y.M., Lawrence, L.M., McGuiggan, J.T., Marsh, P.D., Keevil, C.W. and Leach, S.A. 1998. Extended survival and persistence of *Campylobacter* spp. in water and aquatic biofilms and their detection by immunofluorescent-antibody and -rRNA staining. Appl. Environ. Microbiol. 64: 733–741.

Cabrita, J., Rodrigues, J., Braganca, F., Morgado, C., Pires, I. and Goncalves, A.P. 1992. Prevalence, biotypes, plasmid profile and antimicrobial resistance of *Campylobacter* isolated from wild and domestic animals from northeast Portugal. J. Appl. Bacteriol. 73: 279–285.

CAMPYCHECK. 2003. Improved physiological, immunological and molecular tools for the recovery and identification of emerging *Campylobacteraceae* (CAMPYCHECK): A European Commission Research Project (QLK1 CT 2002 02201). http: //www.campycheck.org/.

Candlish, A.A.G., Pearson, S.M., Aidoo, K.E., Smith, J.E., Kelly, B. and Irvine, H. 2001. A survey of ethnic foods for microbial quality and aflatoxin content. Food Addit. Contam. 18: 129–136.

Cappelier, J.M., Minet, J., Magras, C., Colwell, R.R. and Federighi, M. 1999. Recovery in embryonated eggs of viable but nonculturable *Campylobacter jejuni* cells and maintenance of ability to adhere to HeLa cells after resuscitation. Appl. Environ. Microbiol. 65: 5154–5157.

Casemore, D.P., Jessop, E.G., Douce, D. and Jackson, F.B. 1986. *Cryptosporidium* plus *Campylobacter*: an outbreak in a semi-rural population. J. Hyg. 96: 95–105.

CCDR. 1998. Canadian integrated surveillance report for 1995 on *Salmonella*, *Campylobacter*, pathogenic *E. coli*. Canada Communicable Disease Report 24S5.

CCDR. 2003. *Salmonella*, *Campylobacter*, pathogenic *E. coli* and *Shigella*, from 1996 to 1999. Canada Communicable Disease Report 29S1: 1–32.

CDC. 1983. Campylobacteriosis associated with raw milk consumption – Pennsylvania. MMWR Morb. Mortal Wkly Rep. 32: 337–338, 344.

CDC. 1984. *Campylobacter* outbreak associated with certified raw milk products – California. MMWR Morb. Mortal Wkly Rep. 33: 562.

CDC. 1986. *Campylobacter* outbreak associated with raw milk provided on a dairy tour – California. MMWR Morb Mortal Wkly Rep. 35: 311–312.

CDC. 1991. *Campylobacter* enteritis – New Zealand, 1990. MMWR Morb. Mortal Wkly Rep. 40: 116–117, 123.

CDC. 1998. Outbreak of *Campylobacter* enteritis associated with cross-contamination of food – Oklahoma, 1996. MMWR Morb. Mortal Wkly Rep. 47: 129–131.

CDC. 1999. Outbreak of *Escherichia coli* O157: H7 and *Campylobacter* among attendees of the Washington County Fair, New York, 1999. MMWR Morb Mortal Wkly Rep. 48: 803–805.

CDC. 2002a. Outbreak of *Campylobacter jejuni* infections associated with drinking unpasteurized milk procured through a cow-leasing program – Wisconsin, 2001. MMWR Morb. Mortal Wkly Rep. 51: 548–549.

CDC. 2002b. Preliminary FoodNet data on the incidence of food-borne illnesses – selected sites, United States, 2001. MMWR Morb Mortal Wkly Rep. 51: 325–329.

CDC. 2002c. Surveillance for waterborne disease outbreaks-United States, 1999–2000. MMWR Morb Mortal Wkly Rep. 51: 9.

CDC. 2003. Centers for Disease Control and Prevention: U.S. Food-borne Disease Outbreaks, Annual Listing, 1990–2000. http: //www.cdc.gov/food-borneoutbreaks/us_outb.htm.

CDMR-NI. 2001. Family outbreak of *Campylobacter*. Northern Ireland Communicable Diseases Monthly Report. 10: 5.

CDR. 1998. General outbreaks of food-borne illness in humans, England and Wales: quarterly report (April to June 1998). PHLS Communicable Disease Surveillance Centre.

CDR. 1999a. General outbreaks of food-borne illness in humans, England and Wales: quarterly report (April to June 1999). PHLS Communicable Disease Surveillance Centre.

CDR. 1999b. General outbreaks of food-borne illness in humans, England and Wales: quarterly report (October to December 1998). PHLS Communicable Disease Surveillance Centre.

CDR. 2000a. General outbreaks of food-borne illness in humans, England and Wales: quarterly report (July to September 1999). PHLS Communicable Disease Surveillance Centre.

CDR. 2000b. General outbreaks of food-borne illness in humans, England and Wales: quarterly report (weeks 32–35/2000). PHLS Communicable Disease Surveillance Centre.

CDR. 2001. General outbreaks of food-borne illness in humans, England and Wales: quarterly report (July to September 2000). PHLS Communicable Disease Surveillance Centre.

CDR. 2002a. General outbreaks of food-borne illness in humans, England and Wales: quarterly report (April to June 2002). PHLS Communicable Disease Surveillance Centre.

CDR. 2002b. Outbreaks and incidents of association between human disease and water in England and Wales: 2002. PHLS Communicable Disease Surveillance Centre. http: //www.hpa.org.uk/cdr/pages/enteric. htm#water.

CDR. 2003. General outbreaks of food-borne illness in humans, England and Wales: quarterly report (October to December 2002). PHLS Communicable Disease Surveillance Centre.

Charkowski, A.O., Barak, J.D., Sarreal, C.Z. and Mandrell, R.E. 2002. Differences in growth of *Salmonella enterica* and *Escherichia coli* O157: H7 on alfalfa sprouts. Appl. Environ. Microbiol. 68: 3114–3120.

Chuma, T., Hashimoto, S. and Okamoto, K. 2000. Detection of thermophilic *Campylobacter* from sparrows by mulitplex PCR: the role of sparrows as a source of contamination of broilers with *Campylobacter*. J. Vet. Med. Sci. 62: 1291–1295.

Clark, C.G., Price, L., Ahmed, R., Woodward, D.L., Melito, P.L., Rodgers, F.G., Jamieson, F., Ciebin, B., Li, A. and Ellis, A. 2003. Characterization of waterborne outbreak-associated *Campylobacter jejuni*, Walkerton, Ontario. Emerging Infect. Dis. 9: 1232–1241.

Cohen, D.I., Rouach, T.M. and Rogol, M. 1984. A *Campylobacter jejuni* enteritis outbreak in a military base in Isr. Israel J. Med. Sci. 20: 217–218.

Cooper, I.A. and Slee, K.J. 1971. Human infection by *Vibrio fetus*. Med. J. Aust. 1: 1263–1267.

Costerton, J.W., Stewart, P.S. and Greenberg, E.P. 1999. Bacterial biofilms: a common cause of persistent infections. Science 284: 1318–1322.

Cuk, Z., Annan-Prah, A., Janc, M. and Zajc-Satler, J. 1987. Yoghurt: an unlikely source of *Campylobacter jejuni/coli*. J. Appl. Bacteriol. 63: 201–205.

Daczkowska-Kozon, E. and Brzostek-Nowakowska, J. 2001. *Campylobacter* spp. in waters of three main Western Pomerania water bodies. Int. J. Hyg. Environ. Health. 203: 435–443.

de Boer, E. and Hahne, M. 1990. Cross-contamination with *Campylobacter jejuni* and *Salmonella* spp. from raw chicken products during food preparation. J. Food Protect. 53: 1067–1068.

de Boer, P., Wagenaar, J.A., Achterberg, R.P., Putten, J.P., Schouls, L.M. and Duim, B. 2002. Generation of *Campylobacter jejuni* genetic diversity *in vivo*. Mol. Microbiol. 44: 351–359.

Dekeyser, P., Gossuin-Detrain, M., Butzler, J.P. and Sternon, J. 1972. Acute enteritis due to related vibrio: first positive stool cultures. J. Infect. Dis. 125: 390–392.

Dickins, M.A., Franklin, S., Stefanova, R., Schutze, G.E., Eisenach, K.D., Wesley, I. and Cave, M.D. 2002. Diversity of *Campylobacter* isolates from retail poultry carcasses and from humans as demonstrated by pulsed-field gel electrophoresis. J. Food Protect. 655: 957–962.

Dingle, K.E., Colles, F.M., Wareing, D.R.A., Ure, R., Fox, A.J., Bolton, F.E., Bootsma, H.J., Willems, R.J.L., Urwin, R. and Maiden, M.C.J. 2001. Multilocus sequence typing system for *Campylobacter jejuni*. J. Clin. Microbiol. 39: 14–23.

Djuretic, T., Wall, P.G. and Nichols, G. 1997. General outbreaks of infectious intestinal disease associated with milk and dairy products in England and Wales: 1992 to 1996. Commun Dis Rep CDR Rev. 7: R41–45.

Docherty, L., Adams, M.R., Patel, P. and McFadden, J. 1996. The magnetic immuno-polymerase chain reaction assay for the detection of *Campylobacter* in milk and poultry. Lett. Appl. Microbiol. 22: 288–292.

DOD-GEIS. 1998. DOD-Global Emerging Infectious Disease System. Diarrhoea outbreaks must be taken seriously in the global village. http: //www.geis.ha.osd.mil/GEIS/aboutGEIS/historicaldocs/1998%20-%205Year (StrategyG2c).asp.

Dominguez, C., Gomez, I. and Zumalacarregui, J. 2002. Prevalence of *Salmonella* and *Campylobacter* in retail chicken meat in Spain. Int. J. Food Microbiol. 72: 165–168.

Doyle, M.P. and Schoeni, J.L. 1986. Isolation of *Campylobacter jejuni* from retail mushrooms. Appl. Environ. Microbiol. 51: 449–450.

Duke, L.A., Breathnach, A.S., Jenkins, D.R., Harkis, B.A. and Codd, A.W. 1996. A mixed outbreak of cryptosporidium and *Campylobacter* infection associated with a private water supply. Epidemiol. Infect. 116: 303–308.

DZC. 2001. Annual Report on Zoonoses in Denmark 2000, Ministry of Food, Agriculture and Fisheries. Danish Zoonosis Centre, Copenhagen, Denmark. http: //www.vetinst.dk/high.asp?page_id=138.

Ebright, J.R. and Ryan, L.M. 1984. Acute erosive reactive arthritis associated with *Campylobacter jejuni*-induced colitis. Am. J. Med. 76: 321–323.

EC. 2002. European Commission, Fifth Framework Program. 'CAMPYCHECK': Improved physiological, immunological and molecular tools for the recovery and identification of emerging *Campylobacteriaceae* in the food and water chain (QLK1-CT-2002–02201). http: //dbs.cordis.lu/fep/FP5/FP5_PROJl_search.html.

ENB, S. 2002. An outbreak of gastroenteritis linked to consumption of barbecued chickens. Singapore Ministry of Health, Committee on Epidemic Diseases, Singapore. http: //app.moh.gov.sg/edc/cid_94.asp.

Endtz, H.P., Vliegenthart, J.S., Vandamme, P., Weverink, H.W., van den Braak, N.P., Verbrugh, H.A. and van Belkum, A. 1997. Genotypic diversity of *Campylobacter lari* isolated from mussels and oysters in The Netherlands. Int. J. Food Microbiol. 34: 79–88.

Engberg, J., Gerner-Smidt, P., Scheutz, F., Moller Nielsen, E., On, S.L. and Molbak, K. 1998. Water-borne *Campylobacter jejuni* infection in a Danish town – a 6-week continuous source outbreak. Clin. Microbiol. Infect. 4: 648–656.

Escherich, T. 1886. Beitraege zur Kenntniss der Darmbacterien III Ueber das Vorkommen von Vibrionen im Dramcanal und den Stuhlgaengen der Saeuglinge. Muenchener Medicinische Wochenschrift. 33: 815–835.

Evans, H.S., Madden, P., Douglas, C., Adak, G.K., O'Brien, S.J., Djuretic, T., Wall, P.G. and Stanwell-Smith, R. 1998a. General outbreaks of infectious intestinal disease in England and Wales: 1995 and 1996. Commun. Dis. Publ. Health 1: 165–171.

Evans, M.R., Lane, W., Frost, J.A. and Nylen, G. 1998b. A *Campylobacter* outbreak associated with stir-fried food. Epidemiol. Infect. 121: 275–279.

Evans, M.R., Ribeiro, C.D. and Salmon, R.L. 2003. Hazards of healthy living: bottled water and salad vegetables as risk factors for *Campylobacter* infection. Emerging Infect. Dis. 9: 1219–1225.

Evans, M.R., Roberts, R.J., Ribeiro, C.D., Gardner, D. and Kembrey, D. 1996. A milk-borne *Campylobacter* outbreak following an educational farm visit. Epidemiol. Infect. 117: 457–462.

Evans, S.J. 1992. Introduction and spread of thermophilic campylobacters in broiler flocks. Vet. Rec. 131: 574–576.

Evans, S.J. and Sayers, A.R. 2000. A longitudinal study of *Campylobacter* infection of broiler flocks in Great Britain. Prev. Vet. Med. 46: 209–223.

Fahey, T., Morgan, D., Gunneburg, C., Adak, G.K., Majid, F. and Kaczmarski, E. 1995. An outbreak of *Campylobacter jejuni* enteritis associated with failed milk pasteurization. J. Infect. 31: 137–143.

Fayos, A., Owen, R.J., Hernandez, J., Jones, C. and Lastovica, A. 1993. Molecular subtyping by genome and plasmid analysis of *Campylobacter jejuni* serogroups O1 and O2 (Penner) from sporadic and outbreak cases of human diarrhoea. Epidemiol. Infect. 111: 415–427.

Federighi, M., Magras, C., Pilet, M.F., Woodward, D., Johnson, W., Jugiau, F. and Jouve, J.L. 1999. Incidence of thermotolerant *Campylobacter* in foods assessed by NF ISO 10272 standard: results of a two-year study. Food Microbiol. 16: 195–204.

Fernandez, H. 2000. The Increasing Incidence of Human Campylobacteriosis. Report and Proceedings of a WHO Consultation of Experts. World Health Organization, Copenhagen, Denmark. http: //www.who.int/emc-documents/zoonoses/whocdscsraph20017c.html.

Fernandez, H. and Pison, V. 1996. Isolation of thermotolerant species of *Campylobacter* from commercial chicken livers. Int J Food Microbiol. 29: 75–80.

Fernie, D.S. and Park, R.W. 1977. The isolation and nature of campylobacters (microaerophilic vibrios) from laboratory and wild rodents. J. Med. Microbiol. 10: 325–329.

Fierens, H., Eyers, M., Bastyns, K., Huyghebaert, A. and De Wachter, R. 1993. Rapid and sensitive detection of *Campylobacter* spp. in chicken products by using the polymerase chain reaction (PCR). Med. Fac. Landbouwwet. Univ. Gent. 58 (4b): 1879–1884.

Finch, M.J. and Blake, P.A. 1985. Food-borne outbreaks of campylobacteriosis: the United States experience, 1980–1982. Am. J. Epidemiol. 122: 262–268.

Florkowski, C.M., Ikram, R.B., Crozier, I.M., Ikram, H. and Berry, M.E. 1984. *Campylobacter jejuni* myocarditis. Clin. Cardiol. 7: 558–559.

Flynn, O.M.J., Blair, I.S. and McDowell, D.A. 1994. Prevalence of *Campylobacter* species on fresh retail chicken wings in Northern Ireland. J. Food Protect. 57: 334–336.

Fox, J., Taylor, N. and Penner, J. 1989. Investigation of zoonotically acquired *Campylobacter jejuni* enteritis with serotyping and restriction endonuclease DNA analysis. J. Clin. Microbiol. 27: 2423–2425.

Fricker, C.R. and Park, R.W. 1989. A two-year study of the distribution of 'thermophilic' campylobacters in human, environmental and food samples from the Reading area with particular reference to toxin production and heat-stable serotype. J. Appl. Bacteriol. 66: 477–490.

Friedman, C.R., Neimann, J., Wegener, H.C. and Tauxe, R.V. 2000. Epidemiology of *Campylobacter jejuni* infections in the United States and other industrialized nations. pp. 121–138. In *Campylobacter*. Nachamkin, I., Blaser, M.J. (Eds.). ASM Press, Washington, DC, pp. 121–138.

Frost, J.A., Gillespie, I.A. and O'Brien, S.J. 2002. Public health implications of *Campylobacter* outbreaks in England and Wales, 1995–9: epidemiological and microbiological investigations. Epidemiol. Infect. 128: 111–118.

Frost, J.A., Kramer, J.M. and Gillanders, S.A. 1999. Phage typing of *Campylobacter jejuni* and *Campylobacter coli* and its use as an adjunct to serotyping. Epidemiol. Infect. 123: 47–55.

Frost, J.A., Oza, A.N., Thwaites, R.T. and Rowe, B. 1998. Serotyping scheme for *Campylobacter jejuni* and *Campylobacter coli* based on direct agglutination of heat-stable antigens. J. Clin. Microbiol. 36: 335–339.

FSIS. 1994. Nationwide Beef Microbiological Baseline Data Collection Program (Steers and Heifers): October 1992-September 1993. Food Safety and Inspection Service (FSIS), Washington, DC. http: //www.fsis.usda.gov/OPHS/baseline/contents.htm.

FSIS. 1996a. Nationwide Beef Microbiological Baseline Data Collection Program (Cows and Bulls): December 1993 – November 1994. Food Safety and Inspection Service (FSIS), Washington, DC. http: //www.fsis.usda.gov/OPHS/baseline/contents.htm.

FSIS. 1996b. Nationwide Broiler Chicken Microbiological Baseline Data Collection Program: July 1994-June 1995. Food Safety and Inspection Service (FSIS), Washington, DC. http: //www.fsis.usda.gov/OPHS/baseline/contents.htm.

FSIS. 1996c. Nationwide Ground Beef Microbiological Baseline Data Collection Program. Food Safety and Inspection Service (FSIS), Washington, DC. http: //www.fsis.usda.gov/OPHS/baseline/contents.htm.

FSIS. 1996d. Nationwide Ground Chicken Microbiological Baseline Data Collection Program. Food Safety and Inspection Service (FSIS), Washington, DC. http: //www.fsis.usda.gov/OPHS/baseline/contents.htm.

FSIS. 1996e. Nationwide Ground Turkey Microbiological Baseline Data Collection Program. Food Safety and Inspection Service (FSIS), Washington, DC. http: //www.fsis.usda.gov/OPHS/baseline/contents.htm.

FSIS. 1996f. Nationwide Pork Microbiological Baseline Data Collection Program: April 1995-March 1996. Food Safety and Inspection Service (FSIS), Washington, DC. http: //www.fsis.usda.gov/OPHS/baseline/contents.htm.

FSIS. 1998. Nationwide Young Turkey Microbiological Baseline Data Collection Program: August 1996 – July 1997. Food Safety and Inspection Service (FSIS), Washington, DC. http: //www.fsis.usda.gov/OPHS/baseline/contents.htm.

Furtado, C., Adak, G.K., Stuart, J.M., Wall, P.G., Evans, H.S. and Casemore, D.P. 1998. Outbreaks of waterborne infectious intestinal disease in England and Wales, 1992–5. Epidemiol. Infect. 121: 109–119.

Gaudreau, C. and Michaud, S. 2003. Cluster of erythromycin- and ciprofloxacin-resistant *Campylobacter jejuni* subsp. *jejuni* from 1999 to 2001 in men who have sex with men, Quebec, Canada. Clin. Infect. Dis. 37: 131–136.

Geilhausen, B., Schutt-Gerowitt, H., Aleksic, S., Koenen, R., Mauff, G. and Pulverer, G. 1996. *Campylobacter* and *Salmonella* contaminating fresh chicken meat. Zentralbl. Bakteriol. 284: 241–245.

Gent, R.N., Telford, D.R. and Syed, Q. 1999. An outbreak of *Campylobacter* food poisoning at a university campus. Commun. Dis. Publ. Health 2: 39–42.

Giacoboni, G.I., Itoh, K., Hirayama, K., Takahashi, E. and Mitsuoka, T. 1993. Comparison of fecal *Campylobacter* in calves and cattle of different ages and areas in Japan. J. Vet. Med. Sci. 55: 555–559.

Giesendorf, B.A. and Quint, W.G. 1995. Detection and identification of *Campylobacter* spp. using the polymerase chain reaction. Cell. Mol. Biol. (Noisy-le-grand). 41: 625–638.

Godoy, P., Artigues, A., Nuin, C., Aramburu, J., Perez, M., Dominguez, A. and Salleras, L. 2002. [Outbreak of gastroenteritis caused by *Campylobacter jejuni* transmitted through drinking water]. Med. Clin. (Barc.) 119: 695–698.

Goossens, H., Henocque, G., Kremp, L., Rocque, J., Boury, R., Alanio, G., Vlaes, L., Hemelhof, W., Van den Borre, C., Macart, M. and et al., 1986. Nosocomial outbreak of *Campylobacter jejuni* meningitis in newborn infants. Lancet 2: 146–149.

Gopal, P. and Gill, H. 2000. Oligosaccharides and glycoconjugates in bovine milk and colostrum. Br. J. Nutr. 84: S69-S74.

Gorman, C. 2002. Staying healthy. Playing chicken with our antibiotics. Overtreatment is creating dangerously resistent germs. Time 159: 98, 101.

Grau, F.H. 1988. *Campylobacter jejuni* and *Campylobacter hyointestinalis* in the intestinal tract and on the carcasses of calves and cattle. J. Food Protect. 51: 857–861.

Grennan, B., O'Sullivan, N.A., Fallon, R., Carroll, C., Smith, T., Glennon, M. and Maher, M. 2001. PCR-ELISAs for the detection of *Campylobacter jejuni* and *Campylobacter coli* in poultry samples. Biotechniques 30: 602–606, 608–610.

Gunnarsson, H. and Svedhem, A. 1998. The usefulness of Diffusion-In-Gel-ELISA in clinical practice as illustrated by a *Campylobacter jejuni* outbreak. J. Immunol. Methods 215: 135–144.

Gutteridge, W. and Haworth, E.A. 1994. An outbreak of gastrointestinal illness associated with contamination of the mains supply by river water. Commun. Dis. Rep. CDR Rev. 4: R50–51.

Hahn, A.F. 1998. Guillain–Barré syndrome. Lancet 352: 635–641.

Hald, B., Knudsen, K., Lind, P. and Madsen, M. 2001. Study of the infectivity of saline-stored *Campylobacter jejuni* for day-old chicks. Appl. Environ. Microbiol. 67: 2388–2392.

Hänninen, M.L., Haajanen, H., Pummi, T., Wermundsen, K., Katila, M.L., Sarkkinen, H., Miettinen, I. and Rautelin, H. 2003. Detection and typing of *Campylobacter jejuni* and *Campylobacter coli* and analysis of indicator organisms in three waterborne outbreaks in Finland. Appl. Environ. Microbiol. 69: 1391–1396.

Harrington, C.S., Thomson-Carter, F.M. and Carter, P.E. 1999. Molecular epidemiological investigation of an outbreak of *Campylobacter jejuni* identifies a dominant clonal line within Scottish serotype HS55 populations. Epidemiol. Infect. 122: 367–375.

Harris, N.V., Kimball, T.J., Bennett, P., Johnson, Y., Wakely, D. and Nolan, C.M. 1987. *Campylobacter jejuni* enteritis associated with raw goat's milk. Am. J. Epidemiol. 126: 179–186.

Harris, N.V., Thompson, D., Martin, D.C. and Nolan, C.M. 1986. A survey of *Campylobacter* and other bacterial contaminants of pre-market chicken and retail poultry and meats, King County, Washington. Am J Public Health. 76: 401–406.

Harrison, W.A., Griffith, C.J., Tennant, D. and Peters, A.C. 2001. Incidence of *Campylobacter* and *Salmonella* isolated from retail chicken and associated packaging in South Wales. Lett. Appl. Microbiol. 33: 450–454.

Havelaar, A., de Wit, M., van Pelt, W., van Duynhoven, Y., Voogt, N., de Boer, E., Willems, R., Duim, B., Jacobs-Reitsma, W. and Wagenaar, J. 2000. The Increasing Incidence of Human Campylobacteriosis. Report and Proceedings of a WHO Consultation of Experts. World Health Organization, Copenhagen. http: //www.who. int/emc-documents/zoonoses/whocdscsraph20017c.html.

Health Canada. 1995. Outbreak of *Campylobacter* infection among farm workers: an occupational hazard. Can. Commun. Dis. Rep. 21–17: 153–156.

Health Canada. 2000. Waterborne outbreak of gastroenteritis associated with a contaminated municipal water supply, Walkerton, Ontario, May-June 2000. Can. Commun. Dis. Rep 26: 170–173.

Hernandez, J., Alonso, J.L., Fayos, A., Amoros, I. and Owen, R.J. 1995. Development of a PCR assay combined with a short enrichment culture for detection of *Campylobacter jejuni* in estuarine surface waters. FEMS Microbiol. Lett. 127: 201–206.

Hershkowici, S., Barak, M., Cohen, A. and Montag, J. 1987. An outbreak of *Campylobacter jejuni* infection in a neonatal intensive care unit. J Hosp. Infect. 9: 54–59.

Hervé, F., Warnet, J.F., Branger, C., Vergez, P. and Hervé, J. 1984. An outbreak of *Campylobacter fetus* subspecies *jejuni* enteritis in newborn infants in a maternity hospital. Ann. Pediatr. (Paris) 31: 277–279.

Hill, G.A. and Grimes, D.J. 1984. Seasonal study of a freshwater lake and migratory waterfowl for *Campylobacter jejuni*. Can. J. Microbiol. 30: 845–849.

Hoar, B.R., Atwill, E.R., Elmi, C. and Farver, T.B. 2001. An examination of risk factors associated with beef cattle shedding pathogens of potential zoonotic concern. Epidemiol. Infect. 127: 147–155.

Hood, A.M., Pearson, A.D. and Shahamat, M. 1988. The extent of surface contamination of retailed chickens with *Campylobacter jejuni* serogroups. Epidemiol. Infect. 100: 17–25.

HPA-UK. 2003. Personal communication, S. Meakins, Health Protection Agency, UK.

Hrudey, S.E., Payment, P., Huck, P.M., Gillham, R.W. and Hrudey, E.J. 2003. A fatal waterborne disease epidemic in Walkerton, Ontario: comparison with other waterborne outbreaks in the developed world. Water Sci. Technol. 47: 7–14.

Hudson, S.J., Lightfoot, N.F., Coulson, J.C., Russell, K., Sisson, P.R. and Sobo, A.O. 1991. Jackdaws and magpies as vectors of milkborne human *Campylobacter* infection. Epidemiol. Infect. 107: 363–372.

Hunt, J.M. and Tran, T.T. 1997. Detection of *Campylobacter jejuni* in enriched samples of inoculated foods by the polymerase chain reaction. 111th Annual Meeting, Association of Official Analytical Chemists, Gaithersburg, MD, City.

IASR. 1999. Infectious Agents Surveillance Report: *Campylobacter* enteritis 1995–1998. National Institute of Infectious Disease, Japan. http: //idsc.nih.go.jp/iasr/20/231/tpc231.html.

Iriarte, M.P. and Owen, R.J. 1996. Repetitive and arbitrary primer DNA sequences in PCR-mediated fingerprinting of outbreak and sporadic isolates of *Campylobacter jejuni*. FEMS Immunol. Med Microbiol. 15: 17–22.

Istre, G.R., Blaser, M.J., Shillam, P. and Hopkins, R.S. 1984. *Campylobacter* enteritis associated with undercooked barbecued chicken. Am. J. Publ. Health 74: 1265–1267.

Ito, K., Kubokura, Y., Kaneko, O.K.I., Totake, Y. and Ogawa, M. 1988. Occurrence of *Campylobacter jejuni* in free-living wild birds from Japan. J. Wildl. Dis. 24: 467–470.

Itoh, T., Saito, K., Maruyama, T., Sakai, S., Ohashi, M. and Oka, A. 1980. An outbreak of acute enteritis due to *Campylobacter fetus* subspecies *jejuni* at a nursery school of Tokyo. Microbiol. Immunol. 24: 371–379.

Jackson, C.J., Fox, A.J. and Jones, D.M. 1996. A novel polymerase chain reaction assay for the detection and speciation of thermophilic *Campylobacter* spp. J. Appl. Bacteriol. 81: 467–473.

Jacobs-Reitsma, W. 2000. *Campylobacter* in the food supply. pp. 467–482. In *Campylobacter*. Nachamkin, I., Blaser, M.J., eds. ASM Press, Washington, DC, pp. 467–482.

Jacobs-Reitsma, W.F. 1997. Aspects of epidemiology of *Campylobacter* in poultry. Vet. Q. 19: 113–117.

Jones, D.M., Sutcliffe, E.M. and Curry, A. 1991a. Recovery of viable but non-culturable *Campylobacter jejuni*. J. Gen. Microbiol. 137 (Pt 10): 2477–2482.

Jones, F.T., Axtell, R.C., Rives, D.V., Scheideler, S.E., Tarver, F.R.J., Walker, R.L. and Wineland, M.J. 1991b. A survey of *Campylobacter jejuni* contamination in modern broiler production and processing systems. J. Food Protect. 54: 259–262, 266.

Jones, I.G. and Roworth, M. 1996. An outbreak of *Escherichia coli* O157 and campylobacteriosis associated with contamination of a drinking water supply. Publ. Health 110: 277–282.

Jones, P.H., Willis, A.T., Robinson, D.A., Skirrow, M.B. and Josephs, D.S. 1981. *Campylobacter* enteritis associated with the consumption of free school milk. J. Hyg. 87: 155–162.

Jorgensen, F., Bailey, R., Williams, S., Henderson, P., Wareing, D.R., Bolton, F.J., Frost, J.A., Ward, L. and Humphrey, T.J. 2002. Prevalence and numbers of *Salmonella* and *Campylobacter* spp. on raw, whole chickens in relation to sampling methods. Int. J. Food Microbiol. 76: 151–164.

Josephson, K.L., Gerba, C.P. and Pepper, I.L. 1993. Polymerase chain reaction detection of nonviable bacterial pathogens. Appl. Environ. Microbiol. 59: 3513–3515.

Kalman, M., Szollosi, E., Czermann, B., Zimanyi, M. and Szekeres, S. 2000. Milkborne *Campylobacter* infection in Hungary. J. Food Prot. 63: 1426–1429.

Kapperud, G., Skjerve, E., Bean, N.H., Ostroff, S.M. and Lassen, J. 1992. Risk factors for sporadic *Campylobacter* infections: results of a case–control study in southeastern Norway. J. Clin. Microbiol. 30: 3117–3121.

Karlyshev, A.V., Linton, D., Gregson, N.A., Lastovica, A.J. and Wren, B.W. 2000. Genetic and biochemical evidence of a *Campylobacter jejuni* capsular polysaccharide that accounts for Penner serotype specificity. Mol. Microbiol. 35: 529–541.

Kazwala, R.R., Jiwa, S.F. and Nkya, A.E. 1993. The role of management systems in the epidemiology of thermophilic campylobacters among poultry in eastern zone of Tanzania. Epidemiol. Infect. 110: 273–278.

Kelly, A.F., Martinez-Rodriguez, A., Bovill, R.A. and Mackey, B.M. 2003. Description of a 'Phoenix' phenomenon in the growth of *Campylobacter jejuni* at temperatures close to the minimum for growth. Appl. Environ. Microbiol. 69: 4975–4978.

Kinde, H., Genigeorgis, C.A. and Pappaioanou, M. 1983. Prevalence of *Campylobacter jejuni* in chicken wings. Appl. Environ. Microbiol. 45: 1116–1118.

King, E.O. 1957. Human infections with *Vibrio fetus* and a closely related vibrio. J. Infect. Dis. 101.

King, C.H., Shotts Jr., E.B.,Wooley, R.E. and Porter, K.G. 1988. Survival of coliforms and bacterial pathogens within protozoa during chlorination. Appl. Environ. Microbiol. 54: 3023–3033.

Kirk, M. 2003. *Campylobacter* outbreaks, 2001–2002. Melbourne, Australia. OzFoodNet: Australian Enhanced Food-borne Disease Surveillance. Personal communication.

Kirk, M., Waddell, R.,Dalton, C., Creaser, A. and Rose, N. 1997. A prolonged outbreak of *Campylobacter* infection at a training facility. Commun. Dis. Intell. 21: 57–61.

Klein, B.S., Vergeront, J.M., Blaser, M.J., Edmonds, P., Brenner, D.J., Janssen, D. and Davis, J.P. 1986. *Campylobacter* infection associated with raw milk. An outbreak of gastroenteritis due to *Campylobacter jejuni* and thermotolerant *Campylobacter fetus* subsp *fetus*. JAMA 255: 361–364.

Koenraad, P.M., Jacobs-Reitsma, W.F., Van der Laan, T., Beumer, R.R. and Rombouts, F.M. 1995. Antibiotic susceptibility of *Campylobacter* isolates from sewage and poultry abattoir drain water. Epidemiol. Infect. 115: 475–483.

Koenraad, P.M., Rombouts, F.M. and Notermans, S.H.W. 1997. Epidemiological aspects of thermophilic *Campylobacter* in water-related environments: A review. Water Environ. Res. 69: 52–63.

Korhonen, L.K. and Martikainen, P.J. 1991. Comparison of the survival of *Campylobacter jejuni* and *Campylobacter coli* in culturable form in surface water. Can. J. Microbiol. 37: 530–533.

Korlath, J.A., Osterholm, M.T., Judy, L.A., Forfang, J.C. and Robinson, R.A. 1985. A point-source outbreak of campylobacteriosis associated with consumption of raw milk. J. Infect. Dis. 152: 592–596.

Kornblatt, A.N., Barrett, T., Morris, G.K. and Tosh, F.E. 1985. Epidemiologic and laboratory investigation of an outbreak of *Campylobacter* enteritis associated with raw milk. Am. J. Epidemiol. 122: 884–889.

Korsak, N.,Daube, G., Ghafir, Y., Chahed, A., Jolly, S. and Vindevogel, H. 1998. An efficient sampling technique used to detect four food-borne pathogens on pork and beef carcasses in nine Belgian abattoirs. J. Food Protect. 61: 535–541.

Kramer, J.M., Frost, J.A., Bolton, F.J. and Wareing, D.R. 2000. *Campylobacter* contamination of raw meat and poultry at retail sale: identification of multiple types and comparison with isolates from human infection. J. Food Prot. 63: 1654–1659.

Kulkarni, N.M., Sherikar, A.T., Sherikar, A.A., Tarwate, B.G., Paturkar, A.M. and Murugkar, H.V. 1992. Incidence of *Campylobacter jejuni* in chicken and their carcasses sold at Bombay city. Indian J. Anim. Sci. 62: 1024–1027.

Kumar, A., Agarwal, R.K., Bhilegaonkar, K.N., Shome, B.R. and Bachhil, V.N. 2001. Occurrence of *Campylobacter jejuni* in vegetables. Int. J. Food Microbiol. 67: 153–155.

Kuusi, M., Nuorti, P. and Ruutu, P. 1998. Campylobacteriosis outbreak in Haukipudas, Finland, August 1998. European Programme for Intervention Epidemiology. http: //www.epiet.org/seminar/1998/kuusi.html.

Lammerding, A.M., Garcia, M.M., Mann, E.D., Robinson, Y., Dorward, W.J., Trusscott, R.B. and Tittiger, F. 1988. Prevalence of salmonella and thermophilic *Campylobacter* in fresh pork, beef, veal and poultry in Canada. J. Food Protect. 51: 47–52.

Lamoureux, M., MacKay, A., Messier, S., Fliss, I., Blais, B.W., Holley, R.A. and Simard, R.E. 1997. Detection of *Campylobacter jejuni* in food and poultry viscera using immunomagnetic separation and microtitre hybridization. J. Appl. Microbiol. 83: 641–651.

Lastovica, A.J. and Skirrow, M.B. 2000. Clinical significance of *Campylobacter* and related species other than *Campylobacter jejuni* and *C. coli*. pp. 89–120. In *Campylobacter*. Nachamkin, I., Blaser, M.J. (Eds.). ASM Press, Washington, DC, pp. 89–120.

Lauwers, S., De Boeck, M. and Butzler, J.P. 1978. *Campylobacter* enteritis in Brussels. Lancet 1: 604–605.

Lauwers, S., Hofman, B., Seghers, M., Van Etterijck, R., Van Zeebroeck, A., de Smet, F. and Pierard, D. 1991. An outbreak of *C. upsaliensis* in a day care centre. Microb. Ecol. Health Dis. 4 (Suppl.): S90.

Layton, M.C., Calliste, S.G., Gomez, T.M., Patton, C. and Brooks, S. 1997. A mixed food-borne outbreak with *Salmonella heidelberg* and *Campylobacter jejuni* in a nursing home. Infect Control Hosp. Epidemiol. 18: 115–121.

Lee, A., Smith, S.C. and Coloe, P.J. 1998. Survival and growth of *Campylobacter jejuni* after artificial inoculation onto chicken skin as a function of temperature and packaging conditions. J. Food Protect. 61: 1609–1614.

Lee, S.H., Levy, D.A., Craun, G.F., Beach, M.J. and Calderon, R.L. 2002. Surveillance for waterborne-disease outbreaks — United States, 1999–2000. Morbid. Mortal. Weekly Rep. 51: 1–52.

Lehner, A., Schneck, C., Feierl, G., Pless, P., Deutz, A., Brandl, E. and Wagner, M. 2000. Epidemiologic application of pulsed-field gel electrophoresis to an outbreak of *Campylobacter jejuni* in an Austrian youth centre. Epidemiol. Infect. 125: 13–16.

Levy, A.J. 1946. A gastro-enteritis outbreak probably due to a bovine strain of Vibrio. Yale J. Biol. Med. 18: 243–258.

Lind, L., Sjogren, E., Melby, K. and Kaijser, B. 1996. DNA fingerprinting and serotyping of *Campylobacter jejuni* isolates from epidemic outbreaks. J. Clin. Microbiol. 34: 892–896.

Lind, L., Sjogren, E., Welinder-Olsson, C. and Kaijser, B. 1989. Plasmids and serogroups in *Campylobacter jejuni*. APMIS 97: 1097–1102.

Lindqvist, R., Andersson, Y., de Jong, B. and Norberg, P. 2000. A summary of reported food-borne disease incidents in Sweden, 1992 to 1997. J. Food Prot. 63: 1315–1320.

Lior, H., Woodward, D.L., Edgar, J.A. and LaRoche, L.J. 1981. Serotyping by slide agglutination of *Campylobacter jejuni* and epidemiology. Lancet. 2: 1103–1104.

Llovo, J., Mateo, E., Munoz, A., Urquijo, M., On, S.L. and Fernandez-Astorga, A. 2003. Molecular typing of *Campylobacter jejuni* isolates involved in a neonatal outbreak indicates nosocomial transmission. J. Clin. Microbiol. 41: 3926–3928.

Long, S.M., Adak, G.K., O'Brien, S.J. and Gillespie, I.A. 2002. General outbreaks of infectious intestinal disease linked with salad vegetables and fruit, England and Wales, 1992–2000. Commun. Dis. Publ. Health 5: 101–105.

Luechtefeld, N.A., Blaser, M.J., Reller, L.B. and Wang, W.L. 1980. Isolation of *Campylobacter fetus* subsp. *jejuni* from migratory waterfowl. J. Clin. Microbiol. 12: 406–408.

Luechtefeld, N.W., Cambre, R.C. and Wang, W.L. 1981. Isolation of *Campylobacter fetus* subsp *jejuni* from zoo animals. J. Am. Vet. Med. Assoc. 179: 1119–1122.

Luechtefeld, N.W. and Wang, W.L. 1981. *Campylobacter fetus* subsp. *jejuni* in a turkey processing plant. J. Clin. Microbiol. 13: 266–268.

Luo, N. and Zhang, Q. 2001. Molecular characterization of a cryptic plasmid from *Campylobacter jejuni*. Plasmid 45: 127–133.

McFadyean, J. and Stockman, S. 1913. Great Britain. Report of the Departmental Committee appointed by the Board of Agriculture and Fisheries to inquire into epizootic abortion. Part III. Abortion in sheep. Appendix.

Machado, R.A., Tosin, I. and Leitao, M.F.F. 1994. Occurence of *Salmonella* sp. and *Campylobacter* sp. in chickens during industrial processing. Revista de Microbiol. 25: 239–244.

McKay, D., Fletcher, J., Cooper, P. and Thomson-Carter, F.M. 2001. Comparison of two methods for serotyping *Campylobacter* spp. J. Clin. Microbiol. 39: 1917–1921.

McMahon, M.A. and Wilson, I.G. 2001. The occurrence of enteric pathogens and *Aeromonas* species in organic vegetables. Int. J. Food Microbiol. 70: 155–162.

McMyne, P.M., Penner, J.L., Mathias, R.G., Black, W.A. and Hennessy, J.N. 1982. Serotyping of *Campylobacter jejuni* isolated from sporadic cases and outbreaks in British Columbia. J. Clin. Microbiol. 16: 281–285

Madden, R.H., Moran, L. and Scates, P. 1998. Frequency of occurrence of *Campylobacter* spp. in red meats and poultry in Northern Ireland and their subsequent subtyping using polymerase chain reaction-restriction fragment length polymorphism and the random amplified polymorphic DNA method. J. Appl. Microbiol. 84: 703–708.

Mafu, A.A., Higgins, R., Nadeau, M. and Cousineau, G. 1989. The incidence of *Salmonella*, *Campylobacter*, and *Yersinia enterocolitica* in swine carcasses and the slaughterhouse environment. J. Food Protect. 52: 642–645.

Magistrado, P.A., Garcia, M.M. and Raymundo, A.K. 2001. Isolation and polymerase chain reaction-based detection of *Campylobacter jejuni* and *Campylobacter coli* from poultry in the Philippines. Int J Food Microbiol. 70: 197–206.

Mandrell, R.E. and Brandl, M.T. 2004. *Campylobacter* species and fresh produce: outbreaks, incidence and biology. pp. In Pre-Harvest and Post-Harvest Food Safety: Contemporary Issues and Future Directions. Beier, R., Ziprin, R., Pillai, P., Phillips, T., eds. Blackwell Publishing, Oxford, pp. 59–72.

Manser, P.A. and Dalziel, R.W. 1985. A survey of *Campylobacter* in animals. J. Hyg. 95: 15–21.

Manzano, M., Pipan, C., Botta, G. and Comi, G. 1995a. Comparison of three culture media for recovering *Campylobacter jejuni* and *Campylobacter coli* from poultry skin, liver and meat. Sciences des Ailments. 15: 615–623.

Manzano, M., Pipan, C., Botta, G. and Comi, G. 1995b. Polymerase chain reaction assay for detection of *Campylobacter coli* and *Campylobacter jejuni* in poultry meat. Zentralbl Hyg. Umweltmed. 197: 370–386.

Marinescu, M., Festy, B., Derimay, R. and Megraud, F. 1987. High frequency of isolation of *Campylobacter coli* from poultry meat in France. Eur. J. Clin. Microbiol. 6: 693–695.

Maruyama, S., Tanaka, T., Katsube, Y., Nakanishi, H. and Nukina, M. 1990. Prevalence of thermophilic campylobacters in crows (*Corvus levaillantii*, *Corvus corone*) and serogroups of the isolates. Nippon Juigaku Zasshi 52: 1237–1244.

Matsusaki, S., Katayama, A., Kawaguchi, N., Tanaka, K. and Goto, A. 1982. Incidence of *Campylobacter jejuni/coli* in foodstuffs. J. Food Hyg. Soc. (Japan). 23: 434–437.

Maurer, A.M. and Sturchler, D. 2000. A waterborne outbreak of small round structured virus, *Campylobacter* and *Shigella* co-infections in La Neuveville, Switzerland, 1998. Epidemiol. Infect. 125: 325–332.

MDH. 2000. Minnesota Department of Health 2000 gastroenteritis outbreak summary. Minnesota Department of Health, Minneapolis, MN. www.health.state.mn.us.

MDH. 2003. Campylobacteriosis, 2002. http: //www.health.state.mn.us/divs/idepc/newsletters/dcn/sum02/campylo.html.

Moedema, G.J., Schets, F.M., van de Giessen, A.W. and Havelaar, A.H. 1992. Lack of colonization of 1 day old chicks by viable, non-culturable *Campylobacter jejuni*. J. Appl. Bacteriol. 72: 512–516.

Mégraud, F., Gavinet, A.M. and Camou-Junca, C. 1987. Serogroups and biotypes of human strains of *Campylobacter jejuni* and *Campylobacter coli* isolated in France. Eur. J. Clin. Microbiol. 6: 641–645.

Meinersmann, R.J. 2000. Population genetics and genealogy of *Campylobacter jejuni*. pp. 351–368. In *Campylobacter*. Nachamkin, I., Blaser, M.J., eds. ASM press, Washington, D.C., pp. 351–368.

Melby, K., Gondrosen, B., Gregusson, S., Ribe, H. and Dahl, O.P. 1991. Waterborne campylobacteriosis in northern Norway. Int. J. Food Microbiol. 12: 151–156.

Melby, K.K., Svendby, J.G., Eggebo, T., Holmen, L.A., Andersen, B.M., Lind, L., Sjogren, E. and Kaijser, B. 2000. Outbreak of *Campylobacter* infection in a subartic community. Eur. J. Clin. Microbiol. Infec.t Dis. 19: 542–544.

Mentzing, L.O. 1981. Waterborne outbreaks of *Campylobacter* enteritis in central Sweden. Lancet 2: 352–354.

Merritt, A., Miles, R. and Bates, J. 1999. An outbreak of *Campylobacter* enteritis on an island resort, north Queensland. Commun. Dis Intell. 23: 215–219.

Miettinen, I.T., Zacheus, O., von Bonsdorff, C.H. and Vartiainen, T. 2001. Waterborne epidemics in Finland in 1998–1999. Water Sci. Technol. 43: 67–71.

Millson, M., Bokhout, M., Carlson, J., Spielberg, L., Aldis, R., Borczyk, A. and Lior, H. 1991. An outbreak of *Campylobacter jejuni* gastroenteritis linked to meltwater contamination of a municipal well. Can. J. Publ. Health 82: 27–31.

Moore, J., Caldwell, P. and Millar, B. 2001a. Molecular detection of *Campylobacter* spp. in drinking, recreational and environmental water supplies. Int. J. Hyg. Environ Health 204: 185–189.

Moore, J.E., Caldwell, P.S., Millar, B.C. and Murphy, P.G. 2001b. Occurrence of *Campylobacter* spp. in water in Northern Ireland: implications for public health. Ulster Med. J. 70: 102–107.

Phillips, C.A. 1998. The isolation of *Campylobacter* spp. from modified atmosphere packaged foods. Int. J. Env. Health Res. 8: 215–221.

Porter, I.A. and Reid, T.M. 1980. A milk-borne outbreak of *Campylobacter* infection. J. Hyg. 84: 415–419.

Potera, C. 1996. Biofilms invade microbiology. Science 273: 1795–1797.

Potter, M.E., Blaser, M.J., Sikes, R.K., Kaufmann, A.F. and Wells, J.G. 1983. Human *Campylobacter* infection associated with certified raw milk. Am. J. Epidemiol. 117: 475–483.

Purdy, D., Ash, C.A. and Fricker, C.R. 1996. Polymerase chain reaction assay for the detection of viable *Campylobacter* species from potable and untreated environmental water samples. In: Campylobacters, Helicobacters and Related Organisms. Newell, D.G., Ketley, J.M., Feldman, R.A., eds. Plenum Press, New York, pp. 147–153.

Rasrinaul, L., Suthienkul, O., Echeverria, P.D., Taylor, D.N., Seriwatana, J., Bangtrakulnonth, A. and Lexomboon, U. 1988. Foods as a source of enteropathogens causing childhood diarrhoea in Thailand. Am. J. Trop. Med. Hyg. 39: 97–102.

Raupach, J.C. and Hundy, R.L. 2003. An outbreak of *Campylobacter jejuni* infection among conference delegates. Commun. Dis. Intell. 27: 380–383.

Rautelin, H., Koota, K., von Essen, R., Jahkola, M., Siitonen, A. and Kosunen, T.U. 1990. Waterborne *Campylobacter jejuni* epidemic in a Finnish hospital for rheumatic diseases. Scand. J. Infect. Dis. 22: 321–326.

Refregier-Petton, J., Rose, N., Denis, M. and Salvat, G. 2001. Risk factors for *Campylobacter* spp. contamination in French broiler-chicken flocks at the end of the rearing period. Prev. Vet. Med. 50: 89–100.

Reintjes, R., van Treeck, U., Oehler, M. and Petruschke, N. 1999. Ein Ausbruch von *Campylobacter*-Enteritis in Nordrhein-Westfalen. Epidemiologisches Bull. 43: 317–320.

Rennie, R.P., Strong, D., Taylor, D.E., Salama, S.M., Davidson, C. and Tabor, H. 1994. *Campylobacter fetus* diarrhoea in a Hutterite colony: epidemiological observations and typing of the causative organism. J. Clin. Microbiol. 32: 721–724.

Rhoades, H.E. 1954. The illustration of the morphology of *Vibrio fetus* by electron microscopy. Am. J. Vet. Res. 15: 630–633.

Ribeiro, C.D. and Frost, J.A. 2000. Family clusters of *Campylobacter* infection. Commun. Dis. Publ. Health 3: 274–276.

Riley, L.W. and Finch, M.J. 1985. Results of the first year of national surveillance of *Campylobacter* infections in the United States. J. Infect. Dis. 151: 956–959.

Robinson, D.A. 1981. Infective dose of *Campylobacter jejuni* in milk. Br. Med. J. (Clin. Res. Ed.). 282: 1584.

Robinson, D.A., Edgar, W.J., Gibson, G.L., Matchett, A.A. and Robertson, L. 1979. *Campylobacter* enteritis associated with consumption of unpasteurised milk. Br. Med. J. 1: 1171–1173.

Robinson, D.A. and Jones, D.M. 1981. Milk-borne *Campylobacter* infection. Br. Med. J. (Clin. Res. Ed). 282: 1374–1376.

Rodrigues, L.C., Cowden, J.M., Wheeler, J.G., Sethi, D., Wall, P.G., Cumberland, P., Tompkins, D.S., Hudson, M.J., Roberts, J.A. and Roderick, P.J. 2001. The study of infectious intestinal disease in England: risk factors for cases of infectious intestinal disease with *Campylobacter jejuni* infection. Epidemiol. Infect. 127: 185–193.

Roels, T.H., Wickus, B., Bostrom, H.H., Kazmierczak, J.J., Nicholson, M.A., Kurzynski, T.A. and Davis, J.P. 1998. A food-borne outbreak of *Campylobacter jejuni* (O: 33) infection associated with tuna salad: a rare strain in an unusual vehicle. Epidemiol. Infect. 121: 281–287.

Rogol, M., Sechter, I., Falk, H., Shtark, Y., Alfi, S., Greenberg, Z. and Mizrachi, R. 1983. Waterborne outbreak of *Campylobacter* enteritis. Eur. J. Clin. Microbiol. 2: 588–590.

Rollins, D.M. and Colwell, R.R. 1986. Viable but nonculturable stage of *Campylobacter jejuni* and its role in survival in the natural aquatic environment. Appl. Environ. Microbiol. 52: 531–538.

Roman, G.C. 1995. Tropical neuropathies. Bailliere's Clin. Neurol. 4: 469–487.

Roman, G.C. 1998. Epidemic neuropathy in Cuba: a public health problem related to the Cuban Democracy Act of the United States. Neuroepidemiology 17: 111–115.

Ronveaux, O., Quoilin, S., Van Loock, F., Lheureux, P., Struelens, M. and Butzler, J.P. 2000. A *Campylobacter coli* food-borne outbreak in Belgium. Acta Clin. Belg. 55: 307–311.

Rosef, O., Gondrosen, B., Kapperud, G. and Underdal, B. 1983. Isolation and characterization of *Campylobacter jejuni* and *Campylobacter coli* from domestic and wild mammals in Norway. Appl. Environ. Microbiol. 46: 855–859.

Rosef, O. and Kapperud, G. 1983. House flies (*Musca domestica*) as possible vectors of *Campylobacter fetus* subsp. *jejuni*. Appl. Environ. Microbiol. 45: 381–383.

158

Rosenfield, J.A., Arnold, G.J., Davey, G.R., Archer, R.S. and Woods, W.H. 1985. Serotyping of *Campylobacter jejuni* from an outbreak of enteritis implicating chicken. J. Infect. 11: 159–165.

Ruiz-Palacios, G.M., Cervantes, L.E., Ramos, P., Chavez-Munguia, B. and Newburg, D.S. 2003. *Campylobacter jejuni* binds intestinal H (O) antigen (Fucalpha 1, 2Galbeta 1, 4GlcNAc), and fucosyloligosaccharides of human milk inhibit its binding and infection. J. Biol. Chem. 278: 14112–14120.

Sacks, J.J., Lieb, S., Baldy, L.M., Berta, S., Patton, C.M., White, M.C., Bigler, W.J. and Witte, J.J. 1986. Epidemic campylobacteriosis associated with a community water supply. Am. J. Publ. Health. 76: 424–428.

Sagoo, S.K., Little, C.L. and Mitchell, R.T. 2001. The microbiological examination of ready-to-eat organic vegetables from retail establishments in the United Kingdom. Lett. Appl. Microbiol. 33: 434–439.

Saha, S.K., Saha, S. and Sanyal, S.C. 1991. Recovery of injured *Campylobacter jejuni* cells after animal passage. Appl. Environ. Microbiol. 57: 3388–3389.

Sahin, O., Luo, N., Huang, S. and Zhang, Q. 2003. Effect of *Campylobacter*-specific maternal antibodies on *Campylobacter jejuni* colonization in young chickens. Appl. Environ. Microbiol. 69: 5372–5379.

Sails, A.D., Bolton, F.J., Fox, A.J., Wareing, D.R. and Greenway, D.L. 2002. Detection of *Campylobacter jejuni* and *Campylobacter coli* in environmental waters by PCR enzyme-linked immunosorbent assay. Appl. Environ. Microbiol. 68: 1319–1324.

Sails, A.D., Bolton, F.J., Fox, A.J., Wareing, D.R. and Greenway, D.L.A. 1998. A reverse transcriptase polymerase chain reaction assay for the detection of thermophilic *Campylobacter* spp. Mol. Cell. Probes 12: 317–322.

Sails, A.D., Fox, A.J., Bolton, F.J., Wareing, D.R. and Greenway, D.L. 2003a. A real-time PCR assay for the detection of *Campylobacter jejuni* in foods after enrichment culture. Appl. Environ. Microbiol. 69: 1383–1390.

Sails, A.D., Swaminathan, B. and Fields, P.I. 2003b. Utility of multilocus sequence typing as an epidemiological tool for investigation of outbreaks of gastroenteritis caused by *Campylobacter jejuni*. J. Clin. Microbiol. 41: 4733–4739.

Saito, S., Yatsuyanagi, J., Sato, H., Shiraishi, H. and Amano, K. 2002. [Typings of *Campylobacter jejuni* isolated from patients of 2 outbreak cases by genotypic and phenotypic methods]. Nippon Saikingaku Zasshi. 57: 465–472.

Savill, M., Hudson, A., Devane, M., Garrett, N., Gilpin, B. and Ball, A. 2003. Elucidation of potential transmission routes of *Campylobacter* in New Zealand. Water Sci. Technol. 47: 33–38.

Sebald, M. and Véron, M. 1963. Teneur en bases da l'ADN et classification des vibrions. Annales de L'institut Pasteur (Paris) 105.

Shandera, W.X., Tormey, M.P. and Blaser, M.J. 1992. An outbreak of bacteremic *Campylobacter jejuni* infection. Mt Sinai J. Med. 59: 53–56.

Shane, S.M. 2000. *Campylobacter* infection of commercial poultry. Rev Sci Tech. 19: 376–395.

Shanker, S., Lee, A. and Sorrell, T.C. 1990. Horizontal transmission of *Campylobacter jejuni* amongst broiler chicks: experimental studies. Epidemiol. Infect. 104: 101–110.

Shmilovitz, M., Kretzer, B. and Rotman, N. 1982. *Campylobacter jejuni* as an etiological agent of diarrhoeal diseases in Israel. Isr. J. Med. Sci. 18: 935–940.

Shreeve, J.E., Toszeghy, M., Pattison, M. and Newell, D.G. 2000. Sequential spread of *Campylobacter* infection in a multipen broiler house. Avian Dis. 44: 983–988.

Sierra, M.L., Gonzalez-Fandos, E., Garcia-Lopez, M.L., Garcia-Fernandez, M.C. and Prieto, M. 1995. Prevalence of *Salmonella*, *Yersinia*, *Aeromonas*, *Campylobacter*, and cold growing *Escherichia coli* on freshly dressed lamb carcasses. J. Food Protect. 58: 1183–1185.

Skelly, C. and Weinstein, P. 2003. Pathogen survival trajectories: an eco-environmental approach to the modeling of human campylobacteriosis ecology. Environ. Health Perspect. 111: 19–28.

Skirrow, M.B. 1977. *Campylobacter* enteritis: a 'new' disease. Br. Med. J. 2: 9–11.

Slader, J., Domingue, G., Jorgensen, F., McAlpine, K., Owen, R.J., Bolton, F.J. and Humphrey, T.J. 2002. Impact of transport crate reuse and of catching and processing on *Campylobacter* and Salmonella contamination of broiler chickens. Appl. Environ. Microbiol. 68: 713–719.

SMI. 1999. Swedish surveillance statistics on infectious diseases for 1999. Swedish Institute for Infectious Disease Control, Solna, Sweden. http: //www.smittskyddsinstitutet.se/download/pdf/rapp99.pdf.

Smith, M.V., 2nd and Muldoon, P.J. 1974. *Campylobacter fetus* subspecies *jejuni* (*Vibrio fetus*) from commercially processed poultry. Appl. Microbiol. 27: 995–996.

Sockett, P.N. 1991. Communicable disease associated with milk and dairy products: England and Wales 1987–1989. Communicable Disease Review. 1: R9–R12.

Sopwith, W., Ashton, M., Frost, J.A., Tocque, K., O'Brien, S., Regan, M. and Syed, Q. 2003. Enhanced surveillance of *Campylobacter* infection in the north west of England 1997–1999. J. Infect. 46: 35–45.

Srour, S.F., Rishpon, S., Rubin, L. and Warman, S. 2002. [An outbreak of *Campylobacter jejuni* enteritis after farm visit in Haifa subdistrict]. Harefuah 141: 683–684, 763.

Stalder, H., Isler, R., Stutz, W., Salfinger, M., Lauwers, S. and Vischer, W. 1983. Contribution to the epidemiology of *Campylobacter jejuni*. From asymptomatic excretion by a cow in the cowshed to overt disease in over 500 persons. Schweiz. Med. Wochenschr. 113: 245–249.

Stamnes, O. and Grude, N. 2002. [Outbreak of campylobacteriosis among soccer players]. Tidsskr Nor Laegeforen. 122: 2850–2852.

Stanley, K., Wallace, J., Currie, J., Diggle, P. and Jones, K. 1998a. Seasonal variation of thermophilic campylobacters in lambs at slaughter. J Appl Microbiol. 84: 1111–1116.

Stanley, K.N., Wallace, J.S., Currie, J.E., Diggle, P.J. and Jones, K. 1998b. The seasonal variation of thermophilic campylobacters in beef cattle, dairy cattle and calves. J. Appl. Microbiol. 85: 472–480.

Steffen, W., Mochmann, H., Kontny, I., Richter, U., Werner, U. and el Naeem, O. 1986. [A food-borne infection caused by *Campylobacter jejuni* serotype Lauwers 19]. Zentralbl. Bakteriol. Mikrobiol. Hyg. [B] 183: 28–35.

Stehr-Green, J.K., Nicholls, C., McEwan, S., Payne, A. and Mitchell, P. 1991. Waterborne outbreak of *Campylobacter jejuni* in Christchurch: the importance of a combined epidemiologic and microbiologic investigation. NZ Med. J. 104: 356–358.

Stelzer, W., Mochmann, H., Richter, U. and Dobberkau, H.J. 1989. A study of *Campylobacter jejuni* and *Campylobacter coli* in a river system. Zentralbl Hyg Umweltmed. 189: 20–28.

Stern, N.J., Bailey, J.S., Blankenship, L.C., Cox, N.A. and McHan, F. 1988. Colonization characteristics of *Campylobacter jejuni* in chick ceca. Avian Dis. 32: 330–334.

Stern, N.J., Fedorka-Cray, P., Bailey, J.S., Cox, N.A., Craven, S.E., Hiett, K.L., Musgrove, M.T., Ladley, S., Cosby, D. and Mead, G.C. 2001. Distribution of *Campylobacter* spp. in selected U.S. poultry production and processing operations. J. Food Protect. 64: 1705–1710.

Stern, N.J., Green, S.S., Thaker, N., Krout, D.J. and Chiu, J. 1984. Recovery of *Campylobacter jejuni* from fresh and frozen meat and poultry collected at slaughter. J. Food Protect. 47: 372–374.

Stern, N.J., Hiett, K.L., Alfredsson, G.A., Kristinsson, K.G., Reiersen, J., Hardardottir, H., Briem, H., Gunnarsson, E., Georgsson, F., Lowman, R., Berndtson, E., Lammerding, A.M., Paoli, G.M. and Musgrove, M.T. 2003. *Campylobacter* spp. in Icelandic poultry operations and human disease. Epidemiol. Infect. 130: 23–32.

Stern, N.J., Jones, D.M., Wesley, I.V. and Rollins, D.M. 1994. Colonization of chicks by non-cultureable *Campylobacter* spp. Lett. Appl. Microbiol. 18: 333–336.

Stern, N.J. and Line, J.E. 2000. *Campylobacter*. pp. 1040–1056. In The Microbiological Safety and Quality of Food. Lund, B.M., Baird-Parker, A.C., Gould, G.W., eds. Aspen Publishers, Gaithersburg, MD, pp. 1040–1056.

Stern, N.J., Robach, M.C., Cox, N.A. and Musgrove, M.T. 2002. Effect of drinking water chlorination on *Campylobacter* spp. colonization of broilers. Avian Dis. 46: 401–404.

Stuart, J., Sufi, F., McNulty, C. and Park, P. 1997. Outbreak of *Campylobacter* enteritis in a residential school associated with bird pecked bottle tops. Commun. Dis. Rep. CDR Rev. 7: R38–40.

Studahl, A. and Andersson, Y. 2000. Risk factors for indigenous *Campylobacter* infection: a Swedish case–control study. Epidemiol. Infect. 125: 269–275.

Suarez, J.A., Etienne, M.K. and Leguen, F. 2002. Food-borne outbreak of *Campylobacter* infection from a group conference in North Miami Beach, Miami-Dade County, 2002. Florida Department of Health Epi Month. Rept. 3: 1.

Talhouk, R.S., el-Dana, R.A., Araj, G.F., Barbour, E. and Hashwa, F. 1998. Prevalence, antimicrobial susceptibility and molecular characterization of *Campylobacter* isolates recovered from humans and poultry in Lebanon. J Med Liban. 46: 310–316.

Tam, C.C., O'Brien, S.J., Adak, G.K., Meakins, S.M. and Frost, J.A. 2003. *Campylobacter coli* – an important food-borne pathogen. J. Infect. 47: 28–32.

Tauxe, R.V., Hargrett-Bean, N., Patton, C.M. and Wachsmuth, I.K. 1988. *Campylobacter* isolates in the United States, 1982–1986. MMWR CDC Surveill Summ. 37: 1–13.

Tauxe, R.V., Pegues, D.A. and Hargrett-Bean, N. 1987. *Campylobacter* infections: the emerging national pattern. Am J Public Health. 77: 1219–1221.

Taylor, D.N., McDermott, K.T., Little, J.R., Wells, J.G. and Blaser, M.J. 1983. *Campylobacter* enteritis from untreated water in the Rocky Mountains. Ann. Intern. Med. 99: 38–40.

Taylor, N.S., Ellenberger, M.A., Wu, P.Y. and Fox, J.G. 1989. Diversity of serotypes of *Campylobacter jejuni* and *Campylobacter coli* isolated in laboratory animals. Lab. Anim. Sci. 39: 219–221.

Tholozan, J.L., Cappelier, J.M., Tissier, J.P., Delattre, G. and Federighi, M. 1999. Physiological characterization of viable-but-nonculturable *Campylobacter jejuni* cells. Appl. Environ. Microbiol. 65: 1110–1116.

Thomas, C., Hill, D. and Mabey, M. 2002. Culturability, injury and morphological dynamics of thermophilic *Campylobacter* spp. within a laboratory-based aquatic model system. J. Appl. Microbiol. 92: 433–442.

Thomas, C., Hill, D.J. and Mabey, M. 1999a. Evaluation of the effect of temperature and nutrients on the survival of *Campylobacter* spp. in water microcosms. J. Appl. Microbiol. 86: 1024–1032.

Thomas, C., Hill, D.J. and Mabey, M. 1999b. Morphological changes of synchronized *Campylobacter jejuni* populations during growth in single phase liquid culture. Lett. Appl. Microbiol. 28: 194–198.

Thunberg, R.L., Tran, T.T., Bennett, R.W., Matthews, R.N. and Belay, N. 2002. Microbial evaluation of selected fresh produce obtained at retail markets. J. Food Prot. 65: 677–682.

Thunberg, R.L., Tran, T.T. and Walderhaug, M.O. 2000. Detection of thermophilic *Campylobacter* spp. in blood-free enriched samples of inoculated foods by the polymerase chain reaction. J. Food Prot. 63: 299–303.

TIGR. 2003. The Institute for Genomic Research, Comprehensive Microbial Resource (CMR). http: //www.tigr.org/tigr-scripts/CMR2/CMRHomePage.spl.

Tokumaru, M., Konuma, H., Umesako, M., Konno, S. and Shinagawa, K. 1991. Rates of detection of *Salmonella* and *Campylobacter* in meats in response to the sample size and the infection level of each species. Int. J. Food Microbiol. 13: 41–46.

Trachoo, N. and Frank, J.F. 2002. Effectiveness of chemical sanitizers against *Campylobacter jejuni*-containing biofilms. J. Food Prot. 65: 1117–1121.

Trachoo, N., Frank, J.F. and Stern, N.J. 2002. Survival of *Campylobacter jejuni* in biofilms isolated from chicken houses. J. Food Prot. 65: 1110–1116.

Tsai, Y.L., Palmer, C.J. and Sangermano, L.R. 1993. Detection of *Escherichia coli* in sewage and sludge by polymerase chain reaction. Appl. Environ. Microbiol. 59: 353–357.

UK-PHLS. 2001. Outbreak of *Campylobacter* infection following microbiology conference. CDR Weekly 11: 2.

UK-PHLS. 2002. Enteric Archive 2002: Surveillance of waterborne disease and water quality, January to June 2002. Commun. Dis. Rep. CDR Wkly 11: 6–10.

Uyttendaele, M., de Troy, P. and Debevere, J. 1999. Incidence of *Salmonella*, *Campylobacter jejuni*, *Campylobacter coli*, and *Listeria monocytogenes* in poultry carcasses and different types of poultry products for sale on the Belgian retail market. J. Food Protect. 62: 735–740.

Uyttendaele, M., Schukkink, R., van Gemen, B. and Debevere, J. 1995. Detection of *Campylobacter jejuni* added to foods by using a combined selective enrichment and nucleic acid sequence-based amplification (NASBA). Appl. Environ. Microbiol. 61: 1341–1347.

van de Giessen, A., Mazurier, S.I., Jacobs-Reitsma, W., Jansen, W., Berkers, P., Ritmeester, W. and Wernars, K. 1992. Study on the epidemiology and control of *Campylobacter jejuni* in poultry broiler flocks. Appl. Environ. Microbiol. 58: 1913–1917.

van de Giessen, A.W., Heuvelman, C.J., Abee, T. and Hazeleger, W.C. 1996. Experimental studies on the infectivity of non-culturable forms of *Campylobacter* spp. in chicks and mice. Epidemiol. Infect. 117: 463–470.

van de Giessen, A.W., Tilburg, J.J., Ritmeester, W.S. and van der Plas, J. 1998. Reduction of *Campylobacter* infections in broiler flocks by application of hygiene measures. Epidemiol. Infect. 121: 57–66.

Vanderlinde, P.B., Shay, B. and Murray, J. 1998. Microbiological quality of Australian beef carcass meat and frozen bulk packed beef. J. Food Protect. 61: 437–443.

Vanderlinde, P.B., Shay, B. and Murray, J. 1999. Microbiological status of Australian sheep meat. J. Food Protect. 62: 380–385.

Varslot, M., Resell, J. and Fostad, I.G. 1996. [Water-borne *Campylobacter* infection – probably caused by pink-footed geese. Two outbreaks in Nord-Trondelag, Stjortdal in 1994 and Verdal in 1995]. Tidsskr Nor Laegeforen. 116: 3366–3369.

Vellinga, A. and Butzler, J.-P. 2000. The Increasing Incidence of Human Campylobacteriosis. Report and Proceedings of a WHO Consultation of Experts. World Health Organization, Copenhagen, Denmark. http: //www.who.int/emc-documents/zoonoses/whocdscsraph20017c.html.

Véron, M. and Chatelain, R. 1973. Taxonomic study of the genus *Campylobacter* Sebald and Véron and designation of the neotype strain for the type species, *Campylobacter fetus* (Smith and Taylor) Sebald and Veron. Int. J. Sys. Bacteriol. 23: 122–134.

Vogt, R.L., Sours, H.E., Barrett, T., Feldman, R.A., Dickinson, R.J. and Witherell, L. 1982. *Campylobacter* enteritis associated with contaminated water. Ann. Intern. Med. 96: 292–296.

Waage, A.S., Vardund, T., Lund, V. and Kapperud, G. 1999. Detection of small numbers of *Campylobacter jejuni* and *Campylobacter coli* cells in environmental water, sewage, and food samples by a seminested PCR assay. Appl. Environ. Microbiol. 65: 1636–1643.

161

Wachtel, M.R., Whitehand, L.C. and Mandrell, R.E. 2002a. Association of *Escherichia coli* O157: H7 with preharvest leaf lettuce upon exposure to contaminated irrigation water. J. Food Prot. 65: 18–25.

Wachtel, M.R., Whitehand, L.C. and Mandrell, R.E. 2002b. Prevalence of *Escherichia coli* associated with a cabbage crop inadvertently irrigated with partially treated sewage wastewater. J. Food Prot. 65: 471–475.

Waldenstrom, J., Broman, T., Carlsson, I., Hasselquist, D., Achterberg, R.P., Wagenaar, J.A. and Olsen, B. 2002. Prevalence of *Campylobacter jejuni*, *Campylobacter lari*, and *Campylobacter coli* in different ecological guilds and taxa of migrating birds. Appl. Environ. Microbiol. 68: 5911–5917.

Wang, H., Boyle, E. and Farber, J. 2000. Rapid and specific enzyme immunoassay on hydrophobic grid membrane filter for detection and enumeration of thermophilic *Campylobacter* spp. from milk and chicken rinses. J. Food Prot. 63: 489–494.

Wang, H., Farber, J.M., Malik, N. and Sanders, G. 1999. Improved PCR detection of *Campylobacter jejuni* from chicken rinses by a simple sample preparation procedure. Int. J. Food Microbiol. 52: 39–45.

Wassenaar, T.M., Geilhausen, B. and Newell, D.G. 1998. Evidence of genomic instability in *Campylobacter jejuni* isolated from poultry. Appl. Environ. Microbiol. 64: 1816–1821.

Wassenaar, T.M. and Newell, D.G. 2000. Genotyping of *Campylobacter* spp. Appl. Environ. Microbiol. 66: 1–9.

Wegmuller, B., Luthy, J. and Candrian, U. 1993. Direct polymerase chain reaction detection of *Campylobacter jejuni* and *Campylobacter coli* in raw milk and dairy products. Appl. Environ. Microbiol. 59: 2161–2165.

Wesley, R.D., Swaminathan, B. and Stadelman, W.J. 1983. Isolation and enumeration of *Campylobacter jejuni* from poultry products by a selective enrichment method. Appl. Environ. Microbiol. 46: 1097–1102.

Westling, K. and Evengard, B. 2001. Myocarditis associated with *Campylobacter* infection. Scand. J. Infect. Dis. 33: 877–878.

Wheeler, J.G., Sethi, D., Cowden, J.M., Wall, P.G., Rodrigues, L.C., Tompkins, D.S., Hudson, M.J. and Roderick, P.J. 1999. Study of infectious intestinal disease in England: rates in the community, presenting to general practice, and reported to national surveillance. The Infectious Intestinal Disease Study Executive. Br. Med. J. 318: 1046–1050.

WHO. 1998. Surveillance Programme for Control of Food-borne Infections and Intoxications in Europe, 7th Report. http: //www.bgvv.de/internet/7threport/7threp_ctryreps_fr.htm.

Wight, J.P., Rhodes, P., Chapman, P.A., Lee, S.M. and Finner, P. 1997. Outbreaks of food poisoning in adults due to *Escherichia coli* O111 and *Campylobacter* associated with coach trips to northern France. Epidemiol. Infect. 119: 9–14.

Willis, W.L. and Murray, C. 1997. *Campylobacter jejuni* seasonal recovery observations of retail market broilers. Poultry Sci. 76: 314–317.

Willison, H.J. and O'Hanlon, G.M. 1999. The immunopathogenesis of Miller Fisher syndrome. J. Neuroimmunol. 100: 3–12.

Wilson, I.G. 1997. Inhibition and facilitation of nucleic acid amplification. Appl. Environ. Microbiol. 63: 3741–3751.

Wilson, I.G. 2002. *Salmonella* and *Campylobacter* contamination of raw retail chickens from different producers: a six year survey. Epidemiol. Infect. 129: 635–645.

Wilson, I.G. and Moore, J.E. 1996. Presence of *Salmonella* spp. and *Campylobacter* spp. in shellfish. Epidemiol. Infect. 116: 147–153.

Winkenwerder, W. 1967. [Experimental infection of sheep with Vibrio fetus strains of human and avian origin]. Zentralbl. Veterinarmed. B. 14: 737–745.

Winquist, A.G., Roome, A., Mshar, R., Fiorentino, T., Mshar, P. and Hadler, J. 2001. Outbreak of campylobacteriosis at a senior centre. J. Am. Geriatr Soc. 49: 304–307.

Winters, D.K., O'Leary, A.E. and Slavik, M.F. 1997. Rapid PCR with nested primers for direct detection of *Campylobacter jejuni* in chicken washes. Mol. Cell. Probes. 11: 267–271.

Winters, D.K., O'Leary, A.E. and Slavik, M.F. 1998. Polymerase chain reaction for rapid detection of *Campylobacter jejuni* in artificially contaminated foods. Lett. Appl. Microbiol. 27: 163–167.

Winters, D.K. and Slavik, M.F. 1995. Evaluation of a PCR based assay for specific detection of *Campylobacter jejuni* in chicken washes. Mol. Cell. Probes. 9: 307–310.

Wood, R.C., MacDonald, K.L. and Osterholm, M.T. 1992. *Campylobacter* enteritis outbreaks associated with drinking raw milk during youth activities. A 10-year review of outbreaks in the United States. JAMA 268: 3228–3230.

Woodward, D.L. and Rodgers, F.G. 2002. Identification of *Campylobacter* heat-stable and heat-labile antigens by combining the Penner and Lior serotyping schemes. J. Clin. Microbiol. 40: 741–745.

Wright, E.P., Tillett, H.E., Hague, J.T., Clegg, F.C., Darnel, R., Culshaw, J.A. and Sorrell, J.A. 1983. Milk-borne *Campylobacter* enteritis in a rural area. J. Hyg. (Lond). 91: 227–234.

Yanagisawa, S. 1980. Large outbreak of *Campylobacter* enteritis among schoolchildren. Lancet 2: 153.

Yogasundram, K., Shane, S.M. and Harrington, K.S. 1989. Prevalence of *Campylobacter jejuni* in selected domestic and wild birds in Louisiana. Avian Dis. 33: 664–667.

Young, C.R., Harvey, R., Anderson, R., Nisbet, D. and Stanker, L.H. 2000. Enteric colonization following natural exposure to *Campylobacter* in pigs. Res. Vet. Sci. 68: 75–78.

Yu, L.S., Uknalis, J. and Tu, S.I. 2001. Immunomagnetic separation methods for the isolation of *Campylobacter jejuni* from ground poultry meats. J. Immunol. Methods 256: 11–18.

Yuki, N., Ho, T.W., Tagawa, Y., Koga, M., Li, C.Y., Hirata, K. and Griffin, J.W. 1999. Autoantibodies to GM1b and GalNAc-GD1a: relationship to *Campylobacter jejuni* infection and acute motor axonal neuropathy in China. J. Neurol. Sci. 164: 134–138.

Zhao, C., Ge, B., De Villena, J., Sudler, R., Yeh, E., Zhao, S., White, D.G., Wagner, D. and Meng, J. 2001. Prevalence of *Campylobacter* spp., *Escherichia coli*, and *Salmonella* serovars in retail chicken, turkey, pork, and beef from the Greater Washington, D.C., area. Appl. Environ. Microbiol. 67: 5431–5436.

Zhao, T., Doyle, M.P. and Berg, D.E. 2000. Fate of *Campylobacter jejuni* in butter. J. Food Prot. 63: 120–122.

Chapter 7

Methods for Epidemiological Analysis of *Campylobacter jejuni*

John D. Klena and Michael E. Konkel

Abstract

Identification of *Campylobacter* spp. to the species level is necessary for unraveling aspects of ecology, prevalence and potential dissemination of the bacteria through the environment. Species identification may be achieved by determining expressed characteristics (phenotypes), heritable traits (genotypes) or a combination of both. In order to gain further insight into an isolate's transmission, survival in the environment and virulence potential, methods must be employed that can discriminate dissimilar from similar isolates. Molecular methods capable of enhanced resolution are most effective at determining the differences between isolates. In addition to discriminatory capability, methods should be portable; that is, they should be able to be shared with and reproduced by other laboratories to maximize data exchange. From a practical aspect, the methods should also be relatively inexpensive to set up and maintain, and technically simple. Finally, the methods should have the potential to generate a result in real time. In this review, we discuss three methods, amplified fragment length polymorphism (AFLP), multilocus sequence typing (MLST) and whole genome microarray, used to discriminate between isolates of *C. jejuni* and highlight the application of these methodologies. Our intent is to evaluate the use of each technique as it applies to answering questions regarding the epidemiology and phylogeny of *C. jejuni*.

INTRODUCTION

Campylobacter jejuni is a zoonotic Gram-negative bacterium and is considered a major gastrointestinal pathogen in both the developed and the developing world (Tauxe, 1992, 1997). Consumption of poultry is considered one of the major risk factors with respect to human exposure to the organism (Harris *et al.*, 1986; Kramer *et al.*, 2000). In the last few years it has become apparent that other reservoirs of *C. jejuni* also pose significant risks to humans (Eberhart-Philips *et al.*, 1997; Neiman *et al.*, 2003; Stanley and Jones, 2003). In order to be able to track a specific bacterial pathogen through various environments, microbiologists and epidemiologists require accurate and reproducible data to fingerprint a particular isolate. Specific assays that can generate information capable of discriminating between members of the same species are essential.

OVERVIEW OF PHENOTYPIC METHODS

Phenotypic methods, those based on detecting expressed characteristics such as the carbohydrate or protein antigens found on the organism's cell surface, are commonly used to characterize *C. jejuni* (Lior *et al.*, 1982; Penner and Hennessy, 1980). Other phenotypic methods assess susceptibility to antibiotics (Huysmans and Turnridge, 1997), metabolic requirements (Lior, 1984; On, 1996), sensitivity to bacteriophages (Gibson *et al.*, 1995; Sails *et al.*, 1998), protein profiles (On *et al.*, 1998; Ursing *et al.*, 1994) and fatty ester content (Steele *et al.*, 1998). In general, phenotypic methods are useful in distinguishing between outbreak- and non-outbreak-associated isolates (Neilsen *et al.*, 2000; Wareing *et al.*, 2002), however they lack sufficient discriminatory power and reproducibility to be used for long-term global epidemiology (Olive and Bean, 1999; Wassenaar and Newell, 2000). In addition, due to difficulties in preparing and maintaining reagents (particularly immunological reagents for serotyping), the poor discrimination afforded by biochemical tests and the frequency of genetic exchange within *C. jejuni*, phenotypic methods have fallen out of favor with many microbiologists.

OVERVIEW OF GENETIC METHODS

Genetic methods work under the assumption that if two isolates are identical, then their DNA, in terms of composition of bases, gene order and function, is also identical. The more dissimilar the DNA, the more dissimilar the isolates. Therefore, discrimination between isolates becomes a matter of determining variations at the genetic level.

Single-gene PCR-based methods

The literature is teeming with different methods for the detection and identification of *Campylobacter* to the species level. The single gene approach, in which different species are discriminated from one another based on the nucleotide sequence of one locus (al Rashid *et al.*, 2000; Eyers *et al.*, 1993; Gonzalez *et al.*, 1997; Gorkiewicz *et al.*, 2003; Konkel *et al.*, 1999; Werno *et al.*, 2002) is generally an effective way for determining the species of an unknown *Campylobacter*. A variation of this method is detection of the presence or absence of a specific gene such as *hipO* (Steinhauserova *et al.*, 2001). In some cases, the single gene approach can be used to determine intra-species differences; for *C. jejuni* variation in the flagellin A (*flaA*) gene, determined either by restriction fragment length polymorphisms (RFLP) or nucleotide sequence differences in the short variable region (SVR), have been used to discriminate between isolates (Meinersmann *et al.*, 1997; Nachamkin *et al.*, 1993). However, single gene identification methods are potentially problematic for epidemiological studies as well. For example, intragenomic and intergenomic recombination has been reported to occur within isolates of *C. jejuni* (de Boer *et al.*, 2002; Harrington *et al.*, 1997; Wassenaar *et al.*, 1998). A consequence of this recombination is that the selected marker may appear different between otherwise genetically indistinguishable isolates. Investigators have attempted to circumvent this problem using methods that examine the whole genome.

Macrorestriction profile-pulsed field gel electrophoresis

Strain NCTC 11168 was the first *C. jejuni* isolate to be sequenced in its entirety (Parkhill *et al.*, 2000) and the genome sequence of at least one additional isolate awaits publication (B. Miller, unpublished). It is not likely that genomic sequencing projects will be used on a routine basis for *C. jejuni* epidemiology as the cost to benefit ratio is unfavorable for this type of approach. Several methods, however, attempt to sample information from the entire genome of an isolate in order to determine the level of variation amongst isolates. The 'gold standard' of these methods is macrorestriction profiling (mrp) using pulsed field gel electrophoresis (PFGE) (de Boer *et al.*, 2000; Dickins *et al.*, 2002; Fitzgerald *et al.*, 2001a; Fitzgerald *et al.*, 2001b; Gibson *et al.*, 1995; Hänninen *et al.*, 1999; Hänninen *et al.*, 2001; Ribot *et al.*, 2001; Steinbrueckner *et al.*, 2001; Stephens *et al.*, 1998; Vargas *et al.*, 2003). In general, the restriction enzyme *Sma*I is used to digest genomic DNA from an isolate, yielding four to twenty DNA fragments ranging in size from 40 to 400 kilobase pairs (Olive and Bean, 1999; Wassenaar and Newell, 2000). Better discriminatory power can be achieved using *Kpn*I (Michaud *et al.*, 2001) as it generates a greater number of DNA fragments, however it is more common to combine the results of separate *Sma*I and *Kpn*I analyses for strain discrimination (On, 1998). Intra- and inter- isolate recombination and deletions can also affect the quality of information available from mrp-PFGE (de Boer *et al.*, 2002; Hänninen *et al.*, 1999; Mixter *et al.*, 2003; Wassenaar *et al.*, 1998). Recombination and deletions can result in the loss or gain of a restriction fragment, making genetically indistinguishable isolates appear unique. This has been shown in animal infection studies where mrp-PFGE profiles of an isolate used to inoculate an animal can be vastly different from those recovered from the same animal (de Boer *et al.*, 2002; Hänninen *et al.*, 1999; Mixter *et al.*, 2003; Wassenaar *et al.*, 1998). Thus, this method may not be suitable for global epidemiological studies.

Multilocus enzyme electrophoresis

An alternative to DNA-based whole genome analyses in provided by multilocus enzyme electrophoresis (MLEE). Isolates are characterized by measuring the relative electrophoretic migration of a number of water soluble enzymes (Selander *et al.*, 1986). Non-synonymous substitutions that accumulate at a genetic level are expressed as amino acid changes at the protein level. These changes may lead to an altered electrophoretic mobility of the protein, with each of these proteins indicating a new genetic allele. Synonymous substitutions may also occur, but by definition, these silent mutations will not lead to changes in the primary sequence, and will therefore not result in a new allele. Synonymous mutations will, however, be represented in the nucleotide sequence. It is therefore probable that better discrimination between isolates will be evident using a nucleotide sequencing approach (Urwin and Maiden, 2003). Using a combination of 9 to 11 enzymes, an electrophoretic type (ET) is determined (Selander *et al.*, 1986). The ET is a numerical record representing the combination of alleles of the different enzymes tested. A comparison of ETs is subsequently used to infer phylogenetic relationships amongst isolates (Meinersmann *et al.*, 2002). MLEE was first used to characterize isolates of *C. jejuni* to address the question of whether all *C. jejuni* isolates are equally capable of causing disease in humans, irregardless of their origin (Aeschbacher and Piffaretti, 1989). MLEE has not been greatly used for *C. jejuni* as the technique is somewhat cumbersome, difficult to reproduce and compare between laboratories and can generate untypeable alleles. Recently, however, MLEE has been used to assess the clonality of *C. jejuni* isolates (Matsuda *et al.*, 2003; Meinersmann *et al.*, 2002; Nachamkin *et al.*, 2001).

Three whole genome methods have been described which have been shown to be useful in the ability to discriminate between isolates of *C. jejuni*. Each method has its own advantages and drawbacks, and no one method is ideal. A realistic picture of the epidemiology of *C. jejuni* may be achieved using combinations of AFLP, MLST and microarray, in addition to mrp-PFGE, MLEE or single gene methods.

AMPLIFIED FRAGMENT LENGTH POLYMORPHISM (AFLP) ISOLATE PROFILING

AFLP is a whole genome-fingerprinting method of bacterial genomes that may be applied to phylogenetic studies and microbial typing (Olive and Bean, 1999; Savelkoul *et al.*, 1999; Wassenaar and Newell, 2000). To perform AFLP, purified genomic DNA is digested with two restriction enzymes of different cutting frequencies, generally a six-base pair cutter (e.g. *Bgl*II or *Hin*dIII) and a four-base pair cutter (e.g. *Csp*6I, *Hha*I, or *Mfe*I). Specific adapter oligonucleotides are ligated to the subsequent fragments. Oligonucleotides are constructed in a fashion such that the original restriction enzyme sites are not restored after ligation so that digestion and ligation can take place simultaneously. Selective amplification of adapter-fragment products is achieved using adapter-specific primers in which the 3' end is extended by one to three nucleotides. It is this extension that provides the stringency to the PCR reaction. One of the adapter-specific primers (generally the six-base pair cutter) is labelled, usually with a fluorochrome. Amplified products are resolved through polyacrylamide gels and detected using an automated sequencer.

Duim *et al.* (1999) suggested that isolates that appeared greater than 90% homologous should be considered genetically indistinguishable. In this study, the relationship between isolates is defined as percent homology, a reflection of the number of bands that are shared between two isolates. Others have suggested a cut off level of 95% or above (Lindstedt *et al.*, 2000). Variability of 5–10% should be allowed for and may be explained by differences in PCR efficiency and gel to gel variations (Savelkoul *et al.*, 1999). As with all typing techniques, it is strongly advised to couple at least two methods of discrimination together to create a more robust data set (Manning *et al.*, 2001; Wassenaar and Newell, 2000).

Intra-species fingerprinting using AFLP

AFLP is becoming an increasingly popular method for *C. jejuni* isolate fingerprinting and outbreak identification (de Boer *et al.*, 2002; Desai *et al.*, 2001; Duim *et al.*, 2001; Hänninen *et al.*, 2001; Hein *et al.*, 2003a; Hein *et al.*, 2003b; Manning *et al.*, 2001; Moreno *et al.*, 2002; On and Harrington, 2000; Schouls *et al.*, 2003; Wagenaar *et al.*, 2001). The first applications of AFLP typing to the study of *Campylobacter* spp. were reported in 1999 (Duim *et al.*, 1999; Kokotovic and On, 1999). Kokotovic and On (1999) used the restriction enzymes *Bgl*II and *Csp*6I and a single selective primer CSP61-A to examine the genetic relationships among 50 previously characterized *Campylobacter* isolates. Included in this sample were genetically distinct as well as indistinguishable isolates. Over 60 DNA fragments in the range 35 to 500 bp were resolved in this analysis. Nineteen AFLP profiles from 27 strains of *C. jejuni* and 18 AFLP profiles from 23 strains of *Campylobacter coli* were determined. AFLP results were in good agreement with previously generated two-enzyme mrp-PFGE data.

Similar results were obtained using the combination of restriction enzymes *Hin*dIII and *Hha*I and the specific amplification primers HindA and HhaA to isolates of *C. jejuni* and *C. coli* (Duim *et al.*, 1999). A total of 40 to 70 evenly distributed and highly reproducible bands ranging from 50 to 450 base pairs in size were generated. As shown earlier by Kokotovic and On (1999), isolates that were genetically indistinguishable or highly similar using other methods such as mrp-PFGE were also shown to be genetically homologous, whereas isolates that were non-related were easily identified by AFLP.

Using a third pair of restriction enzymes (*Bgl*II and *Mfe*I), 20–25 DNA fragments in the range of 50 to 500 base pairs per isolate were generated from a collection of 91

C. jejuni isolates (Lindstedt *et al.*, 2000). In this analysis AFLP was directly compared to mrp-PFGE and *fla*-RFLP. Both sporadic and outbreak-associated isolates were tested. Forty different AFLP patterns were determined from the 91 isolates, with outbreak isolates clustering together at a level of 96–100% genetic homology. AFLP was also shown to be more discriminating than *Sma*I-mrp-PFGE and in some instances, indistinguishable isolates based on mrp-PFGE could be distinguished by AFLP. Similarly, 35 Finnish isolates examined by *Sma*I and *Sac*II mrp-PFGE could be subdivided using AFLP (*Hind*III/*Hha*I) (Hänninen *et al.*, 2001). Overall, the mrp-PFGE results were congruent with the AFLP results, though in some cases, AFLP was more discriminating than PFGE.

Studies have shown that there was no evidence for environment-specific strains as determined by AFLP (Duim *et al.*, 1999; Kokotovic and On, 1999). This was further supported in a separate study of 128 isolates of *C. jejuni*, of these 71 were from humans and 57 were from poultry (Duim *et al.*, 2000). Two major groups of isolates were evident at a 40% genetic homology level and each group consisted of human and poultry isolates. No characteristic pattern of strains infecting either humans or poultry was evident. Further, strains linked to Guillain–Barré (GBS) or Miller Fisher (MFS) syndromes showed no specific AFLP profile. In an unrelated study, 18 *C. jejuni* isolates from the Netherlands and Belgium associated with either GBS or MFS were subjected to AFLP analysis (Endtz *et al.*, 2000). Twelve unique fingerprints were characterized, and neither the GBS nor the MFS isolates clustered distinctively from reference serotype strains.

AFLP analyses have also shown evidence for genetically stable strains. Strain 81116 was originally isolated from a human outbreak in the United Kingdom in 1981 (Palmer *et al.*, 1983). Isolates of *C. jejuni* cultured almost 20 years later showed identical AFLP (*Hind*III/*Hha*I) profiles, as well as mrp-PFGE (*Sma*I, *Kpn*I, and *Bam*HI) and *fla*-RFLP profiles (Manning *et al.*, 2001). Clustering of strains with no apparent epidemiological relationship was evident (On and Harrington, 2000). Eleven isolates of Penner serotype O:41 from South Africa, cultured over a span of 17 years, were shown to be indistinguishable by AFLP (Wassenaar *et al.*, 2000). Similarly, five isolates of O:19 were greater than 92% homologous when analysed by AFLP. Two distinctive (geographically and temporally unlinked) outbreaks in Norway were also shown to be caused by the same isolate (Lindstedt *et al.*, 2000). In their study, Hänninen *et al.* (2001) also showed stable strains of *C. jejuni* in Finland that persisted in poultry and humans over time.

Inter-species identification using AFLP

In addition to demonstrating the genetic relationship between isolates within a species, AFLP was able to group isolates into species-specific clusters (Duim *et al.*, 1999). Subsequently, AFLP has been used to investigate the phylogenetic relationship among 138 strains of *Campylobacter* spp. (On and Harrington, 2000). Most of the 16 species examined in this analysis generated between 30 to over 100 bands in the 50 to 500 base pairs region. *Campylobacter helveticus* and *Campylobacter upsaliensis* resulted in 0 to 5 and 6 to 11 bands, respectively. AFLP was able to reveal close genetic relationships between *Campylobacter* strains as well as species and subspecies. One exception appeared to be *Campylobacter fetus* subsp. *fetus* and *Campylobacter fetus* subsp. *venerealis*. Similar results were also obtained by Duim *et al.* (2001), using *Hind*III and *Hha*I (Duim *et al.*, 2001). However, most isolates of *C. helveticus* and *C. upsaliensis* resulted in stable, reproducible AFLP fingerprints and *C. fetus* subsp. *fetus* isolates could be discriminated from *C. fetus* subsp. *venerealis* isolates at a similarity level of >80% (see also Wagenaar *et al.*, 2001). Both studies conclude that while AFLP accurately discriminates species of *Campylobacter*, the phylogeny was not absolutely congruent with other data, such as 16S rDNA analyses (Vandamme, 2000).

MODIFICATIONS OF AFLP

Single-enzyme AFLP (sAFLP)
A variation to the AFLP method described above is sAFLP (Champion *et al.*, 2002; Moreno *et al.*, 2002). Moreno *et al.* (2002) analysed 69 mainly non-epidemiologically related isolates (23 clinical, 43 poultry, three reference isolates) using *Hind*III as the sole enzyme. From the 69 isolates, 66 unique profiles were resolved, clustering into nine groups. While several groups consisted exclusively of either clinical isolates or poultry isolates, three groups were composed of a combination of isolates from different sources. However, sAFLP did not always distinguish between *C. coli* and *C. jejuni*. Champion *et al.* (2002) analysed three outbreaks in southern Wales using mrp-PFGE and sAFLP and found that both methods were equally able to identify the outbreak-related isolates. Therefore, with regards to outbreak investigation, sAFLP may be a useful alternative to mrp-PFGE.

Multiplex endonuclease genotyping approach (MEGA)
A four-restriction endonuclease/two-adapter method has recently been described for use to fingerprint the unicellular protozoan parasite *Trypanosome brucei* (Agbo *et al.*, 2003). The use of additional restriction enzymes increases the number of DNA fragments in the analysis, and thus can increase the resolution of the method. This fine-tuning of AFLP may be useful when comparing genetically indistinguishable isolates that have different phenotypes.

MULTILOCUS SEQUENCE TYPING
Multilocus sequence typing (MLST) is, in principle, similar to MLEE in that neutral genetic variation from multiple sites on the chromosome is noted (Selander *et al.*, 1986). However, instead of assigning alleles based on alterations in enzyme migration through an electric current, they are assigned based on nucleotide sequence changes (Maiden *et al.*, 1998; Urwin and Maiden, 2003). For bacterial populations that have had MLST applied to them, 450 to 500 base pairs of nucleotide sequence taken from seven or eight housekeeping genes has been as informative or more so than 10 to 20 isozymes analysed by MLEE. Three other advantages that MLST has over MLEE is that the samples are portable, data are easily reproducible in any laboratory and it is easily automated (high-throughput).

Genes for MLST analysis are selected based on the following criteria (Urwin and Maiden, 2003): (1) the genes are under stabilizing selection for conservation of metabolic function; that is, they are not under strong selective pressure; (2) the genes are unlinked on the bacterial genome and thus unlikely to be co-inherited in a single genetic event and (3) the genes are present in all members of the species in question. With regards to *C. jejuni*, three MLST schemes have been proposed, although only one is currently used by the MLST website (http://pubmlst.org/campylobacter/). Regardless of the MLST procedure used for *C. jejuni*, DNA sequence from seven housekeeping genes, including some of the flanking DNA are initially amplified using conserved oligonucleotide primers in a PCR. Amplicons generated by PCR are subsequently purified and used as template for cycle-sequencing using primers designed to amplify a 450 to 500 base pair internal fragment of each gene and fluorescently labelled dNTPs. After removal of unincorporated primer, dNTPs and agarose, the nucleotide sequence of the PCR amplicons is determined, the sequence is trimmed to a set length (reflective of accurate reading) and each strand compared to one another. Finished sequences are assigned allele numbers via the internet database and new alleles are given a unique number that reflects the order in which they were discovered. The combination of alleles in any given scheme is known as the allelic profile and allelic profiles are assigned a sequence type (ST). STs, containing genetically related, but not necessarily identical isolates, are grouped together on the basis of allelic differences. Clonal complexes are assigned on the basis of a combination of genetic and microbiological criteria. For *C. jejuni*, isolates differing in up to and including three alleles are considered genetically related. In addition to the internet-based database for *C. jejuni* MLST analysis, a collection of reference isolates representing known *C. jejuni* clonal complexes is also available (Wareing *et al.*, 2003).

The first reported MLST system for *C. jejuni* used the housekeeping genes aspartase A (*aspA*), glutamine synthetase (*glnA*), citrate synthase (*gltA*), serine hydroxymethyltransferase (*glyA*), phosphoglucomutase (*pgm*), transketolase (*tkt*), and ATP synthase α subunit (*uncA*) (Dingle *et al.*, 2001a). One hundred and ninety-four *C. jejuni* isolates from four countries representing a diversity of sources and serotypes were analysed, identifying 155 unique STs and 62 clonal complexes. Extensive intra- and inter- species horizontal gene exchange was evident and based on measurements of the index of association (I_A) (Maynard Smith *et al.*, 1993) for the entire population of isolates and a subset of isolates, a weakly clonal population structure was indicated. A second system using seven different genes (aspartate-semialdehyde dehydrogenase, *asd*; ATP synthase F1α, *atpA* (this partially overlaps the *uncA* region described by Dingle *et al.* (2001a)); D-Alanine-D-Alanine-ligase, *ddlA*; elongation factor TS, *eftS*; fumarate hydratase, *fumC*, NADH dehydrogenase H, *nuoH*; and GTP-binding protein, *yphC*) to characterize isolates of *C. jejuni* (Suerbaum *et al.*, 2001). In this study of primarily human gastroenteric isolates ($n = 33$) from four countries, frequent intra-species recombination within *C. jejuni*, with extensive diversity in STs provided by a small number of polymorphic nucleotides, was evident. Recently, a third MLST system for *C. jejuni* (Manning *et al.*, 2003) was described. This approach made use of only six loci, one of which (*gltA*) was used previously. The other five loci analysed were adenylate kinase (*adk*), dihydroxy acid dehydratase (*ilvD*), malate dehydrogenase (*mdh*), glucose-6-phosphate isomerase (*pgi*) and a putative oxidoreductase, Cj1585c. When compared directly to the MLST method of Dingle *et al.* (2001a), fewer complexes were evident. However, the overall conclusions from this MLST system were consistent with the other two methods, that is, *C. jejuni* consists of a weakly clonal population with frequent intraspecies recombination.

MLST has been used to show an association between the genotype of *C. jejuni* isolates and source (Dingle *et al.*, 2001b; Dingle *et al.*, 2002; Manning *et al.*, 2003). MLST indicated a strong association among some animals and the ST, for instance 16 of 18 ST403 isolates were found to be from pigs (Manning *et al.*, 2003). Similar associations were demonstrated for sheep and bovine isolates. By analysing more than 800 isolates in the MLST database, Dingle *et al.* (2002) showed that several clonal complexes also had an association with an isolate's source and in some cases a particular surface structure, notably a link between the ST-22 complex and Penner serotype 19. In a separate study MLST reaffirmed the association of one Penner type within *C. jejuni* isolates associated with neuropathies (Dingle *et al.*, 2001b). In this study, 25 *C. jejuni* isolates associated with either GBS or MFS were analysed. These isolates were also compared to nearly 800 non-neuropathic isolates in the MLST database. The 25 isolates were diverse, falling into 13 clonal complexes composed of 20 STs, similar to the overall *C. jejuni* diversity in the MLST database. One ST complex, ST-22, was over-represented in neuropathy-associated isolates as compared with the MLST database, and this complex consisted of isolates of Penner serotype 19, previously shown to be relatively homogeneous (Tsang *et al.*, 2001).

Mrp-PFGE, when compared directly to MLST, is able to generate a greater number of subtypes (Sails *et al.*, 2003a,b) but given the utility and portability of the method, and the ease of data transfer and exchange, it is becoming a method of choice for epidemiological investigations (Sails *et al.*, 2003b; Urwin and Maiden, 2003). Additionally, for long-term epidemiological studies as well as phylogenetic studies, mrp-PFGE is not a suitable tool due to recombinations and reassortments within the *C. jejuni* genome (Hänninen *et al.*, 1999; Wassenaar *et al.*, 1998). To increase the discriminatory power of MLST, information from a gene under positive selection, such as the flagellin gene *flaA*, may be used (Sails *et al.*, 2003b; Urwin and Maiden, 2003). In conjunction with the nucleotide sequence from the *flaA* SVR region, the discriminatory power of MLST was comparable to that of *Sma*I mrp-PFGE (Sails *et al.*, 2003b).

WHOLE GENOME DNA MICROARRAY ANALYSES

An offshoot of the *Campylobacter jejuni* NCTC 11168 nucleotide sequencing project has been the development of whole genome DNA microarrays. Whole genome DNA microarrays have been used to compare interstrain and intrastrain-specific variations in bacteria. With regards to *C. jejuni*, microarrays have been used for three purposes, rapid detection and speciation (Keramas *et al.*, 2003; Volokhov *et al.*, 2003), genomotyping (Dorrell *et al.*, 2001; Kim *et al.*, 2002; Leonard *et al.*, 2003; Pearson *et al.*, 2003) and gene expression (Stintzi, 2003). We will focus on the first two applications in this review.

To establish a whole genome DNA microarray, individual open reading frames (ORF) representing each coding region from the bacterial genome, or a portion of the bacterial chromosome of interest (e.g. flagella synthesis and assembly) are amplified using ORF-specific oligonucleotide primers via PCR (Dorrell *et al.*, 2001; Kim *et al.*, 2002; Leonard *et al.*, 2003; Pearson *et al.*, 2003). Alternatively, specific oligonucleotide capture probes may be used instead of PCR amplicons (Keramas *et al.*, 2003; Volokhov *et al.*, 2003). Using currently available chemistries (Call *et al.*, 2003), DNA probes are immobilized onto a solid matrix such as a glass slide in the form of a spot.

DNA, in the form of specific PCR amplicons (Keramas *et al.*, 2003; Volokhov *et al.*, 2003) or randomly primed products generated from genomic DNA (Dorrell *et al.*, 2001; Kim *et al.*, 2002; Leonard *et al.*, 2003; Pearson *et al.*, 2003) are most commonly fluorescently labelled and subsequently hybridized to the probe spots deposited on the microarray. After hybridization and washing steps, arrays are scanned using a laser- of filter-based high resolution scanner (Call *et al.*, 2003) and images subsequently analysed using computer software.

Rapid species identification using DNA microarrays

Numerous applications of modern microbiology rely on the ability to rapidly and reliably identify bacteria to the species level. Detection of a potential pathogen in food and water supplies destined for animal or human consumption are two examples (Call *et al.*, 2003; Kakinuma *et al.*, 2003). One of the distinct advantages whole genome DNA microarrays have over other identification methods is that several genes, and indeed several signature regions within each gene, can be used to confirm species identity (Keramas *et al.*, 2003; Volokhov *et al.*, 2003).

Microarrays have recently been developed for the rapid identification of some or all of the thermotolerant *Campylobacter* spp. One DNA microarray is based on oligonucleotide capture probes targeting either the 16S rDNA, 23S rDNA, Cj0046 or the *hipO* genes, identifying *C. coli* and *C. jejuni* (Keramas *et al.*, 2003). Similarly, a microarray has been developed that identifies *C. coli*, *C. jejuni*, *C. lari* and *C. upsaliensis* (Volokhov *et al.*, 2003) using fluorescently labelled single stranded RNA products synthesized *in vitro* from PCR products. In this microarray, five target genes (*fur*, *glyA*, *cdtABC*, *ceuBC* and *fliY*) were sequenced from representative isolates of the four species and several relatively short (17 to 36 base pairs) oligonucleotide species-specific probes to each of the genes were developed and spotted on to the glass slide. Both Keremas *et al.* (2003) and Volokohov *et al.* (2003) used multiple probes for identification of bacteria to the species level to improve confidence in detection. Rapid identification of the presence of target bacterial species in food and water should now be possible using microarray technology.

Genomotyping

Beyond detecting and identifying the presence of *Campylobacter* spp. in a sample, whole genome DNA microarrays have been used as a means of comparing isolates of *C. jejuni* at the genome level. Hybridization of an isolate's genomic DNA to a DNA microarray is referred to as genomotyping. Dorrell *et al.* (2001) made use of the ordered pUC18 library containing the entire genome of NCTC 11168 to establish a low-cost whole genome DNA microarray. Using the M13 forward and reverse primers located on every vector, PCR amplicons ranging in size from 0.2 to 2 kb were generated. These amplicons were spotted onto a glass slide and the slides were prepared for hybridization. Fluorescently labelled, randomly primed genomic DNA from 11 isolates, including two laboratory-characterized strains, 81116 and 81-176, were individually hybridized to the NCTC 11168 library. Using an alternative fluorescent label, randomly primed genomic DNA from strain NCTC 11168 was used as an internal control. Genes represented by spots in which the ratio of the fluorescence signal intensity between the control and the sample was below an arbitrarily defined value of 0.5 were considered absent or sufficiently divergent to represent new genes. Using this criteria, the most similar isolate to NCTC 11168 differed in only 0.54% of the genome, whereas the two most divergent genomotypes, relative to NCTC 11168, lacked at least 6.95% of the NCTC 11168 genes. To illustrate the potential power of this method, a set of Penner serotype 2 (HS2) isolates were compared to NCTC 11168. While all of the HS2 isolates were identical to NCTC 11168 with respect to their capsular polysaccharide locus, extensive variation was evident elsewhere. One potential problem noted by Dorrell *et al.* (2001) with their microarray results from the type of amplified NCTC 11168 PCR products spotted onto the slides. Only slightly more than 1/3 of the 1651 annotated genes of the NCTC 11168 genome were represented as single-gene fragments. Slightly more than a third had a second gene either upstream or downstream of the target gene. Slightly under 1/3 of the genome contained at least two additional gene fragments. Amplicons containing multiple gene products could lead to false-positive results due to hybridization to the non-target gene.

To correct potentially misleading results originating from more than one gene fragment per PCR product, Leonard *et al.* (2003) designed gene-specific, non-overlapping, nonhomologous primers to 99% of the NCTC 11168 genome as well as each open reading frame in the recently reported virulence plasmid, pVir (Bacon *et al.*, 2000; Bacon *et al.*, 2002). In addition, the interpretation of the signal intensity of each spot was evaluated in a different manner, based on the shape of the signal-ratio distribution of each spot (Kim *et al.*, 2002). Application of this genomotyping array to five sets of isolates representing different epidemiological situations (two outbreaks, one family distribution, two sets of isolates taken from HIV-positive patients over time) showed the value of the method. Genomotyping was able to distinguish between isolates responsible for outbreaks and temporally associated isolates that were not involved in the outbreak. In one outbreak instance, even though other typing techniques failed to distinguish between isolates, the genomotype microarray clearly showed one isolate as distinct. In a similar manner, genomotype microarray analysis demonstrated that at least two distinctive isolates could be cultured from an HIV-positive individual (Leonard *et al.*, 2003). Both the Dorrell *et al.* (2001) and Leonard *et al.* (2003) DNA microarrays revealed similar regions of divergence between isolates. These regions were the flagellar biosynthetic locus, the LOS locus, the capsular polysaccharide locus and a locus encoding restriction-modification systems. In addition, two loci involved in sugar modification and transport, and genes in the virulence plasmid, were also shown to be significantly divergent between isolates tested.

It is important to note that the software and settings used to analyse hybridized microarrays are critical in classifying whether a gene is present or absent (Kim *et al.*, 2002). For instance, presence or absence of a gene is determined by the comparison of the ratio of the signal obtained on the microarray at any given gene for the reference isolate and the isolate tested. A cut-off value must be assigned to group genes as either present or divergent, and this value is usually empirically determined. If the value is not determined for each data set but arbitrarily assigned, then inappropriate interpretations can be made leading to false conclusions.

Recently, a genomotyping microarray has been described that redefines the boundary between what constitutes a conserved and a variable gene (Pearson *et al.*, 2003). Most microarrays empirically determine the fluorescence signal from a reference strain to a similar strain with known deletions in order to calculate what ratio of fluorescence constitutes a conserved gene (Pearson *et al.*, 2003). Rather than using an empirically determined constant ratio, a new algorithm was described that dynamically determines whether a gene is conserved or variable. This algorithm takes into account the global variability of the whole set of conserved genes as well as the individual variability of the fluorescent measurement for each gene. Further, this method allows a prediction of the relative degree of divergence to be made. That is, a variable gene may be defined as partly or entirely deleted from the tested isolate or the gene could have primary sequence divergence from the control isolate to an extent to significantly reduce hybridization. The algorithm described by Pearson *et al.* (2003) can distinguish between small deletions/changes in a segment of the primary sequence and gene deletions. Using this new method to analyses the 1654 NCTC 11168 genome ORFs in 18 *C. jejuni* isolates, divergence ranged from having only 40 genes (2.6%) to 163 genes (10.2%) absent, with a total of 269 genes variable (16.3%) across all isolates (Pearson *et al.*, 2003). Interestingly, seven hypervariable plasticity regions were defined, four of which were previously unrecognized. The largest plasticity region contained a deletion in 24 consecutive genes. Statistical analysis strongly supports the conclusion that these hypervariable regions are non-random, and thus represent regions of the genome that may have been lost or acquired during *C. jejuni* evolution.

SUMMARY

C. jejuni remains an important cause of bacterial gastroenteritis in developed and developing countries. To reach a better understanding of the sources of infection and the routes of transmission, easily accessible, inexpensive and reliable methods for fingerprinting isolates is necessary. While no method alone is adequate for all applications, AFLP, MLST and DNA microarray techniques have been successfully used to distinguish the epidemiological and phylogenic relationships among isolates of *C. jejuni*. These methods make use of increased genetic information reflective of the entire bacterial genome, so the relationships drawn by AFLP, MLST or DNA microarray are less likely to be misinterpreted as a consequence of point mutations or recombinations at a single genetic locus or region of the genome (Savelkoul *et al.*, 1999). Recent studies have shown that genetically defined isolates may acquire deletions, insertions or point mutations affecting their mrp-PFGE profile (de Boer *et al.*, 2002; Mixter *et al.*, 2003). Methods such as mrp-PFGE may be too sensitive for long term global epidemiological studies and creating phylogenies, while AFLP, MLST and DNA microarray are not (de Boer *et al.*, 2002). Increased application of these methods with a concomitant decrease in the setup and operational costs should permit inroads into the ecology of *C. jejuni* in the near future.

Acknowledgments
Work in the laboratory of M.E.K. is supported by grants from the NIH (DK58911) and USDA National Research Initiative Competitive Grants Program (USDA/NRICGP, 99-35201-8579 and 2002-5111001-983).

References
Aeschbacher, M., and Piffaretti, J.-C. 1989. Population genetics of human and animal enteric *Campylobacter* strains. Infect. Immun. 57: 1432–1437.

Agbo, E.C., Duim, B., Majiwa, P.A.O., Büscher, P., Claassen, E., and te Pas, M.F.W. 2003. Multiplex-endonuclease genotyping approach (MEGA): a tool for the fine scale detection of unlinked polymorphic DNA markers. Chromosoma 111: 518–524.

al Rashid, S.T., Dakuna, I., Louie, H., Ng, D., Vandamme, P., Johnson, W., and Chan, V.L. 2000. Identification of *Campylobacter jejuni*, *C. coli*, *C. lari*, *C. upsaliensis*, *Arcobacter butzleri*, and *A. butzleri*-like species based on the *glyA* gene. J. Clin. Microbiol. 38: 1488–1494.

Bacon, D.J., Alm, R.A., Burr, D.H., Hu, L., Kopecko, D.J., Ewing, C.P., Trust, T.J., and Guerry, P. 2000. Involvement of a plasmid in virulence of *Campylobacter jejuni* 81-176. Infect. Immun. 68: 4384–4390.

Bacon, D.J., Alm, R.A., Hu, L., Hickey, T.E., Ewing, C.P., Batchelor, R.A., Trust, T.J., and Guerry, P. 2002. DNA sequence and mutational analyses of the pVir plasmid of *Campylobacter jejuni* 81-176. Infect. Immun. 70: 6242–6250.

Call, D.R., Borucki, M.K., and Loge, F.J. 2003. Detection of bacterial pathogens in environmental samples using DNA microarrays. J. Microbiol. Methods 53: 235–243.

Champion, O.L., Best, E.L., and Frost, J.A. 2002. Comparison of pulsed-field gel electrophoresis and amplified fragment length polymorphism techniques for investigating outbreaks of enteritis due to campylobacters. J. Clin. Microbiol. 40: 2263–2265.

de Boer, P., Duim, B., Rigter, A., van Der Plas, J., Jacobs-Reitsma, W.F., and Wagenaar, J.A. 2000. Computer-assisted analysis and epidemiological value of genotyping methods for *Campylobacter jejuni* and *Campylobacter coli*. J. Clin. Microbiol. 38: 1940–1946.

de Boer, P., Wagenaar, J.A., Achterberg, R.P., Putten, J.P., Schouls, L.M., and Duim, B. 2002. Generation of *Campylobacter jejuni* genetic diversity *in vivo*. Mol. Microbiol. 44: 351–359.

Desai, M., Logan, J.M.J., Frost, J.A., and Stanley, J. 2001. Genome sequence-based fluorescent amplified fragment length polymorphism of *Campylobacter jejuni*, its relationship to serotyping, and its implications for epidemiological analysis. J. Clin. Microbiol. 39: 3823–3829.

Dickins, M.A., Franklin, S., Stefanova, R., Schutze, G.E., Eisenach, K.D., Wesley, I., and Cave, M.D. 2002. Diversity of *Campylobacter* isolates from retail poultry carcasses and from humans as demonstrated by pulsed-field gel electrophoresis. J. Food Prot. 65: 957–962.

Dingle, K.E., Colles, F.M., Wareing, D.R.A., Ure, R., Fox, A.J., Bolton, F.E., Bootsma, H.J., Willems, R.J.L., Urwin, R., and Maiden, M.C.J. 2001a. Multilocus sequence typing system for *Campylobacter jejuni*. J. Clin. Microbiol. 39: 14–23.

Dingle, K.E., van den Braak, N., Colles, F.M., Price, L.J., Woodward, D.L., Rodges, F.G., Endtz, H.P., van Belkum, A., and Maiden, M.C.J. 2001b. Sequence typing confirms that *Campylobacter jejuni* strains associated with Guillain–Barré and Miller Fisher syndromes are of diverse genetic lineage, serotype and flagella type. J. Clin. Microbiol. 39: 3346–3349.

Dingle, K.E., Colles, F.M., Ure, R., Wagenaar, J.A., Duim, B., Bolton, F.J., Fox, A.J., Wareing, D.R., and Maiden, M.C. 2002. Molecular characterization of *Campylobacter jejuni* clones: a basis for epidemiologic investigation. Emerg. Infect. Dis. 8: 949–955.

Dorrell, N., Mangan, J.A., Laing, K.G., Hinds, J., Linton, D., Al-Ghusein, H., Barrell, B.G., Parkhill, J., Stoker, N.G., Karlyshev, A.V., Butcher, P.D., and Wren, B. 2001. Whole genome comparison of *Campylobacter jejuni* human isolates using a low-cost microarray reveals extensive genetic diversity. Genome Res. 11: 1706–1715.

Duim, B., Wassenaar, T.A., Rigter, A., and Wagenaar, J. 1999. High-resolution genotyping of *Campylobacter* strains isolated from poultry and humans with amplified fragment length polymorphism fingerprinting. Appl. Environ. Microbiol. 65: 2369–2375.

Duim, B., Ang, C.W., van Belkum, A., Rigter, A., van Leeuwen, N.W.J., Endtz, H.P., and Wagenaar, J.A. 2000. Amplified fragment length polymorphism analysis of *Campylobacter jejuni* strains isolated from chickens and from patients with gastroenteritis or Guillain–Barré or Miller Fisher syndrome. Appl. Environ. Microbiol. 66: 3917–3923.

Duim, B., Vandamme, P.A.R., Rigter, A., Laevens, S., Dijkstra, J.R., and Wagenaar, J.A. 2001. Differentiation of *Campylobacter* species by AFLP fingerprinting. Microbiology 147: 2729–2737.

Eberhart-Philips, J., Walker, N., Garrett, N., Bell, D., Sinclair, D., Rainger, W., and Bates, M. 1997. Campylobacteriosis in New Zealand: results of a case–control study. J. Epidemiol. Commun. Health 51: 686–691.

Endtz, H.P., Ang, C.W., van den Braak, N., Duim, B., Rigter, A., Price, L.J., Woodward, D.J., Rodgers, F.G., Johnson, W.M., Wagenaar, J.A., Jacobs, B.C., Verbrugh, H.A., and van Belkum, A. 2000. Molecular characterization of *Campylobacter jejuni* from patients with Guillain–Barré and Miller Fisher syndromes. J. Clin. Microbiol. 38: 2297–2301.

Eyers, M., Chapelle, S., Van Camp, G., Goossens, H., and De Wachter, D. 1993. Discrimination among thermophilic *Campylobacter* species by polymerase chain reaction amplification of 23S rRNA gene fragments. J. Clin. Microbiol. 31: 3340–3443.

Fitzgerald, C., Helsel, L.O., Nicholson, M.A., Olsen, S.J., Swerdlow, D.L., Flahart, R., Sexton, J., and Fields, P.I. 2001a. Evaluation of methods for subtyping *Campylobacter jejuni* during an outbreak involving a food handler. J. Clin. Microbiol. 39: 2386–2390.

Fitzgerald, C., Stanley, K., Andrew, S., and Jones, K. 2001b. Use of pulsed field gel electrophoresis and flagellin gene typing in identifying clonal groups of *Campylobacter jejuni* and *Campylobacter coli* in farm and clinical environments. Appl. Environ. Microbiol. 67: 1429–1436.

Gibson, J.R., Fitzgerald, C., and Owen, R.J. 1995. Comparison of PFGE, ribotyping and phage-typing in the epidemiological analysis of *Campylobacter jejuni* serotype HS2 infection. Epidemiol. Infect. 115: 215–225.

Gonzalez, I., Grant, K.A., Richardson, P.T., Park, S.F., and Collins, M.D. 1997. Specific identification of the enteropathogens *Campylobacter jejuni* and *Campylobacter coli* by using a PCR test based on the *ceuE* gene encoding a putative virulence determinant. J. Clin. Microbiol. 35: 759–763.

Gorkiewicz, G., Feierl, G., Schober, C., Dieber, F., Köfer, J., Zechner, R., and Zechner, E.L. 2003. Species-specific identification of campylobacters by partial 16S rRNA gene sequencing. J. Clin. Microbiol. 41: 2537–2546.

Hänninen, M.-L., Hakkinen, M., and Rautelin, H. 1999. Stability of related human and chicken *Campylobacter jejuni* genotypes after passage through chick intestine studied by pulsed-field gel electrophoresis. Appl. Environ. Microbiol. 65: 2272–2275.

Hänninen, M.-L., Perko-Mäkelä, P., Rautelin, H., Duim, B., and Wagenaar, J.A. 2001. Genomic relatedness within five common Finnish *Campylobacter jejuni* pulsed-field gel electrophoresis genotypes studied by

176

amplified fragment length polymorphism analysis, ribotyping, and serotyping. Appl. Environ. Microbiol. 67: 1581–1586.

Harrington, C.S., Thomson-Carter, F.M., and Carter, P.E. 1997. Evidence for recombination in the flagellin locus of *Campylobacter jejuni*: implications for the flagellin gene typing scheme. J. Clin. Microbiol. 35: 2386–2392.

Harris, N.V., Weiss, N.S., and Nolan, C.M. 1986. The role of poultry and meats in the aetiology of *Campylobacter jejuni/coli* enteritis. Am. J. Publ. Health 76: 407–411.

Hein, I., Mach, R.L., Farnleitner, A.H., and Wagner, M. 2003a. Application of single-strand conformation polymorphism and denaturing gradient gel electrophoresis for *fla* sequence typing of *Campylobacter jejuni*. J. Microbiol. Methods 52: 305–313.

Hein, I., Schneck, C., Knogler, M., Feierl, G., Plesss, P., Kofer, J., Achmann, R., and Wagner, M. 2003b. *Campylobacter jejuni* isolated from poultry and humans in Styria, Austria: epidemiology and ciprofloxacin resistance. Epidemiol. Infect. 130: 377–386.

Huysmans, M.B., and Turnridge, J.D. 1997. Disc susceptibility testing for thermophilic campylobacters. Pathology 29: 209–216.

Kakinuma, K., Fukushima, M., and Kawaguchi, R. 2003. Detection and identification of *Escherichia coli*, *Shigella*, and *Salmonella* by microarrays using the *gyrB* gene. Biotechnol. Bioeng. 20: 721–728.

Keramas, G., Bang, D.D., Lund, M., Madsen, M., Rasmussen, S.E., Bunkenborg, H., Telleman, P., and Christensen, C.B.V. 2003. Development of a sensitive DNA microarray suitable for rapid detection of *Campylobacter* spp. Mol. Cell. Probes 17: 187–196.

Kim, C.C., Joyce, E.A., Chan, K., and Falkow, S. 2002. Improved analytical methods for microarray-based genome-composition analysis. Genome Biol. 3: 1–17.

Kokotovic, B., and On, S.L.W. 1999. High-resolution genomic fingerprinting of *Campylobacter jejuni* and *Campylobacter coli* by analysis of amplified fragment length polymorphisms. FEMS Microbiol. Lett. 173: 77–84.

Konkel, M.E., Grey, S.A., Kim, B.J., Garvis, S.G., and Yoon, J. 1999. Identification of the enteropathogens *Campylobacter jejuni* and *Campylobacter coli* based on the *cadF* virulence gene and its product. J. Clin. Microbiol. 37: 510–517.

Kramer, J.M., Frost, J.A., Bolton, F., and Wareing, D.R. 2000. *Campylobacter* contamination of raw meat and poultry at retail sale: identification of multiple types and comparison with isolates from human infection. J. Food Prot. 63: 1654–1659.

Leonard, E.E., 2nd, Takata, T., Blaser, M.J., Falkow, S., Tompkins, L.S., and Gaynor, E.C. 2003. Use of an open-reading frame-specific *Campylobacter jejuni* DNA microarray as a new genotyping tool for studying epidemiologically related isolates. J. Infect. Dis. 187: 691–694.

Lindstedt, B.A., Heir, E., Vardund, T., Melby, K.K., and Kapperud, G. 2000. Comparative fingerprinting analysis of *Campylobacter jejuni* subsp. *jejuni* strains by amplified-fragment length polymorphism genotyping. J. Clin. Microbiol. 38: 3379–3387.

Lior, H., Woodward, D.L., Edgar, J.A., Laroche, L.J., and Gill, P. 1982. Serotyping of *Campylobacter jejuni* by slide agglutination based on heat-labile antigenic factors. J. Clin. Microbiol. 15: 761–768.

Lior, H. 1984. New, extended biotyping scheme for *Campylobacter jejuni, Campylobacter coli*, and '*Campylobacter laridis*'. J. Clin. Microbiol. 20: 636–640.

Maiden, M.C., Bygraves, J.A., Feil, E., Morelli, G., Russell, J.E., Urwin, R., Zhang, Q., Zhou, J., Zurth, K., Caugant, D.A., Feavers, I.M., Achtman, M., and Spratt, B.G. 1998. Multilocus sequence typing: a portable approach to the identification of clones within populations of pathogenic microorganisms. Proc. Natl. Acad. Sci. USA 95: 3140–3145.

Manning, G., Duim, B., Wassenaar, T., Wagenaar, J.A., Ridley, A., and Newell, D.G. 2001. Evidence for a genetically stable strain of *Campylobacter jejuni*. Appl. Environ. Microbiol. 67: 1185–1189.

Manning, G., Dowson, C.G., Bagnall, M.C., Ahmed, i.H., West, M., and Newell, D.G. 2003. Multilocus sequence typing for comparison of veterinary and human isolates of *Campylobacter jejuni*. Appl. Environ. Microbiol. 69: 6370–6379.

Matsuda, M., Kaneko, A., Stanley, T., Millar, B.C., Miyajima, M., Murphy, P.G., and Moore, J.E. 2003. Characterization of urease-positive thermophilic *Campylobacter* subspecies by multilocus enzyme electrophoresis. Appl. Environ. Microbiol. 69: 3308–3310.

Maynard Smith, J., Smith, N.H., O'Rourke, M., and Spratt, B.G. 1993. How clonal are bacteria? Proc. Natl. Acad. Sci. USA 90: 4384–4388.

Meinersmann, R.J., Helsel, L.O., Fields, P.I., and Hiett, K.L. 1997. Discrimination of *Campylobacter jejuni* isolates by *fla* gene sequencing. J. Clin. Microbiol. 35: 2810–2814.

Meinersmann, R.J., Patton, C.M., Evins, G.M., Wachsmuth, I.K., and Fields, P.I. 2002. Genetic diversity and relationships of *Campylobacter* species and subspecies. Int. J. Syst. Evol. Microbiol. 52: 1789–1797.

Michaud, S., Menard, S., Gaudreau, C., and Arbeit, R.D. 2001. Comparison of *Sma* I-defined genotypes of *Campylobacter jejuni* examined by *Kpn* I: a population-based study. J. Med. Microbiol. 50: 1075–1081.

Mixter, P.F., Klena, J.D., Flom, G.A., Siegesmund, A.M., and Konkel, M.E. 2003. *In vivo* tracking of *Campylobacter jejuni* using a novel recombinant expressing green fluorescent protein. Appl. Environ. Microbiol. 69: 2864–2874.

Moreno, Y., Ferrús, M.A., Vanoostende, A., Hernández, M., Montes, R.M., and Hernández, J. 2002. Comparison of 23S polymerase chain reaction–restriction fragment length polymorphism and amplified fragment length polymorphism techniques as typing systems for thermophilic campylobacters. FEMS Microbiol. Lett. 211: 97–103.

Nachamkin, I., Bohachick, K., and Patton, C.M. 1993. Flagellin gene typing of *Campylobacter jejuni* by restriction fragment length polymorphism analysis. J. Clin. Microbiol. 31: 1531–1536.

Nachamkin, I., Engberg, J., Gutacker, M., Meinersman, R.J., Li, C.Y., Arzate, P., Teeple, E., Fussing, V., Ho, T.W., Asbury, A.K., Griffin, J.W., McKhann, G.M., and Piffaretti, J.-C. 2001. Molecular population genetic analysis of *Campylobacter jejuni* HS: 19 associated with Guillain–Barré syndrome and gastroenteritis. J. Infect. Dis. 184: 221–226.

Neilsen, E.M., Engberg, J., Fussing, V., Petersen, L., Brogren, C.-H., and On, S.L.W. 2000. Evaluation of phenotypic and genotypic methods for subtyping *Campylobacter jejuni* isolates from humans, poultry, and cattle. J. Clin. Microbiol. 38: 3800–3810.

Neiman, J., Engberg, J., Mølbak, K., and Wegener, H.C. 2003. A case–control study of risk factors for sporadic *Campylobacter* infections in Denmark. Epidemiol. Infect. 130: 353–366.

Olive, D.M., and Bean, P. 1999. Principles and applications of methods for DNA-based typing of microbial organisms. J. Clin. Microbiol. 37: 1661–1669.

On, S.L.W. 1996. Identification methods for campylobacters, helicobacters, and related organisms. Clin. Microbiol. Rev. 9: 405–422.

On, S.L.W. 1998. *In vitro* genotypic variation of *Campylobacter coli* documented by pulsed-field gel electrophoretic DNA profiling: implications for epidemiological studies. FEMS Microbiol. Lett. 165: 341–346.

On, S.L.W., Atabay, H.I., Corry, J.E.L., Harrington, C.S., and Vandamme, P. 1998. Emended description of *Campylobacter sputorum* and revision of its infrasubspecific (biovar) divisions, including *C. sputorum* biovar *paraureolyticus*, a urease-producing variant from cattke and humans. Int. J. Syst. Bacteriol. 48: 195–206.

On, S.L.W., and Harrington, C.S. 2000. Identification of taxonomic and epidemiological relationships among *Campylobacter* species by numerical analysis of AFLP profiles. FEMS Microbiol. Lett. 193: 161–169.

Palmer, S.R., Gully, P.R., White, J.M., Pearson, A.D., Suckling, W.G., Jones, D.M., Rawes, J.C., and Penner, J.L. 1983. Water-borne outbreak of *Campylobacter* gastroenteritis. Lancet 1: 287–290.

Parkhill, J., Wren, B.W., Mungall, K., Ketley, J.M., Churcher, C., Basham, D., Chillingworth, T., Davies, R.M., Feltwell, T., Holroyd, S., Jagels, K., Karlshev, A.V., Moule, S., Pallen, M.J., Penn, C.W., Quall, M.A., Rajandrean, M.A., Rutherford, K.M., Van Vliet, A.H.M., Whitehead, S., and Barrell, B.G. 2000. The genome sequence of the food-borne pathogen *Campylobacter jejuni* reveals hypervariable sequences. Nature 403: 665–668.

Pearson, B.M., Pin, C., Wright, J., l'Anson, K., Humphrey, T., and Wells, J.M. 2003. Comparative genome analysis of *Campylobacter jejuni* using whole genome DNA microarrays. FEBS Lett. 554: 224–230.

Penner, J.L., and Hennessy, J.N. 1980. Passive hemagglutination technique for serotyping *Campylobacter fetus* subsp. *jejuni* on the basis of soluble heat-stable antigens. J. Clin. Microbiol. 12: 732–737.

Ribot, E.M., Fitzgerald, C., Kubota, K., Swaminathan, B., and Barrett, T.J. 2001. Rapid pulsed-field gel electrophoresis protocol for subtyping of *Campylobacter jejuni*. J. Clin. Microbiol. 39: 1889–1894.

Sails, A.D., Wareing, D.R., Bolton, F.J., Fox, A.J., and Curry, A. 1998. Characterization of 16 *Campylobacter jejuni* and *C. coli* typing bacteriophages. J. Med. Microbiol. 47: 123–128.

Sails, A.D., Swaminathan, B., and Fields, P.I. 2003a. Clonal complexes of *Campylobacter jejuni* identified by multilocus sequence typing correlate with strain associations identified by multilocus enzyme electrophoresis. J. Clin. Microbiol. 41: 4058–4067.

Sails, A.D., Swaminathan, B., and Fields, P.I. 2003b. Utility of multilocus sequence typing as an epidemiological tool for investigation of outbreaks of gastroenteritis caused by *Campylobacter jejuni*. J. Clin. Microbiol. 41: 4733–4739.

Savelkoul, P.H.M., Aarts, H.J.M., de Haas, J., Dijkshoorn, L., Duim, B., Otsen, M., Rademaker, J.L.W., Schouls, L., and Lenstra, J.A. 1999. Amplified-fragment length polymorphism analysis: the state of the art. J. Clin. Microbiol. 37: 3083–3091.

178

Schouls, L.M., Reulen, S., Duim, B., Wagenaar, J.A., Willems, R.J., Dingle, K.E., Colles, F.M., and van Embden, J.D. 2003. Comparative genotyping of *Campylobacter jejuni* by amplified fragment length polymorphism, multilocus sequence typing, and short repeat sequencing: strain diversity, host range, and recombination. J. Clin. Microbiol. 41: 15–26.

Selander, R.K., Caugant, D.A., Ochman, H., Musser, J.M., Gilmour, M.N., and Whittam, T.S. 1986. Methods of multilocus enzyme electrophoresis for bacterial population genetics and systematics. Appl. Environ. Microbiol. 51: 8783–8884.

Stanley, K., and Jones, K. 2003. Cattle and sheep farms as reservoirs of *Campylobacter*. J. Appl. Microbiol. 94: 104S-113S.

Steele, M., McNab, B., Fruhner, L., DeGrandis, S., Woodward, D., and Odumeru, J.A. 1998. Epidemiological typing of *Campylobacter* isolates from meat processing plants by pulsed-field gel electrophoresis, fatty acid profile typing, serotyping, and biotyping. Appl. Environ. Microbiol. 64: 2346–2349.

Steinbrueckner, B., Ruberg, F., and Kist, M. 2001. Bacterial genetic fingerprint: a reliable factor in the study of the epidemiology of human *Campylobacter* enteritis? J. Clin. Microbiol. 39: 4155–4159.

Steinhauserova, I., Ceskova, J., Fojtikova, K., and Obrovska, I. 2001. Identification of thermophilic *Campylobacter* spp. by phenotypic and molecular methods. J. Appl. Microbiol. 90: 470–475.

Stephens, C.P., On, S.L.W., and Gibson, J.A. 1998. An outbreak of infectious hepatitis in commercially reared ostriches associated with *Campylobacter coli* and *Campylobacter jejuni*. Vet. Microbiol. 61: 183–190.

Stintzi, A. 2003. Gene expression profile of *Campylobacter jejuni* in response to growth temperature variation. J. Bacteriol. 185: 2009–2016.

Suerbaum, S., Lohrengel, M., Sonnevend, A., Ruberg, F., and Kist, M. 2001. Allelic diversity and recombination in *Campylobacter jejuni*. J. Bacteriol. 183: 2553–2559.

Tauxe, R.V. 1992. Epidemiology of *Campylobacter jejuni* infections in the United States and other industrialized nations. In: *Campylobacter jejuni*: Current Status and Future Trends. I. Nachamkin, M.J. Blaser and L. Tompkins, eds. Washington, DC. American Society for Microbiology Press, pp. 9–19.

Tauxe, R.V. 1997. Emerging food-borne diseases: an evolving public health challenge. Emerging Infect. Dis. 3: 425–434.

Tsang, R.S., Figueroa, G., Bryden, L., and Ng, L.K. 2001. Flagella as a potential marker for *Campylobacter jejuni* strains associated with Guillain–Barré syndrome. J. Clin. Microbiol. 39: 762–764.

Ursing, J.B., Lior, H., and Owen, R.J. 1994. Proposal of minimal standards for describing new species of the family *Campylobacter*acea. Int. J. Syst. Bacteriol. 44: 842–845.

Urwin, R., and Maiden, M.C.J. 2003. Multilocus sequence typing: a tool for global epidemiology. Trends Microbiol. 11: 479–487.

Vandamme, P. (2000) Taxonomy of the family *Campylobacter*aceae. In: *Campylobacter*. I. Nachamkin and M.J. Blaser, eds. Washington, DC, American Society for Microbiology Press, pp. 3–27.

Vargas, A.C., Costa, M.M., Vainstein, M.H., Kreutz, L.C., and Neves, J.P. 2003. Phenotypic and molecular characterization of bovine *Campylobacter fetus* strains isolated in Brazil. Vet. Microbiol. 93: 121–132.

Volokhov, D., Chizhikov, V., Chumakov, K., and Rasooly, A. 2003. Microarray-based identification of thermophilic *Campylobacter jejuni*, *C. coli*, *C. lari* and *C. upsaliensis*. J. Clin. Microbiol. 41: 4071–4080.

Wagenaar, J.A., van Bergen, M.A.P., Newell, D.G., Grogono-Thomas, R., and Duim, B. 2001. Comparative study using amplified fragment length polymorphism fingerprinting, PCR genotyping, and phenotyping to differentiate *Campylobacter fetus* strains isolated from animals. J. Clin. Microbiol. 39: 2283–2286.

Wareing, D.R., Bolton, F.J., Fox, A.J., Wright, P.A., and Greenway, D.L. 2002. Phenotypic diversity of *Campylobacter* isolates from sporadic cases of human enteritis in the UK. J. Appl. Microbiol. 92: 502–509.

Wareing, D.R.A., Ure, R., Colles, F.M., Bolton, F.M., Fox, A.J., Maiden, M.J., and Dingle, K.E. 2003. Reference isolates for the clonal complexes of *Campylobacter jejuni*. Lett. Appl. Microbiol. 36: 106–110.

Wassenaar, T.A., Fry, B.N., Lastovica, A., Wagenaar, J.A., Coloe, P.J., and Duim, B. 2000. Genetic characterization of *Campylobacter jejuni* O: 41 isolates in relation with Guillain–Barré syndrome. J. Clin. Microbiol. 38: 874–876.

Wassenaar, T.M., Geilhausen, B., and Newell, D.G. 1998. Evidence of genomic instability in *Campylobacter jejuni* isolated from poultry. Appl. Environ. Microbiol. 64: 1816–1821.

Wassenaar, T.M., and Newell, D.G. 2000. Genotyping of *Campylobacter* spp. Appl. Environ. Microbiol. 66: 1–9.

Werno, A., Klena, J.D., Shaw, G.M., and Murdoch, D.R. 2002. Fatal case of *Campylobacter lari* prosthetic joint infection and bacteremia in an immunocompetent patient. J. Clin. Microbiol. 40: 1053–1055.

Chapter 8

Plasmids of *Campylobacter jejuni* 81-176

Joseph C. Larsen and Patricia Guerry

Abstract

The significant phenotypic variation that exists among different *C. jejuni* strains in virulence in animal models, *in vitro* invasiveness, as well as clinical severity of illness leads one to rationalize that genetic elements must contribute to this variation. The major genotypic differences that exist between *C. jejuni* 81-176, a particularly virulent strain of *C. jejuni*, and other *C. jejuni* strains is the presence of two plasmids in 81-176. One of these plasmids, pVir, has been shown contribute to both *in vitro* epithelial cell invasion and the ability to cause disease in the ferret diarrhoea model. This chapter presents a general description of the pVir plasmid and discusses the potential contribution of pVir to the pathogenesis of *C. jejuni* 81-176.

INTRODUCTION

Despite years of research and a complete genome sequence (Parkhill *et al.*, 2000), there is remarkably little understood about the molecular pathogenesis of *C. jejuni*. While several individual virulence determinants have been characterized extensively, most notably flagella and the cytolethal distending toxin, our understanding of the overall disease process remains rudimentary. The genome sequence revealed the presence of a previously unknown capsule, which has subsequently been shown to be the serodeterminant of the Penner serotyping scheme and essential for intestinal cell invasion (Karlyshev *et al.*, 2000; Bacon *et al.*, 2001). However, the genome sequence failed to reveal any specific secretion systems, type III or type IV, that form the basic framework for most other bacterial pathogenesis schemes. The disease caused by *C. jejuni* can be manifest by a range in symptoms, from a mild watery diarrhoea to a dysenteric-like syndrome. These differences could be due to either differences in host response to infection or to differences in virulence among individual strains, or a combination of the two. However, it is clear that there are major differences among strains in their ability to invade intestinal epithelial cells *in vitro*. Our lab has focused on one strain of *C. jejuni*, 81-176, which was originally isolated from an outbreak of diarrhoeal disease in children who consumed raw milk during a visit to a dairy farm (Korlath *et al.*, 1985). The selection of this strain was based on the fact that 81-176 has been fed to human volunteers on two occasions over 28 years apart and caused disease in both instances (Black *et al.*, 1988; Tribble, unpublished), and that the strain invades intestinal epithelial cells at much higher levels than most other strains (Oelschlaeger *et al.*, 1993). Thus, the genome strain, NCTC 11168 invaded INT407 cells at <1% the level of 81-176 (Bacon *et al.*, 2000). Moreover, 81-176 causes disease in the ferret diarrhoea model, but NCTC 11168 is avirulent (Bacon *et al.*, 2000). These significant differences in virulence phenotypes must have a genetic basis.

Microarray hybridizations have indicated that about 5% of the genome of NCTC 11168 is missing in 81-176 (Dorrell *et al.*, 2001), and comparative genomic analyses in our laboratory have confirmed that a number of 11168 genes are deleted in 81-176 and we have identified a limited number of novel chromosomal genes (Guerry, in preparation). However, the most obvious genomic difference between 11168 and 81-176 is the presence of two large plasmids in the latter strain.

Various studies have shown that a significant proportion of *C. jejuni* strains harbour plasmids. Reports have indicated that anywhere from 19% to 53% of strains contain plasmids (Austen *et al.*, 1980; Taylor *et al.*, 1981; Bradbury *et al.*, 1983; Bopp *et al.*, 1985; Bradbury *et al.*, 1985; Sagara *et al.,* 1987). While a majority of plasmids harboured by *C. jejuni* and *C. coli* strains contain genes encoding antibiotic resistance, there are reports of cryptic plasmids (Bopp *et al.*, 1985).

One of the two plasmids found in *C. jejuni* 81-176, pTet, is a conjugative R factor encoding tetracycline resistance (Batchelor *et al.*, submitted), and is likely related to the *tetO* plasmids previously described by Taylor *et al.* (Taylor *et al.*, 1983). The pTet plasmid is 45 kb in size and transfers to other strains of *C. jejuni* at frequencies from 10^{-4} to 10^{-6} per donor cell (Batchellor *et al.*, submitted). Conjugative transfer of pTet is mediated by a type IV secretion system encoded on this plasmid (see below), similar to those found on conjugative plasmids from other bacterial genera. Plasmids related to pTet are ubiquitous in multiple drug-resistant clinical isolates from Thailand (Warawadee, Mason, and Guerry, in preparation). Moreover, loss of pTet had no effect on invasion of intestinal epithelial cells (Bacon *et al.*, 2000).

The second plasmid found in 81-176, designated pVir, is non-conjugative and has been implicated in virulence. This plasmid will be reviewed in depth throughout this chapter.

GENERAL CHARACTERISTICS OF pVir

The pVir plasmid of *C. jejuni* 81-176 is 37,468 nucleotides in length (Bacon *et al.*, 2002). A circular map of pVir is shown in Figure 8.1. The G+C content of the plasmid is 26%, which is significantly lower than the published *C. jejuni* NCTC 11168 chromosome (Parkhill *et al.*, 2000), suggesting that the plasmid may have been acquired by horizontal transfer. The G+C content of pVir was also lower than a number of other previously identified *C. jejuni* plasmids (Luo *et al.*, 2001), with the exception of a plasmid from *Campylobacter hyointestinalis* (Waterman *et al.*, 1993). The pVir plasmid contains 54 predicted ORFs, although 16 of these ORF would be predicted to encode putative proteins of less than 100 amino acids, making their biological relevance suspect. All ORFs were designated Cjp for '*Campylobacter jejuni* plasmid'. The plasmid has very few intergenic regions, utilizing 83% of the plasmid as coding sequence, which is also similar with the coding density of the NCTC 11168 chromosome (Parkhill *et al.*, 2000). Two intergenic non-coding regions exist within pVir, one of 1147 bp present between *cjp28* and *cjp29*. The other non-coding region of 1049 bp exists between *cjp49* and *cjp50*. The intergenic region between *cjp28* and *cj29* contains a repetitive region that is 260 bp in length and is flanked by 52-bp direct repeats. This region may function as a putative origin of replication of the plasmid.

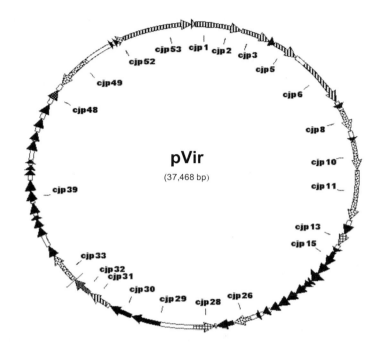

Figure 8.1 Circular map of pVir. Striped arrows represent genes homologous to type IV secretion system genes. Dotted arrows with white backgrounds represent genes with homology to *H. pylori* plasticity zone genes. Grid patterned arrows indicate genes with similarity to proteins of predicted function. Dotted arrows with black backgrounds represent genes with homology to *C. jejuni* NCTC 11168 chromosomal genes. Black arrows represent genes of *C. jejuni* plasmid-specific genes of unknown function. White areas indicate non-coding region.

INCIDENCE OF pVir

Bacon *et al.* determined that *cjp5/virB11* was present on plasmids in 10.3% of isolates obtained from US military personnel in Thailand (Bacon *et al.*, 2000). Although these strains possessed *cjp5/virB11*, it is unknown if the total repertoire of plasmid genes present in these strains was similar to pVir. In the same study, a variety of laboratory strains were also screened and lacked the presence of *cjp5/virB11*. One of these strains, A3249, had caused diarrhoeal disease in a human feeding study, although the symptoms were much less severe than parallel feeding studies with 81-176 (Black *et al.*, 1988). Recently, a DNA microarray was used to examine various genetic profiles among certain *C. jejuni* strains (Leonard *et al.*, 2003). This study also demonstrated the presence of pVir in a significant proportion (31.2%) of strains examined (Leonard *et al.*, 2003). This evidence suggests that pVir or plasmids similar in genetic content exist in nature in a significant proportion of *C. jejuni* strains.

CONTRIBUTION OF pVir TO VIRULENCE OF *C. JEJUNI* 81-176

As discussed in depth below, mutation of several genes present on pVir resulted in reductions in intestinal epithelial cell invasion (Bacon *et al.*, 2000; 2002), and complementation of several of these genes *in trans* restored wild-type levels of invasion (Larsen, Hu, and Guerry, in preparation). One of the non-invasive mutants, in *cjp5/virB11*, was also attenuated in the ferret model of diarrhoeal disease (Bacon *et al.*, 2000). However, electroporation of pVir into NCTC 11168 resulted in no increase in invasion levels (Bacon *et al.*, 2002). This suggested that while the plasmid is required for full invasion of 81-176, there are other unique determinants in 81-176 that are responsible for the higher level of invasion compared with the genome strain. Strain-specific surface structures present in 81-176, such as capsule and LOS, have been shown to effect virulence (Bacon *et al.*, 2001; Guerry *et al.*, 2002) are among the candidates. Nonetheless, there remain numerous unanswered and puzzling questions regarding the role of pVir in virulence of 81-176, as discussed below.

GENES PRESENT ON pVIR WITH HOMOLOGY TO TYPE IV SECRETION SYSTEMS

Type IV secretion systems (TFSS) are bacterial protein secretion systems, which are ancestrally related to conjugation machines (Christie and Vogel 2000; Christie 2001). These secretion systems have been demonstrated to transfer DNA, protein, or nucleoprotein complexes across bacterial membranes. Several essential processes in a variety of species of pathogenic bacteria have been linked to the presence of these systems. Intracellular survival, toxin secretion, and conjugation have been shown to be mediated by TFSS (Christie 2001). Figure 8.2 shows the genetic organization of TFSS from *C. jejuni* 81-176, *Wolinella succinogenes*, *Helicobacter pylori*, and *Agrobacterium tumefaciens*, and Table 8.1 presents the homologies between the various type IV secretion system proteins encoded by these genes. The TFSS present on pVir shares greatest homology to a recently identified TFSS in the ruminant commensal *W. succinogenes* (Baar *et al.*, 2003). Additionally, the TFSS on pVir bears significant homology to two TFSS in *H. pylori*. One of these *H. pylori* TFSS, which was recently identified in an archived stock of the J99 strain, is located

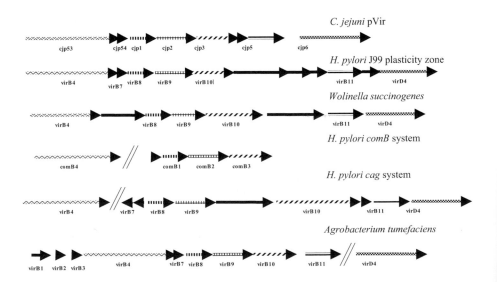

Figure 8.2 Genetic composition of TFSS in a number of bacterial species. Identical patterns indicates orthologous genes, black colour indicates genes not related to TFSS genes or genes without orthologues in the selected organisms. Image is not to scale.

Table 8.1 Comparison of homologies of pVir Type IV secretion proteins with other bacterial species.

pVir TFSS protein	Wolinella succinogenes	H. pylori J99 plasticity zone	H. pylori comB	H. pylori cag	C. jejuni 81-176 pTet
Cjp53/VirB4	40/ 63 (778)	30/51 (833)	31/53 (795)	21/39 (289)	21/38 (779)
Cjp1/VirB8	41/62 (198)	34/57 (223)	35/59 (207)	22/40 (200)	23/40 (176)
Cjp2/VirB9	38/61 (267)	35/51 (224)	32/52 (287)	NS	24/39 (235)
Cjp3/VirB10	41/58 (329)	35/50 (364)	42/59 (228)	35/53 (115)	29/50 (187)
Cjp5/VirB11	59/69 (287)	40/59 (297)	N/A	34/49 (213)	28/48 (314)
Cjp6/VirD4	36/59 (561)	33/55 (613)	N/A	21/41 (465)	22/41 (488)

Results determined by BLASTP analysis. Results presented as percentage identity/similarity (length of amino acid homology). Cjp54/VirB7 was omitted from analysis due to non-significant hits by BLASTP. CLUSTAL analysis had been previously performed with Cjp54 (Bacon et al., 2002). N/A, not applicable; NS, non-significant hit.

within the *H. pylori* plasticity zone and is currently of unknown function (Kersulyte *et al.*, 2003). The other homologous system is the *com* system found in all strains of *H. pylori*, which has been directly implicated in natural competence (Hofreuter *et al.*, 1998, 2001). Both of these TFSS are distinct from the TFSS present on the *cag* pathogenicity island, which is responsible for the translocation of CagA into gastric epithelial cells (Odenbright *et al.*, 2000). It is noteworthy that the second TFSS in *C. jejuni* 81-176 present on the conjugative pTet plasmid appears to be distinct from the pVir encoded TFSS (Batchelor *et al.*, submitted), as shown in Table 8.1.

It has previously been demonstrated that the *comB* locus is directly involved in the natural competence of *H. pylori* strains. The basic structural components of the TFSS, genes designated Cjp54, 1, 2, and 3 (Bacon *et al.*, 2000, 2002), are present on pVir. Mutational analysis of Cjp2 and Cjp3 resulted in significant reductions in intestinal epithelial cell invasion (Bacon *et al.*, 2000; 2002). As with its homologue in the *comB* system of *H. pylori*, mutation of the *cjp3* gene also resulted in diminished natural competence. However, the severity of this reduction was not comparable to the *comB3* mutant phenotype. Additionally, strains of *C. jejuni* that lack pVir are fully competent, suggesting that other chromosomally encoded competence systems exist. These facts, coupled with the recent identification of a type II secretion system mediating competence in *C. jejuni* 81-176 as well as observations in our laboratory, suggests that the contribution of pVir to natural competence is a moderate one (Wiesner *et al.*, 2003). Nonetheless, while modest, this defect appears to be genuine, since the competence defect in the *cjp3* mutants has been complemented *in trans* (Larsen, unpublished). This may reflect a horizontally transferred competence system that has yet to fully acclimate to life in *C. jejuni*.

Cjp54 is an orthologue of VirB7 present in *Agrobacterium tumefaciens* (Goodner *et al.*, 2001; Bacon *et al.*, 2002). This protein is suggested to be an outer membrane lipoprotein that interacts through disulfide bonding to another TFSS structural component, VirB9 (Anderson *et al.*, 1996; Spudich *et al.*, 1996). The presence of a putative lipobox and three Cys residues within its short 42 amino acid sequence suggests that Cjp49 probably fulfils a similar role in *C. jejuni*. Cjp1 is a homologue of ComB1 in *H. pylori* and orthologous to VirB8 in *A. tumefaciens*. This protein is proposed to be an inner membrane protein and has been shown in *A. tumefaciens* to interact with VirB9 and VirB10 (Das *et al.*, 2000). Cross-linking studies performed with wild-type and isogenic mutants of the pVir TFSS suggested that these proteins are able to form high molecular weight complexes, and a Cjp1- and Cjp3-specific interaction was defined (Larsen, unpublished data). This observation suggests that the TFSS structural components most likely interact to form a functional secretion channel in *C. jejuni* 81-176. Cjp2 is a ComB2/VirB9 orthologue, and is predicted to be an outer membrane protein. Cjp3 is an inner membrane protein with a large periplasmic domain and is thought to span the periplasm to interact with Cjp1, and Cjp2, since a similar role has been defined for its orthologue, VirB10, in *A. tumefaciens* (Das *et al.*, 2000).

The other TFSS homologues present on pVir are presumed to provide energy for substrate secretion or assembly of the secretion apparatus (Fullner *et al.*, 1994; Stephens *et al.*, 1995; Dang *et al.*, 1999). Cjp5 shares homology with a family of hexameric ATPases, including *H. pylori* HP0525 and the *A. tumefaciens* VirB11. These proteins are believed to form hexameric rings on the cytoplasmic side of the inner membrane and are thought to provide the energy for substrate secretion (Krause *et al.*, 2000). Recently, Savvies *et al.* proposed a model based on data obtained from the crystal structure of HP0525 (Savvies *et al.*, 2003). This model suggests that the homohexameric form of this protein is dynamic and through the hydrolysis of ATP may provide mechanistic force necessary for assembly of the TFS secreton or substrate secretion (Savvies *et al.*, 2003). Mutational analysis in *A. tumefaciens* has demonstrated that a consensus nucleotide-binding site (Walker A box) was required for virulence (Stephens *et al.*, 1995). A similar motif is seen in Cjp5, with amino acids 150–157 containing (GGTGSGKT). Insertional inactivation of *cjp5* resulted in a severe decrease in the adherence and invasion of intestinal epithelial cells, 13% and 7% of wild-type, respectively (Bacon *et al.*, 2000). Additionally, the *cjp5* mutant was also attenuated in the ferret model of disease, consistent with the low level of invasion of INT407 cells (Bacon *et al.*, 2000). These data suggest that Cjp5 may be the ATPase involved in providing energy necessary for TFS secreton assembly or the secretion of effector proteins.

Two other putative TFSS ATPases are encoded by *cjp6* and *cjp53*. Cjp53 shares homology to the *H. pylori* VirB4 protein. In *A. tumefaciens*, this protein is thought to aid in T-pilus assembly (Dang *et al.*, 1999). Other studies have also shown that the ATPase activity of VirB4 was essential for the transfer of substrates (Berger *et al.*, 1994; Fullner *et al.*, 1994). Mutational inactivation of *virB4* in *Bartonella* demonstrated that this gene was essential in establishing an intraerythrocytic infection (Schulein *et al.*, 2002). Further, in *H. pylori*, mutation of this gene completely abrogated natural competence (Hofreuter *et al.*, 2001). Recently, a signature tagged mutagenesis analysis of *H. pylori* identified *comB4* as being essential for colonization of gerbils (Kavermann *et al.*, 2003). Natural competence was determined not to be essential for colonization, suggesting that this protein may have alternative roles in virulence. In 81-176, however, mutation of the homologue of this gene has no effect on intestinal epithelial cell invasion (Bacon *et al.*, 2002). It will be necessary to determine if mutation of *cjp53*, like *H. pylori comB4*, will lead to a defect in the ability to colonize. The current data suggest Cjp53 may play an as yet undefined role in the TFSS of *C. jejuni* 81-176.

Similarly, Cjp6 is an orthologue of VirD4, and mutation of this gene had no effect on intestinal epithelial cell invasion (Bacon *et al.*, 2002). In *H. pylori*, it was determined that there were VirD4-dependent and independent functions of the TFSS (Selbach *et al.*, 2002). This study demonstrated that secretion of CagA through the TFSS was dependent on VirD4, while the induction of IL-8 was not. Since mutation of *cjp6* had no effect on intestinal epithelial cell invasion, it is possible that other Cjp6-dependent processes have yet to be revealed.

The published data suggest that the TFSS homologues present on pVir comprise a functional secretion channel that is involved in intestinal epithelial cell invasion and natural competence. However, many questions remain. First, mutation of *cjp5* has a dramatic effect on INT407 cell invasion and virulence but has no effect on natural competence (Bacon *et al.*, 2000). This would seem to suggest that this system functions in a bipartite manner, with various components being unique to their specific function. Secondly, the lack of a phenotype associated with mutation of the ATPases (Cjp6, Cjp53) that are essential in other systems is particularly cumbersome. It is unknown at this point if mutation of these genes has an effect on natural competence in 81-176. Further work will be necessary to determine the contribution of these ATPases to the function of the TFSS. Lastly, the presence of any substrate of the TFSS has yet to be defined. Studies in *A. tumefaciens* have suggested that the C-terminal end of TFSS substrates is important for export (Simone *et al.*, 2001). Studies have suggested the presence of a conserved RXR motif present in the

C-terminal 18 amino acids of VirE2, a substrate of the TFSS in *A. tumefaciens* (Simone *et al.*, 2001). However, the only protein that satisfies this criteria encoded on pVir is Cjp34, a protein of unknown function that is predicted to be localized to the inner membrane. This may suggest that other elements, either structural or compositional, contribute to substrate recognition by the TFSS in *C. jejuni*. Identification of substrates of the TFSS will provide key insights into the molecular mechanisms of invasion and interaction of *C. jejuni* with the eukaryotic cell.

GENES HOMOLOGOUS TO *C. JEJUNI* CHROMOSOMAL GENES

Two genes are present on pVir that shared significant homology to genes present on the NCTC 11168 chromosome. One of these genes, *cjp32*, shares significant homology to two genes, *cj0041* (21% identity, 40% similarity) and *cj0320* (22% identity, 43% similarity). The *cjp32* gene encodes a predicted protein of 239 amino acids, which is shorter than the two homologous proteins that are composed of 598 and 276 amino acids, respectively. While Cj0041 does not have significant homology to any known protein, mutation of *cj0041* affected both motility and invasion (Golden *et al.*, 2002). Cj0320 is homologous to FliH, a protein involved in the flagellin export. In *Salmonella*, FliH interacts with another protein, FliI, and inhibits its ATPase activity, both which are needed for flagellar export (Minamino *et al.*, 2000). Both proteins are part of the type III flagellar export apparatus, which bares similarity to type III secretion systems present in a variety of bacterial pathogens (Hueck 1998; Miniamino *et al.*, 2000). FliH homologues are also present within the type III secretion systems of *Yersinia enterocolitica* and *Erwinia amylovora* (Kim *et al.*, 1997; Jackson *et al.*, 2000). In these pathogens, either mutation of the *fliI* or *fliH* homologues resulted in an inhibition of protein secretion. Recently, it was determined that FliI oligomerizes to form a hexameric ring, similar to a TFSS ATPase in *H. pylori*, which is in the same protein family as Cjp5 (Yeo *et al.*, 2000, Claret *et al.*, 2003). Although the type III and type IV protein secretion pathways are clearly distinct, it is possible that a FliI–FliH like interaction could be feasible for Cjp5 and Cjp32. Mutation of *cjp32* results in a drop in intestinal epithelial cell invasion to 34.8% of wild-type, consistent with the possibility that Cjp32 is in involved with protein secretion (Bacon *et al.*, 2002). Whether this protein contributes to the TFSS by regulating Cjp5 or is a soluble substrate of the secretion apparatus remain to be determined. It is also entirely possible that Cjp32 is a component of a distinct type III secretion pathway, although evidence for that notion is weak.

Another pVir gene, *cjp48* has homology to *cj1456c*, a *C. jejuni* gene of unknown function. Mutational analysis of this gene has yet to be performed.

GENES WITH HOMOLOGY TO DNA BINDING/REPLICATION PROTEINS

Several genes on the pVir plasmid encode proteins with significant homology to proteins involved in either DNA replication and/or DNA binding (Bacon *et al.*, 2002). However, it is important to note that none of the mutations in any of these genes effected plasmid replication or segregation, suggesting a distinct role for these proteins (Bacon *et al.*, 2002). Cjp11 possessed homology to DNA topoisomerase I proteins, that are involved in DNA replication. The highest degree of homology was seen with topoisomerases present in the plasticity zones of *H. pylori* (Tomb *et al.*, 1997; Alm *et al.*, 1999; Bacon *et al.*, 2002) that exist in conjunction with their orthologues present elsewhere on the *H. pylori* chromosome, also suggesting that these proteins may function in a capacity separate from DNA replication. Another pVir-encoded protein, Cjp13, has homology with single-stranded DNA-binding proteins. Two other genes, *cjp26* and *cjp28* encode proteins that have homology to proteins involved in plasmid replication. Cjp26 is homologous to ParA, a protein involved in the partitioning of newly replicated plasmid to daughter cells, but a *cjp26* mutant plasmid was stably replicated and maintained suggesting that that this gene is not involved in plasmid replication (Bacon *et al.*, 2002). Interestingly, a *cjp26* homologue is found flanking the region of TFSS genes in *W. succinogenes* (Baar *et al.*, 2003). This may suggest its involvement in the transfer and subsequent integration of this genetic element in *Wolinella*. Cjp26 is also a homologue of HP1000, a gene in the plasticity zone of *H. pylori* 26695 (see below). Cjp28 also shows homology to another replication protein, RepA, but it appears to be a pseudogene, due to the presence of an in-frame stop codon within the coding sequence.

Since mutation of *cjp3* resulted in a modest reduction in natural competence, one begins to wonder if other genes present on pVir are also involved to some degree in natural competence. Likely candidates might include the aforementioned genes.

H. PYLORI PLASTICITY ZONE HOMOLOGS OF UNKNOWN FUNCTION

In addition to *cjp11* and *cjp26*, there are five other genes present on pVir that are homologous to genes present within the plasticity zone of *H. pylori* J99 or 26695. *cjp8*, *cjp10*, *cjp33*, *cjp49* and *cjp52* share homology with genes present in the plasticity zone of either *H. pylori* strain. All these *H. pylori* genes are of unknown function. Mutation of *cjp49*, a homologue of HP1004, resulted in decreases in invasion to 26% of wild-type 81-176 (Bacon *et al.*, 2002). The presence of these orthologous *H. pylori* genes in the pVir plasmid is noteworthy. Since all of the *H. pylori* orthologues present on pVir originate from a region of great interstrain genetic diversity, this may suggest the past horizontal transfer of genes between these two related species or through another species. Plasmid acquisition of DNA sequences has been proposed to be a mechanism for the potentiation of diversity within the plasticity zone of *H. pylori* (Alm *et al.*, 1999; Hofreuter *et al.*, 2002). It is therefore interesting to speculate that the plasticity zone-like region present on pVir might be an area susceptible to rapid genetic change.

PRESENCE OF *CAMPYLOBACTER* PLASMID-SPECIFIC GENES OF UNKNOWN FUNCTION

There are 34 genes of unknown function present on pVir. Sixteen of these putative proteins are predicted to comprised 100 or fewer amino acids and 18 are predicted to be between 100 and 200 amino acids. Mutational analysis of these genes has not been performed and further work will be necessary to determine their function. Two unknown genes, however, did have an effect on the ability of *C. jejuni* to invade intestinal epithelial cells. Forty-seven per cent reduction in invasion was demonstrated with a mutant in *cjp15*, which is predicted to encode a soluble protein of 25.8 kDa (Bacon *et al.*, 2002). Mutation of another gene, *cjp29*, resulted in a decrease in invasion of INT407 cells to levels 15.2% of wild type. Cjp29 is a glutamine rich protein of 44.8 kDa with no significant homology to known proteins. The soluble nature and predicted cytoplasmic localization of both Cjp15 and Cjp29, coupled with their phenotype of reduced invasion, suggest that these proteins may be putative substrates of the TFSS. Further work will be needed to determine if these proteins are secreted or translocated into the eukaryotic cell and what their contribution is to the disease process.

CONCLUDING REMARKS

The possible contribution of plasmids to the pathogenicity of a subset of *C. jejuni* strains is not trivial. While the effect of mutation of the pVir TFS genes on intestinal epithelial cell invasion appears to be indirect, complementation of the decrease in invasion levels has been successful in several mutant backgrounds, suggesting the defects are genuine (Larsen, unpublished results). It seems more likely now that pVir, and in particular its encoded TFSS facilitates some other pathogenic process that aids the organism in its interaction with the human or animal host.

The identification of a TFSS in *C. jejuni* 81-176 provides the opportunity to elucidate molecular mechanisms of pathogenesis, which remain largely unresolved in the field of *Campylobacter* as a whole. With the increasing number of bacterial genome sequences available, it has become clear that the TFSS present on pVir is probably distinct from the *comB* or *cag* pathogenicity island system. The recent identification of cryptic TFSS with high homology to that encoded by pVir in the *H. pylori* J99 plasticity zone and in *W. succinogenes* suggests that this system functions in an as yet undefined capacity. Importantly, aside from the basic type IV secretion components, these organisms lack most of the other genes present on the pVir plasmid. This would seem to indicate that, while these secretion systems may have originated from a common ancestor, they most likely have unique functions and substrates.

References

Alm, R. A., Ling L.-S. L., Moir, D., King, B.L., Brown E.D., Doig, P., Smith, D.R., Noonan, B., Guild, B.C., deJonge, B.L., Carmel, G., Tummino, P.J., Caruso, A., Uria-Nickelsen, M., Mills, D.M., Ives, C., Gibson, R., Merberg, D., Mills, S.D., Jian, Q., Taylor, D.E., Vovis, G.F., and Trust, T. J. 1999. Genomic sequence comparisons of two unrelated isolates of the human gastric pathogen *Helicobacter pylori*. Nature 397: 160–189.

Anderson, L.B., Hertzel, A.V., and Das, A. 1996. *Agrobacterium tumefaciens* VirB7 and VirB9 form a disulphide-linked protein complex. Proc. Natl. Acad. Sci. USA 93: 8889–8894.

Austin, R.A., and Trust, T.J. 1980. Detection of plasmids in the related group of the genus *Campylobacter*. FEMS Microbiol. Lett. 8: 201–204

Baar, C., Eppinger, M., Raddatz, G., Simon, J., Lanz, C., Klimmek, O., Nandakumar, R., Gross, R., Rosinus, A., Keller, H., Jagtap, P., Linke, B., Meyer, F., Lederer, H., and Schuster, S.C. 2003. Complete genome sequence and analysis of *Wolinella succinogenes*. Proc. Natl. Acad. Sci. USA 100: 11690–11695

Bacon, D.J., Alm, R.A., Hu, L., Hickey, T.E., Ewing, C. P., Batchelor, R. A., Trust, T. J., and Guerry P. 2002. DNA sequence and mutational analysis of the pVir plasmid of *Campylobacter jejuni* 81-176. Infect. Immun. 70: 6242–50.

Bacon, D.J., Szymanski, C. M., Burr, D.H., Silver, R. P., Alm, R.A., and Guerry, P. 2001. A phase variable capsule is involved in the virulence of *Campylobacter jejuni*. Mol. Microbiol. 40: 769–778

Bacon, D.J., Alm, R.A., Burr, D. H., Hu, L., Kopecko, D. J., Ewing, C. P., Trust, T. J., and Guerry, P. 2000. Involvement of a plasmid in the virulence of *Campylobacter jejuni* 81-176. Infect. Immun. 68: 4384–4390.

Berger, B.R., and Christie, P. J. 1994. Genetic complementation analysis of the *Agrobacterium tumefaciens virB* operon: *virB2* through *virB11* are essential virulence genes. J. Bacteriol. 176: 3646–3660.

Black, R.E., Levine, M.M., Clements, M.L., Hughes, T.P., and Blaser, M.J. 1988. Experimental *Campylobacter jejuni* infections in humans. J. Infect. Dis. 151: 472–479

Bopp, C.A., Birkness, K.A., Wachsmuth, I.K., and Barrett, T. J. 1985. In vitro antimicrobial susceptibility, plasmid analysis, and serotyping of epidemic-associated *Campylobacter jejuni*. J. Clin. Microbiol. 21: 4–7.

Bradbury, W.C., Marko, M.A., Hennessy, J.N., and Penner, J.L. 1983. Occurrence of plasmid DNA in serologically defined strains of *Campylobacter jejuni* and *Campylobacter coli*. Infect. Immun. 40: 460–463.

Bradbury, W.C., and Munroe, D.L.G., 1985. Occurrence of plasmids and antibiotic resistance among *Campylobacter jejuni* and *Campylobacter coli* isolated from healthy and diarrheic animals. J. Clin. Microbiol. 22: 339–346

Christie, P.J. 2001. Type IV secretion: intracellular transfer of macromolecules by Systems ancestrally related to conjugation machines. Mol. Microbiol. 40: 294–305.

Christie, P.J., and Vogel, J.P. 2000. Bacterial type IV secretion: conjugation systems adapted to deliver effector molecules to host cells. Trends Microbiol. 8: 354–360.

Claret, L., Calder, S.R., Higgins, M., and Hughes, C. 2003. Oligomerization and activation of the FliI ATPase central to bacterial flagellum assembly. Mol. Microbiol. 48: 1349–1355

Dang, T.A., Zhou, X-R, Graf, B., and Christie, P.J. 1999. Dimerization of the *Agrobacterium tumefaciens* VirB4 ATPase and the effect of ATP-binding cassette mutations on the assembly and function of the T-DNA transporter. Mol. Microbiol. 32: 1239–1253.

Das, A., and Xie, Y.H. 2000. The *Agrobacterium* T-DNA transport pore proteins VirB8, VirB9, and VirB10 interact with one another. J Bacteriol. 182: 758–763.

Dorrell, N., Mangan, J.A., Laing, K.G., Hinds, J., Linton, D., Al-Ghusein, H., Barrell, B.G., Parkhill, J., Stoker, N.G., Karlyshev, A.V., Butcher, P.D., and Wren, B.W. 2001. Whole genome comparison of *Campylobacter jejuni* human isolates using a low-cost microarray reveals extensive genetic diversity. Genome Res. 11: 1706–1715.

Fullner, K.J., Stephens, K.M., and Nester, E.W. 1994. An essential virulence protein of *Agrobacterium tumefaciens*. VirB4, requires an intact mononucleotide binding domain to function in transfer of T-DNA. Mol. Gen. Genet. 245: 704–715

Golden, N.J., and Acheson, D.W.K. 2002. Identification of motility and autoagglutination *Campylobacter jejuni* mutants by random transposon mutagenesis. Infect. Immun. 70: 1761–1771.

Goodner, B., Hinkle, G., Gattung, S., Miller, N., Blanchard, M., Qurollo, B., Goldman, B.S., Cao, Y., Askenazi, M., Halling, C., Mullin, L., Houmiel, K., Gordon, J., Vaudin, M., Iartchouck, O., Epp, A., Liu, F., Wollam, C., Allinger, M., Doughty, D., Scott, C., Lappas, C., Marlelz, B., Flanagan, C., Crowell, C., Gurson, J., Lomo, C., Sear, C., Strub, G., Cielo, C., and Slater, S. 2001. Genome sequence of the plant pathogen *Agrobacterium tumefaciens* C58. Science 294: 2323–2328.

Guerry, P., Szymanski, C.M., Prendergrast, M.M., Hickey, T.E., Ewing, C.P., Pattarini, D.L., and Moran, A.P. 2002. Phase variation of *Campylobacter jejuni* 81-176 lipooligosaccharide affects ganglioside mimicry and invasiveness *in vitro*. Infect. Immun. 70: 787–793.

Hofreuter, D., Odenbriet, S., Henke, G., and Haas, R. 1998. Natural competence for DNA transformation in *Helicobacter pylori:* identification and genetic characterization of the comB locus. Mol. Microbiol. 28: 1027–1038.

Hofreuter, D., Odenbreit, S., and Haas, R. 2001. Natural transformation competence in *Helicobacter pylori* is mediated by the basic components of a type IV secretion system. Mol. Microbiol. 41: 379–391.

Hofreuter, D., and Haas, R. 2002. Characterization of two cryptic *Helicobacter pylori* plasmids: a putative source for horizontal gene transfer and gene shuffling. J. Bacteriol. 184: 2755–2766.

Hueck C.J. 1998. Type III protein secretion systems in bacterial pathogens of animals and plants. Microbiol. Mol. Biol. Rev. 62: 379–433.

Jackson, M.W., and Plano, G. 2000. Inteactions between type III secretion apparatus components from *Yersinia pestis* detected using the yeast two-hybrid system. FEMS Microbiol. Lett. 1: 85–90.

Kaverman, H. Burns, BP., Angermuller, K., Odenbriet, S., Fischer, W., Melchers, K., and Hass, R. 2003. Identification and characterization of *Helicobacter pylori* genes essential for gastric colonization. J. Exp. Med. 197: 813–22.

Karlyshev, A.V., Linton, D., Gregson, N.A., Lastovica, A.J., Wren, B.W. 2000. Genetic and biochemical evidence of a *Campylobacter jejuni* capsular polysaccharide that accounts for Penner serotype specificity. 35: 529–541.

Kersulyte. D., B. Velapatino, A. K. Mukhopadhyay, L. Cahuayme, A. Bussalleu, J. Combe, R.H. Giliman, and Berg, D.E. 2003. Cluster of type IV secretion genes in *Helicobacter pylori's* plasticity zone. J. Bacteriol. 185: 3764–3772.

Kim, J.F., Wei, Z.M., and Beer, S.M. 1997. The hrpA and hrpC operons of *Erwinia amylovora* encode components of a type III pathway that secretes harpin. J. Bacteriol. 179: 1690–1697.

Korlath, J.A., Osterholm, M.T., Judy, A., Forgang, J.C., and Robinson, R.A. 1985. A point-source outbreak of campylobacteriosis associated with consumption of raw milk. J. Infect. Dis. 152: 592–596.

Krause, S., M. Barcena, W. Pansegrau, R. Lurz, J. M. Carazo, and Lanka, E. 2000. Sequence-related protein export NTPases encoded by the conjugative transfer region of RP4 and by the *cag* pathogenicity island of *Helicobacter pylori* share similar hexameric structures. Proc. Natl. Acad. Sci. USA 97: 3067–3072

Krause, S., W. Pansegrau, R. Lurz, F. de la Cruz, and.Lanka, E. 2000. Enzymology if type IV macromolecule secretion systems: the conjugative transfer regions of plasmids RP4 and R388 and the *cag* pathogenicity island of *Helicobacter pylori* encode structurally and functionally related nucleoside triphosphate hydrolases. J. Bacteriol. 182: 2761–2770.

Leonard II, E.E., Takata, T., Blaser, M.J., Falkow, S., Tompkins, L.S., and Gaynor, E.C. 2003. Use of an open-reading frame-specific *Campylobacter jejuni* DNA microarray as a new genotyping tool for studying epidemiologically related isolates. J. Infect. Dis. 187: 691–694

Luo, N., and Zhang, Q. 2001. Molecular characterization of a cryptic plasmid from *Campylobacter jejuni*. Plasmid 45: 127–133.

Minamino, T., and Macnab, R.M. 2000. FliH, a soluble component of the type III flagellar export apparatus of *Salmonella* forms a complex with FliI and inhibits its ATPase activity. Mol. Microbiol. 37: 1494–1503.

Oelschlaeger, T.A., Guerry, P., and Kopecko, D. J. 1993. Unusual microtubule-dependent endocytosis mechanisms triggered by *Campylobacter jejuni* and *Citrobacter freundii*. Proc. Natl. Acad. Sci USA. 90: 6884–6888.

Odenbright, S., Puls, J., Sedimaier, B., Gerland, E., Fischer, W., Hass, R. 2000. Translocation of *Helicobacter pylori* CagA into gastric epithelial cells by type IV secretion. Science 287: 1497–1500

Parkhill, J., Wren, B.W., Mungall, K., Ketely, J.M., Churcher, C., Basham, D., Chillingworth, T., Davies, R.M., Fetwell, T., Holroyd, S., Jagels, K., Karlyshev, A.V., Moule, S., Pallen, M.J., Penn, C.W., Quail, M.A., Rajandream, M-A., Rutherford, K.M., van Vliet, A.H.M., Whitehead, S., and Barrell, B.G. 2000. The genome sequence of the food-borne pathogen *Campylobacter jejuni* reveals hypervariable sequences. Nature 403: 665–668.

Sagara, H., Mochizuki, A., Okamura, N., and Nakaya, R. 1987. Antimicrobial resistance of *Campylobacter jejuni* and *Campylobacter coli* with special reference to plasmid profiles of Japanese clinical isolates. Antimicrob. Agents Chemother. 31: 713–719.

Savvides, S.N., H-J. Yeo, M. R. Beck, F. Blaesing, R. Lurz, E. Lanka. R. Buhrdorf, W. Fischer, R. Haas, and G. Waksman. 2003. VirB11 ATPases are dynamic hexameric assemblies: new insights into bacterial type IV secretion. EMBO J. 22: 1969–1980.

Schulein, R., and Dehio, C. 2002. The VirB/VirD4 type IV secretion system of *Bartonella* is essential for establishing intraerythrocytic infection. Mol. Microbiol. 46: 1053–1067

Selbach, M., Moesse, S., Meyer, T.F., and Backert, S. 2002. Functional analysis of the *Helicobacter pylori* cag pathogenicity island reveals both VirD4-CagA dependent and VirD4-CagA independent mechanisms. Infect. Immun. 70: 665–671

Simone, M., McCullen, C.A., Stahl, L.E., and Binns, A.N. 2001. The carboxy-terminus of VirE2 from *Agrobacterium tumefaciens* is required for its transport to host cells by the virB-encoded type IV transport system. Mol. Microbiol. 41: 1283–1293

Spudich, G.M., Fernandez, D., Zhou, X.-R., and Christie, P.J. 1996. Intermolecular disulphide bonds stabilize VirB7 homodimers and VirB7/VirB9 heterodimers during biogenesis of the *Agrobacterium tumefaciens* T-complex transport apparatus. Proc. Natl. Acad. Sci. USA 93: 7512–7517.

Stephens, K.M., Roush, C., and Nester, E.W. 1995. *Agrobacterium tumefaciens* VirB11 protein requires a consensus nucleotide-binding site for function in virulence. J Bacteriol. 177: 27–33

Taylor, D. E., DeGrandis, S.A., Karmali, M.A., and Flemming, P.C. 1981. Transmissible plasmids from *Campylobacter jejuni*. Antimicrob Agents Chemother. 19: 831–835

Taylor, D.E., Garner, R.S., and Allan, B.J. (1983) Characterization of tetracycline resistance plasmids from *Campylobacter jejuni* and *Campylobacter coli*. Antimicrob. Agents Chemother. 24: 930–935

Tomb, J.F., White, O., Kerlavage, A.R., Clayton, R.A., Sutton, G.G., Fleischman, R.D., Ketchum, K.A., Klenck, H.P., Gill, S., Dougherty, B.A., Nelson, K., Quackenbush, J., Zhou, L., Kirkness, E.F., Peterson, S., Loftus, B., Richardson, D., Dodson, R., Khalak, H.G., Glodek, A., McKenney, K., Fitzegerald, L.M., Lee, N., Adams, M.D., Hickey, E.K., Berg, D.E., Gocayne, J.D., Utterback, T.R., Peterson, J.D., Kelley, J.M., Cotton, M.D., Weidman, J.M., Fujii, C., Bowman, C., Watthey, L., Wallin, E., Hayes, W.S., Borodovsky, M., Karp, P.D., Smith, H.O., Fraser, C.M., and Venter, J.C. 1997. The complete genome sequence of the gastric pathogen *Helicobacter pylori*. Nature 388: 539–547

Waterman, S.R., Hackett, J.A., and Manning, P.A. 1993. Characterization of the replicative region of the small cryptic plasmid of *Campylobacter hyointestinalis*. Gene 125: 11–17.

Wiesner, R.S., Hendrixson, D.R., and DiRita, V.J. 2003. Natural transformation of *Campylobacter jejuni* requires the components of a type II secretion system. J. Bacteriol. 185: 5408–5418.

Yeo, H.J., Savvides, S.N., Herr, A.B., Lanka, E., and Waksman, G. 2000. Crystal structure of the hexameric traffic ATPase of the *Helicobacter pylori* type IV secretion system. Mol. Cell. 6: 1461–1472.

192

Chapter 9

Mechanisms of Antimicrobial Resistance in *Campylobacter*

Diane E. Taylor and Dobryan M. Tracz

Abstract
The modes of action of antibiotics and the mechanisms of resistance described in *Campylobacter* species are discussed in the context of antibiotic resistance in general, with primary emphasis on *Campylobacter jejuni*. We also review the frequency of resistance in *C. jejuni*, which appears to be increasing for some antibiotics, particularly tetracycline. In contrast, the frequency of erythromycin resistance in *C. jejuni*, the drug of choice for treatment of serious *Campylobacter* infections, remains low, often < 1%.

INTRODUCTION
Campylobacter jejuni gastroenteritis is an uncomfortable disease to acquire, but it is rarely fatal and is primarily a self-limiting infection. For these reasons, clinical treatment consists principally of oral replacement of fluid and electrolytes lost through vomiting and diarrhoea. Antibiotic treatment is reserved for clinical cases in which the patient is acutely ill with a severe or complicated infection, in systemic infections of immunosuppressed individuals and for control of infections in high risk groups (Skirrow and Blaser, 2000). When antibiotic treatment is administered, erythromycin has traditionally been the drug of choice and continues to be used effectively in clinical treatment of *C. jejuni* enteritis (Nachamkin *et al.*, 2002). Ciprofloxacin may also be used for prophylaxis of travellers' diarrhoea or for treatment of enteric infections. However, the use of these antibiotics in some countries may be limited due to the emergence of drug-resistant strains of *C. jejuni*. In this article we review the prevalence of antibiotic resistance and the mechanisms of resistance identified in *C. jejuni*, in the context of antibiotic resistance in general.

ANTIBIOTICS: MODES OF ACTION AND MECHANISMS OF RESISTANCE
Antibiotics are small, low-weight molecular compounds that are used to kill bacteria or inhibit their growth. Generally, antibiotics have specific targets in the bacterial cell, acting to inhibit processes like protein synthesis, DNA replication and cell wall synthesis. With reference to the treatment of *C. jejuni* gastroenteritis, erythromycin is a protein synthesis inhibitor and ciprofloxacin targets DNA replication. Tetracycline, an antibiotic to which there is now substantial resistance in *C. jejuni*, is an inhibitor of protein synthesis.

There are four main mechanisms of resistance to antibiotics: (1) efflux of the antibiotic from the host cell; (2) modification or destruction of the antibiotic; (3) alteration of the antibiotic target site; and (4) protection of the antibiotic target site (Davies, 1994). These mechanisms are depicted in Figure 9.1. In *C. jejuni*, mutation of the target site is important in resistance to ciprofloxacin and erythromycin, whereas protection of the target is important in resistance to tetracycline.

Fluoroquinolones: ciprofloxacin

Fluoroquinolones are a chemically modified form of the quinolone antibiotic nalidixic acid. They target two bacterial enzymes important in DNA replication, DNA gyrase (type II topoisomerase) and DNA topoisomerase IV. DNA gyrase introduces negative supercoils in the DNA and topoisomerase IV is important in unlinking activity (Drlica, 1999). As both enzymes are essential, the inhibition of their action results in cessation of bacterial growth. Fluoroquinolones trap these enzymes into a complex, prevent DNA replication and effectively poison the cell (Drlica, 1999). The major antibiotic in this class is ciprofloxacin, which is used to treat a wide variety of infections including campylobacteriosis.

Fluoroquinolone resistance – a historical background

Upon their introduction in the 1980s, fluoroquinolones were greeted with widespread enthusiasm from the medical community for treatment of bacterial infections. Their good pharmacological properties contributed to their clinical efficacy in treating gastroenteritis caused by *C. jejuni* and other enteric pathogens (Nachamkin *et al.*, 2000). Although fluoroquinolones were successful in initial clinical trials for treating patients with gastroenteritis caused by *C. jejuni* (Wistrom and Norrby, 1995), resistance quickly emerged.

A study from the Netherlands (Endtz *et al.*, 1991) was the first to demonstrate that the emergence of fluoroquinolone resistance among human clinical isolates of *C. jejuni*

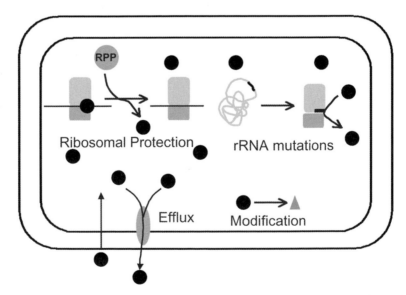

Figure 9.1 Antibiotic resistance mechanisms. The four mechanisms of bacterial resistance are shown. Examples found in *Campylobacter* spp. include the ribosomal protection protein (RPP) Tet (O) which encodes tetracycline (Taylor *et al.*, 1987; Connell *et al.*, 2003) and mutations in rRNA, believed to give resistance to erythromycin in *C. jejuni* (Trieber and Taylor, 2000; Jensen and Aarestrup, 2001) as examples of protection of the antibiotic target. Efflux of antibiotic, which appears to play a fairly minor role in *Campylobacter* spp., and modification of an antibiotic, which are exemplified by chloramphenicol and kanamycin resistance in *Campylobacter* spp., are also shown.

had a temporal relationship with the licensing of fluoroquinolones for use in the animal agriculture industry. In 1987, the fluoroquinolone enrofloxacin was introduced in the Netherlands for treatment of infections in poultry. Resistance profiles of both human and chicken *C. jejuni* isolates prior to the use of enrofloxacin (1982–1985) showed no resistance to ciprofloxacin. In the years after the introduction of enrofloxacin, rates of ciprofloxacin resistance had increased to 11% and 14% for humans and chickens, respectively (Endtz *et al.*, 1991). Further evidence for the link between fluoroquinolone resistance in poultry products and human infections came from studies in the United States, in which molecular subtyping demonstrated ciprofloxacin-resistant *C. jejuni* from human clinical infections and locally purchased retail poultry products were virtually identical (Smith *et al.*, 1999; Smith *et al.*, 2000).

Mechanisms of fluoroquinolone resistance
Resistance to fluoroquinolones is primarily through a mutation in the DNA sequence encoding the *gyrA* subunit of the DNA gyrase enzyme (Drlica, 1999). A single point mutation at Thr-86, Asp-90 or Ala-70 in *gyrA* can result in resistance. A mutation at Thr-86, which is the most common mutation site, is associated with a higher level of resistance to nalidixic acid (64–128 μg/mL) and ciprofloxacin (16–64 μg/mL) than are mutations at Asp-90 or Ala-70 (Wang *et al.*, 1993). Mutants resistant to nalidixic acid arise at a frequency of 10^{-8} in *C. jejuni* showing cross-resistance to fluoroquinolones (Taylor *et al.*, 1985). These mutations can be induced by treating poultry flocks with fluoroquinolones, as McDermott *et al.* (2002) found that fluoroquinolones did not eliminate *C. jejuni* from the intestinal tracts of chickens, but rapidly selected for resistant mutants.

Efflux of ciprofloxacin may be of importance in fluoroquinolone resistance in *C. jejuni*. The *cmeABC* gene operon encodes a multidrug efflux pump that contributes to intrinsic resistance of *C. jejuni* to a number of antibiotics (Lin *et al.*, 2002). The CmeABC efflux system is structurally and functionally similar to that of efflux systems in other Gram-negative bacterial pathogens. Disruption of the CmeABC mediated-efflux results in intercellular accumulation of ciprofloxacin and increases susceptibility to various other antimicrobial agents (Lin *et al.*, 2002). However, recent research suggests that efflux likely plays only a minor role in fluoroquinolone resistance in *C. jejuni* (Payot *et al.*, 2002).

Macrolides: erythromycin
The macrolides, such as erythromycin, are considered to be clinically safe antibiotics that are effective against *C. jejuni* gastroenteritis. Erythromycin is a bacteriostatic agent that acts as a protein synthesis inhibitor. It binds reversibly to the 50S subunit of the 70S bacterial ribosome and causes dissociation of the peptidyl-tRNA, which halts the elongation cycle of protein synthesis at the transpeptidation step (Nakajima, 1999; Trieber and Taylor, 2000).

Mechanisms of erythromycin resistance
Several mechanisms of macrolide resistance are known to occur in bacteria, including *erm*-mediated modification (methylation) of the target site, antibiotic efflux and *ere*-mediated inactivation of the antibiotic (Nakajima, 1999). In Canada, the levels of resistance to erythromycin, the primary drug of choice in treating *C. jejuni* gastroenteritis, are low (Gaudreau and Gilbert, 1998). Because erythromycin-resistant *C. jejuni* isolates cannot be selected for *in vitro*, few studies of the resistance mechanism have been undertaken. A study

by Yan and Taylor (1991) has ruled out the possibility of efflux, antibiotic modification or the action of an RNA methylase. Thus, resistance to erythromycin in *C. jejuni* is most likely due to the alteration of the ribosome target site on the 23S rRNA. These genes have been sequenced in erythromycin-resistant *Campylobacter* spp., and the mutation of an adenine residue in the 23S rRNA has been identified (Trieber and Taylor, 2000), but the type of base substitution is unclear (Jensen and Aarestrup, 2001). The mechanisms of erythromycin resistance in *C. jejuni* and *C. coli* have been reviewed previously (Taylor, 1992; Trieber and Taylor, 2000).

Tetracycline

The tetracyclines are a group of broad spectrum antibiotics that have been used against a wide variety of human pathogens. Chlorotetracycline was the first to be discovered in 1948, followed by both oxytetracycline and tetracycline in 1953 (Chopra and Roberts, 2001). Like many antibiotics, they were isolated from a soil microorganism, *Streptomyces* sp., and have since been chemically modified to improve their clinical properties. These semi-synthetic members of the tetracycline family include doxycycline, an important antibiotic in the clinical treatment of many infections including sexually transmitted diseases, and glycylcycline, the newest tetracycline.

Like the macrolides, the tetracyclines are protein synthesis inhibitors which act at the level of the bacterial ribosome. Initially, tetracycline is accumulated in the bacterial cell by active uptake. Once inside the bacterium, tetracycline binds reversibly to a high-affinity site on the 30S subunit of the 70S ribosome, as well as several other low-affinity sites (Chopra and Roberts, 2001). Binding to the high-affinity site prevents association of the aminoacyl-tRNA (aa-tRNA) with the A site on the ribosome. This stops the elongation step of protein synthesis, in which the aa-tRNA normally decodes the mRNA transcript and delivers the amino acid to the growing polypeptide chain.

Tetracycline resistance

Resistance to tetracycline is mediated by a number of *tet* determinants that can transfer between bacterial genera, including genes for efflux and ribosomal protection (see http://faculty.washington.edu/marilynr) (see Figure 9.1). The mechanism of tetracycline efflux involves membrane proteins such as Tet (A) that actively pump out tetracycline from the bacterial cell. The tetracycline efflux proteins are the best studied of the tetracycline resistance proteins and in Gram-negative bacteria, efflux genes are commonly carried on plasmids (Chopra and Roberts, 2001). However, the mechanism of tetracycline resistance in *C. jejuni* is through ribosomal protection. More specifically, the soluble Tet (O) protein induces a conformational change in the ribosome, which in turn ejects tetracycline from the ribosome and allows protein synthesis to occur (Connell *et al.*, 2003).

Tet (O) and ribosomal protection

In *C. jejuni*, tetracycline resistance is primarily mediated by transmissible plasmids (Taylor *et al.*, 1981), harbouring the *tet* (O) gene (Taylor *et al.*, 1987). Originally, cloning and sequencing of *tet* (O) (GenBank Accession Number M18896) identified a gene of 1917 base pairs (bp) (Manavathu *et al.*, 1988), but recent analysis has corrected the 3′ end of the *tet* (O) sequence, which is now known to be 1920 bp (Connell, 2003). The G+C% content of *tet* (O) is 40% (Manavathu *et al.*, 1988), much higher than that of the *C. jejuni* genome (30%) (Parkhill *et al.*, 2000), leading to speculation that *tet* (O) had its origin in the Gram-positive bacteria (Sougakoff *et al.*, 1987). Originally cloned from the plasmid pUA466, *tet* (*O*) was found to mediate tetracycline resistance up to a minimum inhibitory concentration (MIC) of 64 µg/mL of tetracycline in both *C. jejuni* and *E. coli* JM107 (Taylor, 1986; Taylor *et al.*, 1987). Furthermore, it is believed that the upstream region of *tet* (*O*), which harbours two putative promoters (P1 and P2), is important for full expression of tetracycline resistance (Wang and Taylor, 1991).

Encoded by the *tet* (*O*) gene is the 640 amino acid (~72 kDa) protein, Tet (O), a ribosomal protection protein (RPP) that confers resistance by actively displacing tetracycline from its primary binding site on the ribosome (Trieber *et al.*, 1998; Connell *et al.*, 2002) (see Figure 9.1). This soluble cytoplasmic protein is similar to other ribosomal protection proteins, including Tet (M), Tet (P), Tet (Q), Tet (S), Tet (T), Tet (W) and Otr (A) (Trieber and Taylor, 2000). Of the ribosomal protection proteins involved in tetracycline resistance, Tet (O) and Tet (M) are the best characterized (Chopra and Roberts, 2001), which share approximately 76% sequence similarity (Manavathu *et al.*, 1988). Tet (M) is the most widely distributed RPP and is known to be present in eight Gram-negative bacterial genera and 18 Gram-positive genera (Chopra and Roberts, 2001). For example, 79% of tetracycline-resistant oral bacteria contain *tet* (*M*) (Villedieu *et al.*, 2003). It is noteworthy that no RPPs have been identified in the *Enterobacteriaceae*.

Tet (O) has substantial sequence similarity to the ribosomal elongation factors EF-Tu and EF-G, which are essential in bacterial protein synthesis. The N-terminal region of Tet (O) that has the highest amino acid identity to EF-G and is known to possess GTPase activity (Taylor *et al.*, 1995). The elongation factors interact with the ribosome in a transient manner during the process of protein synthesis. EF-G assists in translocation of the aa-tRNA from the ribosomal A site to the P site. Therefore, Tet (O) may act as a molecular mimic of EF-G and EF-Tu that binds to and causes a conformational change in the ribosome, promoting the release of tetracycline (Connell *et al.*, 2003).

A DNA probe has been used successfully to identify *tet* (*O*) in plasmids and chromosomal DNA from *C. jejuni* (Lee *et al.*, 1994). DNA from all tetracycline-resistant isolates from both human and chicken sources hybridized with the *tet* (*O*) probe. Human clinical isolates of *C. jejuni* had comparable *tet* (*O*) content on their plasmids (47%) and chromosomal DNA (44%), while the majority of *C. jejuni* isolates from chickens carried *tet* (*O*) on plasmids (87%) (Lee *et al.*, 1994). Therefore, plasmids play a major role in the dissemination of tetracycline resistance in *C. jejuni*.

The widespread prevalence of tetracycline resistance in *C. jejuni* human clinical isolates is believed to be due to the use of tetracycline in veterinary medicine for the control of infections and at subtherapeutic levels for growth promotion (Lee *et al.*, 1994; McEwen and Fedorka-Cray, 2002). This effect is similar to that of fluoroquinolone use in animal agriculture. Research by Stuart Levy (Levy *et al.*, 1976; Levy 1978) in the 1970s provided strong evidence that the inclusion of tetracycline in animal feed leads to emergence of tetracycline-resistant *E. coli* in the intestinal flora of humans. Resistant bacterial strains are selected for in the treated animal population and may be transmitted to humans through meat products or alternative routes, such as food in contact with water contaminated by animal faeces (Witte, 1998).

Resistance to other antibiotics

Rates of resistance to chloramphenicol, a protein synthesis inhibitor, are low in *C. jejuni*. The mechanism of chloramphenicol resistance is through the action of the chloramphenicol acetyl transferase (CAT) enzyme, which modifies the antibiotic and renders it ineffective (Wang and Taylor, 1990). Resistance to the aminoglycosides, such as kanamycin, is also rare in *C. jejuni* and involves the modification of the antibiotic by an aminoglycoside phosphotransferase (APH) (Trieber and Taylor, 2000). These resistance determinants are normally carried on plasmids (Tenover *et al.*, 1989).

Frequency of antibiotic resistance in *C. jejuni*

In *C. jejuni*, antibiotic resistance has major clinical implications, as erythromycin is the first line drug in the treatment in severe infections. Ciprofloxacin is commonly used to treat various types of bacterial gastroenteritis. The emergence of multidrug resistance is especially problematic, since it effectively limits the availability of effective antibiotic chemotherapy. Current worldwide antibiotic resistance rates in *C. jejuni* show a diverse response to different antibiotics. However, there are global trends in the increasing incidence of resistance to several antibiotics, such as tetracycline.

Frequency of resistance to fluoroquinolones

Although fluoroquinolones initially had excellent clinical efficacy, the 1990s showed an extraordinary rise in the incidence of ciprofloxacin resistance in *C. jejuni*. Excellent data on the emergence of fluoroquinolone resistance come from Spain, where levels were 0–2% before 1989 and 88% in 1996 (Reina *et al.*, 1994; Ruizi *et al.*, 1998). Studies on resistance in Ontario and Quebec, Canada, have found levels increased to 13% in 1995, up from 0% before 1989 (Karmali *et al.*, 1981; Gaudreau and Gilbert, 1997). The *Campylobacter* Sentinel Surveillance Scheme (CSSS), established in the UK in May, 2000, has found that *Campylobacter* infections acquired abroad were more likely to be resistant to ciprofloxacin than nationally acquired strains (55% to 10%, respectively) (CSSS, 2002).

Frequency of resistance to erythromycin

Although antibiotic resistance rates in a number of bacterial pathogens are increasing annually, the rates of erythromycin resistance in *C. jejuni* have remained very low worldwide. These are encouraging data for clinicians, who can continue to use erythromycin as the first drug of choice in treating *C. jejuni* gastroenteritis. In Alberta Canada, no erythromycin resistance was observed in 1980–81 (Taylor *et al.*, 1986). In a recent study the prevalence of erythromycin-resistant *C. jejuni* in Alberta from 1999 to 2000 was < 1% (Gibreel *et al.*, 2004). Globally, rates range from low levels (0%) in Denmark (Aarestrup *et al.*, 1997) to 10% in Taiwan (Li *et al.*, 1998).

Frequency of resistance to tetracycline

The incidence of tetracycline resistance among human clinical isolates of *C. jejuni* has shown a dramatic upward trend since the 1980s (Table 9.1). In one Canadian study, tetracycline resistance in human clinical isolates of *C. jejuni* increased to 56% from 19% in just over a decade (Gaudreau and Gilbert, 1998). In Alberta, Canada, tetracycline resistance levels were found to be 6.8% and 8.6% in 1980 and 1981, respectively (Taylor *et al.*, 1986), but rose to 50% more recently (1999–2002) (Gibreel *et al.*, 2004).

Globally, there is an increasing trend in the incidence of tetracycline-resistance in *C. jejuni*. In Japan and the United States 55% of clinical isolates are tetracycline resistant (Sagara *et al.*, 1987; Nachamkin, 1994). The extremely high rates of tetracycline resistance among strains of *C. jejuni* have effectively removed tetracycline from the list of antibiotics

Table 9.1 Tetracycline resistance frequencies in *Campylobacter* sp. isolated from human and animal sources worldwide.

Year and country	Frequency of resistance (%)	Organism	Animal source	Reference
1999, United States	85	*C.jejuni*	Raw retail meats	Ge *et al.* (2002)
1994, Taipei	78	*C.jejuni*	Humans	Lee *et al.* (1994)
1995–1997, Canada	56	*C.jejuni*	Humans	Gaudreau and Gilbert (1998)
2000, Ireland	31	*C.jejuni*	Humans	Lucey *et al.* (2002)
1997, Northern Ireland	26	*Campylobacter* sp.	Chickens	Moore *et al.* (2001)
2000, Northern Ireland	19	*Campylobacter* sp.	Humans	Moore *et al.* (2001)
1980–1983, Canada	15	*C.jejuni*	Humans	Lariviere *et al.* (1986)
1997, Denmark	11	*C.jejuni*	Humans	Aarestrup *et al.* (1997)
1981, Canada	9	*C.jejuni*	Humans	Taylor *et al.* (1986)
1994, India	9	*C.jejuni*	Humans	Prasad *et al.* (1994)
1997, Chile	2	*C.jejuni*	Humans	Fernandez *et al.* (2000)

available to treat *C. jejuni* gastroenteritis in Thailand (Engberg *et al.*, 2001), where up to 95% of isolates are tetracycline-resistant (Li *et al.*, 1998). High rates of resistance are also found among poultry, including one study that found all isolates from chickens were tetracycline resistant ($n = 167$) (Lee *et al.*, 1994).

Multi-drug resistance

Strains of *C. jejuni* that are resistant to multiple antibiotics pose a major challenge for effective clinical treatment. This is especially serious in areas of the world where antibiotic resistance in enteric pathogens has reached alarming rates. In Thailand, where *C. jejuni* is resistant to many antibiotics, there is 100% co-resistance between macrolides (azithromycin) and fluoroquinolones (ciprofloxacin) (Hoge *et al.*, 1998). In Taiwan, nalidixic acid-resistant strains were co-resistant to tetracycline and ciprofloxacin at rates of 97% and 66%, respectively (Li *et al.*, 1998).

PLASMIDS IN *CAMPYLOBACTER*: INCIDENCE AND ANTIBIOTIC RESISTANCE

Plasmids are extra-chromosomal genetic elements in bacteria that can carry genes for antibiotic resistance and can undergo inter- and intra-species transfer. In *Campylobacter* organisms, plasmid content varies from 13% to 91%, independent of the source of the isolates (Table 9.2). The majority of plasmids have been found to encode for resistance factors (R factors). On average, 43% of human clinical isolates of *C. jejuni* contain plasmids and high plasmid content has also been described in both animal isolates of *C. jejuni* and *C. coli* (Table 9.2). Isolates with multiple plasmids have been identified in human clinical isolates, and plasmid sizes range from 4.3 to 50 kb (Taylor *et al.*, 1986), with other studies reporting extremely large plasmids (> 100 kb) (Tenover *et al.*, 1985).

In animals, a study by Bradbury and Munroe (1985) found that there was a higher occurrence of plasmids in *C. jejuni* isolated from cattle with a diarrhoeal disease than from healthy animals (31.2% vs. 13.6%, respectively). This suggests a relationship between enteric infections in animals and the presence of plasmids in *C. jejuni*. That same study found that in chickens, the primary source of human *C. jejuni* infections, plasmids were found in 75.8% of isolates (Bradbury and Munroe, 1985).

Plasmids in *C. jejuni* have previously been found to encode tetracycline resistance, giving an MIC of 32 to 128 µg/mL of tetracycline (Taylor *et al.*, 1986). Conjugative transfer of these plasmids has been demonstrated between *Campylobacter* sp., but not to *E. coli*, suggesting their host range is restricted (Taylor *et al.*, 1981). The pTet tetracycline-resistance plasmid is approximately 37 kb (Bacon *et al.*, 2000), although larger tetracycline-resistance plasmids (~100 kb) have been reported (Tenover *et al.*, 1985).

CONCLUSIONS

The low prevalence of erythromycin resistance in *C. jejuni* clinical isolates (often < 1%) and its efficacy in treating the infection make it an ideal choice for chemotherapy, in cases where treatment is indicated, for example in young children or in aged or immunocompromised patients. Tetracycline resistance mediated by Tet (O) continues to rise. The level of dissemination of the Tet (O) determinant and its continued prevalence, in exclusion of other resistance determinants, raises concerns regarding the selective pressures on *C. jejuni* strains in the environment. The prevalence of both erythromycin

Table 9.2 Plasmid content in *Campylobacter* sp.

Organism	Animal source	Plasmid content (%)	Sample size	Reference
C. jejuni	Chickens	91	167	Lee *et al.* (1994)
C. jejuni	Chickens	75	62	Bradbury and Munroe (1985)
C. coli	Humans	70	10	Sagara *et al.* (1987)
C. coli	Pigs	68	71	Bradbury and Munroe (1985)
C. coli	Humans, pigs, chickens	67	6	Taylor and Bryner (1984)
C. jejuni	Humans	52	108	Sagara *et al.* (1987)
C. coli	Unknown	47	17	Bradbury *et al.* (1983)
C. jejuni	Humans	44	41	Lee *et al.* (1994)
C. jejuni	Humans	35	282	Tenover *et al.* (1985)
Campylobacter sp.	Humans	33	30	Austen and Trust (1980)
Campylobacter sp.	Humans	29	242	Lind *et al.* (1989)
C. jejuni	Chickens, pets	29	346	Tenover *et al.* (1985)
C. coli	Humans	29	28	Tenover *et al.* (1985)
C. jejuni	Unknown	28	40	Bradbury *et al.* (1983)
C. jejuni	Humans, cows	25	8	Taylor and Bryner (1984)
C. jejuni	Dairy cows, cattle	24	54	Bradbury and Munroe (1985)
Campylobacter sp.	Humans	21	121	Ambrosio and Lastovica (1983)
C. jejuni	Humans	13	8	Taylor *et al.* (1983)

and tetracycline resistance in human and animal isolates of *C. jejuni* should continue to be monitored.

Acknowledgements
Support for this work was provided by the Natural Sciences and Engineering Research Council of Canada (NSERC) in the form of a grant to D.E.T. and postgraduate scholarship to D.M.T. D.E.T. is an Alberta Heritage Foundation for Medical Research (AHFMR) Scientist and D.M.T. is an AHFMR Student. We thank Sean Connell for Figure 9.1.

References
Aarestrup, F.M., Nielsen, E.M., Madsen, M., and Engberg, J. 1997. Antimicrobial susceptibility patterns of thermophilic *Campylobacter* spp. from humans, pigs, cattle, and broilers in Denmark. Antimicrob. Agents Chemother. 41: 2244–2250.

Ambrosio, R.E., and Lastovica, A.J. 1983. Rapid screening procedure for detection of plasmids *Campylobacter*. S. Afr. J. Sci. 79: 110–111.

Austen, R.A., and Trust, T.J. 1980. Detection of plasmids in the related group of the genus *Campylobacter*. FEMS Microbiol. Lett. 8: 201–204.

Bacon, D.J., Alm, R.A., Hu, L., Hickey, T.E., Ewing, C.P., Batchelor, R.A., Trust, T.J., and Guerry, P. 2000. Involvement of a plasmid in virulence of *Campylobacter jejuni* 81-176. Infect. Immun. 68: 4384–4390.

Bradbury, W.C., Marko, M.A., Hennessy, J.N., and Penner, J.L. 1983. Occurrence of plasmid DNA in serologically defined strains of *Campylobacter jejuni* and *Campylobacter coli*. Infect. Immun. 40: 460–463.

Bradbury, W.C., and Munroe, D.L. 1985. Occurrence of plasmids and antibiotic resistance among *Campylobacter jejuni* and *Campylobacter coli* isolated from healthy and diarrheic animals. J. Clin. Microbiol. 22: 339–346.

Campylobacter Sentinel Surveillance Scheme. 2002. Ciprofloxacin resistance in *Campylobacter jejuni*: case-case analysis as a tool for elucidating risks at home and abroad. J. Antimicrob. Chemother. 50: 561–568.

Chopra, I. and Roberts, M. 2001. Tetracycline antibiotics: mode of action, applications, molecular biology, and epidemiology of bacterial resistance. Microbiol. Mol. Biol. Rev. 65: 232–260.

Connell, S.R. 2003. Interaction of the tetracycline resistance protein Tet (O) with the bacterial ribosome. Medical Microbiology and Immunology. Edmonton, AB, University of Alberta.

Connell, S.R., Trieber, C.A., Dinos, G.P., Einfeldt, E., Taylor, D.E., and Nierhaus, K.H. 2003. Mechanism of Tet (O)-mediated tetracycline resistance. EMBO J. 22: 945–953.

Connell, S.R., Trieber, C.A., Stelzl, U., Einfeldt, E., Taylor, D.E., and Nierhaus, K.H. 2002. The tetracycline resistance protein Tet (O) perturbs the conformation of the ribosomal decoding centre. Mol. Microbiol. 45: 1463–1472.

Davies, J. 1994. Inactivation of antibiotics and the dissemination of resistance genes. Science 264: 375–382.

Drlica, K. 1999. Mechanism of fluoroquinolone action. Curr. Opin. Microbiol. 2: 504–508.

Endtz, H.P., Ruijs, G.J., van Klingeren, B., Jansen, W.H., van der Reyden, T., and Mouton, R.P. 1991. Quinolone resistance in *Campylobacter* isolated from man and poultry following the introduction of fluoroquinolones in veterinary medicine. J. Antimicrob. Chemother. 27: 199–208.

Engberg, J., Aarestrup, F.M., Taylor, D.E., Gerner-Smidt, P., and Nachamkin, I. 2001. Quinolone and macrolide resistance in *Campylobacter jejuni* and *C. coli*: resistance mechanisms and trends in human isolates. Emerging Infect. Dis. 7: 24–34.

Fernandez, H., Mansilla, M., and Gonzalez, V. 2000. Antimicrobial susceptibility of *Campylobacter jejuni* subsp. *jejuni* assessed by E-test and double dilution agar method in Southern Chile. Mem. Inst. Oswaldo Cruz 95: 247–249.

Gaudreau, C., and Gilbert, H. 1997. Comparison of disc diffusion and agar dilution methods for antibiotic susceptibility testing of *Campylobacter jejuni* subsp. *jejuni* and *Campylobacter coli*. J. Antimicrob. Chemother. 39: 707–712.

Gaudreau, C., and Gilbert, H. 1998. Antimicrobial resistance of clinical strains of *Campylobacter jejuni* subsp. *jejuni* isolated from 1985 to 1997 in Quebec, Canada. Antimicrob. Agents Chemother. 42: 2106–2108.

Ge, B., Bodeis, S., Walker, R.D., White, D.G., Zhao, S., McDermott, P.F., and Meng, J. 2002. Comparison of the Etest and agar dilution for *in vitro* antimicrobial susceptibility testing of *Campylobacter*. J. Antimicrob. Chemother. 50: 487–494.

Gibreel, A., Tracz, D.M., Nonaka, L., Ngo, T.M., Connell, S.R., and Taylor, D.E. 2004. Incidence of antibiotic resistance in *Campylobacter jejuni* isolated in Alberta, Canada, from 1999–2002, with special reference to *tet(O)*-mediated tetracycline resistance. Antimicrob. Agents Chemother. 48: 3442–3450.

Hoge, C.W., Gambel, J.M., Srijan, C., Pitarangsi, C., and Echeverria, P. 1998. Trends in antibiotic resistance among diarrhoeal pathogens isolated in Thailand over 15 years. Clin. Infect. Dis. 26: 341–345.

Jensen, L.B. and Aarestrup, F.M. 2001. Macrolide resistance in *Campylobacter coli* of animal origin in Denmark. Antimicrob. Agents Chemother. 45: 371–372.

Karmali, M.A., De Grandis, S., and Fleming, P.C. 1981. Antimicrobial susceptibility of *Campylobacter jejuni* with special reference to resistance patterns of Canadian isolates. Antimicrob. Agents Chemother. 19: 593–597.

Lariviere, L.A., Gaudreau, C.L., and Turgeon, F.F. 1986. Susceptibility of clinical isolates of *Campylobacter jejuni* to twenty-five antimicrobial agents. J. Antimicrob. Chemother. 18: 681–685.

Lee, C.Y., Tai, C.L., Lin, S.C., and Chen, Y.T. 1994. Occurrence of plasmids and tetracycline resistance among *Campylobacter jejuni* and *Campylobacter coli* isolated from whole market chickens and clinical samples. Int. J. Food Microbiol. 24: 161–170.

Lee, L.A., Puhr, N.D., Maloney, E.K., Bean, N.H., and Tauxe, R.V. 1994. Increase in antimicrobial-resistant *Salmonella* infections in the United States, 1989–1990. J. Infect. Dis. 170: 128–134.

Levy, S.B. 1978. Emergence of antibiotic-resistant bacteria in the intestinal flora of farm inhabitants. J. Infect. Dis. 137: 689–690.

Levy, S.B., FitzGerald, G.B., and Macone, A.B. 1976. Changes in intestinal flora of farm personnel after introduction of a tetracycline-supplemented feed on a farm. N. Engl. J. Med. 295: 583–588.

Li, C.C., Chiu, C.H., Wu, J.L., Huang, Y.C., and Lin, T.Y. 1998. Antimicrobial susceptibilities of *Campylobacter jejuni* and *coli* by using E-test in Taiwan. Scand. J. Infect. Dis. 30: 39–42.

Lin, J., Michel, L.O., and Zhang, Q. 2002. CmeABC functions as a multidrug efflux system in *Campylobacter jejuni*. Antimicrob. Agents Chemother. 46: 2124–2131.

Lind, L., Sjogren, E., Welinder-Olsson, C., and Kaijser, B. 1989. Plasmids and serogroups in *Campylobacter jejuni*. APMIS 97: 1097–1102.

Lucey, B., Cryan, B., O'Halloran, F., Wall, P.G., Buckley, T., and Fanning, S. 2002. Trends in antimicrobial susceptibility among isolates of *Campylobacter* species in Ireland and the emergence of resistance to ciprofloxacin. Vet. Rec. 151: 317–20.

Manavathu, E.K., Hiratsuka, K., and Taylor, D.E. 1988. Nucleotide sequence analysis and expression of a tetracycline-resistance gene from *Campylobacter jejuni*. Gene 62: 17–26.

McDermott, P.F., Bodeis, S.M., English, L.L., White, D.G., Walker, R.D., Zhao, S., Simjee, S., and Wagner, D.D. 2002. Ciprofloxacin resistance in *Campylobacter jejuni* evolves rapidly in chickens treated with fluoroquinolones. J. Infect. Dis. 185: 837–840.

McEwen, S.A. and Fedorka-Cray, P.J. 2002. Antimicrobial use and resistance in animals. Clin. Infect. Dis. 34 Suppl 3: S93-S106.

Moore, J.E., Crowe, M., Heaney, N., and Crothers, E. 2001. Antibiotic resistance in *Campylobacter* spp. isolated from human faeces (1980–2000) and foods (1997–2000) in Northern Ireland: an update. J. Antimicrob. Chemother. 48: 455–457.

Nachamkin, I. 1994. Antimicrobial susceptibilities of *Campylobacter jejuni* and *Campylobacter coli* to ciprofloxacin, erythromycin and tetracycline from 1982 to 1992. Med. Microbiol. Lett. 3: 300–305.

Nachamkin, I., Engberg, J., and Aarestrup, F.M. 2000. Diagnosis and antimicrobial susceptibility of *Campylobacter* spp. In: *Campylobacter*, 2nd edition. I. Nachamkin and M.J. Blaser, eds. ASM Press, Washington, D.C. pp. 45–66.

Nachamkin, I., Ung, H., and Li, M. 2002. Increasing fluoroquinolone resistance in *Campylobacter jejuni*, Pennsylvania, USA, 1982–2001. Emerg. Infect. Dis. 8: 1501–1503.

Nakajima, Y. 1999. Mechanisms of bacterial resistance to macrolide antibiotics. J. Infect. Chemother. 5: 61–74.

Parkhill, J., Wren, B.W., Mungall, K., Ketley, J.M., Churcher, C., Basham, D., Chillingworth, T., Davies, R.M., Feltwell, T., Holroyd, S., Jagels, K., Karlyshev, A.V., Moule, S., Pallen, M.J., Penn, C.W., Quail, M.A., Rajandream, M.A., Rutherford, K.M., van Vliet, A.H., Whitehead, S., and Barrell, B.G. 2000. The genome sequence of the food-borne pathogen *Campylobacter jejuni* reveals hypervariable sequences. Nature 403: 665–668.

Payot, S., Cloeckaert, A., and Chaslus-Dancla, E. 2002. Selection and characterization of fluoroquinolone-resistant mutants of *Campylobacter jejuni* using enrofloxacin. Microb. Drug Resist. 8: 335–343.

Prasad, K.N., Mathur, S.K., Dhole, T.N., and Ayyagari, A. 1994. Antimicrobial susceptibility and plasmid analysis of *Campylobacter jejuni* isolated from diarrhoeal patients and healthy chickens in northern India. J. Diarrhoeal. Dis. Res. 12: 270–273.

Reina, J., Ros, M.J., and Serra, A. 1994. Susceptibilities to 10 antimicrobial agents of 1,220 *Campylobacter* strains isolated from 1987 to 1993 from feces of paediatric patients. Antimicrob. Agents Chemother. 38: 2917–2920.

Ruiz, J., Goni, P., Marco, F., Gallardo, F., Mirelis, B., Jimenez De Anta, T., and Vila, J. 1998. Increased resistance to quinolones in *Campylobacter jejuni*: a genetic analysis of *gyrA* gene mutations in quinolone-resistant clinical isolates. Microbiol. Immunol. 42: 223–226.

Sagara, H., Mochizuki, A., Okamura, N., and Nakaya, R.1987. Antimicrobial resistance of *Campylobacter jejuni* and *Campylobacter coli* with special reference to plasmid profiles of Japanese clinical isolates. Antimicrob. Agents Chemother. 31: 713–719.

Skirrow, M.B., and Blaser, M.J. 2000. Clinical aspects of *Campylobacter* infection. In: *Campylobacter*, 2nd Edition. I. Nachamkin and M. J. Blaser, eds. ASM Press: Washington, D.C. pp.69–88.

Smith, K., Bender, J.B., and Osterholm, M.T. 2000. Antimicrobial resistance in animals and relevance to human infections. In: *Campylobacter*, 2nd Edition. I. Nachamkin and M.J. Blaser, eds. ASM Press: Washington, D.C. pp. 483–496.

Smith, K., Besser, J.M., Hedberg, C.W., Leano, F.T., Bender, J.B., Wicklund, J.H., Johnson, B.P., Moore, K.A., and Osterholm, M.T. 1999. Quinolone-resistant *Campylobacter jejuni* infections in Minnesota, 1992–1998. Investigation Team. N. Engl. J. Med. 340: 1525–1532.

Sougakoff, W., Papadopoulou, B., Nordmann, P., and Courvalin, P. 1987. Nucleotide sequence and distribution of gene *tetO* encoding tetracycline resistance in *Campylobacter coli*. FEMS Microbiol. Lett. 44: 153–159.

Taylor, D.E. 1986. Plasmid-mediated tetracycline resistance in *Campylobacter jejuni*: expression in *Escherichia coli* and identification of homology with streptococcal class M determinant. J. Bacteriol. 165: 1037–1039.

Taylor, D.E. 1992. Antimicrobial resistance of *Campylobacter jejuni* and *Campylobacter coli* to tetracycline, chloramphenicol and erythromycin. In: *Campylobacter jejuni*: Current status and future trends. I. Nachamkin, ed. ASM Press: Washington, D.C. pp.74–86.

Taylor, D.E., and Bryner, J.H. 1984. Plasmid content and pathogenicity of *Campylobacter jejuni* and *Campylobacter coli* strains in the pregnant guinea pig model. Am. J. Vet. Res. 45: 2201–2202.

Taylor, D.E., De Grandis, S.A., Karmali, M.A., and Fleming, P.C. 1981. Transmissible plasmids from *Campylobacter jejuni*. Antimicrob. Agents Chemother. 19: 831–835.

Taylor, D.E., Garner, R.S., and Allan, B.J. 1983. Characterization of tetracycline resistance plasmids from *Campylobacter jejuni* and *Campylobacter coli*. Antimicrob. Agents Chemother. 24: 930–935.

Taylor, D.E., Hiratsuka, K., Ray, H., and Manavathu, E.K. 1987. Characterization and expression of a cloned tetracycline resistance determinant from *Campylobacter jejuni* plasmid pUA466. J. Bacteriol. 169: 2984–2989.

Taylor, D.E., Ng, L.K., and Lior, H. 1985. Susceptibility of *Campylobacter* species to nalidixic acid, enoxacin, and other DNA gyrase inhibitors. Antimicrob. Agents Chemother. 28: 708–710.

Taylor, D.E., Chang, N., Garner, R.S., Sherburne, R., and Mueller, L. 1986. Incidence of antibiotic resistance and characterization of plasmids in *Campylobacter jejuni* strains isolated from clinical sources in Alberta, Canada. Can. J. Microbiol. 32: 28–32.

Taylor, D.E., Jerome, L.J., Grewal, J., and Chang, N. 1995. Tet (O), a protein that mediates ribosomal protection to tetracycline, binds, and hydrolyses GTP. Can. J. Microbiol. 41: 965–970.

Tenover, F.C., Gilbert, T., and O'Hara, P. 1989. Nucleotide sequence of a novel kanamycin resistance gene, *aphA*-7, from *Campylobacter jejuni* and comparison to other kanamycin phosphotransferase genes. Plasmid 22: 52–58.

Tenover, F.C., Williams, S., Gordon, K.P., Nolan, C., and Plorde, J.J. 1985. Survey of plasmids and resistance factors in *Campylobacter jejuni* and *Campylobacter coli*. Antimicrob. Agents Chemother. 27: 37–41.

Trieber, C.A., and Taylor, D.E. 2000. Mechanisms of antibiotic resistance in *Campylobacter*. In: *Campylobacter*, 2nd Edition. I. Nachamkin and M. J. Blaser, eds. ASM Press, Washington, DC, pp. 441–454.

Trieber, C.A., Burkhardt, N., Nierhaus, K.H., and Taylor, D.E. 1998. Ribosomal protection from tetracycline mediated by Tet (O): Tet (O) interaction with ribosomes is GTP-dependent. Biol. Chem. 379: 847–855.

Villedieu, A., Diaz-Torres, M.L., Hunt, N., McNab, R., Spratt, D.A., Wilson, M., and Mullany, P. 2003. Prevalence of tetracycline resistance genes in oral bacteria. Antimicrob. Agents Chemother. 47: 878–882.

Wang, Y., Huang, W.M., and Taylor, D.E. 1993. Cloning and nucleotide sequence of the *Campylobacter jejuni* *gyrA* gene and characterization of quinolone resistance mutations. Antimicrob. Agents Chemother. 37: 457–463.

Wang, Y., and Taylor, D. E. 1990. Chloramphenicol resistance in *Campylobacter coli*: nucleotide sequence, expression, and cloning vector construction. Gene 94: 23–28.

Wang, Y., and Taylor, D. E. 1991. A DNA sequence upstream of the *tet* (O) gene is required for full expression of tetracycline resistance. Antimicrob. Agents Chemother. 35: 2020–2025.

Wistrom, J., and Norrby, S.R. 1995. Fluoroquinolones and bacterial enteritis, when and for whom? J. Antimicrob. Chemother. 36: 23–39.

Witte, W. 1998. Medical consequences of antibiotic use in agriculture. Science 279: 996–997.

Yan, W., and Taylor, D.E. 1991. Characterization of erythromycin resistance in *Campylobacter jejuni* and *Campylobacter coli*. Antimicrob. Agents Chemother. 35: 1989–1996.

Chapter 10

Multidrug Efflux Systems in *Campylobacter*

Jun Lin, Masato Akiba and Qijing Zhang

Abstract

Campylobacter jejuni has become increasingly resistant to antimicrobial agents. As a general resistance mechanism, bacterial antimicrobial efflux machinery plays an essential role in the intrinsic and acquired resistance to various antibiotics. Based on the genome sequence of NCTC 11168, *Campylobacter* contains 13 putative multidrug efflux systems, most of which have not been functionally characterized. To date, CmeABC is the only defined antibiotic efflux system in *Campylobacter*, which functions as an energy-dependent efflux pump contributing to *Campylobacter* resistance to antimicrobial agents and adaptation in animal hosts. As exemplified by CmeABC, the expression and function of the efflux systems in *Campylobacter* may be modulated by transcriptional regulation and possible post-translational modification. It is likely that these efflux systems function together to meet the normal physiological needs of *Campylobacter* and facilitate *Campylobacter* adaptation to different environmental conditions including antibiotic treatments. Better understanding of the antibiotic efflux machinery in *Campylobacter* will assist the development of strategies to control the occurrence and spread of antibiotic-resistant *Campylobacter*.

INTRODUCTION

Campylobacter jejuni, the leading bacterial cause of human enteritis in many industrialized countries (Friedman *et al.*, 2000), has become increasingly resistant to antibiotics including fluoroquinolones (Smith *et al.*, 1999; Trieber and Taylor, 2000; Engberg *et al.*, 2001). In general, *Campylobacter* resistance to antibiotics is mediated by three mechanisms, including: (1) synthesis of enzymes (e.g. β-lactamase, dihydrofolate reductase, and acetyltransferase) that modify or inactivate antibiotics; (2) alteration or protection of targets that results in reduced affinity to antibiotics (e.g. mutations in *gyrA* or 23S rRNA genes, and synthesis of protective protein TetO); and (3) active extrusion of drugs from *Campylobacter* cells through efflux transporters (e.g. CmeABC). The detailed information and specific examples related to the first two general mechanisms have been described in a recent review by Trieber and Taylor (2000) and will not be discussed in this chapter. Although specific drug transporters that are involved in the extrusion of a particular drug or class of drugs (e.g. tetracycline efflux protein) have not been reported in *Campylobacter*, multidrug efflux systems, often named multidrug resistance (MDR) pumps, have been recently identified in *Campylobacter* (Lin *et al.*, 2002; Pumbwe and Piddock, 2002). In addition, the genomic sequence of *C. jejuni* NCTC 11168 revealed the presence of multiple putative MDR pumps (Parkhill *et al.*, 2000), although the majority of them have not been functionally characterized. In this chapter, we will summarize the current knowledge on *Campylobacter* MDR systems and discuss future directions for research on the drug efflux machinery in this pathogenic organism.

OVERALL FEATURES OF BACTERIAL MULTIDRUG EFFLUX SYSTEMS

MDR pumps contribute significantly to the intrinsic and acquired resistance to multiple antibiotics in Gram-positive and Gram-negative bacteria as well as in eukaryotic organisms (Reviewed in references Van Bambeke *et al.*, 2000; Putman *et al.*, 2000; Poole, 2001). In bacteria, there are many different types of MDR transporters, which vary in size, structure, and energy source (proton motive force or ATP hydrolysis) (Putman *et al.*, 2000). In general, bacterial antimicrobial efflux transporters belong to five superfamilies, including MFS (major facilitator superfamily), SMR (small multidrug resistance), RND (resistance–nodulation–division), MATE (multidrug and toxic compound extrusion), and ABC (ATP-binding cassette) (Putman *et al.*, 2000). Except ABC type transporters which use the free energy of ATP hydrolysis to extrude drugs out of cells, the other four families utilize the transmembrane electrochemical gradient of protons or sodium ions as energy sources for drug efflux (Putman *et al.*, 2000). These efflux systems are broadly distributed, and a single microorganism can have multiple efflux transporters of different families with overlapping substrate spectra (Putman *et al.*, 2000; Poole, 2001; Paulsen *et al.*, 2001). Comparative genomic analysis revealed that the types and numbers of MDR efflux systems vary greatly among organisms (Paulsen *et al.*, 2001).

Despite the fact that a variety of MDR efflux families exist in bacterial organisms, the RND-type efflux system is a major and important family of drug transporters in Gram-negative bacteria. The RND family transporters are unique to Gram-negative bacteria and are frequently involved in resistance to clinically relevant antibiotics (Poole, 2001). Structurally, a RND type efflux system typically consists of an inner membrane transporter, a periplasmic fusion protein, and an outer membrane protein (Zgurskaya and Nikaido, 2000). These three components function together and form a membrane pump to extrude antibiotics directly from cytoplasm/periplasm to extracellular spaces. Genetically, many of the RND-type MDR efflux systems are encoded by three-gene operons located on bacterial chromosomes (Paulsen *et al.*, 1997; Zgurskaya and Nikaido, 2000). However, some RND-type efflux pumps, such as AcrAB from *Escherichia coli* (Ma *et al.*, 1993), have an outer membrane component that is encoded by a separate gene physically unlinked with the other two members on the bacterial chromosome. More detailed information about molecular properties of various types of bacterial MDR transporters is discussed in an excellent review by Putman *et al.* (2000).

A key feature of these MDR efflux systems, particularly RND-type pumps, is their ability to extrude a broad spectrum of substrates including various clinically relevant antibiotics (Van Bambeke *et al.*, 2000; Putman *et al.*, 2000; Engberg *et al.*, 2001; Poole, 2001). Overexpression of the efflux pumps usually results in a MDR phenotype in bacterial pathogens and is considered a major mechanism of antibiotic resistance in a growing number of pathogenic bacteria (Van Bambeke *et al.*, 2000; Putman *et al.*, 2000; Poole, 2001; Grkovic *et al.*, 2002). Even without overexpression, the MDR efflux pumps work synergistically with other non-efflux resistance mechanisms (such as target mutation) to maintain high levels of antimicrobial resistance in bacteria (Lomovskaya *et al.*, 1999; Oethinger *et al.*, 2000; Wang *et al.*, 2001; Luo *et al.*, 2003). Therefore, the MDR efflux pumps have greatly compromised the effectiveness of antibiotic treatments and posed a serious problem in public health.

The expression of bacterial MDR efflux pumps is usually controlled by transcriptional regulators that either repress or activate the transcription of the MDR efflux genes (Poole, 2001; Grkovic *et al.*, 2002). Many of these regulators are local repressors that directly interact with the promoter regions of MDR efflux genes or operons. For example, repressors AcrR (*E. coli*), QacR (*Staphylococcus aureus*), MtrR (*Neisseria gonorrhoeae*), and MexR (*Pseudomonas aeruginosa*) bind specifically to the upstream sequences of *acrAB*, *qacA*, *mtrCDE*, and *mexAB*, respectively, thereby inhibiting the expression of the corresponding MDR efflux gene (s) (Hagman and Shafer, 1995; Ma *et al.*, 1996; Grkovic *et al.*, 1998; Evans *et al.*, 2001). The operator sequences that interact with the repressor molecules usually consist of inverted repeats (Grkovic *et al.*, 2002). Mutations in the repressors or repressor-binding regions impede the repression and result in overexpression of efflux pumps, which consequently increases bacterial resistance to structurally unrelated antimicrobial agents (Pan and Spratt, 1994; Hagman and Shafer, 1995; Hagman *et al.*, 1995; Poole, *et al.*, 1996; Grkovic *et al.*, 1998; Saito *et al.*, 1999; Wang *et al.*, 2001). Some MDR efflux systems are also controlled by global regulators, such as the *marRAB* regulon in *E. coli* (Alekshun and Levy, 1999b). MarR is a repressor of *marRAB*, while *marA* encodes an activator that not only positively regulates *marRAB* but also activates a variety of genes (including *acrAB*) associated with resistance to antibiotics and oxygen stress (Alekshun and Levy, 1999b; Alekshun *et al.*, 2000). Mutations in *marR* substantially increase the expression of *acrAB* and confer *E. coli* resistance to a variety of antimicrobial agents (Okusu *et al.*, 1996; Alekshun and Levy, 1999a; Kern *et al.*, 2000).

In addition to the mutation-based mechanisms that result in sustained overexpression of MDR efflux pumps in bacteria, the production of some MDR efflux pumps can be conditionally induced by structurally diverse substrates of these pumps or by stress signals (Kato *et al.*, 1992; Ahmed *et al.*, 1994; Ma *et al.*, 1995; Kaatz and Seo, 1995; Grkovic *et al.*, 1998; Brooun *et al.*, 1999; Masuda *et al.*, 2000). This induction is usually due to the direct interaction of the substrates with repressor molecules, which interferes with the binding of repressors to operator DNA and results in increased expression of MDR genes. Two repressors (QacR and AcrR) of such inducible MDR efflux pumps belong to the TetR family of transcriptional regulators, which share a conserved helix–turn–helix DNA-binding motif at their N-terminal regions and have divergent C-terminal sequences that are involved in the binding to inducing compounds (Hillen and Berens, 1994; Grkovic *et al.*, 1998). A recent study by Schumacher *et al.* (2001) has revealed the crystal structures of QacR–drug complexes and identified multiple overlapping drug-binding sites in the C-terminal portion of this repressor protein.

IDENTIFICATION OF MDR PUMP CmeABC IN *CAMPYLOBACTER*

In an early study (Charvalos *et al.*, 1995), Charvalos *et al.* selected MDR *C. jejuni* isolates by *in vitro* passage of the organism on pefloxacin-containing plates. These isolates showed increased resistance to fluoroquinolones (FQs), tetracycline, erythromycin, chloramphenicol, and β-lactams. Accumulation assays found that the resistant isolates showed less accumulation of FQs, suggesting the presence of functional efflux system(s) in these isolates. Although SDS-PAGE identified two putatively overproduced outer membrane proteins in these MDR isolates, the identities of the proteins and their role in efflux of antibiotics were not determined (Charvalos *et al.*, 1995). Completion of *Campylobacter*

ciprofloxacin as an example, at the steady-state reached within 2 min following addition of ciprofloxacin, *cmeB* mutant accumulated about threefold more ciprofloxacin than wild-type 81-176. Addition of carbonyl cyanide *m*-chlorophenylhydrazone (CCCP), a proton conductor, resulted in a very rapid and dramatic increase in cell-associated ciprofloxacin in wild-type 81-176, but not in *cmeB* mutant. At the steady-state following addition of CCCP, the accumulation level of ciprofloxacin was similar between 81-176 and the mutant. These results suggested that CmeABC is the main pump (if not the only one) for the efflux of ciprofloxacin in *Campylobacter*.

The *cmeABC* operon is widely distributed in different *Campylobacter* strains and is constitutively expressed in wild-type strains at a level that can be readily detected with specific antibodies (Lin *et al.*, 2002; Payot *et al.*, 2002; Lin *et al.*, 2003b). Despite extensive evidence demonstrating the contribution of CmeABC to the intrinsic resistance to various antibiotics, the role of CmeABC in the acquired antibiotic resistance in *Campylobacter* has not been well understood, partly because overproduction of CmeABC has not been linked to antibiotic resistance in clinical isolates. In our study (Lin *et al.*, 2002), mutation in *cmeB* resulted in an eight-fold decrease in the MIC of tetracycline in the *cmeB* mutant even though the tetracycline resistance gene *tet (O)* was still present in the mutant, suggesting that *cmeABC* functions synergistically with *tet (O)* to contribute to the acquired resistance to tetracycline. Another study conducted in our laboratory (Luo *et al.*, 2003) demonstrated that CmeABC functioned synergistically with *gyrA* mutations in mediating high levels of resistance to FQs in the *in vivo*-selected clinical isolates that had ciprofloxacin MICs >32 μg/mL. Compared with the wild-type resistant isolates, the *cmeB* mutants showed a marked decrease in their resistance to ciprofloxacin and enrofloxacin, even though the *cmeB* mutants still carried the resistance-associated *gyrA* mutation (Thr-86–Ile). The residual resistance in the *cmeB* mutants of the FQ-resistant isolates was still above the level of the wild-type sensitive isolates but below the level of clinical significance. Thus, without the function of the CmeABC, *gyrA* mutations alone could not maintain the resistant phenotype. Both *gyrA* mutations and the function of CmeABC are absolutely required for maintaining the high level of resistance to FQs. Despite the key role of CmeABC in maintaining the high-level acquired resistance to FQ antibiotics, this efflux pump was not overproduced in the FQ-resistant isolates when compared with the isogenic FQ-sensitive isolates derived in the same study (Luo *et al.*, 2003). This finding contradicts the observation with other Gram-negative bacteria that acquired FQ resistance is often associated with the overproduction of efflux pumps (Poole, 2000), and suggests that CmeABC may have unique regulatory features.

OTHER PUTATIVE MDR TRANSPORTERS IN *CAMPYLOBACTER*

Based on the genome sequence of *C. jejuni* NCTC 11168 (Parkhill *et al.*, 2000) and comparative genomic analysis of MDR transporters (Paulsen *et al.*, 2001), additional putative drug efflux proteins have been identified in *C. jejuni* (Table 10.1). In addition to CmeABC, there is another putative RND-type efflux system in *C. jejuni*. This uncharacterized system is also encoded by a three-gene operon (named *cmeDEF*) composed of *Cj1031*, *Cj1032*, and *Cj1033* (designated *cmeD*, *cmeE*, and *cmeF*, respectively). According to sequence homology to known MDR efflux systems, CmeD, CmeE, and CmeF are predicted to be the outer membrane component, periplasmic fusion protein, and inner membrane RND-type transporter, respectively. The product encoded by *cmeD* (424 aa) shares low, but significant sequence homology to the outer membrane component of bacterial efflux systems, such as HefA (Genebank accession no. AF059041; 25% identity) of *H. pylori* and TolC (X54049; 22% identity) of *E. coli*. *cmeE* encodes 246 aa and shares significant homology with the membrane fusion protein of RND-type efflux systems, such as HefB (AF059041; 35.5% identity) of *H. pylori*. *cmeF* encodes a long ORF of 1005 aa, which shares significant homology with many RND-type drug transporters, such as HefC (AF059041; 37% identity) of *H. pylori*, AcrB (U00734; 25% identity) of *E. coli*. *cmeE* overlaps with *cmeD* and *cmeF*, suggesting that the three genes are likely organized in an operon. The predicted secondary structures of each ORF are also consistent with the known features of typical tripartite efflux systems in bacteria. Interestingly, the three ORFs share limited sequence homology with the corresponding components of the *cmeABC* operon, suggesting the three genes encode a distinct efflux system in *Campylobacter*. PCR analysis using *cmeF*-specific primers conducted in our laboratory indicated that *cmeDEF* is uniformly distributed among *C. jejuni* strains derived from different animal hosts.

To define the function of CmeDEF, our laboratory has recently initiated the work to characterize the *cmeDEF* system. Unlike the finding with CmeABC, insertional mutation in the *cmeF* gene did not result in >2-fold MIC changes in the susceptibility of *Campylobacter* to a wide range of antibiotics, dyes, heavy metals, bile salts, or detergents. RT-PCR analysis indicated that *cmeDEF* was transcribed in wild-type strains. However, immunoblotting using a CmeF-specific antibody demonstrated that CmeF was synthesized at a much lower level than CmeABC in strains 81-176 and NCTC 11168. The low expression level of *cmeDEF* in wild-type strains suggests that this system may not play a major role in the intrinsic resistance of *Campylobacter* to antibiotics, which may partly explain the lack of significant MIC changes in the *cmeF* mutants. At present, it is unknown what conditions can cause the elevated production of CmeDEF and if the overproduction contributes to resistance to antimicrobial agents. Sequence analysis revealed that two inverted repeats occur in the immediate upstream region of the *cmeDEF* operon, suggesting that this system is likely regulated by a transcriptional factor(s). Thus, it is possible that the production of CmeDEF is tightly controlled by repressor proteins in wild-type strains, and it is important to consider the regulation of CmeDEF when defining its function in antimicrobial resistance.

There is another possibility that CmeDEF has intended functions other than antibiotic resistance. To determine the effect of various mutations on *Campylobacter* growth and morphology, we generated single mutants (defective in either CmeABC or CmeDEF) and a double mutant (defective in both CmeABC and CmeDEF) and compared their growth patterns in laboratory media. The mutants and wild-type showed similar growth rates in MH broth. However, in the stationary phase, the double mutant showed a lower survival rate than the single mutants and the wild type, because the viable CFU of the double mutant was consistently 1–2 log units less than those of the wild-type and single mutants in the stationary phase. When different growth phases were examined under microscope, the double mutant transformed into coccoid cells much earlier than the wild-type and the single mutants, suggesting that the double mutant was less fit in the stationary phase. Based on these observations, it is tempting to speculate that CmeDEF along with CmeABC may be involved in the efflux of intracellular toxic compounds and plays an important role in the survival of *Campylobacter* in the stationary phase.

Based on sequence homology, four putative MFS transporters (Cj1375, Cj0035c, Cj1257c, and Cj1687) are present in *C. jejuni* NCTC 11168 (Table 10.1). The corresponding homologues of the four transporters in other bacteria are true MFS-type transporters, mediating resistance to a range of structurally dissimilar drugs including quinolone antimicrobials (Putman *et al.*, 2000). Limited amino acid homology exists among the four putative MFS transporters, suggesting that each of them may have a distinct substrate spectrum. There are also four putative SMR-type transporters in NCTC 11168 (Cj0309c, Cj0310c, Cj1173, and Cj1174; Table 10.1). As the smallest MDR efflux protein, a SMR-type MDR transporter usually function as homooligomeric complexes (Putman *et al.*, 2000). Unlike the majority of SMR transporters found in other bacteria in which SMR transporters are encoded by unlinked genetic loci, the four SMR-type transporters in *Campylobacter* are encoded by gene pairs in two distinct operons (*Cj0309c/Cj0310c* and *Cj1173/Cj1174*, respectively). Each operon encodes two distinct but similar proteins. Similar two-component SMR-type pumps (EbrAB and YkkCD) have been reported in *Bacillus subtilis* (Jack *et al.*, 2000; Masaoka *et al.*, 2000). In each pair, the two members show high sequence similarity, but one member is shorter than the other due to the difference in the C-terminal hydrophilic region (Jack *et al.*, 2000; Masaoka *et al.*, 2000). Both members in each pair (EbrAB or YkkCD) are required for the activity of the MDR pump. The EbrAB system is primarily responsible for the resistance to cationic dyes (e.g. ethidium bromide), while YkkCD extrudes a broad range of substrates including cationic dyes, chloramphenicol, streptomycin, tetracycline as well as phosphonomycin (Jack *et al.*, 2000; Masaoka *et al.*, 2000).

Although there are many ABC transporters encoded on *C. jejuni* genome, there is only one ABC-type transporter (Cj1187c) that appears to be a drug efflux pump (Parkhill *et al.*, 2000). The sequence of *Cj1187c* shares high homology with *arsB* of *E. coli,* which is a membrane subunit of an anion-translocating ATPase that confers resistance to arsenite and antimonite (Bhattacharjee *et al.*, 2000). According to the analysis by Paulsen *et al.* (http://www.membranetransport.org), *Campylobacter* has two putative Na^+-driven multidrug efflux transporters, Cj0560 and Cj0619, which belong to a newly characterized family of transporters termed MATE (Table 10.1). Two characterized MATE-type transporters, NorM from *Vibrio parahaemolyticus* and its *E. coli* homologue YdhE, contribute to resistance to FQs, ethidium bromide, kanamycin, and streptomycin (Morita *et al.*, 1998; Morita *et al.*, 2000). Unlike the transporters in the other families discussed above, the MATE-type transporters have not been extensively characterized even though MATE-type transporters are widely present in various organisms (http://www.membranetransport.org).

ROLE OF MDR EFFLUX SYSTEMS IN *CAMPYLOBACTER* ADAPTATION

One striking feature of CmeABC is its essential role in *C. jejuni* resistance to bile, a group of bactericidal detergents present in the intestinal tracts of animals (Lin *et al.*, 2003b). Detergent-like bile salts kill bacterial cells by destroying the lipid bilayer of membrane (Gunn, 2000). Thus, resistance to bile salts is important for enteric pathogens to survive in the intestinal tract. Although several MDR efflux systems in Gram-negative bacteria were demonstrated to confer resistance to bile salts *in vitro* (Ma *et al.*, 1995; Fralick, 1996; Thanassi *et al.*, 1997; Hagman *et al.*, 1997; Colmer *et al.*, 1998), direct evidence that MDR efflux pumps are required for bacterial adaptation in animal intestines is still limited. Studies conducted in our laboratory have recently demonstrated that CmeABC, by mediating bile resistance, is essential for *Campylobacter* growth in bile-containing media and colonization in animal intestinal tracts (Lin *et al.*, 2003b). Inactivation of *cmeABC* drastically decreased the resistance of *Campylobacter* to various bile salts

(Figure 10.1). Addition of cholate (2 mM) in culture media impaired the *in vitro* growth of the *cmeABC* mutants, but had no effect on the growth of the wild-type strain. When inoculated into chickens, the wild-type strain colonized the birds as early as day 2 post-inoculation with a density as high as 10^7 CFU/g faeces. In contrast, the *cmeABC* mutants failed to colonize any of the inoculated chickens throughout the study. Complementation of the *cmeABC* mutants with a wild-type *cmeABC* allele *in trans* fully restored the *in vitro* growth in bile-containing media and the *in vivo* colonization to the levels of the wild-type strain. Immunoblotting analysis indicated that CmeABC is expressed and immunogenic in chickens experimentally infected with *C. jejuni*. These findings define a key natural function of a multidrug efflux pump in an enteric pathogen and indicate that CmeABC is essential for *Campylobacter* adaptation in animal hosts.

Considering the fact that both CmeABC and CmeDEF are constitutively synthesized in *Campylobacter*, it would be uneconomical to this organism if the two efflux pumps are used strictly for antibiotic resistance. It is very likely that the two pumps have important physiological functions that remain to be discovered. The finding that the *cmeB/cmeF* double mutant does not survive well in the stationary phase suggests that the two efflux pumps are involved in stress tolerance, although the definitive role of each system in the stress response is unknown. It has been proposed that bacterial MDR efflux pumps may function to prevent accumulation of potentially deleterious compounds which may originate inside or outside of bacterial cells (Poole, 2001). Thus, identification of these 'natural' substrates for CmeABC and CmeDEF as well as other putative MDR efflux pumps in *Campylobacter* will greatly facilitate the understanding of how MDR efflux pumps contribute to *Campylobacter* adaptation to various environmental conditions.

REGULATION OF MDR EFFLUX SYSTEMS IN *CAMPYLOBACTER*

Bacterial MDR pumps are tightly regulated by repressors and activators. In many situations, the gene encoding the regulatory protein of a particular efflux system is adjacent to the operon encoding the efflux pump. For example, *mexR*, a gene encoding the repressor for the MexAB-OprM efflux system, is located immediately upstream of the *mexA* gene (Saito *et al.*, 1999; Evans *et al.*, 2001). Based on the presence of inverted repeats in the promoter region of *cmeABC* and *cmeDEF* and the known regulatory features of bacterial MDR pumps, we suspect that CmeABC and CmeDEF are controlled by regulators. Sequence analysis of the *C. jejuni* NCTC 11168 genome indicated that *Cj0368c*, the gene

I   II   III

 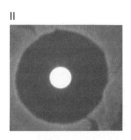 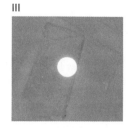

Figure 10.1 Contribution of CmeABC to bile resistance as determined by disk diffusion assay. Each *Campylobacter* culture was incubated with a disk containing 500 μg of cholic acid. The wild-type 21190 is resistant to bile and has no zone of inhibition (I). The *cmeB* mutation greatly increases the sensitivity of 21190 to cholic acid as indicated by a distinct zone of inhibition (II). Complementation of the mutant with a wild-type *cmeABC* allele restores the resistance to cholic acid as indicated by the lack of inhibition (III).

immediately upstream of *cmeA,* encodes a probable transcriptional regulatory protein of 210 aa. The deduced amino acid sequence of *Cj0368c* shares similarity to members of the TetR family of transcriptional repressors of efflux systems (e.g. 29% identity with QacR in 198 aa overlap). Further analysis using MOTIF, a program for searching for protein motifs in multiple databases (http://www.motif.genome.ad.jp), identified a helix–turn–helix DNA binding motif, a signature sequence of the TetR family, at the N-terminal region of Cj0368c (aa 21–64). Together, these observations suggests that *Cj0368c,* named *cmeR,* probably encodes a transcriptional repressor for *cmeABC* and prompted us to examine the role of CmeR in regulating MDR efflux pumps.

To determine if CmeR regulates the expression of *cmeABC,* we constructed an isogenic *cmeR* mutant of 81-176 by insertional mutagenesis (Lin *et al.,* 2003a). Immunoblotting using CmeABC-specific antibodies demonstrated that CmeABC was significantly overproduced in the *cmeR* mutant compared to its parent strain 81-176. Consistent with the overproduction of the efflux pump, the *cmeR* mutant showed increased resistance to multiple antibiotics compared to the wild-type strain. Using gel mobility shift assays, it was shown that the purified recombinant CmeR bound specifically to the promoter region of *cmeABC.* The enhanced expression of CmeABC in the CmeR mutant and the demonstrated binding of CmeR to the promoter region of CmeABC strongly suggest that CmeR serve as a local repressor for CmeABC.

Using *in vitro* stepwise selection from susceptible *C. jejuni* 81-176 on Mueller–Hinton agar plates supplemented with increasing concentration gradients of ciprofloxacin, we obtained a multidrug resistance mutant (CR3e), which showed increased resistance to ampicillin, cefotaxime, erythromycin, and ethidium bromide in addition to ciprofloxacin (Lin *et al.,* 2003a). Immunoblotting also indicated that CmeABC pump was overproduced at a level similar to that of the *cmeR* mutant. Sequence analysis indicated that there was no mutation in the *cmeR* gene in mutant CR3e. However, a single nucleotide deletion occurred between the inverted repeats in the promoter region of *cmeABC,* suggesting that the deletion may affect the binding of CmeR and result in enhanced expression of CmeABC in CR3e. Together, these findings indicate that CmeR functions as a repressor for CmeABC, and the interaction of CmeR with the inverted repeats in the promoter region of *cmeABC* regulates the expression of this drug efflux operon. Thus, mutations in either *cmeR* or the inverted repeat region upstream of *cmeABC* will result in the overproduction of the CmeABC pump.

The insertional mutation in *cmeR* did not affect the level of CmeDEF synthesis as shown by immunoblotting analysis in our laboratory, suggesting that CmeR is not a regulatory protein for CmeDEF. Since there are two inverted repeats in the promoter region of *cmeDEF,* it is very likely that this system is controlled by uncharacterized regulatory protein (s). Analysis of the genomic sequence did not reveal putative regulator genes adjacent to other transporter genes (Table 10.1). However, inverted repeats, typical for binding sites of regulatory proteins, are present upstream of the start codons of every putative transporter genes or operons listed in Table 10.1, suggesting that expression of these putative MDR transporter genes may be controlled by transcriptional factors. Based on the review by Grkovic *et al.* (2002), all of the bacterial MDR transporters identified to date whose expression is under control by transcriptional factors are members of either MFS or RND family. No transcriptional regulation has been reported in any member of other MDR families. It will be interesting to know if the *Campylobacter* MDR transporters are regulated and how the regulation contributes to the adaptation of *Campylobacter* to various environmental conditions including antibiotic treatments.

In addition to the sustained overproduction of CmeABC due to the mutation in CmeR or the promoter region discussed above, there is another possibility that the expression of *cmeABC* is conditionally inducible by its substrates. CmeR shares similarity with QacR and belongs to TetR family of transcriptional regulators. Some substrates of MDR pumps can directly interact with the repressors in the TetR family (e.g. QacR) and block the binding of the repressors to operator DNA, resulting in increased expression of MDR pumps (Ma *et al.*, 1996; Grkovic *et al.*, 1998). As discussed previously (Luo *et al.*, 2003), the FQ-resistant and sensitive *Campylobacter* isolates showed similar levels of production of CmeABC when the isolates were cultured in MH broth without FQ antimicrobials, suggesting that sustained overproduction of CmeABC did not occur in the isolates with high level of resistance to ciprofloxacin. However, this finding did not exclude the possibility that production of CmeABC is inducible in the presence of ciprofloxacin or other substrates. This possibility remains to be examined in future studies.

In addition to transcriptional control, the MDR efflux systems in *Campylobacter* also undergo post-translational modifications, which represents a unique feature of *Campylobacter* MDR pumps. *Campylobacter* has both O- and N-linked glycosylation pathways (Szymanski *et al.*, 2003). The two systems appear to be involved in the glycosylation of multiple soluble and membrane-associated proteins (Wacker *et al.*, 2002; Szymanski *et al.*, 1999; Young *et al.*, 2002). Wacker *et al.* (2002) reported that CmeA (Cj0367c), a periplasmic component of CmeABC MDR efflux system, is modified by the N-linked glycosylation pathway and the modification influences the antigenicity of CmeA. Young *et al.* (2002) identified more than 22 N-linked glycoproteins including CmeE (Cj1032), a putative periplasmic fusion component of the CmeDEF efflux system. At present, it is unknown if the post-translational modification affects the function of CmeABC and CmeDEF.

CONCLUSIONS

In summary, *Campylobacter* contains multiple MDR efflux systems based on genomic sequence analysis (Table 10.1). These MDR efflux systems may function together in *Campylobacter* to meet normal physiological needs, contribute to resistance to structurally diverse antimicrobial agents, and play an important role in the adaptation to different environmental niches. Our understanding of these drug transporters has just begun and is limited, and the majority of these efflux systems remain to be characterized. In particular, the substrates for many of the efflux pumps are unknown. Elucidation of the substrate spectrum of each efflux pump will facilitate the understanding of the function of these systems and their contribution to *Campylobacter* fitness in response to changes in environmental conditions. It should be pointed out that the drug efflux machinery in *Campylobacter* is not designed just for antimicrobial resistance, but also has essential physiological functions.

As the best characterized efflux system in *Campylobacter*, CmeABC has shown to be an important RND-type efflux pump that not only contributes to *Campylobacter* resistance to antibiotics, but is also essential for *in vivo* colonization in animal hosts (Lin *et al.*, 2002; Pumbwe and Piddock, 2002; Luo *et al.*, 2003; Lin *et al.*, 2003b). This pump is controlled by CmeR and possibly by other unknown regulators. Mutation in CmeR or the operator sequence leads to enhanced production of CmeABC. However, the clinical relevance of CmeABC overproduction in the emergence of MDR *Campylobacter* has not been well understood. It is unclear if synergy exists among the MDR pumps and how these systems are coordinately regulated for optimal expression and function in response to different needs. It is also unknown if CmeR affects the expression of other putative drug transporters besides CmeABC or itself is controlled by another global regulator as is the case with AcrR (Grkovic *et al.*, 2002). Use of functional genomics and proteomics will help answer these questions. Our increased understanding of the drug efflux machinery may open new avenues for the development of measures to control *Campylobacter* infection in humans and animal reservoirs. For example, inhibitors that can block the function of CmeABC may not only control antibiotic resistance, but also increase the susceptibility of *C. jejuni* to *in vivo* bile salts, consequently preventing colonization of *Campylobacter* in the animal intestinal tract. Such pump inhibitors may be directly used as novel antimicrobials for therapeutic intervention of *Campylobacter* infection. From the standpoint of vaccine development, the outer membrane components of MDR pumps of Gram-negative bacteria may be exploited as targets of immune interventions to prevent the development of antibiotic-resistant *Campylobacter* and the establishment of infection *in vivo*.

Acknowledgement
This work is supported by grants from the NIH (DK063008) and USDA (00-51110-9741) to Q.Z.

References
Ahmed, M., Borsch, C.M., Taylor, S.S., Vazquez-Laslop, N., and Neyfakh, A.A. 1994. A protein that activates expression of a multidrug efflux transporter upon binding the transporter substrates. J. Biol. Chem. 269: 28506–28513.

Alekshun, M.N., Kim, Y.S., and Levy, S.B. 2000. Mutational analysis of MarR, the negative regulator of marRAB expression in *Escherichia coli*, suggests the presence of two regions required for DNA binding. Mol. Microbiol. 35: 1394–1404.

Alekshun, M.N. and Levy, S.B. 1999a. Alteration of the repressor activity of MarR, the negative regulator of the *Escherichia coli* marRAB locus, by multiple chemicals *in vitro*. J. Bacteriol. 181: 4669–4672.

Alekshun, M.N. and Levy, S.B. 1999b. The *mar* regulon: multiple resistance to antibiotics and other toxic chemicals. Trends Microbiol. 7: 410–413.

Bhattacharjee, H., Zhou, T., Li, J., Gatti, D.L., Walmsley, A.R., and Rosen, B.P. 2000. Structure–function relationships in an anion-translocating ATPase. Biochem. Soc. Trans. 28: 520–526.

Brooun, A., Tomashek, J.J., and Lewis, K. 1999. Purification and ligand binding of EmrR, a regulator of a multidrug transporter. J. Bacteriol. 181: 5131–5133.

Charvalos, E., Tselentis, Y., Hamzehpour, M.M., Kohler, T., and Pechere, J.C. 1995. Evidence for an efflux pump in multidrug-resistant *Campylobacter jejuni*. Antimicrob. Agents Chemother. 39: 2019–2022.

Colmer, J.A., Fralick, J.A., and Hamood, A.N. 1998. Isolation and characterization of a putative multidrug resistance pump from *Vibrio cholerae*. Mol. Microbiol. 27: 63–72.

Engberg, J., Aarestrup, F.M., Taylor, D.E., Gerner-Smidt, P., and Nachamkin, I. 2001. Quinolone and macrolide resistance in *Campylobacter jejuni* and *C. coli*: resistance mechanisms and trends in human isolates. Emerging Infect. Dis. 7: 24–34.

Evans, K., Adewoye, L., and Poole, K. 2001. MexR repressor of the *mexAB-oprM* multidrug efflux operon of *Pseudomonas aeruginosa*: identification of MexR binding sites in the mexA-mexR intergenic region. J. Bacteriol. 183: 807–812.

Fralick, J.A. 1996. Evidence that TolC is required for functioning of the Mar/AcrAB efflux pump of *Escherichia coli*. J. Bacteriol. 178: 5803–5805.

Friedman, C.R., Neimann, J., Wegener, H.C., and Tauxe, R.V. 2000. Epidemiology of *Campylobacter jejuni* Infections in the United States and Other Industrialized Nations. In: *Campylobacter*. I. Nachamkin and M.J. Blaser, eds. ASM Press, Washington, DC, pp. 121–138.

Grkovic, S., Brown, M.H., Roberts, N.J., Paulsen, I.T., and Skurray, R.A. 1998. QacR is a repressor protein that regulates expression of the *Staphylococcus aureus* multidrug efflux pump QacA. J. Biol. Chem. 273: 18665–18673.

Grkovic, S., Brown, M.H., and Skurray, R.A. 2002. Regulation of bacterial drug export systems. Microbiol. Mol. Biol. Rev. 66: 671–701.

Gunn, J.S. 2000. Mechanisms of bacterial resistance and response to bile. Microbes Infect. 2: 907–913.

Hagman, K.E., Lucas, C.E., Balthazar, J.T., Snyder, L., Nilles, M., Judd, R.C., and Shafer, W.M. 1997. The MtrD protein of *Neisseria gonorrhoeae* is a member of the resistance/nodulation/division protein family constituting part of an efflux system. Microbiology 143: 2117–2125.

Hagman, K.E., Pan, W., Spratt, B.G., Balthazar, J.T., Judd, R.C., and Shafer, W.M. 1995. Resistance of *Neisseria gonorrhoeae* to antimicrobial hydrophobic agents is modulated by the *mtrRCDE* efflux system. Microbiology. 141: 611–622.

Hagman, K.E. and Shafer, W.M. 1995. Transcriptional control of the *mtr* efflux system of *Neisseria gonorrhoeae*. J. Bacteriol. 177: 4162–4165.

Hillen, W. and Berens, C. 1994. Mechanisms underlying expression of Tn10 encoded tetracycline resistance. Annu. Rev. Microbiol 48: 345–369.

Jack, D.L., Storms, M.L., Tchieu, J.H., Paulsen, I.T., and Saier, M.H., Jr. 2000. A broad-specificity multidrug efflux pump requiring a pair of homologous SMR-type proteins. J. Bacteriol. 182: 2311–2313.

Johnson, J.M. and Church, G.M. 1999. Alignment and structure prediction of divergent protein families: periplasmic and outer membrane proteins of bacterial efflux pumps. J. Mol. Biol. 287: 695–715.

Kaatz, G.W. and Seo, S.M. 1995. Inducible NorA-mediated multidrug resistance in *Staphylococcus aureus*. Antimicrob. Agents Chemother. 39: 2650–2655.

Kato, S., Nishimura, J., Yufu, Y., Ideguchi, H., Umemura, T., and Nawata, H. 1992. Modulation of expression of multidrug resistance gene (*mdr*-1) by adriamycin. FEBS Lett. 308: 175–178.

Kern, W.V., Oethinger, M., Jellen-Ritter, A.S., and Levy, S.B. 2000. Non-target gene mutations in the development of fluoroquinolone resistance in *Escherichia coli*. Antimicrob. Agents Chemother. 44: 814–820.

Lin, J., Michel, L.O., and Zhang, Q. 2002. CmeABC functions as a multidrug efflux system in *Campylobacter jejuni*. Antimicrob. Agents Chemother. 46: 2124–2131.

Lin, J., Akiba, M., Sahin, O., and Zhang, Q. 2003a. Regulatory mechanisms of multidrug efflux pump CmeABC in *Campylobacter jejuni*. Int. J. Med. Microbiol. 293 (Suppl.): 49 (Abstract).

Lin, J., Sahin, O., Michel, L., and Zhang, Q. 2003b. Critical role of multidrug efflux pump CmeABC in bile resistance and *in vivo* colonization of *Campylobacter jejuni*. Infect. Immun.71: 4250–4259.

Lomovskaya, O., Lee, A., Hoshino, K., Ishida, H., Mistry, A., Warren, M.S., Boyer, E., Chamberland, S., and Lee, V.J. 1999. Use of a genetic approach to evaluate the consequences of inhibition of efflux pumps in *Pseudomonas aeruginosa*. Antimicrob. Agents Chemother. 43: 1340–1346.

Luo, N., Sahin, O., Lin, J., Michel, L.O., and Zhang, Q. 2003. In vivo selection of *Campylobacter* isolates with high levels of fluoroquinolone resistance associated with *gyrA* mutations and the function of the CmeABC efflux pump. Antimicrob. Agents Chemother. 47: 390–394.

Ma, D., Cook, D.N., Alberti, M., Pon,N.G., Nikaido, H., and Hearst, J.E. 1993. Molecular cloning and characterization of *acrA* and *acrE* genes of *Escherichia coli*. J. Bacteriol. 175: 6299–6313.

Ma,D., Cook, D.N., Alberti, M., Pon, N.G., Nikaido, H., and Hearst, J.E. 1995. Genes *acrA* and *acrB* encode a stress-induced efflux system of *Escherichia coli*. Mol. Microbiol. 16: 45–55.

Ma, D., Alberti, M., Lynch, C., Nikaido, H., and Hearst, J.E. 1996. The local repressor AcrR plays a modulating role in the regulation of *acrAB* genes of *Escherichia coli* by global stress signals. Mol. Microbiol. 19: 101–112.

Masaoka, Y., Ueno, Y., Morita, Y., Kuroda, T., Mizushima, T., and Tsuchiya, T. 2000. A two-component multidrug efflux pump, EbrAB, in *Bacillus subtilis*. J. Bacteriol. 182: 2307–2310.

Masuda, N., Sakagawa, E., Ohya, S., Gotoh, N., Tsujimoto, H., and Nishino, T. 2000. Contribution of the MexX–MexY–oprM efflux system to intrinsic resistance in *Pseudomonas aeruginosa*. Antimicrob. Agents Chemother. 44: 2242–2246.

Morita, Y., Kataoka, A., Shiota, S., Mizushima, T., and Tsuchiya, T. 2000. NorM of *Vibrio parahaemolyticus* is an Na (+)-driven multidrug efflux pump. J. Bacteriol. 182: 6694–6697.

Morita, Y., Kodama, K., Shiota, S., Mine, T., Kataoka, A., Mizushima, T., and Tsuchiya, T. 1998. NorM, a putative multidrug efflux protein, of *Vibrio parahaemolyticus* and its homologue in *Escherichia coli*. Antimicrob. Agents Chemother. 42: 1778–1782.

Nakajima, A., Sugimoto, Y., Yoneyama, H., and Nakae, T. 2000. Localization of the outer membrane subunit OprM of resistance- nodulation-cell division family multicomponent efflux pump in *Pseudomonas aeruginosa*. J. Biol. Chem. 275: 30064–30068.

Oethinger, M., Kern, W.V., Jellen-Ritter, A.S., McMurry, L.M., and Levy, S.B. 2000. Ineffectiveness of topoisomerase mutations in mediating clinically significant fluoroquinolone resistance in *Escherichia coli* in the absence of the AcrAB efflux pump. Antimicrob. Agents Chemother. 44: 10–13.

Okusu, H., Ma, D., and Nikaido, H. 1996. AcrAB efflux pump plays a major role in the antibiotic resistance phenotype of *Escherichia coli* multiple-antibiotic-resistance (Mar) mutants. J. Bacteriol. 178: 306–308.

Pan, W. and Spratt, B.G. 1994. Regulation of the permeability of the gonococcal cell envelope by the *mtr* system. Mol. Microbiol. 11: 769–775.

Parkhill, J., Wren, B.W., Mungall, K., Ketley, J.M., Churcher, C., Basham, D., Chillingworth, T.,Davies, R.M., Feltwell, T., Holroyd, S., Jagels, K., Karlyshev, A.V., Moule, S., Pallen, M.J., Penn, C.W., Quail, M.A., Rajandream, M.A., Rutherford, K.M., van Vliet, A.H., Whitehead, S., and Barrell, B.G. 2000. The genome sequence of the food-borne pathogen *Campylobacter jejuni* reveals hypervariable sequences. Nature 403: 665–668.

Paulsen, I.T., Chen, J., Nelson, K.E., and Saier, M.H., Jr. 2001. Comparative genomics of microbial drug efflux systems. J. Mol. Microbiol. Biotechnol. 3: 145–150.

Paulsen, I.T., Park, J.H., Choi, P.S., and Saier, M.H., Jr. 1997. A family of Gram-negative bacterial outer membrane factors that function in the export of proteins, carbohydrates, drugs and heavy metals from Gram-negative bacteria. FEMS Microbiol. Lett. 156: 1–8.

Payot, S., Cloeckaert, A., and Chaslus-Dancla, E. 2002. Selection and characterization of fluoroquinolone-resistant mutants of *Campylobacter jejuni* using enrofloxacin. Microb. Drug Resist. 8: 335–343.

Poole, K. 2000. Efflux-mediated resistance to fluoroquinolones in Gram-negative bacteria. Antimicrob. Agents Chemother. 44: 2233–2241.

Poole, K. 2001. Multidrug resistance in Gram-negative bacteria. Curr Opin. Microbiol 4: 500–508.

Poole, K., Krebes, K., McNally, C., and Neshat, S. 1993. Multiple antibiotic resistance in *Pseudomonas aeruginosa*: evidence for involvement of an efflux operon. J. Bacteriol. 175: 7363–7372.

Poole, K., Tetro, K., Zhao, Q., Neshat, S., Heinrichs, D.E., and Bianco, N. 1996. Expression of the multidrug resistance operon *mexA–mexB–oprM* in *Pseudomonas aeruginosa*: *mexR* encodes a regulator of operon expression. Antimicrob. Agents Chemother. 40: 2021–2028.

Pumbwe, L. and Piddock, L.J. 2002. Identification and molecular characterization of CmeB, a *Campylobacter jejuni* multidrug efflux pump. FEMS Microbiol. Lett. 206: 185–189.

Putman, M., van Veen, H.W., and Konings, W.N. 2000. Molecular properties of bacterial multidrug transporters. Microbiol. Mol. Biol. Rev. 64: 672–693.

Saito, K., Yoneyama, H., and Nakae, T. 1999. *nalB*-type mutations causing the overexpression of the MexAB-OprM efflux pump are located in the mexR gene of the *Pseudomonas aeruginosa* chromosome. FEMS Microbiol. Lett. 179: 67–72.

Schumacher, M.A., Miller, M.C., Grkovic, S., Brown, M.H., Skurray, R.A., and Brennan, R.G. 2001. Structural mechanisms of QacR induction and multidrug recognition. Science 294: 2159–2163.

Smith, K.E., Besser, J.M., Hedberg, C.W., Leano, F.T., Bender, J.B., Wicklund, J.H., Johnson, B.P., Moore, K.A., and Osterholm, M.T. 1999. Quinolone-resistant *Campylobacter jejuni* infections in Minnesota, 1992- 1998. Investigation Team. N. Engl. J. Med. 340: 1525–1532.

Szymanski, C.M., Logan, S.M., Linton, D., and Wren, B.W. 2003. *Campylobacter* – a tale of two protein glycosylation systems. Trends Microbiol. 11: 233–238.

Szymanski, C.M., Yao, R., Ewing, C.P., Trust, T.J., and Guerry, P. 1999. Evidence for a system of general protein glycosylation in *Campylobacter jejuni*. Mol. Microbiol. 32: 1022–1030.

Thanassi, D.G., Cheng, L.W., and Nikaido, H. 1997. Active efflux of bile salts by *Escherichia coli*. J. Bacteriol. 179: 2512–2518.

Trieber, C.A. and Taylor, D.E. 2000. Mechanisms of antibiotic resistance in *Campylobacter*. In: *Campylobacter*. I. Nachamkin and M.J. Blaser, ed. American Society for Microbiology, ASM Press, Washington, DC, pp. 441–454.

Van Bambeke, F., Balzi, E., and Tulkens, P.M. 2000. Antibiotic efflux pumps. Biochem. Pharmacol. 60: 457–470.

Wacker, M., Linton, D., Hitchen, P.G., Nita-Lazar, M., Haslam, S.M., North, S.J., Panico, M., Morris, H.R., Dell, A., Wren, B.W., and Aebi, M. 2002. N-linked glycosylation in *Campylobacter jejuni* and its functional transfer into *E. coli*. Science 298: 1790–1793.

Wang, H., Dzink-Fox, J.L., Chen, M., and Levy, S.B. 2001. Genetic characterization of highly fluoroquinolone-resistant clinical *Escherichia coli* strains from China: role of *acrR* mutations. Antimicrob. Agents Chemother. 45: 1515–1521.

Young, N.M., Brisson, J.R., Kelly, J., Watson, D.C., Tessier, L., Lanthier, P.H., Jarrell, H.C., Cadotte, N., St Michael, F., Aberg, E., and Szymanski, C.M. 2002. Structure of the N-linked glycan present on multiple glycoproteins in the Gram-negative bacterium, *Campylobacter jejuni*. J. Biol. Chem. 277: 42530–42539.

Zgurskaya, H.I. and Nikaido, H. 2000. Multidrug resistance mechanisms: drug efflux across two membranes. Mol. Microbiol. 37: 219–225.

Chapter 11

Genetic Basis for the Variation in the Lipooligosaccharide Outer Core of *Campylobacter jejuni* and Possible Association of Glycosyltransferase Genes with Post-infectious Neuropathies

Michel Gilbert, Peggy C.R. Godschalk, Craig T. Parker, Hubert Ph. Endtz and Warren W. Wakarchuk

Abstract

The lipooligosaccharide (LOS) of *Campylobacter jejuni* displays considerable variation in the structure of its outer core. Microarray and PCR probing studies have shown that there is extensive variation in the gene content of the locus responsible for the biosynthesis of the LOS. DNA sequencing of this locus from multiple strains has demonstrated four other mechanisms that *C. jejuni* uses to vary its LOS outer core: (a) phase variation because of homopolymeric tracts; (b) gene inactivation by the deletion or insertion of a single base (without phase variation); (c) single mutation leading to the inactivation of a glycosyltransferase; and (d) single or multiple mutations leading to glycosyltransferases with different acceptor specificities. These four mechanisms have resulted in 'allelic' glycosyltransferases with potential to phase-vary their expression or modulate their specificity. Alleles representing each of these four mechanisms have been found for some of the outer core glycosyltransferases. Although the various types of alleles have presumably appeared through vertical evolution there is also evidence that horizontal exchange has further contributed to LOS outer core variation. The genetic bases for the variation of LOS outer cores in *C. jejuni* provide a good example of various adaptive evolution strategies used by a mucosal pathogen to modulate the structure of a cell-surface carbohydrate in order to better survive in a host. The role of LOS and ganglioside mimicry in the pathogenesis of the Guillain–Barré and Miller Fisher syndromes is also discussed. Specific LOS biosynthesis genes appear to be associated with these immune-mediated neuropathies.

INTRODUCTION

Cell-surface structures such as the LOS are recognized as antigens by the host and many microorganisms will modulate these structures as a strategy to increase the chances of evading the immune system. Moxon *et al.* (1994) have suggested that bacteria have specific loci that are highly mutable ('contingency' genes) to facilitate the efficient exploration of phenotypic alternatives in order to adapt to the unpredictable aspects of the host environment. The phenotypic variation of bacterial population involves both inter-genomic events, such as recombination, and intra-genomic events such as mutations. A 'contingency' locus can consist of a single gene that shows a high frequency of variation in its expression due to an unstable number of oligonucleotide repeats or bases in a homopolymeric tract. However, the variability of the gene content of specific loci involved in a common pathway is also an example of an adaptive strategy used by a pathogen to deal with a changing and/or hostile environment. Since the recombination events that lead to gene content variability are not uniformly distributed across the genome, this process could be considered 'contingency behaviour'. However, the generation of gene content variability in specific loci clearly involves a different time-scale than the mechanisms that affect the expression of a single gene or operon.

The availability of the complete genome sequence of *C. jejuni* NCTC 11168 has allowed the identification of four loci involved in the biosynthesis of cell-surface carbohydrates (Parkhill *et al.*, 2000; Linton *et al.*, 2001). These four loci constitute nearly 10% of the *C. jejuni* genome. The flagellar modification locus is found next to the flagellin structural genes (*flaA* and *flaB*) and is involved in the biosynthesis of *O*-linked pseudaminic acid and its derivatives (Thibault *et al.*, 2001). The capsular polysaccharide biosynthesis locus was identified by sequence similarity of a group of genes (*kps*) with corresponding genes involved in capsule biosynthesis in *Escherichia coli* (Karlyshev *et al.*, 2000). A third locus contains a large group of carbohydrate biosynthesis genes that were previously proposed to be involved in lipopolysaccharide (LPS) biosynthesis (Fry *et al.*, 1998; Wood *et al.*, 1999). However, this locus was later shown to comprise two adjacent loci which are involved in the synthesis of the LOS core oligosaccharide (Gilbert *et al.*, 2002) and the bacillosamine-containing *N*-linked glycan, respectively (Szymanski *et al.*, 1999; Young *et al.*, 2002). Three of these loci (flagellar modification, capsule and LOS) were shown to be highly divergent by whole genome microarray studies of various *C. jejuni* strains (Dorrell *et al.*, 2001; Leonard *et al.*, 2003; Pearson *et al.*, 2003).

The demonstration by microarray studies of highly divergent loci in *C. jejuni* strains indicates that the sequencing of the whole genome of one, or even a few, strains is not sufficient to gain a comprehensive understanding of the genetic bases for some of the virulence factors present in a pathogen. Consequently, the sequencing of specific loci, such as the one involved in LOS outer core biosynthesis, in multiple strains constitutes a targeted 'small scale' comparative genomics approach that provides a focus on some of the most important pathogenesis mechanisms. The clustering of the genes involved in the biosynthesis of the LOS in *C. jejuni* and the availability of LOS outer core structures in 12 strains provides a good model for such a study on a variable virulence factor. Microarray studies and PCR probing with specific primers can be used to assess the variability of the gene content in the bacterial populations and in various ecological niches. However, these approaches would fail to detect divergent genes that have identical functions and they would consider certain genes as 'conserved' when they have a few mutations that alter significantly their expression or the enzyme activity of their products. The sequencing of the LOS biosynthesis locus in representative strains (i.e. with known LOS outer core structures) has shown that a difference as small as the deletion of a single base in the whole locus can be responsible for the expression of a truncated LOS outer core structure while still mimicking a human ganglioside (Gilbert *et al.*, 2000). The comparison of the LOS biosynthesis locus in these strains has also identified other genetic mechanisms responsible for the variability in the LOS outer core as well as providing information about the evolutionary process that resulted in the differences in gene content.

In addition to being an important cause of gastroenteritis, *C. jejuni* is also the most frequent infectious agent associated with the development of the Guillain–Barré syndrome (GBS), an autoimmune-mediated disorder of the peripheral nervous system that is the most common cause of general paralysis in developed countries since the eradication of poliomyelitis (Jacobs *et al.*, 1998). We will discuss recent data suggesting that specific LOS biosynthesis genes can serve as markers for *C. jejuni* strains isolated from patients with neurological symptoms.

LIPOOLIGOSACCHARIDE OUTER CORE STRUCTURES

Variation in glycoconjugates is caused by the diversity of monosaccharide components and the linkages between them, derivatization with non-carbohydrate moieties and, in some cases, by the sequence of the repeating units. The complete LOS outer core structures have been determined for 12 *C. jejuni* strains (Figure 11.1a–d). Aspinall *et al.* (1993a, 1993b, 1994a, 1995) and Shin *et al.* (1998) determined the LOS outer core structures of eight *C. jejuni* reference strains of the Penner serotyping system. The Penner serotyping system of *C. jejuni* is based on heat-stable antigens and it was originally proposed that the specificity is due to LOS and/or LPS type molecules (Penner *et al.*, 1983; Moran and Penner, 1999). However, recent biochemical and genetic studies have shown that capsular polysaccharides account for Penner serotype specificity (Karlyshev *et al.*, 2000; Bacon *et al.*, 2001). Since the loci responsible for capsule and LOS biosynthesis are distant in the *C. jejuni* genome (Parkhill *et al.*, 2000) and intraspecies gene transfers are known to be frequent in *C. jejuni* (Dingle *et al.*, 2001; Suerbaum *et al.*, 2001), it is possible that strains having the same Penner type could express different LOS outer cores. Consequently, we decided to associate the published LOS outer core structures (Figure 11.1a–d) with the specific strain identification numbers (ATCC, NCTC, etc.) rather than with the Penner types, although the latter are also provided for convenient reference.

In addition to the studies with Penner type strains, Aspinall *et al.* (1994a) also determined the LOS outer structures of two *C. jejuni* strains (OH4382 and OH4384) that were isolated from two siblings who developed the GBS following gastroenteritis. St. Michael *et al.* (2002) determined the LOS outer structure expressed in the 'genome' strain, NCTC 11168 (HS:2), while Hanniffy *et al.* (2001) determined the LOS outer core structure of an HS:2 aerotolerant *C. jejuni* isolate.

The core oligosaccharides of low-molecular-weight LOS of many *C. jejuni* strains have been shown to exhibit molecular mimicry of the carbohydrate moieties of host gangliosides (Figure 11.1a–c). Terminal oligosaccharides identical to those of GM1a, GM2, GM3, GD1a, GD1c, GD3, and GT1a gangliosides have been found in various *C. jejuni* strains. Molecular mimicry of host structures by the saccharide portion of LOS is considered to be a virulence factor of various mucosal pathogens, which may use this strategy to evade the immune response (Moran *et al.*, 1996). The molecular mimicry between *C. jejuni* LOS outer core structures and gangliosides has also been suggested to act as a trigger for autoimmune mechanisms in the development of GBS (Yuki *et al.*, 1992) and this subject is further discussed in a section below.

(a)

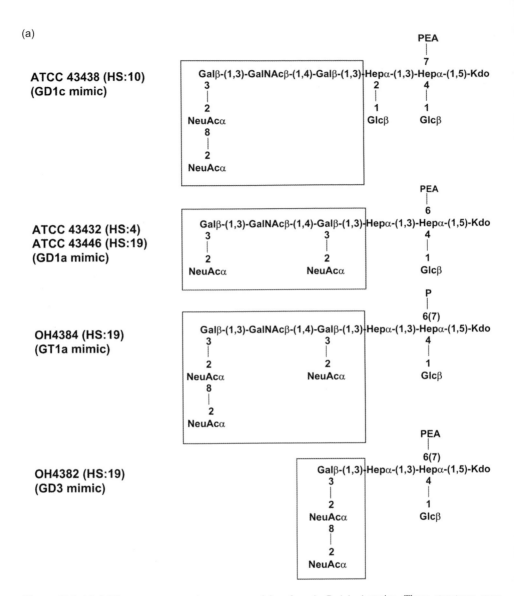

Figure 11.1 (a) LOS outer core structures expressed by class A *C. jejuni* strains. These structures were determined by Shin *et al.* (1998) and Aspinall *et al.* (1993a and 1994a). The abbreviations for the sugar residues and modifications are: Gal for galactose, Glc for glucose, GalNAc for *N*-acetylgalactosamine, NeuAc for *N*-acetylneuraminic (sialic) acid, HEp for L-*glycero*-D-*manno*-heptose, Kdo for 3-deoxy-D-*manno*-2-octulosonic acid, PEA for phosphoethanolamine and P for phosphate. (b) LOS outer core structures expressed by class B *C. jejuni* strains. The structure of ATCC 43456 HS:36 was determined by Aspinall *et al.* (1993a) and the structure of CCUG 10954 HS:23 was determined by Szymanski *et al.* (2003). The abbreviations for the sugar residues and modifications are the same as in (a). (c) LOS outer core structures expressed by class C *C. jejuni* strains. These structures were determined by St. Michael *et al.* (2002), Aspinall *et al.* (1993a and 1993b) and Hanniffy *et al.* (2001). The abbreviations for the sugar residues and modifications are the same as in (a). (d) LOS outer core structure expressed by an unclassified *C. jejuni* strain. This structure was determined by Aspinall *et al.* (1995). The abbreviations for the sugar residues and modifications are the same as in (a) except for QuiNAc which is *N*-acetylated 3-amino-3,6-dideoxyglucose (quinovosamine).

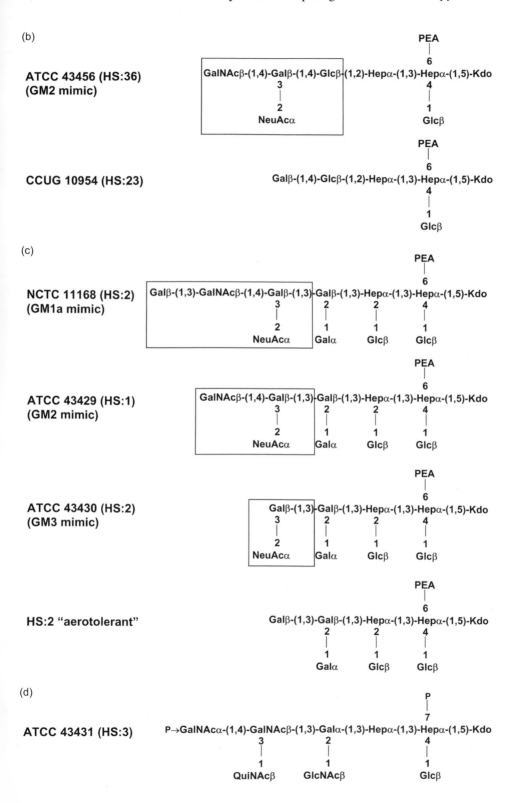

(b)

ATCC 43456 (HS:36)
(GM2 mimic)

CCUG 10954 (HS:23)

(c)

NCTC 11168 (HS:2)
(GM1a mimic)

ATCC 43429 (HS:1)
(GM2 mimic)

ATCC 43430 (HS:2)
(GM3 mimic)

HS:2 "aerotolerant"

(d)

ATCC 43431 (HS:3)

GENE CONTENT VARIATION IN THE LIPOOLIGOSACCHARIDE LOCUS

Sequence analysis of the genome of *C. jejuni* NCTC 11168 has allowed the identification of a cluster of genes involved in LOS biosynthesis (Parkill *et al.*, 2000). This region spans from Cj1131 (*galE*) to Cj1151 (*rfaD* or *gmhD*) but we will focus the discussion on the region between the genes encoding the heptosyltransferases I (Cj1133 or *waaC*) and II (Cj1148 or *waaF*) because it was shown to be the most variable by microarray analysis (Dorrell *et al.*, 2001) and it seems to contain most of the genes responsible for the variable outer core. Another region in the LOS cluster that shows variability in gene content is found between Cj1148 (*waaF*) and Cj1149 (*gmhA*), where one or two genes are inserted in some strains (Millar *et al.*, 2001; GenBank accession numbers AF334961 and AF334762). However, there are currently no data on the function of the variable genes found between Cj1148 and Cj1149.

The identification of the LOS biosynthesis locus in NCTC 11168 has facilitated the cloning of LOS biosynthesis genes from other *C. jejuni* strains (Gilbert *et al.*, 2000; Guerry *et al.*, 2000; Guerry *et al.*, 2002) which resulted in the identification of genes involved in the transfer of galactose, *N*-acetylgalactosamine and sialic acid to the LOS outer core. Another extension of this work was to sequence the corresponding locus from *C. jejuni* strains with known outer core structures in order to find hints about the genetic bases for the variations in structures.

The LOS classes

The sequence of the LOS biosynthesis locus between Cj1133 (*waaC*) and Cj1148 *(waaF)* is available for 20 *C. jejuni* strains (see GenBank accession numbers in the legend of Figure 11.2). The LOS outer core structure is known for 11 of these strains and they express, in total, 10 different structures. There is a total of 37 distinct ORFs found in one or more *C. jejuni* strains. The 20 loci could be grouped in seven classes (A to G, see Figure 11.2) based on the organization of the 37 distinct genes. Although there are 5 ORFs that are common to all classes, the LOS locus shows extensive variability in length (from 9 to 19 ORFs) and in gene content. Some genes are unique to one class (observed in classes B, C, E and G) while some other genes are present in more than one class but not in all. Variability is also observed within a specific class when one considers the sequence diversity of corresponding genes between the various strains. For instance the sequence identity varies from as low as 59% for ORF #6a (CgtB, β-1,3-galactosyltransferase) to above 99% for ORF #13a (WaaF, heptosyltransferase II) when comparing the deduced amino acid sequences of the corresponding genes of the eight class A strains (Gilbert *et al.*, 2002). On the other hand, the level of sequence conservation is very high (99.8% DNA sequence identity) among the three sequenced LOS loci of class C strains (NCTC 11168, ATCC 42429 HS:1 and ATCC 43430 HS:2). Noteworthy is that these three class C strains express different (although related) LOS outer cores (Figure 11.1c). It is possible that the classes that show high sequence identity have evolved more recently than the less conserved classes, although we do not know if the sequenced LOS loci are representative of the total variability present in the *C. jejuni* population.

Recombination and lateral exchange between the classes

The comparison of classes A and B DNA sequences allows us to propose a sequence of evolutionary steps that resulted in class B strains from a class A strain (Figure 11.3). The

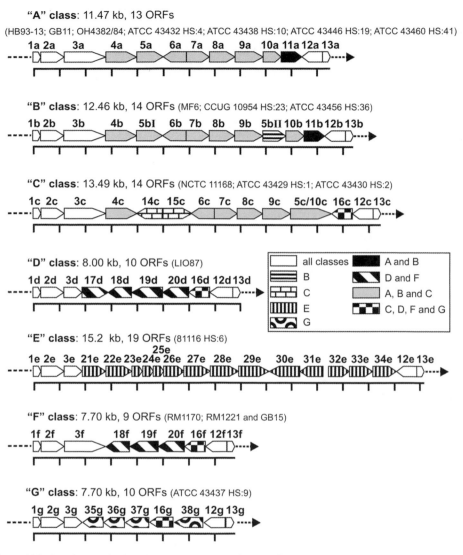

Figure 11.2 Genetic organization of the LOS biosynthesis loci of the different *C. jejuni* strains. The GenBank accession numbers are AY297047 for HB93–13, AY422197 for GB11, AF167345 for OH4382, AF130984 for OH4384, AF215659 for ATCC 43432, AF400048 for ATCC 43438, AF167344 for ATCC 43446, AY044868 for ATCC 43460, AY422196 for MF6, AF401529 for CCUG 10954, AF401528 for ATCC 43456, AL139077 for NCTC 11168, AY044156 for ATCC 43429, AF400047 for ATCC 43430, AF400669 for LIO87, AF343914 and AJ131360 for 81116, AY434498 for RM1170, NC_003912 for RM1221, AY423554 for GB15, and AY436358 for ATCC 43437.

LOS locus of ATCC 43438 (HS:10) is the closest 'A' sequence to the class B sequences and we use it to represent the evolutionary intermediate between classes A and B. The essential difference between classes A and B is the presence of a second *orf5* (*orf5bII*) between *orf9b* and *orf10b* in the latter class (Figure 11.2). We propose that *orf5bII* is the result of duplication of *orf5bI* (Figure 11.3) although horizontal transfer cannot be ruled out. In any case, the DNA sequence of *orf5bII* is very well conserved (99.5% identity)

Gilbert *et al.*

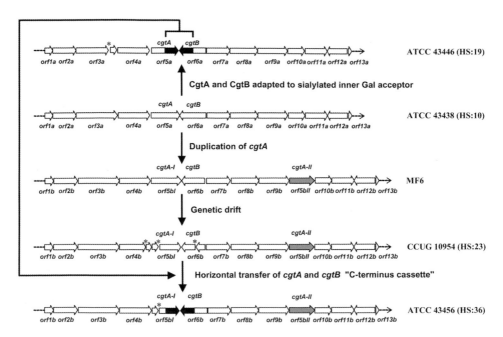

Figure 11.3 Proposed sequence of evolutionary events that resulted in class B strains from a class A strain and in strains that use either sialylated or non-sialylated acceptors. The LOS locus of ATCC 43438 (HS:10) is the closest 'A' sequence to the class B sequences and we use it to represent the evolutionary intermediate between classes A and B. The class B is the result of the duplication of *orf5* (*cgtA*) into *orf5bI* and *orf5bII*. Further variability is due to CgtA and CgtB adapting to use sialylated acceptors and to horizontal exchange events. The * indicates where premature translation stops are observed.

among the three class B strains. Also, the insertion of *orf5bII* between *orf9b* and *orf10b* is very 'precise' since it resulted in the addition of a single base upstream of its start codon and in no base addition after its stop codon (the TGA stop codon of *orf5bII* overlaps with the ATG start codon of *orf10b*). When ignoring *orf5bII*, the DNA sequence identity is very high (96.3%) between the ATCC 43438 (HS:10) and MF6 LOS loci and the alignment does not require the insertion of any major gap. The LOS locus of CCUG 10954 (HS:23) is also very similar to the ones from either ATCC 43438 (HS:10) or MF6 with 96% DNA sequence identity with either one. The LOS locus of ATCC 43456 (HS:36) is slightly more distant from either ATCC 43438 (HS:10) or MF6 with 92.7% and 92.5% identity, respectively. Consequently, we propose that the LOS locus of CCUG 10954 (HS:23) is evolutionary positioned between MF6 and ATCC 43456 (HS:36). The region from nt 4500 to 5700 diverges greatly (less than 70% DNA sequence identity) between ATCC 43456 (HS:36) and either ATCC 43438 (HS:10), CCUG 10954 (HS:23) or MF6, while this region is nearly identical (99% identity) between the last three strains. On the other hand, this region is nearly identical (99% identity) between ATCC 43456 (HS:36) and ATCC 43446 (HS:19) which suggests a lateral transfer event between these two strains. The region from nt 4500 to 5700 spans both *orf5* and *orf6* that encode a β-1,4-*N*-acetylgalactosaminyltrans-ferase and a β-1,3-galactosyltransferase, respectively. Because *orf5* and *orf6* are translated in opposite orientations, the divergence of the region from nt 4500 to 5700 results in a large number of amino acid substitutions in the C-terminus of both the β-1,4-*N*-acetyl-galactosaminyltransferase and the β-1,3-galactosyltransferase. The C-terminus regions of the β-1,4-*N*-acetylgalactosaminyltransferase and the β-1,3-galactosyltransferase are responsible for the specificity for either sialylated or non-sialylated acceptors (Gilbert *et al.*, 2002 and unpublished data) which indicates that the transfer of this 'cassette' between strains will have an impact on the structure of the outer core being synthesized.

Class B may be an evolutionary intermediate between classes A and C. The first step would have been the duplication of *orf5a* into *orf5bI* and *orf5bII*. At least two more recombination events would have been necessary to generate a class C locus by the insertion of *orf14c*, *orf15c* and *orf16c* and the deletion of *orf5bI* and *orf11b*. The three class C loci also have *orf5c* and *orf10c* as an in-frame fusion. It is certainly possible that other evolutionary intermediates exist with different combinations of inserted/deleted ORFs and with *orf5c* and *orf10c* either as separate ORFs or as an in-frame fusion.

There is also an example of the transfer of a whole LOS biosynthesis locus (Gilbert *et al.*, 2004). *C. jejuni* GB11, a strain isolated from a GBS patient, was found to be genetically related to *C. jejuni* NCTC 11168 by various molecular typing and serotyping methods. However, the LOS biosynthesis genes strongly diverged between GB11 and NCTC 11168. The LOS biosynthesis locus of GB11 was found to be nearly identical to the class A LOS locus from the *C. jejuni* ATCC 43446 (HS:19). Analysis of the DNA sequencing data showed that a horizontal exchange event involving at least 14.26 kb has occurred in the LOS biosynthesis locus of GB11 between *galE* (Cj1131c in NCTC 11168) and *gmhA* (Cj1149 in NCTC 11168). Mass spectrometry of the GB11 LOS showed that GB11 expressed an LOS outer core that mimics the carbohydrate portion of the gangliosides GM1a and GD1a, similar to *C. jejuni* ATCC 43446 (HS:19).

THE LOS OUTER CORE BIOSYNTHETIC PATHWAY

There are several studies that aimed directly at determining the functions of specific LOS biosynthesis genes. The functions proposed by sequence homology of both *waaC* (heptosyltransferase I) and *waaF* (heptosyltransferase II) from *Campylobacter* were confirmed by complementation of heptose-deficient strains of *Salmonella typhimurium* (Klena *et al.*, 1998; Oldfield *et al.*, 2002). The heptosyltransferase I is responsible for transferring Hep-I to Kdo while the heptosyltransferase II is transferring Hep-II to Hep-I. Gene-specific inactivation combined with mass spectrometry analysis was used to confirm the function of three genes in a class C strain (MSC57360, the equivalent of ATCC 43429). The inactivation of ORF #7c (a sialyltransferase homologue) and of ORF #9c (*neuC1*, a putative *N*-acetylmannosamine synthetase) resulted in LOS cores with higher electrophoretic mobilities and mass spectrometry confirmed the absence of sialic acid in the LOS core of the *neuC1* mutant (Guerry *et al.*, 2000). The same study also showed that gene-specific inactivation of ORF #5c/10c (a fusion of a β-1,4-*N*-acetylgalactosaminyltrans-ferase, CgtA, and of a CMP-NeuAc synthetase, NeuA) resulted in the loss of GalNAc in the LOS core. A similar approach was used to show that a homologue of ORF #5bII in *C. jejuni* 81-176 is also a β-1,4-*N*-acetylgalactosaminyltransferase (Guerry *et al.*, 2002).

Linton *et al.* (2000a) used gene-specific inactivation and complementation of an *E. coli* NeuB-deficient mutant to show that ORF #8c (*neuB1*, Cj1141) from *C. jejuni* NCTC 11168 is a sialic acid synthase. ORF #6c (Cj1139c) from the same strain was shown to be a β-1,3-galactosyltransferase by gene-specific inactivation (which resulted in the loss of reactivity with cholera toxin) and by *in vitro* assay of the recombinant enzyme expressed in *E. coli* (Linton *et al.*, 2000b). Based on these studies and on comparative data, Linton *et al.* (2001) proposed a model for LOS outer core biosynthesis in *C. jejuni* NCTC 11168.

We propose a biosynthetic pathway (Figure 11.4) for the LOS core in *C. jejuni* OH4384 based on comparative genomics and expression of specific glycosyltransferases that were assayed with synthetic oligosaccharides as acceptors. The *waaC* and *waaF* show high sequence identity with the corresponding genes mentioned above, and we propose that they also add the first and second heptosyl residues in OH4384, respectively. ORF #3a (Cj1135) is homologous to LgtF from *Haemophilus ducreyi* which encodes a β-1,4-glucosyltransferase that transfers glucose to heptose (Filiatrault *et al.*, 2000), and we propose that ORF #3a performs a similar role in OH4384. ORF #4a (Cj1136) is homologous with many bacterial β-1,3-galactosyltransferases and it is a good candidate for transferring the β-1,3-galactosyl residue to Hep-II. However none of these ORF #4a homologues has been shown experimentally to perform exactly this function and this assignment has to be considered speculative.

There is a lot more experimental evidence to support the proposed pathway for the addition of the other outer core residues. We used synthetic fluorescent acceptors to show that CgtA, CgtB and Cst-II are a β-1,4-*N*-acetylgalactosaminyltransferase, a β-1,3-galactosyltransferase and a bifunctional α-2,3-/2,8-sialyltransferase, respectively (Gilbert *et al.*, 2000). The corresponding genes have been inactivated in other *C. jejuni* strains and the observed phenotypes are consistent with these glycosyltransferases being involved in the addition of the proposed residues (Guerry *et al.*, 2000; Linton *et al.*, 2000b; Guerry *et al.*, 2002). We previously proposed that *cst-I* (a gene cloned from OH4384 by activity screening and found downstream of the *prfB* gene, Cj1455, i.e. outside of the LOS locus) was responsible for adding the single sialic acid to the inner galactosyl residue. However, PCR probing showed that *cst-I* was absent in many strains that have a sialic acid on the inner galactosyl residue (M. Gilbert, unpublished), and we now think that it is not involved in LOS biosynthesis. We propose that Cst-II is adding the single sialic acid on the inner galactosyl residue and, since Cst-II is a bifunctional α-2,3-/2,8-sialyltransferase, that it also adds both sialic acid residues to the terminal galactosyl residue. If it is responsible for adding the single α-2,3-sialic acid on the inner galactosyl residue, then it is curious that it does not also add a second α-2,8-sialic at this position. We know that CgtA from OH4384 cannot transfer a β-1,4-*N*-acetylgalactosaminyl residue to a disialylated acceptor

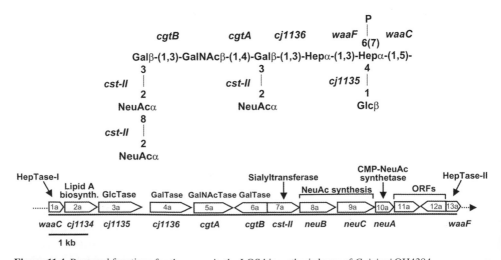

Figure 11.4 Proposed functions for the genes in the LOS biosynthesis locus of *C. jejuni* OH4384.

(Gilbert *et al.*, 2002), which means that the transfer of a second sialic acid on the inner galactosyl residue in OH4384 would result in a truncated structure (GD3 mimic) such as the one observed in OH4382 (Figure 11.1a) where CgtA is inactive because of a frame-shift mutation. It is possible that a small amount of the GD3 mimic is actually present in OH4384 but was undetected because the GT1a mimic is the main outer core structure that is synthesized. An explanation for the preferential expression of the GT1a mimic could be that CgtA adds the β-1,4-*N*-acetylgalactosaminyl residue as soon as the first sialic acid is added to the inner galactosyl residue which promotes the synthesis of the longer LOS outer core. In other words, CgtA would act 'faster' than Cst-II in the competition for this acceptor.

The proposed LOS core biosynthesis pathway in OH4384 (Figure 11.4) involves seven glycosyltransferases that account for seven of the thirteen ORFs found in the LOS locus of this strain. ORF #12a is also proposed to be a glycosyltransferase based on sequence homology, but its role in LOS biosynthesis is not clear. Since this ORF is present in all of the classes (Figure 11.2), it is possible that it plays a role in other strains but not in OH4384. Two of the remaining ORFs are involved in sialic acid biosynthesis: ORF #8a and #9a are homologous with NeuB (sialic acid synthase) and NeuC (UDP-GlcNAc-2-epimerase/ManNAc synthase), respectively. ORF #10a is homologous with NeuA from various organisms and is responsible for the synthesis of CMP-sialic acid, the donor used by the sialyltransferase. ORF #2a is homologous with various acyltransferases proposed to be involved in lipid A biosynthesis. Finally, ORF #11a is homologous with various bacterial acetyltransferases but acetylation of the LOS structure has not been reported. However this modification could have been undetected since it is easily lost during extraction procedures.

MECHANISMS GENERATING LOS CORE VARIATION

The differences in the gene content of the LOS loci provide a basis for differences in LOS outer cores among *C. jejuni* strains. However gene contents cannot always be directly correlated with the presence of sugar residues or modifications in outer core structures. The various LOS biosynthesis classes are the results of multiple recombination events and it is likely that some of the genes are remnants that are not used by some of the strains, which could result in 'excess' genes in some of the classes. As mentioned above, ORF #12 is present in all classes but it has no clear role in OH4384 and it may be functional in only one (or some) of the classes. On the other hand, the same set of homologous genes is able to result in different LOS core structures. For instance, there are at least four different outer cores expressed by the eight class A strains while the three class C strains express three different outer cores. The expression of different outer cores by the same set of homologous genes is the result of multiple mechanisms that are used by *C. jejuni* to turn on or off a gene or to modulate the substrate specificities of the glycosyltransferases. We give detailed examples of these various mechanisms below.

Phase variation due to homopolymeric tracts

Based on the whole-genome sequence, Parkhill *et al.* (2000) showed that short nucleotide homopolymeric runs of variable lengths are commonly found in genes involved in the biosynthesis of *C. jejuni* carbohydrates. Homopolymeric tracts are the simplest motifs of the short sequence DNA repeats that are proposed to be involved in a mechanism resulting in high-frequency on-off switching of the associated gene. A slipped-strand mispairing mechanism would generate variations in homopolymeric tract lengths during replication and would result in a mixture of translationally in-frame and out-of-frame variants of a gene (Levinson and Gutman, 1987). In some cases, the repeats are located upstream of the coding region of a gene and variation in their number will influence the transcriptional status. Phase-variation due to short sequence DNA repeats have been observed in virulence genes (including LOS biosynthesis genes) in *Haemophilus influenzae* (Hood *et al.*, 1996), *Helicobacter pylori* (Saunders *et al.*, 1998), *Neisseria meningitidis* and *Neisseria gonorrhoeae* (Kahler and Stevens, 1998).

Linton *et al.* (2000b) studied, in detail, the variation of a homopolymeric G tract in ORF #6c (β-1,3-galactosyltransferase) from *C. jejuni* NCTC 11168. They observed either 8 or 9 G tracts that resulted in the expression of either a GM1a or a GM2 ganglioside mimic in *C. jejuni* NCTC 11168. Guerry *et al.* (2002) observed variation in the homopolymeric G tract of CgtA (corresponding to ORF #5bII in Figure 11.2) from *C. jejuni* 81-176. They observed from 9 to 12 G tracts with 10 G corresponding to the full-length translation of CgtA and GM2 mimicry while the other variants resulted in GM3 mimicry. Homopolymeric G-tracts are found in many other LOS biosynthesis genes (nine in total, see Table 11.1) but they were not studied as thoroughly as far as determining the frequency of each variant and the phenotype being expressed by each of them. Some of the G-tracts are found in all members of a class (for instance ORF #5bII and ORF #6c in classes B and C, respectively) while others are unique to one member of a class (for instance ORF #5a in ATCC 43438 and ORF #6b in CCUG 10954). However, it is interesting to note that six of the class A strains lack homopolymeric G-tracts longer than five bases in their LOS locus (HB93–13, GB11, OH4382, OH4384, ATCC 43432 HS:4 and ATCC 43446 HS:19).

So far, there is a single example of a phase-variable homopolymeric tract that is not a G-tract in *C. jejuni* LOS loci: a phase-variable homopolymeric A-tract is found in the *cst-II* gene (ORF #7b) of *C. jejuni* MF6 (GenBank No. AY422196). We observed a mixture of 14 and 15 A tracts in the *cst-II* gene of MF6, with 15 A bases resulting in an in-frame variant.

Gene inactivation by the deletion or insertion of a single base (without phase variation)

There are two examples of glycosyltransferase genes that are found as inactive versions in some of the strains due to non phase-variable frame-shift mutations. The first example is ORF #5a in OH4382 where a stretch of seven A bases, rather than eight, results in the expression of a severely truncated CgtA (29 amino acids). The LOS outer core structure of OH4382 is consistent with the absence of β-1,4-*N*-acetylgalactosaminyltransferase activity since it is truncated at the inner galactosyl residue (Figure 11.1a).

Another example of where a non phase-variable frame-shift mutation has occurred is in some variants of ORF #3. ORF#3 is proposed to encode a one-domain β-glucosyltransferase in families D, E and G and a two-domain β-glucosyltransferase in classes A, B, C and F (Table 11.2). A premature translation termination occurs in ORF #3d of LIO87 because of a missing T base which results in the expression of a severely truncated enzyme (22 amino acids). However the LOS outer core structure of LIO87 is unknown and we cannot interpret the impact of this mutation. On the other hand, a missing A base in six class A strains (HB93–13, GB11, OH4382, OH4384, ATCC 43432 and ATCC 43446) results in the expression of a 418-amino acid protein which is truncated in the second domain (the full-length two-domain enzyme is 515 amino acids). The LOS outer core structure

Table 11.1 Location of homopolymeric G-tracts in the LOS biosynthesis locus of the various *C. jejuni* strains.

Gene	Strain	Class	Proposed function	G-tract length	Translation of the most frequent variant
orf5	ATCC 43438 (HS:10)	A	β-1,4-GalNAc transferase	9	In frame
orf5-11	CCUG 10954 (HS:23)	B	β-1,4-GalNAc transferase	9/**10**[a]	In frame
	ATCC 43456 (HS:36)	B	β-1,4-GalNAc transferase	9/**10**	In frame
	MF6	B	β-1,4-GalNAc transferase	10/**11**	Out of frame
orf6	CCUG 10954 (HS:23)	B	β-1,3-Gal transferase	8/**9**	Out of frame
	ATCC 43429 (HS:1)	C	β-1,3-Gal transferase	9	Out of frame
	ATCC 43430 (HS:2)	C	β-1,3-Gal transferase	9	Out of frame
	NCTC 11168	C	β-1,3-Gal transferase	**8**/9[b]	In frame
orf7	CCUG 10954 (HS:23)	B	α-2,3-/α-2,8-sialyltransferase	9/**10**	Out of frame
orf11	ATCC 43438 (HS:10)	A	Putative acetyltransferase	10/**11**	Out of frame
	ATCC 43460 (HS:41)	A	Putative acetyltransferase	8	Out of frame
	MF6	B	Putative acetyltransferase	10/**11**	Out of frame
orf16	ATCC 43429 (HS:1)	C	unknown	8/**9**	In frame
	ATCC 43430 (HS:2)	C	unknown	8/**9**	In frame
	NCTC 11168	C	unknown	**8**/9[c]	Out of frame
	LIO87	D	unknown	8/**9**	In frame
	GB15	F	unknown	7/**8**	Out of frame
	RM1170	F	unknown	8 [d]	Out of frame
	RM1221	F	unknown	9[d]	In frame
	ATCC 43437 (HS:9)	G	unknown	8[d]	Out of frame
orf20	LIO87	D	Putative glycosyltransferase	**9**/10	In frame
	GB15	F	Putative glycosyltransferase	9/**10**	Out of frame
	RM1170	F	Putative glycosyltransferase	10[d]	Out of frame
	RM1221	F	Putative glycosyltransferase	9[d]	In frame
orf23	81116	E	Unknown	8[d]	In frame
orf25	81116	E	Unknown	10[d]	Out of frame

[a]The number in bold indicates the most frequent variant when heterogeneity was observed.
[b]Linton *et al.* (2000b).
[c]The Sanger Centre, http://www.sanger.ac.uk/Projects/C_jejuni/
[d]Data on heterogeneity are not available.

Table 11.2 Functional alleles of the glucosytransferase gene (*orf3*) present in various *C. jejuni* strains.

Number of domains	Allele	Strains	Proposed activity	Other features
One	1	81116[a]	Glc-β1,4 to Hep-I	
		ATCC 43431 (HS:3)		
		ATCC 43437 (HS:9)[a]		Premature translational stop (not phase-variable)
One	2	LIO87[a]	Inactive	
Two	3	ATCC 43429 (HS:1)	Glc- β1,4 to Hep-I and Glc- β1,2 to Hep-II	
		ATCC 43430 (HS:2)		
		ATCC 43438 (HS:10)		
		CCUG 10954 (HS:23)		
		ATCC 43456 (HS:36)		
		ATCC 43460 (HS:41)[a]		
		NCTC 11168		
		MF6[a]		
		GB15[a]		
		RM1170[a]		
		RM1221[a]		
Two	4	ATCC 43432 (HS:4)	Glc- β1,4 to Hep-I	Premature translational stop in the second domain (not phase-variable)
		ATCC 43446 (HS:19)		
		OH4382		
		OH4384		
		GB11[a]		
		HB93–13[a]		

[a]The LOS outer core structures of these strains have not been determined. The proposed activity is based on sequence similarity with the corresponding gene in other strains but has not been confirmed by the presence of the proposed linkage.

is known for four (OH4382, OH4384, ATCC 43432 and ATCC 43446) of the strains with this frame-shift mutation and they are all missing the β-1,2-glucose on Hep-II (see Figure 11.1a). Consequently, we suggested that the second domain of ORF #3 is a β-1,2-glucosyltransferase (Gilbert *et al.*, 2002).

Single mutation leading to the inactivation of a glycosyltransferase

A third mechanism of on/off regulation of a glycosyltransferase is found in *C. jejuni* ATCC 43430 (HS:2). This *C. jejuni* strain has an LOS locus almost identical to NCTC 11168 (only eight base differences over 13.49 kb), but expresses a different (truncated) LOS outer core (Figure 11.1c). Six of the eight base differences cause frame shift changes in ORF#6c (β-1,3-galactosyltransferase) and in ORF #16c (unknown function, 5 bases are missing in NCTC 11168). One of the base differences causes a silent mutation in ORF #16c while the last base difference causes an amino acid change (Cys92Tyr, NCTC 11168 → ATCC 43430) in ORF #5c/10c (β-1,4-*N*-acetylgalactosaminyltransferase/CMP-NeuAc synthetase natural fusion). ORF #5c/10c from both ATCC 43430 (HS:2) and NCTC 11168 were expressed in *E. coli* and their activity was assayed *in vitro* (Gilbert *et al.*, 2002). Both versions were found to have similar CMP-NeuAc synthetase activity while only the NCTC 11168 version had β-1,4-*N*-acetylgalactosaminyltransferase activity. Consequently the Cys92Tyr has resulted in an inactive glycosyltransferase that allows ATCC 43430 (HS:2) to express a different ganglioside mimic (GM3) than NCTC 11168 (GM1a).

Homologous glycosyltransferases with different acceptor specificities

Although the β-1,4-*N*-acetylgalactosaminyltransferase alleles (CgtA, CgtA-I and CgtA-II) from the classes A, B and C are clearly homologous, the level of conservation among them is only 34% (Gilbert *et al.*, 2002). The acceptor preference was found to vary significantly with the ATCC 43438 (HS:10) version using only a non-sialylated acceptor, the version from OH4384 using only a monosialylated acceptor while the version from ATCC 43456 (HS:36) and NCTC 11168 are able to use both a monosialylated and a disialylated acceptor. In most cases the acceptor specificity correlates with the natural acceptor. For example, ATCC 43438 (HS:10) has no sialic acid on the inner Gal residue of the LOS outer core whereas OH4384, ATCC 43456 HS:36 and NCTC 11168 have a single sialic acid on the inner Gal residue (Figure 11.1a–c). The ability of the β-1,4-*N*-acetylgalactosaminyltrans-ferase from ATCC 43456 (HS:36) and NCTC 11168 to use a disialylated acceptor could seem superfluous since these strains express LOS outer cores with a single sialic acid on the inner Gal residue. However, it is possible that this allele plays a role in *C. jejuni* strains that have not been characterized yet.

Another example of mutations leading to different acceptor specifities is provided by ORF #7a, which was named Cst-II when we cloned it from *C. jejuni* OH4384 (Gilbert *et al.*, 2000). We use this designation for all of the versions from classes A and B. Guerry *et al.* (2000) showed that ORF #7 from MSC57360 (an equivalent of ATCC 43429 HS:1) is responsible for transferring the α-2,3-sialic acid and named this glycosyltransferase Cst-III, a designation that we use for ORF #7c from class C strains. An alignment of the deduced protein sequences of 11 ORF #7 versions from the A, B and C classes gave 50% identity (Gilbert *et al.*, 2002). However, when the classes A and B versions only are aligned together, the level of protein sequence identity rises to 92% while the three class

C versions are 100% identical between themselves. Pair-wise alignments between Cst-III and each variant of Cst-II gave 52% protein sequence identity on average.

Six versions of Cst-II were expressed in *E. coli* and their α-2,3- and -2,8-sialyltransferase activities were assayed *in vitro* (Gilbert *et al.*, 2002). Four versions (OH4382/84, ATCC 43438 HS:10, CCUG 10954 HS:23 and ATCC 43460 HS:41) were found to be bifunctional (both α-2,3- and α-2,8-sialyltransferase activities) and two versions (ATCC 43432 HS:4 and ATCC 43446 HS:19) were found to have only the α-2,3-sialyltransferase activity. An alignment of the amino acid sequences of the various Cst-II versions indicated that only three residues (Asn51, Leu54 and Ile269) were specific for the bifunctional Cst-II versions. Site-directed mutagenesis was used to determine which of these residues were essential for bifunctional sialyltransferase activity. An Asn51Thr substitution in Cst-II from OH4384 completely abolished the α-2,8-sialyltransferase activity. The opposite substitution (Thr51Asn) in the monofunctional Cst-II from ATCC 43446 conferred it the ability to perform both activities (α-2,3- and α-2,8-sialyltransferase). The other two residues (Leu54 and Ile269) unique to bifunctional Cst-II variants as well as the very variable residue #53 were found to affect the relative ratios of α-2,3- and α-2,8-sialyltransferase activities. Only Asn51 was found to be absolutely essential for α-2,8-sialyltransferase activity.

Although Cst-III (present in class C strains) also has Asn51, it seems to have only α-2,3-sialyltransferase activity (monofunctional) as observed in the LOS outercore structures and from *in vitro* assays. Since Cst-II and Cst-III have diverged significantly, it is not too surprising that the presence of Asn51 in Cst-III is not sufficient to confer it α-2,8-sialyltransferase activity. The low level of protein sequence conservation between Cst-II and Cst-III might be a consequence of adaptation to significantly different acceptor environments: in classes A and B the acceptor β-1,3-galactosyl residue is attached to an unsubstituted sugar residue (either heptose or glucose, see Figure 1.1a and b) while in class C the acceptor β-1,3-galactosyl is attached to a galactosyl residue that is substituted with an α-1,2-galactosyl residue (see Figure 11.1c).

Role of 'allelic' glycosyltransferases in LOS core variation

The four mechanisms described above result in homologous glycosyltransferases having different impacts on the LOS outer core structures in the different classes and sometimes between different strains from the same class. We list the different variants of ORF #3, ORF #5 and ORF #7 in Tables 11.2, 11.3 and 11.4, respectively, to give examples of the extent of variability that exists for some the LOS biosynthesis genes. By definition, alleles can designate any variant of a gene and in the case of multiple loci sequence typing schemes, they would be used to designate any DNA sequence variants even if the difference is a silent mutation. However we use the term 'functional' alleles to designate variants that have different activities as determined by activity measurement using model compounds, gene-specific inactivation or correlation between the presence of variants and specific structures in the LOS outer core. The amino acid sequence identity between alleles having different functions can be as high as 99% (for instance CgtA alleles 3 and 4, Table 11.3) while the sequence identity of alleles having the same functions can be as low as 81% (for instance the variants of ORF #3 allele 3, Table 11.2). We also consider as distinct functional alleles two variants that differ by the presence or absence of a phase-variable homopolymeric tract since such two variants differ in their ability to undergo on/off regulation. Lateral gene exchanges are known to be frequent in *C. jejuni*, which means that functional allele shuffling can result in extensive LOS core variation.

Table 11.3 Functional alleles of the β-1,4-*N*-acetylgalactosaminyltransferase gene (*orf5*, gene names: *cgtA*, *cgtA-I* and *cgtA-II*) present in various *C. jejuni* strains.

Gene	Allele	Strains	Proposed activity	Other features
cgtA	1	ATCC 43438 (HS:10)	The acceptor is a non-sialylated galactosyl residue	
cgtA-I		MF6[a]		
cgtA	2	ATCC 43432 (HS:4)	The acceptor is a monosialylated galactosyl residue	
		ATCC 43446 (HS:19)		
		ATCC 43460 (HS:41)[a]		
		OH4384		
		GB11[a]		
cgtA	3	ATCC 43429 (HS:1)	The acceptor can be a monosialylated or a disialylated galactosyl residue	Fused with CMP-NeuAc synthetase
		NCTC 11168		
cgtA	4	ATCC 43430 (HS:2)	Inactive (Cys-92 to Tyr mutation)	Fused with CMP-NeuAc synthetase
cgtA	5	OH4382	Inactive (single base deletion)	
		HB93-13[a]		
cgtA-I		CCUG 10954 (HS:23)		
		ATCC 43456 (HS:36)		
cgtA-II	6	CCUG 10954 (HS:23)	The acceptor can be a monosialylated or a disialylated galactosyl residue	Phase-variable (G-tract)
		ATCC 43456 (HS:36)		
		MF6[a]		

[a]The LOS outer core structures of these strains have not been determined. The proposed activity is based on sequence similarity with the corresponding gene in other strains but has not been confirmed by the presence of the proposed linkage.

Table 11.4 Functional alleles of the sialyltransferase gene (*orf7*, gene names: *cst-II* and *cst-III*) present in various *C. jejuni* strains,

Gene	Allele	Strains	Proposed activity	Other features
cst-II	1	ATCC 43432 (HS:4) ATCC 43446 (HS:19) ATCC 43456 (HS:36) GB11[a]	α-2,3- to Gal	
cst-II	2	ATCC 43438 (HS:10) ATCC 43460 (HS:41)[a] OH4382 OH4384 HB93-13[a]	α-2,3- to Gal and α-2,8- to NeuAc	
cst-II	3	CCUG 10954 (HS:23)	α-2,3- to Gal and α-2,8- to NeuAc	Phase-variable (G-tract)
cst-II	4	MF6[a]	α-2,3- to Gal and α-2,8- to NeuAc	Phase-variable (A-tract)
cst-III	5	ATCC 43429 (HS:1) ATCC 43430 (HS:2) NCTC 11168	α-2,3- to Gal	The acceptor Gal residue is attached to a Gal residue that is substituted with a Galα-1,2- residue

[a]The LOS outer core structures of these strains have not been determined. The proposed activity is based on sequence similarity with the corresponding gene in other strains but not has not been confirmed by the presence of the proposed linkage.

GUILLAIN–BARRÉ SYNDROME AND LOS BIOSYNTHESIS LOCI

The Guillain–Barré syndrome

The Guillain–Barré syndrome is the most common cause of acute neuromuscular paralysis worldwide, with a global incidence rate of 1–2 cases per 100 000 people per year (Hughes and Rees, 1997). GBS is an immune-mediated polyneuropathy characterized by an acute progressive, symmetrical motor weakness of the extremities and loss of tendon reflexes. Sensory and autonomous disturbances may also be present (Hughes, 1990). The degree of paralysis varies widely and up to one-third of the patients need artificial respiration (Winer *et al.*, 1988a). Paralysis gradually resolves several weeks after the onset of symptoms, but recovery is not always complete. Depending on the type of predominant nerve fibre damage, GBS can be divided into demyelinating and axonal forms. Acute inflammatory demyelinating polyneuropathy (AIDP) is the most frequent form of GBS in Northern America and Europe (Rees *et al.*, 1995a), whereas acute motor axonal neuropathy (AMAN) and acute motor-sensory axonal neuropathy (AMSAN) occur most often in China, Japan and Mexico (Ramos-Alvarez *et al.*,1969; McKhann *et al.*, 1991; Sobue *et al.*, 1997). The Miller Fisher syndrome is a rare variant of GBS, which is characterized by ophthalmoplegia, limb ataxia and areflexia in the absence of limb weakness (Fisher, 1956).

Treatment of GBS patients primarily involves general supportive medical care. Plasma exchange and the intravenous administration of immunoglobulins have been shown to have a beneficial effect on the speed of recovery and residual functional deficits (Guillain–Barré Study Group, 1985; van der Meché and Schmitz, 1992). Despite treatment the mortality rate is approximately 5% and up to 20% of the patients suffers from disabling residual deficits (van Koningsveld *et al.*, 2000).

In approximately two-third of GBS patients, an infectious illness is reported in the weeks before the onset of neurological symptoms (Hughes and Rees, 1997; van Koningsveld *et al.*, 2000). *C. jejuni* has been identified as the most frequent triggering infectious agent, preceding GBS in 14–80% of all cases (Jacobs *et al.*, 1998; Kuroki *et al.*, 1991; Rees *et al.*, 1995a; Willison and O'Hanlon, 1999; van Koningsveld *et al.*, 2001). However, it is estimated that only approximately 0.3–1 per 1000 *Campylobacter* infections are followed by GBS (Allos, 1997; McCarthy *et al.*, 1999). A preceding *C. jejuni* infection is associated with a severe, pure motor form of GBS and poor prognosis (Rees *et al.*, 1995b; Jacobs *et al.*, 1996). Other microorganisms significantly associated with the development of GBS are cytomegalovirus, Epstein–Barr virus, *Mycoplasma pneumoniae* and *Haemophilus influenzae* (Winer *et al.*, 1988b; Jacobs *et al.*, 1998).

The role of ganglioside mimicry in GBS

Although the pathogenesis of GBS has not been elucidated, studies have provided evidence that molecular mimicry plays an important role in triggering GBS (Yuki, 2001a). Molecular mimicry refers to a structural resemblance of microbial antigens with host structures (Damian, 1964). This resemblance induces a cross-reactive antibody and/or T-cell response leading to the neurological symptoms. As mentioned above, the LOS of many *C. jejuni* strains has been shown to exhibit molecular mimicry with gangliosides in peripheral nerves (Moran, 1997). Gangliosides are membrane glycolipids that are highly enriched in nerve tissue. They are composed of a ceramide tail that is inserted into the membrane and a highly variable oligosaccharide part that contains one or more sialic acid molecules. In most GBS patients, high titres of anti-ganglioside antibodies are present in the acute phase serum (Ilyas *et al.*, 1988; Quarles and Weiss, 1999). Antibody reactivity against numerous gangliosides has been detected in GBS patients and the specificity ranges widely between patients. The specificity of the cross-reactive antibody response is associated with the clinical features of GBS. The presence of antibodies against GM1a, GD1a, GalNAc-GD1a and GM1b is associated with pure motor GBS, whereas anti-GQ1b reactivity is present in most patients with MFS and GBS with oculomotor symptoms and ataxia (Chiba *et al.*, 1993; Jacobs *et al.*, 1996; Ho *et al.*, 1999; Ang *et al.*, 1999; Hao *et al.*, 1999; Willison and O'Hanlon, 1999). This phenomenon may be explained by the relative abundance of a certain ganglioside in specific nerves or nerve structures (Kusunoki *et al.*, 1993; Chiba *et al.*, 1993). The pattern of anti-ganglioside reactivity is also associated with the type of antecedent infection. Anti-GM1a and anti-GD1a reactivity have been linked to a preceding *C. jejuni* infection, although this relationship has not been found in all studies (Hughes *et al.*, 1999). Infection with CMV is associated with anti-GM2 reactivity and *M. pneumoniae* infection with antibodies against galactocerebroside (GalC) (Kusunoki *et al.*, 1995; Irie *et al.*, 1996; Jacobs *et al.*, 1997).

It has been demonstrated that anti-ganglioside antibodies of *C. jejuni*-infected GBS patients cross-react with *C. jejuni* LOS (Yuki, 1997). Immunization of animals with *C. jejuni* LOS leads to a cross-reactive anti-ganglioside antibody response (Ritter *et al.*, 1996; Wirguin *et al.*, 1997; Ang *et al.*, 2000a and 2000b). Furthermore, it has been shown that the structure of the *C. jejuni* LOS corresponds with the specificity of the anti-ganglioside antibodies, both in GBS/MFS patients and in immunized animals (Ang *et al.*, 2002a and 2002b). However, *C. jejuni* strains isolated from patients with enteritis without GBS/MFS ('enteritis-only') can also harbour ganglioside-like structures in their LOS (Nachamkin *et al.*, 1999), indicating that ganglioside mimicry *per se* is not sufficient to induce GBS. Additional factors, both pathogen- and host-related, may also be necessary for the induction of a neuropathogenic immune response, indicating the importance of animal disease models to further elucidate the pathogenesis of *C. jejuni*-associated GBS and to fulfil Koch's postulates. It has been possible to induce an anti-ganglioside antibody response in various animal models after injection with *C. jejuni* LOS (Ritter *et al.*, 1996; Wirguin *et al.*, 1997; Goodyear *et al.*, 1999; Ang *et al.*, 2000a and 2000b). Yuki *et al.* (2001b) described high anti-ganglioside antibody titres and acute flaccid paralysis in rabbits after immunization with purified GM1a ganglioside. There is only one report, which remains unconfirmed, of neuropathy development in chickens after a challenge with whole *C. jejuni* bacteria (Li *et al.*, 1996).

C. jejuni strains associated with Guillain–Barré syndrome

Because GBS is a rare disease and stool cultures are often negative at the onset of neurological symptoms, the number of GBS-associated *C. jejuni* isolates available for study is limited. Several GBS/MFS-associated *C. jejuni* collections from different geographical areas have been extensively studied by various phenotyping and genotyping methods (Nishimura *et al.*, 1997; Endtz *et al.*, 2000; Engberg *et al.*, 2001). So far, it has not been possible to identify a factor common to all GBS-associated strains. However, in some parts of the world certain Penner serotypes are over-represented among isolates from GBS cases. In GBS-associated *C. jejuni* strains from Japan, Mexico and the United States, a predominance of clonally related serotype HS:19 strains is observed (Kuroki *et al.*, 1993; Fujimoto *et al.*, 1997; Nachamkin *et al.*, 2003). In South-Africa, most strains isolated from GBS patients are serotype HS:41, and are also clonally related (Lastovica *et al.*, 1997; Wassenaar *et al.*, 2000). In contrast, Dutch GBS/MFS-associated *C. jejuni* strains are genetically heterogeneous and the serotype distribution resembles that of the general *C. jejuni* population in the Netherlands (Endtz *et al.*, 2000). Comparison of a worldwide non-HS:19 collection of GBS-associated and enteritis-only strains by Engberg *et al.* (2001) also showed heterogeneity and did not reveal a GBS-specific marker.

Since the first report of a ganglioside mimic in the LOS of a GBS-associated *C. jejuni* strain in 1993 (Yuki *et al.*, 1993), a great variety of ganglioside-like structures have been identified in the LOS of GBS/MFS-associated strains. Mass spectrometry has revealed the presence of GM1a, GD3, GD1a and GT1a mimics in GBS-associated strains and of GD3 mimics in MFS-associated strains (Yuki *et al.*, 1994; Aspinall *et al.*, 1994a and 1994b; Salloway *et al.*, 1996; Prendergast *et al.*, 1998). Serological studies have confirmed and extended these findings. However, as the specificity of serological assays varies they are not suitable to determine the exact chemical structure of the LOS. Thus, serum reactivity for a specific ganglioside such as GM1a should be interpreted as the presence of a GM1a-like structure in the LOS, but not as a confirmation of the presence of the complete GM1a glycan portion. Only a few studies have compared the presence of ganglioside-like structures in the LOS between substantial collections of GBS/MFS-associated strains and enteritis-only strains. Ang *et al.* (2002b) analysed a group of Dutch non-clonal *C. jejuni* strains with several serological techniques and found that strains from GBS/MFS patients more frequently express ganglioside mimics than strains from uncomplicated enteritis patients. GM1a-like structures were detected in 85% of GBS-associated isolates and in 57% of enteritis-only strains, whereas GQ1b-like structures were expressed by 100% of MFS-associated strains and by only 9% of enteritis-only strains. Nachamkin *et al.* (2002) found that GBS-associated strains, both HS:19 and non-HS:19, were strongly associated with the expression of GD1a-like mimicry when compared to enteritis-only isolates.

LOS biosynthesis genes and GBS

Analysis of the genes that are involved in the biosynthesis of ganglioside mimics may help to further elucidate the pathogenesis of GBS and may eventually lead to the development of molecular markers for the identification of potentially neuropathogenic strains. The biosynthesis of ganglioside mimics in GBS-associated *C. jejuni* strains (OH4382 and OH4384) was first studied in 2000 (Gilbert *et al.*, 2000). We demonstrated that disialyl groups, which are present in the ganglioside mimics of most strains isolated from patients with MFS or GBS with oculomotor symptoms, are attached to the outer core by a bifunctional sialyltransferase encoded by the *cst-II* gene. In accordance with these data, van Belkum *et al.* (2001) found that the *cst-II* gene was present in all tested *C. jejuni* strains that expressed a GQ1b-like structure and in only half of the strains that lacked a GQ1b-like mimic. Nachamkin *et al.* (2002) found a strong association between the simultaneous presence of three LOS biosynthesis genes, *cst-II, cgtA* and *cgtB*, and GBS-associated *C. jejuni* isolates. Very recently, the analysis of the LOS biosynthesis loci in a collection of Dutch GBS/MFS-associated and enteritis-only strains revealed that specific classes of the LOS locus are associated with GBS and MFS (Godschalk *et al.*, 2003). The class A locus was detected in 60% of the GBS-associated strains versus 14% of the control strains. The class A locus is also present in clonal HS:19 and HS:41 strains, two serotypes that are over-represented among GBS strains from certain geographical areas. This finding suggests that it is not the Penner (HS) serotype, but rather the structure of the LOS biosynthesis locus that is associated with GBS. Further evidence in support of the association between GBS and a class A LOS locus is provided by the observation that GB11, a GBS-associated HS:2 strain, is genetically very closely related to the enteritis-only class C strain NCTC 11168, but contains a class A ('HS:19') LOS locus, which it has probably acquired through horizontal transfer (Gilbert *et al.*, 2004).

The presence of a class B locus was found to be significantly associated with MFS and the expression of GQ1b-like structures (Godschalk *et al.*, 2003). It is possible that class B loci contain a specific characteristic that enables a strain to trigger MFS. However, only a few MFS-associated isolates were studied and analysis of additional strains is needed to confirm the association between a class B locus and MFS. The presence of GQ1b-like structures was only detected in classes A and B strains and was more common in class B strains. This suggests that *cst-II* with bifunctional sialyltransferase activity is only present in strains with either class A or B and may be more common in class B strains. GM1a-like structures were predominantly expressed by strains with a class A locus, but were also found in strains with other LOS loci. In most strains with a class D or E LOS locus, including a few GBS-associated strains, ganglioside mimics could not be detected with the available serological methods, which may be due to the lack of a sialyltransferase-encoding gene in these LOS loci. Other types of mimicry, either *Campylobacter* related or not, may have triggered GBS in these cases.

Using several mechanisms, including phase variation, *C. jejuni* strains can differentially express a variety of core structures. The relative amounts of these structures may vary between strains, even if they have the same class of LOS locus. One may therefore hypothesize that high levels of a ganglioside mimic are crucial for triggering GBS. So far, regulatory mechanisms for the expression of ganglioside mimics have not been identified. It is possible that the interaction between host and pathogen in the gut plays an important regulatory role in the expression of ganglioside mimics (Preston *et al.*, 1996).

Both the class A and B LOS loci were found to be associated with neuropathy. Genes or DNA sequence polymorphisms that are unique to these loci may be important in the pathogenesis of GBS and MFS. A more detailed analysis of LOS biosynthesis genes in GBS/MFS-associated strains, including the construction of specific gene knock-out mutants, is needed to gain further insight.

DISCUSSION

Various molecular typing studies have shown that *C. jejuni* generates extensive genetic diversity through intra- and interspecies recombination (Dingle *et al.*, 2001; Suerbaum *et al.*, 2001). Variation in genetic composition provides a bacterial population with a genome plasticity that may increase the adaptation potential and potentially the survival of a pathogen. Lateral exchange has created the various LOS classes and vertical evolution resulted in various alleles. Later horizontal exchange events have created new combinations of alleles. Some of these combinations may have improved the ability of the 'recipient' strain to survive in the host which allowed them to proliferate and remain present in the environment. An example of an outer core structure that could result from 'shuffling' of known alleles is the tetrasialylated GQ1b mimic. Some *C. jejuni* strains were reported to express a GQ1b epitope in their LOS outer core based on antibody probing (van Belkum *et al.*, 2001). Although the expression of an 'authentic' GQ1b mimic requires confirmation by structural analysis, the combination of a bifunctional sialyltransferase (Cst-II alleles 2, 3 or 4 in Table 11.4) and of a β-1,4-*N*-acetylgalactosaminyltransferase able to use a disialylated acceptor (CgtA alleles 3 or 6 in Table 11.3) could result in the expression of a tetrasialylated LOS outer core.

The genetic bases for LOS structure diversity has also been studied extensively in *Neisseria gonorrhoeae* and *Neisseria meningitidis* (Yang and Gotschlich, 1996; Kahler and Stephens, 1998; Jennings *et al.*, 1999; Zhu *et al.*, 2002). Three genetic loci (*lgt-1*, *lgt-2* and *lgt-3*) were reported to encode the glycosyltransferases responsible for biosynthesis of the LOS oligosaccharide chains. Differences in gene contents could explain many of the differences observed in LOS outer cores between various *N. meningitidis* immunotypes. Phase-variation mechanisms were also found to play roles in LOS structural heterogeneity. For instance, both *lgtA* and *lgtC* have variable G-tracts which will modulate the expression of the structure (s) of the α-chain (the oligosaccharide attached to heptose I) since these two genes encode glycosyltransferases that compete for the same acceptor (a galactosyl residue).

Another interesting example of multiple mechanisms used to modulate the activity of a gene involved in lipopolysaccharide biosynthesis is provided by the α-1,2-fucosyltransferase (FutC) of *H. pylori* (Wang *et al.*, 2000). There is a deletion in the promoter region of the *futC* gene in some *H. pylori* isolates that results in an off-status. The *H. pylori futC* gene also contains a homopolymeric C tract as well as imperfect TAA repeats in the middle of the gene and these elements mediate slipped-strand mispairing. There are two elements, a Shine–Dalgarno-like sequence and a stem loop structure, regulating translational frameshifting in the *futC* gene and this mechanism allows the expression of a full-length FutC by promoting a –1 reading frame slipping in some *H. pylori* isolates that would otherwise express a truncated FutC. All these regulating elements result in *H. pylori* isolates having various levels of FutC activity which result in variability in the fucosylated antigens being synthesized.

The availability of variable cell-surface structures should be advantageous to a pathogen by enabling the organism to evade the immune system. The five different mechanisms used to vary the LOS outer core structure can be effective over various time scales (Figure 11.5). The different gene complements are probably a result of vertical evolution as well as of lateral gene transfers and it is the mechanism that has taken place over the longest period in the evolution of the LOS locus. The second 'least frequent' mechanism is the accumulation of mutations resulting in different acceptor specificities. It is likely that many neutral mutations have to accumulate before a new useful phenotype appears. For instance the amino acid sequence identity between alleles 1 (active on non-sialylated acceptor) and 2 (active on monosialylated acceptor) of CgtA (β-1,4-*N*-acetyl-galactosaminyltransferase) is only 72%. On the other hand, a one amino acid substitution can change Cst-II from a mono- to bifunctional sialyltransferase (and vice versa) and it is possible that such mutations could occur during the course of an infection or an outbreak.

A single amino acid substitution can also lead to the inactivation of a glycosyltransferase, an effect that is likely to be more frequent than the acquisition of a new acceptor specificity. However, we have only one example of such a mutation resulting in variation in *C. jejuni* LOS outer core (the Cys92Tyr substitution in CgtA of ATCC 43430, which results in the expression of a GM3 mimic rather than a GM1a mimic). Interestingly, a similar amino acid substitution (Cys437Tyr) was found to cause the conversion of *Shigella flexneri* serotype 2a to serotype Y by inactivating the glucosyltransferase (GtrII) responsible for adding a glucosyl residue on a specific rhamnose residue, which gives rise to the serotype

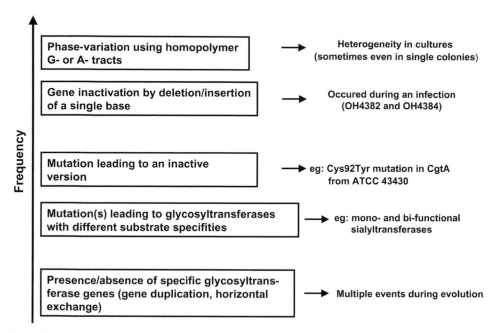

Figure 11.5 Relative frequency of the mechanisms that result in LOS outer core variation in *C. jejuni.*

2a epitope (Chen *et al.*, 2003). This serotype conversion was observed by comparing *S. flexneri* isolates recovered from the same patient over a period of six weeks. Since *S. flexneri* infection results in serotype-specific immunity it was proposed that serotype conversion could enhance the survival of the bacteria in the patient.

Phase-variation with homopolymeric tracts is the mechanism that has the most chance to occur during an infection since it was shown to create a high level of heterogeneity, sometimes even in single colonies. However, no study has demonstrated directly that phase variation provides an advantage during the course of a *C. jejuni* infection. Non phase-variable single base frame-shift mutations are more stable than the homopolymeric tracts since there are no report of reversion during subcultures. However, the 'one-base difference' between *C. jejuni* OH4382 and OH4384 occurred during the course of an infection since these two strains were isolated from siblings (Aspinall *et al.*, 1994b). This might be the direct result of selective pressure to change the LOS outer core during the course of infection.

Comparative genomics of the LOS biosynthesis locus in 20 *C. jejuni* strains combined with gene-specific inactivation studies and *in vitro* assays with model compounds have elucidated five mechanisms responsible for the generation of variability in the LOS outer core structure. However, the LOS outer core structure is known for 12 strains only, with the LOS biosynthesis locus sequence being available for 11 of these strains. It will be important to have more LOS outer structures available, particularly for classes D to G as well as for yet uncharacterized classes. These LOS outer core structures would help to determine the functions of the ORFs with currently unknown functions. Although many of these ORFs show homology with putative glycosyltransferases, bioinformatics analyses often fail to suggest the specific function (i.e. the sugar residue that is transferred and the linkage specificity).

Another application of these LOS biosynthesis studies is to determine if specific classes or genes are associated with ecological niches, animal reservoirs, infectivity potential and post-infectious complications. The association of *cst-II* with the expression of GQ1b-like epitope (van Belkum *et al.*, 2001) and the high frequency of class A loci in *C. jejuni* strains isolated from GBS patients (Godschalk *et al.*, 2003) are two examples of the relevance of these probing studies. Since *C. jejuni* seems to be part of the normal flora of birds and poultry is a major source of infection for humans, there is an incentive to determine if the strains causing gastroenteritis and the bird commensal strains are the same. It is possible, for instance, that ganglioside-mimicking strains are more pathogenic to humans while other types of outer core are more successful for chicken colonization. In addition to probing for class-specific genes it will be important to determine if the other mechanisms responsible for LOS outer core variation also play a critical role in bird colonization, human infection or post-infectious complications.

Acknowledgements

We would like to thank Drs. Alex van Belkum and Stéphane Bernatchez for helpful discussion. This work was supported by the NRC Genome Sciences and Health Related Research Initiative and a research grant from the Human Frontier Science Program.

References

Allos, B.M. 1997. Association between *Campylobacter* infection and Guillain–Barré syndrome. J. Infect. Dis. 176 (Suppl. 2): S125–128.

Ang, C.W., Yuki, N., Jacobs, B.C., Koga, M., van Doorn, P.A., Schmitz, P.I.M., and van Der Meché, F.G.A. 1999. Rapidly progressive, predominantly motor Guillain–Barré syndrome with anti-GalNAc-GD1a antibodies. Neurology 53: 2122–2127.

Ang, C.W., de Klerk, M.A., Jacobs, B.C., Laman, J.D., van der Meché, F.G.A., van den Braak, N., Endtz, H.P., and van Doorn, P.A. 2000a. *Campylobacter jejuni* lipopolysaccharides from Guillain–Barré and Miller Fisher patients induce anti-GM1 and anti-GQ1b antibodies in rabbits. J. Neurol. 247 (Suppl. 3): 153.

Ang, C.W., Endtz, H.Ph., Jacobs, B.C., Laman, J.D., de Klerk, M.A., van der Meché, F.G.A., and van Doorn, P.A. 2000b. *Campylobacter jejuni* lipopolysaccharides from Guillain–Barré syndrome patients induce IgG anti-GM1 antibodies in rabbits. J. Neuroimmunol. 104: 133–138.

Ang, C.W., Noordzij, P.G., de Klerk, M.A., Endtz, H.P., van Doorn, P.A., and Laman, J.D. 2002a. Ganglioside mimicry of *Campylobacter jejuni* lipopolysaccharides determines antiganglioside specificity in rabbits. Infect. Immun. 70: 5081–5085.

Ang, C.W., Laman, J.D., Willison, H.J., Wagner, E.R., Endtz, H.P., de Klerk, M.A., Tio-Gillen, A.P., van den Braak, N., Jacobs, B.C., and van Doorn, P.A. 2002b. Structure of *Campylobacter jejuni* lipopolysaccharides determines antiganglioside specificity and clinical features of Guillain–Barré and Miller Fisher patients. Infect. Immun. 70: 1202–1208.

Aspinall, G.O., McDonald, A.G., Raju, T.S., Pang, H., Moran, A.P., and Penner, J.L. 1993a. Chemical structures of the core regions of *Campylobacter jejuni* serotypes O: 1, O: 4, O: 23, and O: 36 lipooligosaccharides. Eur. J. Biochem. 213: 1017–1027.

Aspinall, G.O., MaDonald, A.G., Raju, T.S., Pang, H., Kurjanczyk, L.A., Penner, J.L., and Moran, A.P. 1993b. Chemical structure of the core region of *Campylobacter jejuni* serotype O: 2 lipopolysaccharide. Eur. J. Biochem. 213: 1029–1937.

Aspinall, G.O., McDonald, A.G., Pang, H., Kurjanczyk, L.A., and Penner, J.L. 1994a. Lipopolysaccharides of *Campylobacter jejuni* serotype O: 19: structures of core oligosaccharide regions from the serostrain and two bacterial isolates from patients with the Guillain–Barré syndrome. Biochemistry 33: 241–249.

Aspinall, G.O., Fujimoto, S., McDonald, A.G., Pang, H., Kurjanczyk, L.A., and Penner, J.L. 1994b. Lipopolysaccharides from *Campylobacter jejuni* associated with Guillain–Barré syndrome patients mimic human gangliosides in structure. Infect. Immun. 62: 2122–2125.

Aspinall, G.O., Lynch, C.M., Pang, H., Shaver, R.T., and Moran, A.P. 1995. Chemical structures of the core region of *Campylobacter jejuni* O: 3 lipopolysaccharide and an associated polysaccharide. Eur. J. Biochem. 231: 570–578.

Bacon, D.J., Szymanski, C.M., Burr, D.H., Silver, R.P., Alm, R.A., and Guerry, P. 2001. A phase-variable capsule is involved in virulence of *Campylobacter jejuni* 81-176. Mol. Microbiol. 40: 769–777.

Chen, J.-H., Hsu, W.-B., Chiou, C.-S., and Chen, C.-M. 2003. Conversion of *Shigella flexneri* serotype 2a to serotype Y in a shigellosis patient due to a single amino acid substitution in the protein product of the bacterial glucosyltransferase *gtrII* gene. FEMS Microbiol. Lett. 224: 277–283.

Chiba, A., Kusunoki, S., Obata, H., Machinami, R., and Kanazawa, I. 1993. Serum anti-GQ1b IgG antibody is associated with ophthalmoplegia in Miller Fisher syndrome and Guillain–Barré syndrome: clinical and immunohistochemical studies. Neurology 43: 1911–1917.

Damian, R.T. 1964. Molecular mimicry: antigen sharing by parasite and host and its consequences. Am. Nat. 98: 129–149.

Dingle, K.E., Colles, F.M., Wareing, D.R.A., Ure, R., Fox, A.J., Bolton, F.E., Bootsma, H.J., Willems, R.J.L., Urwin, R., and Maiden, M.C.J. 2001. Multilocus sequence typing system for *Campylobacter jejuni*. J. Clin. Microbiol. 39: 14–23.

Dorrell, N., Mangan, J.A., Laing, K.G., Hinds, J., Linton, D., Al-Ghusein, H., Barrell, B.G., Parkhill, J., Stoker, N.G., Karlyshev, A.V., Butcher, P.D., and Wren, B.W. 2001. Whole genome comparison of *Campylobacter jejuni* human isolates using a low-cost microarray reveals extensive genetic diversity. Genome Res. 11: 1706–1715.

Endtz, H.P., Ang, C.W., van den Braak, N., Duim, B., Rigter, A., Price, L.J., Woodward, D.L., Rodgers, F.G., Johnson, W.M., Wagenaar, J.A., Jacobs, B.C., Verbrugh, H.A., and van Belkum, A. 2000. Molecular characterization of *Campylobacter jejuni* from patients with Guillain–Barré and Miller Fisher syndromes. J. Clin. Microbiol. 38: 2297–2301.

Engberg, J., Nachamkin, I., Fussing, V., McKhann, G.M., Griffin, J.W., Piffaretti, J.C., Nielsen, E.M., and Gerner-Smidt, P. 2001. Absence of clonality of *Campylobacter jejuni* in serotypes other than HS: 19 associated with Guillain–Barré syndrome and gastroenteritis. J. Infect. Dis. 184: 215–220.

Filiatrault, M.J., Gibson, B.W., Schilling, B., Sun, S., Munson Jr, R.S., and Campagnari, A.A. 2000. Construction and characterization of *Haemophilus ducreyi* lipooligosaccharide (LOS) mutants defective in expression of heptosyltransferase III and β-1,4-glucosyltransferase: identification of LOS glycoforms containing lactosamine repeats. Infect. Immun. 68: 3352–3361.

Fisher, M. 1956. An unusual variant of acute idiopathic polyneuritis (syndrome of ophthalmoplegia, ataxia and areflexia). N. Engl. J. Med. 225: 57–65.

Fry, B.N., Korolik, V., ten Brinke, J.A., Pennings, M.T.T., Zalm, R., Teunis, B.J.J., Coloe, P.J., and van der Zeijst, B.A.M. 1998. The lipopolysaccharide biosynthesis locus of *Campylobacter jejuni* 81116. Microbiology 144: 2049–2061.

Fujimoto, S., Allos, B.M., Misawa, N., Patton, C.M., and Blaser, M.J. 1997. Restriction fragment length polymorphism analysis and random amplified polymorphic DNA analysis of *Campylobacter jejuni* strains isolated from patients with Guillain–Barré syndrome. J. Infect. Dis. 176: 1105–1108.

Gilbert, M., Brisson, J.-R., Karwaski, M.-F., Michniewicz, J., Cunningham, A.-M., Wu, Y., Young, N.M., and Wakachuk, W.W. 2000. Biosynthesis of ganglioside mimics in *Campylobacter jejuni* OH4384: identification of the glycosyltransferase genes, enzymatic synthesis of model compounds and characterization of nanomole amounts by 600-MHz ^{1}H and ^{13}C NMR analysis. J. Biol. Chem. 275: 3896–3906.

Gilbert, M., Karwaski, M.-F., Bernatchez, S., Young, N.M., Taboada, E., Michniewicz, J., Cunningham, A.-M., and Wakarchuk, W.W. 2002. The genetic bases for the variation in the lipo-oligosaccharide of the mucosal pathogen, *Campylobacter jejuni*: biosynthesis of sialylated ganglioside mimics in the core oligosaccharide. J. Biol. Chem. 277: 327–337.

Gilbert, M., Godschalk, P.C.R., Karwaski, M.-F., Ang, C.W., van Belkum, A., Li, J., Wakarchuk, W.W., and Endtz, H.P. 2004. Evidence for the acquisition by horizontal exchange of the lipooligosaccharide biosynthesis locus in *Campylobacter jejuni* GB11, a strain isolated from a Guillain–Barré syndrome patient. Infect. Immun. 72: 1162–1165.

Godschalk, P., van Belkum, A., Gilbert, M., Ang, W., Jacobs, B., Lastovica, A., van den Braak, N., Wakarchuk, W., Verbrugh, H., and Endtz, H. 2003. Specific classes of the *Campylobacter jejuni* lipo-oligosaccharide gene locus are associated with the Guillain–Barré and Miller Fisher syndromes. Int. J. Med. Microbiol. 293 (S35): 35.

Goodyear, C. S., O'Hanlon, G.M., Plomp, J.J., Wagner, E.R., Morrison, I., Veitch, J., Cochrane, L., Bullens, R.W.M., Molenaar, P.C., Conner, J., and Willison, H.J. 1999. Monoclonal antibodies raised against Guillain–Barré syndrome-associated *Campylobacter jejuni* lipopolysaccharides react with neuronal gangliosides and paralyse muscle–nerve preparations. J. Clin. Invest. 104: 697–708.

Guerry, P., Ewing, C.P., Hickey, T.E., Prendergast, M.M., and Moran A.P. 2000. Sialylation of lipooligosaccharide cores affects immunogenicity and serum resistance of *Campylobacter jejuni*. Infect. Immun. 68: 6656–6662.

Guerry, P., Szymanski, C.M., Prendergast, M.M., Hickey, T.E., Ewing, C.P., Pattarini, D.L., and Moran, A.P. 2002. Phase variation of *Campylobacter jejuni* 81-176 lipooligosaccharide affects ganglioside mimicry and invasiveness *in vitro*. Infect. Immun. 70: 787–793.

Guillain–Barré Syndrome Study Group 1985. Plasmapheresis and acute Guillain–Barré syndrome. Neurology 35: 1096–1104.

Hanniffy, O.M., Shashkov, A.S., Moran, A.P., Senchenkova, S.N., and Savage, A.V. 2001. Chemical structure of the core oligosaccharide of aerotolerant *Campylobacter jejuni* O: 2 lipopolysaccharide. Carbohydr. Res. 330: 223–229.

Hao, Q., Saida, T., Yoshino, H., Kuroki, S., Nukina, M., and Saida, K. 1999. Anti-GalNAc-GD1a antibody-associated Guillain–Barré syndrome with a predominantly distal weakness without cranial nerve impairment and sensory disturbance. Ann. Neurol. 45: 758–768.

Ho, T.W., Willison, H.J., Nachamkin, I., Li, C.Y., Veitch, J., Ung, H., Wang, G.R., Liu, R.C., Maizeblath, D.R., Asbury, A.K., Griffin, J.W., and McKhann, G.M. 1999. Anti-GD1a antibody is associated with axonal but not demyelinating forms of Guillain–Barré syndrome. Ann. Neurol. 45: 168–173.

Hood, D.W., Deadman, M.E., Jennings, M.P., Bisercic, M., Fleischmann, R.D., Venter, J.C., and Moxon, E.R. 1996. DNA repeats identify novel virulence genes in *Haemophilus influenzae*. Proc. Natl. Acad. Sci. USA 93: 11121–11125.

Hughes, R.A.C. 1990. Guillain–Barré Syndrome. Springer-Verlag, London.

Hughes, R.A.C., and Rees, J.H. 1997. Clinical and epidemiologic features of Guillain–Barré syndrome. J. Infect. Dis. 176 (Suppl. 2): S92–98.

Hughes, R.A.C., Hadden, R.D.M., Gregson, N.A., and Smith, K.J. 1999. Pathogenesis of Guillain–Barré syndrome. J. Neuroimmunol. 100: 74–97.

Ilyas, A. A., Willison, H.J., Quarles, R.H., Jungalwala, F.B., Maizeblath, D.R., Trapp, B.D., Griffin, D.E., Griffin, J.W., and McKhann, G.M. 1988. Serum antibodies to gangliosides in Guillain–Barré syndrome. Ann. Neurol. 23: 440–447.

Irie, S., Saito, T., Nakamura, K., Kanazawa, N., Ogino, M., Nukazawa, T., Ito, H., Tamai, Y., and Kowa, H. 1996. Association of anti-GM2 antibodies in Guillain–Barré syndrome with acute cytomegalovirus infection. J. Neuroimmunol. 68: 19–26.

Jacobs, B.C., van Doorn, P.A., Schmitz, P.I.M., Tio-Gillen, A.P., Herbrink, P., Visser, L.H., Hooijkaas, H., and van der Meché, F.G.A. 1996. *Campylobacter jejuni* infections and anti-GM1 antibodies in Guillain–Barré syndrome. Ann. Neurol. 40: 181–187.

Jacobs, B.C., van Doorn, P.A., Groeneveld, J.H., Tio-Gillen, A.P., and van der Meche, F.G. 1997. Cytomegalovirus infections and anti-GM2 antibodies in Guillain–Barré syndrome. J. Neurol. Neurosurg. Psychiatry 62: 641–643.

Jacobs, B.C., Rothbarth, P.H., van der Meché, F.G.A., Herbrink, P., Schmitz, P.I.M., de Klerk, M.A., and van Doorn, P.A. 1998. The spectrum of antecedent infections in Guillain–Barré syndrome. Neurology 51: 1110–1115.

Jennings, M.P., Srikhanta, Y.N., Moxon, E.R., Kramer, M., Poolman, J.T., Kuipers, B., and van der Ley, P. 1999. The genetic basis of the phase variation repertoire of lipopolysaccharide immunotypes in *Neisseria meningitidis*. Microbiology 45: 3013–3021.

Kahler, C.M., and Stephens, D.S. 1998. Genetic basis for biosynthesis, structure, and function of meningococcal lipooligosaccharide (endotoxin). Crit. Rev. Microbiol. 24: 281–334.

Karlyshev, A.V., Linton, D., Gregson, N.A., Lastovica, A.J., and Wren, B.W. 2000. Genetic and biochemical evidence of a *Campylobacter jejuni* capsular polysaccharide that accounts for Penner serotype specificity. Mol. Microbiol. 35: 529–541.

Klena, J.D., Grey, S.A., and Konkel, M.E. 1998. Cloning, sequencing, and characterization of the lipopolysaccharide biosynthetic enzyme heptosyltransferase I gene (*waaC*) from *Campylobacter jejuni* and *Campylobacter coli*. Gene 222: 177–185.

Kuroki, S., Haruta, T., Yoshioka, M., Kobayashi, Y., Nukina, M., and Nakanishi, H. 1991. Guillain–Barré syndrome associated with *Campylobacter* infection. Pediatr. Infect. Dis. J. 10: 149–151.

Kuroki, S., Saida, T., Nukina, M., Haruta, T., Yoshioka, M., Kobayashi, Y., and Nakanishi, H. 1993. *Campylobacter jejuni* strains from patients with Guillain–Barré syndrome belong mostly to Penner serogroup 19 and contain beta-*N*-acetylglucosamine residues. Ann. Neurol. 33: 243–247.

Kusunoki, S., Chiba, A., Tai, T., and Kanazawa, I. 1993. Localization of GM1 and GD1b antigens in the human peripheral nervous system. Muscle Nerve 16: 752–756.

Kusunoki, S., Chiba, A., Hitoshi, S., Takizawa, H., and Kanazawa, I. 1995. Anti-Gal-C antibody in autoimmune neuropathies subsequent to mycoplasma infection. Muscle Nerve 18: 409–413.

Lastovica, A.J., Goddard, E.A., and Argent, A.C. 1997. Guillain–Barré syndrome in South Africa associated with *Campylobacter jejuni* O: 41 strains. J. Infect. Dis. 176 (Suppl. 2): S139–143.

Leonard, E.E., Takata, T., Blaser, M.J., Falkow, S., Tompkins, L.S., and Gaynor, E.C. 2003. Use of an open-reading frame-specific *Campylobacter jejuni* DNA microarray as a new genotyping tool for studying epidemiologically related isolates. J. Infect. Dis. 187: 691–694.

Levinson, G., and Gutman, G.A. 1987. Slipped-strand mispairing: a major mechanism for DNA sequence evolution. Mol. Biol. Evol. 4: 203–221.

Li, C.Y., Xue, P., Tian, W.Q., Liu, R.C., and Yang, C. 1996. Experimental *Campylobacter jejuni* infection in the chicken: an animal model of axonal Guillain–Barré syndrome. J. Neurol. Neurosurg. Psychiatry 61: 279–284.

Linton, D., Karlyshev, A.V., Hitchen, P.G., Morris, H.R., Dell, A., Gregson, N.A., and Wren, B.W. 2000a. Multiple *N*-acetyl neuraminic acid synthetase (*neuB*) genes in *Campylobacter jejuni*: identification and characterization of the gene involved in sialylation of lipo-oligosaccharide. Mol. Microbiol. 35: 1120–1134.

Linton, D., Gilbert, M., Hitchen, P.G., Dell, A., Morris, H.R., Wakarchuk, W.W. Gregson, N.A., and Wren, B.W. 2000b. Phase variation of a β-1,3 galactosyltransferase involved in generation of the ganglioside GM$_1$-like lipo-oligosaccharide of *Campylobacter jejuni*. Mol. Microbiol. 37: 501–514.

Linton, D., Karlyshev, A.V., and Wren, B.W. 2001. Deciphering *Campylobacter jejuni* cell surface interactions from the genome sequence. Curr. Opinion Microbiol. 4: 35–40.

McCarthy, N., Andersson, Y., Jormanainen, V., Gustavsson, O., and Giesecke, J. 1999. The risk of Guillain–Barré syndrome following infection with *Campylobacter jejuni*. Epidemiol. Infect. 122: 15–17.

McKhann, G.M., Maizeblath, D.R., Ho, T., Griffin, J.W., Li, C. Y., Bai, A.Y., Wu, H.S., Yei, Q.F., Zhang, W. C., Zhaori, Z., Jiang, Z., and Asbury, A.K. 1991. Clinical and electrophysiological aspects of acute paralytic disease of children and young adults in northern China. Lancet 338: 593–597.

Millar, L.A., Oldfield, N.J., and Ketley, J.M. 2001. Genetic variation of the lipooligosaccharide biosynthesis region of *Campylobacter jejuni*. Int. J. Med. Microbiol. 291 (S31): 82.

Moran, A.P., Prendergast, M.M., and Appelmelk, B.J. 1996. Molecular mimicry of host structures by bacterial lipopolysaccharides and its contribution to disease. FEMS Immunol. Med. Microbiol. 16: 105–115.

Moran, A.P. 1997. Structure and conserved characteristics of *Campylobacter jejuni* lipopolysaccharides. J. Infect. Dis. 176 (Suppl. 2): S115–121.

Moran, A.P., and Penner, J.L. 1999. Serotyping of *Campylobacter jejuni* based on heat-stable antigens: relevance, molecular basis and implications in pathogenesis. J. Appl. Microbiol. 86: 361–377.

Moxon, E.R., Rainey, P.B., Nowak, M.A., and Lenski, R.E. 1994. Adaptative evolution of highly mutable loci in pathogenic bacteria. Cur. Biol. 4: 24–33.

Nachamkin, I., Ung, H., Moran, A.P., Yoo, D., Prendergast, M.M., Nicholson, M.A., Sheikh, K., Ho, T., Asbury, A.K., McKhann, G.M., and Griffin, J.W. 1999. Ganglioside GM1 mimicry in *Campylobacter* strains from sporadic infections in the United States. J. Infect. Dis. 179: 1183–1189.

Nachamkin, I., Liu, J., Li, M., Ung, H., Moran, A.P., Prendergast, M.M., and Sheikh, K. 2002. *Campylobacter jejuni* from patients with Guillain–Barré syndrome preferentially expresses a GD$_{1a}$-like epitope. Infect. Immun.70: 5299–5303.

Nachamkin, I., Lobato, C., Arzate, P., Gonzalez, A., Rodriguez, P., Garcia, A., Cordero, L.M., Perea, L.G., Perez, J.C., Ribera, M., Fitzgerald, C., Maizeblath, D., Pinto, M.R., Asbury, A.K., and McKhann, G.M. 2003. A prospective study of Guillain–Barré syndrome in Mexico City, 1996–2002. Int. J. Med. Microbiol. 293 (S35): 36.

Nishimura, M., Nukina, M., Kuroki, S., Obayashi, H., Ohta, M., Ma, J.J., Saida, T., and Uchiyama, T. 1997. Characterization of *Campylobacter jejuni* isolates from patients with Guillain–Barré syndrome. J. Neurol. Sci. 153: 91–99.

Oldfield, N.J., Moran, A.P., Millar, L.A., Prendergast, M.M., and Ketley, J.M. 2002. Characterization of the *Campylobacter jejuni* heptosyltransferase II gene, *waaF*, provides evidence that extracellular polysaccharide is lipid A core independent. J. Bacteriol. 184: 2100–2107.

Parkhill, J., Wren, B.W., Mungall, K., Ketley, J.M., Churcher, C., Basham, D., Chillingworth, T., Davies, R.M., Feltwell, T., Holroyd, S., Jagels, K., Karlyshev, A.V., Moule, S., Pallen, M.J., Penn, C.W., Quail, M.A., Rajandream, M.-A., Rutherford, K.M., van Vliet, A.H.M., Whitehead, S., and Barrell, B.G. 2000. The genome sequence of the food-borne pathogen *Campylobacter jejuni* reveals hypervariable sequences. Nature 403: 665–669.

Pearson, B.M., Pin, C., Wright, J., I'Anson, K., Humphrey, T., and Wells, J.M. 2003. Comparative genome analysis of *Campylobacter jejuni* using whole genome DNA microarrays. FEBS Lett. 554: 224–230.

Penner, J.L., Hennessy, J.N., and Congi, R.V. 1983. Serotyping of *Campylobacter jejuni* and *Campylobacter coli* on the basis of thermostable antigens. Eur. J. Clin. Microbiol. 2: 378–383.

Prendergast, M.M., Lastovica, A.J., and Moran, A.P. 1998. Lipopolysaccharides from *Campylobacter jejuni* O: 41 strains associated with Guillain–Barré syndrome exhibit mimicry of GM1 ganglioside. Infect. Immun. 66: 3649–3655.

Preston, A., Mandrell, R.E., Gibson, B.W., and Apicella, M.A. 1996. The lipooligosaccharides of pathogenic Gram-negative bacteria. Crit. Rev. Microbiol. 22: 139–180.

Quarles, R.H., and Weiss, M.D. 1999. Autoantibodies associated with peripheral neuropathy. Muscle Nerve 22: 800–822.

Ramos-Alvarez, M., Bessudo, L., and Sabin, A. B. 1969. Paralytic syndromes associated with noninflammatory cytoplasmic or nuclear neuronopathy. Acute paralytic disease in Mexican children, neuropathologically distinguishable from Landry–Guillain–Barré syndrome. JAMA 207: 1481–1492.

Rees, J.H., Soudain, S.E., Gregson, N.A., and Hughes, R.A.C. 1995a. *Campylobacter jejuni* infection and Guillain–Barré syndrome. N. Engl. J. Med. 333: 1374–1379.

Rees, J.H., Gregson, N.A., and Hughes, R.A.C.1995b. Anti-ganglioside GM1 antibodies in Guillain–Barré syndrome and their relationship to *Campylobacter jejuni* infection. Ann. Neurol. 38: 809–816.

Ritter, G., Fortunato, S.R., Cohen, L., Noguchi, Y., Bernard, E.M., Stockert, E., and Old, L.J. 1996. Induction of antibodies reactive with GM2 ganglioside after immunization with lipopolysaccharides from *Campylobacter jejuni*. Int. J. Cancer 66: 184–190.

Salloway, S., Mermel, L.A., Seamans, M., Aspinall, G.O., Nam Shin, J.E., Kurjanczyk, L.A., and Penner, J.L. 1996. Miller Fisher syndrome associated with *Campylobacter jejuni* bearing lipopolysaccharide molecules that mimic human ganglioside GD$_3$. Infect. Immun. 64: 2945–2949.

Saunders, N.J., Peden, J.F., Hood, D.W., and Moxon, E.R. 1998. Simple sequence repeats in the *Helicobacter pylori* genome. Mol. Microbiol. 27: 1091–1098.

Shin, J.E.N., Ackloo, S., Mainkar, A.S., Monteiro, M.A., Pang, H., Penner, J.L., and Aspinall, G.O. 1998. Lipo-oligosaccharides of *Campylobacter jejuni* serotype O: 10. Structures of core oligosaccharide regions from a bacterial isolate from a patient with the Miller Fisher syndrome and from the serotype reference strain. Carbohydr. Res. 305: 223–232.

Sobue, G., Li, M., Terao, S., Aoki, S., Ichimura, M., Ieda, T., Doyu, M., Yasuda, T., Hashizume, Y., and Mitsuma, T. 1997. Axonal pathology in Japanese Guillain–Barré syndrome: a study of 15 autopsied cases. Neurology 48: 1694–1700.

St. Michael, F., Szymanski, C.M., Li, J., Chan, K.H., Khieu, N.H., Larocque, S., Wakarchuk, W.W., Brisson, J.R., and Monteiro, M.A. 2002. The structures of the lipooligosaccharide and capsule polysaccharide of *Campylobacter jejuni* genome sequenced strain NCTC 11168. Eur. J. Biochem. 269: 5119–5136.

Suerbaum, S., Lohrengel, M., Sonnevend, A., Ruberg, F., and Kist, M. 2001. Allelic diversity and recombination in *Campylobacter jejuni*. J. Bacteriol. 183: 2553–2559.

Szymanski, C.M., Yao, R., Ewing, C.P., Trust, T.J., and Guerry, P. 1999. Evidence for a system of general protein glycosylation in *Campylobacter jejuni*. Mol. Microbiol. 32: 1022–1030.

Szymanski, C.M., St. Michael, F., Jarrell, H.C., Li, J., Gilbert, M., Larocque, S., Vinogradov, E., and Brisson, J.R. 2003. Detection of conserved *N*-linked glycans and phase variable lipo-oligosaccharides and capsules from *Campylobacter* cells by mass spectrometry and high resolution magic angle spinning NMR spectroscopy. J. Biol. Chem. 278: 24509–24520.

Thibault, P., Logan, S.M., Kelly, J.F., Brisson, J.-R., Ewing, C.P., Trust, T.J., and Guerry, P. 2001. Identification of the carbohydrate moieties and glycosylation motifs in *Campylobacter jejuni* flagellin. J. Biol. Chem. 276: 34862–34870.

van Belkum, A., van den Braak, N., Godschalk, P., Ang, W., Jacobs, B., Gilbert, M., Wakarchuk, W., Verbrugh, H., and Endtz, H. 2001. A *Campylobacter jejuni* gene associated with immune-mediated neuropathy. Nature Med. 7: 752–753.

van der Meche, F.G., and Schmitz, P.I. 1992. A randomized trial comparing intravenous immune globulin and plasma exchange in Guillain–Barré syndrome. Dutch Guillain–Barré Study Group. N. Engl. J. Med. 326: 1123–1129.

van Koningsveld, R., van Doorn, P.A., Schmitz, P.I.M., Ang, C.W., and van der Meche, F.G.A. 2000. Mild forms of Guillain–Barré syndrome in an epidemiologic survey in The Netherlands. Neurology 54: 620–625.

van Koningsveld, R., Rico, R., Gerstenbluth, I., Schmitz, P.I.M., Ang, C.W., Merkies, I.S.J., Jacobs, B.C., Halabi, Y., Endtz, H.P., van der Meché, F.G.A., and van Doorn, P.A. 2001. Gastroenteritis-associated Guillain–Barré syndrome on the Caribbean island Curaçao. Neurology 56: 1467–1472.

Wang, G., Ge, Z., Rasko, D.A., and Taylor, D.E. 2000. Lewis antigens in *Helicobacter pylori*: biosynthesis and phase variation. Mol. Microbiol. 36: 1187–1196.

Wassenaar, T.M., Fry, B.N., Lastovica, A.J., Wagenaar, J.A., Coloe, P.J., and Duim, B. 2000. Genetic characterization of *Campylobacter jejuni* O: 41 isolates in relation with Guillain–Barré syndrome. J. Clin. Microbiol. 38: 874–876.

Willison, H.J., and O'Hanlon, G.M. 1999. The immunopathogenesis of Miller Fisher syndrome. J. Neuroimmunol. 100: 3–12.

Winer, J.B., Hughes, R.A.C., and Osmond, C. 1988a. A prospective study of acute idiopathic neuropathy. I. Clinical features and their prognostic value J. Neurol. Neurosurg. Psychiatry 51: 605–612.

Winer, J.B., Hughes, R.A.C., Anderson, M.J., Jones, D.M., Kangro, H., and Watkins, R.P. 1988b. A prospective study of acute idiopathic neuropathy. II. Antecedent events. J. Neurol. Neurosurg. Psychiatry 51: 613–618.

Wirguin, I., Briani, C., Suturkova-Milosevic, L., Fisher, T., Della-Latta, P., Chalif, P., and Latov, N. 1997. Induction of anti-GM1 ganglioside antibodies by *Campylobacter jejuni* lipopolysaccharides. J. Neuroimmunol. 78: 138–142.

Wood, A.C., Oldfield, N.J., O'Dwyer, C.A., and Ketley, J.M. 1999. Cloning, mutation and distribution of a putative lipopolysaccharide biosynthesis locus in *Campylobacter jejuni*. Microbiology 145: 379–388.

Yang, Q.L., and Gotschlich, E.C. 1996. Variation of gonococcal lipooligosaccharide structure is due to alterations in poly-G tracts in *lgt* genes encoding glycosyltransferases. J. Exp. Med. 183: 323–327.

Young, N.M., Brisson, J.-R., Kelly, J., Watson, D.C., Tessier, L., Lanthier, P.H., Jarrell, H.C., Cadotte, N., St. Michael, F., Aberg, E., and Szymanski, C.M. 2002. Structure of the *N*-linked glycan present on multiple glycoproteins in the Gram-negative bacterium, *Campylobacter jejuni*. J. Biol. Chem. 277: 42530–42539.

Yuki, N., Handa, S., Taki, T., Kasama, T., Takahashi, M., Sito, K., and Miyatake, T. 1992. Cross-reactive antigen between nervous tissue and a bacterium elicits Guillain–Barré syndrome: molecular mimicry between ganglioside GM1 and lipopolysaccharide from Penner's serotype 19 of *Campylobacter jejuni*. Biomed. Res. 13: 451–453.

Yuki, N., Taki, T., Inagaki, F., Kasama, T., Takahashi, M., Saito, K., Handa, S., and Miyatake, T. 1993. A bacterium lipopolysaccharide that elicits Guillain–Barré syndrome has a GM1 ganglioside-like structure. J. Exp. Med.178: 1771–1775.

Yuki, N., Taki, T., Takahashi, M., Saito, K., Tai, T., Miyatake, T., and Handa, S. 1994. Penner's serotype 4 of *Campylobacter jejuni* has a lipopolysaccharide that bears a GM1 ganglioside epitope as well as one that bears a GD1a epitope. Infect. Immun. 62: 2101–2103.

Yuki, N. 1997. Molecular mimicry between gangliosides and lipopolysaccharides of *Campylobacter jejuni* isolated from patients with Guillain–Barré syndrome and Miller Fisher syndrome. J. Infect. Dis.176 (Suppl. 2): S150–153.

Yuki, N. 2001a. Infectious origins of, and molecular mimicry in, Guillain–Barré and Fisher syndromes. Lancet Infect. Dis. 1: 29–37.

Yuki, N., Yamada, M., Koga, M., Odaka, M., Susuki, K., Tagawa, Y., Ueda, S., Kasama, T., Ohnishi, A., Hayashi, S., Takahashi, H., Kamijo, M., and Hirata, K. 2001b. Animal model of axonal Guillain–Barré syndrome induced by sensitization with GM1 ganglioside. Ann. Neurol. 49: 712–720.

Zhu, P., Klutch, M.J., Bash, M.C., Tsang, R.S.W., Ng, L.-K., and Tsai, C.-M. 2002. Genetic diversity of three *lgt* loci for biosynthesis of lipooligosaccharide (LOS) in *Neisseria* species. Microbiology 148: 1833–1844.

Chapter 12

The Polysaccharide Capsule of *Campylobacter jejuni*

A.V. Karlyshev, O.L. Champion, G.W.P. Joshua and B.W. Wren

Abstract

The discovery of the *Campylobacter jejuni* capsular polysaccharide (CPS) is one of the most important recent advances in the study of the biology of this pathogen. The CPS consists of repeating oligosaccharide units attached to a phospholipid, and it is not chemically linked to lipooligosaccharide (LOS). In addition, it is CPS and not LOS that is associated with Penner serotypes. The CPS structures seem to be relatively stable for particular *C. jejuni* strains maintained in laboratory conditions, although some variant structures may be detected. However, among different *C. jejuni* strains the CPS structures are highly variable and a number of genetic mechanisms responsible for such variation have been uncovered. CPS is an important virulence factor and may also be essential for increased survival of *C. jejuni* in the environment.

INTRODUCTION

Campylobacter jejuni is the most frequent bacterial gastrointestinal pathogen in man, yet it does not grow aerobically and human to human transition is rare. Therefore, one may wonder 'how can an organism of such limited hardiness and growth capabilities be responsible for an ever-increasing level of human food borne disease' (Solomon and Hoover, 1999). Better understanding of the biology of the pathogen is required to resolve this paradox.

One important feature of *C. jejuni* is the production of a variety of cell-surface located glycoconjugants, including glycoproteins and glycolipids. There are two types of glycolipids produced by *C. jejuni*: lipo-oligosaccharide (LOS) and capsular polysaccharide (CPS). LOS has been extensively investigated over the last decade. The chemical structures of LOSs extracted from different strains (Moran *et al.*, 2000) and genetic mechanisms of its variation have been reported (Gilbert *et al.*, 2002) (Linton *et al.*, 2000) (see Chapter 11). In contrast, the role of the recently discovered CPS is less well studied. However, the data available suggest that this extracellular structure may play an important role both in pathogenesis and in the survival of *Campylobacter* in the environment.

The *Campylobacter* glycolipids, which were referred to as lipopolysaccharides (LPSs) in many previous publications, have been under investigation since the development of the Penner serotyping scheme (Penner and Hennessy, 1980). Although a number of alternative typing techniques have been recently reported (e.g. Shi *et al.*, 2002), this remains the major typing scheme used to distinguish *C. jejuni* strains worldwide. The scheme is based on indirect agglutination of heat-stable antigens bound to erythrocytes with a set of typing antisera raised against sero-reference strains. Analysis of the glycolipids present in the extracts used for typing revealed both LOS and, in some cases, also a 'high molecular weight lipopolysaccharide' (HMW LPS). Despite the lack of experimental evidence, the latter was considered to be derived from LOS via the attachment of a long polysaccharide chain (Moran *et al.*, 2000). However, some earlier studies had provided indirect evidence that these could be separate molecules. For example, whereas LOS was readily stained with silver, the HMW LPS could only be detected via Western blotting (Mandatori and Penner, 1989). If the HMW LPS were derived from LOS, both molecules would be stainable with silver. Aspinal *et al.* (1995) reported that HMW LPS from O:3 strain of *C. jejuni* is not linked to LOS. However, for years this finding was considered to be an exceptional case. Despite the lack of experimental evidence it was presumed that in other strains of *C. jejuni* the HMW LPS was associated with LOS (Moran *et al.*, 2000). Subsequent genetic and biochemical evidence, however, has proved that this was not the case.

DISCOVERY OF CPS AND A CAPSULE

For many years, the presence of capsule remained unnoticed. The possible existence of a *Campylobacter* CPS was first suggested by Chart and co-workers in 1996 (Chart *et al.*, 1996b), who identified a glycolipid loosely associated with bacterial cell surface. This material was extractable during mild treatment of a bacterial cell suspension at 50°C. Loose association with the bacterial cell surface is a typical feature of CPS from a wide range of bacterial species (Vann and Freese, 1994).

Genetic evidence for the production of a CPS was not obtained until 1999. Whilst sample-sequencing a random pUC18-based library of the *C. jejuni* NCTC 11168 genome (Karlyshev *et al.*, 1999), we isolated several sequences with strong similarity to *kps* genes involved in CPS biosynthesis in many bacterial species, including *Escherichia coli*. Subsequently, a large gene cluster involved in CPS biosynthesis was identified in the complete genome sequence of NCTC 11168 (Parkhill *et al.*, 2000) (Figure 12.1). The discovery of a CPS produced by various strains of *Campylobacter* (Karlyshev *et al.*, 2000) supported the predictions based on the genome sequencing data.

Several lines of evidence corroborated the presence of a *C. jejuni* polysaccharide capsule. We found that mutations in some of the *kps* genes in all *C. jejuni* strains tested resulted in the loss of the HMW LPS bands (Figure 12.2) detectable via Western blotting (Karlyshev *et al.*, 2000). Another important observation was that the CPS is not chemically linked to LOS and is the main component in the Penner serotyping scheme (Karlyshev *et al.*, 2000). It consists of a large number of repeating units and has a highly heterogenic structure among different strains.

The results of treatment of the *C. jejuni* CPSs with phospholipases (PLs) clearly indicated the presence of phospholipid moieties similar to those found in type II/III CPS from other bacteria (Whitfield and Roberts, 1999). However, the reason for the complete disappearance of *C. jejuni* CPSs on Western blots after such treatment was not clear (Karlyshev *et al.*, 2000). Our subsequent work using the cationic dye Alcian Blue demonstrated that polysaccharide released after PL treatment could be detected directly on gels (Figure 12.3), indicating that the phospholipid moiety was crucial for binding to polyvinylidene difluoride (PVDF) membranes (Karlyshev and Wren, 2001). These experiments also demonstrated that, whereas phospholipase treatment reduces the size of a CPS, it has no effect on LOS (Figure 12.3, lanes 1 and 2), thus confirming

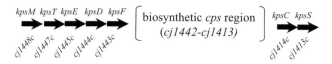

Figure 12.1 The capsular polysaccharide gene cluster of strain NCTC 11168. The cluster contains a central variable biosynthetic region flanked by two sets of conserved *kps* genes with high similarity to the genes involved in transport and assembly of capsular polysaccharides in *E. coli* and other bacteria (Karlyshev *et al.*, 2000).

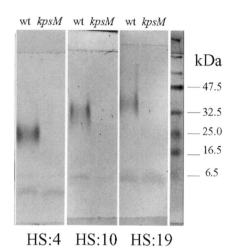

Figure 12.2 The effect of the mutation of *kpsM* on the production of CPS in *C. jejuni* strains NCTC 12561 (HS:4), NCTC 12509 (HS:10) and NCTC 12517 (HS:19). Western blot analysis of proteinase K-treated cell lysates using homologous Penner serotyping antisera (Karlyshev *et al.*, 2000).

that these molecules are not related structurally. We also found that in certain growth conditions the CPS can be detached from the cell surface resulting in the production of extracellular polysaccharide (EPS). Finally, our characterization of the CPS culminated in the visualization by electron microscopy of capsular material stabilized by Alcian blue (Figure 12.4) surrounding the *C. jejuni* cell (Karlyshev *et al.*, 2001).

More recently, Oldfield *et al.* (2002) have shown that a deep LOS core *waaF* mutant showed changes in LOS, but not the 'HMW LPS' supporting data that the HMW LPS is likely to be due to another polysaccharide, the CPS. Subsequently Bacon and colleagues (Bacon *et al.*, 2001) confirmed our data on the genetics and biochemistry of the CPS. The authors also found that, compared to wild-type strain, the *kpsM* mutant of 81-176 strain revealed reduced adhesion and invasion of epithelial cells, as well as reduced virulence in a ferret diarrhoeal model (Bacon *et al.*, 2001). However, the exact mechanism of such biological effects remains to be elucidated. In addition, we have shown that the capsule is important for the colonization of chicks (Karlyshev *et al.*, unpublished observation).

GENETICS OF CPS BIOSYNTHESIS
The finding that CPS is a major Penner serotyping antigen implied similarity in the organization of the gene clusters among strains belonging to the same serotype and preliminary PCR analysis confirmed this hypothesis (Karlyshev, unpublished observation).

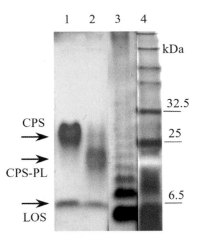

Figure 12.3 Use of Alcian Blue dye for direct in-gel staining of a polysaccharide moiety (CPS-PL) released by the CPS of strain G1 after treatment with phospholipase (Karlyshev and Wren, 2001). Note that treatment did not affect LOS. LPS from *S. enterica* serovar Typhimurium is included for comparison. Lane 1, proteinase K-treated 50°C saline extract, containing CPS with minor amounts of LOS; lane 2, same material as lane 1 after phospholipase treatment; lane 3, LPS from *S. enterica* serovar Typhimurium; lane 4, molecular mass markers.

The PCR analysis was confirmed by hybridization analysis using a *C. jejuni* DNA microarray. Five of the strains investigated in the microarray study had the same Penner serotype (HS:02) and revealed identical patterns of hybridization in the CPS loci (Dorrell *et al.*, 2001). However, although the microarray technique using large DNA fragments can provide some general indications as to the presence or absence of certain genes in various strains, it has significant limitations. For example, some genomic rearrangements (e.g. gene fusions) will not necessarily result in changes in hybridization pattern. Structural variability of CPS has recently been uncovered by DNA sequencing studies (Karlyshev *et al.*, in press). Using a long range PCR technique, we amplified and sequenced *cps* regions from four *C. jejuni* strains belonging to different serotypes, and compared them with the *cps* region from strain NCTC 11168. Comparative analysis revealed variation in the organization of the *cps* regions studied (Figure 12.5). However, despite this variation, some genes in various *cps* clusters were found to be almost identical (these genes are linked by bars in Figure 12.5). In particular, in strains NCTC 11168 (HS:1), 81-176 (HS:23/36), NCTC12517 (HS:19) and G1 (HS:1) the *kpsC*-proximal genes appear to be most highly conserved. One can assume that these genes encode enzymes required for common steps in the biosynthesis of the respective CPSs of these strains.

Another group of highly conserved genes found in strains NCTC 11168 (HS:1), 81-176 (HS:23/36) and 176.83 (HS:41) includes *hddC*, *gmhA2* and *hddA*, which are responsible for the biosynthesis of heptose residues (Valvano *et al.*, 2002) found in the CPS of these strains (see below). Owing to the very high similarity of these heptose biosynthesis genes to ones present in the LOS biosynthesis gene cluster, these genes may have been acquired recently via interstrain genetic exchange. Campylobacters are naturally transformable and, significantly, the possibility of a genetic exchange between campylobacters *in vivo* has recently been experimentally confirmed (Boer *et al.*, 2002).

Substantial variation has been observed in the small number of *C. jejuni cps* gene clusters that have been investigated. Given that CPS forms the structural basis for the

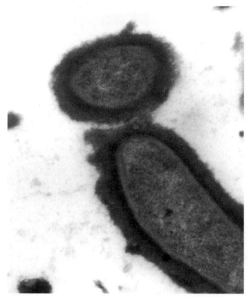

Figure 12.4 EM visualization of *C. jejuni* capsule stabilized with Alcian Blue dye (Karlyshev *et al.*, 2001). The capsule is visible as a thick amorphous material surrounding the bacterial cells.

Penner serotyping system, the large number of Penner serotypes that have been described leads one to assume that there is a lot more variation in the pattern of CPS gene cluster organization still to be described.

CPS structure

The standard Penner serotyping procedure allows detection of as many as 66 different serotypes of *C. jejuni* (McKay *et al.*, 2001). The number of different CPS structures should even be higher given that over 20% of strains remain untypable due to the limited number of antisera available (Oza *et al.*, 2002). The structures of the repeating units of CPS extracted from some *Campylobacter* strains have been determined and found to be extremely diverse (Figure 12.6). One interesting observation is that, as predicted by the presence of heptose biosynthesis genes in the CPS cluster, some CPSs contain heptose residues. The heptose residues found in polysaccharides of other pathogenic bacteria are thought to be important for their virulence (Kneidinger *et al.*, 2002).

Also present are amidated glucuronic acid residues (GlcA6-NGro) in the CPS of strains NCTC12517 (HS:19) and NCTC 11168 (HS:2) (Figure 12.6), which correlates with the presence of the *udg* gene (Figure 12.5), encoding uridyl-glucose dehydrogenase, required for the biosynthesis of glucuronic acid (GlcA). GlcA is a common component of other group II/III capsular polysaccharides found in some strains of *E. coli* (Sussman, 1997). Glucuronic and mannuronic acid residues were also found in the exopolysaccharide produced by *Pseudomonas aeruginosa* strains associated with cystic fibrosis (May and Chakrabarty, 1994). Interestingly, none of the reported CPS structures of *C. jejuni* contained free glucuronic acid residues, all being in an amidated form. However, one cannot exclude the possibility that due to genetic variation, some of the *C. jejuni* CPS molecules will not carry a modifying group and retain a negative charge.

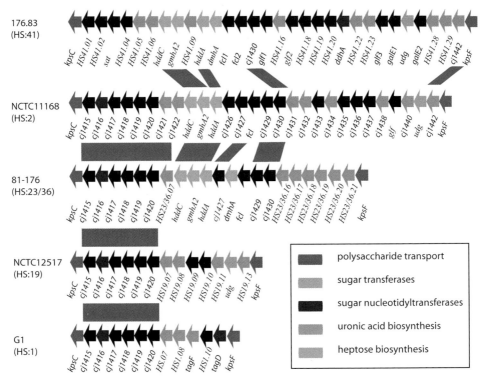

Figure 12.5 Comparison of the genomic *cps* biosynthesis regions found in different *C. jejuni* strains. Outlined bars indicate genes encoding products with >80% identity with respective counterparts from NCTC 11168 strain. Likely functions of each open reading frame are shown; black arrows indicate genes of unknown function.

Strain 81116 (HS:6) was found to produce two different CPS molecules with unrelated structures (Figure 12.6). One possibility is that there could be two different *cps* gene clusters present in this strain. However, according to our data (Karlyshev *et al.*, 2000) no CPS can be detected in the 81116 (*kpsM::kan^r*) mutant, suggesting that the same *kpsM* gene, encoding a component of a CPS transporter, may be involved in the biosynthesis of both polysaccharides. The genetic origin of this phenomenon remains to be elucidated.

CONCLUSION AND FUTURE DIRECTIONS

Extensive study of *C. jejuni* in recent years, assisted by the availability of the complete genomic sequence, has resulted in a better understanding of the virulence mechanisms of this important pathogen. However, many questions remain unanswered. One conundrum is the apparent extent to which bacteria that appear to be so vulnerable when grown under laboratory conditions can survive in the environment. It is also unclear why *C. jejuni* causes gastrointestinal diseases in humans but is completely avirulent in the avian host.

The discovery of the polysaccharide capsule produced by *C. jejuni* has had a significant impact on our understanding of the biology of the bacterium. However, more work is required for a proper understanding of its biological functions. There is some evidence indicating a role for CPS as a virulence factor; for example, a non-capsulated mutant of 81-176 was found to have increased serum sensitivity (Bacon *et al.*, 2001). The CPS, especially in its EPS form, may be important for resistance to desiccation (Roberson and

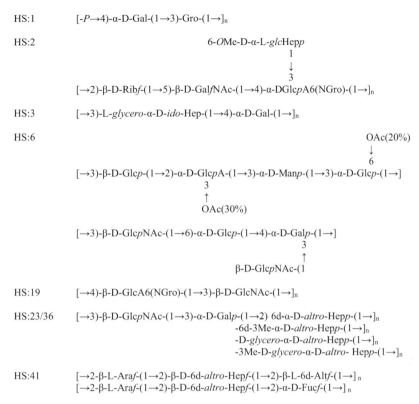

HS:1   [-P→4)-α-D-Gal-(1→3)-Gro-(1→]$_n$

HS:2

6-OMe-D-α-L-*glc*Hep$p$
1
↓
3

[→2)-β-D-Rib$f$-(1→5)-β-D-Gal$f$NAc-(1→4)-α-DGlc$p$A6(NGro)-(1→]$_n$

HS:3   [→3)-L-*glycero*-α-D-*ido*-Hep-(1→4)-α-D-Gal-(1→]$_n$

HS:6

OAc(20%)
↓
6

[→3)-β-D-Glc$p$-(1→2)-α-D-Glc$p$A-(1→3)-α-D-Man$p$-(1→3)-α-D-Glc$p$-(1→]
3
↑
OAc(30%)

[→3)-β-D-Glc$p$NAc-(1→6)-α-D-Glc$p$-(1→4)-α-D-Gal$p$-(1→]
3
↑
β-D-Glc$p$NAc-(1

HS:19   [→4)-β-D-GlcA6(NGro)-(1→3)-β-D-GlcNAc-(1→]$_n$

HS:23/36   [→3)-β-D-Glc$p$NAc-(1→3)-α-D-Gal$p$-(1→2) 6d-α-D-*altro*-Hep$p$-(1→]$_n$
-6d-3Me-α-D-*altro*-Hep$p$-(1→]$_n$
-D-*glycero*-α-D-*altro*-Hep$p$-(1→]$_n$
-3Me-D-*glycero*-α-D-*altro*- Hep$p$-(1→]$_n$

HS:41   [→2-β-L-Ara$f$-(1→2)-β-D-6d-*altro*-Hep$f$-(1→2)-β-L-6d-Alt$f$-(1→]$_n$
[→2-β-L-Ara$f$-(1→2)-β-D-6d-*altro*-Hep$f$-(1→2)-α-D-Fuc$f$-(1→]$_n$

Figure 12.6 The known *C. jejuni* CPS structures: HS:1 (Moran *et al.*, 2000), HS:2 (NCTC 11168) (Michael *et al.*, 2002), HS:3 (Aspinall *et al.*, 1995), HS:6 (81116) (Muldoon *et al.*, 2002), HS:19 (Aspinall *et al.*, 1994), HS:23/36 (Aspinall *et al.*, 1992). Abbreviations for sugar residues: Alt, altrose; Fuc, fucose; Gal, galactose; GalNac, N-acetyl-galactosamine; Gro, glycerol; Glc, glucuronic acid, Hep, heptose; Man, mannose.

Firestone, 1992) and in biofilm formation. In a biofilm bacteria may also be protected against oxidative and other stresses. Our finding that in certain conditions the capsular material can be released from bacterial cell surface in the form of EPS via cleavage of the phospholipid bond may indicate the existence of an as yet unknown mechanism of CPS processing.

At present there are no data regarding regulation of CPS biosynthesis. One obvious factor which may affect CPS production is temperature. However, in preliminary experiments, we could not detect temperature dependence in CPS production (Karlyshev, unpublished observation). The possibility remains, therefore, that CPS production is regulated by as yet unidentified factor (s). This is supported by the observation that the biosynthesis of *C. jejuni* 'HMW LPS' (now known as CPS) may be induced upon colonization of chickens (Chart *et al.*, 1996a).

Another area of interest is the possible role of CPS in the development of Guillain–Barré syndrome (GBS). Although an autoimmune response against ganglioside-like structures present in LOS is considered to be the main contributing factor to the development of the disease, the possibility that uronic acid-like structures present in CPS could also contribute to the disease cannot be excluded (Yuki, 1997). Association of GBS syndrome with specific serotypes has been reported (e.g. Lastovica *et al.*, 1997). Due to the link between serotypes and CPS, it would be interesting to ascertain whether or not there is a correlation between these CPS structures and GBS and if the specific combination of CPS and LOS structure is significant.

Our data have revealed multiple mechanisms of genetic variation in the *cps* region among different strains. Some of the genes in the *cps* regions found in different strains are almost identical suggesting interstrain genetic exchange. Other mechanisms of genetic variation in the *cps* region include gene fusion, deletion, duplication and phase variation. The availability of genetic variation generated by different mechanisms that result in the production of CPS of different structures may be essential for evasion of host immune response, as well as for adaptation of bacteria to adverse environmental conditions.

In some cases the genetic variation in the *cps* region should result in changes of the CPS structure, which may ultimately lead to serotype changes. Interestingly, however, the serotypes of the strains maintained *in vitro* appear to be extremely stable and no serotype changes could be detected even after a large number of passages *in vivo* and *in vitro* (Nielsen *et al.*, 2001; Manning *et al.*, 2001). One possible explanation for this apparent contradiction is that although variant cells with a different CPS structure may indeed be present in the population, the proportion of these cells may remain low without a selective advantage. Selective pressure may be imposed *in vitro*, when a particular CPS structure is advantageous for survival under certain conditions (e.g. through biofilm formation or selection of cells with increased tolerance to desiccation) or *in vivo,* when a host has developed an immune response to a particular CPS. Chickens, therefore, may not be the best model for *in vivo* studies on the immune response against CPS due to their poor immune response to some polysaccharide antigens (Jeurissen *et al.*, 1998). Consequently, there may be no selective pressure in chickens for the survival of bacteria producing any particular CPS. Therefore, in animals with a strong immune response to CPS the frequency of serotype changes may be much higher. A change in *C. jejuni* serotype has been detected in an infected patient (Mills *et al.*, 1992), which may correlate with the more effective CPS-specific immune response mounted by humans.

A recent development on the structural analysis of the CPS has been the application of high resolution magic angle spinning (HR-MAS) NMR to examine CPS directly from campylobacter cells (Szymanski *et al.*, 2003). HR-MAS NMR examination of growth from individual colonies of *C. jejuni* NCTC 11168 demonstrated different forms of the capsular glycan modified by phase-variable genes (Szymanski *et al.*, 2003). These variants show different staining patterns on deoxycholate-PAGE and reactivity with immune sera. One of the identified modifications was a novel -OP=O (NH$_2$)OMe phosphoramide, not observed previously in nature (Szymanski *et al.*, 2003). This sensitive method also exhibited the feasibility of complementing the Penner serotyping system.

Another implication of the recent *C. jejuni* CPS studies relates to serotype designation. The designation 'O' (e.g. O:41) relates to LPS which is inappropriate. It would therefore be more appropriate to use either 'K' (e.g. K:41) as in *E. coli* capsular antigen designation or HS (e.g. HS:41) for heat-stable, which has become more commonly used in recent publications as the serotype designation in *C. jejuni.*

We believe that our studies on *C. jejuni* CPS will lead to a better understanding of its genetics, structure and function (s). This may in turn result in the development of improved intervention strategies to reduce the burden of *Campylobacter*-associated disease.

Acknowledgements

We acknowledge the BBSRC and the Leverhulme Trust for financial support.

References

Aspinall, G.O., McDonald, A.G., and Pang, H. 1992. Structures of the O chains from lipopolysaccharides of *Campylobacter jejuni* serotypes O: 23 and O: 36. Carbohydrate Res. 231: 13–30.

Aspinall, G.O., McDonald, A.G., and Pang, H. 1994. Lipopolysaccharides of *Campylobacter jejuni* serotype O: 19: structures of O antigen chains from the serostrain and two bacterial isolates from patients with the Guillain–Barré syndrome. Biochemistry 33: 250–255.

Aspinall, G.O., Lynch, C.M., Pang, H., Shaver, R.T., and Moran, A.P. 1995. Chemical structures of the core region of *Campylobacter jejuni* O: 3 lipopolysaccharide and an associated polysaccharide. Eur. J. Biochem. 231: 570–578.

Bacon, D.J., Szymanski, C.M., Burr, D.H., Silver, R.P., Alm, R.A., and Guerry, P. 2001. A phase-variable capsule is involved in virulence of *Campylobacter jejuni* 81-176. Mol. Microbiol. 40: 769–777.

Boer, P., Wagenaar, J.A., Achterberg, R.P., Putten, J.P., Schouls, L.M., and Duim, B. 2002. Generation of *Campylobacter jejuni* genetic diversity *in vivo*. Mol. Microbiol. 44: 351–359.

Chart, H., Conway, D., Frost, J.A., and Rowe, B. 1996a. Outer membrane characteristics of *Campylobacter jejuni* grown in chickens. FEMS Microbiol. Lett. 145: 469–472.

Chart, H., Frost, J.A., Oza, A., Thwaites, R., Gillanders, S., and Rowe, B. 1996b. Heat-stable serotyping antigens expressed by strains of *Campylobacter jejuni* are probably capsular and not long-chain lipopolysaccharide. J. Appl. Bacteriol. 81: 635–640.

Dorrell, N., Mangan, J.A., Laing, K.G., Hinds, J., Linton, D., Al-Ghusein, H., Barrell, B.G., Parkhill, J., Stoker, N.G., Karlyshev, A.V., Butcher, P.D., and Wren, B.W. 2001. Whole genome comparison of *Campylobacter jejuni* human isolates using a low-cost microarray reveals extensive genetic diversity. Genome Res. 11: 1706–1715.

Gilbert, M., Karwaski, M.F., Bernatchez, S., Young, N.M., Taboada, E., Michniewicz, J., Cunningham, A.M., and Wakarchuk, W.W. 2002. The genetic bases for the variation in the lipo-oligosaccharide of the mucosal pathogen, *Campylobacter jejuni*. Biosynthesis of sialylated ganglioside mimics in the core oligosaccharide. J. Biol. Chem. 277: 327–337.

Jeurissen, S.H., Janse, E.M., van Rooijen, N., and Claassen, E. 1998. Inadequate anti-polysaccharide antibody responses in the chicken. Immunobiology 198: 385–395.

Karlyshev, A.V., Henderson, J., Ketley, J.M., and Wren, B.W. 1999. Procedure for the investigation of bacterial genomes: random shot-gun cloning, sample sequencing and mutagenesis of *Campylobacter jejuni*. Biotechniques 26: 50–52, 54, 56.

Karlyshev, A.V., Linton, D., Gregson, N.A., Lastovica, A.J., and Wren, B.W. 2000. Genetic and biochemical evidence of a *Campylobacter jejuni* capsular polysaccharide that accounts for Penner serotype specificity. Mol. Microbiol. 35: 529–541.

Karlyshev, A.V., McCrossan, M.V., and Wren, B.W. 2001. Demonstration of polysaccharide capsule in *Campylobacter jejuni* using electron microscopy. Infect. Immun. 69: 5921–5924.

Karlyshev, A.V., and Wren, B.W. 2001. Detection and initial characterization of novel capsular polysaccharide among diverse *Campylobacter jejuni* strains using alcian blue dye. J. Clin. Microbiol. 39: 279–284.

Kneidinger, B., Marolda, C., Graninger, M., Zamyatina, A., McArthur, F., Kosma, P., Valvano, M.A., and Messner, P. 2002. Biosynthesis pathway of ADP-L-glycero-beta-D-manno-heptose in *Escherichia coli*. J. Bacteriol. 184: 363–369.

Lastovica, A.J., Goddard, E.A., and Argent, A.C. 1997. Guillain–Barré syndrome in South Africa associated with *Campylobacter jejuni* O: 41 strains. J. Infect. Dis. 176 Suppl 2: S139–143.

Linton, D., Gilbert, M., Hitchen, P.G., Dell, A., Morris, H.R., Wakarchuk, W.W., Gregson, N.A., and Wren, B.W. 2000. Phase variation of a beta-1,3 galactosyltransferase involved in generation of the ganglioside GM1-like lipo-oligosaccharide of *Campylobacter jejuni*. Mol. Microbiol. 37: 501–514.

Mandatori, R., and Penner, J.L. 1989. Structural and antigenic properties of *Campylobacter coli* lipopolysaccharides. Infect. Immun. 57: 3506–3511.

Manning, G., Duim, B., Wassenaar, T., Wagenaar, J.A., Ridley, A., and Newell, D.G. 2001. Evidence for a genetically stable strain of *Campylobacter jejuni*. Appl. Environ. Microbiol. 67: 1185–1189.

May, T.B., and Chakrabarty, A.M. 1994. Isolation and assay of *Pseudomonas aeruginosa* alginate. Methods Enzymol. 235: 295–304.

McKay, D., Fletcher, J., Cooper, P., and Thomson-Carter, F.M. 2001. Comparison of two methods for serotyping *Campylobacter* spp. J. Clin. Microbiol. 39: 1917–1921.

Michael, F.S., Szymanski, C.M., Li, J., Chan, K.H., Khieu, N.H., Larocque, S., Wakarchuk, W.W., Brisson, J.R., and Monteiro, M.A. 2002. The structures of the lipooligosaccharide and capsule polysaccharide of *Campylobacter jejuni* genome sequenced strain NCTC 11168. Eur. J. Biochem. 269: 5119–5136.

Mills, S.D., Kuzniar, B., Shames, B., Kurjanczyk, L.A., and Penner, J.L. 1992. Variation of the O antigen of *Campylobacter jejuni in vivo*. J. Med. Microbiol. 36: 215–219.

Moran, A.P., Penner, J.L., and Aspinall, G.O. 2000. *Campylobacter* polysaccharides. In *Campylobacter*. Nachamkin, I. and Blaser, M.J., eds. Washington: ASM Press, pp. 241–257.

Muldoon, J., Shashkov, A.S., Moran, A.P., Ferris, J.A., Senchenkova, S.N., and Savage, A.V. 2002. Structures of two polysaccharides of *Campylobacter jejuni* 81116. Carbohydrate Res. 337: 2223–2229.

Nielsen, E.M., Engberg, J., and Fussing, V. 2001. Genotypic and serotypic stability of *Campylobacter jejuni* strains during *in vitro* and *in vivo* passage. Int. J. Med. Microbiol. 291: 379–385.

Oldfield NJ, Moran AP, Millar LA, Prendergast MM, Ketley JM. 2002 Characterization of the *Campylobacter jejuni* heptosyltransferase II gene, waaF, provides genetic evidence that extracellular polysaccharide is lipid A core independent. J. Bacteriol.184: 2100–7

Oza, A.N., Thwaites, R.T., Wareing, D.R., Bolton, F.J., and Frost, J.A. 2002. Detection of heat-stable antigens of *Campylobacter jejuni* and *C. coli* by direct agglutination and passive hemagglutination. J. Clin. Microbiol. 40: 996–1000.

Parkhill, J., Wren, B.W., Mungall, K., Ketley, J.M., Churcher, C., Basham, D., Chillingworth, T., Davies, R.M., Feltwell, T., Holroyd, S., Jagels, K., Karlyshev, A.V., Moule, S., Pallen, M.J., Penn, C.W., Quail, M.A., Rajandream, M.A., Rutherford, K.M., van Vliet, A.H., Whitehead, S., and Barrell, B.G. 2000. The genome sequence of the food-borne pathogen *Campylobacter jejuni* reveals hypervariable sequences. Nature 403: 665–668.

Penner, J.L., and Hennessy, J.N. 1980. Passive hemagglutination technique for serotyping *Campylobacter fetus* subsp. *jejuni* on the basis of soluble heat-stable antigens. J. Clin. Microbiol. 12: 732–737.

Roberson, E.B., and Firestone, M.K. 1992. Relationship between desiccation and exopolysaccharide production in a soil *Pseudomonas* sp. Appl. Environ. Microbiol. 58: 1284–1291.

Shi, F., Chen, Y.Y., Wassenaar, T.M., Woods, W.H., Coloe, P.J., and Fry, B.N. 2002. Development and application of a new scheme for typing *Campylobacter jejuni* and *Campylobacter coli* by PCR-based restriction fragment length polymorphism analysis. J. Clin. Microbiol. 40: 1791–1797.

Solomon, E.B., and Hoover, D.G. 1999. *Campylobacter jejuni*: A bacterial paradox. J. Food Safety 19: 121–136.

Sussman, M. 1997. Capsules of *Escherichia coli*. In *Escherichia coli*: Mechanisms of Virulence. Sussman, M., ed. Cambridge: Cambridge University Press, pp. 114–143.

Szymanski CM, Michael FS, Jarrell HC, Li J, Gilbert M, Larocque S, Vinogradov E, Brisson JR. 2003. Detection of conserved N-linked glycans and phase-variable lipooligosaccharides and capsules from campylobacter cells by mass spectrometry and high resolution magic angle spinning NMR spectroscopy. J. Biol. Chem. 2003 278: 24509–20

Valvano, M.A., Messner, P., and Kosma, P. 2002. Novel pathways for biosynthesis of nucleotide-activated glycero-manno-heptose precursors of bacterial glycoproteins and cell surface polysaccharides. Microbiology 148: 1979–1989.

Vann, W.F., and Freese, S.J. 1994. Purification of *Escherichia coli* K antigens. Methods Enzymol. 235: 304–311.

Whitfield, C., and Roberts, I.S. 1999. Structure, assembly and regulation of expression of capsules in *Escherichia coli*. Mol. Microbiol. 31: 1307–1319.

Yuki, N. 1997. Molecular mimicry between gangliosides and lipopolysaccharides of *Campylobacter jejuni* isolated from patients with Guillain–Barré syndrome and Miller Fisher syndrome. J. Infect. Dis. 176 Suppl 2: S150–153.

Chapter 13

Protein Glycosylation in *Campylobacter*

Christine M. Szymanski, Scarlett Goon, Brenda Allan and
Patricia Guerry

Abstract

Glycoproteins are ubiquitous in eukaryotes where it is estimated that more than half of all proteins are glycosylated (Apweiler *et al.*, 1999). The significance of co-translational modification of proteins with sugars is well known. Both *N*- and *O*-linked sugar modifications influence multiple biological processes and changes in their biosynthetic pathways have been implicated in several diseases. It is now known that bacteria also glycosylate their proteins. Many *O*-linked systems have been described including flagellar modification with pseudaminic acid derivatives in *Campylobacter* and *Helicobacter* organisms. Recently, the first report of an *N*-linked bacterial protein glycosylation pathway has been described in *Campylobacter jejuni*. An overview of the literature describing these two protein glycosylation pathways in campylobacters is presented. Further studies describing the commonality and significance of bacterial protein glycosylation is also discussed.

INTRODUCTION

Protein glycosylation is a common mechanism of modifying proteins in eukaryotes. Glycans can be attached to proteins through the hydroxyl groups of serines or threonines to form an *O*-linkage. Alternatively, glycans can be attached to amino groups of asparagines in the conserved sequon, Asn-X-Ser/Thr, where X is any amino acid except for proline, to form an *N*-linkage. It is now well recognized that bacteria also glycosylate their proteins (for reviews see Benz and Schmidt, 2002; Messner and Schäffer, 2003; Power and Jennings, 2003). Campylobacters are unique in that they possess both *N*- and *O*-linked protein glycosylation pathways. This is not surprising since genome sequencing of *C. jejuni* NCTC 11168 revealed a small densely packed genome largely devoted to carbohydrate biosynthesis (Parkhill *et al.*, 2000). The presence of two characterized glycosylation pathways makes campylobacters an ideal model system for understanding bacterial protein glycosylation and for comparative studies with the respective eukaryotic systems.

N-LINKED GLYCOSYLATION OF *CAMPYLOBACTER* PROTEINS

A novel cluster of carbohydrate biosynthetic genes was identified in *C. jejuni* 81-176 and first established to play a role in protein glycosylation in the Guerry laboratory (Szymanski *et al.*, 1999). Preliminary observations demonstrated that mutagenesis of the genes from this locus altered the reactivity of multiple *C. jejuni* proteins with both rabbit and human immune sera (Szymanski *et al.*, 1999). Chemical deglycosylation of various protein fractions gave similar results suggesting that the immunoreactive epitopes were carbohydrates on glycoproteins. The genes from this locus were therefore named *pgl* for protein glycosylation (*pglA–pglG*) and are summarized in Figure 13.1. Studies by Linton *et al.* (2002) extended the locus (*pglH–pglJ*) and identified two glycoproteins, PEB3 (major antigenic protein) and CgpA (*Campylobacter* glycoprotein A) that were modified by this pathway. They further showed that these and other unidentified glycoproteins bound to the GalNAc-specific lectin, soybean agglutinin (SBA). Subsequent work by Young *et al.* (2002) identified nearly 30 *C. jejuni* glycoproteins (Table 13.1), including PEB3 and CgpA, and provided the first demonstration that the glycan was linked to proteins through asparagine residues in the same sequon used by eukaryotes. The complete structure of the *N*-linked glycan was determined to be a heptasaccharide with the structure: GalNAc-α1,4-GalNAc-α1,4-[Glcβ1,3-]GalNAc-α1,4-GalNAc-α1,4-GalNAc-α1,3-Bac-β1,*N*-Asn-Xaa, where Bac is bacillosamine, 2,4-diacetamido-2,4,6-trideoxyglucose (Figure 13.2, Young *et al.*, 2002). Wacker *et al.* (2002) confirmed the heptasaccharide structure in both PEB3 and an additional glycoprotein, AcrA (putative multidrug efflux pump subunit). They also provided the first demonstration that the *Campylobacter pgl* gene cluster alone can function in *E. coli* to glycosylate recombinant PEB3 and AcrA. These findings were recently reviewed by Szymanski *et al.* (2003a). Subsequently, it was demonstrated for the first time that *N*-linked glycans could be readily detected directly from intact bacterial cells by high-resolution magic angle spinning NMR (HR-MAS NMR) (Szymanski *et al.*, 2003b). HR-MAS NMR has thus become a powerful tool to examine *N*-linked glycan conservation and assembly. It was using this method that the conservation of the

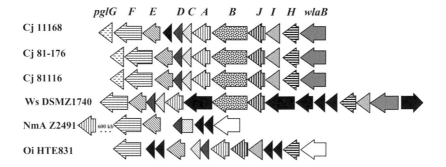

Figure 13.1 The commonality of *pgl* gene homologues in bacteria. The *pgl* genes are conserved in all campylobacters examined including *C. jejuni* NCTC 11168 (NC002163, Cj1119c-Cj1130c), 81-176 (AF108897), 81116 (Y11648) and *C. coli* VC167 (Guerry, unpublished). Recently, genome sequencing identified a similar locus in *Wolinella succinogenes* DSMZ 1740 (NC005090, WS0036-WS0054) that consisted of all *pgl* gene homologues, including *pglB*, required for *N*-linked glycosylation. Note that the *pglG* homologue is located at another position on the chromosome of *W. succinogenes*. The *pglB* homologue, is missing from both the *Neisseria meningitidis* group A Z2491 (NC003116, NMA0048–NMA0643) and *Oceanobacillus iheyensis* HTE831 (NC004193, OB2896-OB2885) clusters. *Bacillus* species including *B. subtilis* (NC000964), *B. fragilis* (NC003228) and *B. halodurans* (NC002570) also contain *pgl* gene homologues similar to those shown for *O. iheyensis* (see Table 13.2). In addition, all clusters exhibit a transporter homologue, *wlaB*, adjacent to the sugar biosynthetic genes that may be involved in the movement of the sugars across the membrane. The putative transporters (grey arrows) show homology to ABC transporters in the bacteria synthesizing *N*-linked glycans while the putative transporters in the other clusters (white arrows) show homology to O-antigen flippases. Black arrows represent gene insertions. The proposed function of the encoded Pgl enzymes, levels of homology and additional homologues are summarized in Table 13.2.

Table 13.1 Putative *N*-linked glycoproteins in *Campylobacter jejuni.*[a]

Cj no.	Predicted function
0114[b]	**Probable periplasmic protein**
0143c	Periplasmic solute binding protein for ABC transport system
0175c	Putative iron uptake ABC transporter/ periplasmic iron binding protein[c]
0200c	**Probable periplasmic protein**
0289c	**Major antigenic peptide (PEB3)**[d]
0367c	Putative membrane fusion component of efflux system (AcrA)/lipoprotein[e]
0376	Probable periplasmic protein
0415	Putative oxidoreductase subunit
0420	Probable periplasmic protein
0493	Translation elongation factor EF-G (FusA)
0511	Probable secreted proteinase
0694	Probable periplasmic protein
0779	Thioredoxin peroxidase (Tpx)
0835c	Aconitate hydratase (AcnB)
0843c	Putative secreted transglycosylase
0906c	Probable periplasmic protein
0944c	Probable periplasmic protein
0998c	Probable periplasmic protein
1018c	Periplasmic branched-chain amino acid ABC transporter (LivK)
1032	Probable membrane fusion component of efflux system
1214c	Hypothetical protein
1221	60-kDa chaperonin (Cpn60; GroEL)
1345c	Probable periplasmic protein
1444c	Putative capsule polysaccharide exporter periplasmic protein (KpsD)
1496c	Probable periplasmic protein
1565c	Paralysed flagellum protein (PflA)
1643	Putative periplasmic protein
1670c	**Probable periplasmic protein (CgpA)**[f]

[a]Containing the sequon: Asn-Xaa-Ser/Thr and reacting with SBA and/or showing changes in *pgl* mutants. These proteins were identified by Young *et al.* (2002) unless otherwise noted.
[b]Confirmed glycoproteins are shown in bold.
[c]Previously observed by Szymanski and Guerry (unpublished).
[d]Also described by Linton *et al.* (2000) and Wacker *et al.* (2002).
[e]Identified by Wacker *et al.* (2002).
[f]Also identified by Linton *et al.* (2002).

N-linked glycan structure was demonstrated in other strains of *C. jejuni* as well as the related species, *C. coli* (Szymanski *et al.*, 2003b; Szymanski *et al.*, 2003c). These findings support early analysis by Southern blots (Szymanski *et al.*, 1999) and recent microarray analyses (Dorrell *et al.*, 2001; Leonard *et al.*, 2003), demonstrating that the *pgl* locus is present in multiple isolates. In addition, gene sequence comparisons have demonstrated that the locus is highly conserved and lacking contingency genes that are found in all other *Campylobacter* surface exposed glycan structures such as capsule, lipooligosaccharide and flagellin (Parkhill *et al.*, 2000).

Commonality in sugar modifications

Amino and deoxy sugars such as pseudaminic acid and bacillosamine are unusual sugars that are commonly used by bacteria in their glycoproteins, lipopolysaccharides and capsules. The full characterization of the *Campylobacter O*- and *N*-linked glycosylation pathways should provide a useful model for deciphering the function of the enzyme

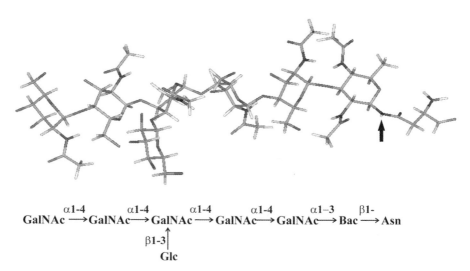

Figure 13.2 *C. jejuni* N-linked glycan assembled by the Pgl enzymes. A model of the heptasaccharide N-linked (arrow) to asparagine is shown with the corresponding structure below.

homologues involved in similar pathways in other bacteria. It was recently demonstrated that the N-linked glycan in *C. jejuni* is attached to multiple glycoproteins through 2,4-diacetamido-2,4,6-trideoxyglucose. The O-linked glycan in *Neisseria meningitidis* can also be attached to pilin with a similar, if not identical, diamino sugar: 2,4-diacetamido-2,4,6-trideoxyhexose (Stimson *et al.*, 1995). Therefore, it is not surprising that *N. meningitidis* contains several *pgl* gene homologues. One of the first diamino sugars ever isolated was N-acetylbacillosamine from *Bacillus subtilis* (Sharon and Jeanloz, 1960) and several *Bacillus* species also contain *pgl* gene homologues that resemble those found in *Campylobacter* and *Neisseria* (Table 13.2). However, it is interesting that the *C. jejuni* NCTC 11168 *pgl* gene cluster is more closely related to the *pgl* cluster found in the extremely halotolerant and alkaliphilic deep-sea organism, *Oceanobacillus iheyensis* rather than to the neisserial *pgl* cluster (Figure 13.1 and Table 13.2). It is also interesting to note that a large majority of these gene homologues are clustered together and show co-linearity in all three classes of bacteria suggesting that this pathway has been acquired by horizontal gene transfer. The dehydratases, aminotransferases and acetyltransferases are required for the formation of both the pseudaminic and bacillosamine type sugars and therefore show homologues in organisms that synthesize either sugar. The *pglB* homologue, that is required for N-linked glycosylation, is missing from both the *Neisseria* and *Bacillus* clusters. This would be expected in *Neisseria* since the pilin glycan is O-linked rather than N-linked. The structures modified by the *pgl* locus in the *Bacillus* species remain to be determined. However, until recently, only homologues of the sugar biosynthetic pathway have been identified, leaving *Campylobacter* organisms as the only bacteria capable of attaching these sugars through an N-linkage to the protein (i.e. containing a *pglB* homologue). As additional genome sequencing projects are completed, more homologues of the Pgl pathway are being identified. This includes two other proteobacteria, *Wolinella succinogenes* (Baar *et al.*, 2003) and *Desulfovibrio desulfuricans* (Szymanski *et al.*, 2003a), which have been demonstrated to contain *pglB* homologues and thus also have the potential for N-linked glycosylation (see Table 13.2 and below). These observations and the fact that other

Table 13.2 Homologues of the *C. jejuni* NCTC 11168 *pgl* genes.

Cj[1]	Nm	Bf	Bs	Bh	Oc	Ws	Dd	Other homologues	Putative function
pglG						52*		*Vibrio acfB*/signal transducing protein (3)*	Unknown
pglF	*pglD* 74	36	11	82	84	139	96	*Clostridium* (92) *Yersinia* (82)	Dehydratase for Bac[2]
pglE *wlaJ*	*pglC* 22	74	68		79	121	85	Multiple homologues Unique to *Campylobacter*	AminoTF[3] for Bac Unknown
pglD	*pglB* 27	16	19		29	30	22	*Escherichia neuD* (19)	acetylTF for Bac
pglC	*pglB* 53	63	61	28	62	48	52	Undecaprenyl-P- glycosyl-1-P- TFs	Bac attachment to lipid
pglA	*pglA* 34	6	7	10	31	72	25	*Coxiella* glycosylTF (35)	GalNAcα1–3 TF
pglB						103	17	*Methanobacterium* (14) STT3 Archaea/eukaryotes	N-linked oligoTF
pglJ		6		29	28	63	18	*Campylobacter pglH* (20)	GalNAcα1–4 TF
pglI	*lgtA* 7	7	8	16	11	7	16	*Streptococcus* β1-3-Glc-TF CpsVII (17)	Glcβ1–3 TF
pglH	5	7	5	8	13	39	11	*Campylobacter pglJ* (23)	GalNAcα1–4 TF

[1]Cj, *Campylobacter jejuni*; Nm, *Neisseria meningitidis*; Bf, *Bacteroides fragilis*; Bs = *Bacillus subtilis*; Bh, *Bacillus halodurans*; Oc, *Oceanobacillus iheyensis*; Ws, *Wolinella succinogenes*; Dd, *Desulfovibrio desulfuricans*

[2]Bac, bacillosamine.

[3]TF, transferase.

*BlastX E-value (10⁻⁸) indicated numerically.

organisms, such as *Streptococcus sanguis* which has been reported to also produce an *N*-linked glycoprotein (Erickson and Herzberg, 1993), suggest that *N*-linked glycosylation will become a more common method of bacterial protein modification than was previously believed.

N-linked biosynthetic pathway

The key enzyme involved in *N*-linked transfer of the oligosaccharides to protein is PglB. This enzyme belongs to the STT3 family of oligosaccharyltransferases commonly found in eukaryotes and archaea. STT3 homologues are large proteins (700–970 amino acids) with 10–12 transmembrane domains and a conserved C-terminal catalytic domain (WWDYG) shown to be essential for activity (Yan and Lennarz, 2002). The *Campylobacter* PglB was the first bacterial homologue to be identified (Szymanski *et al.*, 1999). In *Campylobacter*, substitution of tryptophan 458 and aspartic acid 459 residues with alanine (W458A, D459A) in the [457]WWDYG[462] domain of PglB caused loss of protein glycosylation demonstrating that this domain is conserved and essential not only in eukaryotes and archaea, but also in bacteria (Wacker *et al.*, 2002). PglB homologues have now been identified in the genomes of the soil and human intestinal sulphate-reducing bacterium, *D. desulfuricans* (Szymanski *et al.*, 2003a) and in the non-pathogenic host-associated bacterium, *W. succinogenes* (Baar *et al.*, 2003). The bacterial STT3 homologues are smaller, yet all contain the essential domain: *C. jejuni* (713 amino acids, WWDYG); *W. succinogenes* (707 amino acids, WWDYG) and *D. desulfuricans* (704 amino acids, WWDWG).

The additional Pgl enzymes are involved in sugar biosynthesis and assembly and their putative functions are listed in Table 13.2. A detailed description of the *N*-linked biosynthetic pathway has been recently described (Szymanski *et al.*, 2003a). The definitive role of each enzyme is currently under investigation (Amber *et al.*, 2003; Bernatchez *et al.*, 2003).

Biological role of *N*-linked glycosylation in *Campylobacter*

Immune response during human infection

We previously demonstrated that the glycan plays a large role in immunity during infection by masking the primary amino acid epitopes (Szymanski *et al.*, 1999). The glycoproteins affected by *pgl* mutation demonstrated changes in antigenicity with hyperimmune rabbit antiserum. In addition, these glycosyl epitopes were reported to be recognized during human enteric infection. An immunoblot of total membrane proteins of wild-type 81-176, *pglB*, *pglE* and the complement, *pglE* (pCS101) is shown in Figure 13.3. The membrane proteins were detected with HS:23 typing serum (Figure 13.3B) or with serum from a human volunteer who had been experimentally infected with 81-176 (Figure 13.3C). This individual developed diarrhoea and was subsequently shown to be immune to rechallenge (Tribble, unpublished). Immunodetection with HS:23 typing serum showed a result similar to that previously described with HS:36 serum with these mutants (Szymanski *et al.*, 1999). Thus, numerous proteins that are immunoreactive in wild-type 81-176, particularly at apparent M_r 15–61 kDa, lose immunoreactivity in both mutants. Immunodetection with antiserum from the human volunteer showed a more extensive loss of reactivity including proteins in the M_r range of 15–26 kDa. Immunoreactivity was restored when *pglE* was complemented *in trans* (Figure 13.3B and C, lane 4).

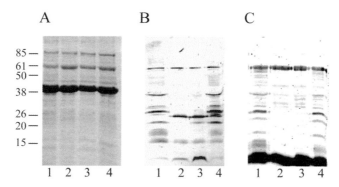

Figure 13.3 Western blot comparison of rabbit and human sera reactivity to *C. jejuni* 81-176 and *pgl* mutant membrane glycoproteins. (A) Coomassie stain of total membrane proteins separated on 12.5% SDS-PAGE to demonstrate similarities in protein profiles. (B and C) Immunoblots of the same preparations detected with HS:23 serotyping serum and serum from the human feeding study, respectively. Lane 1, 81-176 wild-type; lane 2, *pglB*; lane 3, *pglE* and lane 4, *pglE* (pCS101). Molecular masses are shown in kDa.

Poultry colonization

N-linked glycosylation clearly also plays an important role in *C. jejuni*–host cell interactions. Mutants in the *pgl* pathway from *C. jejuni* 81-176 show reduced levels of adherence and invasion in INT407 cells and colonization of the intestinal tracts of mice (Szymanski *et al.*, 2002). The influence of protein glycosylation on the ability of *C. jejuni* NCTC 11168 to colonize poultry was tested in a day-old chick model (Carrillo *et al.*, 2004) and is shown in Table 13.3. Leghorn chicks were randomly assigned into groups of 24 or 25 birds and provided with feed and water *ad libitum*. All birds were orally challenged with 10^9 CFU of *C. jejuni* wild-type and mutants in a 0.5 mL volume of normal saline. Colonization of the birds was monitored by culturing cloacal swabs on Karmali medium and growing under microaerophilic conditions. Birds were maintained for seven days after challenge and were then euthanized and the caeca cultured. Inactivation of either *pglB* (Young *et al.*, 2002) or *pglH* (constructed with pEAp27 containing Cj1127c–Cj1133c with the Km resistance cassette inserted in a non-polar orientation into the *Mlu*I restriction site of *pglH*) resulted in loss of colonization in chicks (Table 13.3). The demonstration that protein glycosylation is important in colonization of chicks is consistent with the previous observation in mice (Szymanski *et al.*, 2002).

Common role for N-linked protein glycosylation

Since inactivation of the Pgl pathway affects so many proteins (Table 13.2), one would expect multiple pleiotropic effects. However, the conservation of the *N*-linked pathway in *C. jejuni* and *C. coli* and the multitude of functionally diverse secreted proteins modified by this pathway suggest a common role for this modification in *Campylobacter*. In eukaryotes, glycoproteins are required to exhibit *N*-linked saccharides of minimum size for proper folding, secretion and function (Imperiali and O'Connor, 1999; Helenius and Aebi, 2001; Dempski, Jr and Imperiali, 2002). However, in *C. jejuni* proteins such as PEB3 have been isolated with and without the modification in approximately equal amounts (Young *et al.*, 2002). It remains to be determined whether the unmodified proteins are also active or non-functional intermediates. Also, not all glycoprotein sequons in *C*.

Szymanski *et al.*

Table 13.3 Chick colonization studies with *C. jejuni* NCTC 11168 and *pgl* mutants.

Inoculum	1 day[1] swab	3 day swab	5 day swab	7 day swab	7 day caecal culture
PBS[2]	0/25[3]	0/25	0/25	0/25	0/25
Wild type	17/24	15/24	17/24	19/24	24/24
pglB–	1/24	0/24	0/24	0/24	0/24
pglH–	3/24	0/24	0/24	0/24	0/24

[1]Days post challenge.
[2]Phosphate-buffered saline control.
[3]Number of birds positive for *C. jejuni*.

jejuni are modified suggesting that glycosylation may be influenced by protein folding and conformation in the periplasm possibly requiring a similar Asx-turn as that described for eukaryotes (Imperiali and Hendrickson, 1995). Furthermore, not all *C. jejuni* sequon containing proteins are glycosylated suggesting that proteins destined to be glycosylated may be separately directed towards the complex. In contrast to *Mycobacterium tuberculosis* (Herrmann *et al.*, 1996), the *C. jejuni* *N*-linked glycan does not appear to be protective against proteolytic cleavage since AcrA can be cleaved by trypsin at the residue adjacent to the modified asparagine (Wacker *et al.*, 2002). However, proteolytic protection under native protein conditions has not yet been examined. In addition, the glycan structure is not variable as observed with other *Campylobacter* surface polysaccharides or with the variable pilin terminal galactose affected by *N. meningitidis pglA* and the corresponding *N. gonorrhoeae pgtA* (Jennings *et al.*, 1998; Banerjee *et al.*, 2002). But, it is interesting to note that there is variability in the number of heptasaccharides attached to each individual protein and in the state of modification in *Campylobacter*. The ability to vary protein glycosylation has also been described for *M. tuberculosis* where the extent of protein glycosylation has been shown to modulate immune responses (Horn *et al.*, 1999; Romain *et al.*, 1999). The role of the *N*-linked glycan in the *C. jejuni* protein glycosylation pathway awaits further elucidation.

O-LINKED GLYCOSYLATION OF *CAMPYLOBACTER* FLAGELLIN
There have been indications for quite some time that *Campylobacter* flagellins are posttranslationally modified. *N*-terminal amino acid sequencing from internal peptides of flagellins from *C. jejuni* strain VC74 (Logan and Trust, unpublished) and *C. coli* VC167 (Logan *et al.*, 1989) revealed the presence of numerous blocked residues. Subsequent DNA sequence analysis of flagellin genes revealed that the blocked amino acids were all serine residues (Logan *et al.*, 1989). Alm *et al.* (1992) observed that the apparent mass of FlaA from *C. coli* VC167 changed when it was expressed in different strains of *Campylobacter*, again suggesting not only posttranslational modification but also variability in modifications among different strains. Doig *et al.* (1996) provided the first chemical evidence for glycosylation by demonstration that purified flagellins were reactive with periodate.

Chemical analyses of the glycans on flagellin
Subsequent and more recent direct chemical analyses of *Campylobacter* flagellins have confirmed that these proteins are heavily glycosylated with *O*-linked glycans. In the three *Campylobacter* strains studied thus far, *C. jejuni* NCTC 11168, *C. jejuni* 81-176 and *C. coli* VC167 T2, the carbohydrate moieties contribute approximately 6000 Da

to the mass of each flagellin, or approximately 10% of the total mass (Thibault *et al.*, 2001). The modifications were originally thought to include a sialic acid (Neu5Ac) residue based on lectin-binding activity with the sialic acid-specific lectin *Limax flavus* agglutinin (Doig *et al.*, 1996) and the similarity of some of the genes involved in flagellin glycosylation to sialic acid biosynthetic genes (Guerry *et al.*, 1996). However, further analysis by mass spectrometry and nuclear magnetic resonance spectroscopy showed that the major modification on two of the *Campylobacter* strains, 81-176 and VC167, is 5,7-diacetamido-3,5,7,9-tetradeoxy-L-*glycero*-L-*manno*-nonulosonic acid (or pseudaminic acid, Pse5Ac7Ac), a nine-carbon sugar that bears great resemblance to sialic acid (Figure 13.4) (Thibault *et al.*, 2001; Logan *et al.*, 2002). In addition to Pse5Ac7Ac, Figure 13.5 shows the structures of additional glycosyl residues found on the flagellin of 81-176: an acetamidino form of pseudaminic acid (PseAm), an acetylated version (Pse5Ac7AcOAc), and one which bears dihydroxyproprionyl groups (Pse5Pr7Pr).

Similar to 81-176, flagellin from *C. coli* VC167 is also glycosylated with Pse5Ac7Ac and an acetamidino form of pseudaminic acid. However, this acetamidino modification is structurally and immunologically distinct from that found on 81-176 (Logan *et al.*, 2002). Confirmation of the acetamidino structure present on 81-176 and VC167 has been difficult due to its minor presence on flagellin. However, the difference between the two forms is believed to be the location of the acetamidino group on either the C-5 or C-7 position of pseudaminic acid (Figure 13.5). For 81-176, it appears that the acetamidino group is present on the C-5 position (Pse5Am7Ac), whereas for VC167, it is present at the C-7 position (Pse5Ac7Am). In addition to the acetamidino modification, VC167 flagellin is also glycosylated with two groups of mass 432 Da and 431 Da (Logan *et al.*, 2002). Although further confirmation is necessary, these two masses may correspond to Pse5Ac7Ac and PseAm, respectively, extended by a deoxypentose residue. A summary of the different modifications present on the two strains of *Campylobacter* is summarized in Table 13.4.

Biosynthesis of pseudaminic acids

Although pseudaminic acid has been reported in several bacteria, including *Pseudomonas aeruginosa*, *Shigella boydii* and *Sinorhizobium fredii*, there are few data on its biosynthesis (Gil-Serrano *et al.*, 1999; Rocchetta *et al.*, 1999; Luneberg *et al.*, 2000). In the case of *Campylobacter*, most of the genes known to affect flagella glycosylation map near the flagellin structural genes. A comparison of this region in 81-176, VC167 and the genome strain NCTC 11168 indicated that there are similarities that likely lead to the biosynthesis of Pse5Ac7Ac, as well as major genetic differences that reflect divergent pathways of PseAm biosynthesis, as well as strain-specific variations (Logan *et al.*, 2002). The most striking genetic feature among these strains is the difference in complexity. The genome

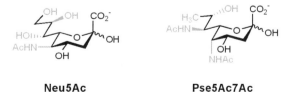

Neu5Ac   **Pse5Ac7Ac**

Figure 13.4 The structures of Neu5Ac and Pse5Ac7Ac. Structural differences between the two residues are indicated in grey.

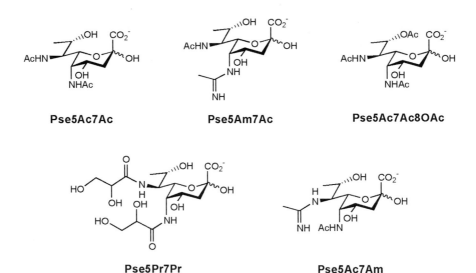

Figure 13.5 The five forms of pseudaminic acid residues found on *C. jejuni* 81-176 and/or *C. coli* VC167.

Table 13.4 Flagellar modifications in *C. jejuni* 81-176 and *C. coli* VC167

Mass/structure	81-176	VC167
315[1]/PseAm[2]	+	+
316/Pse5Ac7Ac	+	+
358/Pse5Ac7AcOAc	+	−
408/Pse5Pr7Pr	+	−
430/PseAm2 + deoxypentose	−	+
431/Pse5Ac7Ac + deoxypentose	−	+

[1]Mass in Da.
[2]Structure is currently unconfirmed, but believed to be Pse5Ac7Am for 81-176 and Pse5Am7Ac for VC167.

strain NCTC 11168 is the most complex, while 81-176 is missing 19 genes found in NCTC 11168 and VC167 is intermediate in complexity between 81-176 and NCTC 11168 (Guerry *et al.*, 1996; Parkhill *et al.*, 2000; Thibault *et al.*, 2001; Logan *et al.*, 2002). Based on mutational analysis of potential genes involved in flagellin glycosylation in the two simpler systems, 81-176 and VC167, putative pathways for the biosynthesis of Pse5Ac7Ac and PseAm have been proposed and are summarized in Figure 13.6.

Campylobacter strains 81-176 and VC167 appear to use a common pathway in the biosynthesis of Pse5Ac7Ac, the major modification found on flagellin (Figure 13.6). Biosynthesis of this sugar begins with UDP-GlcNAc and involves the gene Cj1293/*pseB*. The predicted 81-176 PseB protein is 63% identical and 76% similar to the *Helicobacter pylori* HP0840 protein, over 328 amino acids. This *H. pylori* protein has been shown to be a C_6-dehydratase/C_4-reductase that converts UDP-GlcNAc to UDP-QuiNAc via a UDP-4-keto-6-methyl-GlcNAc intermediate (Creuzenet *et al.*, 2000). Schirm *et al.* (2003) recently showed that flagellins of *H. pylori* are also glycosylated, although a role for HP0840 in glycosylation of *Helicobacter* flagellins has not been reported. Functional enzyme assay analysis of the Cj1293 product showed similar activity to HP0840 (Goon, unpublished). The conversion of UDP-QuiNAc to Pse5Ac7Ac is less well understood (see Figure 13.6).

However, other genes that appear to be involved in Pse5Ac7Ac biosynthesis include the so-called 'third set' of Neu5Ac genes, Cj1317/*neuB3* and Cj1311/*neuA3*. Based on sequence homology to Neu5Ac enzymes, it is hypothesized that Cj1317 condenses an altrose precursor with PEP to give Pse5Ac7Ac and Cj1311 likely converts Pse5Ac7Ac to the donor substrate CMP-Pse5Ac7Ac.

Although the pathway to Pse5Ac7Ac appears conserved, the pathways for synthesis of PseAm in 81-176 and VC167 are completely distinct, reflecting the structural differences between the forms of PseAm found in these two strains. In 81-176, mutation of Cj1316 or *pseA* resulted in flagellins that were exclusively glycosylated with Pse5Ac7Ac, suggesting that PseAm is likely synthesized from CMP-Pse5Ac7Ac via the gene product of *pseA* (Thibault *et al.*, 2001). The *pseA* gene in VC167 appears to be a pseudogene, and mutation of *pseA* in VC167 did not alter flagellin modifications. Instead, in VC167, PseAm is synthesized via a pathway that includes the genes annotated as the 'second set' of sialic acid genes in NCTC 11168 (Parkhill *et al.*, 2000). Since some of these genes were previously identified in VC167 and called *ptmA* and *ptmB* (for posttranslational modification), this nomenclature has been retained (Guerry *et al.*, 1996). These *ptm* genes are found in VC167 and NCTC 11168, but not in 81-176 and are responsible for synthesizing the alternative acetamidino sugar. To date, six genes have been confirmed to be in the *ptm* pathway, *ptmA* through *ptm F* (Guerry *et al.*, 1996; Logan *et al.*, 2002), and two of these resemble sialic acid biosynthetic genes. Flagellins from all the VC167 *ptm* mutants have been shown by

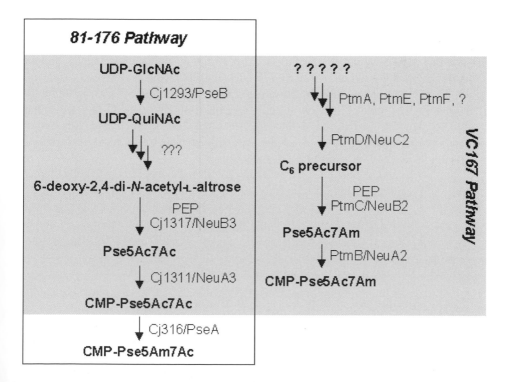

Figure 13.6 Putative biosynthetic pathways for flagellin modifications in 81-176 (shown in the empty box) and VC167 (shaded in grey). The common pathway to Pse5Ac7Ac is called the Pse pathway (shown in the empty box and shaded in grey). The independent pathway to Pse5Ac7Am in VC167 is called the Ptm pathway.

mass spectrometry to have lost all PseAm residues and the corresponding Ser/Thr residues are instead modified with Pse5Ac7Ac (Logan *et al.*, 2002).

There are also some interesting strain-specific differences. Thus, flagellins from all six *ptm* mutants of VC167 have lost the glycan of mass 316 Da, which is the PseAm residue as well as the glycan of mass 431 Da. Although the structure bearing the latter mass has not been determined because of the low levels of this glycan, the mass and the genetics suggest that the 431 Da glycan may be a PseAm sugar further modified with a deoxypentose unit. Similarly, mutation of *pseB*/Cj1293 in VC167 results in loss of both the 315 Da glycan, which is Pse5Ac7Ac, and the 430 Da glycan. This again suggests that the 430 Da glycan is Pse5Ac7Ac extended by a deoxypentose.

It is also interesting that the genome strain, NCTC 11168, has both the *ptm* pathway and *pseA*, and should, therefore, be able to synthesize both forms of PseAm. It would be interesting to determine if the basic modifications are varied in this genetically more complex glycosylation system.

Glycosylation is required for flagella assembly

There are a number of lines of evidence suggesting that glycosylation is essential for assembly of flagella filaments in *C. jejuni*. Mutants in *neuB3*, *neuA3* and *pseB* of several *Campylobacter* strains that encode or probably encode enzymes in the biosynthetic pathway of the modifications, are non-motile (Linton *et al.*, 2000; Thibault *et al.*, 2001; Goon *et al.*, 2003). In addition, mutation of a number of other genes of unknown function, but termed *maf* (for motility associated factor) in this region also result in a non-motile phentoype (Karlyshev *et al.*, 2002).

Comparative genetic analysis of 81-176 and VC167 has, however, provided direct evidence that glycosylation of flagellin is required for assembly of a flagella filament (Goon *et al.*, 2003). A *pseB* mutant in 81-176 was non-motile and synthesized a truncated flagella filament consistent with the absence of FlaA (Goon *et al.*, 2003). Examination of whole cells by immunoblot with an anti-flagellin antibody revealed that the mutant accumulated flagellin of a reduced M_r that was consistent with the absence of carbohydrate modifications. Furthermore, analysis of the intracellular flagellin by mass spectrometry confirmed it lacked any carbohydrate modification.

In contrast, a *pseB* mutant in VC167 was fully motile but synthesized a flagella filament composed of flagellins modified exclusively with PseAm and the related modification of mass 431 Da (Logan *et al.*, 2002; Goon *et al.*, 2003). Mutants in any of the *ptm* genes in VC167, as mentioned above, resulted in flagellins modified with only Pse5Ac7Ac and the related 430 Da group. However, a VC167 double mutant in both a *ptm* gene and *pseB*, key genes in each of the pathways shown in Figure 13.6, was non-motile and accumulated small amounts of unglycosylated flagellin intracellularly (Goon *et al.*, 2003).

Biological significance of flagella modifications

The major biological significance of flagellin modifications is the observation, summarized above, that *Campylobacter* flagellins must be glycosylated for assembly into filaments. This is similar to the case of *Caulobacter crescentus*, another polarly flagellated organism (Johnson *et al.*, 1983; LeClerc *et al.*, 1998), although there is no information on the chemical nature of the modifications in this organism. At least for *C. coli* VC167, modification with either Pse5Ac7Ac or PseAm appears sufficient for filament assembly, but filaments composed of flagellins modified exclusively with PseAm are physically different from those with either Pse5Ac7Ac or a mixture of the modifications (Goon *et al.*, 2003). Thus, flagellins from a VC167 *pseB* mutant can be solubilized in 1% SDS, unlike wild-type flagellins or flagellins from VC167 mutants in the *ptm* pathway (Goon *et al.*, 2003). This is the first suggestion that glycosyl modifications contribute to the physical nature of the flagella filament. Additionally, the modifications confer serospecificity on the flagella filament (Logan *et al.*, 2002), and, because of their surface exposure (Power *et al.*, 1994) may affect the interactions of *Campylobacter* organisms with their host.

CONCLUSIONS

A large volume of information has rapidly accumulated over the past 5 years on both the *O*- and *N*-linked protein glycosylation pathways in *Campylobacter*. Not only do we now know that both pathways exist, studies have identified many of the genes involved in protein modification, the sugar structures synthesized by these gene products and the proteins targeted by both pathways. Furthermore, these pathways are similar to the respective pathways in eukaryotes, gene homologues of both pathways exist in other bacteria, and the synthesized sugars share common biosynthetic precursors. Thus, investigation of these biosynthetic pathways may have broader relevance and application. Yet, many unanswered questions still remain. Why does *Campylobacter* dedicate a relatively large amount of its limited genetic repertoire to two pathways of protein glycosylation? Why are these genes so highly conserved through evolution? Where are these enzymes located within the cell? Why are so many proteins glycosylated by the *N*-linked pathway? Are additional proteins modified by either of these pathways? Why are only certain proteins modified? Are proteins modified co- or post-translationally? How are the sites to be modified recognized? What is the general role of *N*-linked glycosylation? And, how are these pathways regulated and related to other cell processes? The discovery of *O*- and *N*-linked glycosylation pathways in *Campylobacter* organisms is both exciting and fruitful. Further studies may provide a better understanding of how *Campylobacter* interacts with the host, may identify potential vaccine targets and may provide novel enzymes for glycoengineering. Following the surge of the 'omics' era, we approach a new time of systems biology, where the understanding of protein–protein interactions and the fine-tuning mechanisms of protein function through protein modification will provide new information on protein trafficking, folding, function and interaction within complex systems. Thus, the study of protein glycosylation in *Campylobacter* provides a unique model for understanding the role and relevance of protein modification by carbohydrates.

Acknowledgements

We would like to thank Erika Aberg and Peter Lau for technical assistance, Jean-Robert Brisson for *N*-glycan modelling and Martin Young for support. Funding to CMS was from the NRC Genomics and Health Initiate, to BA from the Alberta Poultry Producers and OMAF Food Safety Research Program, and to PG from NIAID (1 RO1 AI43559).

References

Alm, R.A., Guerry, P., Power, M.E., and Trust, T.J. 1992. Variation in antigenicity and molecular weight of *Campylobacter coli* VC167 flagellin in different genetic backgrounds. J. Bacteriol. 174: 4230–4238.

Amber, S., Alaimo, C., Feldman, M., Nita-Lazar, M., Wacker, M., Dorrell, N., Linton, D., Wren, B., and Aebi, M. 2003. *N*-linked protein glycosylation in *Campylobacter jejuni*: Analyzing the glycosylation machinery. Int. J. Med. Microbiol. 293: 63.

Apweiler R, Hermjakob H, and Sharon N. 1999. On the frequency of protein glycosylation, as deduced from analysis of the SWISS-PROT database. Biochim. Biophys. Acta 1473: 4–8.

Banerjee, A., Wang, R., Supernavage, S. L., Ghosh, S.K., Parker, J., Ganesh, N.F., Wang, P.G., Gulati, S., and Rice P.A. 2002. Implications of phase variation of a gene (*pgtA*) encoding a pilin galactosyl transferase in gonococcal pathogenesis. J. Exp. Med. 196: 147–162.

Baar, C., Eppinger, M., Raddatz, G., Simon, J., Lanz, C., Klimmek, O., Nandakumar, R., Gross, R., Rosinus, A., Keller, H., Jagtap, P., Linke, B., Meyer, F., Lederer, H., and Schuster, S.C. 2003. Complete genome sequence and analysis of *Wolinella succinogenes*. Proc. Natl. Acad. Sci. U.S.A. 2003 100: 11690–11695.

Benz, I. and Schmidt, M.A. 2002. Never say never again: protein glycosylation in pathogenic bacteria. Mol. Microbiol. 45: 267–276.

Bernatchez, S., Wakarchuk, W.W., and Young, N.M. 2003. The UDP-glucose 4-epimerase (GalE; CJ1131c) of the *pgl* (*wla*) locus in *Campylobacter jejuni* NCTC 11168 is really a UDP-N-acetylglucosaminyl 4-epimerase. Int. J. Med. Microbiol. 293: 129.

Carrillo, C.D., Taboada, E., Nash, J.H., Lanthier, P., Kelly, J., Lau, P.C., Verhulp, R., Mykytczuk, O., Sy, J., Findlay, W.A., Amoaka, K., Gomis, S., Willson, P., Austin, J.W., Potter, A., Babiuk, L., Allan, B., and Syzmanski, C.M. 2004. Genome-wide expression analyses of *Campylobacter jejuni* NCTC 11168 reveals coordinate regulation of motility and virulence by *flhA*. J. Biol. Chem. 279: 20327–20338.

Creuzenet, C., Schur, M.J. Li, J. Wakarchuk, W.W., and Lam, J.S. 2000. FlaA1, a new bifunctional UDP-GlcNAc C_6 dehydratase C_4 reductase from Helicobacter pylori. J. Biol. Chem. 275: 34873–34880.

Dempski, R.E. Jr. and Imperiali, B. 2002. Oligosaccharyl transferase: gatekeeper to the secretory pathway. Curr. Opin. Chem. Biol. 6: 844–850.

Doig, P., Kinsella, N., Guerry, P., and Trust, T.J. 1996. Characterization of a posttranslational modification of *Campylobacter* flagellin: identification of a sero-specific glycosyl moiety. Mol. Microbiol. 19: 379–387.

Dorrell, N., Mangan, J.A., Laing, K.G., Hinds, J., Linton, D., Al-Ghusein, H., Barrell B.G., Parkhill J., Stoker N.G., Karlyshev A.V., Butcher P.D., and Wren B.W. 2001. Whole genome comparison of *Campylobacter jejuni* human isolates using a low-cost microarray reveals extensive genetic diversity. Genome Res. 11: 1706–1715.

Erickson, P.R. and Herzberg, M.C. 1993. Evidence for the covalent linkage of carbohydrate polymers to a glycoprotein from *Streptococcus sanguis*. J. Biol. Chem. 268: 23780–23783.

Gil-Serrano, A.M., Rodriguez-Carvajal, M.A., Tejero-Mateo, P., Espartero, J.L., Mendendez, M., Corzo, J., Ruiz-Sainz, J.E., and Buendia-Claveria, A.M. 1999. Structural determination of a 5-acetamido-3,5,7,9-tetradeoxy-7-(3-hydroxybutyramido)-L-glycero-L-manno-nonulosonic acid-containing homopolysaccharide isolated from *Sinorhizobium fredii* HH103. Biochem. J. 342: 527–535.

Goon, S., Kelly, J.F., Logan, S.M., Ewing, C.P., and Guerry, P. 2003. Pseudaminic acid, the major modification on *Campylobacter* flagellin, is synthesized from UDP-GlcNAc via Cj1293. Mol. Microbiol. 50: 659–671.

Guerry, P., Doig, P., Alm, R.A., Burr, D.H., Kinsella, N. and Trust, T.J. 1996. Identification and characterization of genes required for post-translational modification of *Campylobacter coli* VC167 flagellin. Mol. Microbiol. 19: 369–378.

Helenius, A., and Aebi, M. 2001. Intracellular functions of *N*-linked glycans. Science 291: 2364–2369.

Herrmann, J.L., O'Gaora, P., Gallagher, A., Thole, J.E., and Young, D.B. 1996. Bacterial glycoproteins: a link between glycosylation and proteolytic cleavage of a 19kDa antigen from *Mycobacterium tuberculosis*. EMBO J. 15: 3547–3554.

Horn, C., Namane, A., Pescher, P., Riviere, M., Romain, F., Puzo, G., Barzu, O., and Marchal, G. 1999. Decreased capacity of recombinant 45/47-kDa molecules (Apa) of *Mycobacterium tuberculosis* to stimulate T lymphocyte responses related to changes in their mannosylation pattern. J. Biol. Chem. 274: 32023–32030.

Imperiali, B. and Hendrickson, T.L. 1995. Asparagine-linked glycosylation: specificity and function of oligosaccharyl transferase. Bioorg. Med. Chem. 3: 1565–1578.

Imperiali, B. and O'Connor, S.E. 1999. Effect of *N*-linked glycosylation on glycopeptide and glycoprotein structure. Curr. Opin. Chem. Biol. 3: 643–649.

Jennings, M.P., Virji, M., Evans, D., Foster, V., Srikhanta, Y.N., Steeghs, L., van der Ley, P., and Moxon E.R. 1998. Identification of a novel gene involved in pilin glycosylation in *Neisseria meningitidis*. Mol. Microbiol. 29: 975–984.

Johnson, R. C., Ferber, D. M. and Ely, B. 1983. Synthesis and assembly of flagellar components by *Caulobacter crescentus* motility mutants. J. Bacteriol. 154: 1137–1144.

Karlyshev, A.V., Linton, D., Gregson, N.A., and Wren, B.W. 2002. A novel paralogous gene family involved in phase variable flagella-mediated motility in *Campylobacter jejuni*. Microbiology 148: 473–480.

LeClerc, G., Wang, S. P., and Ely, B. 1998. A new class of *Caulobacter crescentus* flagellar genes. J. Bacteriol. 180: 5010–5019.

Leonard, E.E. 2nd, Takata, T., Blaser, M.J., Falkow, S., Tompkins, L.S., and Gaynor, E.C. 2003. Use of an open-reading frame-specific *Campylobacter jejuni* DNA microarray as a new genotyping tool for studying epidemiologically related isolates. J. Infect. Dis. 187: 691–694.

Linton, D., Gilbert, M. Hitchen, P.G., Dell, A., Morris, H.R. Wararchuk, W.W., Gregson, N.A., and Wren, B.W. 2000. Phase variation of a β-1,3 galactosyltransferase involved in generation of the ganglioside GM1-like lipo-oligosaccharide of *Campylobacter jejuni*. Mol. Microbiol. 37: 501–514.

Linton, D., Allan, E., Karlyshev, A.V., Cronshaw, A.D., and Wren, B.W. 2002. Identification of *N*-acetylgalactosamine-containing glycoproteins PEB3 and CgpA in *Campylobacter jejuni*. Mol. Microbiol. 43: 497–508.

Logan, S.M., Trust, T.J., and Guerry, P. 1989. Evidence for posttranslational modification and gene duplication of *Campylobacter* flagellin. J. Bacteriol. 171: 3031–3038.

Logan, S.M., Kelly, J.F., Thibault, P., Ewing, C.P., and Guerry, P. 2002. Structural heterogeneity of carbohydrate modifications affects serospecificity of *Campylobacter* flagellins. Mol. Microbiol. 46: 587–597.

Luneberg, E., Zetzmann, N., Alber, D. Knirel, Y.A., Kooistra, O., Zahringer, U., and Frosch, M. 2000. Cloning and functional characterization of a 30kb gene locus required for lipopolysaccharide biosynthesis in *Legionella pneumophilia*. Int. J. Med. Microbiol. 290: 37–49.

Messner, P. and Schäffer, C. 2003. Prokaryotic glycoproteins. In: Progress in the chemistry of organic natural products. Herz, W. *et al.*, eds. Springer-Verlag, Vienna, pp. 51–124.

Parkhill, J., Wren, B.W., Mungall, K., Ketley, J.M., Churcher, C., Basham, D., Chillingworth, T., *et al.*, 2000. The genome sequence of the food-borne pathogen *Campylobacter jejuni* reveals hypervariable tracts. Nature 403: 665–668.

Power, M. E., Guerry, P., McCubbin, W. D., Kay, C. M. and Trust, T. J. 1994. Structural and antigenic characteristics of *Campylobacter coli* FlaA flagellin. J. Bacteriol. 176: 3303–3313.

Power, P.M. and Jennings, M.P. 2003. The genetics of glycosylation in Gram-negative bacteria. FEMS Microbiol. Lett. 218: 211–222.

Rocchetta, H.L., Burrows, L.L., and Lam, J.S. 1999. Genetics of O-antigen biosynthesis in *Pseudomonas aeruginosa*. Microbiol. Mol. Biol. Rev. 63: 523–553.

Romain, F., Horn, C., Pescher, P., Namane, A., Riviere, M., Puzo, G., Barzu, O., and Marchal, G. 1999. Deglycosylation of the 45/47-kilodalton antigen complex of *Mycobacterium tuberculosis* decreases its capacity to elicit *in vivo* or *in vitro* cellular immune responses. Infect. Immun. 67: 5567–5572.

Schirm, M., Soo, E., Aubry, A., Austin, J., Thibault, P., and Logan, S. 2003. Structural, genetic and functional characterization of the flagellin glycosylation process in *Helicobacter pylori*. Mol. Microbiol. 48: 1579–1592.

Sharon, N. and Jeanloz, R.W. 1960. The diaminohexose component of a polysaccharide isolated from *Bacillus subtilis*. J. Biol. Chem. 235: 1–5.

Stimson, E., Virji, M., Makepeace, K., Dell, A., Morris, H.R., Payne, G., Saunders, J.R., Jennings, M.P., Barker, S., Panico, M., *et al.* Meningococcal pilin: a glycoprotein substituted with digalactosyl 2,4-diacetamido-2,4,6-trideoxyhexose. Mol. Microbiol. 17: 1201–1214.

Szymanski, C.M., Yao, R., Ewing, C.P., Trust, T.J., and Guerry, P. 1999. Evidence for a system of general protein glycosylation in *Campylobacter jejuni*. Mol. Microbiol. 32: 1022–1030.

Szymanski, C.M., Burr, D.H., and Guerry, P. 2002. *Campylobacter* protein glycosylation affects host cell interactions. Infect. Immun. 70: 2242–2244.

Szymanski, C. M., Logan, S.M., Linton, D., and Wren, B.W. 2003a. *Campylobacter* – a tale of two protein glycosylation systems. Trends Microbiol. 11: 233–238.

Szymanski, C.M., Michael, F.S., Jarrell, H.C., Li, J., Gilbert, M., Larocque, S., Vinogradov, E., and Brisson, J.R. 2003b. Detection of conserved *N*-linked glycans and phase-variable lipooligosaccharides and capsules from *Campylobacter* cells by mass spectrometry and high-resolution magic angle spinning NMR spectroscopy. J. Biol. Chem. 278: 24509–24520.

Szymanski, C., Cadotte, N., Lau, P., Aberg, E., Watson, D., Brisson, J.-R., and Jarrell, H. 2003c. Characterization of the *N*-linked protein glycosylation pathway in *Campylobacter* using high-resolution magic angle spinning NMR. Int. J. Med. Micro. 293: 79.

Thibault, P., Logan, S.M., Kelly, J.F., Brisson, J-R., Trust, T.J., and Guerry, P. 2001. Identification of the carbohydrate moieties and glycosylation motifs in *Campylobacter jejuni* flagellin. J. Biol. Chem. 276: 34862–34870.

Wacker, M., Linton, D., Hitchen, P.G., Nita-Lazar, M., Haslam, S.M., North, S.J., Panico, M., Morris, H.R., Dell, A., Wren, B.W., and Aebi, M. 2002. *N*-linked glycosylation in *Campylobacter jejuni* and its functional transfer into *E. coli*. Science 298: 1790–1793. Yan, Q. and Lennarz, W.J. 2002. Studies on the function of oligosaccharyl transferase subunits. Stt3p is directly involved in the glycosylation process. J. Biol. Chem. 277: 47692–47700.

Young, N.M., Brisson, J.R., Kelly, J., Watson, D.C., Tessier, L., Lanthier, P.H., Jarrell, H.C., Cadotte, N., St. Michael, F., Aberg, E., and Szymanski, C.M. 2002. Structure of the *N*-linked glycan present on multiple glycoproteins in the Gram-negative bacterium, *Campylobacter jejuni*. J. Biol. Chem. 277: 42530–42539.

Chapter 14

Metabolism, Electron Transport and Bioenergetics of *Campylobacter jejuni*: Implications for Understanding Life in the Gut and Survival in the Environment

David J. Kelly

Abstract

Understanding the ability of *Campylobacter jejuni* to survive in the food chain and the environment, but to be a commensal in the avian gut and a pathogen in humans, will require a detailed knowledge of its metabolism and electron transport. This chapter focuses on aspects of carbon, nitrogen and electron flow which are relevant to developing such knowledge. It appears that *C. jejuni* is unable to use hexose sugars as carbon sources because of the absence of the key glycolytic enzyme 6-phosphofructokinase, yet the presence of a typical catabolic pyruvate kinase suggests some catabolic role for the lower part of the Embden–Meyerhof pathway, in addition to a major function in gluconeogenesis. The major carbon and nitrogen sources likely to be used by *C. jejuni in vivo* are amino acids, and it possesses several enzymes for the amino acid deamination. Serine catabolism is especially significant. A major prediction from the genome sequence is an unexpected complexity in the electron transport chains of *C. jejuni*, with a wide variety of electron donors and alternative electron acceptors to oxygen capable of being utilized. This underlines a hitherto unappreciated metabolic versatility in this bacterium, which may contribute to its ability to occupy diverse niches. Nevertheless, *C. jejuni* appears to be unable to grow under strictly anaerobic conditions, due to the use of an oxygen-dependent ribonucleotide reductase for DNA synthesis.

INTRODUCTION: THE NEED TO UNDERSTAND METABOLISM IN RELATION TO GROWTH, SURVIVAL, COLONISATION AND PATHOGENICITY

The relentless increase in the incidence of gastroenteritis in recent years due to enteric campylobacters, particularly *Campylobacter jejuni*, has re-focused attention on the physiological and metabolic properties of this group of bacteria. Despite the importance of *C. jejuni* as a food-borne pathogen, its physiology has not been studied in as much detail as other enteric pathogens. It can be a difficult organism to culture, has complex nutritional requirements and is microaerophilic, and this combination of features has undoubtedly hindered progress in elucidating many, rather basic, aspects of its physiology and metabolism. Nevertheless, progress in such studies is crucial if we are to understand growth of the organism in animal and human hosts, what carbon sources it is using *in vivo*, its survival in the environment and the food chain, and its mechanisms of pathogenicity. Physiological studies may also give leads in identifying potential targets for controlling or preventing the growth of *C. jejuni*.

The determination of the complete genome sequence of strain 11168 (Parkhill *et al.*, 2000) has provided some major new insights into the biology of *C. jejuni*, not least in terms of its metabolism, but it is important that predictions from the genome sequence are tested experimentally. It should now be possible to discover which metabolic pathways or enzyme activities are essential in the different environments that *C. jejuni* is capable of surviving or growing in and this will give us insights into the strategies it uses for successful colonization. While global transcriptomic and proteomic approaches are crucial, so too is an appreciation of the *enzymatic functions* of many of the gene products which will be subject to changes in abundance under different growth conditions. For a small-genome pathogen with apparently relatively few complex regulatory circuits, a fascinating challenge is to understand how *C. jejuni* can be a commensal in the avian gastrointestinal (GI) tract, a pathogen in the human GI tract and a persistent survivor in the environment and the food chain. In these different situations, the precise types of carbon and nitrogen sources used and the metabolic repertoire of the bacteria may be significantly altered. Yet these crucial changes are very poorly understood at present. In this chapter, I will review current knowledge on some of the metabolic aspects of *C. jejuni* physiology, with emphasis on those features of carbon, nitrogen and electron flow which are likely to be of importance in understanding growth in the environment and *in vivo*.

EXPERIMENTAL APPROACHES TO METABOLIC STUDIES IN *C. JEJUNI*

Growth of *C. jejuni* requires a complex medium, often supplemented with undefined blood products such as serum. In addition, the bacterium is a microaerophile with specific gas atmosphere requirements of both low oxygen and high carbon dioxide. For metabolic studies, these features can sometimes be frustrating, and conventional metabolic and substrate utilization studies (using simple defined media and radioisotopes for example) can be difficult to achieve. Assays of enzyme activity in cell-free extracts are important to perform and generally not a problem with campylobacters, but other approaches can be particularly useful. Nuclear magnetic resonance (NMR) spectroscopy, in particular, can be used to investigate both individual enzyme reactions and metabolic pathways. NMR spectroscopy also provides a fast and efficient method for the study of labelling patterns, which is non-invasive. It enables compounds to be identified without chemical isolation or purification and, therefore, several measurements can be taken from the same sample in a non-destructive manner. This enables studies to be carried out with suspensions of intact, viable cells and enzymic processes can be followed *in situ* without any cell disruption. Because of the exceptional resolution of NMR spectroscopy, a single spectrum can discriminate between individual carbon atoms within the same molecule and simultaneously provide the analysis of metabolic flux into several divergent pathways. Metabolic studies with campylobacters, for example, have employed both one- and two-dimensional NMR techniques (Mendz *et al.*, 1997; Smith *et al.*, 1999; Sellars *et al.*, 2002). However, NMR analysis is relatively insensitive and should ideally be performed in concert with conventional assays of enzyme activity to provide additional confirmation of the operation of metabolic pathways.

THE MAJOR ROUTES OF CARBON METABOLISM

Glycolysis, gluconeogenesis and anaplerotic reactions

Although it is widely accepted that *C. jejuni* neither oxidizes nor ferments glucose (Smibert, 1984), genes encoding enzymes of the Embden–Meyerhof (EM) and pentose phosphate pathways are present in the genome (Parkhill *et al.*, 2000). Most of the reactions of the Embden–Meyerhof pathway are reversible *in vivo*, with the notable exceptions of the 6-phosphofructokinase and pyruvate kinase (PYK) reactions, both of which constitute key control points in many organisms. *C. jejuni* lacks a gene that could encode a 6-phosphofructokinase, which may account for the inability of the bacterium to catabolize glucose. However, a gene encoding fructose-1,6-biphosphatase is present, which strongly suggests that the Embden–Meyerhof pathway functions in gluconeogenesis. Nevertheless, a homologue of PYK (Cj0392c) is also present in the genome (Parkhill *et al.*, 2000). This is surprising since PYK catalyses the physiologically irreversible conversion of phosphoenolpyruvate (PEP) and ADP to pyruvate and ATP at the final stage of the glycolytic pathway. PYK is thus a catabolic enzyme which should not be required if the only function of the EM pathway in *C. jejuni* was in gluconeogenesis.

    C. jejuni is predicted to possess a complete citric acid cycle (CAC) (Parkhill *et al.*, 2000; Kelly, 2001) and net synthesis of oxaloacetate (OAA) from pyruvate or PEP is required to permit the CAC to fulfil both biosynthetic and energy-generating roles. *C. jejuni* possesses homologues of anaplerotic enzymes that may function in this capacity, the activities of some of which have been detected by NMR studies (Mendz *et al.*, 1997), as follows:

(i)   Pyruvate carboxylase (PYC) is encoded by the *pycA* (Cj1037c) and *pycB* (Cj0933c) homologues, which correspond to the biotin carboxylase and biotin carboxyl carrier subunits, respectively, of the two-subunit type of carboxylase (Goss *et al.*, 1981). PYC is a biotin-containing enzyme that fulfils an important anaplerotic function in many organisms by catalysing the ATP-dependent carboxylation of pyruvate to yield OAA.

(ii)   Phosphoenolpyruvate carboxykinase (PCK) provides the PEP required for gluconeogenesis. PCK catalyses the ATP-dependent decarboxylation of OAA to PEP. *C. jejuni* has a homologue of the *Escherichia coli pckA* gene (Cj0932c) and, interestingly, it is contiguous with the *pycB* homologue.

(iii)   Besides PYC and PCK, the only other predicted decarboxylating/carboxylating enzyme located at the metabolic node around pyruvate is malate oxidoreductase [malic enzyme (MEZ)]. MEZ catalyses the oxidative decarboxylation of malate to pyruvate coupled with NAD (P) reduction. Cj1287c encodes a probable malate oxidoreductase homologous to that of *Bacillus stearothermophilus* (Parkhill *et al.*, 2000).

    In order to ascertain the role of these enzymes, insertion mutants in *pycA*, *pycB*, *pyk* and *mez* were generated by Velayudhan and Kelly (2002). However, this could not be achieved for *pckA*, indicating that PCK could be an essential enzyme in *C. jejuni*. The lack of PEP synthase and pyruvate orthophosphate dikinase activities confirmed a unique role for PCK in PEP synthesis. The *pycA* mutant was able to grow normally in complex media, but was unable to grow in defined media with pyruvate or lactate as major carbon sources, thus indicating an important role for PYC in anaplerosis. The activity of PYC was strictly dependent on the presence of ATP, pyruvate, bicarbonate and Mg^{2+}, although the divalent cation requirement could be fulfilled by Co^{2+} and poorly by Mn^{2+}. PEP could not substitute for pyruvate. ATP could not be replaced with ADP or AMP. In the presence of ATP, both AMP and ADP acted as potent inhibitors. Incubation of the cell extract with avidin completely abolished PYC activity. PYC activity was not affected by acetyl-CoA. Although insensitive to α-ketoglutarate, the enzyme was significantly inhibited by glutamate and aspartate. Michaelis–Menten kinetics were followed when the initial velocity was plotted against varying pyruvate and bicarbonate concentrations and the determined K_m values were; pyruvate, 0.24 mM, ATP, 1.76 mM, bicarbonate, 0.64 mM. Sequence and biochemical data indicate that the *C. jejuni* PYC is a member of the $\alpha_4\beta_4$, acetyl-CoA-independent class of PYCs, with a 65.8 kDa subunit containing the biotin moiety.

Whilst growth of the *mez* mutant was comparable to the wild-type, a *pyk* mutant displayed a slightly decreased growth rate in complex media. Nevertheless, the *mez* and *pyk* mutants were able to grow with pyruvate, lactate and malate as carbon sources in defined media (Velayudhan and Kelly 2002). PYK was present in cell extracts at a much higher specific activity (>800 nmol min^{-1} mg^{-1} protein) than PYC or PCK (< 65 nmol min^{-1} mg^{-1} protein), was activated by fructose 1,6-bisphosphate and displayed other regulatory properties strongly indicative of a catabolic role. It was concluded that PYK may function in the catabolism of unidentified substrates which are metabolized through PEP. Thus, the lower part of the EM pathway in *C. jejuni* may have a catabolic role in some situations (e.g. during *in vivo* growth?) even though the absence of 6-phosphofructokinase does not allow the catabolism of sugars. These results highlight how little we know about the nature of the carbon sources available to *C. jejuni* in the gut – a variety of approaches (transcriptomic, proteomic and metabolic) will be needed to address this problem. In view of the high K_m of MEZ for malate (≈ 9 mM) and the lack of a growth phenotype of the *mez* mutant, MEZ seems to have only a minor anaplerotic role in *C. jejuni* (Velayudhan and Kelly, 2002).

The citric acid cycle – relationship to microaerophily?
The genome sequence indicates that *C. jejuni* has the capacity to operate a complete oxidative citric-acid cycle, with all of the key enzymes present including obvious homologues of succinate dehydrogenase, the α and β subunits of succinyl-CoA synthetase (SCS) and an NAD-linked malate dehydrogenase (MDH), which are absent in its close relative *H. pylori* (Kelly and Hughes, 2001). However, there are some other enzymes that are normally restricted to obligately anaerobic bacteria or which are otherwise somewhat unusual.

The entry of carbon from pyruvate into the cycle requires its oxidative decarboxylation to acetyl-CoA which, in most aerobic bacteria, is carried out by the pyruvate dehydrogenase multienzyme complex, using NAD as an electron acceptor. However, in many obligate anaerobes a flavodoxin- or ferredoxin-dependent pyruvate:acceptor oxidoreductase (POR) catalyses this reaction. Daucher and Kreig (1995) showed that many species and strains of *Campylobacter* possess this latter type of enzyme, which could be detected using benzyl or methyl viologen as an artificial electron acceptor in cell-free extracts. However, conflicting data presented by Mendz *et al.* (1997) suggested the use of a conventional NAD-dependent pyruvate dehydrogenase, based on NMR assays in which

NAD, lipoate or thiamine pyrophosphate (TPP) stimulated pyruvate decarboxylation. Strain differences maybe responsible here, but the genome sequence of *C. jejuni* 11168 at least, clearly reveals the presence of genes for the oxidoreductase type of enzyme and not for pyruvate dehydrogenase (Parkhill *et al.*, 2000). The significance of this is that the acceptor:oxidoreductases are iron–sulphur cluster-containing enzymes, which are usually very sensitive to inactivation by molecular oxygen or reactive oxygen species. This has been clearly shown in the related *H. pylori* by the oxygen lability of the purified pyruvate:flavodoxin oxidoreductase (Hughes *et al.*, 1995). In addition, both *H. pylori* and *C. jejuni* contain a related oxygen-labile citric acid cycle enzyme, 2-oxoglutarate:acceptor oxidoreductase (OOR; Hughes *et al.*, 1998; Kelly, 2001) which replaces the function of the 2-oxoglutarate dehydrogenase multienzyme complex found in conventional aerobes. *Campylobacter jejuni* is a true microaerophile which exhibits oxygen-dependent growth but which is unable to grow at normal atmospheric oxygen tensions, and it has been proposed that the presence of these proteins might contribute to the microaerophilic phenotype of both *Helicobacter* and *Campylobacter* (Kelly, 1998, 2001). However, there is an interesting difference between the POR of *C. jejuni* and *H. pylori* in that the former enzyme is a large single subunit protein containing four separate functional domains that are present as separate gene products in *H. pylori*. In contrast, the OORs of both bacteria are very similar four subunit enzymes. The situation in *C. jejuni* and the fact that there is only a rather low overall sequence similarity between the POR and OOR subunit sequences in *H. pylori*, indicates that these two enzymes have evolved independently. More work is needed on the properties and comparative molecular biology of these enzymes in *C. jejuni* to determine their importance and role in contributing to microaerophilic growth.

Amino acid metabolism and nitrogen assimilation

The main carbon sources used by *C. jejuni in vivo* are likely to be amino acids, based on the fact that the bacterium is unable to metabolize exogenous sugars but does have the transport and enzymatic capacity for amino acid catabolism (Parkhill *et al.*, 2000).

There are essentially four enzymatic amino acid catabolic mechanisms that might be operative, depending on the particular chemistry of the amino acid side-chain: (i) deamination (oxidative or non-oxidative), (ii) transamination via an oxo-acid acceptor like 2-oxoglutarate or oxaloacetate (iii) β-elimination of ammonia via dehydratases or amino acid:ammonia lyases (iv) decarboxylation. These four mechanisms must provide intermediates that can readily feed into central metabolic pathways, and some amino acids might require a combination of enzymes to satisfy this requirement. In *C. jejuni* 11168, based on scrutiny of the available genome sequence and a consideration of the most likely catabolic pathways in which these enzymes participate, there are enzymes that would be predicted to allow the complete catabolism of only aspartate, asparagine, glutamate, glutamine, serine and proline. This is a very narrow range of amino acids but it accords exactly with the amino acid utilization data of Leach *et al. (*1997) and, interestingly, aspartate, glutamate, proline and serine are the most abundant amino acids in chicken excreta (Parsons *et al.*, 1982). Cj0021c encodes a potential fumarylacetoacetate hydrolase which is the final enzyme in tyrosine catabolism in some bacteria, but the other enzymes of this complex pathway appear to be absent. No amino acid catabolic decarboxylases are present, but the three other principal mechanistic routes are represented. Asparagine might be deaminated to aspartate by Cj0029 (a predicted asparaginase) and there are two possible aspartate or glutamate transaminases (Cj0150 and Cj0762). A key aspartate catabolic enzyme is likely be encoded by Cj0087, a predicted aspartate:ammonia lyase (aspartase) which would directly form fumarate by β-elimination of ammonia. Proline could be oxidized to glutamate via a predicted bifunctional proline dehydrogenase PutA (Cj1503) with Δ^1-pyrroline-5-carboxylate (P5C) as an intermediate. The genome sequence reveals a two-gene operon (Cj1624c and Cj1625c) encoding homologues of the *E. coli* serine dehydratase (SdaA) and serine transporter SdaC, for L-serine transport and catabolism. The hydropathy profile of SdaC shows that it contains 10 or 11 potential membrane-spanning segments and its overall topology is similar to the *E. coli* SdaC protein and other members of the hydroxy/aromatic amino acid permease (HAAAP) family. Glutamate metabolism is dominated by the glutamine synthase (GS) and glutamine:2-oxoglutarate amino-transferase (GOGAT) system; an obvious glutamate dehydrogenase is absent. GS/GOGAT is thus the major route by which nitrogen is incorporated into cellular amino acids. 2-Oxoglutarate is a crucial intermediate in *C. jejuni* amino acid metabolism, as it takes part in transamination reactions as well as being the primary amino-group acceptor in the interconversion of glutamine and glutamate.

Some biochemical studies have been carried out on amino acid utilization in both batch and continuous cultures (Karmali *et al.*, 1986; Westfall *et al.*, 1986; Leach *et al.*, 1997; Mendz *et al.*, 1997). In complex media, the most heavily depleted amino acids were serine, aspartate, glutamate and proline (Leach *et al.*, 1997). In continuous culture there was a strong growth rate effect on the utilization of different amino acids, with aspartate and serine being metabolized preferentially at high dilution (growth) rates, with a switch to glutamine and proline at lower dilution rates (Leach *et al.*, 1997).

L-Serine seems to be a particularly favoured amino acid for catabolism. *C. jejuni* possesses an active serine dehydratase (deaminase) which converts serine to pyruvate and ammonia (Leach *et al.*, 1997; Mendz *et al.*, 1997). The pyruvate can be rapidly and efficiently metabolized via POR and the citric-acid cycle, which may explain why this amino acid is so well utilized. Most serine dehydratases from aerobic bacteria are pyridoxal 5'-phosphate (PLP)-dependent enzymes. Recent results (Velayudhan *et al.*, 2004) have shown that the *sdaA* gene in *C. jejuni* encodes an L-serine dehydratase that is devoid of PLP, and is instead an oxygen-labile iron–sulphur enzyme. The enzyme lacks the consensus sequence centred on a lysyl residue to which PLP binds via a Schiff base (Ogawa *et al.*, 1989), in addition to lacking the glycine-rich region reported for all PLP-dependent enzymes (Marcaeu *et al.*, 1988). Moreover, the PLP-reactive carbonyl reagents phenylhydrazine and hydroxylamine did not significantly affect enzyme activity. The absorption spectrum of the purified protein was, however, typical of that for an iron–sulphur protein, with broad peaks at 300–350 nm and 420 nm (Velayudhan *et al.*, 2004). Consistent with this, the *C. jejuni* SdaA sequence contains four conserved cysteine residues, which could coordinate a [4Fe–4S] cluster. Moreover, both the spectral features and enzyme activity were lost upon exposure to air but activity could be restored by treatment with ferrous iron and a reducing agent (Velayudhan *et al.*, 2004). These properties are similar to those exhibited by the L-serine dehydratases of many anaerobes, e.g. *Peptostreptococcus asaccharolyticus,* which are PLP-independent, iron–sulphur-containing enzymes (Grabowski *et al.*, 1993). They form a family of enzymes which share mechanistic similarities with the dehydration of citrate by aconitase (Grabowski *et al.*, 1993) but they are generally highly specific for L-serine deamination. The employment of an oxygen-labile dehydratase for serine catabolism is another example of the use by *C. jejuni* of a type of enzyme more commonly found in anaerobic bacteria than aerobes.

Mutation of *sdaA* resulted in a ~10-fold reduction in the activity of serine dehydratase in cell-free extracts. The residual activity was found to be due to the PLP-dependent biosynthetic L-threonine dehydratase (IlvA; Cj0828c), but this activity was unable to support growth of the mutant on L-serine in a minimal medium, and there was no evidence from ^{1}H-NMR spectroscopy for significant utilization of L-serine by intact cells (Velayudhan *et al.*, 2004). However, growth of the *sdaA* mutant in a complex medium was indistinguishable from that of the parent strain. This indicates that an inability to deaminate serine does not result in a growth disadvantage when a larger array of carbon sources in addition to amino acids are available. Most significantly, however, it was found that there was a complete lack of colonization of 2-week old chickens by the isogenic *sdaA* mutant compared to its wild-type colonization proficient parent strain, as evidenced by a repeated lack of recovery of viable bacteria from inoculated chickens by either enrichment from cloacal swabs over a 6-week period or by caecal contents counts at the end of the experiment (Velayudhan *et al.*, 2004). This clearly shows that serine dehydratase encoded by *sdaA* is an essential colonization factor for *C. jejuni*. Moreover, the data suggest that a source of serine is readily available to the bacteria and that its catabolism via SdaA is a crucial determinant of the ability of the bacteria to grow in the chicken caecum. These results are surprising in view of the potentially diverse types of alternative carbon/nitrogen/ energy sources which would be present in this environment, and indicate a high degree of selectivity in the types of amino acid utilized by *C. jejuni* in this niche.

ELECTRON TRANSPORT CHAINS

Overview

An appreciation of the role and complexity of electron transport systems in bacteria is essential in understanding the ability to grow under different environmental conditions, as it is the design of the electron transport chain that determines the degree of metabolic flexibility a bacterium possesses in terms of the variety of electron donors and acceptors which can be employed to support growth. In this context, some relevant characteristics of bacterial respiratory chains are (i) the degree of branching at both 'dehydrogenase' and 'reductase' ends, (ii) the use of alternative electron acceptors in addition to molecular oxygen, (iii) the presence of a variety of types of cytochromes and, often, more than one type of quinone, (iv) 'cross-talk' between electron transport pathways, optimizing the possibility of each reductant being paired with a wide choice of oxidants, and (v) the degree to which each electron transport chain contributes to concomitant proton translocation and energy transduction (Kelly *et al.*, 2001). Early work on campylobacters led to the identification of a rich complement of cytochromes (Harvey and Lascelles, 1980; Lascelles and Calder, 1985), and the detailed studies of Hoffman and Goodman (1982) and Carlone and Lascelles (1982) indicated that the respiratory chain of *C. jejuni* was complex. This work can now be interpreted in terms of the genome sequence of *C. jejuni* strain 11168 (Parkhill *et al.*, 2000). Figure 14.1 shows a partial reconstruction of the major predicted electron transport pathways as deduced from the genome sequence and biochemical evidence (Sellars *et al.*, 2002). The respiratory chain in *C. jejuni* is clearly highly branched and is significantly more complex than might be expected for such a small genome pathogen.

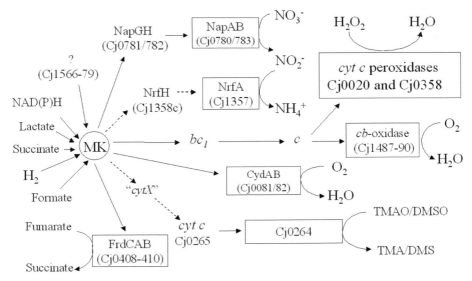

Figure 14.1 Some predicted electron transport chains in *C. jejuni*. Integral membrane oxidoreductases on the electron donor side of the menaquinone pool include an NDH-1 like complex (Cj1566–1579), the electron donor to which is unknown, hydrogenase, formate dehydrogenase and succinate dehydrogenase. Peripherally associated oxidoreductases include (among several others) malate:quinone oxidoreductase and a D-lactate dehydrogenase. Reducing equivalents are transferred to the sole quinone, menaquinone (MK) in the lipid bilayer of the inner membrane. Menaquinol reduces the trimeric cytochrome bc_1 complex, which in turn reduces periplasmic cytochrome *c*. Cytochrome *c* is reoxidized by one of the terminal oxidases, a *cb*-type cytochrome *c* oxidase. A separate quinol oxidase is also present, with some similarity to the *E. coli* CydAB complex. Cytochrome *c* may also be reoxidized by hydrogen peroxide in the periplasm through the activity of two separate cytochrome *c* peroxidases. Several alternative reductases are predicted in *C. jejuni*. Fumarate reductase (FrdABC) catalyses electron transfer from menaquinol to fumarate as terminal acceptor. Periplasmic nitrate, nitrite and TMAO/DMSO reductases are also present. Solid lines indicate experimentally established or highly likely routes of electron transport, while dotted lines indicate uncertainty as to the exact route, possibly with the participation of unidentified additional redox proteins (e.g. 'cyt X'). Figure modified from Sellars *et al.* (2002).

Electron donors

Hydrogen

Uptake hydrogenase mediates the transfer of electrons from hydrogen to the quinone pool. The enzyme is membrane bound and consists of three polypeptides, HydA, HydB, and HydC. *C. jejuni* contains a set of structural genes for a NiFe-type uptake hydrogenase, as well as accessory genes for nickel incorporation, but no functional studies have yet been done on these genes. *C. jejuni* has been shown to possess hydrogenase activity in membrane fractions (Carlone and Lascelles, 1982; Hoffman and Goodman, 1982). Hydrogen could be a very important source of electrons for the growth of campylobacters in the gut, as many obligate anaerobes in the intestine produce hydrogen from redox reactions associated with fermentation. It would be of great interest to know if hydrogenase activity is essential for or contributes to the colonization of the avian gut by *C. jejuni*. Recent studies with *H. pylori* have shown that hydrogenase mutants do not colonize a mouse model of infection and microelectrode measurements have shown that hydrogen is present in the mouse stomach at concentrations much greater than the K_m value of the hydrogenase enzyme (Olson and Maier, 2002).

Formate

Next to hydrogen, formate is an exceptionally good electron donor, as it has a low redox potential (-420 mV). Moreover, formate is produced by anaerobes in the gut from mixed acid type fermentation reactions, so it is available to an organism like *C. jejuni*, which has the enzymatic capacity for formate oxidation. Formate-dependent oxygen respiration in *C. jejuni* was originally demonstrated and characterized by Hoffman and Goodman (1982) and Carlone and Lascelles (1982). The respiratory activities determined with membrane vesicles were 50–100 times greater with formate or hydrogen as substrates when compared with the rates achieved with other electron donors such as succinate, lactate, malate or NADH, suggesting that the former are indeed excellent electron donors (Hoffman and Goodman, 1982), and indicating the presence of an active formate dehydrogenase. In *E. coli*, formate dehydrogenase-N (formate: quinone oxidoreductase-N) catalyses transmembrane proton translocation from the cytoplasm to the periplasm, and consists of three subunits, a large selenomolybdoprotein containing the catalytic site (FdnG), a smaller iron–sulphur protein (FdnH) and a *b*-type cytochrome (FdnI) (Gennis and Stewart, 1996). The genome sequence of *C. jejuni* 11168 (Parkhill *et al.*, 2000) reveals an operon encoding putative formate dehydrogenase subunits (*fdhA-D*/Cj1511c–Cj1508c). *C. jejuni* *fdhA* (Cj1511c) encodes a large 104 kDa selenocysteine containing molybdoprotein equivalent to the *E. coli* 110 kDa FdnG (?) subunit. *fdhB* (Cj1510c) encodes a 24 kDa iron–sulphur subunit equivalent to the *E. coli* 32 kDa FdnH (?) subunit. *fdhC* (Cj1509c) encodes a 35 kDa cytochrome *b* subunit equivalent to the *E. coli* 20 kDa FdnI (?) subunit. In addition to these three subunits, both *E. coli* and *C. jejuni* encode an FdhD protein (29 kDa) required for activity of the formate dehydrogenase enzyme complex (Berg *et al.*, 1991). Molecular and functional studies are clearly warranted on the *C. jejuni* enzyme. In particular, it would be of great interest to know if *fdh* mutants are deficient in colonization of chickens or other animal models, as this would provide evidence for the utilization and importance of formate as an *in vivo* electron donor.

NAD (P)H

The function of NADH in bioenergetics is to act as the major electron donor for oxidative phosphorylation, via interaction with a proton-translocating quinone oxidoreductase (Complex I or NDH-1), while the role of NADPH is as a source of electrons for biosynthetic reactions. NDH-1 is a large enzyme with a multitude of redox centres and is most commonly made up of 14 different subunits in bacteria (Friedrich, 1998; Yagi *et al.*, 1998). Electrons are passed from NADH, via 1 FMN and 9 Fe–S centres in NDH-1, to ubiquinone/menaquinone in the respiratory chain. Four protons are translocated across the membrane for every two electrons transferred by NDH-1. The genome sequence of *C. jejuni* 11168 shows that the bacterium contains a cluster of genes (Cj1566–1579) encoding a potential NADH-quinone oxidoreductase of the NDH-1 type. However, an examination of the deduced proteins encoded by this gene cluster suggests that the complex may not actually oxidize NADH, because of the lack of the NQO1 (NuoE) and NQO2 (NuoF) subunits (Smith *et al.*, 2000). These subunits are thought to be essential components for the function of the NDH-1 complex. NuoF binds NADH, and also possesses a bound FMN and an Fe–S centre. NuoE, G, B and I, have cysteine residues that probably coordinate Fe–S clusters (Smith *et al.*, 2000). In the place of the expected *nuoE* and *nuoF* genes, *C. jejuni* (and also *H. pylori*) has ORFs of unknown function. Cj1574c from *C. jejuni* encodes a protein 230 aa in length (26 kDa) which has 26.5% identity to the *H. pylori* equivalent

(HP1265) which encodes a protein 328 aa in length (37 kDa). Cj1575c from *C. jejuni* encodes a protein 75 aa in length (8.7 kDa) with 52.0% identity to the *H. pylori* equivalent (HP1264) which encodes a protein 76 aa in length (8.9 kDa) (Tomb *et al.*, 1997; Parkhill *et al.*, 2000). Significantly, these proteins do not contain any obvious NAD (P)H-binding motif.

There are essentially two possibilities regarding the function of this complex in *C. jejuni*. The first is that electrons from NAD (P)H are transferred to it via an intermediate protein that interacts with the NQO1 and NQO2 replacements (Finel, 1998; Smith *et al.*, 2000). Alternatively, coupling of NADPH with the respiratory chain may not occur via the NDH-1 homologue at all, but through an alternative quinone reductase which may or may not be proton-translocating. This implies that an as-yet unidentified electron donor interacts with the Cj1566–1579 complex, and that its function is very different to most bacterial Complex I systems. Obviously, the nature of the electron donor and whether this type of enzyme is proton translocating is an important unresolved question for understanding the bioenergetics of *C. jejuni*.

Lactate, malate and succinate
Several types of organic acid appear to be able to act as electron donors in *C. jejuni*. Hoffman and Goodman (1982) demonstrated the activity of a lactate dehydrogenase in oxygen-linked respiration. A H^+/O ratio of 2.12 was found for lactate in whole cell proton-pulse assays (Hoffman and Goodman, 1982). Analysis of the genome sequence has yet to produce a locus for a membrane-associated lactate dehydrogenase enzyme in *C. jejuni*, although an annotated L-lactate dehydrogenase (Cj1167) is present, which is probably a fermentative enzyme. Malate can act as a direct electron donor to the quinone pool due to the presence of a flavoprotein type malate oxidoreductase (Mqo). Analysis of the *C. jejuni* genome reveals a putative oxidoreductase (Cj0393c) with 24.6% identity to *E. coli* malate:quinone oxidoreductase and 49.3% identity to the closely related Mqo of *H. pylori*, indicating that both bacteria use a similar enzyme for malate oxidation (Kather *et al.*, 2000). This is also consistent with the data of Hoffman and Goodman (1982), who showed that malate stimulated respiration in *C. jejuni*, producing a H^+/O ratio of 2.0.

In *E. coli*, the interconversion of fumarate and succinate can be carried out by two related enzymes; succinate dehydrogenase, which is expressed under aerobic conditions, or fumarate reductase, which is induced under anaerobiosis (Spiro and Guest, 1991). Succinate dehydrogenase (succinate:quinone oxidoreductase) catalyses electron transfer from succinate to ubiquinone, where succinate is oxidized to fumarate as part of the tricarboxylic acid cycle. As noted above, *C. jejuni* appears to have a complete citric-acid cycle with genes encoding clearly identifiable homologues of all of the conventional enzymes (Parkhill *et al.*, 2000), including separate *sdh* and *frd* operons. The *C. jejuni* genome reveals an *sdhABC* operon (Parkhill *et al.*, 2000). *sdhA* (Cj0437) encodes a 66 kDa protein with 37.3% identity to *E. coli* SdhA and contains an FAD binding domain. *sdhB* (Cj0438) encodes a 35.8 kDa putative iron–sulphur protein. *sdhC* (Cj0439) encodes a 31.4 kDa putative succinate dehydrogenase subunit C which is a probable functional equivalent to *E. coli* succinate dehydrogenase subunits C and D.

Terminal oxidases for oxygen-dependent respiration
Most bacteria possess at least two terminal oxidases, often a quinol oxidase and a cytochrome *c* oxidase (Poole and Cook, 2000), with distinct catalytic properties and control

of expression. Genome analysis of *C. jejuni* indicates the absence of any homologues for an *aa$_3$*-type cytochrome *c* oxidase. Whereas only one terminal oxidase, a *cb* (*cbb$_3$*)-type cytochrome *c*-oxidase has been found to terminate the respiratory chain of *Helicobacter pylori* (Nagata *et al.*, 1996), *Campylobacter jejuni* has two terminal oxidases (Parkhill *et al.*, 2000). This confirms the earlier spectroscopic work of Carlone and Lascelles (1982) and Hoffman and Goodman (1982). One oxidase is a *cb*-type cytochrome *c*-oxidase, very similar to that described in *H. pylori*. Subunit I (*ccoN*/Cj1490c) has a molecular weight of 56 kDa and contains the haem–copper oxidase catalytic centre and a copper B binding region signature. Subunit II (*ccoO*/Cj1489c) has a molecular weight of 25 kDa and contains a cytochrome *c* family haem-binding site signature. Subunit III (*ccoP*/Cj1487c) has a molecular weight of 31 kDa and contains two haem C binding motifs. Subunit IV (*ccoQ*/Cj1488c) has a molecular weight of 10.4 kDa with 35.1% identity to the corresponding *H. pylori* subunit. No functional studies have yet been reported on this oxidase. Carlone and Lascelles (1982) demonstrated an alternative terminal oxidase in *C. jejuni* and analysis of the genome sequence reveals the presence of a *cydAB*-like operon encoding subunits I and II. Subunit I (58 kDa) is encoded by *cydA* (Cj0081), and has 45.7% identity to *E. coli* cytochrome *bd* oxidase subunit I. Subunit II (42 kDa) is encoded by *cydB* (Cj0082) and is similar to the *E. coli* subunit II with 27.7% identity (Parkhill *et al.*, 2000). The relative physiological roles of the two oxidases in *C. jejuni* are as yet unknown, but the possession of a *cb*-type oxidase is likely to endow the bacterium with the ability to continue respiration at extremely low oxygen concentrations, as might be found in the niches which it occupies in the gut of animals or humans.

Hydrogen peroxide as an electron acceptor

Hydrogen peroxide is produced as a by-product in the reduction of molecular oxygen to water. It is a partially reduced species that is toxic to the cell. Hydrogen peroxide can be degraded to H_2O and O_2 by the cytoplasmic enzyme catalase, but in the periplasm, H_2O_2 can be broken down to water alone by a periplasmic cytochrome *c* peroxidase (CCP), with the requirement of reduced cytochrome *c* as an electron donor. Studies of CCP in a variety of bacteria have revealed that the enzyme has a molecular weight of 34–37 kDa and contains two haeme C, one high potential and one low potential. The high potential haem is the source of the second electron for H_2O_2 reduction and the low potential haem acts as a peroxidatic centre. Analysis of the *C. jejuni* genome reveals two separate genes encoding putative cytochrome *c* peroxidase homologues (Parkhill *et al.*, 2000). Cj0020 and Cj0358 are 34 kDa and 37 kDa respectively with both proteins containing two cytochrome *c* haem-binding site signatures. Cj0020 is similar to the cytochrome *c* peroxidase of *Nitrosomonas europaea* and *Pseudomonas aeruginosa* with 46% and 42% identity respectively (Parkhill *et al.*, 2000). Cj0358 is similar to the *H. pylori* putative cytochrome *c* peroxidase homologue (HP1461) with 58.5% identity, and to the cytochrome *c* peroxidase of *Pseudomonas aeruginosa* with 48.9% identity (Parkhill *et al.*, 2000). The physiological roles and regulation of expression of the peroxidases have yet to be studied. However, cytochrome *c* peroxidase has been studied in the related microaerophile *Campylobacter mucosalis*, where the oxidation of formate leads to the generation of periplasmic H_2O_2. This is reduced by CCP, using electrons from cytochrome c_{553} that has been reduced by the bc_1 complex. Thus, removal of H_2O_2 from the periplasm leads to net proton extrusion and energy conservation (Goodhew *et al.*, 1988).

Respiration using alternative electron acceptors to oxygen

Fumarate as electron acceptor

In the absence of oxygen, fumarate can be used as an alternative terminal electron acceptor for the proton-translocating electron transport chain, and is important in ATP generation in many anaerobic bacteria. Although the presence of fumarate reductase provides evidence of anaerobic-type respiration, Veron *et al.* (1981) reported anaerobic growth with fumarate to be absent in *C. jejuni*. Yet cytochrome *b*-linked fumarate reductase activity was demonstrated in the membranes of both *C. fetus* and *C. jejuni* (Carlone and Lascelles, 1982; Harvey and Lascelles, 1980) and *C. fetus* is fully capable of anaerobic growth with fumarate (Veron *et al.*, 1981). Analysis of the *C. jejuni* genome reveals a *frdABC* operon, encoding subunits similar to the thoroughly investigated fumarate reductase of *W. succinogenes* and to that in *H. pylori* which, like *C. jejuni*, is apparently unable to grow anaerobically with fumarate. Why then is *C. jejuni* apparently unable to grow by anaerobic respiration of fumarate? Sellars *et al.* (2002) found that although growth with fumarate (and other electron acceptors such as nitrate, nitrite and TMAO; see below) was insignificant under strictly anaerobic conditions, electron acceptor-dependent growth was possible under severely oxygen-limited conditions. Their results indicated that some oxygen requiring metabolic reaction (s) prevented anaerobic growth. There are several possibilities for this, but as *C. jejuni* only contains genes for an oxygen-requiring class I type of ribonucleotide reductase, it was suggested that the inability to synthesize DNA anaerobically is the most likely explanation. Consistent with this, cells incubated anaerobically with electron acceptors did not divide properly but formed filaments analogous to those seen after treatment of aerobic cells with the RNR inhibitor hydroxyurea (Sellars *et al.*, 2002). Thus, *C. jejuni* can use alternative electron acceptors in energy conserving reactions, but only if some oxygen is present to satisfy the requirement for deoxyribonucleotide production. A similar explanation for *H. pylori* is likely, as it too only contains the class I RNR genes. This conclusion is directly relevant for understanding growth of *C. jejuni* in the gut, as it implies that in the niches occupied by the bacterium, small amounts of oxygen must be present for continued viability. Whether in these severely oxygen-limited niches alternative electron acceptors (like fumarate) are being used in addition to or even instead of molecular oxygen will require studies with the relevant mutants in animal models.

Because the active site of fumarate reduction is on the cytoplasmic side of the membrane, if quinol oxidation releases protons into the periplasm, fumarate reductase will be an electrogenic enzyme in *C. jejuni* and will contribute to the generation of a proton-motive force. However, this is not the case for the other reductases discussed below, which are all periplasmic in location.

Nitrate as electron acceptor

There are three types of bacterial nitrate reductases (i) soluble, cytoplasmic, assimilatory nitrate reductases (NAS); (ii) membrane-associated 'respiratory' nitrate reductases (NAR); (iii) soluble, periplasmic, 'dissimilatory' nitrate reductases (NAP) (Potter *et al.*, 2001). The NAP class of nitrate reductase is a two-subunit complex, located in the periplasm, which is coupled to quinol oxidation via a membrane anchored tetrahaem cytochrome. *C. jejuni* is predicted to possess a periplasmic NAP enzyme, encoded by a *napAGHBD* operon (Parkhill *et al.*, 2000; Kelly 2001; Sellars *et al.*, 2002). Nitrate reductase activity has been demonstrated in intact cells by nitrite accumulation assays (Sellars *et al.*, 2002).

As with fumarate, nitrate-dependent growth in *C. jejuni* can only be demonstrated under oxygen-limited, not strictly anaerobic conditions (Sellars *et al.*, 2002).

NapA is a ~90 kDa catalytic subunit which binds a bis-molybdenum guanosine dinucleoside (MGD) co-factor and a [4Fe–4S] cluster. NapB is a ~16 kDa electron transfer subunit which in other bacteria binds two *c*-type haems (Richardson *et al.*, 2001). NapD (13 kDa) is proposed to be involved in maturation of NapA prior to export to the periplasm (Berks *et al.*, 1995; Potter and Cole, 1999). The *nap* operon of *C. jejuni* does not contain *napC*, encoding the NapC subunit. NapC acts as an electron donor to the NapAB complex, a role which was shown to be essential in the function of nitrate reductases in many bacteria, including *E. coli* and *Paracoccus pantotrophus* (Berks *et al.*, 1995; Potter and Cole, 1999). Analysis of the *C. jejuni* genome shows that it does encode a NapC homologue (Cj1358c), but it is upstream of the *nrfA* nitrite reductase gene. The operonal location of Cj1358c implies that it is part of the nitrite reductase system, and thus may not be involved in electron transfer to the nitrate reductase. A potential role of Cj1358c in nitrite reduction is suggested by studies on the *nrf* operon in *Wolinella succinogenes,* which encodes a NapC-like subunit (NrfH) similar to the product of Cj1358c (Simon *et al.*, 2000). NrfH is the mediator between the quinone pool and the cytochrome *c* nitrite reductase (NrfA) of *W. succinogenes* (Simon *et al.*, 2000). This leaves the question of how the NapAB complex is coupled to menaquinol oxidation. One possibility, depicted in Figure 14.2, is that the NapG and NapH subunits could function in this role (Sellars *et al.*, 2002). NapG (27 kDa) is predicted to bind up to four Fe–S clusters. NapH (30 kDa) is predicted to be an integral membrane protein with four transmembrane helices with both the N- and C-terminus on the cytoplasmic side of the membrane. Because of the periplasmic location of Nap, a proton-motive force can only be generated at the level of the primary dehydrogenase and not from quinol oxidation.

There is now some evidence that nitrate respiration may play a significant role in the growth of human and animal pathogens *in vivo*. In *Mycobacterium bovis*, membrane-bound nitrate reductase (Nar) activity has been shown to contribute to virulence (Weber *et al.*, 2000), as *nar* mutants were unable to colonize a mouse animal model. In a wide range of Gram-negative bacterial pathogens, including *C. jejuni*, periplasmic nitrate reductases are more commonly present than Nar-type enzymes (Potter *et al.*, 2001). Nitrate concentrations in human body fluids are in the range 10–50 μM (Potter *et al.*, 2001). Significantly, it has been shown that in *E. coli* (which has both Nar and Nap) the Nap enzyme has a much higher affinity for nitrate compared to the membrane-bound Nar (Potter *et al.*, 2001), making Nap ideally suited to a role in scavenging the low nitrate concentrations encountered *in vivo*. As yet, it is not known if the *C. jejuni* Nap has any *in vivo* role; it may also have other functions such as in survival in foods or the environment where nitrate is present, or in nitrogen assimilation in combination with nitrite reductase, which produces ammonia.

Nitrite as electron acceptor

The nitrite-reducing enzyme present in *C. jejuni* is cytochrome *c* nitrite reductase (NrfA; Cj1537), which is the terminal enzyme in the six-electron dissimilatory reduction of nitrate to ammonia:

$$NO_2^- + 8H^+ + 6e^- \rightarrow NH_4^+ + 2H_2O$$

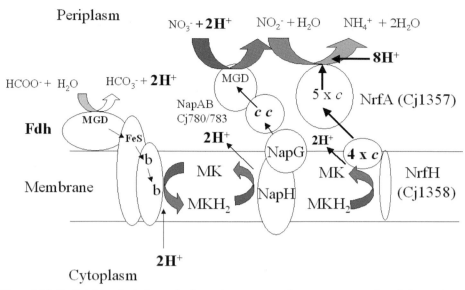

Figure 14.2 Predicted topological organization and consequences for energy conservation of electron transport chains to nitrate and nitrite in *C. jejuni*. Formate is depicted as the electron donor to the quinone pool, through the action of formate dehydrogenase (Fdh). It is thought that NapGH form a quinol oxidase complex and pass electrons to the nitrate reductase structural proteins, NapAB. The NrfH homologue Cj1358 is believed to be the electron donor to NrfA, the nitrite reductase. With nitrate or nitrite as electron acceptors, the terminal reductases are both predicted to be periplasmic, so that reduction of the electron acceptor and quinol oxidation occur on the same (periplasmic) side of the membrane. In these cases a proton-motive force can only be generated at the level of the primary dehydrogenase.

This is a periplasmic enzyme which was detected in intact cells of *C. jejuni* using methyl viologen as an electron donor (Sellars *et al.*, 2002). Three basic elements are required to form a functional periplasmic nitrite reductase complex (Einsle *et al.* (2000); (i) the NrfA enzyme; (ii) a system to oxidize menaquinol and transport electrons to NrfA; (iii) a special haem lyase needed for the covalent attachment of the active site haem group. The cytochrome *c* nitrite reductases (NrfA) are pentahaem enzymes with a molecular mass of 55–65 kDa. NrfA acts as a homodimer, with each monomer presumed to be functional and to act independently. The active site is located at haem 1, with a Ca^{2+} ion in close proximity (Einsle *et al.*, 2000). There are two distinct systems for the transfer of electrons resulting from menaquinol oxidation to NrfA, the NrfH tetrahaem protein described by Simon *et al.* (2000) for *Wolinella succinogenes*, and the NrfBCD system of, for example, *E. coli* K-12, in which the NrfA is connected to the quinol oxidase NrfCD via the soluble pentahaem cytochrome *c* NrfB (Hussain *et al.*, 1994).

The organization of the gene for the nitrite-reducing enzyme in *C. jejuni* (Cj1357c) and the upstream gene encoding a putative NapC/NirT-type cytochrome *c* (Cj1358c), indicates similarity to the *Wolinella succinogenes* nitrite-reducing system in which NrfA accepts electrons from a membrane anchored tetrahaem cytochrome, NrfH (Simon *et al.*, 2000). NrfH is a tetrahaem cytochrome *c* acting as a quinol oxidase to receive electrons from the quinone pool and interacts with the positively charged environment around haem 2 of the NrfA enzyme (Einsle *et al.*, 2000). NrfA contains an unusual CXXCK motif for the haem 1 binding site, instead of the CXXCH motif found in many haem-containing enzymes. This substitution in NrfA may explain the requirement for a modified haem lyase for covalent attachment of the active site haem The *C. jejuni* NrfA homologue, Cj1357c, is slightly larger at 69 kDa when compared to the *W. succinogenes* NrfA (58 kDa), and contains a novel ligation at haem 1, with a CXXCH motif instead of CXXCK. This would explain the absence in *C. jejuni* of genes encoding haem lyases and assembly proteins that are found in other bacteria (Einsle *et al.*, 2000). Although the functional consequences of this substitution are unknown at present, it is possible it could have important implications for the catalytic activity of the protein. A possible mechanism of coupled nitrate and nitrite reduction in *C. jejuni* is shown in Figure 14.2.

S- or N-oxides as electron acceptors

The structurally related compounds trimethylamine-N-oxide (TMAO) and dimethylsulphoxide (DMSO) may appear to be somewhat esoteric or unusual electron acceptors, but are in fact widely distributed in many environments, particularly water; TMAO is an excretory product of some fish and DMSO is present in some algae. The reduction of TMAO and DMSO may be carried out by two different reductase enzymes as in *E. coli* (Gennis and Stewart, 1996), or reduction of both TMAO and DMSO may be carried out by a single reductase as in the Dor system of *Rhodobacter* (Kelly *et al.*, 1988; Knablein *et al.*, 1996). In either case the reduction is a two-electron transfer process:

$$(CH_3)_2SO + 2H^+ + 2e^- \rightarrow (CH_3)_2S + H_2O$$
$$(CH_3)_3NO + 2H^+ + 2e^- \rightarrow (CH_3)_3N + H_2O$$

The *torCAD* operon in *E. coli* encodes the TMAO reductase complex, consisting of a large periplasmic molybdoprotein (TorA, 94 kDa), a membrane anchored periplasmic cytochrome *c* (TorC, 43 kDa) and a presumed intrinsic membrane subunit (TorD, 22 kDa) (Gennis and Stewart, 1996). The DMSO reductase complex contains three subunits, consisting of a large extrinsic membrane-bound molybdoprotein, facing the cytoplasm (DmsA, 82.6 kDa), a smaller iron–sulphur subunit (DmsB, 23.6 kDa), and an intrinsic membrane subunit (DmsC, 22.6 kDa) (Bilous and Weiner, 1988; Weiner *et al.*, 1988; Gennis and Stewart, 1996).

Both TMAO and DMSO reductase activities can be measured in intact cells of *C. jejuni* using methyl viologen linked assays and TMAO- and DMSO-dependent growth under oxygen limited conditions has been demonstrated (Sellars *et al.*, 2002). Analysis of the *C. jejuni* genome reveals a gene encoding a 93 kDa molybdoprotein DorA/TorA homologue (Cj264) containing conserved residues for the binding of an MGD co-factor and an N-terminal motif (RRXFL/IK), similar to the 'twin-arginine' translocase recognition motif (S/TRRXFLK), characteristic of extracytoplasmic proteins containing complex redox co-factors (Berks, 1996). The Cj264c gene was mutated, and the mutant was found to deficient in both TMAO and DMSO reductase activity, indicating that a single enzyme is responsible (Sellars *et al.*, 2002). In addition there is an upstream gene encoding a 22 kDa monohaem *c* type cytochrome (Cj265c) with similarity to the C-terminus of the membrane-anchored pentahaem *c* type cytochrome TorC of *E. coli* (Gon *et al.*, 2001; Sellars *et al.*, 2002). TorC contains a C-terminal extension (TorC (C)) with an additional haem-binding site, responsible for the electron transfer to TorA. The tetrahaem N-terminus (TorC (N)) is involved in the binding of TorC to TorA (Gon *et al.*, 2001). The monohaem *c* type cytochrome (Cj265c) of *C. jejuni,* may be involved in electron transport to the DorA/TorA homologue (Cj264c), as with the C-terminal part of the TorC

Kelly

subunit in *E. coli*, although additional redox proteins are likely to be involved in electron transfer from quinol to Cj265. The fact that *C. jejuni*, a small genome pathogen, possesses a functional electron transport chain to TMAO/DMSO underlines its surprising metabolic versatility, and indicates a possible role for these electron acceptors in aiding survival outside the host.

CONCLUSIONS

In order to understand fully why campylobacters can survive in the environment, behave as commensals in the avian gut and yet be pathogens in the human gut, we need to investigate and appreciate the metabolism and physiology of these bacteria. This task has been greatly facilitated recently by the availability of the complete genome sequence of *C. jejuni* 11168 and by better techniques for growing these bacteria in defined gas atmospheres and for probing their metabolism with techniques like NMR spectroscopy. A major feature that is emerging is an unexpected metabolic versatility, which is particularly reflected in the complexity of the electron transport chains in *C. jejuni*. Despite a genome size of only 1.7 Mb, it is clear that this bacterium has retained the enzymatic equipment to be flexible with respect to the variety of electron donors and acceptors it can use, which is surely a function of its ability to occupy so many niches. Major unanswered questions exist about exactly what sources of carbon, nitrogen, sulphur, etc, the bacteria are using during growth *in vivo*. These could turn out to be quite specific (like serine) and could be useful in identifying a means to control the growth of *C. jejuni* in animal reservoirs for example. In the future, the application of both conventional biochemical techniques and post-genomic technologies will be required to fully elucidate the metabolism of this important pathogen.

Acknowledgements

Work in my laboratory on *Campylobacter jejuni* has been supported by grants and studentships from the UK Biotechnology and Biological Sciences Research Council.

References

Berg, B.L., Li, J., Heider, J. and Stewart, V. 1991. Nitrate-inducible formate dehydrogenase in *Escherichia coli* K-12. I. Nucletide sequence of the *fdnGHI* operon and evidence that opal (UGA) encodes selenocysteine. J. Biol. Chem. 266: 22380–22385.

Berks, B.C., Richardson, D.J., Reilly, A., Willis, A.C. and Ferguson, S.J. 1995. The *napEDABC* gene cluster encoding the periplasmic nitrate reductase system of *Thiosphaera pantotropha*. Biochem. J. 309: 983–992

Berks, B.C. 1996. A common export pathway for proteins binding complex redox cofactors? Mol. Microbiol. 22: 393–404

Bilous, P.T. and Weiner, J.H. 1988. Molecular cloning and expression of the *Escherichia coli* DMSO reductase operon. J. Bacteriol. 170: 1511–1518

Carlone, G.M. and Lascelles, J. 1982. Aerobic and anaerobic respiratory systems in *Campylobacter fetus* subsp. *jejuni* grown in atmospheres containing hydrogen. J. Bacteriol. 152: 306–314

Daucher, J.A. and Kreig, N.R. 1995. Pyruvate-ferredoxin oxidoreductase in *Campylobacter* species. Can. J. Microbiol. 41: 198–201.

Einsle, O., Stach, P., Messerschmidt, A., Simon, J., Kröger, A., Huber, R. and Kroneck, M.H. 2000. Cytochrome *c* nitrite reductase from *Wolinella succinogenes*. J. Biol. Chem. 275: 39608–39616

Finel, M. 1998. Does NADH play a central role in energy metabolism in *Helicobacter pylori*? Trends Biochem. Sci. 23: 412–414

Friedrich, T. 1998. The NADH: ubiquinone oxidoreductase (complex I) from *Escherichia coli*. Biochim. Biophys. Acta 1364: 134–146

Gennis, R.B. and Stewart, V. 1996. Respiration. In: *Escherichia coli* and *Salmonella typhimurium*. Neidhardt F.C. ed, ASM Press, Washington D.C. 2nd edition, Vol. 1, pp. 217–261

Gon, S., Guidici-Orticoni, M.T., Mejean, V. and Iobbi-Nivol, C. 2001. Electron transfer and binding of the *c*-type cytochrome TorC to the trimethylamine N-oxide reductase in *Escherichia coli*. J. Biol. Chem. 276: 11545–11551

Goodhew, C.F., ElKurdi, A.B. and Pettigrew, G.W. 1988. The microaerophilic respiration of *Campylobacter mucosalis*. Biochim. Biophys. Acta 933: 114–123.

Goss, J.A., Cohen, N.D. and Utter, M.F. 1981. Characterization of the subunit structure of pyruvate carboxylase from *Pseudomonas citronellolis*. J. Biol. Chem. 256: 11819–11825.

Grabowski, R., Hofmeister, A.E.M. and Buckel, W. 1993. Bacterial L-serine dehydratases: a new family of enzymes containing iron–sulfur clusters. Trends Biochem. Sci. 18: 297–300.

Harvey, S. and Lascelles, J. 1980. Respiration systems and cytochromes in *Campylobacter fetus* subsp. *intestinalis*. J. Bacteriol. 144: 917–922

Hoffman, P.S. and Goodman, T.G. 1982. Respiratory physiology and energy conservation efficiency of *Campylobacter jejuni*. J. Bacteriol. 150: 319–326

Hughes, N.J., Chalk, P.A., Clayton, C.L. and Kelly, D.J. 1995. Identification of carboxylation enzymes and characterization of a novel four-subunit pyruvate: flavodoxin oxidoreductase from *Helicobacter pylori*. J. Bacteriol. 177: 3953–3959

Hughes, N.J., Clayton, C.L., Chalk, P.A. and Kelly, D.J. 1998. *Helicobacter pylori porCDAB* and *oorDABC* genes encode distinct pyruvate: flavodoxin and 2-oxoglutarate: acceptor oxidoreductases which mediate electron transport to NADP. J. Bacteriol. 180: 1119–1128

Hussain, H., Grove, J., Griffiths, L., Busby, S. and Cole, J. 1994. A 7-gene operon essential for formate-dependent nitrite reduction to ammonia by enteric bacteria. Mol. Microbiol. 12: 153–163

Kather, B., Stingl, K., van der Rest, M.E., Altendorf, K. and Molenaar, D. 2000. Another unusual type of citric acid cycle enzyme in *Helicobacter pylori:* malate: quinone oxidoreductase. J. Bacteriol. 182: 3204–3209

Karmali, M.A., Roscoe, M. and Fleming, P.C. 1986. Modified ammonia electrode method to investigate D-asparagine breakdown by *Campylobacter* strains. J. Clin. Microbiol. 23, 743–747.

Kelly, D.J. 1998. The physiology and metabolism of the human gastric pathogen *Helicobacter pylori*. Adv. Microb. Physiol. 40: 137–189

Kelly, D.J. 2001. The physiology and metabolism of *Campylobacter jejuni* and *Helicobacter pylori*. J. Appl. Microbiol. 90: 16S-24S

Kelly, D.J. and Hughes, N.J. 2001 The citric acid cycle and fatty acid biosynthesis. In: *Helicobacter pylori*: Physiology and Genetics. Mobley, H.L.T., Mendz, G.L. and Hazell, S.L. eds. ASM Press, Washington, DC., pp. 135–146

Kelly, D.J., Hughes, N.J. and Poole, R.K. 2001. Microaerobic physiology: Aerobic respiration, Anaerobic respiration, and carbon dioxide metabolism. In: *Helicobacter pylori*: Physiology and Genetics. Mobley, H.L.T., Mendz, G.L. and Hazell, S.L. eds. ASM Press, Washington, D. C., pp 113–124.

Kelly, D.J., Richardson, D.J., Ferguson, S.J. and Jackson, J.B. 1988. Isolation of transposon Tn5 insertion mutants of *Rhodobacter capsulatus* unable to reduce trimethylamine-N-oxide and dimethylsulphoxide. Arch. Microbiol. 150: 138–144

Knablein, J., Mann, K., Ehlert, S., Fonstein, M., Huber, R. and Schneider, F. 1996. Isolation, cloning, sequence analysis and localization of the operon encoding dimethyl sulfoxide/trimethylamine N-oxide reductase from *Rhodobacter capsulatus*. J. Mol. Biol. 263: 40–52

Lascelles, J. and Calder, K. 1985. Participation of cytochromes in some oxidation-reduction systems in *Campylobacter fetus*. J. Bacteriol. 164: 401–409

Leach, S., Harvey, P. and Wait, R. 1997. Changes with growth rate in the membrane lipid composition of and amino acid utilization by continuous cultures of *Campylobacter jejuni*. J. Appl. Microbiol. 82: 631–640

Marceau, M., Lewis, S.D. and Shafer, J.A. 1988. The glycine-rich region of *Escherichia coli* D-serine dehydratase. Altered interaction with pyridoxal 5′-phosphate produced by substitution of aspartic acid for glycine. J. Biol. Chem. 263: 16934–16941

Mendz, G.L., Ball, G. E., and Meek, D.J. 1997. Pyruvate metabolism in *Campylobacter* spp. Biochim. Biophys. Acta. 1334: 291–302.

Nagata, K., Tsukita, S., Tamura, T. and Sone, N. 1996. A *cb*-type cytochrome-*c* oxidase terminates the respiratory chain in *Helicobacter pylori*. Microbiology 142: 1757–1763

Ogawa, H., Konishi, K., and Fujioka, M. 1989. The peptide sequences near the bound pyridoxal phosphate are conserved in serine dehydratase from rat liver and threonine dehydratases from yeast and *Escherichia coli*. Biochim. Biophys. Acta. 139: 139–141.

Olson, J.W. and Maier, R.J. 2002. Molecular hydrogen as an energy source for *Helicobacter pylori*. Science 298: 1788–1790.

Parkhill, J., Wren, B.W., Mungall, K., Ketley, J.M., Churcher, C., Basham, D., Chillingworth, T., Davies, R.M., Feltwell, T., Holroyd, S., Jagels, K., Karlyshev, A.V., Moule, S., Pallen, M.J., Penn, C.W., Quail, M.A., Rajandream, M-A, Rutherford, K.M., van Vliet, A.H.M., Whitehead, S. and Barrell, B.G. 2000. The genome sequence of the food-borne pathogen *Campylobacter jejuni* reveals hypervariable sequences. Nature 43: 665–668

Parsons, C.M., Potter, L.M. and Brown, R.D. Jr. 1982. Effects of dietry protein and intestinal microflora on excretion of amino acids in poultry. Poultry Sci. 61: 939–946.

Poole, R.K. and Cook, G.M. 2000. Redundancy of aerobic respiratory chains in bacteria: Routes, reasons and regulation. Adv. Microb. Physiol. 42: 165–224.

Potter, L., Angove, H., Richardson, D. and Cole, J. 2001. Nitrate reduction in the periplasm of Gram-negative bacteria. Adv. Microb. Physiol. 45: 51–112

Potter, L.C. and Cole J.A. (1999) Essential roles for the products of the *napABCD* genes, but not *napFGH*, in periplasmic nitrate reduction by *Escherichia coli* K-12. Biochem. J. 344: 69–67

Richardson, D.J., Berks, B.C., Russell, D.A., Spiro, S. and Taylor, C.J. 2001 Functional, biochemical and genetic diversity of prokaryotic nitrate reductases. Cell. Mol. Life Sci. 58: 165–178

Sellars, M.J., Hall, S,J. and Kelly, D.J. 2002. Growth of *Campylobacter jejuni* supported by respiration of fumarate, nitrate, nitrite, trimethylamine-N-oxide or dimethylsulfoxide requires oxygen. J. Bacteriol. 184: 4187–4196.

Simon, J., Gross, R., Einsle, O., Kroneck, P.M.H., Kröger, A. and Klimmek, O. 2000. A NapC/NirT-type cytochrome *c* (NrfH) is the mediator between the quinone pool and the cytochrome *c* nitrite reductase of *Wolinella succinogenes*. Mol. Microbiol. 35: 686–696

Smibert, R.M. 1984. Genus *Campylobacter* Sebald and Véron 1963 In: Bergey's Manual of Systematic Bacteriology, Vol. 1. N.R. Krieg and J.G. Holt, eds. Williams and Wilkins, Baltimore, MD, pp. 111–116

Smith, M.A., Finel, M., Korolik, V. and Mendz, G.L. 2000. Characteristics of the aerobic respiratory chains of the microaerophiles *Campylobacter jejuni* and *Helicobacter pylori*. Arch. Microbiol. 174: 1–10

Smith, M.A., Mendz, G.L., Jorgensen, M.A. and Hazell, S.L. 1999. Fumarate metabolism and the microaerophily of *Campylobacter jejuni*. Int. J. Biochem. Cell. Biol. 31: 961–975.

Spiro, S. and Guest, J.R. 1991. Adaptive responses to oxygen limitation in *Escherichia coli*. Trends Biochem. Sci. 16: 310–314

Tomb, J-F, White, O., Kerlavage, A.R., Clayton, R.A., Sutton, G.G., Fleishmann, R.D., Ketchum, K.A., Klenk, H.P., Gill, S., Dougherty, B., Nelson, K., Quakenbush, J., Zhou L, Kirkness, E.F., Peterson, S., Loftus, B., Richardson, D., Dodson, R., Khalak, H.G., Glodek, A., McKenney, K., Fitzegerald, L.M., Lee, N., Adams, M.D., Hickey, E., Berg, D.E., Gocayne, J.D., Utterback, T.R., Peterson, J.D., Kelley, J.M., Cotton, M.D., Weidman, J.M., Fujii, C., Bowman, C., Watthey, L., Wallin, E., Hayes, W.S., Borodovsky, M., Karp, P.D., Smith, H.O., Fraser, C.M. and Venter, J.C. 1997. The complete genome sequence of the gastric pathogen *Helicobacter pylori*. Nature. 388: 539–547

Velayudhan, J and Kelly, D.J. 2002. Analysis of gluconeogenic and anaplerotic enzymes in *Campylobacter jejuni*: an essential role for phosphoenolpyruvate carboxykinase. Microbiology. 48: 685–694.

Velayudhan, J., Jones, M.A., Barrow, P.A. and Kelly, D.J. 2004. L-serine catabolism via an oxygen-labile L-serine dehydratase (SdaA) is essential for colonization of the avian gut by *Campylobacter jejuni*. Infect. Immun. 72: 260–268.

Veron, M., Lenvoise-Furet, A. and Beaune, P. (1981) Anaerobic respiration of fumarate as a differential test between *Campylobacter fetus* and *Campylobacter jejuni*. Curr. Microbiol. 6: 349–354

Weber, I., Fritz, C., Ruttkowski, S., Kreft, A. and Bange, F.C. 2000. Anaerobic nitrate reductase (narGHJI) activity of *Mycobacterium bovis* BCG *in vitro* and its contribution to virulence in immunodeficient mice. Mol. Microbiol. 35: 1017–1025

Weiner, J.H., MacIsaac, D.P., Bishop, R.E. and Bilous, P.T. 1988. Purification and properties of *Escherichia coli* DMSO reductase, an iron–sulfur molybdoenzyme with broad substrate specificity. J. Bacteriol. 170: 1505–1510

Westfall, H.N., Rollins, D.M. and Weiss, E. 1986. Substrate utilization by *Campylobacter jejuni* and *Campylobacter coli*. Appl. Env. Microbiol. 52: 700–705.

Yagi, T., Yano, T., Di Bernardo, S. and Matsuno-Yagi, A. 1998. Prokaryotic complex I (NDH-1), an overview. Biochim. Biophys. Acta 1364: 125–133

Chapter 15

Iron Transport and Regulation

Karl G. Wooldridge and Arnoud H.M. van Vliet

Abstract

Given the small size of the genome of *Campylobacter jejuni*, a surprisingly large number of its genes are implicated in iron scavenging, metabolism, storage and regulation. This is likely to reflect the central role of iron in the host-pathogen relationship. *C. jejuni* is known to utilize haem iron and ferric siderophores, both of which are likely to be available during intestinal colonization and infection. Genomic data suggest that there are additional iron uptake pathways in *C. jejuni*: these include a ferrous iron transporter and receptors for additional, uncharacterized iron sources. Furthermore, there is genetic heterogeneity among *C. jejuni* isolates: some pathways are common to all isolates, while other pathways are restricted to a subset of strains. Two iron-responsive regulatory circuits in *C. jejuni* are responsible for regulation of iron homeostasis, and for protection against oxidative stress, respectively.

INTRODUCTION

In order for any pathogen to colonize a particular niche within its host it must be able to obtain sufficient nutrients to grow and proliferate. Iron in particular is known to be central to the host-pathogen relationship (Weinberg, 1978; Finkelstein *et al.*, 1983; Wooldridge and Williams, 1993). It is a cofactor of many enzymes involved in essential cellular processes, including electron transfer and DNA synthesis. In addition to its essential biological roles, it is also toxic because of its participation in Haber–Weiss–Fenton chemistry in which reactive oxygen species are liberated. These molecules can cause extensive damage to biological systems (Touati, 2000). Living organisms, from bacteria to mammals and birds, have evolved finely balanced systems for iron acquisition and storage to provide iron for their metabolism, without allowing the concentration of free iron to reach dangerous levels.

In mammals and birds most of the iron is sequestered within cells, where it may be complexed as a cofactor of various proteins, or stored in the iron-storage protein ferritin (Aisen *et al.*, 2001). Since iron is a nutrient required by all cells and tissues, it must be transported in significant amounts via the circulatory system. Within the blood, lymph and tissue fluid most of the extracellular iron is complexed with the iron-binding glycoprotein transferrin. This protein, which is normally only 30% saturated, serves to sequester iron in a form that is non-toxic. It is taken up into cells by a process of receptor-mediated endocytosis, and iron is released upon acidification of the endosome. Iron-free (*apo-*) transferrin is returned to the cell surface and released owing to a reduced affinity of *apo-* transferrin for the transferrin receptor (Aisen *et al.*, 2001). Mucosal secretions contain the related protein lactoferrin, which has no known role in iron transport but binds iron avidly making it unavailable to many bacteria. Unlike transferrin, which loses its affinity for iron in mildly acidic conditions, lactoferrin retains a high affinity for ferric iron at pH values as low as 3.5 (Abdallah and El Hage Chahine, 2000).

The mechanisms utilized by *C. jejuni* to obtain sufficient iron for metabolism and growth, as well as regulatory systems that sense iron levels and control the expression of genes necessary for iron acquisition and protection against oxidative damage, are beginning to be elucidated. Sequencing and annotation of the complete genome of one strain of *C. jejuni* (Parkhill *et al.*, 2000) has facilitated the identification of many genes with a potential role in iron acquisition and regulation, complementing experimental work in which a number of iron scavenging systems and regulatory circuits have been investigated (van Vliet *et al.*, 2002a). It is likely that iron acquisition and regulation is similar in the two closely related species *C. jejuni* and *C. coli* but there may also be important differences. Most of the experimental data available come from studies of *C. jejuni* but evidence from *C. coli* is also discussed. Figure 15.1 and Table 15.1 give an overview of all putative iron acquisition systems of *C. jejuni,* as identified by genome sequence analyses as well as functional analyses as further discussed in this chapter.

AVAILABILITY OF IRON IN THE INTESTINE

Since the intestinal mucosa is the principal site of colonization by *C. jejuni*, it is important to consider the iron status of this environment, which is complex and incompletely understood. Iron can exist in either ferrous (Fe^{2+}) or ferric (Fe^{3+}) forms, depending on the pH and redox conditions of the environment. Under oxidizing conditions, uncomplexed soluble ferrous iron is rapidly oxidized to the ferric form, which forms insoluble precipitates that are largely unavailable to biological systems. The redox environment of the gut, however, is variable, and oxygen levels are likely to be significantly lower than atmospheric levels. Furthermore, most of the available iron is likely to be complexed with other molecules that will affect both its oxidation state as well as its bioavailability.

There are two potential sources of iron in the intestine: the mucosa and food. Secretions of the intestine, like other mucosal secretions, contain lactoferrin, which will bind to the majority of the available ferric iron making it unavailable to most bacteria. Furthermore, it has been shown that lactoferrin secretion can be increased in response to infection by some enteric pathogens (Qadri *et al.*, 2002). Intestinal cells and, during inflammation, blood cells may also be released into the intestine and may be a potential source of haem iron. In food, iron is available as both haem iron (principally in myoglobin and haemoglobin) and non-haem iron (Lombard *et al.*, 1997). Non-haem ferric iron is likely to be reduced to the ferrous form in the very low pH environment of the stomach. Experimental and clinical observations have supported a role for reduction of iron in the stomach by gastric juice (Bezwoda *et al.*, 1978). Furthermore, ascorbic acid, which is concentrated in gastric juice (Rathbone *et al.*, 1989), is capable of forming monomeric complexes with ferrous iron (Conrad and Schade, 1968) and the bile salt taurocholate also binds ferrous iron avidly (Sanyal *et al.*, 1990). Such interactions may inhibit the oxidation and polymerization of iron after it leaves the low pH environment of the stomach.

Intestinal microflora produce iron-binding siderophores (discussed below), which compete with lactoferrin and bind ferric iron with high affinity. Thus, the potential sources of iron in the intestinal environment include ferric iron, bound to either lactoferrin or microbial siderophores, ferrous iron complexed with small molecules, and haem iron. Gram-negative bacteria have been shown to utilize all of these sources of iron and here we will discuss the evidence for utilization of these sources by *C. jejuni*.

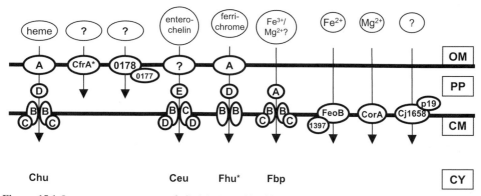

Figure 15.1 Iron transport systems of *C. jejuni*, as identified through comparative genome analysis and experimental data. For all systems their known or expected substrate is indicated wherever possible. Question marks indicate that the substrate is currently unknown. Systems marked with an asterisk are not present in all strains tested. OM, Outer membrane; CM, cytoplasmic membrane; PP, periplasm; CY, cytoplasm.

IRON ACQUISITION

Utilization of ferrous iron

Many bacteria are known to utilize ferrous iron and there are both high- and low-affinity pathways for ferrous iron uptake. The low-affinity pathway is non-saturable and has not been extensively characterized. A high-affinity ferrous iron transport system in *E. coli* is composed of two proteins, FeoA and FeoB, which are expressed under anaerobic conditions (Kammler *et al.*, 1993). The function of the FeoA protein is unknown but the FeoB protein is a cytoplasmic membrane protein with homology to ATPases. Homologues of *feoB* are present in many bacteria including *Salmonella enterica* serovar Typhimurium (Tsolis *et al.*, 1996) and *Helicobacter pylori* (Velayudhan *et al.*, 2000), and have been suggested to be one of the early archetypes of iron transporters (Hantke, 2003). The effect of *feoB* mutations varies among enteric and gastric pathogens. A *feoB* mutant of *S.* Typhimurium was not attenuated in oral or intraperitoneal infections of mice, although it was outcompeted by wild-type organisms in competitive intestinal colonization experiments (Tsolis *et al.*, 1996). By contrast, an *E. coli* K-12 *feo* mutant was deficient in murine colonization (Stojiljkovic *et al.*, 1993), and an *H. pylori feoB* mutant was unable to colonize the gastric mucosa of experimentally challenged mice (Velayudhan *et al.*, 2000). These data suggest that Feo-mediated ferrous iron uptake makes an important contribution to iron acquisition by *H. pylori* in the low pH environment of the stomach, and to *E. coli* in the intestine, but is of more marginal importance for intestinal colonization by *S.* Typhimurium, in which alternative (ferric) iron uptake pathways are in place. The genome of *C. jejuni* NCTC 11168 contains a homologue of FeoB (Cj1398) with 53% identity to its *H. pylori* counterpart. An additional gene (Cj1397), which is upstream of *feoB* and is probably part of the same operon, has weak identity to *E. coli feoA*. It is thus likely that *C. jejuni* is capable of utilizing ferrous iron via the Feo system, but this remains to be determined experimentally.

   In addition to the Feo system, *H. pylori* is capable of transporting ferrous iron via the CorA magnesium transporter (Hantke, 1997; Wainwright *et al.*, 2001; Pfeiffer *et al.*, 2002). The physiological relevance of this is unknown since under normal extracellular conditions the magnesium ion concentration is several orders of magnitude higher than

Table 15.1 Iron-related proteins of *C. jejuni*

	Number[b]	Localization[a]	MM (kDa)	Function
Transport-related				
CfrA	Cj0755	OM	77.6	Outer membrane receptor (ligand unknown)
ChuA	Cj1614	OM	80.9	Haemin receptor
ChuB	Cj1615	CM	35.6	Permease component of ABC transport system (putative haem transport)
ChuC	Cj1616	CM	29.5	Nucleotide-binding component of ABC transport system (putative haem transport)
ChuD	Cj1617	PP	30.1	Periplasmic binding protein component of ABC transport system (putative haem transport)
CeuB	Cj1352	CM	35.5	Permease component of ABC transport system (enterochelin transport)
CeuC	Cj1353	CM	34.9	Permease component of ABC transport system (enterochelin transport)
CeuD	Cj1354	CM	28.6	Nucleotide-binding component of ABC transport system (enterochelin transport)
CeuE	Cj1355	PP	36.3	Periplasmic binding protein component of ABC transport system (enterochelin transport)
_[c]	Cj0177	CM?	32.5	Lipoprotein, similar to PhuW from haemin uptake locus of *P. aeruginosa*.
_[c]	Cj0178	OM	84.1	Probable TonB-dependent receptor with homology to transferrin-binding proteins
_[c]	Cj0444	OM	72.9	Probable TonB-dependent receptor (pseudogene in *C. jejuni* NCTC 11168)
FhuA	_[d]	OM	83.8	Probable TonB-dependent receptor with homology to *E. coli* ferrichrome receptor
FhuB	_[d]	CM	66.1	Permease component of ABC transport system (putative ferrichrome transport)
FhuD	_[d]	PP	41.1	Periplasmic binding protein component of ABC transport system (putative ferrichrome transport)
FeoA	Cj1397	CM	8.4	Ferrous iron transport
FeoB	Cj1398	CM	69.0	
FbpA	Cj0175c	PP	37.5	Putative periplasmic binding protein component of ABC transport system
FbpB	Cj0174c	CM	62.4	Putative permease of ABC transport system
FbpC	Cj0173c	CM	34.8	Putative ATPase component of ABC transport system

TonB1	Cj0181	CM	28.1	TonB proteins transduce energy from the cytoplasmic membrane to outer membrane receptors. ExbB and ExbD are accessory proteins that serve to stabilise TonB. TonB3 is required for uptake of haemin, enterochelin and ferrichrome in *C. coli*.
ExbB1	Cj0179	CM	28.4	
ExbD1	Cj0180	CM	15.5	
TonB2	Cj1630	CM	26.3	
ExbB2	Cj1628	CM	15.8	
ExbD2	Cj1629	CM	14.7	
TonB3	Cj0753c	CM	26.4	
Storage				
Cft	Cj0612c	CY	19.5	Iron storage and protection from oxidative stress
Dps	Cj1534c	CY	17.3	Iron storage and protection from oxidative stress
Regulation				
Fur	Cj0400	CY	18.1	Repressor of many iron-regulated genes
PerR	Cj0322	CY	15.9	Iron-dependent repression of peroxide stress-related genes
Other				
p19	Cj1659	PP	19.6	Unknown

[a]Either experimentally determined or based on amino acid homology. Abbreviations: OM, outer membrane; CM, cytoplasmic membrane; PP, periplasm; CY, cytoplasm (see also Figure 15.1).

[b]Number of ORFs in *C. jejuni* NCTC 11168 genome

[c]No current designation.

[d]Not present in genomic sequence of *C. jejuni* NCTC 11168.

the iron concentration, and abolishes iron transport via CorA (Smith and Maguire, 1998). CorA has been reported to transport iron under low magnesium conditions (Hantke, 1997), however, and such conditions are known to exist inside eukaryotic cells. There is a homologue of CorA in strain NCTC 11168 (Cj0726c) and, since some *C. jejuni* strains are known to invade eukaryotic cells both *in vitro* and during infection (Wooldridge and Ketley, 1997), it is possible that CorA-mediated iron transport is important for intracellular survival during *C. jejuni* infection.

Siderophore-mediated iron scavenging

Many bacteria secrete low molecular weight iron-binding molecules called siderophores. These compounds, which are frequently either catechols or hydroxamates, are secreted from the bacterial cell, bind iron with extremely high affinity, and are then taken back up into the cell in an energy-dependent process via high-affinity outer membrane receptors (Braun *et al.*, 1998). The energy required for the transport of siderophores across the outer membrane is derived from the proton motive force (PMF) across the cytoplasmic membrane and is transduced via the TonB protein complex (Braun and Braun, 2002). This complex comprises the TonB protein and two accessory proteins: ExbB and ExbD (Braun and Braun, 2002). The complex is associated with the cytoplasmic membrane and is thought to span the periplasm allowing the TonB protein to interact with the outer membrane receptors (Hannavy *et al.*, 1990). Siderophore-dependent iron transport across the cytoplasmic membrane requires an additional set of proteins that constitute an ABC transport system. A periplasmic binding protein (PBP) binds the ferric siderophore complex and delivers it to a permease complex consisting of one or two integral membrane proteins, each in association with a peripheral membrane protein with ATPase activity. ABC transport systems involved in iron uptake by Gram-negative bacteria constitute a distinct subset of these ubiquitous transport systems (Linton and Higgins, 1998). Uptake of iron across the outer and cytoplasmic membranes via TonB-dependent and ABC transport systems, respectively, is illustrated in Figure 15.2.

Despite one early report of the production of uncharacterized siderophores by some strains of *C. jejuni* (Field *et al.*, 1986), it appears that most strains do not synthesize siderophores (Baig *et al.*, 1986; Field *et al.*, 1986; Pickett *et al.*, 1992). Furthermore, no genes encoding potential siderophore biosynthesis genes have been identified in the genome of *C. jejuni* NCTC 11168 (Parkhill *et al.*, 2000). The ability to utilize exogenous siderophores produced by other microorganisms, however, has been demonstrated convincingly. *C. jejuni* have been shown to utilize the catechol siderophore enterochelin and the hydroxamate siderophore ferrichrome (Field *et al.*, 1986). The hydroxamate siderophores aerobactin, desferrioxamine-B and rhodotorulic acid, on the other hand, were unable to supply iron to *C. jejuni* in feeding assays (Field *et al.*, 1986).

Enterochelin is produced by a wide variety of bacteria from the family Enterobacteriaceae. This compound is likely to be produced in significant amounts by resident microflora of the mammalian and avian small and large intestines, and thus may be a significant source of iron to *C. jejuni* during colonization and infection. Genes encoding components of the enterochelin uptake system have been identified in both *C. jejuni* (Park and Richardson, 1995) and *C. coli* (Richardson and Park, 1995). The *ceuBCDE* operon encodes components of an ABC transport system: the *ceuB* and *ceuC* genes both encode integral membrane proteins, which presumably form the permease complex in the cytoplasmic membrane; *ceuD* encodes the ATP-binding protein component and *ceuE*

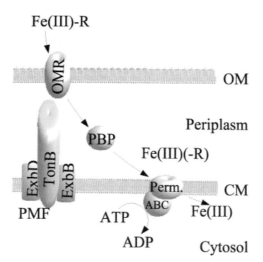

Fe(III)-R

OMR

OM

Periplasm

PBP

Fe(III)(-R)

ExhD TonB ExbB

Perm. CM

ABC Fe(III)

PMF ATP

ADP Cytosol

Figure 15.2 TonB- and ABC-dependent transport of iron compounds across the Gram-negative bacterial envelope. Ferric iron, complexed with siderophores or haem [Fe (III)-R], is bound by a specific outer membrane receptor (OMR). Transport across the outer membrane (OM) requires energy derived from the proton motive force (PMF) across the cytoplasmic membrane (CM) and transduced via the periplasm-spanning protein TonB. The cytoplasmic membrane proteins ExbB and ExbD serve to stabilize the TonB protein. After transfer across the outer membrane the iron complex is bound by a periplasmic binding protein (PBP), which delivers it to a permease complex in the cytoplasmic membrane. This complex consists of one or two integral membrane permease protein(s) (Perm.) and one or two ATP-binding cassette-containing protein(s) (ABC). Transport across the cytoplasmic membrane requires the hydrolysis of ATP.

encodes the periplasmic-binding protein (Figure 15.1) (Richardson and Park, 1995). Mutation studies using *C. coli* have shown that all of these genes are required for efficient uptake of enterochelin, at least for this *Campylobacter* species (Richardson and Park, 1995). No outer membrane receptor for enterochelin has been identified but a *C. coli tonB* mutant has been shown to be unable to utilize enterochelin (Guerry *et al.*, 1997), suggesting that there is an unidentified TonB-dependent outer membrane receptor for this ligand. While enterochelin may be an important source of iron during intestinal colonization it is not essential for successful colonization of the avian gut by *C. coli* since a *ceuE* mutant was not impaired in chick colonization experiments (Cawthraw *et al.*, 1996).

Enteric microorganisms are not known to produce ferrichrome, which is produced by certain soil fungi. It is possible that the ferrichrome uptake system in *C. jejuni* is also capable of transporting related but unidentified hydroxamate siderophores produced by enteric microorganisms. Alternatively, ferrichrome may be an important source of iron to *C. jejuni* outside of the host, for example during periods of survival in faecal material, which may become colonized by fungi. Recently, genes with homology to the *fhuABD* genes of *E. coli*, which encode the outer membrane receptor, the integral membrane permease and the periplasmic-binding protein components, respectively, of a ferrichrome uptake system, have been identified in a proportion of *C. jejuni* strains (Figure 15.1) (Galindo *et al.*, 2001). A gene encoding a putative ATP-binding protein component of this system has not been identified and a role for these genes in ferrichrome transport in *C. jejuni* remains to be demonstrated.

Utilization of haem compounds

Haem is an important potential source of iron to enteric microorganisms and may be available from food sources or as a result of inflammation of the intestine. A number of bacteria are known to utilize this compound as free haem, as part of the haemoglobin molecule, or in complexes with the haem-binding proteins haemopexin or albumin. These include *Vibrio cholerae* (Henderson and Payne, 1994), *Shigella dysenteriae* (Mills and Payne, 1997), enterohaemorrhagic *E. coli* (Torres and Payne, 1997), *Pseudomonas aeruginosa* (Ochsner *et al.*, 2000), *Neisseria* species (Stojiljkovic *et al.*, 1996), *Yersinia* species (Hornung *et al.*, 1996) and *Haemophilus influenzae* (Cope *et al.*, 1995). *C. jejuni* have been shown to utilize haem, haem bound to the acute-phase protein haemopexin, haemoglobin and haemoglobin complexed with haptoglobin (Pickett *et al.*, 1992).

There appear to be at least three distinct mechanisms of bacterial haem acquisition (Lee, 1995; Ochsner *et al.*, 2000). The most widespread mechanism is similar to the TonB-dependent siderophore transporters. These systems, which have been described in *P. aeruginosa* (Ochsner *et al.*, 2000), *Yersinia pestis* (Hornung *et al.*, 1996) and *Y. enterocholitica* (Stojiljkovic and Hantke, 1992) are composed of an outer membrane haem receptor, a TonB homologue and its accessory proteins ExbB and ExbD, and a periplasmic-binding protein-dependent system. The second category consists of an outer membrane receptor and a haem-binding protein that is secreted via a dedicated type I protein secretion system. Haem uptake systems of this type have been described in *Serratia marcescens* (Binet and Wandersman, 1996; Ghigo *et al.*, 1997) and *P. aeruginosa* (Ochsner *et al.*, 2000). A third mechanism for haem uptake that is found in a number of bacteria including *Haemophilus influenzae* is mediated by a haem-binding outer membrane lipoprotein (Hanson *et al.*, 1992; Reidl and Mekalanos, 1996).

A *C. jejuni* 70 kDa outer membrane protein with homology to TonB-dependent receptors was previously shown to be required for utilization of haem compounds (Pickett *et al.*, 1992; van Vliet *et al.*, 1998; Rock *et al.*, 2001). This gene, designated *chuA* (Cj1614), is adjacent to the *chuBCD* genes (Cj1615-Cj1617) that are predicted to encode components of an ABC transport system (Figure 15.1). Surprisingly, while a *chuA* mutant is deficient in haemin utilization, mutants in the *chuBCD* genes have no apparent defect (Rock *et al.*, 2001). This is reminiscent of the TonB- and ABC-dependent haem uptake systems in *V. cholerae* (Occhino *et al.*, 1998) and *Bradyrhizobium japonicum* (Nienaber *et al.*, 2001), both of which have operons containing genes encoding components of TonB-dependent and ABC transport systems in which the outer membrane receptor and TonB are absolutely required for haem uptake, but the ABC transport components are dispensable, at least under the experimental conditions used. It is possible that additional systems may compensate for the loss of the ABC transport components of these haem transport systems.

In addition to the *chuABCD* haem uptake system in *C. jejuni* there is a gene (Cj0178) (Figure 15.1) encoding a putative outer membrane receptor with significant homology to the genes encoding both the HasA and HasR proteins of *P. aeruginosa*. These proteins constitute a haem uptake system of the second category described above in which HasA is a secreted haem-binding protein and HasR is an outer membrane receptor (Ochsner *et al.*, 2000). Interestingly, this gene is adjacent to a gene (Cj0177) with homology to a *P. aeruginosa* gene of unknown function, *phuW*, which encodes a lipoprotein and is part of the TonB-dependent haem-uptake locus of this organism (Ochsner *et al.*, 2000). Cj0178, however, has higher homology to the bacterial transferrin and lactoferrin-binding proteins (discussed below) and the true role of this gene awaits mutational analysis. It appears that the genome of *C. jejuni* NCTC 11168 does not have any functional homologues of the *H. influenzae hpbA* gene, which encodes the haem-binding lipoprotein of the third type of haem uptake system described above (Hanson *et al.*, 1992).

Some bacteria are capable of increasing the available haem pool by elaboration of haemolysins. These toxins can lyse erythrocytes and other cells thus liberating haem compounds that can be subsequently used as a source of iron. *C. jejuni* has been shown to be haemolytic and two genes have been demonstrated to encode proteins with haemolytic activity. A gene that was subsequently shown to be the periplasmic binding protein component of the enterochelin uptake system, and named *ceuE,* conferred a haemolytic phenotype when cloned in *E. coli* (Park and Richardson, 1995). It is not known, however, if CeuE functions as a haemolysin in *C. jejuni*, and its periplasmic location argues against this possibility. Interestingly, CeuE is a lipoprotein (Richardson and Park, 1995): an unusual property for a periplasmic binding protein. It is possible, therefore, that it has a second role as a membrane-bound haemolysin. An outer membrane phospholipase A (PldA) of *C. coli* has been shown to be the major haemolytic determinant of at least one strain of this organism (Grant *et al.*, 1997). A *pldA* homologue is present in the genome of *C. jejuni* NCTC 11168 (Cj1351) (Parkhill *et al.*, 2000), where it is upstream of the *ceu* operon, although it is probably transcribed separately.

Another way in which intestinal pathogens can increase the supply of haem compounds is through tissue invasion leading to inflammation. Many strains of *C. jejuni* are known to invade human cells and campylobacter infection is often accompanied by a pronounced inflammatory response (Ketley, 1997; Wooldridge and Ketley, 1997). This results in release of erythrocytes and other cells into the intestinal lumen where, especially in the presence of haemolysins, they constitute an additional source of haem compounds.

Utilization of host iron-binding glycoproteins

A number of Gram-negative bacterial pathogens are capable of deriving iron directly from the host iron-binding glycoproteins transferrin and lactoferrin: pathogenic *Neisseria* species (Mickelsen *et al.*, 1982; Lee and Schryvers, 1988), *H. influenzae* (Schryvers, 1989), *Pasteurella haemolytica* (Ogunnariwo and Schryvers, 1990); *Actinobacillus pleuropneumoniae* (Gerlach *et al.*, 1992) and *Bordetella pertusis* (Menozzi *et al.*, 1991) are all able to use transferrin and/or lactoferrin as a sole iron source. Iron uptake from these molecules is dependent on highly specific receptors that are typically composed of two proteins. The first component is an essential, outer-membrane-embedded protein with homology to TonB-dependent siderophore receptors. An accessory protein interacts with the first component and is required for efficient binding of the host glycoprotein. These receptors display a high degree of host specificity and typically will only bind transferrin or lactoferrin from a particular host species. *H. pylori* has also been shown to utilize human lactoferrin (Husson *et al.*, 1993; Velayudhan *et al.*, 2000). An iron-regulated 70 kDa protein has been purified from the outer membrane of *H. pylori* that binds human lactoferrin but not human transferrin or bovine lactoferrin (Dhaenens *et al.*, 1997). However, utilization of lactoferrin and transferrin by *H. pylori* remains controversial (Husson *et al.*, 1993; Dhaenens *et al.*, 1999; Velayudhan *et al.*, 2000). This may be a strain-dependent phenomenon, and requires further investigation and identification of the gene encoding the receptor protein.

In the genome of *C. jejuni* NCTC 11168 there are no significant homologues of the accessory proteins, but one gene (Cj0178) has significant homology with the TonB-dependent outer membrane components of transferrin- and lactoferrin-binding proteins. When transferrin or lactoferrin were supplied as a sole iron source to several strains of *C. jejuni*, however, they did not support growth (Pickett *et al.*, 1992) suggesting that *C. jejuni* is unable to utilize these potential iron sources.

TonB and associated genes in *C. jejuni*

Since a *tonB* homologue identified in *C. coli* was demonstrated to be essential for utilization of enterochelin, ferrichrome and haem iron (Guerry *et al.*, 1997), it was surprising to discover that the genome of *C. jejuni* NCTC 11168 contained two additional *tonB* homologues: Cj0181 (*tonB1*) and Cj1630 (*tonB2*), in addition to the presumed orthologue of the gene identified in *C. coli*, which was designated *tonB3* (Cj0753c). Multiple paralogues of the *tonB*-associated genes *exbB* and *exbD* are also present in the genome of *C. jejuni* NCTC 11168. The *tonB3* gene is not coupled to any *exb* genes but *tonB1* and *tonB2* are both apparently transcriptionally coupled to a pair of *exbB*/*exbD* genes (Cj0179/Cj0180 and Cj1628/Cj1629, respectively). A third pair of *exbB*/*exbD* genes is not coupled to a *tonB* gene but rather is adjacent to an ATP synthase operon.

The presence of multiple *tonB* genes is not unprecedented. In *V. cholerae* a haem uptake locus contains an additional set of *tonB*/*exbB*/*exbD* genes (Occhino *et al.*, 1998). Both *V. cholerae tonB* genes can mediate transport of haemin and the siderophores vibriobactin and ferrichrome. Only *tonB1*, however, can mediate uptake of the siderophore shizokinen, and *tonB2* is required for uptake of enterochelin (Seliger *et al.*, 2001). Furthermore, *tonB1* was shown to be more efficient in haemin utilization, especially under conditions of high osmolarity (Seliger *et al.*, 2001).

It is not known whether the multiple *tonB* and *exbB*/*exbD* genes found in *C. jejuni* are also present in the *C. coli* strain in which the *tonB3* orthologue was shown to be required for uptake of haemin and siderophores: the reported phenotype may be specific for this *Campylobacter* species. It is possible that the multiple *tonB* and *exbB*/*exbD* genes in *C. jejuni* have overlapping as well as specific roles but this remains to be investigated.

Other *C. jejuni* genes with a potential role in iron uptake

A number of *C. jejuni* genes are known to have homology with bacterial iron uptake systems but are not associated with any known iron transport pathway. The *cfrA* gene (Cj0755) (Figure 15.1) shows significant homology to a number of TonB-dependent siderophore receptor genes. It was originally described in *C. coli* (Guerry *et al.*, 1997) and is adjacent to a *tonB* gene (*tonB3*; Cj0753c, see above). A *C. coli cfrA* knockout strain did not have any defect in utilization of enterochelin, ferrichrome or haem. It is likely, therefore, that the CfrA protein is the receptor for an as yet unidentified siderophore. Hybridization experiments showed that *cfrA* was present in all *C. coli* strains tested but in only a subset of *C. jejuni* strains (Guerry *et al.*, 1997).

Another gene with extensive homology to TonB-dependent outer membrane receptors (Cj0444) is a pseudogene in the genome of strain NCTC 11168 (Parkhill *et al.*, 2000). It is possible, however, that in other strains this gene encodes a TonB-dependent receptor for an unidentified siderophore.

Upstream of Cj0177 and Cj0178, and transcribed in the opposite direction, there is another set of genes (Cj0173c–Cj0176c) that are homologous to genes encoding an unusual ferric iron transporter originally described in *S. marcescens* (Zimmermann *et al.*, 1989). Such systems are reminiscent of ABC transporter-dependent ferric iron transport systems, but they lack an outer membrane receptor and are TonB-independent. These genes, tentatively called *fbpA-C* (van Vliet *et al.*, 2002a), encode the periplasmic-binding component, the inner membrane permease component and the nucleotide-binding components, respectively, of an ABC transport system. Recently it has been suggested that these ABC-transport systems represent manganese transporters, which would explain the lack of associated outer membrane receptor (Kehres and Maguire, 2003).

An iron-regulated gene (Cj1659) encoding a 19 kDa periplasmic protein, p19, has been described (Janvier *et al.*, 1998; van Vliet *et al.*, 1998). A gene (Cj1658) encoding a putative membrane protein precedes Cj1659 and both genes have presumed orthologues in an iron uptake pathogenicity island of *Y. pestis* (Carniel, 2001). An orthologue of Cj1659 in the magnetotactic bacterium MV-1 is essential for the formation of Fe_3O_4-containing magnetosomes, and it is suggested to function in iron acquisition using a mechanism similar to the ferroxidase/ferric iron transport system of *Sacharomyces cerevisiae* (B.L. Dubbels and D.A. Bazylinski, personal communication). The genes in *C. jejuni*, however, have not been demonstrated to have any role in iron transport.

IRON STORAGE

When iron is abundant it is advantageous to store some iron against future iron starvation. Iron in excess, however, can be toxic, so bacteria require strategies to store iron in a non-toxic form. Two distantly related bacterial proteins are capable of iron storage: the non-haem-containing ferritins, which are also present in eukaryotic cells, and the haem-containing bacterioferritins (Andrews, 1998). Both proteins form hollow spherical multimers in which in excess of 2000 iron atoms can be stored in the form of a ferric-hydroxyphosphate core (Andrews, 1998). In *H. pylori*, the ferritin protein Pfr was shown to be important for iron storage since a *pfr* mutant had a reduced ability to build up intracellular stores of iron (Bereswill *et al.*, 1998) which, in the parental strain, were shown to be mobilized and to support growth under conditions of restricted iron availability (Waidner *et al.*, 2002). The *pfr* mutant was also more susceptible to iron toxicity, especially under low pH conditions, and was unable to colonize the gastric mucosa in a Mongolian gerbil model (Waidner *et al.*, 2002).

The genome of *C. jejuni* contains both a ferritin gene, *cft* (Cj0612c) and a putative bacterioferritin gene, *dps* (Cj1534c). The Cft protein has been shown to contain iron and the growth of a *cft* mutant was inhibited under low iron conditions, indicating that this protein is a functional iron storage protein (Wai *et al.*, 1996). The *cft* mutant was also more sensitive to oxidative stress than its parent (Wai *et al.*, 1996). The *C. jejuni* bacterioferritin gene is homologous to the neutrophil-activating protein (NAP) of *H. pylori*. NAP has been shown to bind iron (Tonello *et al.*, 1999) and it has also been reported to be a DNA-binding protein with similarities to the Dps family of DNA-binding proteins (Almiron *et al.*, 1992). Expression of NAP is repressed under conditions of iron starvation, an effect at least partly regulated by the Fur protein (see below) (Cooksley *et al.*, 2003). An *H. pylori* *dps* mutant was shown to be more sensitive to oxygen than the wild-type strain (Tonello *et al.*, 1999). The *C. jejuni* Dps protein has recently been cloned from strain 81-176, and was constitutively expressed, able to bind up to 40 atoms of iron per monomer, but did not appear to bind DNA (Ishikawa *et al.*, 2003). An isogenic *C. jejuni* *dps* mutant was more sensitive to H_2O_2; an effect that was abrogated by chelation of free ferric iron (Ishikawa *et al.*, 2003). These data suggest that *C. jejuni* bacterioferritin/Dps has a role in protection against oxidative stress when intracelluar iron levels are high.

IRON-RESPONSIVE GENETIC REGULATION

The levels of iron encountered by *C. jejuni* in the host gut will be different than those encountered during periods outside the host. Iron levels in the intestinal mucosa will also vary in response to alimentation. Invasive *C. jejuni* will encounter different levels of iron in the submucosa or within cells after cellular invasion, and availability of host iron will vary in response to inflammation, in which additional haem iron may become available but levels of free ferric iron may be reduced by increased secretion of lactoferrin. *C. jejuni* must obtain sufficient iron to meet its needs but also avoid the consequences of iron overload and oxidative damage that may result from the production of reactive oxygen species (Touati, 2000). In order to respond to changes in iron levels *C. jejuni*, like other bacteria, is able to regulate expression of genes involved in iron acquisition, storage and defence against oxidative stress (Field *et al.*, 1986; Wooldridge *et al.*, 1994; Chan *et al.*, 1995; van Vliet *et al.*, 1998; van Vliet *et al.*, 1999). This is achieved by the iron-responsive repressor proteins Fur and PerR.

The Fur repressor

Many Gram-negative bacteria achieve iron-responsive genetic regulation via the Fur repressor protein (Escolar *et al.*, 1999). When intracellular ferrous iron is present in excess it is bound by the Fur protein and the Fur–Fe^{2+} complex then binds to specific operator sequences, referred to as Fur-boxes, which are positioned upstream of iron responsive genes. A functional *C. jejuni fur* homologue has been identified (Wooldridge *et al.*, 1994; Chan *et al.*, 1995) and shown to regulate a number of genes in response to iron levels including the outer membrane receptors CfrA and ChuA, the periplasmic binding proteins CeuE and ChuD, and the periplasmic protein p19 (van Vliet *et al.*, 1998). The consensus Fur-binding sequence that has been determined previously for *E. coli* Fur-regulated genes (Escolar *et al.*, 1999) was recently reinterpreted for *B. subtilis* (Baichoo and Helmann, 2002), *E. coli* (Lavrrar and McIntosh, 2003), *H. pylori* (Delany *et al.*, 2001; van Vliet *et al.*, 2002b; van Vliet *et al.*, 2003) and *C. jejuni* (van Vliet *et al.*, 2002a). The *C. jejuni* consensus (six trimers of 5′-NAT of which at least 4 trimers should match and where N is any nucleotide) identified at least one Fur box in each known Fur-regulated gene in *C. jejuni* and, additionally, identified Fur boxes upstream of a number of other genes with a known or presumed role in iron uptake or storage (van Vliet *et al.*, 2002a). These included the genes encoding the ferrous iron transporter FeoB, the periplasmic binding protein CeuB, the outer membrane receptors Cj0178 and FhuA, one of the ExbB proteins and the putative bacterioferritin gene Cj1534c (van Vliet *et al.*, 2002a). The region directly upstream of the *fur* gene itself contains sequences with similarity to Fur boxes suggesting that *fur* may be autoregulated as it is in *E. coli* (Wooldridge *et al.*, 1994; Chan *et al.*, 1995). The *fur* gene in *C. jejuni* strain 81116, however, does not seem to have its own promoter (van Vliet *et al.*, 2000) and is probably expressed from non-iron-regulated promoters of two genes upstream of *fur* (Chan *et al.*, 1995; van Vliet *et al.*, 2000). It is still possible that Fur-binding upstream of the *fur* gene may modulate expression of the *fur* gene but experimental evidence for this is lacking.

The PerR repressor

By comparing protein expression in wild-type and *fur* mutant cells under conditions of high and low iron availability two cytosolic proteins; catalase (KatA) and alkyl hydroperoxide reductase (AhpC), were shown to be iron repressible by a Fur-independent mechanism (van Vliet *et al.*, 1998). These proteins, encoded by *katA* (Cj1385) and *ahpC* (Cj0334), are involved in defence against hydrogen peroxide and hydroperoxides, respectively (Grant and Park, 1995; Baillon *et al.*, 1999; van Vliet *et al.*, 1999). Upstream of *ahpC*, and transcribed divergently, a third gene involved in oxidative stress defence, *fdxA*, was shown to be iron-induced (van Vliet *et al.*, 2001). The genome of *C. jejuni* NCTC 11168 was subsequently shown to contain a homologue of the *perR* gene of *Bacillus subtilis* (van Vliet *et al.*, 1999). This gene is a homologue of *fur* and is responsible for iron- and manganese-dependent regulation of *ahpC*, *katA* and a number of other genes in *B. subtilis* (Bsat *et al.*, 1998). *C. jejuni perR* mutants were subsequently shown to express very high levels of both catalase and alkyl hydroperoxide reductase and to be highly resistant to peroxide stress (van Vliet *et al.*, 1999). PerR repressors have been described in only a small number of bacteria and the *C. jejuni* PerR protein is the only repressor of this type to be described in a Gram-negative bacterium (Mongkolsuk and Helmann, 2002). Like Fur repressors, PerR repressors also recognize a specific sequence upstream of their target genes although the consensus sequence has only been confirmed in *B. subtilis* (Herbig and Helmann, 2001). Identification of PerR-binding sites is complicated by their similarity to Fur-binding sequences. In *C. jejuni*, however, putative PerR-binding sequences have been identified upstream of the *katA* and *ahpC* genes, and several other genes with a role in oxidative stress defence also contain putative Fur or PerR boxes including the thioredoxin reductase gene (*trxB*; Cj0146c), the superoxide dismutase gene *sodB* (Cj0169), the bacterioferritin homologue (Cj1534c), the cytochrome c peroxidase gene *ccp* (Cj0358) and the *perR* gene itself (Baillon *et al.*, 1999; van Vliet *et al.*, 2000; van Vliet *et al.*, 2001; van Vliet *et al.*, 2002a).

The Fur and PerR regulons both respond to iron levels and have similar recognition sequences. The Fur regulon is composed mainly of genes with a role in iron homeostasis, while the two confirmed PerR-regulated genes as well as other genes with putative PerR boxes are involved in oxidative stress defence. There is at least some overlap between these systems, however, since the *katA* gene was still partially regulated by Fur in a *perR* mutant. Conversely, expression of Fur-regulated genes was not affected in a *perR* mutant (van Vliet *et al.*, 1999). It is possible, therefore, that PerR boxes are a subset of Fur boxes and that Fur recognizes both Fur and PerR- boxes, but PerR only recognizes PerR boxes. This arrangement, together with regulated expression of both repressors, will allow the fine-tuning of expression levels of iron-regulated genes under different environmental conditions (Fuangthong *et al.*, 2002; Mongkolsuk and Helmann, 2002; van Vliet *et al.*, 2002a).

CONCLUSIONS AND FUTURE DIRECTIONS

We have described the potential sources of iron that are likely to be available to *C. jejuni* during intestinal colonization and infection. Of these, *C. jejuni* has been shown to be able to utilize haem compounds and some siderophores. In addition, clues from the genome sequence indicate that other iron sources may also be important. These include ferrous iron and possibly additional, unidentified siderophores. Perhaps surprisingly, it appears that *C. jejuni* is not able to utilize lactoferrin, which is a source of iron for many mucosal pathogens including the closely related pathogen *H. pylori* (Dhaenens *et al.*, 1997; Velayudhan *et al.*, 2000).

C. jejuni is a highly heterogeneous species and it has been suggested that there is a core set of genes present in all strains and other, contingency genes, which are not essential under all conditions and are only found in a subset of *C. jejuni* strains (Dorrell *et al.*, 2001). The *fhuABD* genes and the *cfrA* gene, for example, are only present in some isolates of *C. jejuni* (Guerry *et al.*, 1997; Dorrell *et al.*, 2001; Galindo *et al.*, 2001); it is possible that other strain-restricted iron uptake systems remain to be discovered.

The presumed roles of a number of systems that have been inferred from the genome remain to be demonstrated experimentally. These include the FeoB-mediated ferrous iron uptake system, the putative ferrichrome transport system mediated by the products of the *fhuABC* genes and the putative ferric transport system mediated by the products of the *fbpABC* genes. The role of other genes with a presumed role in iron uptake will be more difficult to determine since there is no obvious iron source to test. These systems include the *cfrA* siderophore receptor homologue and the transferrin-binding protein homologue Cj0178.

Comparative genome analysis, facilitated by the recent release of the genome sequence of the enteric *Helicobacter* species *H. hepaticus* (Suerbaum *et al.*, 2003), could also give new insights into the role of the different iron-acquisition and iron-responsive regulatory systems. *C. jejuni*, *H. hepaticus* and *H. pylori* share the Fur regulator, but the *H. hepaticus* genome encodes both a PerR-like regulator (absent in *H. pylori*) (van Vliet *et al.*, 1999) and a NikR nickel-responsive regulator (absent in *C. jejuni*) (van Vliet *et al.*, 2002c). Similarly, *H. hepaticus* contains homologues of the *C. jejuni* OM receptors CfrA and ChuA and of the *H. pylori* FrpB OM receptor. The elucidation of the roles of these genes in any of the three species may lead to a better understanding of the diverse roles of these systems in enteric colonization by *C. jejuni*.

It is possible that there are redundant as well as exclusive roles for the multiple *tonB* and *exbB*/*exbD* genes present in the *C. jejuni* genome. The roles of each gene needs to be determined by constructing mutants in individual as well as combinations of these genes, and testing the mutants for their ability to utilize iron via the known iron transport pathways. Similarly, the precise mechanisms by which Fur and PerR cooperate, and the full complement of genes that are regulated by these repressors in *C. jejuni*, await further experimental evidence.

A direct role in pathogenesis or colonization for any of the iron transport systems described has yet to be demonstrated, and this is largely due to the lack of a good model for *C. jejuni* pathogenesis (Newell, 2001). A role for the *fur* gene has been demonstrated in chick colonization (van Vliet *et al.*, 2002a) and this model may provide clues as to the relative importance of iron transport systems during intestinal colonization by *C. jejuni*.

Note added in proof

The expression profile of *C. jejuni* under iron limitation and in a *fur* mutant have been published recently (Palyada *et al.* 2004. J. Bacteriol. 186: 4714–4729; and Holmes *et al.* 2005. Microbiology 151: 243–257).

Acknowledgements

We thank Brad Dubbels and Dennis Bazylinski for useful comments and Julian Ketley, Charles Penn, Peter Williams and our many other colleagues for their intellectual input over the years. Research on *Campylobacter jejuni* iron metabolism by the authors has been supported by grants from the Wellcome Trust.

References

Abdallah, F.B. and El Hage Chahine, J.M. 2000. Transferrins: iron release from lactoferrin. J. Mol. Biol. 303: 255–266.

Aisen, P., Enns, C. and Wessling-Resnick, M. 2001. Chemistry and biology of eukaryotic iron metabolism. Int. J. Biochem. Cell Biol. 33: 940–959.

Almiron, M., Link, A.J., Furlong, D. and Kolter, R. 1992. A novel DNA-binding protein with regulatory and protective roles in starved *Escherichia coli*. Genes Dev. 6: 2646–2654.

Andrews, S.C. 1998. Iron storage in bacteria. Adv. Microb. Physiol. 40: 281–351.

Baichoo, N. and Helmann, J.D. 2002. Recognition of DNA by Fur: a reinterpretation of the Fur box consensus sequence. J Bacteriol 184: 5826–5832.

Baig, B.H., Wachsmuth, I.K. and Morris, G.K. 1986. Utilization of exogenous siderophores by *Campylobacter* species. J. Clin. Microbiol. 23: 431–433.

Baillon, M.L.A., van Vliet, A.H.M., Ketley, J.M., Constantinidou, C. and Penn, C.W. 1999. An iron-regulated alkyl hydroperoxide reductase (AhpC) confers aerotolerance and oxidative stress resistance to the microaerophilic pathogen *Campylobacter jejuni*. J. Bacteriol. 181: 4798–4804.

Bereswill, S., Waidner, U., Odenbreit, S., Lichte, F., Fassbinder, F., Bode, G. and Kist, M. 1998. Structural, functional and mutational analysis of the *pfr* gene encoding a ferritin from *Helicobacter pylori*. Microbiology 144: 2505–2516.

Bezwoda, W., Charlton, R., Bothwell, T., Torrance, J. and Mayet, F. 1978. The importance of gastric hydrochloric acid in the absorption of nonheme food iron. J. Lab. Clin. Med. 92: 108–116.

Binet, R. and Wandersman, C. 1996. Cloning of the *Serratia marcescens hasF* gene encoding the Has ABC exporter outer membrane component: a TolC analogue. Mol. Microbiol. 22: 265–273.

Braun, V., Hantke, K. and Köster, W. 1998. Bacterial iron transport: mechanisms, genetics, and regulation. Met. Ions Biol. Syst. 35: 67–145.

Braun, V. and Braun, M. 2002. Active transport of iron and siderophore antibiotics. Curr. Opin. Microbiol. 5: 194–201.

Bsat, N., Herbig, A., Casillas-Martinez, L., Setlow, P. and Helmann, J.D. 1998. *Bacillus subtilis* contains multiple Fur homologues: identification of the iron uptake (Fur) and peroxide regulon (PerR) repressors. Mol. Microbiol. 29: 189–198.

Carniel, E. 2001. The *Yersinia* high-pathogenicity island: an iron-uptake island. Microbes Infect. 3: 561–569.

Cawthraw, S., Park, S.F., Wren, B.W., Ketley, J.M., Ayling, R. and Newell, D.G. 1996. Usefulness of the chick colonization model to investigate potential colonization factors of campylobacters. In: *Campylobacter*, *Helicobacter* and Related Organisms. D.G. Newell, J.M. Ketley, and R.A. Feldman, eds. Plenum Press, New York, pp. 649–652.

Chan, V.L., Louie, H. and Bingham, H.L. 1995. Cloning and transcription regulation of the ferric uptake regulatory gene of *Campylobacter jejuni* TGH9011. Gene 164: 25–31.

Conrad, M.E. and Schade, S.G. 1968. Ascorbic acid chelates in iron absorption: a role for hydrochloric acid and bile. Gastroenterology 55: 35–45.

Cooksley, C., Jenks, P.J., Green, A., Cockayne, A., Logan, R.P. and Hardie, K.R. 2003. NapA protects *Helicobacter pylori* from oxidative stress damage, and its production is influenced by the ferric uptake regulator. J. Med. Microbiol 52: 461–469.

Cope, L.D., Yogev, R., Muller-Eberhard, U. and Hansen, E.J. 1995. A gene cluster involved in the utilization of both free heme and heme: hemopexin by *Haemophilus influenzae* type b. J. Bacteriol. 177: 2644–2653.

Delany, I., Spohn, G., Rappuoli, R. and Scarlato, V. 2001. The Fur repressor controls transcription of iron-activated and -repressed genes in *Helicobacter pylori*. Mol. Microbiol. 42: 1297–1309.

Dhaenens, L., Szczebara, F. and Husson, M.O. 1997. Identification, characterization, and immunogenicity of the lactoferrin-binding protein from *Helicobacter pylori*. Infect. Immun. 65: 514–518.

Dhaenens, L., Szczebara, F., Van Nieuwenhuyse, S. and Husson, M.O. 1999. Comparison of iron uptake in different *Helicobacter* species. Res. Microbiol. 150: 475–481.

Dorrell, N., Mangan, J.A., Laing, K.G., Hinds, J., Linton, D., Al-Ghusein, H., Barrell, B.G., Parkhill, J., Stoker, N.G., Karlyshev, A.V., Butcher, P.D. and Wren, B.W. 2001. Whole genome comparison of *Campylobacter jejuni* human isolates using a low-cost microarray reveals extensive genetic diversity. Genome Res. 11: 1706–1715.

Escolar, L., Perez-Martin, J. and de Lorenzo, V. 1999. Opening the iron-box: transcriptional metalloregulation by the Fur protein. J. Bacteriol. 181: 6223–6229.

Field, L.H., Headley, V.L., Payne, S.M. and Berry, L.J. 1986. Influence of iron on growth, morphology, outer membrane protein composition, and synthesis of siderophores in *Campylobacter jejuni*. Infect. Immun. 54: 126–132.

Finkelstein, R.A., Sciortino, C.V. and McIntosh, M.A. 1983. Role of iron in microbe-host interactions. Rev. Infect. Dis. 5: S759–777.

Fuangthong, M., Herbig, A.F., Bsat, N. and Helmann, J.D. 2002. Regulation of the *Bacillus subtilis fur* and *perR* genes by PerR: not all members of the PerR regulon are peroxide inducible. J. Bacteriol. 184: 3276–3286.

Galindo, M.A., Day, W.A., Raphael, B.H. and Joens, L.A. 2001. Cloning and characterization of a *Campylobacter jejuni* iron-uptake operon. Curr. Microbiol. 42: 139–143.

Gerlach, G.F., Anderson, C., Potter, A.A., Klashinsky, S. and Willson, P.J. 1992. Cloning and expression of a transferrin-binding protein from *Actinobacillus pleuropneumoniae*. Infect. Immun. 60: 892–898.

Ghigo, J., Letoffe, S. and Wandersman, C. 1997. A new type of hemophore-dependent heme acquisition system of *Serratia marcescens* reconstituted in *Escherichia coli*. J. Bacteriol. 179: 3572–3579.

Grant, K.A. and Park, S.F. 1995. Molecular characterization of *katA* from *Campylobacter jejuni* and generation of a catalase-deficient mutant of *Campylobacter coli* by interspecific allelic exchange. Microbiology 141: 1369–1376.

Grant, K.A., Belandia, I.U., Dekker, N., Richardson, P.T. and Park, S.F. 1997. Molecular characterization of *pldA*, the structural gene for a phospholipase A from *Campylobacter coli*, and its contribution to cell-associated hemolysis. Infect. Immun. 65: 1172–1180.

Guerry, P., Perez-Casal, J., Yao, R., McVeigh, A. and Trust, T.J. 1997. A genetic locus involved in iron utilization unique to some *Campylobacter* strains. J. Bacteriol. 179: 3997–4002.

Hannavy, K., Barr, G.C., Dorman, C.J., Adamson, J., Mazengera, L.R., Gallagher, M.P., Evans, J.S., Levine, B.A., Trayer, I.P. and Higgins, C.F. 1990. TonB protein of *Salmonella typhimurium*. A model for signal transduction between membranes. J. Mol. Biol. 216: 897–910.

Hanson, M., Slaughter, C. and Hansen, E. 1992. The *hbpA* gene of *Haemophilus influenzae* type b encodes a heme-binding lipoprotein conserved among heme-dependent *Haemophilus* species. Infect. Immun. 60: 2257–2266.

Hantke, K. 1997. Ferrous iron uptake by a magnesium transport system is toxic for *Escherichia coli* and *Salmonella typhimurium*. J. Bacteriol. 179: 6201–6204.

Hantke, K. 2003. Is the bacterial ferrous iron transporter FeoB a living fossil? Trends Microbiol 11: 192–195.

Henderson, D. and Payne, S. 1994. Characterization of the *Vibrio cholerae* outer membrane heme transport protein HutA: sequence of the gene, regulation of expression, and homology to the family of TonB-dependent proteins. J. Bacteriol. 176: 3269–3277.

Herbig, A.F. and Helmann, J.D. 2001. Roles of metal ions and hydrogen peroxide in modulating the interaction of the *Bacillus subtilis* PerR peroxide regulon repressor with operator DNA. Mol. Microbiol. 41: 849–859.

Hornung, J.M., Jones, H.A. and Perry, R.D. 1996. The hmu locus of *Yersinia pestis* is essential for utilization of free haemin and haem–protein complexes as iron sources. Mol. Microbiol. 20: 725–739.

Husson, M.O., Legrand, D., Spik, G. and Leclerc, H. 1993. Iron acquisition by *Helicobacter pylori*: importance of human lactoferrin. Infect. Immun. 61: 2694–2697.

Ishikawa, T., Mizunoe, Y., Kawabata, S., Takade, A., Harada, M., Wai, S.N., and Yoshida, S. 2003. The iron-binding protein Dps confers hydrogen peroxide stress resistance to *Campylobacter jejuni*. J. Bacteriol. 185: 1010–1017.

Janvier, B., Constantinidou, C., Aucher, P., Marshall, Z.V., Penn, C.W. and Fauchere, J.L. 1998. Characterization and gene sequencing of a 19-kDa periplasmic protein of *Campylobacter jejuni/coli*. Res. Microbiol. 149: 95–107.

Kammler, M., Schon, C. and Hantke, K. 1993. Characterization of the ferrous iron uptake system of *Escherichia coli*. J. Bacteriol. 175: 6212–6219.

Kehres, D.G., and Maguire, M.E. 2003. Emerging themes in manganese transport, biochemistry and pathogenesis in bacteria. FEMS Microbiol. Rev. 27: 263–290.

Ketley, J.M. 1997. Pathogenesis of enteric infection by *Campylobacter*. Microbiology 143: 5–21.

Lavrrar, J.L., and McIntosh, M.A. 2003. Architecture of a fur binding site: a comparative analysis. J. Bacteriol. 185: 2194–2202.

Lee, B.C. and Schryvers, A.B. 1988. Specificity of the lactoferrin and transferrin receptors in *Neisseria gonorrhoeae*. Mol. Microbiol. 2: 827–829.

Lee, B.C. 1995. Quelling the red menace: haem capture by bacteria. Mol. Microbiol. 18: 383–390.

Linton, K.J. and Higgins, C.F. 1998. The *Escherichia coli* ATP-binding cassette (ABC) proteins. Mol. Microbiol. 28: 5–13.

Lombard, M., Chua, E. and O'Toole, P. 1997. Regulation of intestinal non-haem iron absorption. Gut 40: 435–439.

Menozzi, F.D., Gantiez, C. and Locht, C. 1991. Identification and purification of transferrin- and lactoferrin-binding proteins of *Bordetella pertussis* and *Bordetella bronchiseptica*. Infect. Immun. 59: 3982–3988.

Mickelsen, P.A., Blackman, E. and Sparling, P.F. 1982. Ability of *Neisseria gonorrhoeae*, *Neisseria meningitidis*, and commensal *Neisseria* species to obtain iron from lactoferrin. Infect. Immun. 35: 915–920.

Mills, M. and Payne, S.M. 1997. Identification of *shuA*, the gene encoding the heme receptor of *Shigella dysenteriae*, and analysis of invasion and intracellular multiplication of a *shuA* mutant. Infect. Immun. 65: 5358–5363.

Mongkolsuk, S. and Helmann, J.D. 2002. Regulation of inducible peroxide stress responses. Mol. Microbiol. 45: 9–15.

Newell, D.G. 2001. Animal models of *Campylobacter jejuni* colonization and disease and the lessons to be learned from similar *Helicobacter pylori* models. J. Appl. Microbiol. 90: 57S-67S.

Nienaber, A., Hennecke, H. and Fischer, H.M. 2001. Discovery of a haem uptake system in the soil bacterium *Bradyrhizobium japonicum*. Mol. Microbiol. 41: 787–800.

Occhino, D.A., Wyckoff, E.E., Henderson, D.P., Wrona, T.J. and Payne, S.M. 1998. *Vibrio cholerae* iron transport: haem transport genes are linked to one of two sets of *tonB*, *exbB*, *exbD* genes. Mol. Microbiol. 29: 1493–1507.

Ochsner, U.A., Johnson, Z. and Vasil, M.L. 2000. Genetics and regulation of two distinct haem-uptake systems, *phu* and *has*, in *Pseudomonas aeruginosa*. Microbiology 146: 185–198.

Ogunnariwo, J.A. and Schryvers, A.B. 1990. Iron acquisition in *Pasteurella haemolytica*: expression and identification of a bovine-specific transferrin receptor. Infect. Immun. 58: 2091–2097.

Park, S.F. and Richardson, P.T. 1995. Molecular characterization of a *Campylobacter jejuni* lipoprotein with homology to periplasmic siderophore-binding proteins. J. Bacteriol. 177: 2259–2264.

Parkhill, J., Wren, B.W., Mungall, K., Ketley, J.M., Churcher, C., Basham, D., Chillingworth, T., Davies, R.M., Feltwell, T., Holroyd, S., Jagels, K., Karlyshev, A.V., Moule, S., Pallen, M.J., Penn, C.W., Quail, M.A., Rajandream, M.A., Rutherford, K.M., van Vliet, A.H.M., Whitehead, S. and Barrell, B.G. 2000. The genome sequence of the food-borne pathogen *Campylobacter jejuni* reveals hypervariable sequences. Nature 403: 665–668.

Pfeiffer, J., Guhl, J., Waidner, B., Kist, M. and Bereswill, S. 2002. Magnesium uptake by CorA is essential for viability of the gastric pathogen *Helicobacter pylori*. Infect. Immun. 70: 3930–3934.

Pickett, C.L., Auffenberg, T., Pesci, E.C., Sheen, V.L. and Jusuf, S.S. 1992. Iron acquisition and hemolysin production by *Campylobacter jejuni*. Infect. Immun. 60: 3872–3877.

Qadri, F., Raqib, R., Ahmed, F., Rahman, T., Wenneras, C., Das, S.K., Alam, N.H., Mathan, M.M. and Svennerholm, A.M. 2002. Increased levels of inflammatory mediators in children and adults infected with *Vibrio cholerae* O1 and O139. Clin. Diagn. Lab. Immunol. 9: 221–229.

Rathbone, B.J., Johnson, A.W., Wyatt, J.I., Kelleher, J., Heatley, R.V. and Losowsky, M.S. 1989. Ascorbic acid: a factor concentrated in human gastric juice. Clin. Sci. (Lond.) 76: 237–241.

Reidl, J. and Mekalanos, J. 1996. Lipoprotein e (P4) is essential for hemin uptake by *Haemophilus influenzae*. J. Exp. Med. 183: 621–629.

Richardson, P.T. and Park, S.F. 1995. Enterochelin acquisition in *Campylobacter coli*: characterization of components of a binding-protein-dependent transport system. Microbiology 141: 3181–3191.

Rock, J.D., van Vliet, A.H.M. and Ketley, J.M. 2001. Haemin uptake in *Campylobacter jejuni*. Int. J. Med. Microbiol. 291 (S31): 125.

Sanyal, A.J., Hirsch, J.I. and Moore, E.W. 1990. Premicellar taurocholate avidly binds ferrous (Fe++) iron: a potential physiologic role for bile salts in iron absorption. J. Lab. Clin. Med. 116: 76–86.

Schryvers, A.B. 1989. Identification of the transferrin- and lactoferrin-binding proteins in *Haemophilus influenzae*. J. Med. Microbiol. 29: 121–130.

Seliger, S.S., Mey, A.R., Valle, A.M. and Payne, S.M. 2001. The two TonB systems of *Vibrio cholerae*: redundant and specific functions. Mol. Microbiol. 39: 801–812.

Smith, R.L. and Maguire, M.E. 1998. Microbial magnesium transport: unusual transporters searching for identity. Mol. Microbiol. 28: 217–226.

Stojiljkovic, I. and Hantke, K. 1992. Hemin uptake system of *Yersinia enterocolitica*: similarities with other TonB-dependent systems in gram-negative bacteria. EMBO J. 11: 4359–4367.

Stojiljkovic, I., Cobeljic, M. and Hantke, K. 1993. *Escherichia coli* K-12 ferrous iron uptake mutants are impaired in their ability to colonize the mouse intestine. FEMS Microbiol. Lett. 108: 111–115.

Stojiljkovic, I., Larson, J., Hwa, V., Anic, S. and So, M. 1996. HmbR outer membrane receptors of pathogenic *Neisseria* spp.: iron- regulated, haemoglobin-binding proteins with a high level of primary structure conservation. J. Bacteriol. 178: 4670–4678.

Suerbaum, S., Josenhans, C., Sterzenbach, T., Drescher, B., Brandt, P., Bell, M., Droge, M., Fartmann, B., Fischer, H.P., Ge, Z., Horster, A., Holland, R., Klein, K., Konig, J., Macko, L., Mendz, G.L., Nyakatura, G., Schauer, D.B., Shen, Z., Weber, J., Frosch, M. and Fox, J.G. 2003. The complete genome sequence of the carcinogenic bacterium *Helicobacter hepaticus*. Proc. Natl. Acad. Sci. USA 100: 7901–7906.

Tonello, F., Dundon, W.G., Satin, B., Molinari, M., Tognon, G., Grandi, G., Del Giudice, G., Rappuoli, R. and Montecucco, C. 1999. The *Helicobacter pylori* neutrophil-activating protein is an iron-binding protein with dodecameric structure. Mol. Microbiol. 34: 238–246.

Torres, A.G. and Payne, S.M. 1997. Haem iron-transport system in enterohaemorrhagic *Escherichia coli* O157: H7. Mol. Microbiol. 23: 825–833.

Touati, D. 2000. Iron and oxidative stress in bacteria. Arch. Biochem. Biophys. 373: 1–6.

Tsolis, R.M., Baumler, A.J., Heffron, F. and Stojiljkovic, I. 1996. Contribution of TonB- and Feo-mediated iron uptake to growth of *Salmonella typhimurium* in the mouse. Infect. Immun. 64: 4549–4556.

van Vliet, A.H.M., Wooldridge, K.G. and Ketley, J.M. 1998. Iron-responsive gene regulation in a *Campylobacter jejuni fur* mutant. J. Bacteriol. 180: 5291–5298.

van Vliet, A.H.M., Baillon, M.L., Penn, C.W. and Ketley, J.M. 1999. *Campylobacter jejuni* contains two Fur homologues: characterization of iron-responsive regulation of peroxide stress defense genes by the PerR repressor. J. Bacteriol. 181: 6371–6376.

van Vliet, A.H.M., Rock, J.D., Madeleine, L.N. and Ketley, J.M. 2000. The iron-responsive regulator Fur of *Campylobacter jejuni* is expressed from two separate promoters. FEMS Microbiol. Lett. 188: 115–118.

van Vliet, A.H.M., Baillon, M.L.A., Penn, C.W. and Ketley, J.M. 2001. The iron-induced ferredoxin FdxA of *Campylobacter jejuni* is involved in aerotolerance. FEMS Microbiol. Lett. 196: 189–193.

van Vliet, A.H.M., Ketley, J.M., Park, S.F. and Penn, C.W. 2002a. The role of iron in *Campylobacter* gene regulation, metabolism and oxidative stress defense. FEMS Microbiol. Rev. 26: 173–186.

van Vliet, A.H.M., Stoof, J., Vlasblom, R., Wainwright, S.A., Hughes, N.J., Kelly, D.J., Bereswill, S., Bijlsma, J.J.E., Hoogenboezem, T., Vandenbroucke-Grauls, C.M.J.E., Kist, M., Kuipers, E.J. and Kusters, J.G. 2002b. The role of the Ferric Uptake Regulator (Fur) in regulation of *Helicobacter pylori* iron uptake. Helicobacter 7: 237–244.

van Vliet, A.H.M., Poppelaars, S.W., Davies, B.J., Stoof, J., Bereswill, S., Kist, M., Kuipers, E.J., Penn, C.W. and Kusters, J.G. 2002c. NikR mediates nickel-responsive transcriptional induction of urease expression in *Helicobacter pylori*. Infect. Immun. 70: 2846–2852.

van Vliet, A.H.M., Stoof, J., Poppelaars, S.W., Bereswill, S., Homuth, G., Kist, M., Kuipers, E.J. and Kusters, J.G. 2003. Differential regulation of amidase- and formamidase-mediated ammonia production by the *Helicobacter pylori* Fur repressor. J. Biol. Chem. 278: 9052–9057.

Velayudhan, J., Hughes, N.J., McColm, A.A., Bagshaw, J., Clayton, C.L., Andrews, S.C. and Kelly, D.J. 2000. Iron acquisition and virulence in *Helicobacter pylori*: a major role for FeoB, a high-affinity ferrous iron transporter. Mol. Microbiol. 37: 274–286.

Wai, S.N., Nakayama, K., Umene, K., Moriya, T. and Amako, K. 1996. Construction of a ferritin-deficient mutant of *Campylobacter jejuni*: contribution of ferritin to iron storage and protection against oxidative stress. Mol. Microbiol. 20: 1127–1134.

Waidner, B., Greiner, S., Odenbreit, S., Kavermann, H., Velayudhan, J., Stahler, F., Guhl, J., Bisse, E., van Vliet, A.H., Andrews, S.C., Kusters, J.G., Kelly, D.J., Haas, R., Kist, M. and Bereswill, S. 2002. Essential role of ferritin Pfr in *Helicobacter pylori* iron metabolism and gastric colonization. Infect. Immun. 70: 3923–3929.

Wainwright, S.A., Velayudhan, J. and Kelly, D.J. 2001. The magnesium transporter CorA catalyses low-affinity iron uptake in *Helicobacter pylori*. Int. J. Med. Microbiol. 291 (S31): 100.

Weinberg, E.D. 1978. Iron and infection. Microbiol. Rev. 42: 45–66.

Wooldridge, K.G. and Williams, P.H. 1993. Iron uptake mechanisms of pathogenic bacteria. FEMS Microbiol. Rev. 12: 325–348.

Wooldridge, K.G., Williams, P.H. and Ketley, J.M. 1994. Iron-responsive genetic regulation in *Campylobacter jejuni*: cloning and characterization of a *fur* homologue. J. Bacteriol. 176: 5852–5856.

Wooldridge, K.G. and Ketley, J.M. 1997. *Campylobacter*–host cell interactions. Trends Microbiol. 5: 96–102.

Zimmermann, L., Angerer, A. and Braun, V. 1989. Mechanistically novel iron (III) transport system in *Serratia marcescens*. J. Bacteriol. 171: 238–243.

Chapter 16

Campylobacter jejuni Stress Responses During Survival in the Food Chain and Colonization

Simon F. Park

Abstract

Campylobacter jejuni is probably the most ubiquitous bacterial pathogen in the food chain and is the leading cause of bacterial food borne diarrhoeal disease throughout the world. It can be isolated from a wide variety of environments including farms, surface waters, foods and the intestinal tracts of various animals. In particular, *C. jejuni* is widely distributed in the intestinal tract of poultry. The organism is able to survive exposure to a multitude of inimical conditions whilst in the food chain, and also during infection, and its ability to tolerate these is fundamental to the continuation of the contamination cycle. The following discussion will focus on the stress responses elicited by this pathogen, and how these correlate with its ability to survive in the food chain and to tolerate the non-immune defence mechanisms encountered in the human host.

INTRODUCTION

Campylobacter jejuni is probably the most ubiquitous bacterial pathogen in the food chain today and also one of the commonest pathogens found in the environment. Levels of contamination on poultry products may be as high as 10^5 CFU per carcass (Jacobs-Reitsma, 2000) and consequently the poultry products within a supermarket may easily harbour a total population of up to 10^7 organisms. Furthermore, an individual chicken may carry a population of 10^9 viable organisms per gram in the gastrointestinal tract. Given the fact that 95% of the poultry flock in the UK is contaminated with campylobacters (Evans, 1997) and that 800 million broilers are raised every year, the number of this pathogen entering the food chain is immense. High rates of faecal shedding of *C. jejuni* by domesticated and wild animals (Wesley *et al.*, 2000) also results in large numbers of the organisms being continually excreted into the environment. Accordingly, campylobacters can be isolated from river water, wastewaters (Koenraad *et al.*, 1997), and even sand on bathing beaches (Bolton *et al.*, 1999). The organism is also commonly found in the home, and following the preparation of meals prepared from raw chicken, *C. jejuni* can become widely disseminated at hand and food contact surfaces within the kitchen (Cogan *et al.*, 1999) The frequent isolation of *C. jejuni* at many points in the food chain suggests that it is able to tolerate a variety of different inimical conditions. Since adaptability is a crucial characteristic of bacteria that are able to survive in a wide variety of environmental conditions this review will focus on the adaptive responses of *C. jejuni* to the different environments that it may encounter in the food chain. In this context, it follows the journey of this bacterium during colonization of poultry, from the farm into the kitchen and finally during infection of the human host.

STRESS RESPONSES OF IMPORTANCE AT THE POULTRY FARM

Whilst other vectors for the transmission for campylobacters do exist, such as pets, unpasteurized milk and water (Freidman *et al.*, 2000), for the purpose of this review, the entry of *C. jejuni* into the food chain will begin at the poultry farm. Whilst the primary environment for *C. jejuni* is the digestive tract of poultry and other warm-blooded animals such wild birds and rodents, it can also isolated from the air, flies, water feeders, feed, farmers' boots, chicken feathers and faeces in poultry units (Berndtson *et al.*, 1996; Stern *et al.*, 1997). It is thought that broiler flocks initially become contaminated from environmental sources. In one documented case, colonization resulted from the transfer of the organism from cattle to broilers via the farmer's footwear (van der Geissen *et al.*, 1998). Survival and transmission in the farm environment may entail a number of different responses some of which, for example, desiccation, will be discussed later. However, since survival at non-growth temperatures, starvation and survival in water, and resistance to oxygen and oxidative stress, are likely to be of primary importance in the farm environment, these will be discussed here.

Survival at non-growth temperatures

Whilst *C. jejuni* has an optimum growth temperature of 42°C, the lower temperature limit for the growth of *C. jejuni* generally lies between 30°C and 36°C and this places a definitive limitation on the ability of the organism to multiply outside of a warm-blooded host. The ability of many bacteria to replicate at temperatures far below that required for optimum growth is associated with the production of characteristic cold shock proteins. In *E. coli* for example, a temperature downshift from 37°C to 10°C induces the expression of proteins which are members of a family of nine cold shock inducible proteins. Among these is the major cold shock protein CspA whose production can reach more than 10% of cellular protein synthesis (Goldstein *et al.*, 1990) and which is thought to act as an RNA chaperone to block the formation of secondary structures in mRNA (Yamanaka *et al.*, 1998). The profile of protein expression in *C. jejuni*, however, is similar at 37°C, 20°C and 4°C and the induction of specific cold-shock proteins has not been observed (Hazeleger *et al.*, 1998). The failure of *C. jejuni* to elicit the characteristic cold shock response is reflected by the absence of any genes encoding members of the CspA family of cold-shock proteins in the genome sequence (Parkhill *et al.*, 2000), and may explain its inability to grow at low temperatures. The absence of this wide-spread response may also explain the observation that unlike other micro-organisms, which show a gradual reduction in growth rate near the minimal growth temperature, *C. jejuni* shows a dramatic and sudden growth rate decline near the lower temperature limit (Hazeleger *et al.*, 1998). Nevertheless *C. jejuni* is still metabolically active at temperatures far below its minimal growth temperature and is still motile, and therefore able to move to favourable environments at temperatures as low as 4°C (Hazeleger *et al.*, 1998).

Survival in water: starvation and hypoosmotic stress

Although growth is not possible in feeding water and surface water, the survival of *C. jejuni* in these environments is likely to play a vital role in maintaining the contamination cycle in poultry farms. Survival in water is dependent on a number of environmental parameters such as temperature, the presence of oxygen, sunlight and the availability of nutrients. For example, the rate of loss of viability in river water, a microcosm containing nutrients, is markedly less than that observed in deionized water (Thomas *et al.*, 1999) and occurs more rapidly at higher temperatures and when the pathogen is exposed to light (Obiri-Danso *et al.*, 2001;Thomas *et al.*, 2002). The issue of survival in water is further complicated by the fact that when exposed to starvation in water or saline, *C. jejuni* becomes unable to colonize chicks long before it actually looses culturability in the laboratory (Newell *et al.*, 1985, Hald *et al.*, 2001). In this state the cells may be considered to be 'culturable but not infectious' (Hald *et al.*, 2001).

Bacteria generally have distinct responses to starvation depending upon which individual nutrient, for example carbon, phosphorous or nitrogen, has become depleted (Spector and Foster, 1993). The response of *C. jejuni* to starvation is very poorly understood and although protein synthesis has been shown to occur during the first hours of starvation (Cappelier *et al.*, 2000), it is not clear whether this is the result of a specific induced stress response or is just a continuation of non-adaptive protein synthesis. In the absence of detailed experimental evidence, however, it is possible to gain some information about the likely starvation response from the genome sequence.

In other bacteria starvation and/or entry into the stationary phase elicits a programme of profound structural and physiological changes which result in increased resistance to heat shock, oxidative, osmotic and acid stress. For many food-borne pathogens, this adaptive process has an important bearing on the ability of organisms to survive the physical challenges encountered in the food chain (Rees *et al.*, 1995). The central regulator for these changes in many Gram-negative bacteria is the alternative σ-factor RpoS, which, accordingly, is critical for the survival of the bacterial cell following exposure to many types of environmental stresses (Loewen *et al.*, 1998). An analysis of the *C. jejuni* NCTC 11168 genome sequence (Parkhill *et al.*, 2000), however, indicates that RpoS is absent from this organism. Furthermore, stationary phase cultures of *C. jejuni* do not elicit the induced cross-protection seen in many bacteria and in fact the reverse of this situation is observed as cells from this growth phase are actually more sensitive to heat stress (Figure 16.1), oxidative stress, and acid than cells taken from the exponential growth phase (Kelly *et al.*, 2000; Murphy *et al.*, 2003).

RpoS is responsible for the regulation of many stationary phase-inducible genes expressed in stationary phase. However, the expression of certain genes is also repressed in this growth stage and the CsrA/CsrB (carbon storage regulator) forms a global regulatory system that controls the expression of some of these repressed genes at the post-transcriptional level in many bacteria (Romeo, 1998). In *E. coli* for example, glycogen catabolism, gluconeogenesis, glycolysis, motility and adherence are modulated by CsrA, which acts by binding to and destabilizing specific mRNAs. CsrB, is a non-coding RNA molecule that forms a large globular ribonucleoprotein complex with CsrA and antagonises its action. CsrA may even activate certain exponential-phase metabolic pathways and also serves as a repressor of biofilm formation and an activator of biofilm dispersal (Jackson *et al.*, 2002). CsrA also controls genome-wide changes in gene expression in *Salmonella enterica* serovar Typhimurium, but in this case, it also regulates invasion genes (Lawhon *et al.*, 2003). *C. jejuni* possesses a homologue of CsrA (Cj1103) and given the lack of other key starvation and stationary phase regulators, it is possible that it plays a key role in modulating gene expression in *C. jejuni* in response to starvation and stationary phase.

The stringent response occurs in many bacteria and involves the coordinated regulation

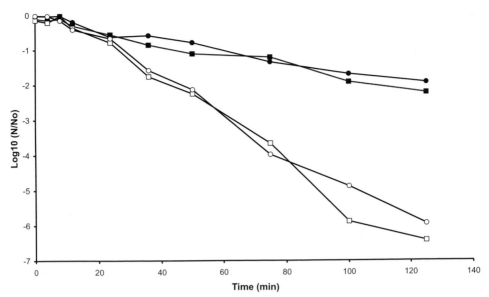

Figure 16.1 Effect of growth phase on the resistance of *C. jejuni* to heat shock. *C. jejuni* NCTC 11168 (□, ■) and *C. jejuni* 81116 (○, ●) were grown in Mueller–Hinton Broth to exponential phase (■, ●) and to stationary phase (48 hour) (□, ○) and samples heated to 50°C and viability measured by plate counting.

of numerous cellular activities (Goldman and Jakubowski, 1990) in response to amino acid and carbon-energy source starvation. The gaunosine nucleotide (p)ppGpp is considered to be central to this response and a number of enzymes involved in the metabolism of this molecule are now well characterized including ppGpp synthase I (RelA), (p)ppGpp 3′-pyrophosphohydrolase (SpoT), pppGpp 5′-phosphohydrolase (Gpp), and nucleoside 5′-diphosphate kinase (Ndk). Genes encoding these enzymes are present in the genome sequence but as of yet, there is no experimental evidence to indicate that *C. jejuni* exhibits a true stringent response. Notably, *Helicobacter pylori*, like *C. jejuni*, has the enzymatic machinery for (p)ppGpp metabolism but guanosine polyphosphate levels do not rise as a result of amino acid starvation and as such the organisms does not exhibit a stringent response (Scoarughi *et al.*, 1999).

In *E. coli* and other enteric bacteria, the well-characterized cAMP–CRP system allows the use of alternative carbon sources. Levels of cAMP rise during carbon starvation (Botsford and Drexler, 1978) and this allows CRP to act as an activator of approximately two thirds of the carbon-starvation-responsive genes in *E. coli* (Marschall and Hengge-Aronis, 1995). However, glucose-linked catabolite repression, as it functions in many bacteria, probably does not operate in *C. jejuni* since glucose is not utilized as a carbon source and may not, therefore, be transported into the cell for this purpose. Consequently, whilst the genome encodes a peptide with homology to the CRP/FNR (Cj0466) family of regulatory proteins the exact function of this protein is not yet known.

Many bacteria synthesize storage polymers when faced with starvation and these play roles during periods of starvation and stress. *E. coli* and various other bacteria, for example, are able to synthesize glycogen when a specific nutrient is limiting but when carbon is in excess. During carbon limitation glycogen can then be metabolized and used as a carbon or energy source. Similarly, massive accumulations of polyphosphate, containing hundreds of phosphate residues linked by high-energy phosphoanhydride bonds are generated in response to amino acid starvation in *E. coli* (Kuroda *et al.*, 1997). Whilst the exact function of this polymer is not yet known, it is essential for the induction of RpoS-mediated gene expression in the stationary phase in *E. coli* (Shiba *et al.*, 1997) and may serve as an important energy and phosphate store in starved cells of *H. pylori* (Nilsson *et al.*, 2002). *C. jejuni* seems unable to produce stores of glycogen since genes encoding enzymes for this pathway are not represented in the genome sequence. However, the bacterium is able to produce polyphosphate (S.F. Park, unpublished), although the exact function of this polymer is not yet known.

Oxidative stress

Whilst *C. jejuni* is considered to be microaerophilic, exposure to atmospheric oxygen is an inevitable consequence of its contamination cycle and leads to the formation of reactive oxygen intermediates (ROIs), including superoxide radicals. If these highly reactive agents are not neutralized then lethal damage to nucleic acids, proteins and membranes may ensue. Consequently, as a response against distinct ROIs excesses, many prokaryotic cells are able to induce the synthesis of anti-oxidant enzymes and these play an important role in the survival of these bacteria during exposure to air.

A number of enzymes, including superoxide dismutase (SOD) (Purdy and Park, 1994; Pesci *et al.*, 1994; Purdy *et al.*, 1999), alkyl hydroperoxide reductase (Ahp) (Baillon *et al.*, 1999) and catalase (KatA) (Grant and Park, 1995), play roles in the oxidative defence system of campylobacters. The deleterious effect of exposure to superoxide radicals are counteracted by the activity of SOD and both *C. jejuni* and *C. coli* possesses a single iron co-factored SOD (Purdy and Park, 1994; Pesci *et al.*, 1994). This enzyme plays a key role in the defence against oxidative stress and aerotolerance as SOD-deficient *Campylobacter* mutants are less able to survive in air, in food, during freezing and also colonize chicks poorly (Purdy *et al.*, 1999; Stead and Park, 2000). Iron-regulated proteins, including AhpC (Baillon *et al.*, 1999) and ferredoxin (van Vliet *et al.*, 2001) also contribute to aerotolerance.

The recognition and response to oxidative stress in many enteric Gram-negative bacteria is mediated by the key regulators SoxRS and OxyR, which recognize superoxide and peroxide respectively. These are not present in *C. jejuni* but an alternative regulator termed PerR, has been identified (van Vliet *et al.*, 1999). This bears considerable similarity to PerR from *Bacillus subtilis* which acts as a negative regulator of the peroxide responsive regulon (Bsat *et al.*, 1998), and mediates at least part of the response to oxidative stress response in *C. jejuni* as it has been shown to negatively regulate both AhpC and catalase expression (van Vliet *et al.*, 1999).

Hypoosmotic stress

Introduction of the bacterial cell into water leads to a massive influx of water due to osmosis. Unless the bacterium is able to react quickly to prevent or reverse this process, lysis will rapidly ensue. Campylobacters are able to survive in water for extended periods yet little is known about the mechanisms by which they counter hypoosmotic stress. In many bacteria, the response to hypo-osmotic shock involves both solute and water efflux. Mechanosensitive or stretch activated channels are the primary routes for this expulsion and these are often activated following the transition from high to low osmolarity (Levina *et al.*, 1999). In *E. coli* three types of mechanosensitive channel have been identified MscM (M for mini), MscS (for small) and MscL (for large) (Berrier *et al.*, 1996). Functional homologues of mechanosensitive channels have been identified in many Gram-negative and Gram-Positive bacteria (Moe *et al.*, 1998) and two other genes associated with MscS activity, *kefA* and *yggB* are also widely distributed (Levina *et al.*, 1999). No experimental evidence concerning the molecular response of *C. jejuni* to low osmolarity has been published. However, *H. pylori*, which is closely related to *C. jejuni* possesses a homologue of an MscL channel (Kloda and Matinac, 2001). Other than the protein Cj0238, which exhibits weak similarity to this, there are no obvious genes which could encode alternative mechanosensitive channels in *C. jejuni*. Certain bacteria also possess aquaporins and as these are specific channels that allow the rapid influx/efflux of water they are also thought to play an essential role in maintenance of turgor during fluctuations in osmolarity (Calamita, 2000). Again there is no obvious homologue for this class of protein in *C. jejuni*.

COLONISATION OF POULTRY

The gastrointestinal tract of certain warm-blooded animals is thought to be the natural habitat of *Campylobacter* species. In particular, birds, and especially poultry, are regarded as the primary reservoir for *C. jejuni* and the organism is generally considered to be a commensal in these hosts. When poultry is experimentally inoculated with campylobacters, colonization is found to occur largely in the caeca, large intestine and cloaca and is generally restricted to the intestinal mucus layer in the crypts of the intestinal epithelium at these locations (Beery *et al.*, 1988). To reach this environment the organism must survive the stresses imposed by the upper gastrointestinal tract such as the pH of the stomach and exposure to bile salts. The response of *C. jejuni* to these environments will be discussed later. The large intestine and caeca represent highly specialised environments and *C. jejuni* has evolved specialised strategies which allow it to exploit these restricted ecological niches. For example, the optimal growth temperature of the organism (42°C) mirrors that of the avian gut and this differs considerably from that encountered in the mammalian gut (37°C). The ability of *C. jejuni* to become established in the gastrointestinal tract of chickens is believed to involve binding of the bacterium to its surface cells. The outer membrane protein CadF, which facilitates the binding of campylobacters to fibronectin, is thought to be responsible for this since a deficient mutant is incapable of colonizing the caecum of newly hatched chicks (Ziprin *et al.*, 1999). Other facets of the physiology of campylobacters also point to the fact that the organism has evolved to occupy the mucous layer of the gastrointestinal tract. Firstly, their microaerophilic nature is probably a reflection of the concentration of oxygen encountered in this niche. Secondly, competition for essential nutrients amongst the bacterial community within the gastrointestinal is likely to be intense and it appears that campylobacters have evolved mechanisms to allow them to compete successfully in this environment. Iron is such an essential nutrient, which will have limited availability within the gut. As a consequence campylobacters have evolved mechanisms to acquire iron from both the host and the gastrointestinal flora (van Vliet *et al.*, 2002; see Chapter 15).

Anaerobiosis

When *C. jejuni* is present in the mucus layer of the chicken gastrointestinal tract, it is likely to be exposed to a microaerobic atmosphere optimal for growth. However, if the organism becomes detached, it may enter the lumen of the colon which is strictly anaerobic. *C.*

jejuni is generally recognized as being a microaerophile and thus would not be expected to replicate in this environment. However, the discovery of genes encoding a number of terminal reductases which could potentially allow the use of a wide range of alternative electron acceptors to oxygen such as fumarate, nitrate, and nitrite has raised the possibility that under specific conditions *C. jejuni* could grow in the absence of oxygen. *C. jejuni*, however, has been shown not to grow under strictly anaerobic conditions with any of the above electron acceptors but these compounds did increase growth in environments in which the rate of oxygen transfer was severely limited (Sellars *et al.*, 2002; Chapter 14). Such environments may be encountered during colonization and consequently these alternative respiratory pathways may contribute to energy conservation under oxygen-limited conditions *in vivo*. Whilst the response of *E. coli* to oxygen limiting conditions is well known and is mediated via Fnr and Arc signal sensing mechanisms (Sawers, 1999), whether *C. jejuni* senses changes in oxygen concentration and/or redox potential is not yet known. However it has been reported to be aerotactic (Hazeleger *et al.*, 1998) and may possess aerotaxis receptor proteins (Marchant *et al.*, 2002; Chapter 18, this volume).

Uric acid resistance

The urine of birds enters the cloaca and is moved by reverse peristaltic action into the colon. Here it comes into contact with an epithelial surface that can change the composition of the urine (Braun, 1999). The urine contains high concentrations of uric acid which is the chief nitrogenous waste of birds. In most avian species characterized to date, this region of the gastrointestinal tract is populated by bacterial species that are able to tolerate and degrade uric acid (Mead, 1989). Whilst *C. jejuni* occupies essentially the same environment it is not known how it responds to uric acid. Free radicals derived from uric acid have been shown to actively release nitric oxide (Skinner *et al.*, 1998) and given that this is bactericidal it could have an impact on *Campylobacter* survival.

POULTRY PROCESSING AND RETAIL

Generally, *C. jejuni* is transported to the poultry processing plants via colonized poultry either inside the gastrointestinal tract or on feathers. The carcasses are dipped into a scald tank of hot water, which helps to loosen the feathers and then moved to mechanical plucking machines. For most of this process the organism is protected from external stresses since it resides in the gastrointestinal tract. Nevertheless, the organism can be isolated from scald water at temperatures of 60°C and feather plucking machines (Wempe *et al.*, 1983). Once the head is removed an eviscerating machine removes the intestines and internal organs. The carcasses are then washed using potable water and chilled down to 4°C by cold air jets or cold water sprays. Following a period of maturation (8–10 hours) the birds are trussed and bagged for sale.

Chilling and freezing

Chilling is generally considered to promote the survival of *C. jejuni* (Chan *et al.*, 2001). Nevertheless, exposure to low temperature will generate stress and *C. jejuni* must be able to respond to this in order to survive. For example, proteins are known to precipitate in cold (Nicholson and Scholtz, 1996) and cell membrane fluidity decreases. As mentioned earlier *C. jejuni* does not elicit the characteristic cold-shock response seen in many other bacteria and does not possess members of the CspA family of cold-shock proteins.

Recently, evidence for a link between osmoprotection and cyrotolerance has emerged. In this context, the synthetic pathway for the osmotic compatible solute trehalose is induced on exposure to cold in *E. coli*, and is also essential for viability at low temperatures (Kandror *et al.*, 2002). Similarly, glycine betaine transport, whilst osmotically regulated, is also cryo-activated in the food-borne pathogen *Listeria monocytogenes* (Mendum and Smith, 2002) and this compatible solute also confers cyrotolerance (Ko *et al.*, 1994). *E. coli* also responds to cold shock by increasing the level of spermidine acetylation (Tabor, 1968) and mutants which lack spermidine acetyl transferase lyse in the presence of excess sperimidine at 7°C (Limsuwun and Jones, 2000). In contrast to other food-borne pathogens, little is known of the response of *C. jejuni* to chilling. It obviously is able to tolerate this process and even chlorination of chilling water has little effect on the survival of bacteria bound to chicken skin (Yang *et al.*, 2001). Nevertheless, an analysis of the genome sequences suggests that the organism does not possess any of the cryotolerance mechanisms described above and consequently alternative mechanisms may operate. Also since substantial variability in the resistance of different strains to chilling has been reported (Chan *et al.*, 2001) it is possible that some cryotolerance mechanisms are strain specific.

If poultry products are frozen, *C. jejuni* rapidly loses viability (Humphrey and Cruickshank, 1985). However, campylobacters can still be isolated from frozen meats and poultry products (Fernandez and Pison, 1996; Humphrey and Cruickshank, 1985). Several factors, including ice nucleation and dehydration, have been implicated in the freeze-induced injury of bacterial cells. More recently a study carried out with yeast has shown that reactive oxygen intermediates are generated during freeze-thawing, and that these contribute to lethal damage to the cell (Park *et al.*, 1998). SOD-deficient mutants of *C. jejuni* are highly susceptible to freezing and thawing and since the resistance of these cells to freeze-thaw was restored by freezing in the absence of oxygen, it is likely that, similar to yeast cells, superoxide radicals are generated during this process and that SOD is important in the resistance of campylobacters to these damaging agents (Stead and Park 2000).

SURVIVAL IN THE KITCHEN

Survival of *C. jejuni* in the kitchen may entail survival at low temperatures, exposure to oxidative stress, exposure to heat, desiccation, and exposure to biocides. Furthermore, since cross-contamination in the kitchen is also a significant contributory factor for infection (Cogan *et al.*, 1999), survival on fingertips may also be important. Survival times can range from 1 to 4 minutes if dried or up to 1 hour if the fingers are also contaminated with blood products (Coates *et al.*, 1987).

Response to heating

Although thermophilic in growth requirement, campylobacters are sensitive to heat and readily inactivated by pasteurization treatments and domestic cooking processes. The physiological events that take place during adaptation to elevated temperature are well documented from studies with other bacteria but far from resolved for campylobacters. Bacterial cells exposed to temperatures above that which is optimal for growth generally respond by eliciting a heat shock response involving the synthesis of proteins able to act as chaperones or ATP-dependent proteases (Arsene *et al.*, 2000). At least twenty-four proteins are preferentially synthesized by *C. jejuni* immediately following heat shock (Konkel *et al.*,

1998) and some have been identified as GroELS, DnaJ, DnaK and Lon protease (Konkel *et al.*, 1998; Thies *et al.*, 1998: Thies *et al.*, 1999ab). The recent finding, that *dnaJ* mutants are severely retarded in growth at 46°C and also unable to colonize chickens suggests that the heat shock response plays an important role in both thermotolerance and colonization (Konkel *et al.*, 1998). Whilst it is clear that campylobacters are able to elicit a heat shock response similar to that observed in other bacteria, the regulatory mechanisms governing this response have not yet been studied in detail. There are potentially three alternative regulatory systems controlling the induction of the heat shock response in *C. jejuni*. The RacRS regulon, previously characterized as a two component regulatory system, is required for the differential expression of proteins at 37°C and 42°C (Bras *et al.*, 1999) and homologues of HrcA and HspR, negative regulators of the heat shock response in *B. subtilis* (Schulz and Schuman, 1996) and *Streptomyces coelicolour* (Bucca *et al.*, 1997), respectively, are also present. Furthermore, putative consensus recognition sequences for HrcA binding have been identified upstream of the *groESL* and *dnaK* operons of *C. jejuni* (Thies *et al.*, 1999ab) providing further support for a regulatory role for the HcrA orthologue.

Response to desiccation and hyperosmotic stress

Campylobacters are much less tolerant of desiccation than other bacterial food-borne pathogens (Doyle and Roman, 1982) and because the stresses encountered during desiccation are intimately linked with osmotic stress, this may be a reflection of their limited capacity to respond to elevated osmolarities. In the absence of any empirical information relating to osmoregulation, the genome sequence is the only source for information on the osmoregulatory capacity of *C. jejuni*.

The most rapid response to osmotic up-shock in many bacteria is the accumulation of potassium which is brought about by the activation of low- and high-affinity transport systems. *C. jejuni* contains at least one potassium transporter, encoded by two adjacent genes (Cj1283 and Cj1284), which resembles the KtrAB transporter previously characterized in *Vibrio alginolyticus* (Nakamura *et al.*, 1998). It is not known, however, whether this system plays any role in osmoregulatory potassium transport. The Kdp system is a high-affinity ATP-driven potassium transport system, originally characterized in *Enterobacteriaceae*, which is involved in potassium transport during osmotic upshock. Four membrane-bound subunits form the catalytic subunit (KdpB), the potassium translocating subunit (KdpA), a stabilizing peptide (KdpF) and an unknown function (KbpC). At the promoter-distal end of the operon are the *kdpDE* genes which encode a sensor kinase and response regulator responsible for the osmotically inducible nature of the operon (Altendorf *et al.*, 1992) Whilst the *C. jejuni* genome contains genes encoding the KdpB (Cj0677) catalytic subunit other anticipated Kdp components are not expressed. For example, Cj0679 encodes a truncated copy of KdpD, lacking the C-terminal two-component histidine kinase domain, whilst KdpA (Cj0676) and KdpC (Cj0678) are pseudogenes. Furthermore, KdpE and KdpF both appear to be completely absent (Figure 16.2). The exact role of this seemingly redundant system in *C. jejuni* NCTC 11168 (Parkhill *et al.*, 2000), is not yet known but it is possible that the operon is intact in other strains.

Long term resistance to osmotic stress in other bacteria has been correlated with mechanisms for the synthesis or transport of compatible solutes (Kempf and Bremer, 1998). It is clear from an analysis of the *C. jejuni* genome sequence (Parkhill *et al.*, 2000) that whilst this organism may possess a low-affinity transporter for proline and

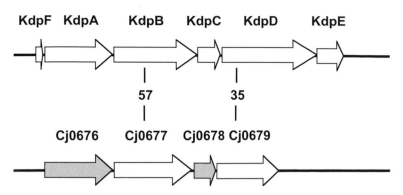

Figure 16.2 Genome organization of the *E. coli* Kdp operon with the partial Kdp operon from *C. jejuni*. Gene annotation is shown. The spacing between ORFs is not to scale. Pseudogenes are represented by shading. Where available identity between the two sequences is given.

glycine betaine similar to ProP (Cairney *et al.*, 1985), it does not possess any previously characterized high-affinity transporters for known compatible solutes such as the ProU system for betaine transport (Stirling *et al.*, 1989) In addition, *C. jejuni* lacks the capacity to synthesize compatible solutes by known pathways and accordingly does not contain the osmoregulatory betaine (Lamark *et al.*, 1991) and trehalose (Strom and Kaasen, 1993) biosynthetic pathways seen in other bacteria. Therefore, the limited capacity for the accumulation of compatible solutes may explain the apparent sensitivity of campylobacters to osmotic stress and desiccation.

THE HUMAN HOST

The acid and nitrosative stress response

Following ingestion *C. jejuni* is immediately exposed to the low pH of the stomach and its ability to survive in this environment is essential for infection. Compared to other food-borne pathogens *C. jejuni* appears to be unusually sensitive to low pH (Cuk *et al.*, 1987). However, acid tolerance may be strain dependent as some strains appear to have increased acid resistance compared with commonly used strains such as 81116 and NCTC 11351 (Murphy *et al.*, 2003). One such acid-tolerant strain, CI120, exhibits an adaptive tolerance response to acid, which requires *de novo* protein synthesis, and which can be induced by sublethal exposure to acid and aerobic conditions (Murphy *et al.*, 2003). However, it is not clear at present whether this is a specific response to acid or aerobic conditions or whether it is an aspect of a general stress response. In this context, it should be noted that proteins induced during exposure to alkaline pH have been identified as heat shock proteins (Wu *et al.*, 1994).

Exposure to the bactericidal effects of nitric oxide may also be a consequence of entry into the stomach. The reduction of food-borne nitrate to nitrite in the mouth has been associated with the oral microbial flora on the posterior tongue (Duncan *et al.*, 1995; Li *et al.*, 1997). Once nitrite enters the stomach, it encounters HCl, and this results in the chemical generation of nitric oxide, which is bactericidal. In the stomach, this mechanism is thought to provide a powerful defence against gut pathogens (Dykhuizen *et al.*, 1996; Duncan *et al.*, 1995). For example, *H. pylori*, which is closely related to campylobacters, is killed rapidly by nitrosative stress (Dykhuizen *et al.,* 1998). Bacteria utilize flavohaemoglobins or single domain globins as a defence system against nitrosative stress (Poole and Hughes, 2000) and *C. jejuni* possesses a protein (Cj1586; also called Cgb) which is a homologue of the single domain globins. The expression of Cgb is induced by nitric oxide and as deficient mutants also display increased sensitivity towards this agents, the protein plays a clear role in the defence of *C. jejuni* against nitrosative stress (Elvers *et al.*, 2004).

Resistance to bile and hyperosmotic stress in small intestine
After egress from the stomach, *C. jejuni* will be exposed to bile salts, low oxygen, hyperosmolarity and iron limitation in the small intestine. The response to low oxygen and hyperosomolarity has been outlined in detail above, and mechanisms for maintaining iron homeostasis are discussed in detail elsewhere (vanVliet *et al.*, 2002; see Chapter 15).

Bile comprises a mixture of bactericidal detergents, and resistance to this agent is important during infection as reduced capacity to tolerate bile has been correlated with a loss of colonization in a number of enteric pathogens (Lacroix *et al.*, 1996; Bina and Mekalonaos, 2001). Multidrug efflux transporters contribute to the intrinsic resistance of bacteria to many antimicrobial agents and have been implicated as a protective mechanism against bile salts (Zgurskaya, 2002). A multidrug efflux pump, designated CmeABC, has recently been characterized in *C. jejuni*. Since inactivation of the system results in a 4000-fold decrease in the MIC for bile (Lin *et al.*, 2002), it obviously plays an important role in the resistance of *C. jejuni* to this agent and, consequently, it is likely to play an important role during colonization.

The chyme that floods into the intestine from the stomach is not typically hyperosmotic, but as its macromolecular components are digested the osmolarity may increase dramatically. An osmolarity gradient from the tip to the base of villi has also been detected with the tip having an osmolarity of 700 milliosmoles per kg H_2O (Hallback *et al.*, 1978). In this environment the response to hyperosmotic stress as outlined above may be important.

Invasion of epithelial cells
The exact mechanism by which *C. jejuni* causes illness is not yet fully established. Certain strains are known to be able to invade intestinal epithelial cells, and this has been suggested as a possible disease-causing process (Wooldridge and Ketley, 1997). It is not known whether *C. jejuni* recognizes contact or invasion of epithelial cells and whether this triggers the expression of genes required for survival in these environments. However, it has recently been shown that it is able to synthesize a novel set of proteins upon co-culture with epithelial cells. These secreted proteins have been collectively referred to as *Campylobacter* invasion antigens (Cia) and are induced in response to bile salts and also other unidentified signals (Rivera-Amill *et al.*, 1999, 2001).

Since enteroinvasive bacteria directly activate expression of NO synthase and NO production in human colonic epithelial cells, it is thought that NO and/or its redox products are an important component of the defence of intestinal cells against microbial infection (Witthoft *et al.*, 1998). Accordingly, enhanced NO synthesis is also apparent in patients with infective gastroenteritis and this has been attributed to the increased activity of the L-arginine/NO pathway (Forte *et al.*, 1999). If the single domain globin Cgb described above is involved in NO resistance and forms part of a response to nitrosative stress this may have a role during invasion. Superoxide dismutase is also thought to contribute to the survival of *C. jejuni* in epithelial cells as SOD-deficient mutants were 12 times more sensitive to host killing than the isogenic wild-type strain (Pesci *et al.*, 1994). Furthermore, there is a potential interaction of superoxide with nitric oxide generated by the inducible nitric oxide synthase and accordingly, the two agents agent may interact to produce highly toxic products, including nitrogen dioxide and peroxynitrite (Fang, 1997). In *Salmonella enterica* serovar Typhimurium SOD protects periplasmic or inner membrane targets by reducing superoxide levels and thereby limiting one of the precursors of peroxynitrite and limiting its formation in phagocytes (DeGroote *et al.*, 1997). It is likely then that the iron-SOD and Cgb may play an important role in the response by campylobacters that reduces levels of superoxide and NO, both of which are bactericidal, and which also limits the formation of peroxynitrite.

Quorum sensing

The control of gene expression in response to cell density allows individual bacteria of numerous species to communicate and coordinate actions through the production and detection of extracellular signalling compounds (Bassler, 1999). Such systems, which allow cell-cell communication, also play important roles in virulence and the formation of complex communities of bacteria. Through this mechanism, pathogens are able to synchronise production of specific proteins needed for infection processes. This process can occur via the sensing of acyl homoserine lactone derivatives as is the case for many Gram-negative bacteria (Fuqua *et al.*, 2001) or via the production of signalling peptides in Gram-positive organisms (Kleerebezem *et al.*, 1997). More recently, another quorum-sensing pathway that is widespread in both Gram-positive and Gram-negative bacteria has been characterized (Surrette *et al.*, 1999). This is based on LuxS, which generates the signalling compound 4-hydroxy-5-methyl-3 (2H)-furanone (Schauder *et al.*, 2001). Whilst there is some debate whether all bacteria that possess LuxS use the pathway as a cell to cell signalling compound (Winzer *et al.*, 2002ab) it has been implicated in a number of cell signalling roles (Ohtani *et al.*, 2002; Stevenson and Babb, 2002). Whilst *C. jejuni* does not contain any gene predicted to encode an acyl homoserine lactone synthase and thus, is unlikely to produce acyl homoserine lactone based signalling molecules, it does contain LuxS and generates the cognate signalling molecule (Elvers and Park, 2002). However, the role of this signalling pathway is not yet clear.

ALTERNATIVE MECHANISMS MEDIATING SURVIVAL AND STRESS RESISTANCE

From the above discussion it is apparent that *C. jejuni* does not possess many of the adaptive responses that have been established for model organisms such as *E. coli* and *B. subtilis.* The unusual sensitivity of these organisms is a striking feature compared to other food-borne pathogens and seems to reflect this apparent lack of adaptive mechanisms and a minimal capacity for recognizing and responding to environmental stress. However, since the physiology of *C. jejuni* is still poorly understood and the survival strategies that operate in food and the environment are not yet known, it is possible that alternative survival strategies do exist.

Phase variable gene expression and polymorphism

C. jejuni appears to be an unusually diverse bacterial species as evidenced by variation in virulence (Everest *et al.*, 1992; Pickett *et al.*, 1996; Harvey *et al.*, 1999; Purdy *et al.*,

2000); differences in the genetic content of strains (Dorrell *et al.*, 2001) and variability in tolerance of strains to low temperature (Chan *et al.*, 2001) and acid (Murphy *et al.*, 2003). The relative proportion of *Campylobacter* subtypes can also change during poultry processing and it possible that this process selects for the most tolerant forms or variants (Newell *et al.*, 2001).

Cells of *C. coli* UA585 and other *C. jejuni* strains (Figure 16.3; S.F.P., unpublished data) were taken from stationary phase and assessed for resistance to various environmental stresses, including exposure to 0.1% sodium deoxycholate, 0.2 M NaCl, 150 μM methyl viologen and selection on a defined media. For every condition tested, only a small fraction of the original population was able to survive to grow and form colonies. When single colonies, derived from each of these surviving populations, were subcultured under non-selective conditions and re-exposed to the same stress, the cells showed a marked increase in survival (Figure 16.3). Since these colonies had been passaged in the absence of environmental stress, the increase in resistance cannot be due to phenotypic adaptation and must, therefore, be due to the survival of a subpopulation of cells having a particular genetic composition that increases their resistance to the particular stress. Recently, a similar mechanism has also been shown to modulate the expression of *Campylobacter* virulence. Within any population of *C. jejuni* cells, only a small subset produces a phase variable capsule which promotes serum resistance, invasion and virulence (Bacon *et al.*, 2001). The basis of such variation may be evident in the genome sequence of *C. jejuni*. The presence of 25 actively polymorphic homonucleotide tracts in the genome of strain NCTC 11168 (Parkhill *et al.*, 2000) suggests that differential expression of some genes can be mediated by mutation in these hypervariable tracts. The genome also contains numerous pseudogenes containing one frameshift mutation at potentially polymorphic regions. It is interesting to note that a potentially variable region (G9) can be found in the coding sequence of KdpA (Cj0676). Since this is a component of the osmoregulatory potassium

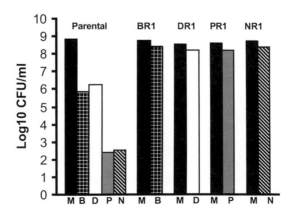

Figure 16.3 Survival of *C. coli* UA585 and variants under different selective growth conditions. Cells of the parental strain (Parental) UA585 were plated on to Mueller–Hinton agar (M), MHA containing 0.1% sodium deoxycholate (B), a defined media (D), MHA containing 150 μM methyl viologen (P) and MHA containing 0.2 M NaCl (N) and survival determined. One survivor from each condition was then subcultured (under non-selective conditions) and then plated onto MHA (M) and the same selective condition it originally survived (B, D, P or N). BR1, deoxycholate resistant survivor: DR1 variant able to grow on defined media: PR1 variant able to grow in the presence of 150 μM methyl viologen: NR1 variant able to grow in the presence of 0.2 M NaCl.

Table 16.1 *C. jejuni* genes encoding proteins that are potentially secreted by the TAT apparatus

Protein	Predicted function	N-terminal sequence*
Cj1513c	Periplasmic protein	VKN**RREFLK**KSAFALGAAGVLGSTTLALAKDEERKDLVKG
Cj0379c	Unknown	MLITPEKLYKQ**RRNFLK**LGAGALISSSVLASKLSALNFTSDT
Cj0005c	Molybdenum containing oxidoreductase	MKQNDQKEN**RRDFLK**NIGLGLFGISVLSNFSFENFLGSKALA
Cj1186c	Ubiquinol-cytochrome C reductase iron–sulfur subunit	MATSES**RRSFMG**FAFGSVAAVGGVFSLVAMKKTWDPLPSVK
Cj0414	Oxidoreductase subunit	MQDNIID**RRSFFK**LGLLGGSVVAASTIGGGAVLKAAELTHSH
Cj1516	Periplasmic oxidoreductase	MN**RRNFLK**FNALTLASMGVAYANPMHDMHMHKNHSINHDL
Cj0145	Unknown	ME**RRLFLK**GSALGSMVAFFASSNLSAAMLKDKDLLGFKAVS
Cj0780	Nitrate reductase	MN**RRDFIK**NTAIASAASVAGLSVPSSMLGAQEEDWKWDKAV
Cj0781	Ferredoxin	MKG**RREFFV**SAFKAACLCTGGGFLANLALKADDNYALRPPG
Cj1511c	Formate dehydrogenase large subunit	MSSVGENIKLT**RRSFLK**MAALSSLATPLLARSETLREASADEL
Cj0264c	Molybdopterin-containing oxidoreductase	MLD**RRKFLK**IGASLSALPLIPSLSAGKTVEASKVSLGLVKNGE
Cj0437	Succinate dehydrogenase flavoprotein subunit	MGEFS**RRDFI K**TACISVGALAASSSGVYALDDSSKMDKDINL

*Potential twin-arginine recognition motifs (consensus RRXFLK) are underlined and in bold. Additional TAT-dependent proteins harbouring less conserved signal sequences may also exist.

transporter Kdp, it suggests that the expression of this activity may be phase variable to some extent. The presence of these hypervariable sequences provides the organism with a mechanism for generating diversity via phenotypic selection. For example, switching in certain genes may mean that only a limited number of cells in any population would survive a given stress, or be able to grow in a particular environment, but once selection had been applied these cells would survive to perpetuate the resistant phenotype.

Twin-arginine transporter

A novel secretion pathway, originally discovered in plants, has recently been characterized in bacteria (Bogsch *et al.*, 1998). This novel system is capable of exporting folded proteins carrying a special twin arginine (RR) signal peptide (Blaudeck *et al.*, 2001) and consequently is designated TAT (twin-arginine translocation). In *Pseudomonas aeruginosa,* the expression of several virulence factors and a number proteins involved in iron-uptake, anaerobic respiration, osmotic stress defence, motility, and biofilm formation depend on this system (Ochsner *et al.*, 2002) A possible twin arginine translocase exists in *C. jejuni* (Cj0578c; Cj0579c; Cj1176c) and a number of proteins also contain the signal motif (Table 16.1) recognized by this transporter. Consequently, the *Campylobacter* TAT system may also play a similar role in the stress response of this organism.

CONCLUSION

One of the most intriguing aspects of *Campylobacter* research is the apparent sensitivity of the organism when assessed in the laboratory and this contrasts with the persistence of the organism in the food chain and the incidence of infection. The information that we have derived for these pathogens suggests that they lack many of the key adaptive mechanisms, such as those mediated by RpoS that have been established in model organisms such as *E. coli.* The publication of the genome sequence seems to highlight this and it is now clear that *C. jejuni* has a minimal regulatory capacity compared to many other bacteria. It is possible, however, that the organism has adopted alternative survival strategies and genetic polymorphisms and phase-variation may be central to this.

References

Altendorf, K., Siebers, A., and Epstein, W. 1992. The KDP ATPase of *Escherichia coli*. Ann. N.Y. Acad. Sci. 30: 671: 228–243.

Arsene F., Tomoyasu, T., and Bukau, B. 2000. The heat shock response of *Escherichia coli*. Int. J. Food Microbiol. 55: 3–9.

Bacon, D.J., Szymanski C.M., Burr, D.H., Silver, R.P., Alm, R.A., and Guerry, P. 2001. A phase-variable capsule is involved in virulence of *Campylobacter jejuni* 81-176. Mol. Microbiol. 40: 769–777.

Baillon, M.L., van Vliet, A.H., Ketley, J.M., Constantinidou, C., and Penn, C.W. 1999. An iron-regulated alkyl hydroperoxide reductase (AhpC) confers aerotolerance and oxidative stress resistance to the microaerophilic pathogen *Campylobacter jejuni*. J. Bacteriol. 181: 4798–4804.

Bassler, B.L. 1999. How bacteria talk to each other: regulation of gene expression by quorum sensing. Curr. Opin. Microbiol. 2: 582–587.

Beery, J.T., Hugdahl, M.B., and Doyle, M.P. 1988. Colonization of gastrointestinal tracts of chicks by *Campylobacter jejuni*. Appl. Environ. Microbiol. 54: 2365–2370.

Berndtson, E., Danielsson-Tham M.L., and Engvall, A. 1996. *Campylobacter* incidence on a chicken farm and the spread of *Campylobacter* during the slaughter process. Int. J. Food Microbiol. 32: 35–47.

Berrier, C., Besnard, M., Ajouz, B., Coulombe, A., and Ghazi, A. 1996. Multiple mechanosensitive ion channels from *Escherichia coli*, activated at different thresholds of applied pressure. J. Membr. Biol. 151: 175–187.

Bina J.E., and Mekalanos, J.J. 2001 *Vibrio cholerae tolC* is required for bile resistance and colonization. Infect. Immun. 69: 4681–4685.

Blaudeck, N., Sprenger, G.A., Freudl, R., and Wiegert, T. 2001. Specificity of signal peptide recognition in TAT-dependent bacterial protein translocation. J. Bacteriol. 183: 604–610.

Bogsch, E.G., Sargent, F., Stanley, N.R., Berks, B.C., Robinson, C., and Palmer, T. 1998. An essential component of a novel bacterial protein export system with homologues in plastids and mitochondria. J. Biol. Chem. 273: 18003–18006.

Bolton, F.J., Surman, S.B., Martin, K., Wareing, D.R., and Humphrey, T.J. 1999. Presence of *Campylobacter* and *Salmonella* in sand from bathing beaches. Epidemiol. Infect. 122: 7–13.

Botsford, J.L., and Drexler, M. 1978. The cyclic $3',5'$-adenosine monophosphate receptor protein and regulation of cyclic $3',5'$-adenosine monophosphate synthesis in *Escherichia coli*. Mol. Gen. Genet. 165: 47–56.

Bras, A.M., Chatterjee, S., Wren B.W., Newell, D.G. and Ketley, J.M. 1999. A novel *Campylobacter jejuni* two-component regulatory system important for temperature-dependent growth and colonization. J. Bacteriol. 181: 3298–3302.

Braun, E.J. 1999. Integration of renal and gastrointestinal function. J. Exp. Zool. 283: 495–499.

Bsat, N., Herbig, A., Casillas-Martinez, L., Setlow, P., and Helmann, J.D. 1998. *Bacillus subtilis* contains multiple Fur homologues: identification of the iron uptake (Fur) and peroxide regulon (PerR) repressors. Mol. Microbiol. 29: 189–198.

Bucca, G., Hindle, Z., and Smith, C.P. 1997 Regulation of the *dnaK* operon of *Streptomyces coelicolour* A3 (2) is governed by HspR, an autoregulatory repressor protein. J. Bacteriol. 179: 5999–6004.

Cairney, J., Booth, I.R., and Higgins, C.F. 1985. *Salmonella typhimurium proP* gene encodes a transport system for the osmoprotectant betaine. J. Bacteriol. 164: 1218–1223.

Calamita, G. 2000. The *Escherichia coli* aquaporin-Z water channel. Mol. Microbiol. 37: 254–262.

Cappelier, J.M., Rossero, A., and Federighi, M. 2000. Demonstration of a protein synthesis in starved *Campylobacter jejuni* cells. Int. J. Food Microbiol. 55: 63–67.

Chan, K.F., Le Tran, H., Kanenaka, R.Y., and Kathariou, S. 2001. Survival of clinical and poultry-derived isolates of *Campylobacter jejuni* at a low temperature (4 degrees C) Appl. Environ. Microbiol. 67: 4186–4191.

Coates, D., Hutchinson, D.N., and Bolton, F.J. 1987. Survival of thermophilic campylobacters on fingertips and their elimination by washing and disinfection. Epidemiol. Infect. 99: 265–274.

Cogan, T.A., Bloomfield, S.F., and Humphrey, T.J. 1999. The effectiveness of hygiene procedures for prevention of cross-contamination from chicken carcases in the domestic kitchen. Lett. Appl. Microbiol. 29: 354–358.

Cuk, Z., Annan-Prah, A., Janc, M., and Zajc-Satler, J. 1987. Yoghurt: an unlikely source of *Campylobacter jejuni/coli*. J. Appl. Bacteriol. 63: 201–205.

DeGroote, M.A., Ochsner, U. A., Shiloh, M.A., Nathan, C., McCord, J.M., Dinauer, M.C., Libby, S.J., Vazquez-Torres, A., Xu, Y., and Fang, F. 1997. Periplasmic superoxide dismutase protects *Salmonella* from products of phagocyte NADPH-oxidase and nitric oxide synthase. Proc. Natl. Acad. Sci. USA 94: 13997–14001.

Dorrell, N., Mangan, J.A., Laing, K.G., Hinds, J., Linton, D., Al-Ghusein, H., Barrell, B.G., Parkhill, J., Stoker, N.G., Karlyshev, A.V., Butcher, P.D., and Wren, B.W. 2001. Whole genome comparison of *Campylobacter jejuni* human isolates using a low-cost microarray reveals extensive genetic diversity Genome. Res. 11: 1706–1715.

Doyle, M.P., and Roman, D.J. 1982. Response of *Campylobacter jejuni* to sodium chloride. Appl. Environ. Microbiol. 43: 561–565.

Duncan, C., Dougall, H., Johnston, P., Green, S., Brogan, R., Leifert, C., Smith, L., Golden, M., and Benjamin, N. 1995. Chemical generation of nitric oxide in the mouth from the enterosalivary circulation of dietary nitrate. Nat. Med. 1: 546–551.

Dykhuizen, R.S., Frazer, R., Duncan, C., Smith, C.C., Golden, M., Benjamin N., and Leifert, C. 1996. Antimicrobial effect of acidified nitrite on gut pathogens: importance of dietary nitrate in host defense. Antimicrob. Agents Chemother. 40: 1422–1425.

Dykhuizen, R.S., Fraser A., McKenzie, H., Golden, M., Leifert, C., and Benjamin N. 1998. *Helicobacter pylori* is killed by nitrite under acidic conditions. Gut. 42: 334–337.

Elvers, K.T., and Park, S.F. 2002. Quorum sensing in *Campylobacter jejuni*: detection of a *luxS* encoded signaling molecule. Microbiology 148: 1475–1481.

Elvers, K.T., Wu, G., Gilberthorpe, N.J., Poole, R.K., and Park, S.F. 2004. Role of an inducible single-domain hemoglobin in mediating resistance to nitric oxide and nitrosative stress in *Campylobacter jejuni* and *Campylobacter coli*. J. Bacteriol. 186: 5332–5341.

Evans, S.A. 1997. Epidemiological studies of *Salmonella* and *Campylobacter* in poultry, PhD thesis, University of London, London, UK.

Everest, P.H., Goossens, H., Butzler, J.P., Lloyd, D., Knutton, S., Ketley, J.M., and Williams, P.H. 1992. Differentiated Caco-2 cells as a model for enteric invasion by *Campylobacter jejuni* and *C. coli*. J. Med. Microbiol. 37: 319–325.

Fang, F.C. 1997. Mechanisms of nitric oxide-related antimicrobial activity. J. Clin. Invest. 99: 2818–2825.

Fernandez, H., and Pison, V. 1996. Isolation of thermotolerant species of *Campylobacter* from commercial chicken livers. Int. J. Food Microbiol. 29: 75–80.

Forte, P., Dykhuizen, R.S., Milne, E., McKenzie, A., Smith, C.C., and Benjamin, N. 1999. Nitric oxide synthesis in patients with infective gastroenteritis. Gut 45: 355–361.

Friedman, C.R., Neimann, J., Wegener, H.C., and Tauxe, R.V. 2000 Epidemiology of *Campylobacter jejuni* infections in the United States and other industrialized nations. In: *Campylobacter* 2nd Edition. I. Nachamkin and M.J. Blaser, eds. ASM Press, Washington, DC, pp. 121–138.

Fuqua, C., Parsek, M.R., and Greenberg, E.P. 2001. Regulation of gene expression by cell-to-cell communication: acyl-homoserine lactonequorum sensing. Annu. Rev. Genet. 35: 439–468.

Goldman, E., and H. Jakubowski. 1990. Uncharged tRNA, protein synthesis, and the bacterial stringent response. Mol. Microbiol. 4: 2035–2040.

Goldstein, J., Pollitt, N.S., and Inouye, M. 1990. Major cold shock protein of *Escherichia coli*. Proc. Natl. Acad. Sci. USA 87: 283–287.

Grant, K.A., and Park, S.F. 1995. Molecular characterization of *katA* from *Campylobacter jejuni* and generation of a catalase-deficient mutant of *Campylobacter coli* by intraspecific allelic exchange. Microbiology. 141: 1369–1376.

Hald, B., Knudsen, K., Lind, P., and Madsen, M. 2001. Study of the infectivity of saline-stored *Campylobacter jejuni* for day-old chicks. Appl. Environ. Microbiol. 67: 2388–2392.

Hallback, D.A., Hulten, L., Jodal, M., Lindhagen, J., and Lundgren, O. 1978. Evidence for the existence of a countercurrent exchanger in the small intestine in man. Gastroenterology 74: 683–690.

Harvey, P., and Leach S. 2002. Acid adaptation in *Campylobacter jejuni*: a mechanism for survival in food production and the host. Poster presentation at the 150th Ordinary meeting of the Society for General Microbiology, University of Warwick.

Harvey, P., Battle, T., and Leach, S. 1999. Different invasion phenotypes of *Campylobacter* isolates in Caco-2 cell monolayers. J. Med. Microbiol. 48: 461–469.

Hazeleger, W.C., Wouters, J.A., Rombouts, F. M., and Abee, T. 1998. Physiological activity of *Campylobacter jejuni* far below the minimal growth temperature. Appl. Environ. Microbiol. 164: 3917–3922.

Humphrey, T.J., and Cruickshank, J. G. 1985. Antibiotic and deoxycholate resistance in *Campylobacter jejuni* following freezing or heating. J. Appl. Bacteriol. 59: 65–71.

Jackson, D.W., Suzuki, K., Oakford, L., Simecka, J.W., Hart, M.E., and Romeo T. 2002. Biofilm formation and dispersal under the influence of the global regulator CsrA of *Escherichia coli*. J. Bacteriol 184: 290–301.

Jacobs-Reitsma, W. 2000. *Campylobacter* in the food supply. In: *Campylobacter,* 2nd edition. I. Nachamkin and M.J. Blaser, eds. ASM Press, Washington, DC. pp. 467–481.

Kandror, O., DeLeon, A., and Goldberg, A.L. 2002. Trehalose synthesis is induced upon exposure of *Escherichia coli* to cold and is essential for viability at low temperatures. Proc. Natl. Acad. Sci. USA 99: 9727–9732.

Kelly, A.F. Park, S.F., Bovill, R., and Mackey, B.M. 2000. The survival of *Campylobacter jejuni* and *C. coli* during stationary phase: evidence for the absence of a phenotypic stationary phase response in *C. jejuni*. Appl. Environ. Microbiol. 67: 2248–2254.

Kempf, B., and Bremer, E. 1998. Uptake and synthesis of compatible solutes as microbial stress responses to high-osmolality environments. Arch. Microbiol. 170: 319–330.

Kleerebezem M., Quadri, L.E., Kuipers, O.P., and de Vos, W.M. 1997. Quorum sensing by peptide pheromones and two-component signal-transduction systems in Gram-positive bacteria. Mol. Microbiol. 24: 895–904.

Kloda, A., and Martinac, B. 2001. Structural and functional differences between two homologous mechanosensitive channels of *Methanococcus jannaschii*. EMBO J. 20: 1888–1896.

Ko, R., Smith, L.T., and Smith, G.M. 1994. Glycine betaine confers enhanced osmotolerance and cryotolerance on *Listeria monocytogenes*. J. Bacteriol. 176: 426–431.

Koenraad, P.M.F.J., Rombouts, F.M., and Notermans, S.H.W. 1997. Epidemiological aspects of thermophilic *Campylobacter* in water-related environments. Water Env. Res. 69: 52–63.

Konkel, M.E., Kim, B.J., Klena, J.D., Young, C.R., and Ziprin, R. 1998. Characterization of the thermal stress response of *Campylobacter jejuni*. Infect. Immun. 66: 3666–3672.

Kuroda A., Murphy, H., Cashel, M., and Kornberg, A. 1997. Guanosine tetra- and pentaphosphate promote accumulation of inorganic polyphosphate in *Escherichia coli*. J. Biol. Chem. 272: 21240–21243.

Lacroix, F.J., Cloeckaert, A., Grepinet, O., Pinault, C., Popoff, M.Y., Waxin, H., and Pardon, P. 1996. *Salmonella typhimurium acrB*-like gene: identification and role in resistance to biliary salts and detergents and in murine infection. FEMS. Microbiol. Lett. 135: 161–167.

Lamark, T., Kaasen, I., Eshoo, M.W., Falkenberg, P., McDougall, J., and Strom, A.R. 1991. DNA sequence and analysis of the bet genes encoding the osmoregulatory choline-glycine betaine pathway of *Escherichia coli*. Mol. Microbiol. 5: 1049–1064.

Lawhon, S.D., Frye, J.G., Suyemoto, M., Porwollik, S., McClelland, M., and Altier, C. 2003. Global regulation by CsrA in *Salmonella typhimurium*. Mol. Microbiol. 48: 1633–1645.

Levina, N., Totemeyer, S., Stokes, N.R., Louis, P., Jones, M.A., and Booth, I.R. 1999. Protection of *Escherichia coli* cells against extreme turgor by activation of MscS and MscL mechanosensitive channels: identification of genes required for MscS activity. EMBO J.18: 1730–1737.

Li, H., Duncan, C., Townend, J., Killham, K., Smith, L.M., Johnston, P., Dykhuizen, R., Kelly, D., Golden, M., Benjamin, N., and Leifert, C. 1997. Nitrate-reducing bacteria on rat tongues. Appl. Environ. Microbiol. 63: 924–930.

Limsuwun, K., and Jones, P.G. 2000. Spermidine acetyltransferase is required to prevent spermidine toxicity at low temperatures in *Escherichia coli*. J. Bacteriol. 182: 5373–5380.

Lin, J., Michel, L.O., and Zhang, Q. 2002. CmeABC Functions as a multidrug efflux system in *Campylobacter jejuni*. Antimicrob. Agents. Chemother. 46: 2124–2131.

Loewen, P.C., Hu, B., Strutinsky, J. and Sparling, R. (1998) Regulation in the *rpoS* regulon of *Escherichia coli*. Can. J. Microbiol. 44: 707–717.

Marchant, J., Wren, B., and Ketley J.M. 2002. Exploiting genome sequence: predictions for mechanisms of *Campylobacter* chemotaxis. Trends Microbiol. 10: 155–159.

Marschall, C., and Hengge-Aronis, R. 1995. Regulatory characteristics and promote ranalysis of *csiE*, a stationary phase-inducible gene under the control of σ^s and the cAMP–CRP complex in *Escherichia coli*. Mol. Microbiol. 18: 175–184.

Mead, G.C. 1989. Microbes of the avian cecum: types present and substrates utilized. J. Exp. Zool. Suppl. 3: 48–54.

Mendum, M.L., and Smith, L.T. 2002. Characterization of glycine betaine porter I from *Listeria monocytogenes* and its roles in salt and chill tolerance. Appl. Environ. Microbiol. 68: 813–819.

Moe, P.C., Blount, P., and Kung, C. 1998. Functional and structural conservation in the mechanosensitive channel MscL implicates elements crucial for mechanosensation. Mol. Microbiol. 28: 583–592.

Murphy, C., Carroll, C., and Jordan, K.N. 2003. Induction of an adaptive tolerance response in the food-borne pathogen *Campylobacter jejuni*. FEMS Microbiol. Lett. 223: 89–93.

Nakamura, T., Yuda, R., Unemoto, T., and Bakker, E.P. 1998. KtrAB, a new type of bacterial K^+-uptake system from *Vibrio alginolyticus*. J. Bacteriol. 180: 3491–3494.

327

Newell, D.G.,. McBride, H., Saunders F., Dehele, Y., and Pearson, A.D. 1985. The virulence of clinical and environmental isolates of *Campylobacter jejuni*. J. Hyg. Camb. 94: 45–54.

Newell, D.G., Shreeve, J.E., Toszeghy, M., Domingue, G., Bull, S., Humphrey, T., and Mead, G. 2001. Changes in the carriage of *Campylobacter* strains by poultry carcasses during processing in abattoirs. Appl. Environ. Microbiol. 67: 2636–2640.

Nicholson, E.M., and Scholtz, J.M. 1996. Conformational stability of the *Escherichia coli* HPr protein: test of the linear extrapolation method and a thermodynamic characterization of cold denaturation. Biochemistry. 35: 11369–11378.

Nilsson, H.O., Blom, J., Abu-Al-Soud, W., Ljungh, A.A., Andersen, L.P., and Wadstrom T. 2002. Effect of cold starvation, acid stress, and nutrients on metabolic activity of *Helicobacter pylori*. Appl. Environ. Microbiol. 68: 11–19.

Ochsner, U.A., Snyder, A., Vasil, A.I., and Vasil, M.L. 2002. Effects of the twin-arginine translocase on secretion of virulence factors, stress response, and pathogenesis. Proc. Natl. Acad. Sci. USA 99: 8312–8317.

Obiri-Danso, K., Paul, N., and Jones, K. 2001. The effects of UVB and temperature on the survival of natural populations and pure cultures of *Campylobacter jejuni, Camp. coli, Camp. lari* and urease-positive thermophilic campylobacters (UPTC) in surface waters. J. Appl. Microbiol. 90: 256–267.

Ohtani, K., Hayashi, H., and Shimizu, T. 2002. The *luxS* gene is involved in cell-cell signalling for toxin production in *Clostridium perfringens*. Mol. Microbiol. 44: 171–179.

Park, J.I., C.M. Grant, M.J. Davies, and I.W. Dawes.1998. The cytoplasmic Cu,Zn superoxide dismutase of *Saccharomyces cerevisiae* is required for resistance to freeze-thaw stress: generation of free radicals during freezing and thawing. J. Biol. Chem. 273: 22921–22928.

Parkhill, J., Wren, B.W., Mungall, K., Ketley, J.M., Churcher, C., Basham, D., Chillingworth, T., Davies, R.M., Feltwell, T., Holroyd, S., Jagels, K., Karlyshev, A.V., Moule, S., Pallen, M.J., Penn, C.W., Quail, M.A., Rajandream, M.A., Rutherford, K.M., van Vliet, A.H., Whitehead, S., and Barrell B.G. 2000. The genome sequence of the food-borne pathogen *Campylobacter jejuni* reveals hypervariable sequences. Nature. 403: 665–668.

Pesci, E.C., Cottle, D.L., and Picket, C. L. 1994. Genetic, enzymatic, and pathogenic studies of the iron superoxide dismutase of *Campylobacter jejuni*. Infect. Immun. 62: 2687–2695.

Pickett, C.L., Pesci, E.C., Cottle, D.L., Russell, G., Erdem, A.N. and Zeytin, H. 1996. Prevalence of cytolethal distending toxin production in *Campylobacter jejuni* and relatedness of *Campylobacter* sp. *cdtB* gene. Infect. Immun. 64: 2070–2078.

Poole R.K., and Hughes, M.N. 2000. New functions for the ancient globin family: bacterial responses to nitric oxide and nitrosative stress. Mol. Microbiol. 36: 775–783.

Purdy, D., and Park, S.F. 1994. Cloning, nucleotide sequence, and characterization of a gene encoding superoxide dismutase from *Campylobacter jejuni* and *Campylobacter coli*. Microbiology 140: 1203–1208.

Purdy, D., Cawthraw S., Dickinson, J.H., Newell, D.G. and Park, S.F. 1999. Generation of a superoxide dismutase-deficient mutant of *Campylobacter coli*: evidence for the significance of SOD in *Campylobacter* survival and colonization. Appl. Environ. Microbiol.65: 2540–2546.

Purdy, D., Buswell, C.M., Hodgson, A.E., McAlpine, K., Henderson, I., and Leach, S. A. 2000. Characterization of cytolethal distending toxin (CDT) mutants of *Campylobacter jejuni*. J. Med. Microbiol. 49: 473–479.

Rees, C.E., Dodd, C.E., Gibson, P.T., Booth, I.R. and Stewart G. S.A.B. 1995. The significance of bacteria in stationary phase to food microbiology. Int. J. Food. Microbiol. 28: 263–275.

Rivera-Amill, V., Kim, B.J., Seshu, J., and Konkel, M.E. 2001. Secretion of the virulence-associated *Campylobacter* invasion antigens from *Campylobacter jejuni* requires a stimulatory signal. J. Infect. Dis. 183: 1607–1616.

Rivera-Amill, V., Konkel, M.E. 1999. Secretion of *Campylobacter jejuni* Cia proteins is contact dependent. Adv. Exp. Med. Biol. 473: 225–229.

Romeo, T. 1998. Global regulation by the small RNA-binding protein CsrA and the non-coding RNA molecule CsrB. Mol. Microbiol. 29: 1321–1330.

Sawers, G. 1999. The aerobic/anaerobic interface. Curr. Opin. Microbiol. 2: 181–187.

Schauder, S., Shokat, K., Surette, M.G., and Bassler, B.L. 2001. The LuxS family of bacterial autoinducers: biosynthesis of a novel quorum-sensing signal molecule. Mol. Microbiol. 41: 463–476.

Schulz, A., and Schumann, W. 1996 *hrcA*, the first gene of the *Bacillus subtilis dnaK* operon encodes a negative regulator of class I heat shock genes. J. Bacteriol. 178: 1088–1093.

Scoarughi, G.L.,. Cimmino, C., and Donini, P. 1999. *Helicobacter pylori*: a eubacterium lacking the stringent response. J. Bacteriol. 181: 552–555.

Sellars, M.J., Hall, S.J., and Kelly, D.J. 2002. Growth of *Campylobacter jejuni* supported by respiration of fumarate, nitrate, nitrite, trimethylamine-N-oxide, or dimethyl sulfoxide requires oxygen. J. Bacteriol. 184: 4187–4196.

Shiba, T., Tsutsumi, K., Yano, H., Ihara, Y., Kameda, A., Tanaka, K., Takahashi, H. Munekata, M., Rao, N.N., and Kornberg, A. 1997. Polyphosphate and the induction of rpoS expression. Proc. Natl. Acad. Sci. USA 94: 11210–11215.

Skinner, K.A., White, C.R., Patel, R., Tan, S., Barnes, S., Kirk, M., Darley-Usmar, V., and Parks, D.A. 1998. Nitrosation of uric acid by peroxynitrite. Formation of a vasoactive nitric oxide donor. J. Biol. Chem. 273: 24491–24497.

Spector, M.P., and Foster, J.W. 1993 Starvation response of *Salmonella typhimurium* In: Starvation in Bacteria. S. Kjelleberg, ed. Plenum Press, New York, pp. 201–224.

Stead, D., and Park, S.F. 2000 Roles of Fe superoxide dismutase and catalase in resistance of *Campylobacter coli* to freeze-thaw stress. Appl. Environ. Microbiol. 66: 3110–3112.

Stern, N.J., Myszewski, M.A., Barnhart, H.M., and Dreesen D.W. 1997. Flagellin A gene restriction fragment length polymorphism patterns of *Campylobacter* spp. isolates from broiler production sources. Avian Dis. 41: 899–905.

Stevenson, B., and Babb, K. 2002. LuxS-mediated quorum sensing in *Borrelia burgdorferi*, the Lyme disease spirochete. Infect. Immun. 70: 4099–40105.

Stirling, D.A., Hulton, C.S., Waddell, L., Park, S.F., Stewart, G.S., Booth, I.R., and Higgins C.F. 1989. Molecular characterization of the *proU* loci of *Salmonella typhimurium* and *Escherichia coli* encoding osmoregulated glycine betaine transport systems. Mol. Microbiol. 3: 1025–1038.

Strom, A.R., and Kaasen, I. 1993 Trehalose metabolism in *Escherichia coli*: stress protection and stress regulation of gene expression. Mol. Microbiol. 8: 205–210.

Surette, M.G., Miller, M.B., and Bassler, B.L. 1999 Quorum sensing in *Escherichia coli, Salmonella typhimurium*, and *Vibrio harveyi*: a new family of genes responsible for autoinducer production. Proc. Natl. Acad. Sci. USA 96: 1639–1644.

Tabor, C.W. 1968. The effects of temperature on the acetylation of spermidine. Biochem. Biophys. Res. Commun. 30: 339–342.

Thies, F.L.,. Hartung H.-P, and Giegerich, G. 1998. Cloning and expression of the *Campylobacter jejuni lon* gene. FEMS Microbiol. Lett. 165: 329–334.

Thies, F.L., Karch, H., Hartung, H.-P., and Giegerich, G. 1999a. Cloning and expression of the *dnaK* gene of *Campylobacter jejuni* and antigenicity of heat shock protein 70. Infect. Immun.67: 1194–1200.

Thies, F.L., Weishaupt, A. Karch, H., Hartung, H.-P. and Giegerich, G. 1999b. Cloning, sequencing and molecular analysis of the *Campylobacter jejuni groESL* bicistronic operon. Microbiology 145, 89–98.

Thomas, C., Hill, D.J., and Mabey, M. 1999. Evaluation of the effect of temperature and nutrients on the survival of *Campylobacter* spp. in water microcosms. J. Appl. Microbiol. 186: 1024–1032.

Thomas C., Hill, D., and Mabey, M 2002. Culturability, injury and morphological dynamics of thermophilic *Campylobacter* spp. within a laboratory-based aquatic model system and following exposure to sunlight. J. Appl. Microbiol. 92: 433–442.

van de Giessen, A.W., Tilburg, J.J., Ritmeester, W.S., and van der Plas, J. 1998. Reduction of *Campylobacter* infections in broiler flocks by application of hygiene measures. Epidemiol. Infect.121: 57–66.

van Vliet, A.H., Baillon, M.L., Penn, C.W., and Ketley, J.M. 1999. *Campylobacter jejuni* contains two Fur homologues: characterization of iron-responsive regulation of peroxide stress defense genes by the PerR repressor. J. Bacteriol. 181: 6371–6376.

van Vliet, A.H., Baillon, M.A., Penn, C.W., and Ketley, J.M. 2001. The iron-induced ferredoxin FdxA of *Campylobacter jejuni* is involved in aerotolerance. FEMS. Microbiol. Lett. 196: 189–193.

van Vliet, A.H., Ketley, J.M., Park, S.F., and Penn, C.W. 2002. The role of iron in *Campylobacter* gene regulation, metabolism and oxidative stress defense. FEMS. Microbiol. Rev. 26: 173–186.

Wempe, J.M., Genigeorgis, C.A., Farver, T.B., and Yusufu, H.I.T. 1983. Prevalence of *Campylobacter jejuni* in two California chicken processing plants. Appl. Environ. Microbiol. 45: 355–359.

Wesley, I.V., Wells, S.J., Harmon, K.M., Green, A., Schroeder-Tucker, L., Glover, M., and Siddique, I. 2000. Fecal shedding of *Campylobacter* and *Arcobacter* spp. in dairy cattle. Appl. Environ. Microbiol. 66: 1994–2000.

Winzer, K., Hardie, K.R., Williams, P. 2002a Bacterial cell-to-cell communication: sorry, can't talk now – gone to lunch! Curr. Opin. Microbiol. 5: 216–222.

Winzer, K., Hardie, K.R., Burgess, N., Doherty, N., Kirke, D., Holden, M.T., Linforth, R., Maizeell, K.A., Taylor, A.J., Hill, P.J., and Williams, P. 2002b. LuxS: its role in central metabolism and the *in vitro* synthesis of 4-hydroxy-5-methyl-3 (2H)-furanone. Microbiology. 2002b 148: 909–922.

Witthoft, T., Eckmann, L., Kim, J.M., and Kagnoff, M.F. 1998. Enteroinvasive bacteria directly activate expression of iNOS and NO production in human colon epithelial cells. Am. J. Physiol. 275: G564–571.

Wooldridge, K.G., and J.M. Ketley. 1997. *Campylobacter* host cell interactions. Trends Microbiol. 5: 96–102.

Wu, Y.L., Lee, L.H., Rollins, D.M., and Ching, W.M. 1994. Heat shock- and alkaline pH-induced proteins of *Campylobacter jejuni*: characterization and immunological properties. Infect. Immun. 62: 4256–4260.

Yamanaka, K., Fang, L., and Inouye, M. 1998. The CspA family in *Escherichia coli*: multiple gene duplication for stress adaptation. Mol. Microbiol. 27: 247–255.

Yang, H., Li, Y., and Johnson, M.G. 2001. Survival and death of *Salmonella typhimurium* and *Campylobacter jejuni* in processing water and on chicken skin during poultry scalding and chilling. J. Food. Prot. 64: 770–776.

Zgurskaya, H.I. 2002. Molecular analysis of efflux pump-based antibiotic resistance. Int. J. Med. Microbiol. 292: 95–105.

Ziprin, R.L., Young, C.R., Stanker, L.H., Hume, M.E., and Konkel, M.E. 1999. The absence of cecal colonization of chicks by a mutant of *Campylobacter jejuni* not expressing bacterial fibronectin-binding protein. Avian Dis. 43: 586–589.

Chapter 17

Motility

Aparna Jagannathan and Charles Penn

Abstract

A key feature of many pathogenic bacteria is motility by means of flagella. Motility and flagellar synthesis have been implicated by several classical studies in intestinal colonization and virulence, and *C. jejuni* also depends on flagellar motility for its pathogenicity. With the involvement of more than 40 flagellar structural and regulatory genes, the flagellar system includes a type III secretion system necessary for flagellar assembly. It differs significantly however from the *E. coli/Salmonella* paradigm of regulation of flagellar gene expression. The regulation of expression of these genes is probably influenced primarily by cell cycle events, but in addition flagellar expression appears to be linked to flagellin glycosylation and to be potentially phase-variable. The association of these phenomena with virulence modulation in *C. jejuni* is underinvestigated.

INTRODUCTION

Motility is perhaps the best-established virulence factor of *Campylobacter jejuni* and its close relative *C. coli*. It has also proved to be a vital weapon in the armoury of many other enteric bacterial pathogens, and new angles on the properties of bacterial flagellins as potential pro-inflammatory or immunomodulatory proteins are currently emerging. Research on the role of motility in *Campylobacter* virulence has been extensive over the past 15 years or so, and much of the earlier work has been thoroughly reviewed elsewhere (Guerry *et al.*, 1992; Guerry *et al.*, 2000; Nuijten *et al.*, 1992). This review will therefore consider more recent work in this area, against the background of new developments in the understanding of bacterial motility and flagellins, and will also provide an update on the molecular genetics and biogenesis of the flagella of *C. jejuni*.

THE FLAGELLINS

The flagella of *C. jejuni* are of the 'complex' type (i.e. are composed, unlike those of *E. coli*, of more than one flagellin monomer in the filament, which constitutes the major part of the flagellar structure). The flagellins were among the first antigens of *C. jejuni* and *C. coli* to be characterized (Logan *et al.*, 1987; Newell *et al.*, 1984), proving to be major antigens recognized immunologically during infection and a significant heat-labile (HL) antigenic factor in the Lior serotyping scheme (Lior *et al.*, 1982). This scheme, eventually defining more than 50 serotypes, was initially believed to exploit the flagellar antigen as one of the serodeterminants contributing to antigenic diversity among strains of *C. jejuni*, although the surface components imparting thermolabile specificity were later shown to be different from the flagella in some serotypes (Alm *et al.*, 1991). Antigenic characterization also indicated heterogeneity in *Campylobacter* flagellins within strains, with a capability to alternate or phase-switch between types (Harris *et al.*, 1987; Logan *et al.*, 1987). A region of the protein involved in antigenicity was identified in preliminary structural studies (Power *et al.*, 1994). *C. jejuni* has the capacity spontaneously to undergo bi-directional transition between flagellated and aflagellate phenotypes which suggests its ability to modulate its motility phenotype through variable expression of flagella (Caldwell *et al.*, 1985). Furthermore, it is now clear that post-translational modification of flagellins by glycosylation is also potentially phase-variable (see below) and likely to affect flagellar function. This ability to modulate flagellar expression and function is perhaps compatible with the necessity for many pathogenic bacteria to adapt to the environmental changes they encounter as they progress through the various stages of pathogenesis, e.g. colonization followed by invasion of epithelial cells. In practical terms, it should be noted that the strain NCTC 11168, the first subject for whole genome sequencing of *C. jejuni*, is probably subject to flagellar phase switching of some kind in that many laboratory stocks of the organism, while displaying intact full-length flagella, appear phenotypically non-motile. Their motility is however restored by passage *in vivo* through the avian digestive tract (Jones *et al.*, 2003), and can also be selected for *in vitro* using soft agar 'motility' media (Karlyshev *et al.*, 2002). The flagellins were the subject of some of the earliest molecular genetic experimentation in these organisms, in which flagellin genes were disrupted leading to loss of motility and of virulence (Guerry *et al.*, 1991; Wassenaar *et al.*, 1991). Studies focused primarily on the flagellin genes *flaA* and *flaB*, isolated from genomic libraries by the use of an oligonucleotide probe designed on the basis of flagellin sequence from *C. coli* (Guerry *et al.*, 1991), and by antibody probing of an expression library from *C. jejuni* (Nuijten *et al.*, 1990). Genetic investigation showed that for both organisms, there were two flagellin genes arranged in tandem, the upstream gene *flaA* being dependent on a σ^{28} promoter and the downstream gene *flaB* on a σ^{54} promoter (Guerry *et al.*, 1990; Nuijten *et al.*, 1990). In both organisms, the DNA and deduced amino acid sequences of the two genes were highly similar although not identical, most of the variation being found in the central region of the protein sequence. Mutants in *flaA* were poorly motile and possessed only short truncated flagella, while mutants in *flaB* possessed near-normal flagella and motility but in some experiments showed some decrease in pathogenic potential (Guerry *et al.*, 1991; Wassenaar *et al.*, 1993). The conclusion was that FlaA is the major flagellin while FlaB is a minor component of the flagella, possibly not expressed at all under some conditions (Wassenaar *et al.*, 1993), and the exact function of which remains uncertain. One possibility is that FlaB is a component of the proximal part of the flagellar filament while FlaA forms the bulk of the distal part of the structure, as is believed to occur in *Helicobacter pylori* (Kostrzynska *et al.*, 1991), but the similarity of the two proteins, which are antigenically near-identical, has precluded definitive testing of this hypothesis in *C. jejuni*. Although both FlaA and FlaB subunits are required for a fully active flagellar filament in wild-type cells, at the regulatory level, a variant showing expression of *flaB* forming full-length flagella composed of FlaB, under selective pressure *in vivo*, was also observed suggesting a predominant transcription of one gene at a time (Wassenaar *et al.*, 1994). Investigation of the regulation of expression of the two flagellins, and the possibility that recombination between the two sequences (Alm *et al.*, 1993; Harrington *et al.*, 1997) occurs, or that compensatory over-expression of FlaB may occur when FlaA is deleted, indicates that there are complexities in flagellin expression that are not yet fully understood.

ROLE OF FLAGELLAR GENES AND FLAGELLINS AS MARKERS IN EPIDEMIOLOGY

In past decades, phenotypic techniques like biotyping, serotyping and phagetyping have served to identify and distinguish *C. jejuni* subtypes although the epidemiological data provided have not always been unequivocal. Genotyping on the other hand is a better alternative in providing more precise information on the genetic diversity of *Campylobacter*, with the use of techniques like ribotyping, PFGE (pulsed field gel electrophoresis) and PCR based methods such as MLST (multilocus sequence typing), most of which provide information about the basic genetic framework of a strain rather than loci that may be subject to variation and selective pressure. Such methods can usefully be complemented by methods such as flagellin gene typing, which do focus on genes that are highly variable and subject to selection and potentially to horizontal transfer, but may effectively identify related strains in 'short term' epidemiological investigations of potential outbreaks. Techniques that have been employed for flagellin gene typing include RFLP (restriction fragment length polymorphism), DGGE (denaturing gradient gel electrophoresis) and sequence analysis of the *flaA* gene. Recently, the application of SSCP (single strand conformation polymorphism) analysis on the 3' variable (SVR) region (see below) of *flaA* provided distinct patterns for different strains, indicating its use as a primary subtyping method (Hein *et al.*, 2003).

The first detailed report of flagellin gene or *fla* typing by PCR-RFLP analysis was by Nachamkin and colleagues (Nachamkin *et al.*, 1993). This approach relies on the specific amplification of the flagellin-encoding *flaA* gene followed by restriction enzyme digestion, and is able to distinguish strains within a given serotype. Additionally, *fla* typing was shown to correlate with both HL and O (heat-stable antigen) serotypes in a study wherein 404 *Campylobacter* strains were determined to belong to 83 distinct flagellar types (Nachamkin *et al.*, 1996). However, there are limitations to this typing method as the flagellin gene typing data do not necessarily reveal consistent differences between strains belonging to different species and between strains belonging to a single species, which may exhibit conserved flagellin profiles (Santesteban *et al.*, 1996). Presence of these conserved flagellin profiles across different genotypes could be due to horizontal gene transfer. Furthermore, reports of recombination of *fla* sequences within and between flagellin loci imply that such an intra/intergenomic recombination may not allow a flagellin gene typing method to efficiently differentiate *Campylobacter* populations for epidemiological purposes (Harrington *et al.*, 1997). Studies and exploitation of the *fla* typing approach have however progressed further with comparisons of the entire *flaA* sequence from a panel of 15 strains, revealing the presence of two highly variable regions, a short variable region (SVR) of 150 bp around bases 450 to 600 of the gene, and another region around bases 700–1450 out of the total of about 1764 nucleotides (Meinersmann *et al.*, 1997). Further, SVR analysis of 20 isolates from *C. jejuni* outbreaks demonstrated considerable variation, suggesting that flagellin genotyping is an effective method for discriminating between strains, and therefore can complement or replace serotyping methods. Several studies using flagellin gene typing for identification and typing of campylobacters have been conducted. Fla typing by RFLP involving a combination of both flagellin genes, *flaA* and *flaB* from each strain may also be informative, allowing specific subtyping of *C. jejuni* isolates (Petersen and Newell, 2001). Other studies include flagellin gene PCR-RFLP analysis of a 702 bp PCR amplified portion of the flagellin gene *flaA* that revealed 22 types from isolated human and chick strains of *C. jejuni* (Steinhauserova *et al.*, 2002).

The flagellin gene has also been proposed as a marker to identify Guillain–Barré syndrome (GBS)-related strains by means of a particular variant of the SVR of *flaA* (Tsang *et al.*, 2001). In addition, the study of diverse *C. jejuni* genotypes defined by MLST and *flaA* SVR nucleotide sequences and serotypes that reflect capsular and flagellar antigens demonstrated an association with the development of neuropathy in GBS patients (Dingle *et al.*, 2001). Thus flagellin typing is a useful tool aiding in both the epidemiological study of campylobacters and as a marker for strains associated with GBS.

Finally, the flagellins have potential as immunogens of significance in vaccine development. In a study by Lee and others (Lee *et al.*, 1999), it was observed that both intranasal and oral vaccination with a truncated recombinant flagellin subunit vaccine elicited a protective immune response against *C. jejuni* as revealed by molecular, immunological and animal model methods.

THE ROLE OF MOTILITY AND FLAGELLAR EXPRESSION IN VIRULENCE

It should be noted that intestinal colonization *per se*, in animal models that do not exhibit real pathology (most do not), is not strictly a manifestation of virulence; however a process analogous to intestinal colonization (without pathology) in animal models is probably an essential step in pathogenesis in humans, and it is a reasonable assumption that organisms that cannot colonize the intestine will exhibit reduced virulence in humans. A connection between colonization potential and flagellum-based motility has long been observed in *C. jejuni*. As is the case for many motile intestinal bacterial pathogens, motility, or at least the possession of flagella, was found to be an important factor in early studies of undefined variants or mutants in animal colonization experiments (Caldwell *et al.*, 1985; Logan *et al.*, 1989). Experiments in which the colonization potential of defined flagellar mutants was compared with that of wild-type parental strains provided the first concrete demonstration of the importance of flagellar genes in the disease process. Knock-out mutants in the two major flagellins were tested in animal models for colonization capability, and while FlaA appeared essential for colonization in rabbits (Pavlovskis *et al.*, 1991) and chicks (Nachamkin *et al.*, 1993), such a role could not be demonstrated for FlaB (Wassenaar *et al.*, 1993).

In vitro models of virulence have also been used in similar experiments, mainly based on adherence to and invasion of cultured cell lines. In one study, *flaA* was shown to be essential for invasion while *flaB* was not, but it was also clear that other factors in addition to motility were required for invasion. Furthermore, it was shown that the FlaA flagellin *per se* does not appear to be an adhesin (Wassenaar *et al.*, 1991). Experiments in a rabbit ligated ileal loop model, which is an imperfect but indicative model for pathogenesis *in vivo*, also showed that a spontaneous and undefined non-motile variant was incapable of eliciting the same degree of host reaction (fluid accumulation, tissue damage) as wild-type strains (Everest *et al.*, 1993). Several *in vitro* studies have demonstrated the role of flagella in host cell internalization. One such study (Grant *et al.*, 1993) showed that both *flaA* and *flaB* mutants adhered to cultured epithelial cells as readily as the wild-type. However they were deficient in their ability to enter epithelial cells.

As well as the designed mutants in the flagellins referred to above, mutants made by various methods have also been extensively screened for motility and in some cases for virulence-related attributes. Thus among mutants obtained by insertional mutagenesis of library clones followed by recombination back into the *Campylobacter* chromosome, non-invasive mutants were selected using a tissue culture model and some of these proved to be non-motile (Yao *et al.*, 1994). An interesting outcome from this work was the discovery of the *pflA* gene, a mutant which showed a flagellate yet non-motile phenotype and a reduced ability to invade INT407 cells; the exact function of the gene in motility has not yet been defined. The protein encoded by *pflA* falls within the family of tetratricopeptide repeat proteins (Andrade *et al.*, 2001), suggesting that protein–protein interaction may be involved in its function.

The effective use of transposon-based mutagenesis methods was not fully realized until the introduction of an Himar 1 transposase-based (Golden *et al.*, 2000) *in vitro* transposon mutagenesis approach. Another *in vitro* approach that was partially successful was based on Tn552 (Colegio *et al.*, 2001). Further development of a Himar 1-based method led to an improved *in vitro* system (Hendrixson *et al.*, 2001). In all of these developments in mutagenesis methodology, and in exploration of the mutants obtained (Golden and Acheson, 2002; Yao *et al.*, 1997), it is striking that motility mutants were always isolated and these have proved highly illuminating in studies of the genetics of motility and its significance in pathogenesis. In the interesting study by Golden and Acheson (2002) a correlation has been demonstrated between flagellar expression and autoagglutination. Experiments using an autoagglutination assay showed that twenty transposition-derived motility mutants were unable to autoagglutinate. Although most of these mutants lacking an autoagglutination phenotype showed a negligible or reduced level of FlaA expression, four mutants expressing wild-type levels of FlaA also demonstrated an inability to autoagglutinate, indicating that FlaA may not be the sole requirement for autoagglutination, a phenomenon that is likely to reflect surface properties of the organism and hence may be related to its pathogenicity. Indeed it has been shown independently that autoagglutination is strongly correlated with expression of flagella (Misawa and Blaser, 2000).

Other aspects of the potential of flagellar expression in determining virulence include first, the suggestion that parts of the flagellar 'secreton', the only complete and proven protein secretion mechanism in most strains (other than the autotransporter or 'type V' route (Henderson *et al.*, 2000)), might be responsible for extracellular secretion of virulence factor proteins. Such a proposal was first put forward for *Yersinia enterocolitica* (Young *et al.*, 1999) and it is an attractive hypothesis that such a mechanism might operate in *C. jejuni*, where proteins potentially involved in virulence are known to be secreted extracellularly (Konkel *et al.*, 1999) but the route and mechanism for this are not known. Extracellular secretion of effector proteins by pathogenic Gram-negative bacteria is a widespread virulence mechanism. Second, Gram-negative bacterial flagellins *per se* are increasingly prominent in the literature as pro-inflammatory mediators (Gewirtz *et al.*, 2001; Liaudet *et al.*, 2003; Moors *et al.*, 2001; Zhou *et al.*, 2003), not least via the Toll-like receptor 5 pathway. In the absence of any great diversity of potential virulence factors in *C. jejuni* with potential to stimulate intestinal pathology, it is tempting to speculate that flagellin might have such an effect in *Campylobacter* infections of humans.

THE FLAGELLAR GENE COMPLEMENT IN *C. JEJUNI*

As in enteric group bacteria, the *C. jejuni* flagellum comprises three structural elements: basal body, hook and filament. The majority of the genes involved in biogenesis of this complex structure, a total of about 40, are conserved with those of the well-defined paradigm based on research in the 'enteric' group organisms *E. coli* and *Salmonella enterica* serovar Typhimurium (Macnab, 2003) (hereafter referred to as the *E. coli* model). However, the

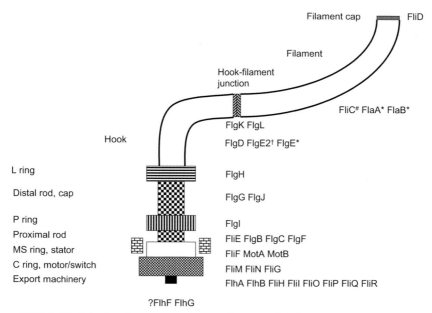

Figure 17.1 Model of the flagellar gene complement of *C. jejuni*. The known structural elements of bacterial flagellar assembly are detailed on the left of the diagram with associated flagellar proteins on the right. The proteins FlhDC, FliA, FlgM, FliK, FlgR and σ^{54} which are not shown are known to have a role in the transcriptional regulation of flagellar genes. Also not shown are FlgA, FlgN, FliJ, FliS and FliT which are believed to act as chaperones. Genes encoding proteins labelled with '#' along with FlhDC, FliK, FlgA, FlgN and FliJ have not been found in the *C. jejuni* genome. The proteins labelled with '*' are not present in the *E. coli* genome, and in addition in the latter organism σ^{54} is not involved in flagellar expression and there is no equivalent to FlgS. FlgE2 corresponds to FlgE of *E. coli* and the gene annotated in *C. jejuni* to encode FlgE has an unknown role in flagellar biosynthesis.

regulatory genes and the flagellar gene hierarchy clearly differ between the enterics and *C. jejuni*, a notable difference being the lack of the *flhDC* master regulator in *C. jejuni*: in the peritrichous organisms *E. coli* and *Salmonella enterica*, the *flhDC* transcriptional activators activate the flagellar regulon. Perusal of the flagellar gene complement as defined by the genome annotation (Parkhill *et al.*, 2000) shows several differences both from the enteric paradigm (Macnab, 2003) and from the situation in the closely related *Helicobacter pylori* (Josenhans *et al.*, 2002). These differences are summarized in Table 17.1.

In other polar flagellate bacteria, such as *Vibrio cholerae* (Prouty *et al.*, 2001), *Vibrio parahaemolyticus* (Kim and McCarter, 2000), *Pseudomonas aeruginosa* (Arora *et al.*, 1997) and *Caulobacter crescentus* (Wu and Newton, 1997), studied so far, flagellar gene regulation is accomplished by FlgR-like transcriptional activators, which are required for flagellar gene expression under the control of the alternative sigma factor σ^{54}. It should be emphasized that in many polar flagellate bacteria, this is one of the fundamental differences from the enteric group paradigm, where there is no known involvement of σ^{54} in flagellar gene expression. The presence of only a single σ^{54}-activator protein of *C. jejuni* is in pronounced contrast to the multiple activators of the other organisms listed above. The *C. jejuni* flagellar system shares similarities to that of *H. pylori* wherein a single transcriptional activator FlgR along with σ^{54} and σ^{28} all play a role in regulation of flagellar gene expression (Spohn and Scarlato, 1999). More recent studies on the flagellar model in *H. pylori* reveal a flagellar organization with three hierarchical levels of gene

Table 17.1 Flagellar gene orthologues and novel motility genes in *C. jejuni*, *H. pylori* and *E. coli*.

Gene	Location or function of protein	Present or absent in		
		C. jejuni[5]	*H. pylori*[5]	*E. coli*
flgA	Periplasmic chaperone	–	–	+
flgF	Proximal rod	+ (*flgG2*)	+ (*flgG2*)	+
flgJ	Peptidoglycan hydrolase and rod cap	–	–	+
flgL	Hook-filament junction protein 3	– (*flaD*)	+	+
flgM	Anti-σ[28] factor	+[6]	+	+
flgN	Export chaperone	–	–	+
flhDC	Master regulator operon	–	–	+
fliJ	Chaperone	–	–	+
fliK	Hook-length sensor	–	–	+
fliO	Possible type III export component	–	–	+
fliT	Chaperone	–	–	+
fliY Cj0059c	Motor-switch protein	+	HP1030	–
rpoN Cj0670	σ[54]	+	HP0714	Not flagellar regulon
flaC Cj0720	Flagellin-like gene	+	–	–
flgR Cj1024c	σ[54] activator	+	HP0703	–
flaB Cj1338c	Flagellin	+	HP0115	–
flgE Cj0043	'2nd' flgE	+	HP0908	–
flaG Cj0547	Flagellin-like	+	HP0751	–
flaG' Cj1313	Flagellin-like	+	HP0327	–
flhF Cj0064c	FtsY-like	+	HP1035	–
flhG Cj0063c	MinD-like	+	HP1034	–
pflA Cj1565	Paralysed flagella protein	+	HP1274	–
Cj0848	FlhB-like C-terminal domain	+	HP1575	–
Cj0041[1]	Unknown	+	HP0906	–
Cj1075[1]	Unknown conserved hypothetical	+	HP1377	–
Cj1497c[1]	Unknown	+	–	–
Cj0062c[2]	Unknown	+	HP1033	–
Cj0248[2]	Unknown	+	–	–
Cj0883c[2]	Transcriptional regulator, upstream of *flhA*	+	–	–
Cj1026c[2]	Probable lipoprotein	+	HP0837	–
Cj1189c[2]	Probable aerotaxis receptor protein	+	–	+
Cj1190c[2]	MCP protein	+	HP0599	+
Cj1336[3] *maf5*	Motility accessory factor (MAF)	+	HP0114	–
Cj1318[4]	MAF gene	+	HP0114	–
Cj0390[4]	Tetratricopeptide repeat protein	+	HP1479	–
Cj1293[4]	Putative sugar dehydrogenase-flagellin modification protein	+	HP0840	–

[1-4]Genes in which insertional mutants were reported to be non-motile, by [1]Golden and Acheson (2002), [2]Hendrixson *et al.* (2001), [3]Karlyshev *et al.* (2002), [4]Colegio *et al.* (2001).
[5]Gene names in parentheses denote annotated name in *C. jejuni* or *H. pylori*, or equivalent gene in *H. pylori*. See text for details.
[6]Probably present but not annotated. See text for details.

expression, class I with genes encoding the basal body and export apparatus including *flgR*, class II with σ[54]-dependent genes and class III with σ[28]-dependent genes; in addition a small number of genes have both σ[54] and σ[28] promoters (Josenhans *et al.*, 2002). This organization may provide a useful model for the flagellar regulon in *C. jejuni*. However, the regulatory control of FlaA by FlgR in *C. jejuni* appears to differ from that of *H.*

pylori where in the latter, FlgR represses FlaA gene expression. No such role of FlgR as a repressor was observed in *C. jejuni*, where a *flgR* knock-out mutant did not exhibit increased expression of FlaA, but rather the reverse (Jagannathan *et al.*, 2001); however the transcriptional analysis of such mutants has not yet been reported.

Some of the individual gene differences between the *E. coli* paradigm and *C. jejuni* merit further comment. First, the annotated *flgG2* and *flaD* of the genome sequence appear to correspond to *flgF* and *flgL* respectively of *E. coli*. Second, while sequence similarity suggests that Cj1464 is a homologue of the experimentally defined *flgM* of *H. pylori*, no experimental evidence for this has yet been published. One reason why Cj1464 was not recognized as a putative *flgM* at the time of the genome annotation is that it encodes a very small predicted protein of only 65 amino acids, and although the C-terminal 32 amino acids (more than half the protein) show 45% identity with the corresponding region of the *E. coli* FlgM, the statistical significance of this similarity in search algorithm outputs is below the default cut-offs for identification of orthologues. Significantly, several of the other unidentified genes, *fliJ*, *fliT* and *flgN*, also encode small proteins and may remain unrecognized for similar reasons. Third, an unexplained duplication of the flagellar hook protein gene *flgE* deserves comment. Although one copy of this gene, Cj0043, is located downstream of *flgD*, as in the *E. coli* genome, and appears likely to be transcribed along with *flgD*, there is no evidence that this gene is involved in flagellar hook construction and, for example, there are no reports of transposon mutants in this gene affecting motility. Instead the gene annotated *flgE2* (Cj1729c), which is located separately on the chromosome very close to the origin of replication, appears to be the functional gene: its orthologue in *C. coli* has been shown experimentally to be essential for flagellar expression and motility (Kinsella *et al.*, 1997). The protein encoded by Cj1729c is the largest known FlgE protein at over 90 kDa. Fourth, a second *flhB*-like gene, Cj0848c, is also present in the genome and the predicted protein product closely resembles the C-terminal part of the full-length FlhB protein encoded by Cj0335. The role of the Cj0848c protein is unknown, and knockout mutants, either in the wild type or in combination with insertionally inactivated *flhB*, showed no difference in their flagellar phenotype from that of the parental strain (Nahid Kamal, personal communication). Yet this small *flhB*-like gene is conserved in *H. pylori*, suggesting a conserved function between the two species.

GENES PROVEN AND IMPLICATED TO BE INVOLVED IN FLAGELLIN GLYCOSYLATION

The *C. jejuni* genome sequencing project (Parkhill *et al.*, 2000) revealed a large cluster of about 50 genes running from Cj1493 to Cj1342, including those encoding the flagellins but no other structural flagellar genes, many of which were thought at the time of the genome annotation to be involved in flagellar modification/glycosylation. This supposition was based on the occurrence in the cluster of several putative sialic acid biosynthesis genes, *neuA* and *neuB*, as well as the flagellin modification genes *ptmA* and *ptmB*, and putative acyl- and amino- transferases and acyl carrier proteins likely to be involved in sugar synthesis and polymerization. The genes concerned and their functions have recently been reviewed (Szymanski *et al.*, 2003) in light of detailed research of some aspects of this flagellin-specific O-glycosylation pathway as well as the second, 'general' N-glycosylation pathway also characterized in *C. jejuni* (Szymanski *et al.*, 1999; Szymanski *et al.*, 2002). Interestingly, the genome sequence revealed first, that the majority of genes in the flagellin modification cluster were not shared with *H. pylori*; and second, that several of the 'hypervariable' sequence regions detected during the genome sequencing work, where variable numbers of single nucleotide repeats in homopolymeric sequence runs were discovered in different sequencing reads, were in genes located in this cluster. Eight of the 25 genomic hypervariable regions were present in this flagellin modification region. These observations suggest first, that the flagellin glycosylation system or mechanism may not be shared with the closely related *H. pylori*, suggesting that the phenotype conferred may be uniquely important in the pathogenesis of *Campylobacter*; and second that flagellin glycosylation may be subject to high frequency phase switching due to variation in the polymorphic sequences, and hence may be involved in immune evasion or other rapid adaptations to conditions during the process of infection. Another unique feature of the flagellin modification region is the presence of multiple copies of two novel families of paralogous genes, the *maf* or '1318' family typified by Cj1318 (*maf1*) (Karlyshev *et al.*, 2002) with seven copies within the region, and the '617' family typified by Cj0617, with four additional copies within this flagellin modification region (see below). Two of the *maf* genes and all of the 617 family genes contain hypervariable repeats. The functions of these genes remain unknown.

Molecular genetics of motility

The roles of flagella in chemotaxis and motility are essential for the survival in the ecological niches of many flagellate bacteria. Flagella are highly complex bacterial organelles requiring coordinated and ordered expression of 40 or more genes for their synthesis and function. Unusually, the flagellar 'structural protein' genes in the *C. jejuni* genome are often located singly or in small dispersed clusters, scattered throughout the chromosome and often without a logical operon structure comprising functionally related genes. The organization of transcription of these gene clusters is often unclear. Their locations, genomic context and transcriptional polarities are summarized in Table 17.2.

Since the discovery of the two flagellin genes *flaA* and *flaB* in *C. coli* and *C. jejuni*, several other flagellar genes have been experimentally characterized and shown to be involved in motility, including *flgE*, *flgFG*, *pflA*, *flhA*, *flhB* and *fliN*. The *flgE* gene (corresponding to the *flgE2/Cj1729c* gene described in the *C. jejuni* genome annotation (Parkhill *et al.*, 2000)) encoding the functional hook protein was identified by phenotypic characterization of a *flgE* mutant of *C. coli*. This mutant was defective in production of hook and flagellar filaments and associated with reduced extracellular flagellin expression but an intracellular accumulation of flagellin (Kinsella *et al.*, 1997). Subsequent studies on the *flgE* gene of *C. jejuni* (again corresponding to *flgE2/Cj729c*) showed that it exhibited variability in its sequence and in the antigenicity of the encoded FlgE protein among different strains, suggesting that it might be under selective pressure from the immune system of colonized hosts (Luneberg *et al.*, 1998).

The genes *flgF* and *flgG* in *C. jejuni*, functioning as a single operon, are thought to participate in flagellar biogenesis; *flgF*, annotated as *flgG2* (Parkhill *et al.*, 2000), and *flgG* encode the basal body rod proteins (Chan *et al.*, 1998) as observed in the *E. coli* flagellar system (Macnab, 2003). Characterization of *flhA* (formerly designated *flbA* (Miller *et al.*, 1993)) mutants suggested a role for *flhA* in flagellar biogenesis as unlike the wild-type parent strain, *flhA* mutants were unable to synthesize flagellin (Miller *et al.*, 1993). Another study of the *flhA* gene in *C. coli* demonstrated that a reversible frameshift mutation in the *flhA* gene confers phase variability to flagellin gene expression (Park *et al.*, 2000), although the polymorphic homopolymeric repeat sequence involved is absent from the gene sequence

Table 17.2 Genomic context of flagellar genes in *C. jejuni* NCTC 11168

ORF numbers	Annotated designations of flagellar genes and clusters and flanking genes (arrows indicate direction of transcription)								
Cj0039–Cj0044	*typA* ↓	*Cj0040* ↑	*Cj0041* ↑	*flgD* ↑	*flgE* ↑	*Cj0044c* ↓			
Cj0067–Cj0058	*Cj0067* ↓	*aroQ* ↑	*folk* ↑	*flFh* ↑	*flhG* ↑	*Cj0062c* ↑	*fliA* ↑	*fliM* ↑	*Cj0058* ↓
Cj0193c–Cj0196c	*tig* ↓	*folE* ↑	*fliI* ↑	*purF* ↓	*dxs*				
Cj0317–Cj0321	*hisC* ↓	*fliF* ↑	*fliG* ↑	*fliH* ↑					
Cj0334–Cj0338	*ahpC* ↓	*flhB* ↑	*motB* ↑	*motA* ↑	*polA* ↑				
Cj0350–Cj0352	*Cj0350* ↑	*fliN* ↑	*Cj0352* ↑						
Cj0529c–Cj0523	*Cj0529c* ↑	*flgB* ↑	*flgC* ↑	*fliE* ↑	*pbpB* ↑				
Cj0546–Cj0550	*Cj054* ↑	*ftaG* ↑	*fliD* ↑	*fliS* ↑	*Cj0550* ↑				
Cj0669–Cj0671	*Cj066* ↑	*rpoN* ↑	*dcuB* ↑						
Cj0688–Cj0686	*pta* ↓	*flgH* ↑	*gcpE* ↓						
Cj0696–Cj0699	*ftsZ* ↑	*flgG2* ↑	*flgG* ↑	*glnA* ↓					

Cj0721c–Cj0719c	*Cj0721c* ↑	*ftaC* ↑	*Cj0719c* ↑						
Cj0821–Cj0819	*gluM* ↓	*ftiP* ↑	*Cj0819* ↓						
Cj0849c–Cj0847	*Cj0849c* ↑	*Cj0848c* ↑	*psd* ↓						
Cj0888c–Cj0881c	*Cj0888* ↑	*tRNA-gly* ↓	*tRNA-leu* ↓	*flaD* ↑	*ftsK* ↑	*rpsO* ↓	*Cj0883c* ↑	*flhA* ↑	*Cj0881c* ↑
Cj1025c–Cj1023c	*Cj1025* ↑	*flgR* ↑	*asd* ↑						
Cj1180c–Cj1178c	*Cj1180c* ↑	*ftiR* ↑	*Cj1178c* ↑						
Cj1330–Cj1333	*Cj1330* ↑	*ptm* ↑	*ptmA* ↑	*Cj1333* ↑					
Cj1340c–Cj1337	*Cj1340c* ↑	*fla* ↑	*flaB* ↑	*Cj1337* ↓					
Cj1407c–Cj1409	*Cj1407* ↓	*ftiL* ↑	*acpS* ↑						
Cj1461–Cj1467	*Cj1461* ↑	*flgI* ↑	*Cj1463* ↑	*flgM* ↑	*Cj1465* ↑	*flgK* ↑			
Cj1674–Cj1676	*Cj167* ↑	*ftiQ* ↑	*Cj1676* ↑						
Cj1731c–Cj1728c	*Cj1331* ↑	*flgE* ↑	*Cj1328c* ↑						

in *C. jejuni*. Mutations in the gene *flhB* also abolished the presence of flagella, however a small minority of cells possessed truncated flagella (Matz *et al.*, 2002); no explanation for this phenomenon is currently available. Expression of any surface FlaA was eliminated in both *flhA* and *flhB* mutants and both genes appear to participate in flagellar biogenesis (Golden and Acheson, 2002).

In more recent studies, numerous flagellar mutants were isolated by random transposon mutagenesis (see above) (Colegio *et al.*, 2001; Golden and Acheson, 2002; Hendrixson *et al.*, 2001). In total at least 25 mutants with altered motility were isolated and genes affected included some of those affected in mutants previously characterized as well as some novel and even unknown genes (see Table 17.1). Some of these motility mutants lost the ability to invade INT407 cells despite increased adherence (Golden and Acheson, 2002) demonstrating the functional aspect of flagellar genes in determining virulence potential. Finally, studies with genes of the motility accessory factor (*maf*) family typified by Cj1318 suggest a role in flagellar mediated motility (Karlyshev *et al.*, 2002). The genes *maf1* and *maf4* contain intragenic polyG sequences and the gene maf5 is involved in flagellar formation; insertional mutants in *maf5*, which are non-motile, were partially restored to motility by phase variation in *maf1*. An insertional mutant in Cj1318 (*maf1*) generated by the use of Tn552 was also shown to be non-motile (Colegio *et al.*, 2001).

On the regulatory front, the *C. jejuni* flagellar regulon is controlled in large part by a transcriptional activator, FlgR and two alternative sigma factors σ^{54} and σ^{28} that regulate transcription of the flagellar genes. Knockout mutants that lacked the *rpoN* gene encoding the sigma factor σ^{54} showed complete absence of flagella and flagellin expression (Jagannathan *et al.*, 2001). The role of the σ^{54} form of RNA polymerase in pathogens with small genomes is likely to aid in transcriptional modulation through interaction with enhancer binding proteins or activator proteins (Studholme and Buck, 2000). The σ^{54} enhancer-binding activator FlgR belongs to the NtrC group of response regulators and is believed to be part of a two-component signal transduction system (Jagannathan *et al.*, 2001). Mutants in *flgR* showed absence of flagella and of flagellin expression indicating a functional role in flagellar gene regulation (Jagannathan *et al.*, 2001). Gene homologues of *flgR* are known to control or regulate flagellar gene expression in several organisms besides *C. jejuni* including *C. crescentus*, *H. pylori*, *P. aeruginosa*, *V. cholerae* and *V. parahaemolyticus* (see above). The cognate sensor of FlgR, which based on its homology with the experimentally demonstrated candidate HP0244 is likely to be Cj0793 (Beier and Frank, 2000), may transfer phosphate to a conserved aspartate residue in the amino-terminus of FlgR rendering it capable of activating σ^{54}-dependent transcription. The FlrC homologue of FlgS in *V. cholerae* has also been proved to demonstrate such a mechanism (Prouty *et al.*, 2001).

A characteristic feature of many flagellate bacteria, including members of the *Enterobacteriaceae*, is the involvement of alternative sigma factor σ^{28} in the expression of flagella. In *C. jejuni*, non-motile *fliA* (σ^{28}) mutants revealed truncated flagella and partial flagellin expression with a negligible level of *flaA* transcripts (Hendrixson *et al.*, 2001; Jagannathan *et al.*, 2001). The expression of the rod, hook, basal body and flagellin genes under the control of the two alternative sigma factors of *C. jejuni* σ^{28} and σ^{54} (and FlgR) as observed in other polar flagellate bacteria appears central to the organization and ordered expression of *C. jejuni* flagellar genes. This suggests a model of *C. jejuni* flagellar gene transcription as a hierarchy of MS ring/switch and export genes, followed by basal body-hook genes and finally filament-associated genes. The expression of the flagellins FlaA and FlaB that constitute the filament may thus be controlled by several factors including not only σ^{54} and σ^{28} but potentially others also.

An association between chemotaxis and virulence has long been recognized in bacteria (see Chapter 18). In *C. jejuni*, mutants of the *cheA* chemotaxis gene that encodes for a histidine kinase have been isolated by transposon mutagenesis (Colegio *et al.*, 2001; Golden and Acheson, 2002; Hendrixson *et al.*, 2001). Mutations in *cheA* result in an invasion defective phenotype with reduced motility that interestingly exhibited reduced *flaA* expression. On the other hand, CheY mutants showed reduced motility with an increased adherence and invasion capacity compared with the wild-type cells (Golden and Acheson, 2002; Yao *et al.*, 1997); however, such mutants were unable to colonize mice (Yao *et al.*, 1997). The FlaA levels also appeared to be unaffected. Two other proteins CetA and CetB were shown to mediate motility and aer-like energy taxis (Hendrixson *et al.*, 2001). In this study, deletion of these genes resulted in *C. jejuni* with a reduced motility phenotype. The authors postulated a bipartite nature of a signalling apparatus wherein CetB recognizes energy poor environments, and interacts with CetA to transduce signals to the chemotaxis proteins, altering flagellar rotation and directing motility to access new environments.

Flagellin glycosylation

In bacteria glycosylation of proteins has only recently been fully recognized and accepted as a parallel with the processes that occur in eukaryotic cells (Szymanski *et al.*, 2003; see Chapter 13). Glycosylation is now known to occur in a variety of proteins involved in pathogenesis such as S-layer proteins, pili, adhesins and flagella; the role and mechanism of glycosylation in bacteria is however not completely understood. Over the past decade, it has become apparent that glycosylation of the flagellin subunits plays a key role in antigenic specificity and expression of the flagellar structure in *Campylobacter*. Glycosylation studies in *C. jejuni* began with the identification of post-translationally modified serine residues in the flagellin gene, the most likely presumptive modification of the serine residues being phosphorylation (Logan *et al.*, 1989). Heterogeneity in the flagellar proteins was later confirmed with the study of reactivity of flagellins in both *C. jejuni* and *C. coli* to carbohydrate-specific periodate oxidation and staining, and demonstration of lectin-binding properties, implying the presence of sialic acid-containing structures (Doig *et al.*, 1996). Analyses of flagellins by SDS-PAGE indicates a discrepancy in mobility in comparison with the molecular weight predicted from the DNA sequence of the flagellin genes, and this discrepancy can be attributed to the additional molecular mass of glycosyl residues attached to the flagellin. Furthermore, the presence of a carbohydrate moiety on flagellin was shown to contribute to its serospecificity (Doig *et al.*, 1996), and some of the genes involved in glycosylation were also defined (Guerry *et al.*, 1996). Recently, definitive information about the nature of the O-linked glycosyl modification was obtained by application of mass spectrometry and NMR spectroscopy to flagellin, leading to a detailed structural analysis of its glycosylation (Thibault *et al.*, 2001). The flagellins have about 19 modified serine and threonine residues, mainly situated in the central part of the molecule, with a substantial amount of the O-linked nine-carbon sugar pseudaminic acid found in the glycosylated moieties. The sequence of this central region of the flagellin is less conserved than the amino and carboxy termini, which are believed to be involved in assembly of the flagellum by non-covalent interaction of subunits, and the central part is also believed to be exposed on the surface of the assembled flagellum, hence being potentially exposed to the environment and involved in antigenicity of the structure. Recent studies have identified genes such as *ptmA*, *ptmB*, *neuB2* and *neuB3* as playing a key role in glycosylation of flagellin. Both genes *ptmA* and *ptmB* affected flagellin antigenicity as shown by investigation of specific mutants in these genes; *ptmB* affected flagellin apparent molecular weight (Guerry *et al.*, 1996). Regarding *neuB* involvement, in one strain a *neuB2* mutant showed reduced apparent molecular weight of flagellin, while altered glycosylation affecting subunit interactions of flagellins could be a possibility in the *neuB3* mutant which has a non-motile and aflagellate phenotype (Linton *et al.*, 2000).

LINKS BETWEEN FLAGELLIN GLYCOSYLATION AND FLAGELLAR STRUCTURAL BIOGENESIS

Several observations suggest that although the flagellin glycosylation genes are not generally homologous with known genes for flagellar structural proteins, there is an intimate connection between flagellin glycosylation and the process of flagellar assembly. First, it is likely that glycosylation of the flagellins, like that of other proteins, occurs in association with export of the proteins from the cytoplasm, or at least in association with the export machinery. Since the known route for flagellin export is via the central channel through the flagellar basal body, it can be assumed that the flagellin glycosylation system is also associated with this structure. However nothing is currently known about the organization or mechanism of this process. Second, a number of the genes (in addition to *flaA* and *flaB*) which have been mutagenized either by random methods or by design, and which lead to a non-motile and/or aflagellate phenotype, are located in the flagellin modification region of the chromosome. These are Cj1293, a putative GlcNAc epimerase/dehydratase, Cj1313, a putative acetyltransferase, the *neuB3* gene Cj1317, and Cj1318 and Cj1337 (*maf1* and *maf5* – see above). Current knowledge of the structural protein gene expression cascade, modelled on the *E. coli/Salmonella* paradigm, provides no clues as to how these modifications to the flagellin glycosylation process might interfere with flagellar structural biogenesis. Indeed, it is likely that control at the level of translation of structural genes may play a large part in modulation of flagellar biogenesis by glycosylation-associated factors.

CONCLUSIONS

Motility is well established as a factor in colonization and probably also in virulence of *Campylobacter* spp. The large majority of protein structural genes expected to be involved in flagellar biogenesis have now been identified in the *C. jejuni* genome, although a small number of 'missing' regulatory genes and chaperones remain to be found or explained away. Glycosylation of flagellins is now well proven and increasingly well understood; however its relationship with control of flagellar biogenesis, and the role of phase switching in hypervariable genes associated with glycosylation and leading to phase variation in flagellar expression, require further exploration.

Acknowledgements
We thank the Darwin Trust of Edinburgh for funding a studentship to A.J. and the BBSRC for financial support.

References

Alm, R.A., Guerry, P., Power, M.E., Lior, H., and Trust, T.J. 1991. Analysis of the role of flagella in the heat-labile Lior serotyping scheme of thermophilic campylobacters by mutant allele exchange. J. Clin. Microbiol. 29: 2438–2445.

Alm, R.A., Guerry, P., and Trust, T.J. 1993. Significance of duplicated flagellin genes in *Campylobacter*. J. Mol. Biol. 230: 359–363.

Andrade, M.A., Perez-Iratxeta, C., and Ponting, C.P. 2001. Protein repeats: structures, functions, and evolution. J. Struct. Biol. 134: 117–131.

Arora, S.K., Ritchings, B.W., Almira, E.C., Lory, S., and Ramphal, R. 1997. A transcriptional activator, FleQ, regulates mucin adhesion and flagellar gene expression in *Pseudomonas aeruginosa* in a cascade manner. J. Bacteriol. 179: 5574–5581.

Beier, D., and Frank, R. 2000. Molecular characterization of two-component systems of *Helicobacter pylori*. J. Bacteriol. 182: 2068–2076.

Caldwell, M.B., Guerry, P., Lee, E.C., Burans, J.P., and Walker, R.I. 1985. Reversible expression of flagella in *Campylobacter jejuni*. Infect. Immun. 50: 941–943.

Chan, V.L., Louie, H., and Joe, A. 1998. Expression of the *flgFG* operon of *Campylobacter jejuni* in *Escherichia coli* yields an extra fusion protein. Gene. 225: 131–141.

Colegio, O.R., Griffin, T.J.T., Grindley, N.D., and Galan, J.E. 2001. *In vitro* transposition system for efficient generation of random mutants of *Campylobacter jejuni*. J. Bacteriol. 183: 2384–2388.

Dingle, K.E., Van Den Braak, N., Colles, F.M., Price, L.J., Woodward, D.L., Rodgers, F.G., Endtz, H.P., Van Belkum, A., and Maiden, M.C. 2001. Sequence typing confirms that *Campylobacter jejuni* strains associated with Guillain–Barré and Miller Fisher syndromes are of diverse genetic lineage, serotype, and flagella type. J. Clin. Microbiol. 39: 3346–3349.

Doig, P., Kinsella, N., Guerry, P., and Trust, T.J. 1996. Characterization of a post-translational modification of *Campylobacter* flagellin: identification of a sero-specific glycosyl moiety. Mol. Microbiol. 19: 379–387.

Everest, P.H., Goossens, H., Sibbons, P., Lloyd, D.R., Knutton, S., Leece, R., Ketley, J.M., and Williams, P.H. 1993. Pathological changes in the rabbit ileal loop model caused by *Campylobacter jejuni* from human colitis. J. Med. Microbiol. 38: 316–321.

Gewirtz, A.T., Simon, P.O., Jr., Schmitt, C.K., Taylor, L.J., Hagedorn, C.H., O'Brien, A.D., Neish, A.S., and Madara, J.L. 2001. *Salmonella typhimurium* translocates flagellin across intestinal epithelia, inducing a proinflammatory response. J. Clin. Invest. 107: 99–109.

Golden, N.J., Camilli, A., and Acheson, D.W. 2000. Random transposon mutagenesis of *Campylobacter jejuni*. Infect. Immun. 68: 5450–5453.

Golden, N.J., and Acheson, D.W. 2002. Identification of motility and autoagglutination *Campylobacter jejuni* mutants by random transposon mutagenesis. Infect. Immun. 70: 1761–1771.

Grant, C.C., Konkel, M.E., Cieplak, W., Jr., and Tompkins, L.S. 1993. Role of flagella in adherence, internalization, and translocation of *Campylobacter jejuni* in nonpolarized and polarized epithelial cell cultures. Infect. Immun. 61: 1764–1771.

Guerry, P., Logan, S.M., Thornton, S., and Trust, T.J. 1990. Genomic organization and expression of *Campylobacter* flagellin genes. J. Bacteriol. 172: 1853–1860.

Guerry, P., Alm, R.A., Power, M.E., Logan, S.M., and Trust, T.J. 1991. Role of two flagellin genes in *Campylobacter* motility. J. Bacteriol. 173: 4757–4764.

Guerry, P., Alm, R.A., Power, M.E., and Trust, T.J. 1992. Molecular and structural analysis of *Campylobacter* flagellin. In *Campylobacter jejuni*. Current Status and Future Trends. Nachamkin, I., Blaser, M.J. and Tompkins, L.S., eds. Washington, DC: American Society for Microbiology, pp. 267–281.

Guerry, P., Doig, P., Alm, R.A., Burr, D.H., Kinsella, N., and Trust, T.J. 1996. Identification and characterization of genes required for post-translational modification of *Campylobacter coli* VC167 flagellin. Mol. Microbiol. 19: 369–378.

Guerry, P., Alm, R.A., Szymanski, C.M., and Trust, T.J. 2000. Structure, function and antigenicity of *Campylobacter* flagella. In *Campylobacter*. Nachamkin, I. and Blaser, M.J., eds. Washington, DC: American Society for Microbiology, pp. 405–421.

Harrington, C.S., Thomson-Carter, F.M., and Carter, P.E. 1997. Evidence for recombination in the flagellin locus of *Campylobacter jejuni*: implications for the flagellin gene typing scheme. J. Clin. Microbiol. 35: 2386–2392.

Harris, L.A., Logan, S.M., Guerry, P., and Trust, T.J. 1987 Antigenic variation of *Campylobacter* flagella. J. Bacteriol. 169: 5066–5071.

Hein, I., Mach, R.L., Farnleitner, A.H., and Wagner, M. 2003. Application of single-strand conformation polymorphism and denaturing gradient gel electrophoresis for *fla* sequence typing of *Campylobacter jejuni*. J. Microbiol. Methods 52: 305–313.

Henderson, I.R., Cappello, R., and Nataro, J.P. 2000. Autotransporter proteins, evolution and redefining protein secretion: response. Trends. Microbiol. 8: 534–535.

Hendrixson, D.R., Akerley, B.J., and DiRita, V.J. 2001. Transposon mutagenesis of *Campylobacter jejuni* identifies a bipartite energy taxis system required for motility. Mol. Microbiol. 40: 214–224.

Jagannathan, A., Constantinidou, C., and Penn, C.W. 2001. Roles of *rpoN*, *fliA*, and *flgR* in expression of flagella in *Campylobacter jejuni*. J. Bacteriol. 183: 2937–2942.

Jones, M., Marston, K., Woodall, C., Maskell, D., Karlyshev, A., Linton, D., Wren, B., and Kelly, D. 2003. *In vivo* and *in vitro* selection of *C. jejuni* NCTC 11168: Development of a rational model for physiological studies of *Campylobacter* colonization. Int. J. of Med. Microbiol. 293: 70.

Josenhans, C., Niehus, E., Amersbach, S., Horster, A., Betz, C., Drescher, B., Hughes, K.T., and Suerbaum, S. 2002. Functional characterization of the antagonistic flagellar late regulators FliA and FlgM of *Helicobacter pylori* and their effects on the *H. pylori* transcriptome. Mol. Microbiol. 43: 307–322.

Karlyshev, A.V., Linton, D., Gregson, N.A., and Wren, B.W. 2002. A novel paralogous gene family involved in phase-variable flagella-mediated motility in *Campylobacter jejuni*. Microbiology 148: 473–480.

Kim, Y.K., and McCarter, L.L. 2000. Analysis of the polar flagellar gene system of *Vibrio parahaemolyticus*. J. Bacteriol. 182: 3693–3704.

Kinsella, N., Guerry, P., Cooney, J., and Trust, T.J. 1997. The flgE gene of *Campylobacter coli* is under the control of the alternative sigma factor sigma54. J. Bacteriol. 179: 4647–4653.

Konkel, M.E., Kim, B.J., Rivera-Amill, V., and Garvis, S.G. 1999. Bacterial secreted proteins are required for the internalization of *Campylobacter jejuni* into cultured mammalian cells. Mol. Microbiol. 32: 691–701.

Kostrzynska, M., Betts, J.D., Austin, J.W., and Trust, T.J. 1991. Identification, characterization, and spatial localization of two flagellin species in *Helicobacter pylori* flagella. J. Bacteriol. 173: 937–946.

Lee, L.H., Burg, E., 3rd, Baqar, S., Bourgeois, A.L., Burr, D.H., Ewing, C.P., Trust, T.J., and Guerry, P. 1999. Evaluation of a truncated recombinant flagellin subunit vaccine against *Campylobacter jejuni*. Infect. Immun. 67: 5799–5805.

Liaudet, L., Szabo, C., Evgenov, O.V., Murthy, K.G., Pacher, P., Virag, L., Mabley, J.G., Marton, A., Soriano, F.G., Kirov, M.Y., Bjertnaes, L.J., and Salzman, A.L. 2003. Flagellin from gram-negative bacteria is a potent mediator of acute pulmonary inflammation in sepsis. Shock 19: 131–137.

Linton, D., Karlyshev, A.V., Hitchen, P.G., Morris, H.R., Dell, A., Gregson, N.A., and Wren, B.W. 2000. Multiple N-acetyl neuraminic acid synthetase (neuB) genes in *Campylobacter jejuni*: identification and characterization of the gene involved in sialylation of lipo-oligosaccharide. Mol. Microbiol. 35: 1120–1134.

Lior, H., Woodward, D.L., Edgar, J.A., Laroche, L.J., and Gill, P. 1982. Serotyping of *Campylobacter jejuni* by slide agglutination based on heat-labile antigenic factors. J. Clin. Microbiol. 15: 761–768.

Logan, S.M., Harris, L.A., and Trust, T.J. 1987. Isolation and characterization of *Campylobacter* flagellins. J. Bacteriol. 169: 5072–5077.

Logan, S.M., Trust, T.J., and Guerry, P. 1989. Evidence for posttranslational modification and gene duplication of *Campylobacter* flagellin. J. Bacteriol. 171: 3031–3038.

Luneberg, E., Glenn-Calvo, E., Hartmann, M., Bar, W., and Frosch, M. 1998. The central, surface-exposed region of the flagellar hook protein FlgE of *Campylobacter jejuni* shows hypervariability among strains. J. Bacteriol. 180: 3711–3714.

Macnab, R.M. 2003. How Bacteria Assemble Flagella. Annu. Rev. Microbiol. 57: 77–100

Matz, C., van Vliet, A.H., Ketley, J.M., and Penn, C.W. 2002. Mutational and transcriptional analysis of the *Campylobacter jejuni* flagellar biosynthesis gene *flhB*. Microbiology 148: 1679–1685.

Meinersmann, R.J., Helsel, L.O., Fields, P.I., and Hiett, K.L. 1997. Discrimination of *Campylobacter jejuni* isolates by *fla* gene sequencing. J. Clin. Microbiol. 35: 2810–2814.

Miller, S., Pesci, E.C., and Pickett, C.L. 1993. A *Campylobacter jejuni* homologue of the LcrD/FlbF family of proteins is necessary for flagellar biogenesis. Infect. Immun. 61: 2930–2936.

Misawa, N., and Blaser, M.J. 2000. Detection and characterization of autoagglutination activity by *Campylobacter jejuni*. Infect. Immun. 68: 6168–6175.

Moors, M.A., Li, L., and Mizel, S.B. 2001. Activation of interleukin-1 receptor-associated kinase by gram-negative flagellin. Infect. Immun. 69: 4424–4429.

Nachamkin, I., Yang, X.H., and Stern, N.J. 1993. Role of *Campylobacter jejuni* flagella as colonization factors for three-day-old chicks: analysis with flagellar mutants. Appl. Environ. Microbiol. 59: 1269–1273.

Nachamkin, I., Ung, H., and Patton, C.M. 1996. Analysis of HL and O serotypes of *Campylobacter* strains by the flagellin gene typing system. J. Clin. Microbiol. 34: 277–281.

Newell, D.G., McBride, H., and Pearson, A.D. 1984. The identification of outer membrane proteins and flagella of *Campylobacter jejuni*. J. Gen. Microbiol. 130 (Pt 5): 1201–1208.

Nuijten, P.J., van Asten, F.J., Gaastra, W., and van der Zeijst, B.A. 1990. Structural and functional analysis of two *Campylobacter jejuni* flagellin genes. J. Biol. Chem. 265: 17798–17804.

Nuijten, P.J., Wassenaar, T.M., Newell, D.G., and van der Zeijst, B.A. 1992. Molecular characterization and analysis of *Campylobacter jejuni* flagellin genes and proteins. In *Campylobacter jejuni*. Current status and future trends. Nachamkin, I., Blaser, M.J. and Tompkins, L.S., eds. Washington DC: American Society for Microbiology, pp. 282–296.

Park, S.F., Purdy, D., and Leach, S. 2000. Localized reversible frameshift mutation in the *flhA* gene confers phase variability to flagellin gene expression in *Campylobacter coli*. J. Bacteriol. 182: 207–210.

Parkhill, J., Wren, B.W., Mungall, K., Ketley, J.M., Churcher, C., Basham, D., Chillingworth, T., Davies, R.M., Feltwell, T., Holroyd, S., Jagels, K., Karlyshev, A.V., Moule, S., Pallen, M.J., Penn, C.W., Quail, M.A., Rajandream, M.A., Rutherford, K.M., van Vliet, A.H., Whitehead, S., and Barrell, B.G. 2000. The genome sequence of the food-borne pathogen *Campylobacter jejuni* reveals hypervariable sequences. Nature 403: 665–668.

Pavlovskis, O.R., Rollins, D.M., Haberberger, R.L., Jr., Green, A.E., Habash, L., Strocko, S., and Walker, R.I. 1991. Significance of flagella in colonization resistance of rabbits immunized with *Campylobacter* spp. Infect. Immun. 59: 2259–2264.

Petersen, L., and Newell, D.G. 2001. The ability of Fla-typing schemes to discriminate between strains of *Campylobacter jejuni*. J. Appl. Microbiol. 91: 217–224.

Power, M.E., Guerry, P., McCubbin, W.D., Kay, C.M., and Trust, T.J. 1994. Structural and antigenic characteristics of *Campylobacter coli* FlaA flagellin. J. Bacteriol. 176: 3303–3313.

Prouty, M.G., Correa, N.E., and Klose, K.E. 2001. The novel sigma54- and sigma28-dependent flagellar gene transcription hierarchy of *Vibrio cholerae*. Mol. Microbiol. 39: 1595–1609.

Santesteban, E., Gibson, J., and Owen, R.J. 1996. Flagellin gene profiling of *Campylobacter jejuni* heat-stable serotype 1 and 4 complex. Res. Microbiol. 147: 641–649.

Spohn, G., and Scarlato, V. 1999. Motility of *Helicobacter pylori* is coordinately regulated by the transcriptional activator FlgR, an NtrC homologue. J. Bacteriol. 181: 593–599.

Steinhauserova, I., Ceskova, J., and Nebola, M. 2002. PCR/restriction fragment length polymorphism (RFLP) typing of human and poultry *Campylobacter jejuni* strains. Lett. Appl. Microbiol. 34: 354–358.

Studholme, D.J., and Buck, M. 2000. The biology of enhancer-dependent transcriptional regulation in bacteria: insights from genome sequences. FEMS. Microbiol. Lett. 186: 1–9.

Szymanski, C.M., Yao, R., Ewing, C.P., Trust, T.J., and Guerry, P. 1999. Evidence for a system of general protein glycosylation in *Campylobacter jejuni*. Mol. Microbiol. 32: 1022–1030.

Szymanski, C.M., Burr, D.H., and Guerry, P. 2002. *Campylobacter* protein glycosylation affects host cell interactions. Infect. Immun. 70: 2242–2244.

Szymanski, C.M., Logan, S.M., Linton, D., and Wren, B.W. 2003. *Campylobacter* – a tale of two protein glycosylation systems. Trends. Microbiol. 11: 233–238.

Thibault, P., Logan, S.M., Kelly, J.F., Brisson, J.R., Ewing, C.P., Trust, T.J., and Guerry, P. 2001. Identification of the carbohydrate moieties and glycosylation motifs in *Campylobacter jejuni* flagellin. J. Biol. Chem. 276: 34862–34870.

Tsang, R.S., Figueroa, G., Bryden, L., and Ng, L. 2001. Flagella as a potential marker for *Campylobacter jejuni* strains associated with Guillain–Barré syndrome. J. Clin. Microbiol. 39: 762–764.

Wassenaar, T.M., Bleumink-Pluym, N.M., and van der Zeijst, B.A. 1991. Inactivation of *Campylobacter jejuni* flagellin genes by homologous recombination demonstrates that *flaA* but not *flaB* is required for invasion. EMBO J. 10: 2055–2061.

Wassenaar, T.M., van der Zeijst, B.A., Ayling, R., and Newell, D.G. 1993. Colonization of chicks by motility mutants of *Campylobacter jejuni* demonstrates the importance of flagellin A expression. J. Gen. Microbiol. 139 (Pt 6): 1171–1175.

Wassenaar, T.M., Bleumink-Pluym, N.M., Newell, D.G., Nuijten, P.J., and van der Zeijst, B.A. 1994. Differential flagellin expression in a *flaA flaB+* mutant of *Campylobacter jejuni*. Infect. Immun. 62: 3901–3906.

Wu, J., and Newton, A. 1997. Regulation of the *Caulobacter* gene hierarchy; not just for motility. Mol. Microbiol. 24: 233–239.

Yao, R., Burr, D.H., Doig, P., Trust, T.J., Niu, H., and Guerry, P. 1994. Isolation of motile and non-motile insertional mutants of *Campylobacter jejuni*: the role of motility in adherence and invasion of eukaryotic cells. Mol. Microbiol. 14: 883–893.

Yao, R., Burr, D.H., and Guerry, P. 1997. CheY-mediated modulation of *Campylobacter jejuni* virulence. Mol. Microbiol. 23: 1021–1031.

Young, G.M., Schmiel, D.H., and Miller, V.L. 1999. A new pathway for the secretion of virulence factors by bacteria: the flagellar export apparatus functions as a protein-secretion system. Proc. Natl. Acad. Sci. USA 96: 6456–6461.

Zhou, X., Giron, J.A., Torres, A.G., Crawford, J.A., Negrete, E., Vogel, S.N., and Kaper, J.B. 2003. Flagellin of enteropathogenic *Escherichia coli* stimulates interleukin-8 production in T84 cells. Infect. Immun. 71: 2120–2129.

the caecal crypts (Lee *et al.*, 1986). Thus, chemotaxis towards mucin combined with efficient motility in a viscous milieu would enable the bacteria to remain localized within the mucus (Szymanski *et al.*, 1995). The importance of chemotaxis in colonization was first demonstrated using chemically mutagenized motile but non-chemotactic strains that were unable to colonize the intestinal tract of suckling mice (Takata *et al.*, 1992). The more recent use of a defined chemotaxis mutant further supports the importance of chemotaxis in colonization by *C. jejuni* and also highlighted a wider role for the system in adhesion and invasion of host cells (Yao *et al.*, 1997).

Given that campylobacters are specialized colonizers of the animal intestinal tract, a detailed understanding of the mechanisms involved in intestinal colonization is essential if we are to improve our understanding of the biology of this organism. In this chapter we describe the components of the chemosensory pathway in *C. jejuni* and develop a model for *Campylobacter* chemotaxis based on comparisons with systems in *E. coli* and other bacteria.

E. COLI-BASED MODEL FOR CHEMOSENSORY SIGNAL TRANSDUCTION

The chemotactic signal transduction pathway has been well studied in *E. coli* and *Salmonella* Typhimurium (for detailed reviews see Eisenbach, 1996; Falke *et al.*, 1997; Bren and Eisenbach 2000). The chemotaxis systems of other bacteria have also been studied but in less detail and these include *Bacillus subtilis*, *Rhodobacter sphaeroides* and *Sinorhizobium meliloti* (Armitage, 1999; Sonenshein *et al.*, 2001). The central tenet of the pathway appears to be highly conserved in motile bacteria irrespective of type of flagella and the mode of locomotion they employ (Bren and Eisenbach, 2000). These basic steps are the assessment of environmental chemical status, through ligand binding to clustered transmembrane MCP (methyl-accepting chemotaxis proteins) receptors, transduction of the signal to a histidine protein kinase (HK), CheA, which in turn passes the phosphate signal to a response regulatory protein, CheY, that interacts with the flagellar motor. The level of CheY phosphorylation governs the direction of rotation of the flagellar motor and in *E. coli* clockwise rotation induces tumbling motion and counterclockwise rotation induces straight-line motion. By control of the switching between tumbles (clockwise flagellar rotation) and directed runs (counterclockwise flagellar rotation), the bacterial cell is able to rapidly move towards chemoattractants and away from chemorepellants. Other components of the pathway are involved in signal modulation and adaptation. It has also been suggested that at least some chemotaxis pathway components, such as CheA, interact with components of other sensory and metabolic pathways to modulate responses to intracellular changes (Alexandre and Zhulin, 2001). Such potential interactions add another level of intricacy to an already sophisticated system.

Chemosensory complex

The first step in the pathway is the binding of chemotactic ligands such as sugars or small peptide binding proteins to docking domains of the MCP receptors. Five such receptors have been characterized in *E. coli*: the aspartate receptor, Tar; the serine receptor, Tsr; the ribose/galactose receptor, Trg; the dipeptide receptor, Tap; and the redox potential sensor, Aer (Bren and Eisenbach, 2000). MCPs are large transmembrane proteins with spatially and functionally defined periplasmic, transmembrane and cytoplasmic domains (Stock and Levit 2000). Transmembrane helices connect via a linker region to the cytoplasmic signalling domain with a structure of two coiled-coil antiparallel helices, which, as a homodimer, are supercoiled into four α helical bundles (Djordjevic and Stock, 1998; Kim *et al.*, 1999). One hairpin turn of dimer antiparallel helices forms a conserved CheW (see below) binding domain of approximately 30 amino acids. The cytoplasmic signalling domain also contains an exposed conserved region rich in glutamate and glutamine residues that are subject to methylation and demethylation by CheR and CheB (see below).

Not all receptors are equally represented in the cell membrane, Tsr and Tar are about 10 times more abundant than Tap and Trg. Homodimeric receptor molecules associate into trimers, which can consist of three identical or mixed homodimers. These then further aggregate into higher order receptor clusters that exhibit strong cooperativity in signalling and adaptation. Receptors can form mixed clusters independent of the types of receptors present. Each receptor cluster forms a signalling organelle-like structure consisting of several thousand individual homodimers approximately 100 nm across in the polar region of the *E. coli* cell (Grebe and Stock, 1998; Lybarger and Maddock, 2000; Ames *et al.*, 2002; Falke, 2002).

The MCP receptors are a part of a sensory complex of proteins that also contains two molecules of a scaffold protein CheW and two molecules of the HK, CheA. CheW has two conserved docking domains, one to allow binding of CheW to the MCP and the other to CheA. CheW is thought to be instrumental in the formation of ternary MCP–CheW–CheA complexes as well as further clustering of such ternary complexes (Liu and Parkinson 1989; Ames and Parkinson 1994; Liu *et al.*, 1997; Boukhvalova *et al.*, 2002a; Boukhvalova *et al.*, 2002b). The CheA HK is bound to CheW and receives the signal from the sensory domain of the MCP. It is not known if CheW is directly involved in signal transduction or if the signal is passed directly from the MCP to CheA (Bourret and Stock, 2002). Binding of the chemoattractant to the periplasmic domain of the receptor initiates a cascade of signal transduction events (Ottemann *et al.*, 1999; Peach *et al.*, 2002). The transduction of the signal from ligand–MCP to CheA results in inhibition of the autophosphorylation of CheA, which therefore leads to reduced levels of phosphorylated CheY.

CheA exists as a homodimer of 71 kDa subunits with two symmetric active sites that utilize ATP and Mg^{2+} ions for reversible autophosphorylation of the His48 residue (Swanson *et al.*, 1993; Surette *et al.*, 1996). Each CheA molecule consists of two distinct functional domains, P1, an amino-terminal phosphotransfer domain containing the His48 residue, and P2, a docking domain for CheY and CheB, which must compete for access to the binding site. The phosphate is then passed from His48 of CheA to Asp57 of CheY. The phosphate can also be passed to CheB, which is involved in the adaptation response. A shorter form of CheA, $CheA_S$, is also expressed in motile cells (Parkinson and Kofoid, 1992); both versions of the protein are present in cells. It is produced from an alternative translation start signal and possesses the catalytic and docking domains as well as the carboxy terminal identical to $CheA_L$ but the amino terminal carrying His48 is absent. $CheA_S$ is able to phosphorylate His48 of $CheA_L$ efficiently *in vitro* (Levit *et al.*, 1996) but it is unknown if phosphorylation occurs *in vivo*. $CheA_S$ does play a role in signal modulation through interaction with CheZ phosphatase (see below).

Receptor responses and adaptation

Methylation and demethylation of chemosensory receptor molecules control the level of response to chemotactic stimuli and the rate of receptor adaptation. CheR methyltransferase and CheB methylesterase interact with conserved cytoplasmic domains of the MCP receptors to control the level of receptor methylation and consequently the rate of autophosphorylation of CheA. CheR is a two-domain protein with the active methylation site at the amino-terminus and a docking, carboxy-terminal domain. In *E. coli*, only the high-abundance MCP receptors, such as Tar and Tsr appear to have a conserved docking site for CheR. Thus, the methyltransferase activity of CheR bound to high abundance receptor can be extended to neighbouring low abundance receptor in the mixed receptor complex (Djordjevic and Stock, 1998; Lybarger and Maddock, 1999; Wu *et al.*, 1996). CheR catalyses transfer of a methyl group from S-adenosyl-L-methionine (AdoMet) to four to six glutamyl residues on the MCP receptor leading to enhanced auto-phosphorylation of CheA and ultimately increased tumbling behaviour (Ninfa *et al.*, 1991).

CheB is a two-domain protein with a conserved amino-terminal response receiver (RR) domain homologous to CheY that is a subject to phosphorylation on the conserved aspartate residue by CheA. The carboxy-terminal methylesterase domain of CheB demethylates the MCP residues methylated by CheR and has amidase activity catalysing conversion of glutamine residues to glutamate. Phosphorylation of the RR domain increases methylesterase activity up to 100-fold compared with unphosphorylated CheB. Demethylation of MCP receptors leads to inhibition of CheA autophosphorylation and therefore a reduction in phospho-CheY (Ninfa *et al.*, 1991; Falke *et al.*, 1997). The CheY-like RR domain of CheB appears to regulate methylesterase activity, but is not required for methylesterase function (Borczuk *et al.*, 1986). The methylesterase and the regulatory domains are connected by a flexible linker and although the isolated domains lacking the linker can associate *in vitro*, the regulatory activity of the N-terminal domain is reduced compared with native protein (Anand and Stock, 2002).

The balance between the response to stimulus and adaptation to existing conditions in bacteria is remarkable; the single cell can sense and respond to changes in environmental levels of chemicals whether they are chemoattractants or repellants in a matter of milliseconds. In a similar timeframe, bacteria can adapt to local chemical levels and bring the sensory apparatus back to a pre-stimulus state, ready to respond to any further changes. Bacteria can produce a similar response to small changes in chemical concentration over five orders of magnitude using the same receptor complexes. It appears that the clustering of different MCP receptor types at the cell poles allows for logarithmic amplification of the signal from the single (or several) ligand-occupied receptor molecules due to cooperativity in the cluster, so that a strong chemotactic response can be produced in relatively low ligand concentrations. This works not only for stimulation or suppression of CheA kinase activity, but also for adaptation via CheR and CheB, which need only to bind to a high abundance receptor in order to also exert their function on neighbouring receptor molecules (Bourret and Stock, 2002; Gestwicki and Kiessling, 2002; Jasuja *et al.*, 1999).

Linking the chemosensory complex to flagellar rotational changes

Reduced chemoattractant binding to receptor MCP stimulates CheA autophosphorylation and therefore increased levels of phosphorylated CheY. Following diffusion from the chemosensory complex, phospho-CheY binds to FliM in the flagellar motor. Increased levels of phospho-CheY promote flagellar rotational switching, leading to increased tumbling and hence directional change (Bren and Eisenbach, 2000).

The response regulator CheY is a 14 kDa protein that is able to undergo significant conformational changes upon phosphorylation of residue Asp57 (Sanders *et al.*, 1989).

The active site, a typical aspartate kinase domain, is formed by Asp57 and two additional Asp residues (Asp12 and Asp13, involved in binding of Mg^{2+} ions) as well as Lys109 and Thr87. Changes in conformation due to phosphorylation status are transferred to the C-terminal docking domain leading to non-phosphorylated CheY having greater affinity for CheA and phospho-CheY having greater affinity for FliM and the phosphatase CheZ (Blat and Eisenbach, 1996; McEvoy *et al.*, 1999; Sola *et al.*, 2000). Transfer of phosphate to the CheY active site from CheA induces spatial re-positioning in CheY (Lee *et al.*, 2001) and promotes undocking from CheA and allows docking to FliM. CheY has overlapping binding sites for CheA, CheZ and FliM so that the binding of these proteins to CheY is mutually exclusive. It has also been shown that CheY is subject to acetylation on residues 92 and 109 mediated by acetyl-CoA synthetase and that acetylation may be involved in reaction to repellents (Barak and Eisenbach, 2001).

Interactions between phospho-CheY with the amino-terminal region of FliM and changes in rotational direction are complex and not well understood. In *E. coli*, CheY binds to FliM and the proportion of phosphorylated molecules will determine whether the tumbling motion will be induced (Bren and Eisenbach, 2000). It appears that phosphorylated CheY molecules need to occupy at least 70% of the available FliM molecules for a change in rotational direction to occur. Therefore, multiple phosphorylated CheY binding events induce a tumbling motion by switching the counterclockwise flagella rotation to clockwise.

In *E. coli* the CheZ phosphatase modulates the auto-phosphatase activity of CheY. CheZ is thought to have at least two domains with the C-terminus involved in binding to CheY (Boesch *et al.*, 2000; Zhao *et al.*, 2002). The presence of phospho-CheY stimulates oligomerization of dimeric CheZ, which has greater affinity to CheY and increased phosphatase activity. The exact mechanism of CheZ action is as yet unknown, but it was suggested (Zhao *et al.*, 2002) that it directly participates in the catalysis of CheY dephosphorylation by interaction with the active site of CheY. It has also been shown that CheZ can form a complex with CheA$_S$ under reducing conditions and may further increase the rate of dephosphorylation of phospho-CheY (Wang and Matsumura, 1996). The exact nature of CheZ–CheA$_S$ interactions and the interaction of this complex with the chemosensory protein complex is not known.

Alternative chemotaxis systems

Chemotaxis pathways of a number of other bacteria have been at least partially characterized and even though the basic backbone of sensor–CheA–CheW–CheY is a common principle in the chemotaxis systems, there are marked differences that relate to shape, type of flagella and motility, host and habitat of different bacteria (Josenhans and Suerbaum, 2002; Armitage, 1999). Some bacteria, such as *Pseudomonas aeruginosa*, *R. sphaeroides, Helicobacter pylori* and *Vibrio cholerae*, encode more than one set of chemotaxis genes (Heidelberg *et al.*, 2000; Martin *et al.*, 2001; Ferrandez *et al.*, 2002; Armitage and Schmitt, 1997; Pittman *et al.*, 2001). Separate *che* operons may be involved in different types of motility such as directed or twitching motility, or free movement in the environment and movement towards host tissues during infection.

Not all bacteria carry chemotaxis genes found in *E. coli*. For example, the chemotactic pathway of *B. subtilis* encodes at least three proteins not found in *E. coli* and *Salmonella* spp: CheC, CheD and CheV. Proteins CheC and CheD are thought to be involved in adaptation (Kirby *et al.*, 2001) and the CheV protein consists of CheW and RR domains. In

B. subtilis CheW and CheV-W are in part functionally redundant, since CheV-W can partly substitute for the lack of CheW (Rosario *et al.,* 1994). In fact, the close *Campylobacter* relative, *H. pylori* (Pittman *et al.,* 2001), along with many other very diverse bacteria, such as *Borrelia burdorferi* (Fraser *et al.,* 1997), *R. sphaeroides* (Shah *et al.,* 2000), *S. meliloti* (Armitage and Schmitt, 1997) and *B. subtilis* (Rosario *et al.,* 1994), encode more than one CheY and/or proteins with RR domains, which may act as a 'phosphate sink' (Armitage, 1999). It has been recently suggested that CheZ in *E. coli* may be a recent adaptation from free living state to host colonization/infection (Schmitt, 2002).

The chemotactic pathway in *Campylobacter*

Chemotactic motility is an important factor in intestinal colonization by *C. jejuni*. The derivation of the complete genome sequence has enabled us to rapidly identify the components that make up the chemotactic pathway in *C. jejuni*. Given the limited genetic approaches available with *C. jejuni* such a complete description of the pathway components would not have been feasible within a reasonable time frame. *In silico* analysis of the genome sequence of *C. jejuni* NCTC 11168 (Parkhill *et al.,* 2000) reveals the presence of orthologues of chemotaxis and aerotaxis genes (Table 18.1, Figure 18.1). As in other bacteria, regulatory components of the chemotaxis pathway in *C. jejuni* respond to signals detected by a range of signal binding receptor proteins. These receptors transduce the recognition of a ligand into a signal that is fed into the Che protein network.

Chemosensory complex of the *C. jejuni* pathway

As in other chemotactic bacteria, the chemosensory complex that senses and transduces a signal to CheY is produced in *C. jejuni* (Table 18.1, Figure 18.1). Several potential chemotaxis receptors with periplasmic domains that may bind ligands are encoded by *C. jejuni* (Parkhill *et al.,* 2000). In addition, genes encoding the scaffold protein CheW and a protein with a CheA kinase domain are also present. Again, as in other chemotactic bacteria, there are differences in *C. jejuni* to the model characterized in *E. coli* (Bren and Eisenbach, 2000). Genome analysis has highlighted that, in addition to *cheW*, there is a *cheV* gene that encodes a CheW domain coupled to a RR domain. Moreover, the CheA kinase produced by *C. jejuni* also contains a RR domain and hence is designated CheAY here, although annotated as CheA (Parkhill *et al.,* 2000). The genes *cheV*, *cheAY* and *cheW* are located next to one another and the genes flanking the three *che* genes in the apparent operon do not appear to be associated with chemotaxis. No other genes thought to be involved in chemotaxis, including orthologues of *E. coli cheZ*, *B. subtilis cheC* and *cheD* and *Caulobacter crescentus cheD* and *cheL* (unknown function), are present in the *C. jejuni* genome (Table 18.1).

Chemoreceptors

The *C. jejuni* NCTC 11168 genome encodes 10 proteins that contain the conserved cytoplasmic signalling domain found in chemotaxis receptors (designated Tlp for *t*ransducer-*l*ike *p*rotein) and two aerotaxis (Aer) receptor proteins (summarized in Table 18.1). Two of these genes (*tlp9* and *aer2*) have been found to be required for energy taxis and were named *cet*B and *cet*A for *C*ampylobacter *e*nergy *t*axis, respectively (Hendrixson *et al.,* 2001). Although the remainder are likely to be involved in sensing and thus important in how *C. jejuni* monitors the environment, it is not possible to determine

Table 18.1 Table summarizing the structure analysis of the candidate signal transduction, chemoreceptor and aerotaxis genes found in the *C. jejuni* genome.

(a) Signal transduction

Transducer details			Transducer domain features			
Name	Genome ORF	Length (amino acid residues)	Response receiver domain	Histidine kinase domain	Methylation domain	W domain
CheY	Cj1118	130	Yes	No	No	No
CheAY	Cj0284	769	Yes	Yes	No	No
CheV	Cj0285	318	Yes	No	No	Yes
CheW	Cj0283	173	No	Yes	No	Yes
CheR	Cj0925	262	No	No	Transferase	No
CheB	Cj0924	184	No	No	Esterase	No

(b) Chemoreceptor and aerotaxis

Receptor details			Receptor domain features			
Name	Genome ORF	Length (amino acid residues)	Genomic repeat structures*	Transmembrane domains†	Methylation sites#	Group type
Tlp1	Cj1506c	700		Yes (2)	Yes	A
Tlp2	Cj0144	659	Repeat 1	Yes (2)	Yes	A
Tlp3	Cj1564	662	Repeat 1	Yes (2)	Yes	A
Tlp4	Cj0262c	665	Repeat 1	Yes (2)	Yes	A
Tlp5	Cj0246c	375		No	No	C
Tlp6	Cj0448c	365		No	No	C
Tlp7‡	Cj0951/0952	526		Yes (2)	No	A
Tlp8	Cj1110c	429		No	No	C
CetA	Cj1190c	459		Yes (1)	Yes	B
Tlp10	Cj0019c	592		Yes (2)	No	A
Aer1	Cj1191c	164		No	No	Aer
CetB	Cj1189c	165		No	No	Aer

†Transmembrane domains were predicted using Tmpred. Number of domains in parentheses.
#Methylation sites are based on the *E. coli* consensus of Glx-Gl̲x̲-X-X-Ala-Ser/Thr (where Glx can be a glutamine or glutamate residue, and the methylated residue is underlined).
‡Tlp7 may include two coding sequences and assumes fusion of Cj0951 and Cj0952 (see text for details).
*Repeat structure named as in Parkhill *et al.* (2000).

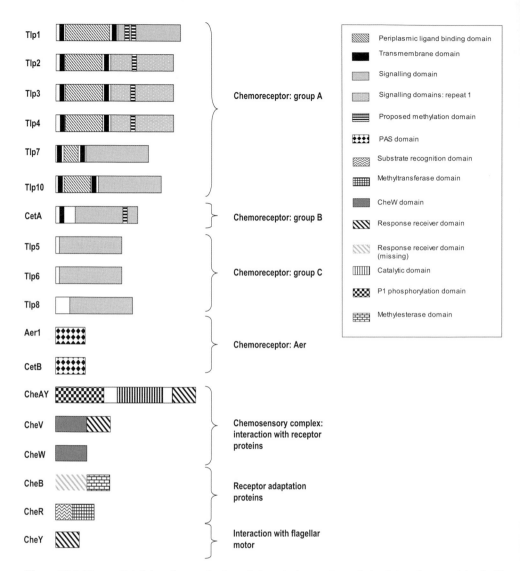

Figure 18.1 The predicted domain organization of chemotaxis receptor and signal transducer proteins in *C. jejuni*. The chemotaxis proteins are arranged in groups to highlight roles and structure–function relatedness. Shared domain structures are highlighted and identified in the key. Tlp7 is depicted assuming fusion of Cj0951 and Cj0952.

which receptors are directly involved in chemotaxis, as such receptors may be involved in other systems such as respiration, virulence or quorum sensing (Josenhans and Suerbaum, 2002; Armitage and Schmitt 1997). The genes encoding the receptor proteins do not map closely together and are not generally arranged into probable single transcriptional units. The distances between ORFs suggests that only *tlp*5, 6 and 10 are co transcribed with other genes, these being genes of unknown function (*tlp*5 and 6) and a gene encoding cytochrome C551 peroxidase (*tlp*10). In addition, *aer*1, *cetA* and *cetB* are adjacent to each other, with *cet*AB being possibly co-transcribed.

Sequence analysis, including prediction of transmembrane and coiled-coil domains, enables the classification of the *C. jejuni* receptors into several groups (Table 18.1, Figure 18.1). One group (A: Tlp1, Tlp2, Tlp3, Tlp4 and Tlp10) have a similar structure to the well-characterized MCPs of *E. coli*, and to family A of transducers in *Halobacterium salinarum* (Zhang *et al.,* 1996). The receptors contain domains consistent with membrane-spanning proteins. The C-terminal regions are homologous to the cytoplasmic signalling domains of other chemoreceptors while the N-terminal regions are less conserved and thus are likely to contain the ligand-binding domains (Table 18.1, Figure 18.1). The *C. jejuni* genome contains few repeats (Parkhill *et al.,* 2000); however, the cytoplasmic signalling domains of Tlp2, 3 and 4 form the three identical genome copies of repeat 1, which begin at the second transmembrane domain (Figure 18.1). The presence of transmembrane domains and a periplasmic ligand-binding domain means the group A receptors are likely to sense ligands external to the cell (Figure 18.2). There is a degree of similarity between some of the ligand-binding domains of this group of Tlps and of the *H. pylori* receptor proteins, TlpA and TlpC, perhaps reflecting the detection of similar signals. Tlp7 (Cj0951) contains a MCP signal domain but no transmembrane regions or ligand-binding domain. However, the ORF may continue upstream of Cj0951 (Parkhill *et al.,* 2000) into Cj0952, which shows similarity to other MCP proteins and contains transmembrane domains. Although a pseudogene in NCTC 11168, the deduced protein of Cj0951–Cj0952 would also be in receptor group A (Figure 18.1).

The second group (B) contains one receptor homologue (CetB) that is similar to family B of transducers described in *H. salinarum* (Zhang *et al.,* 1996). The predicted structure is of a cytoplasmic protein anchored to the membrane by a single N-terminal transmembrane region (Table 18.1, Figure 18.2). In the absence of a specific ligand-binding domain, CetB interacts with the aerotaxis-like protein, CetA (Hendrixson *et al.,* 2001) (see below).

The third group (C: Tlp 5,6,8) is likely to contain receptor proteins that detect cytoplasmic signals and are similar to family C of transducers in *H. salinarum* (Zhang *et al.,* 1996). The receptors are homologous to the C-terminal signalling domains of a variety of chemoreceptors with one (Tlp8) having additional unique N-terminal sequence (Table 18.1, Figure 18.1). These cytoplasmic chemotaxis-like receptor proteins may sense the internal physiological state of the cell, which may have a role in *C. jejuni* chemotaxis (Figure 18.2).

A final group contains two homologues of cytoplasmic Aer proteins. Aer1 and CetA are similar along most of their length to the N-terminal region of *E. coli* Aer and thus supports a role in redox sensing (Bibikov *et al.,* 1997; Taylor *et al.,* 1999). Moreover, as in *E. coli* Aer, Aer1 and CetA contain PAS domains and probably bind the cofactor flavin adenine dinucleotide (FAD) (Bibikov *et al.,* 1997; Repik *et al.,* 2000), further supporting a role in sensing oxygen-related changes in redox potential. PAS domains are involved in signal transduction by cytoplasmic proteins, possibly involving a second transducer protein and PAS has also been associated with protein–protein interactions in some systems (Taylor and Zhulin, 1999). The presence of redox sensing proteins may enable the location of optimal oxygen concentrations for microaerophilic *C. jejuni*. Transposon mutants of *cet*A and *cet*B disrupted energy taxis (Hendrixson *et al.,* 2001). Therefore, it was proposed that campylobacters sense energy levels through CetB and transduce the signal into the

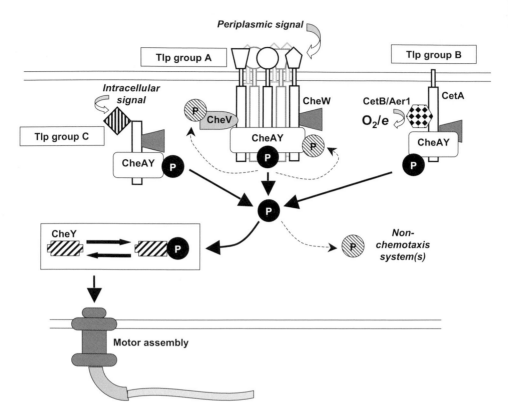

Figure 18.2 A model of chemotaxis signal transduction pathways in *C. jejuni*. Three pathways are proposed to be associated with different chemosensory complexes focused on the structure-based grouping of predicted chemoreceptors. Group A receptors, possibly arranged in large array clusters of mixed receptor specificity and positioned at the cell poles, are predicted to sense periplasmic signals. Binding of ligands to periplasmic domains of group A receptors might involve accessory periplasmic ligand-binding proteins (not shown). For the group B receptor CetA, a chemotactic response to redox potential changes is mediated by interaction with CetB. The group C receptors might play a role in chemotactic responses to intracellular signals, possibly via other cytoplasmic ligand-binding proteins (not shown). Some receptors, for example Aer1, might transduce signals into other non-chemotaxis systems. Signal loss, for example decreasing binding of attractant to the receptor, is transduced to CheAY, CheW and possibly CheV (shown with group A receptors only) in the chemosensory complex, activating the CheAY kinase domain. Following autophosphorylation of CheAY, the phosphate is transferred to CheY and phospho-CheY then binds to FliM on the flagellum motor. Sufficient binding of phospho-CheY to the flagellum motor leads to reversal from counter-clockwise to clockwise rotation (not shown). System adaptation, which resets the signalling properties of the receptor, is presumed to occur via reversible methylation by CheB and CheR (not shown). The CheAY kinase domain might also pass phosphate to other, non-CheY, response receiver (RR) domains (broken arrows). Direct or indirect routing of phosphate to the RR domain of CheV or CheAY may act as a 'phosphate sink' (see text). Alternatively, phosphorylation of CheV might have a role in system adaptation and involve the regulation of CheB methylesterase activity via the RR domain of CheV (see text).

chemotaxis pathway via CetA (Figure 18.2). The mutation of *aer*1 did not affect motility or energy taxis and thus the role for this receptor remains unknown.

The CheW scaffolding protein

Given that *C. jejuni* CheW shares many conserved residues throughout the full length of the protein with CheW proteins from other bacteria, it is likely to act as an essential structural component of the *C. jejuni* ternary complex (Table 18.1, Figure 18.1). *C. jejuni* CheW does, however, have three insertions when compared with the *E. coli* sequence, two of six amino acids and one of seven amino acids. Two of these are located in the N-terminal region of the protein, and are shared by *H. pylori* (Tomb *et al.*, 1997). These differences may possibly reflect the different domain content of CheAY and/or the presence of CheV in the complex.

All of the *C. jejuni* chemoreceptor homologues share the highly conserved cytoplasmic signalling domain. Since this domain is believed to interact physically with CheW, the presence of this sequence suggests that all of the receptors form a ternary complex (Le Moual and Koshland, 1996). In Tlp7, however, this region is less well conserved, perhaps as *tlp*7 (Cj0951–Cj0952) is a non-functional pseudogene. Alternatively, if strains exist where Tlp7 is produced, it may not form a ternary complex or may interact with another component rather than CheW to form a different ternary complex. It is possible that Tlp7 could associate with CheV, rather than CheW and this interaction could be part of an alternative pathway not controlling swimming behaviour.

The CheA kinase

The histidine protein kinase CheA forms another component of the sensory complex. In *C. jejuni* only one gene encodes a polypeptide with a CheA domain and this is fused to a C-terminal RR domain (33% identical to CheY) (Table 18.1, Figure 18.1). Designated here as CheAY, but annotated (Parkhill *et al.*, 2000) as *cheY*, it shows identity over its entire length to the fused CheAY proteins of other bacteria (*H. pylori*, 60% identity; *Rhodospirillum centenum*, 38% identity; *Myxococcus xanthus* FrzE, 27% identity). *C. jejuni* CheAY contains the known functional domains present in CheA homologues from a range of other bacterial species, including the P1 phosphorylation domain, the site of phosphorylation (His48) and residues that surround this site (i.e. Gly52, His67 and Glu70; all numbers as in *E. coli* CheA) (Dutta *et al.*, 1999). Various other residues are shared throughout the domain and the presence of an internal conserved methionine residue (Met98) suggests that $CheA_S$ (Djordjevic and Stock, 1998) could also feature in *C. jejuni*. The catalytic kinase domain, including the N, G1, F and G2 boxes (Dutta *et al.*, 1999), which is a highly conserved region between CheA proteins, is present and conserved in *C. jejuni*. The receptor-docking region, which is essential for receptor–CheW interaction, is again conserved in *C. jejuni* CheA. In contrast, the P2 RR docking domain, and the flexible linkers flanking it, do not show any conserved regions in *C. jejuni*, but this region does not appear to be generally conserved between different CheA proteins. The CheAY RR domain is found at the C-terminus (Figure 18.1) and appears to be fully functional (see below).

It seems probable therefore that *C. jejuni* CheA has a similar function to other CheA proteins in chemotaxis pathways already studied, i.e. it is a histidine kinase, located as part of the ternary complex, that autophosphorylates (with kinase activity modulated by the receptor) and passes the phosphate group onto CheY (Figure 18.2). It is likely that the less conserved P2 domain does interact with CheY as CheAY is the only protein to contain CheA domains. However, given the lack of a CheB RR domain and the presence of RR domains in CheAY and CheV, the P2 domain in CheAY may reflect the unique aspects of the *C. jejuni* chemotaxis system. Whether CheA is capable of phosphorylating the Che-associated RR domains in *C. jejuni* is not known, but certainly all of the RR domains appear, from sequence analysis at least, able to be phosphorylated by a histidine kinase.

The CheV protein

In addition to CheW, the *C. jejuni* genome contains a second gene that encodes a protein with a N-terminal region with identity to CheW but which is fused to a C-terminal RR domain (Table 18.1, Figure 18.1). This gene therefore encodes a CheV protein similar to that found in *H. pylori* (41% identity) and *B. subtilis* (23% identity). A comparison of the predicted CheV amino acid sequences from *C. jejuni*, *H. pylori* and *B. subtilis* shows that they are very similar, with several highly conserved clusters, as well as many shared identical or similar residues spaced throughout the whole sequence. As there is a high degree of conservation between the CheV 'W domain' and CheW proteins, it is likely that CheV is also incorporated into the ternary chemosensory complex (Figure 18.2). Such a possibility raises the question of the relative ratios (Levit *et al.*, 1998) of each of the components in the *C. jejuni* sensory complex, that is receptor:CheW:CheAY:CheV. Alternatively, although complexes may be mixed (Figure 18.2), containing both CheV and CheW, it is possible that complexes are CheW or CheV specific. The model whereby the sensory complex forms large, possibly polar, mixed receptor clusters (Maddock and Shapiro, 1993) further complicates the issue, as does the possibility of differential expression of *cheV* and *cheW*, although the genes do appear to be under the same transcriptional control in *C. jejuni*. In addition, the phosphorylation state of the CheV RR domain may influence incorporation of CheV into the complex.

Responses and adaptation

Unlike *H. pylori* (Pittman *et al.*, 2001) and similar to *E. coli* (Bren and Eisenbach 2000), the *C. jejuni* genome encodes both a CheR and a CheB (Table 18.1, Figure 18.1). The *cheB* and *cheR* genes are separated by 15 bp of sequence and are probably co-transcribed; the orientation and gap between the flanking genes, *rpiB* and *pebC* suggest that only *cheR* and *cheB* are co-transcribed. With the production of CheR and CheB one would expect that at least some of the *C. jejuni* chemoreceptors can be methylated. Analysis has highlighted the presence of possible near-consensus methylation sites in Tlp1, 2, 3, 4 and 9 (Le Moual and Koshland, 1996) (Table 18.1). In contrast no sequences sufficiently close to the consensus were found in Tlp5, 6, 7, 8, and 10 indicating that these chemoreceptors are not likely to be methylated (Table 18.1).

The predicted *C. jejuni* CheB polypeptide is less than half the length of other CheB proteins and shows amino acid sequence similarity along its whole length to only the C-terminal portion of other CheB proteins as it does not contain a RR domain. Hence, *C. jejuni* CheB consists of the methylesterase domain only and this structure was, until recently, unique to *C. jejuni* (Table 18.1, Figure 18.1). Whole genome sequence analysis indicates that proteins with a methylesterase only domain architecture are produced by several bacteria, including the close relatives, *Wolinella succinogenes* (Baar *et al.*, 2003) and *H. hepaticus* (Suerbaum *et al.*, 2003). The sequence of the *C. jejuni* methylesterase domain itself is similar to other CheB sequences (e.g. 25% identity to *E. coli* CheB). All of the residues known to be necessary for methylesterase function are present in *C. jejuni* CheB (West *et al.*, 1995), including the catalytic triad (Ser164, His190 and Asp286 as in *E. coli*), residues implicated in stabilizing or correctly positioning the active site (Met283, Thr165, Asp261, Ser259, Val260, Asn254) and the nucleotide binding fold consensus sequences (GXSGG and GXGXXG), and in the appropriate positions.

C. jejuni CheR (Table 18.1) shows highest identity to CheR from *B. subtilis* (18% identity), than to CheRs from *S.* Typhimurium (15% identity) and *E. coli* (14% identity).

This is a lower level of similarity than for some of the other chemotaxis components, but there are shared residues throughout the CheR sequences, indicating similarity of both of the domains (Figure 18.1). Residues associated with substrate (S-adenosylmethionine; AdoMet) positioning (Ile155 and Val232, numbers according to *E. coli* CheR) and hydrogen bond formation with AdoMet (Asp154, Ala38 and Asn212) are present (Djordjevic and Stock, 1997). The presence of these residues suggests that *C. jejuni* CheR has a similar methyltransferase function to its *S.* Typhimurium equivalent.

In *E. coli*, the CheB C-terminal methylesterase domain is active only when the N-terminal RR domain is phosphorylated by phospho-CheA. Phosphorylation of CheB provides a link between the ligand binding state of the chemoreceptors and the activity of CheB, thus allowing effective adaptation by methylation. In other CheBs, the unphosphorylated RR domain inhibits the methylesterase activity by packing against the C-terminal domain, directly inhibiting access to the active site (Djordjevic and Stock 1998). The absence of the RR domain in *C. jejuni* (Table 18.1, Figure 18.1) might therefore be predicted to relieve the inhibition of methylesterase activity, and indeed, the isolated *S.* Typhimurium 21 kDa C-terminal domain retains catalytic activity and specificity for the transmembrane chemoreceptors (West *et al.*, 1995). Therefore, a likely prediction would be that in *C. jejuni* CheB and CheR are both constitutively active, suggesting a functionally redundant constant rate of CheR and CheB methylation/demethylation and an absence of adaptation.

Given the apparent absence of CheB control, there may be some other mechanism for regulating the activity of CheR and CheB, thus providing the necessary link with the chemotaxis signal in *C. jejuni*. One possibility is that the regulation is built into the chemoreceptor itself and the C-terminal domain of CheB alone could be enough to allow wild-type chemotaxis (Lupas and Stock, 1989); in practice, however, it is not (Djordjevic and Stock, 1998). An alternative possibility is that one of the extra Che-associated RR domains could serve to inhibit methylesterase activity when phosphorylated. The structure of the catalytic domain of CheB does present many potential targets for inhibitory interactions (Lupas and Stock, 1989). In *B. subtilis*, the CheV RR domain is thought to have a role in methylation (Rosario *et al.*, 1994). CheV could either inhibit CheB directly, as part of the ternary complex (Figure 18.2), or it could simply be involved in causing phosphorylation-dependent changes in the conformation of the ternary complex that regulate CheB access to the methylation sites.

Linking to flagellar rotational changes

The *cheY* gene is located adjacent to the LOS biosynthesis gene cluster (Parkhill *et al.*, 2000) and at the start of an operon where no other gene is likely to be involved in chemotaxis. The CheY amino acid sequence shows all the characteristics of a response regulator, as well as regions of homology generally conserved amongst CheYs (Table 18.1, Figure 18.1). In particular Asp57, the site of phosphorylation, is present, and also Asp12, Asp13, Thr87, Tyr106 and Lys109, which form the active site catalysing phosphorylation in CheY (all numbers as in *E. coli*). The conservation of these residues suggests that the regulator functions in the same way as CheY regulators from other bacterial species, that is when phosphorylated by CheA it interacts with the flagellar motor to reverse the direction of flagellar rotation (Figure 18.2). The role of CheY in chemotaxis has been confirmed experimentally as mutation of *cheY* resulted in the loss of the ability to swarm on semi-solid media (Yao *et al.*, 1997; Marchant *et al.*, 1998). As in *E. coli*, CheY is thought to interact with the flagellar motor to cause clockwise rotation (Marchant *et al.*, 1998). Moreover, computerized video tracking analysis of a *cheY* mutant verified that the fully motile bacterial cells were straight swimming and not able to change direction (Marchant, Wren and Ketley, unpublished results) and therefore loss of CheY leaves the cell unable to reverse the direction of flagellar rotation. Further analysis of the *cheY* mutant (Yao *et al.*, 1997) provided support for the role of chemotaxis in colonization and confirmed previous observations with undefined chemically mutagenized motile but non-chemotactic mutants that were unable to colonize the intestinal tract of suckling mice (Takata *et al.*, 1992). Using a mouse colonization model (Yao *et al.*, 1997) the mutation of *cheY* was found to significantly affect intestinal colonization. When compared to wild-type in a ferret model, the *cheY* mutant showed reduced disease symptoms, including less mucus diarrhoea and an absence of dehydration (Yao *et al.*, 1997). Amelioration of symptoms was thought to reflect reduced colonization of the ferret intestine. Assessment in the ferret model of a *cheY* diploid strain with a normal chemotactic phenotype was also found to affect the severity of disease (Yao *et al.*, 1997). It was proposed that in a *cheY* merodiploid excess CheY reduced virulence by disturbing interactions with host epithelial cells.

RR domain interactions

Three chemotaxis-related genes encode polypeptides with RR domains that have a high level of identity to those found in CheY homologues (Table 18.1, Figure 18.1). The first is CheY itself and the others are CheAY, which has a RR domain fused to a CheA kinase domain, and CheV, which has a CheW-like region fused to a RR domain. There is good experimental evidence that CheY is the only component of the chemotaxis pathway to interact with the motor (Yao *et al.*, 1997; Marchant *et al.*, 1998). Therefore (at least under the conditions tested), all tactic responses must be integrated at a point higher in the pathway (possibly CheA, as in *E. coli*) and pass through CheY (Figure 18.2).

If CheY interacts only with the motor, this raises the question as to the function(s) of the other RR domains in CheAY and CheV. A different role for the RR domains in CheAY and CheV is supported by the fact that CheY is less similar to the CheAY-RR and CheV-RR than to CheY proteins from other bacteria. Multiple CheY domains are often found in species that lack the CheZ phosphatase. *S. meliloti*, for example, also lacks CheZ and has two proteins containing CheY domains. CheY2 interacts with the motor to influence flagellar rotation. CheY1 may act as a 'phosphate sink' (Armitage, 1999), enabling phosphate groups to be passed from CheY2 back to CheA, and then to CheY1 (Armitage and Schmitt, 1997; Sourjik and Schmitt, 1998). A similar situation has been proposed for *R. sphaeroides* (Shah *et al.*, 2000) that has four *cheY* genes. In *S. meliloti* and *R. sphaeroides*, the CheYs are separate proteins, but it is possible that in *C. jejuni* the CheY domain fused to a CheA could also fulfil such a function (Figure 18.2).

The function of the CheV protein is not completely understood. CheV has been studied in *B. subtilis* (Rosario *et al.*, 1994), and genes encoding amino acid sequences similar to CheV are present in a variety of bacteria including *Vibrio* spp. (for example, *V. cholerae* (Heidelberg *et al.*, 2000) and *V. vulnificus* (Chen *et al.*, 2003)), *P. syringae* (Buell *et al.*, 2003) and *Clostridium acetobutylicum* (Nolling *et al.*, 2001) as well as the close relatives to *C. jejuni*, *Helicobacter* spp. (*H. hepaticus* (Suerbaum *et al.*, 2003) and *H. pylori* (Tomb *et al.*, 1997)) and *Wolinella succinogenes* (Baar *et al.*, 2003). CheV can be phosphorylated by CheA (Kirby *et al.*, 2000) and thus could act as a phosphate sink for CheY and CheB. CheV can partly substitute for CheW in *B. subtilis*, but not for CheY (Rosario *et al.*,

1994) (Figure 18.2). The CheY domain of *B. subtilis* CheV is not thought to interact with the flagellar motor and no evidence has been found for CheV acting as a sink in *B. subtilis* (Kirby *et al.,* 2000). Strains lacking the RR domain of CheV are defective in MCP methylation and methanol release assays however, so the CheV-RR domain may have a role in adaptation (Rosario *et al.,* 1994) (Figure 18.2). Given the lack of a RR domain in the *C. jejuni* CheB, CheV is unlikely to act as a sink for CheB, but this does not preclude an involvement for CheV in adaptation, regulating CheB activity.

CONCLUSIONS

A range of deduced chemotaxis protein orthologues was identified from the genome sequence of *C. jejuni* NCTC 11168 with the corresponding genes scattered throughout the genome (Table 18.1, Figure 18.1). The gene content and the nature of the domains encoded by these chemotaxis genes can be used to predict a role for the chemotaxis proteins expressed in *C. jejuni*. A general scheme illustrating the possible chemoreceptor signal transduction pathways discussed in this review that are present in *C. jejuni* is summarized diagrammatically in Figure 18.2.

C. *jejuni* is predicted to have four membrane-spanning MCP chemotaxis receptors, two membrane-spanning homologues which may or may not be involved in chemotaxis, one membrane-bound chemoreceptor predicted to mediate aerotactic responses, and three cytoplasmic receptors, which may sense the internal physiological and metabolic state of the cell (Figure 18.2). All of these proteins except Tlp7 appear to be capable of forming a ternary complex and therefore of signalling into the chemotaxis pathway to affect flagellar rotation and cause a behavioural response (Figure 18.2). Tlp7 has a slightly altered conserved domain, suggesting that it may not form a ternary complex with CheW.

The rest of the signalling pathway (Figure 18.2) has elements that are similar to *E. coli*. CheAY and CheW appear to form a ternary complex with the chemoreceptors, functioning to phosphorylate CheY, which then interacts with the flagellar motor to cause clockwise flagellar rotation. But there are some important departures from the *E. coli* paradigm that affect the adaptation and termination of the chemotactic response in *C. jejuni*. Dephosphorylation of CheY leading to the decoupling of CheY from the flagellum motor may involve the additional RR domain(s) present in *C. jejuni* (Table 18.1, Figure 18.1). There is an additional RR domain fused to the CheA domain of CheAY. This possibly functions as a phosphate sink, to drain the pool of phosphate from CheAY and CheY, and terminate the negative response (Figure 18.2). There is evidence that methylation/demethylation-based adaptation of the chemoreceptors also occurs in *C. jejuni*. CheR and CheB probably act to methylate and demethylate at least some of the receptors (Table 18.1) but the nature of the control mechanism regulating this process is not clear in *C. jejuni*. Although the *C. jejuni* CheB lacks the regulatory RR domain, there is a CheV component that consists of a CheW domain and a RR domain which may have a role in adaptation (Table 18.1, Figure 18.1). CheV is probably involved in at least some of the ternary complexes and CheV-W is probably at least partly functionally redundant with CheW. The CheV-RR domain could be involved in methylation-dependent adaptation by altering the conformation of the ternary complex to regulate the access of CheB (and/or CheR) to the methylation sites or by inhibiting the CheB methylesterase when it is phosphorylated. Alternatively, there could be a methylation-independent adaptation mechanism that causes a phosphorylation-dependent conformation change in the chemoreceptor that alters CheA kinase activity.

A wider role for components of the chemotaxis pathway in *C. jejuni* pathogenesis was highlighted by the analysis of the *cheY* gene (Yao *et al.*, 1997). A *che*Y mutant was found to show enhanced adhesion and invasion of host cells and the mutation ameliorated disease symptoms in infection using a ferret model. The latter probably reflected a defect in colonization due to an inability to express chemotaxis towards mucin. More intriguing was the phenotype of a motile, swarming *che*Y merodiploid. Here a presumably higher level of expression of CheY led to poor adhesion and invasion to host cells *in vitro*, while *in vivo* the presence of an extra *cheY* resulted in the ability to colonize the ferret intestine, but not cause disease. Excess CheY protein in a *che*Y merodiploid may have affected virulence by disturbing interactions with host epithelial cells in the intestine. These observations may therefore highlight a regulatory link between chemotactic motility and other virulence systems in *C. jejuni*.

Chemotactic motility is central to the intestinal lifestyle of *C. jejuni* and, by extension, an essential prerequisite to pathogenesis in human disease. The chemoreceptor signal transduction pathway may even have input into the expression of other virulence determinants in campylobacters. The model illustrated in Figure 18.2 is speculative and almost entirely based on the exploitation of genomic sequence data. However, it does form a reasonable foundation upon which we are basing an experimental investigation of chemotaxis in *C. jejuni*. A thorough understanding of the chemotaxis system in campylobacters will be important in addressing the problem of the intestinal colonization of poultry and livestock animals and the initiation of disease in man.

Acknowledgments

J.M.K. and V.K. would like to thank the Wellcome Trust for a short-term international travel fellowship to V.K. J.M.K. would also like to thank the Biotechnology and Biological Sciences Research Council for support.

References

Alexandre, G., and Zhulin, I.B. 2001. More than one way to sense chemicals. J. Bacteriol. 183: 4681–4686.

Ames, P., and Parkinson, J.S. 1994. Constitutively signaling fragments of Tsr, the *Escherichia coli* serine chemoreceptor. J. Bacteriol. 176: 6340–6348.

Ames, P., Studdert, C.A., Reiser, R.H., and Parkinson, J.S. 2002. Collaborative signaling by mixed chemoreceptor teams in *Escherichia coli*. Proc. Natl. Acad. Sci. USA 99: 7060–7065.

Anand, G.S., and Stock, A.M. 2002. Kinetic basis for the stimulatory effect of phosphorylation on the methylesterase activity of CheB. Biochemistry 41: 6752–6760.

Armitage, J.P. 1999. Bacterial tactic responses. Adv. Microb. Physiol. 41: 229–289.

Armitage, J.P., and Schmitt, R. 1997. Bacterial chemotaxis: *Rhodobacter sphaeroides* and *Sinorhizobium meliloti* – variations on a theme? Microbiology 143 (Pt 12): 3671–3682.

Baar, C., Eppinger, M., Raddatz, G., Simon, J., Lanz, C., Klimmek, O., Nandakumar, R., Gross, R., Rosinus, A., Keller, H., *et al.*, 2003. Complete genome sequence and analysis of *Wolinella succinogenes*. Proc. Natl. Acad. Sci. USA 100: 11690–11695.

Barak, R., and Eisenbach, M. 2001. Acetylation of the response regulator, CheY, is involved in bacterial chemotaxis. Mol. Microbiol. 40: 731–743.

Bibikov, S.I., Biran, R., Rudd, K.E., and Parkinson, J.S. 1997. A signal transducer for aerotaxis in *Escherichia coli*. J. Bacteriol. 179: 4075–4079.

Blat, Y., and Eisenbach, M. 1996. Conserved C-terminus of the phosphatase CheZ is a binding domain for the chemotactic response regulator CheY. Biochemistry 35: 5679–5683.

Boesch, K.C., Silversmith, R.E., and Bourret, R.B. 2000. Isolation and characterization of nonchemotactic CheZ mutants of *Escherichia coli*. J. Bacteriol. 182: 3544–3552.

Borczuk, A., Staub, A., and Stock, J. 1986. Demethylation of bacterial chemoreceptors is inhibited by attractant stimuli in the complete absence of the regulatory domain of the demethylating enzyme. Biochem. Biophys. Res. Commun. 141: 918–923.

Boukhvalova, M., VanBruggen, R., and Stewart, R.C. 2002a. CheA kinase and chemoreceptor interaction surfaces on CheW. J. Biol. Chem. 277: 23596–23603.

Boukhvalova, M.S., Dahlquist, F.W., and Stewart, R.C. 2002b. CheW binding interactions with CheA and Tar. Importance for chemotaxis signaling in *Escherichia coli*. J. Biol. Chem. 277: 22251–22259.

Bourret, R.B., and Stock, A.M. 2002. Molecular information processing: lessons from bacterial chemotaxis. J. Biol. Chem. 277: 9625–9628.

Bren, A., and Eisenbach, M. 2000. How signals are heard during bacterial chemotaxis: protein–protein interactions in sensory signal propagation. J. Bacteriol. 182: 6865–6873.

Buell, C.R., Joardar, V., Lindeberg, M., Selengut, J., Paulsen, I.T., Gwinn, M.L., Dodson, R.J., Deboy, R.T., Durkin, A.S., Kolonay, J.F., *et al.*, 2003. The complete genome sequence of the Arabidopsis and tomato pathogen *Pseudomonas syringae* pv. tomato DC3000. Proc. Natl. Acad. Sci. USA 100: 10181–10186.

Chen, C.Y., Wu, K.M., Chang, Y.C., Chang, C.H., Tsai, H.C., Liao, T.L., Liu, Y.M., Chen, H.J., Shen, A.B., Li, J.C., *et al.*, 2003. Comparative genome analysis of *Vibrio vulnificus*, a marine pathogen. Genome Res. 13: 2577–2587.

Djordjevic, S., and Stock, A.M. 1997. Crystal structure of the chemotaxis receptor methyltransferase CheR suggests a conserved structural motif for binding S-adenosylmethionine. Structure 5: 545–558.

Djordjevic, S., and Stock, A.M. 1998. Structural analysis of bacterial chemotaxis proteins: components of a dynamic signaling system. J. Struct. Biol.124: 189–200.

Dutta, R., Qin, L., and Inouye, M. 1999. Histidine kinases: diversity of domain organization. Mol. Microbiol. 34: 633–640.

Eisenbach, M. 1996. Control of bacterial chemotaxis. Mol .Microbiol .20: 903–910.

Falke, J.J. 2002. Cooperativity between bacterial chemotaxis receptors. Proc. Natl. Acad. Sci. USA 99: 6530–6532.

Falke, J.J., Bass, R.B., Butler, S.L., Chervitz, S.A., and Danielson, M.A. 1997. The two-component signaling pathway of bacterial chemotaxis: a molecular view of signal transduction by receptors, kinases, and adaptation enzymes. Annu. Rev .Cell. Dev. Biol. 13: 457–512.

Ferrandez, A., Hawkins, A.C., Summerfield, D.T., and Harwood, C.S. 2002. Cluster II *che* genes from *Pseudomonas aeruginosa* are required for an optimal chemotactic response. J. Bacteriol. 184: 4374–4383.

Fraser, C.M., Casjens, S., Huang, W.M., Sutton, G.G., Clayton, R., Lathigra, R., White, O., Ketchum, K.A., Dodson, R., Hickey, E.K., *et al.*, 1997. Genomic sequence of a Lyme disease spirochaete, *Borrelia burgdorferi*. Nature 390: 580–586.

Gestwicki, J.E., and Kiessling, L.L. 2002. Inter-receptor communication through arrays of bacterial chemoreceptors. Nature 415: 81–84.

Grebe, T.W., and Stock, J. 1998. Bacterial chemotaxis: the five sensors of a bacterium. Curr. Biol. 8: R154–157.

Heidelberg, J.F., Eisen, J.A., Nelson, W.C., Clayton, R.A., Gwinn, M.L., Dodson, R.J., Haft, D.H., Hickey, E.K., Peterson, J.D., Umayam, L., *et al.*, 2000. DNA sequence of both chromosomes of the cholera pathogen *Vibrio cholerae*. Nature 406: 477–483.

Hendrixson, D.R., Akerley, B.J., and Dirita, V.J. 2001. Transposon mutagenesis of *Campylobacter jejuni* identifies a bipartite energy taxis system required for motility. Mol. Microbiol. 40: 214–224.

Hugdahl, M.B., Beery, J.T., and Doyle, M.P. 1988. Chemotactic behaviour of *Campylobacter jejuni*. Infect. Immun. 56: 1560–1566.

Jasuja, R., Lin, Y., Trentham, D.R., and Khan, S. 1999. Response tuning in bacterial chemotaxis. Proc. Natl. Acad. Sci. USA 96: 11346–11351.

Josenhans, C., and Suerbaum, S. 2002. The role of motility as a virulence factor in bacteria. Int. J. Med. Microbiol. 291: 605–614.

Kim, K.K., Yokota, H., and Kim, S.H. 1999. Four-helical-bundle structure of the cytoplasmic domain of a serine chemotaxis receptor. Nature 400: 787–792.

Kirby, J.R., Kristich, C.J., Saulmon, M.M., Zimmer, M.A., Garrity, L.F., Zhulin, I.B., and Ordal, G.W. 2001. CheC is related to the family of flagellar switch proteins and acts independently from CheD to control chemotaxis in *Bacillus subtilis*. Mol. Microbiol. 42: 573–585.

Kirby, J.R., Niewold, T.B., Maloy, S., and Ordal, G.W. 2000. CheB is required for behavioural responses to negative stimuli during chemotaxis in *Bacillus subtilis*. Mol. Microbiol. 35: 44–57.

Le Moual, H., and Koshland, D.E., Jr. 1996. Molecular evolution of the C-terminal cytoplasmic domain of a superfamily of bacterial receptors involved in taxis. J. Mol. Biol. 261: 568–585.

Lee, A., O'Rourke, J.L., Barrington, P.J., and Trust, T.J. 1986. Mucus colonization as a determinant of pathogenicity in intestinal infection by *Campylobacter jejuni*: a mouse cecal model. Infect. Immun. 51: 536–546.

Lee, S.Y., Cho, H.S., Pelton, J.G., Yan, D., Berry, E.A., and Wemmer, D.E. 2001. Crystal structure of activated CheY. Comparison with other activated receiver domains. J. Biol. Chem. 276: 16425–16431.

Levit, M., Liu, Y., Surette, M., and Stock, J. 1996. Active site interference and asymmetric activation in the chemotaxis protein histidine kinase CheA. J. Biol. Chem. 271: 32057–32063.

Levit, M.N., Liu, Y., and Stock, J.B. 1998. Stimulus response coupling in bacterial chemotaxis: receptor dimers in signalling arrays. Mol. Microbiol. 30: 459–466.

Liu, J.D., and Parkinson, J.S. 1989. Role of CheW protein in coupling membrane receptors to the intracellular signaling system of bacterial chemotaxis. Proc. Natl. Acad. Sci. USA 86: 8703–8707.

Liu, Y., Levit, M., Lurz, R., Surette, M.G., and Stock, J.B. 1997. Receptor-mediated protein kinase activation and the mechanism of transmembrane signaling in bacterial chemotaxis. EMBO J. 16: 7231–7240.

Lupas, A., and Stock, J. 1989. Phosphorylation of an N-terminal regulatory domain activates the CheB methylesterase in bacterial chemotaxis. J. Biol. Chem. 264: 17337–17342.

Lybarger, S.R., and Maddock, J.R. 1999. Clustering of the chemoreceptor complex in *Escherichia coli* is independent of the methyltransferase CheR and the methylesterase CheB. J. Bacteriol. 181: 5527–5529.

Lybarger, S.R., and Maddock, J.R. 2000. Differences in the polar clustering of the high- and low-abundance chemoreceptors of *Escherichia coli*. Proc. Natl. Acad. Sci. USA 97: 8057–8062.

Maddock, J.R., and Shapiro, L. 1993. Polar location of the chemoreceptor complex in the *Escherichia coli* cell. Science 259: 1717–1723.

Marchant, J., Henderson, J., Wren, B., and Ketley, J. 1998. Role of the *cheY* gene in the chemotaxis of *Campylobacter jejuni*. In: *Campylobacter*, Helicobacter and Related Organisms. (eds. A. Lastovica, D. Newell, and E. Lastovica), pp. 306–311. University of Cape Town, Cape Town.

Martin, A.C., Wadhams, G.H., and Armitage, J.P. 2001. The roles of the multiple CheW and CheA homologues in chemotaxis and in chemoreceptor localization in *Rhodobacter sphaeroides*. Mol. Microbiol. 40: 1261–1272.

McEvoy, M.M., Bren, A., Eisenbach, M., and Dahlquist, F.W. 1999. Identification of the binding interfaces on CheY for two of its targets, the phosphatase CheZ and the flagellar switch protein fliM. J. Mol. Biol. 289: 1423–1433.

Ninfa, E.G., Stock, A., Mowbray, S., and Stock, J. 1991. Reconstitution of the bacterial chemotaxis signal transduction system from purified components. J. Biol. Chem. 266: 9764–9770.

Nolling, J., Breton, G., Omelchenko, M.V., Makarova, K.S., Zeng, Q., Gibson, R., Lee, H.M., Dubois, J., Qiu, D., Hitti, J., *et al.*, 2001. Genome sequence and comparative analysis of the solvent-producing bacterium *Clostridium acetobutylicum*. J. Bacteriol. 183: 4823–4838.

Ottemann, K.M., Xiao, W., Shin, Y.K., and Koshland, D.E. 1999. A piston model for transmembrane signaling of the aspartate receptor. Science 285: 1751–1754.

Parkhill, J., Wren, B.W., Mungall, K., Ketley, J.M., Churcher, C., Basham, D., Chillingworth, T., Davies, R.M., Feltwell, T., Holroyd, S., *et al.*, 2000. The genome sequence of the food-borne pathogen *Campylobacter jejuni* reveals hypervariable sequences. Nature 403: 665–668.

Parkinson, J.S., and Kofoid, E.C. 1992. Communication modules in bacterial signaling proteins. Annu. Rev. Genet. 26: 71–112.

Peach, M.L., Hazelbauer, G.L., and Lybrand, T.P. 2002. Modeling the transmembrane domain of bacterial chemoreceptors. Protein Sci. 11: 912–923.

Pittman, M.S., Goodwin, M., and Kelly, D.J. 2001. Chemotaxis in the human gastric pathogen *Helicobacter pylori*: different roles for CheW and the three CheV paralogues, and evidence for CheV2 phosphorylation. Microbiology 147: 2493–2504.

Repik, A., Rebbapragada, A., Johnson, M.S., Haznedar, J.O., Zhulin, I.B., and Taylor, B.L. 2000. PAS domain residues involved in signal transduction by the Aer redox sensor of *Escherichia coli*. Mol. Microbiol. 36: 806–816.

Rosario, M.M., Fredrick, K.L., Ordal, G.W., and Helmann, J.D. 1994. Chemotaxis in *Bacillus subtilis* requires either of two functionally redundant CheW homologues. J .Bacteriol. 176: 2736–2739.

Sanders, D.A., Gillece-Castro, B.L., Stock, A.M., Burlingame, A.L., and Koshland, D.E. 1989. Identification of the site of phosphorylation of the chemotaxis response regulator protein, CheY. J. Biol. Chem. 264: 21770–21778.

Schmitt, R. 2002. Sinorhizobial chemotaxis: a departure from the enterobacterial paradigm. Microbiology 148: 627–631.

Shah, D.S., Porter, S.L., Harris, D.C., Wadhams, G.H., Hamblin, P.A., and Armitage, J.P. 2000. Identification of a fourth *cheY* gene in *Rhodobacter sphaeroides* and interspecies interaction within the bacterial chemotaxis signal transduction pathway. Mol. Microbiol. 35: 101–112.

Sola, M., Lopez-Hernandez, E., Cronet, P., Lacroix, E., Serrano, L., Coll, M., and Parraga, A. 2000. Towards understanding a molecular switch mechanism: thermodynamic and crystallographic studies of the signal transduction protein CheY. J. Mol. Biol. 303: 213–225.

Sonenshein, A., JA, H., and RM, L. 2001. *Bacillus subtilis* and its Closest Relatives: from Genes to Cells. ASM Press, Washington DC, pp. 646.

Sourjik, V., and Schmitt, R. 1998. Phosphotransfer between CheA, CheY1, and CheY2 in the chemotaxis signal transduction chain of *Rhizobium meliloti*. Biochemistry 37: 2327–2335.

Stock, J., and Levit, M. 2000. Signal transduction: hair brains in bacterial chemotaxis. Curr .Biol. 10: R11–14.

Suerbaum, S., Josenhans, C., Sterzenbach, T., Drescher, B., Brandt, P., Bell, M., Droge, M., Fartmann, B., Fischer, H.-P., Ge, Z., *et al.*, 2003. The complete genome sequence of the carcinogenic bacterium *Helicobacter hepaticus*. Proc. Natl. Acad. Sci. USA 100: 7901–7906.

Surette, M.G., Levit, M., Liu, Y., Lukat, G., Ninfa, E.G., Ninfa, A., and Stock, J.B. 1996. Dimerization is required for the activity of the protein histidine kinase CheA that mediates signal transduction in bacterial chemotaxis. J. Biol. Chem. 271: 939–945.

Swanson, R.V., Schuster, S.C., and Simon, M.I. 1993. Expression of CheA fragments which define domains encoding kinase, phosphotransfer, and CheY binding activities. Biochemistry 32: 7623–7629.

Szymanski, C.M., King, M., Haardt, M., and Armstrong, G.D. 1995. *Campylobacter jejuni* motility and invasion of Caco-2 cells. Infect. Immun. 63: 4295–4300.

Takata, T., Fujimoto, S., and Amako, K. 1992. Isolation of nonchemotactic mutants of *Campylobacter jejuni* and their colonization of the mouse intestinal tract. Infect. Immun. 60: 3596–3600.

Taylor, B.L., and Zhulin, I.B. 1999. PAS domains: internal sensors of oxygen, redox potential, and light. Microbiol. Mol. Biol. Rev. 63: 479–506.

Taylor, B.L., Zhulin, I.B., and Johnson, M.S. 1999. Aerotaxis and other energy-sensing behaviour in bacteria. Annu. Rev. Microbiol. 53: 103–128.

Tomb, J.F., White, O., Kerlavage, A.R., Clayton, R.A., Sutton, G.G., Fleischmann, R.D., Ketchum, K.A., Klenk, H.P., Gill, S., Dougherty, B.A., *et al.*, 1997. The complete genome sequence of the gastric pathogen *Helicobacter pylori*. Nature 388: 539–547.

Wang, H., and Matsumura, P. 1996. Characterization of the CheAS/CheZ complex: a specific interaction resulting in enhanced dephosphorylating activity on CheY-phosphate. Mol. Microbiol. 19: 695–703.

West, A.H., Martinez-Hackert, E., and Stock, A.M. 1995. Crystal structure of the catalytic domain of the chemotaxis receptor methylesterase, CheB. J. Mol. Biol. 250: 276–290.

Wu, J., Li, J., Li, G., Long, D.G., and Weis, R.M. 1996. The receptor binding site for the methyltransferase of bacterial chemotaxis is distinct from the sites of methylation. Biochemistry 35: 4984–4993.

Yao, R., Burr, D.H., and Guerry, P. 1997. CheY-mediated modulation of *Campylobacter jejuni* virulence. Mol. Microbiol. 23: 1021–1031.

Zhang, W., Brooun, A., McCandless, J., Banda, P., and Alam, M. 1996. Signal transduction in the archaeon *Halobacterium salinarium* is processed through three subfamilies of 13 soluble and membrane-bound transducer proteins. Proc. Natl. Acad. Sci. USA 93: 4649–4654.

Zhao, R., Collins, E.J., Bourret, R.B., and Silversmith, R.E. 2002. Structure and catalytic mechanism of the *E. coli* chemotaxis phosphatase CheZ. Nat. Struct. Biol. 9: 570–575.

Chapter 19

Invasion

Lan Hu and Dennis J. Kopecko

Abstract

Campylobacter is the leading cause of food-borne bacterial enteritis worldwide. The results of intestinal biopsies of patients, infected primates and other experimentally infected model animals, together with experimental infection of cultured human intestinal epithelial cells have clearly demonstrated that *C. jejuni* invade the intestine. Further, these collective data emphasize the importance of bacterial invasiveness as a virulence factor for *Campylobacter* pathogenesis. Following passage through the stomach, *Campylobacter* adhere to and invade colonic epithelial cells, trigger signal transduction events that induce host cytoskeletal rearrangements and bacterial uptake, induce interleukin 8 production, and cause colitis. The molecular components of the invasion process (es) are just beginning to be characterized.

INTRODUCTION

Bacterial diseases of the gastrointestinal tract typically result from a complex set of interactions between the offending bacteria and the host. In the process of evolution, humans have developed a variety of innate mechanisms to protect themselves from pathogenic organisms (e.g. mucous layer, epithelial barrier, gastric acidity). At the same time, bacteria have evolved pathogenic attributes to circumvent these innate host defences. Many bacterial pathogens are known to develop specific interactions with host mucosal surfaces that result in disease initiation [e.g. *Salmonella* spp., *Shigella* spp., *Neisseria gonorrhoeae*, enteropathogenic *Escherichia coli* (EPEC)]. These pathogens typically exploit host cell machinery, which contributes to their pathogenic ability. For example, upon establishing a unique association with human intestinal epithelial cells, *Salmonella*, *Shigella*, and EPEC secrete effector proteins via type III secretion pathways into the eukaryotic host cell which initiate host signal transduction events that lead to cytoskeletal rearrangements and eventual intimate colonization (i.e. for EPEC) or internalization (i.e. for *Salmonella* and *Shigella*) of the adjacent pathogen. Current evidence has revealed that *Campylobacter* also interact intimately with host cells during disease pathogenesis.

Invasion into the epithelial mucosa is now considered to be an essential virulence mechanism of several pathogenic enteric bacteria including *Salmonella, Shigella*, enteroinvasive *E. coli*, *Yersinia* spp., and *Listeria monocytogenes* (Falkow *et al.*, 1992; Finlay and Falkow, 1988; Robins-Browne *et al.*, 1985; Sansonetti, 1991; Takeuchi, 1967; Takeuchi *et al.*, 1967). The results of intestinal biopsies of patients (van Spreeuwel *et al.*, 1985), infected primates (Russell *et al.*, 1993) and other experimentally infected model animals (Babakhani and Joens, 1993; Doig *et al.*, 1996; Newell and Pearson, 1984; Yao *et al.*, 1997), together with experimental infection of cultured human intestinal epithelial cells (Ketley, 1997; De Melo *et al.*, 1989; Konkel and Joens, 1989; Oelschaeger *et al.*, 1993), have clearly demonstrated that *C. jejuni* invade the cells of the intestinal tract. Further, these collective data emphasize the importance of bacterial invasiveness as a virulence factor for *Campylobacter* pathogenesis. Recent studies have begun to reveal the molecular mechanisms of this invasion process (es) (reviewed in Kopecko *et al.*, 2001).

Cultured eukaryotic cell invasion assay techniques have become standard experimental tools for the study of bacterial internalization mechanisms. These adherence/invasion assay procedures allow one to quantitate over time the number of bacteria attached to the surface of or internalized into cultured host cells (Elsinghorst, 1994; Finlay and Falkow, 1988; Kihlstrom and Nilsson, 1997). In combination with the use of specific biochemical inhibitors, bacterial genetic mutants or various microscopic techniques, these assays can reveal both the bacterial requirements (e.g. nascent protein synthesis) or host cell requisites (e.g. cytoskeletal elements or signal transduction pathways) of these processes, as well as the adaptation of the host in response to the invading pathogen (Galan and Sansonetti, 1996). Early work by Bukholm and Kapperud (1987) suggested that *Campylobacter* spp. invade cultured HEp-2 and A549 cells, but only when coinfected with other enteropathogenic bacteria, such as *Salmonella* or *Shigella*. However, De Melo *et al.* (1989) clearly demonstrated that *C. jejuni* alone could invade human tracheal epithelial HEp-2 cells, and showed that invasion was enhanced in the presence of mucin (de Melo and Pechere, 1988). In 1989, Konkel and Joens found that several *Campylobacter* species were both adherent to and invasive in HEp-2 cells (Konkel and Joens, 1989) and also in the human colonic Caco-2 cell line (Konkel *et al.*, 1992c). Enhanced levels of HEp-2 cell invasion occurred when *C. jejuni* were coinfected with enteroviruses (Konkel and Joens, 1990). In these early studies, relatively low levels of *Campylobacter* invasion were observed (i.e. generally < 0.01% of the inoculum entered host cells within 2–3 h). However, later studies revealed that certain *C. jejuni* strains can invade human embryonic intestinal (INT407) cells at levels of 100- to >1000-fold higher than observed earlier and at efficiencies of 1 to >5% of the inoculum entering host cells within 2–3 h (Konkel and Joens, 1989; Oelschlaeger *et al.*, 1993). Thus, evidence is abundant to demonstrate that *Campylobacter* spp. invade the intestinal mucosa, and that this step is an important component of pathogenesis. Our current conception of the mechanism (s) of *Campylobacter* invasion is a composite developed from *in vivo* animal studies and *in vitro* cell culture invasion assays, together with data from light, fluorescence, and electron microscopic studies.

Pathogenic microorganisms utilize a variety of different molecular strategies to subvert host cell machinery to enable these pathogens to invade susceptible host cells (Falkow *et al.*, 1992; Finlay and Falkow, 1988; Goldberg and Sansonetti, 1993; Isberg and Falkow, 1985; Isberg and Van Nhieu, 1994; Misawa and Blaser, 1988), as exemplified below. Certain viruses and *Chlamydia psittaci* bind to receptors on the host cell surface and are internalized by the process of receptor-mediated endocytosis (Goldstein *et al.*, 1979; Helenius *et al.*, 1980; Matlin *et al.*, 1981), which does not require the overt involvement of host cell microtubules (MTs) or microfilaments (MFs). Some pathogens utilize MF-dependent entry processes which mimic phagocytosis [i.e. Fc- or C3-receptor mediated uptake (Griffin *et al.*, 1976)], but utilize different bacterial ligands and host receptors to zipper the host membrane tightly around the pathogen [e.g. *Chlamydia trachomatis* (Byrne and Moulder, 1978) and the *Yersinia enterocolitica inv* pathway (Isberg and Falkow, 1985; Miller *et al.*, 1988)]. Some *Chlamydia* serovars then transit along MTs within the host to the perinuclear region where pathogen maturation apparently occurs (Clausen *et al.*, 1997). Upon contact with epithelial cells, invasive *Salmonella* or *Shigella* spp. secrete diffusible effector proteins which translocate into the host cell. These effectors trigger a signal transduction cascade that induces a host actin-based cytoskeletal reorganization which leads to MF-dependent, macropinocytotic membrane engulfment of these pathogens (Clerc and Sansonetti, 1987; Finlay and Falkow, 1988; 1989; Galan *et al.*, 1992; Goldberg and Sansonetti, 1993; Huang *et al.*, 1998). Somewhat different, *Listeria monocytogenes* recognize E-cadherin as a receptor and enter host cells via a MF-dependent process (Mengaud *et al.*, 1996). Interestingly, once inside a host cell and released from the endosome, cytoplasmic *Listeria* (Cossart and Kocks, 1994), *Shigella* (Goldberg and Sansonetti, 1993), *Rickettsia* (Heizen *et al.*, 1993), and vaccinia virus (Cudmore *et al.*, 1995) nucleate actin to form 'tails' to move within the host cell or to adjacent host cells (i.e. promotes intercellular spread of the pathogen). In contrast to a molecular understanding of many of the processes described above, some microorganisms enter cells via, as yet, less characterized mechanisms that require MTs and/or MFs [e.g. *N. gonorrhoeae* (Makino *et al.*, 1991; Richardson and Sadoff, 1988); *Citrobacter freundii* and *C. jejuni* (Oelschlaeger *et al.*, 1993)]. *C. jejuni* have been variously reported to have different host cytoskeletal requirements for invasion, dependent upon the bacterial strain and host cell utilized (Fauchere *et al.*, 1989; Konkel and Joens, 1989; Konkel *et al.*, 1992b; Oelschlaeger *et al.*, 1993). This has created confusion akin to that caused by finding divergent host cell cytoskeletal requirements for invasion by different serovars of *Chlamydia* (Byrne and Moulder, 1978; Majeed *et al.*, 1993; Majeed and Kihlstrom, 1991). Only a few *C. jejuni* strains have been studied in any detail for molecular mechanisms of invasion, but the results suggest that *Campylobacter* may encode either MF-dependent or MT-dependent pathways for host invasion (some strains may encode both mechanisms). Alternatively, a more complex single uptake mechanism involving both MTs and MFs, proposed previously (Oelschlaeger *et al.*, 1993), may be involved in *C. jejuni* invasion. Recent mechanistic studies have begun to unravel the molecular events involved in *Campylobacter* internalization into the gut mucosa.

The pathogenesis of *Campylobacter* gastroenteritis is a complex process in which bacterial adherence, protein secretion, and invasion of host mucosal surfaces are likely to be essential early steps. In this chapter, we will focus on the invasion into host cells of *C. jejuni* (i.e. the key agent of *Campylobacter*-related human intestinal illness).

ADHERENCE – THE FIRST STEP OF INVASION

Humans and animals typically acquire campylobacters in contaminated food or water. Following passage through the stomach, these organisms colonize the ileum and colon where they can interfere with the normal secretory or absorptive capacity of the intestine. *C. jejuni* have been found to be part of the normal gut flora in many animals including avians, rodents, and dogs. In mice, *C. jejuni* colonize the mucous layer and crypts of the intestinal mucosa, mainly in the colon and caecum. In other animals, different parts of the gastrointestinal tract may be preferentially colonized, depending upon the ecological microenvironment (e.g. oxygen tension, pH, and presence of host receptors) (Lee *et al.*, 1986). The first visible interaction between a pathogenic enteric organism and its host entails colonization of the mucus barrier and then specific attachment to the mucosal cell surface.

Bacterial pathogens have numerous ways of attaching themselves to host cells; most pathogens encode multiple different adherence functions. Bacterial adherence is most typically due to a specific interaction between molecules on the bacterial surface (i.e. adhesins) and molecules on the host surface (i.e. receptors). Two common structures that bacteria use to attach to host cells are pili (or fimbriae), and afimbrial adhesins. As is the case for other intestinal pathogens, the ability of *C. jejuni* to colonize the gastrointestinal tract by binding to epithelial cells has been proposed to be essential for disease production. Fauchere *et al.* (1986) initially reported that *C. jejuni* isolated from individuals with fever and diarrhoea exhibited much greater binding to epithelial cells than strains isolated from asymptomatic individuals, but did not examine specific adherence mechanisms.

Motility and chemotaxis appear important for adherence of *Campylobacter*. Microbes may express alternative cellular adherence mechanisms depending upon the environmental conditions encountered at different host surfaces. In addition to bacterial colonization of the mucus layer and adherence to the mucosal cell surface, invasive bacteria may also exhibit a unique, transient interaction at the host cell surface prior to internalization.

Functional flagella are essential for the internalization of *C. jejuni in vitro* and for colonization of the mouse intestine (Alm *et al.*, 1993; Grant *et al.*, 1993; Wassenaar *et al.*, 1991; Yanagawa *et al.*, 1994; Yao *et al.*, 1994). Aflagellate, non-motile variants lose invasion ability (Yao *et al.*, 1994). Furthermore, a paralysed flagellar mutant with a defined mutation in *pflA*, but containing a full-length flagellum comprising mainly FlaA, bound to but exhibited markedly reduced levels of invasion into cultured cells. Nonflagellated mutants are impaired in their adherence to host cells. which is enhanced by bacterial centrifugation onto host cells (Wassenaar *et al.*, 1991). However, centrifugation of Fla⁻ *C. jejuni* onto a host cell monolayer did not restore invasion ability, indicating that motility does not simply function to make bacteria contact host cells. It remains undetermined if the flagellum plays some additional role (e.g. as an invasion-specific ligand or protein secretion system) in the entry process. Alternatively, motile *C. jejuni* may seek 'preferred invasion' host membrane sites (e.g. due to basolateral sequestering of receptors or to a numerical limitation of host receptors), or bacterial invasion ligands may be located polarly on the bacterial surface, requiring motility to initiate effective host cell contact.

Numerous studies have been performed to identify and characterize potential *Campylobacter* adhesins that mediate the organism's attachment to host cells. Possible reported adhesins include flagella, outer membrane proteins (OMPs), and lipopolysaccharides (LPS) (Fauchere *et al.*, 1989; McSweegan *et al.*, 1987; McSweegan and Walker, 1986). De Melo and Pechere (1990) identified four *C. jejuni* proteins with apparent molecular masses of 28, 32, 36, and 42 kDa that bind to HEp-2 cells. A *C. jejuni* gene encoding a 28 kDa protein, termed PEB1, has been identified as a conserved antigen in *C. jejuni* and *C. coli* strains and proposed to be an adhesin (Pei *et al.*, 1991 and 1998; Pei and Blaser, 1993). PEB1 shares homology with a periplasmic binding protein involved in nutrient acquisition (Garvis *et al.*, 1996). In other studies using a similar technique, it was found that HeLa cells bound 26 and 30 kDa proteins present in outer membrane extracts of an invasive strain of *C. jejuni* (Fauchere *et al.*, 1986; 1989). Antiserum raised against these HeLa cell-bound proteins inhibited the adhesion of homologous *C. jejuni* to this cell line, whereas antibodies against an unrelated 90 kDa OMP were without effect (Fauchere *et al.*, 1986; 1989). Many pathogenic microorganisms are capable of binding to components of the extracellular matrix such as fibronectin, laminin, vitronectin, and collagen. *C. jejuni* has also been reported to bind to extracellular matrix components (Ketley, 1995; Konkel *et al.*, 1997; Konkel *et al.*, 1999a). Scanning electron microscopy of infected INT407 cells first indicated that *C. jejuni* binds to fibronectin. This suspected binding was confirmed and is mediated by a 37 kDa OMP (CadF) which is conserved among *C. jejuni* isolates (Konkel *et al.*, 1999a). An anti-CadF antibody reduced the binding of two *C. jejuni* clinical isolates to immobilized Fn by greater than 50% (Monteville *et al.*, 2003). A surface-exposed lipoprotein, JlpA, has also been reported to be a *C. jejuni* adhesin (Jin *et al.*, 2001). JlpA is a 42.3 kDa protein which plays a role in adherence of *C. jejuni* to HEp-2 epithelial cells. Thus, a number of *Campylobacter* OMPs have been associated with binding to various eukaryotic cells.

Lipopolysaccharides (e.g. possessing an O-antigen) and lipooligosaccharides (LOS) are the major surface antigens of Gram-negative bacteria and play an important role in the interaction of these bacteria with their host and or the environment. These surface polysaccharide molecules in *C. jejuni* also serve as host mucosal adherence factors. The high molecular weight glycan capsule produced by some *C. jejuni* strains is required for cell invasion *in vitro* and full virulence in the ferret animal model (Bacon *et al.*, 2001). Thus, *C. jejuni* exhibit numerous host cell adhesins, but the relative importance of each and their recognition ligands on the host cell are not yet defined. Understanding their specific involvement in disease pathogenesis is complicated by the presence of multiple adherence factors.

BACTERIAL PROTEIN SECRETION

To provoke disease, invasive microorganisms must express products that bind host cell receptors and facilitate subsequent internalization, as well as their survival in this changed environment. In order to adjust to the metabolic needs of a new environment and upregulate virulence functions, *Salmonella* undergo gross changes in protein expression during growth in the intracellular environment, with >30 proteins induced and ~100 proteins repressed relative to growth in the absence of host cells (Buchmeier and Heffron, 1990). Similarly, new proteins are induced in campylobacters upon contact with both viable and non-viable host cells. These newly synthesized proteins were detectable within 30 min and a subset of these proteins are induced by released host cell components (Konkel *et al.*, 1993; Konkel *et al.*, 1999b; Rivera-Amill *et al.*, 2001). Non-viable campylobacters are still able to attach to host cells indicating that *de novo* bacterial protein synthesis is not required for bacterial adherence to eukaryotic cells (Konkel and Cieplak, 1992). In contrast, *Campylobacter* invasion does require nascent bacterial protein synthesis (Konkel and Cieplak, 1992; Oelschlaeger *et al.*, 1993), whereas blocking host cell protein synthesis with cycloheximide does not inhibit invasion by *C. jejuni* over a several hour period (Oelschlaeger *et al.*, 1993), suggesting that the eukaryotic requisites for invasion are present during normal host cell growth. Rabbit antisera raised against host cell-cultivated *C. jejuni* recognized nine new *C. jejuni* proteins, relative to media-grown bacteria. Furthermore, invasion was reduced by 98% when conducted in the presence of this antisera, indicating a role for one or more of these proteins in invasion (Konkel *et al.*, 1993). The antisera was utilized to clone a gene termed *ciaB* (i.e. *Campylobacter* invasion antigen B) that encodes a protein of ~73 kDa (Konkel *et al.*, 1999b). Mutants in *ciaB* are noninvasive for INT407 cells and somehow are blocked in the secretion of at least eight *C. jejuni* F38011 proteins when in the presence of INT407 cells. Furthermore, this CiaB protein was shown by immunofluorescence microscopy to translocate into infected INT407 cells. The components of the Cia secretion apparatus are still unknown. Proteins homologous or analogous with a classical type III secretory apparatus have not been found in the genome of *C. jejuni* NCTC 11168. However, it has been reported (Rivera-Amill *et al.*, 2001) that the Cia proteins may be secreted from the *C. jejuni* flagellar apparatus in a manner similar to virulence factors of *Yersinia* (Young *et al.*, 1999).

Type IV secretion systems are versatile transporters of proteins or nucleic acids, and can secrete substrates across the bacterial envelope and direct them into target cells. Type IV systems are known to export three types of substrates: DNA conjugation intermediates, the multisubunit pertussis toxin (PT), and monomeric proteins. Type IV systems have been shown to contribute to the virulence of several pathogens including *Bordetella pertussis*, *Brucella abortus*, *Brucella suis*, *Helicobacter pylori* (Christie and Vogel, 2000) and *Legionella pneumophila* (Vogel and Isberg, 1999). The *H. pylori* CagA protein is directly exported into the cytoplasm of gastric cells via type IV secretion. CagA is a 125–145 kDa protein that is tyrosine-phosphorylated after export by host cell kinases (Christie and Vogel, 2000).

A plasmid has been found to be intimately involved in the invasion process of *C. jejuni* 81-176 (Bacon *et al.*, 2000; 2002). DNA sequence analysis of this plasmid, pVir, revealed that it consists of 37,468 nucleotides and contains 54 open reading frames (ORFs). Seven ORFs of pVir are homologous with VirB4, VirB7, VirB8, VirB9, VirB10, VirB11 and VirD4 of type IV secretion systems of *A. tumefaciens*. Site-specific mutation of four of these putative type IV secretion genes in *C. jejuni* resulted in a significant decrease in both invasion of intestinal epithelial cells *in vitro*, as well as a reduction in natural transformation frequency. One of the mutants, a *virB11* homologue, was tested in the ferret diarrhoea model and caused significantly less disease than wild-type 81-176 (Bacon *et al.*, 2000). Thus, some *C. jejuni* strains may utilize the flagellar secretion system or a type IV secretory apparatus to secrete invasion effectors into the host cells.

HOST CYTOSKELETAL REQUIREMENTS

Bacterial internalization has typically been observed to involve rearrangement of host cytoskeletal structures resulting in endocytosis of the pathogen. The cytoskeleton of eukaryotic cells is a complex array of proteins, the most prominent of which are actin and tubulin, which, respectively, constitute MFs and MTs. These filamentous structures, together with intermediate filaments, are involved in both cellular and subcellular movements, in the determination of host cell shape and, not surprisingly, in bacterial invasion. Most invasive enteric organisms including *Salmonella*, *Shigella*, *Listeria*, *Yersinia* (Kihlstrom and Nilsson, 1977; Isberg *et al.*, 1987; Miller *et al.*, 1988; Finlay and Falkow, 1988; 1989; 1990; Elsinghorst *et al.*, 1989; Finlay *et al.*, 1991) have been found to trigger largely MF-dependent entry pathways.

Campylobacter internalization has been variously reported to require MFs (deMelo *et al.*, 1989; Konkel and Joens, 1989; Konkel *et al.*, 1992b), MTs (Oelschlaeger *et al.*, 1993; Hu and Kopecko, 1999), both MFs and MTs (Oelschlaeger *et al.*, 1993; Biswas *et al.*, 2003; Monteville *et al.*, 2003), or neither (Russell and Blaker, 1994), depending upon the host cell type, the methods employed (often times different and lacking appropriate controls), and the *C. jejuni* strain studied. Only a few *C. jejuni* strains have been studied in any detail for invasion mechanism. The available data, though limited, suggest that *Campylobacter* may encode separate MF-dependent (Konkel *et al.*, 1992b) or MT-dependent (Oelschlaeger *et al.*, 1993; Hu and Kopecko, 1999) pathways for host invasion; some strains (e.g. *C. jejuni* VC84) may encode both MT- and MF-dependent mechanisms (Oelschlaeger *et al.*, 1993; Biswas *et al.*, 2003; Monteville *et al.*, 2003).

Oelschlaeger *et al.* (1993) demonstrated that the entry of *C. jejuni* 81-176 into cultured human intestinal INT407 cells required intact MTs, but was unaffected by the specific depolymerization of actin filaments. Evidence for the strict MT-dependence of this process was strengthened by the inhibitory behaviour of a series of MT depolymerizing agents that act at different MT sites. Thus, colchicine, demecolcine, nocodazole, vincristine or vinblastine all caused a >90% reduction in 81-176 invasion, whereas concurrent *Salmonella* serovar Typhi (*S. typhi*) Ty2 uptake was not reduced by MT depolymerization but, by contrast, was totally blocked by cytochalasin D, which depolymerizes actin filaments (Oelschlaeger *et al.*, 1993; Hu and Kopecko, 1999). Recent studies have verified and extended this strict MT-dependence to *C. jejuni* 81-176 entry into Caco-2 cells (Hu and Kopecko, 1999). Thus, strain 81-176 enters host cells via a novel mechanism that requires host cell MTs but not actin filaments. This finding has been readily reproducible in our assays. Immunofluorescence (IF) microscopy of differentially labelled, time-course-infected INT407 cells has revealed a close association of *C. jejuni* 81-176 with microtubules during the entire invasion process (Hu and Kopecko, 1999). Early in infection, *C. jejuni* are observed interacting with the host cell at the tips of finger-like protrusions of the host cell membrane, being extended by one or a few bundled MTs (Kopecko *et al.*, 2001; Hu and Kopecko, 1999). These early membrane protrusions, initially detected by IF microscopy, have now been observed by EM (L. Hu *et al.*, unpublished). The appearance of these structures suggests that initiation of invasion involves localized reorganization of the host cytoskeleton (i.e. depolymerization of cortical actin filaments and the formation of MT-based membrane projections) in response to a signal (s) from *C. jejuni*, as similar membrane protrusions are not triggered by either the non-invasive RY213 mutant of 81-176 nor by *S. typhi* in control studies (Hu and Kopecko, 1999). Konkel *et al.* (1999b) have reported that *C. jejuni* strain F38011 synthesizes and secretes a minimum of eight novel proteins upon contact with host cells; mutation in one of these genes, *ciaB*, blocks invasion ability. Strain 81-176 also secretes similar proteins (M. Konkel, pers. commun.), which apparently serve as effectors of host cell signalling, cytoskeletal reorganization and bacterial uptake. During invasion, 81-176 cells have been observed to colocalize with MTs and to move, over time, from the periphery of the host cell to the perinuclear region (Hu and Kopecko, 1999). The perinuclear migration of internalized *C. jejuni* was first reported following IF microscopy of several different strains (Konkel *et al.*, 1992b). Separate studies have shown that *C. jejuni* strains can translocate across a polarized monolayer, both paracellularly and transcellularly, without affecting monolayer integrity, implying that bacterial exocytosis occurs basolaterally (Konkel *et al.*, 1992c). EM studies have shown for several different strains that *C. jejuni* remains within a membrane-bound vacuole during passage through the host cell (Oelschlaeger *et al.*, 1993; Konkel *et al.*, 1992c). Thus, movement of the endosomal vacuole containing *C. jejuni* apparently occurs along MTs. IF microscopic studies have confirmed that *C. jejuni* are colocalized with both MTs and dynein (an MT-based motor protein) throughout the invasion process (Hu and Kopecko, 1999). Further, orthovanadate, an inhibitor of dynein activity, was observed to reduce *C. jejuni* (but not control *S. typhi*) invasion into epithelial cells significantly, indicating a role for this MT minus-end directed molecular motor in the *C. jejuni* invasion process. Although Konkel *et al.* (1992b) have reported that *C. jejuni* strain M129 causes an accumulation of actin at the point of bacteria–host cell contact, no actin accumulation was observed with strain 81-176 (Hu and Kopecko, 1999). Interestingly, internalization of 81-176 requires polymerized MTs, but uptake is not inhibited by taxol, which stabilizes MTs (Oelschlaeger *et al.*, 1993). Finally, it is difficult to interpret the significance of a recent study, utilizing starvation cell culture conditions, which resulted in the finding of both MF and MT involvement in *C. jejuni* 81-176 invasion (Monteville *et al.*, 2003).

INVASION EFFICIENCY AND AFFECTING FACTORS

The invasion efficiency for cultured human cells varies greatly with different *C. jejuni* strains (Everest *et al.*, 1992; Konkel., *et al.*, 1992b, 1992c; Oelschlaeger *et al.*; Hu and Kopecko, 1999). Generally, 0.5% to >5.0% of the inoculum is typically internalized into host cells during efficient *C. jejuni* entry. Recent clinical isolates which are highly motile tend to be more invasion proficient (Everest *et al.*, 1992). Invasion ability is adversely affected by *C. jejuni* passage over time and selection for motile organisms helps maintain *Campylobacter* in an efficient invasive mode. Whether some *C. jejuni* strains lack invasion determinants or have lost virulence gene expression with passage remain important unanswered questions.

The same *Campylobacter* strain can vary in invasion efficiency for different host cell lines and this dependence appears to be host cell line-specific (Konkel *et al.*, 1992a; Oelschlaeger *et al.*, 1993). *C. jejuni* and *C. coli* can invade many cell lines (e.g. HeLa, INT407, HCT-8, Caco-2, HEp-2, HT29, A498, Vero, CHO-k1 and MDCK). However, *Campylobacter* appears to be most effectively internalized by cell lines of human origin (de Melo and Pechere, 1988; Konkel *et al.*, 1992a; Oelschlaeger *et al.*, 1993; Hu and Kopecko, 1999). Because of their ease of cultivation (i.e. HEp-2 or HeLa) or intestinal origin (i.e. INT407, Caco-2, HT29), these human epithelial cell lines have typically been used to study *Campylobacter* invasion. Semi-confluent monolayers of INT407 or Caco-2 cells appear optimally susceptible to *Campylobacter* invasion. Young, but fully confluent monolayers are reduced several-fold in invasion susceptibility relative to semi-confluent monolayers of the same host cells. Invasion of mid-log-phase *C. jejuni* into seven day-old differentiated Caco-2 cells is reduced approximately 10 times relative to entry into young, semi-confluent Caco-2 cells (Hu and Kopecko, 1999). Together with electron microscopic (EM) studies showing typical *C. jejuni* invasion at or within host cell junctions (Oelschlaeger *et al.*, 1993; Konkel, 1992c), this observation indicates that the as-yet-unidentified host invasion receptors of enterocytes could be sequestered basolaterally during differentiation. Further, EM studies of differentiated Caco-2 cells have revealed that only ~5% of the monolayer is heavily infected with 7–20 bacteria per host cell, and these infected cells appear M-cell-like (L. Hu and D. Kopecko, unpublished). Although preliminary, this finding suggests that M cells could be a major portal of mucosal entry for *C. jejuni*, as observed previously in the rabbit intestinal model of *C. jejuni* infection (Walker *et al.*, 1988). Kinetic studies of *C. jejuni* 81-176 entry into INT407 cells, analysed over a range of starting multiplicities of infection (MOI) from 0.02 to 20 000 bacteria per epithelial cell, have revealed several important properties of the invasion process (Hu and Kopecko, 1999). The efficiency of internalization (i.e. the proportion of the inoculum that invades host cells within a two hr period) is highest at an MOI of 0.02 and decreases steadily at higher MOIs, presumably owing to autoagglutination of strain 81-176 at higher densities. By contrast, the total number of internalized *C. jejuni* colony-forming units (CFUs) increases gradually from an MOI of 0.02 to a peak at an MOI of 200, and then decreases slightly thereafter. Thus, the optimal invasion efficiency occurs at the lowest MOI, but the maximal number of internalized *C. jejuni* occurs at an MOI of 200, where the efficiency of invasion is slightly lower. The finding that invasion efficiency is highest at the lowest MOI (0.02) suggests that this organism is a highly efficient solitary invader;

i.e. a single *C. jejuni* can induce its own uptake into host cells. This contrasts sharply with *S. typhi* invasion (Huang *et al.*, 1998); invasion efficiency is suboptimal at lower and higher MOIs, but reaches a broad optimum at an MOI = 40. However, the maximal number of internalized *S. typhi* is achieved at MOIs of 40 and above, suggesting that *S. typhi* invasion is a cooperative process which requires that the host cell 'accumulate' signalling events from many bacteria to trigger entry of a few bacteria. Although *S. typhi* or *Shigella* invade all available INT407 monolayer cells within one hr at an MOI of 40 (Huang *et al.*, 1998), the percentage of *C. jejuni*-infected INT407 cells reaches only 70% after two hr at an MOI of 200, suggesting that *C. jejuni* entry may be host cell-cycle dependent (Hu and Kopecko, 1999). Direct staining of bacteria in infected INT407 cells has revealed that approximately two bacteria are internalized per-infected host cell, and this number did not increase even when the MOI was increased to 20,000. These observations demonstrate that *C. jejuni* invasion occurs via a kinetically saturable process that strictly limits the number of bacteria that can enter host cells (Hu and Kopecko, 1999; 2000).

SIGNAL TRANSDUCTION

Host cell signalling pathways are known to be induced by virtually all invasive enteric bacteria including *S. typhimurium* (Bliska *et al.*, 1993; Falkow *et al.*, 1992; Galan *et al.*, 1992) and *Shigella flexneri* (Goldberg and Sansonetti, 1993). *Campylobacter* also induces the host cell's signalling pathways (Wooldridge *et al.*, 1996; Hu and Kopecko, 2000; Kopecko *et al.*, 2001).

The host 'invasion' receptor(s) is probably situated in membrane invaginations (i.e. caveolae or coated-pits). Inhibitors that interfere with maturation of coated pits (e.g. g-strophantin and monodansylcadaverine) or membrane caveolae (i.e. filipin III) have been found to block host cell entry of several *C. jejuni* strains including 81-176 (Konkel *et al.*, 1992b; Oelschlaeger *et al.*, 1993; Wooldridge *et al.*, 1996; Hu and Kopecko, 2000; Biswas *et al.*, 2000), suggesting that the invasion receptor (s) likely reside in these host cell membrane structures. It is unclear if filipin III, g-strophantin and monodansylcadaverine block formation of the same structures (e.g. receptors in caveolae) or if both coated and non-coated membrane invaginations are involved in *C. jejuni* entry (Hu and Kopecko, 2000).

Common signal transduction pathways in eukaryotic cells are initiated via activation of membrane-associated receptor protein kinases, which results in specific protein phosphorylation events, thereby activating host proteins. Recent studies involving the use of protein kinase inhibitors genistein and staurosporine have shown that inhibition of protein tyrosine kinases markedly reduced invasion by many *C. jejuni* strains, including N82 and 81-176 (Biswas *et al.*, 2000; Hu and Kopecko, 2000; Wooldridge *et al.*, 1996). Further studies using immunoblotting with anti-phosphotyrosine antibody of proteins from infected versus uninfected cells have demonstrated that at least nine presumed host proteins are specifically tyrosine-phosphorylated during invasion by *C. jejuni* 81-176 (Hu and Kopecko, 1999b; 2000). One of these proteins has been identified as phosphoinositol 3-kinase, through the use of specific inhibitors wortmannin and LY294002 and by Western immunoblotting analyses (Hu and Kopecko, 1999b). The heterotrimeric G proteins of the Giα subfamily have also been implicated in the host signalling events necessary for entry of *C. jejuni* strain N82, but no cytoskeletal requirements have yet been reported for this strain (Wooldridge *et al.*, 1996). Finally, host cells normally respond to transient increases in intracellular free Ca^{2+} levels by rearranging the cytoskeleton or by upregulating specific nuclear transcription. Preliminary studies using extracellular versus intracellular Ca^{2+} chelators have demonstrated that *C. jejuni* 81-176 triggers an invasion-essential, specific release of Ca^{2+} from intracellular stores (Hu and Kopecko, 1999b).

INTRACELLULAR SURVIVAL AND INTESTINAL TRANSLOCATION

Once internalized, *C. jejuni* do not appear to encode functions that allow escape from the vacuole and these bacteria undergo, at best, very limited intracellular replication during the first 8–10 hours post infection. EM studies have revealed that *C. jejuni* move intracellularly within vacuoles to the basolateral membrane before exocytosis (Konkel *et al.*, 1992c; L. Hu *et al.*, unpublished data). Reinfection of adjacent cells apparently occurs basolaterally, with presumed increased efficiency owing to suspected basolateral sequestering of host 'invasion' receptors. All strains of *C. jejuni* produce cytolethal-distending toxin (CDT), a nuclease that results in cell-cycle arrest and host DNA damage (Pickett *et al.*, 2000; Lara-Tejero and Galan, 2000). Separately, *C. jejuni* invasion of the mucosa or CDT alone triggers basolateral IL-8 release from the epithelium. *In vitro* data suggest that *C. jejuni* can survive within macrophages/monocytes for several days, which might allow for some localized *Campylobacter* dissemination to the draining lymph nodes. The proinflammatory cytokine IL-8 triggers an influx of PMNs from the lamina propria, which prevent further spread of *C. jejuni* (Autenrieth *et al.*, 1995). The resulting focal necrosis of the epithelium (i.e. dysentery) could result additively from the local inflammatory response plus host cell death caused by CDT (Kopecko *et al.*, 2001).

Bacterial translocation entails the movement of viable bacteria across the gastrointestinal epithelial barrier, where further extraintestinal dissemination can occur. For example, after gut mucosal translocation, *Salmonella typhi* are disseminated via macrophages/monocytes to various extraintestinal sites during typhoid fever pathogenesis. Although animal models are useful for studying intestinal translocation, polarized epithelial cell lines provide a simple and controlled experimental alternative to animal models. Polarized human colonic carcinoma (Caco-2) cells, with differentiated apical and basolateral surfaces separated by tight junctions, express several markers characteristic of normal small intestinal cells and have a well-defined brush border (Finlay and Falkow, 1990). Apical addition of *S. typhimurium* to polarized cells was initially shown to be followed by bacterial endocytosis and passage through the host cell to the basolateral domain. This transcytosis process occurred in the presence of tight cell junctions during the first few hours, but caused a significant decrease in monolayer electrical resistance by 6 h (Finlay and Falkow, 1990). Campylobacters also have been observed to translocate across a tight epithelial cell monolayer (Everest *et al.*, 1992; Grant *et al.*, 1993; Konkel *et al.*, 1992c). *C. jejuni* can penetrate from the apical to the basolateral surface of polarized Caco-2 cells without disrupting transepithelial electrical resistance. Bacteria were found within vacuoles inside Caco-2 cells, and presumably pass through the monolayer while remaining within a vacuole. Translocated bacteria were observed below the cell monolayer less than 1 hour after inoculation at the apical surface and continued to translocate for at least 6 h. Electron microscopic studies also indicate that some *Campylobacter* isolates transcytose between epithelial cells without undergoing invasion (Everest *et al.*, 1992; Konkel *et al.*, 1992c). Thus, it appears that *Campylobacter* may cross polarized epithelial cells via both transcellular and paracellular routes. Unlike *Salmonella*, *C. jejuni* do not cause a loss in transepithelial electrical resistance during the first six hours, indicating that the tight junctions between cells are not disrupted (Konkel *et al.*, 1992c).

Grant *et al.* (1993) reported that either *C. jejuni* motility or the product of the *flaA* gene is essential for the bacterium to cross polarized monolayers, since *flaA⁻ flaB⁻* (Mot⁻) and *flaA⁻ flaB⁺* (Mot⁻) mutants were unable to cross the cell barrier. The bacterial protein synthesis inhibitor, chloramphenicol, also reduced monolayer translocation of *C. jejuni* (Konkel *et al.*, 1992c), suggesting that bacterial endocytosis is an important part of transcytosis. No other requirements have been defined.

SUMMARY

Clinical infections, experimental infections in humans and animals, and *in vitro* analyses of adherence/invasion in cultured human cells have now clearly demonstrated that cell invasiveness is a necessary step in *Campylobacter*-induced inflammatory diarrhoea. The current understanding of the host cell signalling events and structural rearrangements triggered by *C. jejuni* now provides the molecular framework on which to build a detailed molecular cell biological model for bacterial internalization, transcytosis and eventually for disease pathogenesis (see Kopecko *et al.*, 2001).

References

Alm, R.A., Guerry, P., and Trust, T.J. 1993. The *Campylobacter* sigma 54 *flaB* flagellin promoter is subject to environmental regulation. J. Bacteriol. 175: 4448–4455.

Autenrieth, I.B., Schwarzkopf, A., Ewald, J.H., Karch, H., and Lissner, R. 1995. Bactericidal properties of *Campylobacter jejuni* specific immunoglobulin M antibodies in commercial immunoglobulin preparations. Antimicrob. Agents Chemother. 39: 1965–1969.

Babakhani, F.K., and Joens, L.A. 1993. Primary swine intestinal cells as a model for studying *Campylobacter jejuni* invasiveness. Infect. Immun. 61: 2723–2726.

Bacon, D.J., Alm, R.A., Burr, D.H., Hu, L., Kopecko, D.J., Ewing, C.P., Trust, T.J., and Guerry, P. 2000. Involvement of a plasmid in virulence of *Campylobacter jejuni* 81-176. Infect. Immun. 68: 4384–4390.

Bacon D.J, Szymanski C.M, Burr D.H, Silver R.P, Alm R.A, and Guerry P. 2001. A phase-variable capsule is involved in virulence of *Campylobacter jejuni* 81-176. Mol. Microbiol. 40 (3): 769–777.

Bacon, D.J., Alm, R.A., Hu, L., Hickey, T.E., Ewing, C.P., Batchelor, R.A., Trust, T.J., and Guerry, P. 2002. DNA sequence and mutational analysis of the pVir plasmid of *Campylobacter jejuni* 81-176. Infect. Immun. 70: 6242–6250.

Biswas, D., Itoh, K., and Sasakawa, C. 2000. Uptake pathways of clinical and healthy animal isolates of *Campylobacter jejuni* into INT407 cells. FEMS Immunol. Med. Microbiol. 29: 203–211.

Biswas, D., Itoh, K., and Sasakawa, C. 2003. Role of microfilaments and microtubules in the invasion of INT407 cells by *Campylobacter jejuni*. Microbiol. Immunol. 47: 469–473.

Bliska, J.B., Galan, J.E. and Falkow, S. 1993. Signal transduction in the mammalian cell during bacterial attachment and entry. Cell. 73: 903–920.

Buchmeier, N.A., and Heffron, F. 1990. Induction of *Salmonella* stress proteins upon infection of macrophages. Science 248: 730–732.

Bukholm, G., and Kapperud, G. 1987. Expression of *Campylobacter jejuni* invasiveness in cell cultures coinfected with other bacteria. Infect. Immun. 55: 2816–2821.

Byrne, G.I., and Moulder, J.W. 1978. Parasite-specified phagocytosis of *Chlamydia psittaci* and *Chlamydia trachomatis* by L and HeLa cells. Infect. Immun. 19: 598–606.

Christie P.J, and Vogel J.P. 2000. Bacterial type IV secretion: conjugation systems adapted to deliver effector molecules to host cells. Trends Microbiol. 8: 354–360.

Clausen, J.D., Christiansen, G., Holst, H.U., and Birkelund, S. 1997. *Chlamydia trachomatis* utilizes the host cell microtubule network during early events of infection. Mol. Microbiol. 25: 441–449.

Clerc, P., and Sansonetti, P.J. 1987. Entry of *Shigella flexneri* into HeLa cells: evidence for directed phagocytosis involving actin polymerization and myosin accumulation. Infect. Immun. 55: 2681–2688.

Cossart, P., and Kocks, C. 1994. The actin-based motility of the facultative intracellular pathogen *Listeria monocytogenes*. Mol. Microbiol. 13: 395–402.

Cover, T.L., Perez-Perez, G.I., and Blaser, M.J. 1990. Evaluation of cytotoxic activity in fecal filtrates from patients with *Campylobacter jejuni* or *Campylobacter coli* enteritis. FEMS Microbiol Lett. 58: 301–304.

Cudmore, S., Cossart, P., Griffiths, G., and Way, M. 1995. Actin-based motility of vaccinia virus. Nature 378: 636–638.

de Melo, M.A., and Pechere, J.C. 1988. Effect of mucin on *Campylobacter jejuni* association and invasion on HEp- 2 cells. Microb. Pathogen. 5: 71–76.

de Melo, M.A., Gabbiani, G., and Pechere, J.C. 1989. Cellular events and intracellular survival of *Campylobacter jejuni* during infection of HEp-2 cells. Infect. Immun. 57: 2214–2222.

de Melo, M.A., and Pechere, J.C. 1990. Identification of *Campylobacter jejuni* surface proteins that bind to eucaryotic cells *in vitro*. Infect. Immun. 58: 1749–1756.

Doig, P., Kinsella, N., Guerry, P., and Trust, T.J. 1996. Characterization of a post-translational modification of *Campylobacter* flagellin: identification of a sero-specific glycosyl moiety. Mol. Microbiol. 19: 379–387.

Elsinghorst, E. A. 1994. Measurement of invasion by gentamicin resistance. Methods Enzymol. 236: 405–420.

Elsinghorst, E.A., Baron, L.S., and Kopecko, D.J. 1989. Penetration of human intestinal epithelial cells by *Salmonella*: molecular cloning and expression of *Salmonella typhi* invasion determinants in *Escherichia coli*. Proc. Natl. Acad. Sci. USA 86: 5173–5177.

Everest, P.H., Goossens, H., Butzler, J.P., Lloyd, D., Knutton, S., Ketley, J.M., and Williams, P.H. 1992. Differentiated Caco-2 cells as a model for enteric invasion by *Campylobacter jejuni* and *C. coli*. J. Med. Microbiol. 37: 319–325.

Falkow, S., Isberg, R.R., and Portnoy, D.A. 1992. The interaction of bacteria with mammalian cells. Annu. Rev. Cell. Biol. 8: 333–363.

Fauchere, J., Kervella, M., Rosenau, A., Mohanna, K., and Veron, M. 1989. Adhesion to HeLa cells of *Campylobacter jejuni* and *C. coli* outer membrane components. Res. Microbiol. 140: 379–392.

Fauchere, J.L., Rosenau, A., Veron, M., Moyen, E.N., Richard, S., and Pfister, A. 1986. Association with HeLa cells of *Campylobacter jejuni* and *Campylobacter coli* isolated from human feces. Infect. Immun. 54: 283–287.

Finlay, B.B., and Falkow, S. 1988. Comparison of the invasion strategies used by *Salmonella cholerae-suis*, *Shigella flexneri* and *Yersinia enterocolitica* to enter cultured animal cells: endosome acidification is not required for bacterial invasion or intracellular replication. Biochimie 70: 1089–1099.

Finlay, B.B., and Falkow, S. 1989. Common themes in microbial pathogenicity. Microbiol. Rev. 53: 210–230.

Finlay, B.B., and Falkow, S. 1990. *Salmonella* interactions with polarized human intestinal Caco-2 epithelial cells. J. Infect. Dis. 162: 1096–1106.

Finlay, B.B., Ruschkowski, S, and Dedhar, S. 1991. Cytoskeletal rearrangements accompanying *Salmonella* entry into epithelial cells. J. Cell. Sci. 99: 283–296.

Galan, J.E., Pace, J., and Hayman, M.J. 1992. Involvement of the epidermal growth factor receptor in the invasion of cultured mammalian cells by *Salmonella typhimurium*. Nature. 357: 588–589.

Galan, J.E. and Sansonetti, P.J. 1996. Molecular and cellular bases of *Salmonella* and *Shigella* interactions with host cells. In: *Escherichia coli* and *Salmonella*: Cellular and Molecular Biology (Vol. 2) Neidhardt, F.C. *et al.*, eds. ASM Press, Washington DC, pp. 2757–2773.

Garvis, S.G., Puzon, G.J., and Konkel, M.E. 1996. Molecular characterization of a *Campylobacter jejuni* 29-kilodalton periplasmic binding protein. Infect. Immun. 64: 3537–3543.

Goldberg, M.B., and Sansonetti, P.J. 1993. *Shigella* subversion of the cellular cytoskeleton: a strategy for epithelial colonization. Infect. Immun. 61: 4941–4946.

Goldstein, J.L., Anderson, R.G., and Brown, M.S. 1979. Coated pits, coated vesicles, and receptor-mediated endocytosis. Nature 279: 679–685.

Grant, C.C., Konkel, M.E., Cieplak, W. Jr., and Tompkins, L.S. 1993. Role of flagella in adherence, internalization, and translocation of *Campylobacter jejuni* in nonpolarized and polarized epithelial cell cultures. Infect. Immun. 61: 1764–1771.

Griffin, F.M. Jr., Griffin, J.A. and Silverstein, S.C. 1976. Studies on the mechanism of phagocytosis. II. The interaction of macrophages with anti-immunoglobulin IgG-coated bone marrow-derived lymphocytes. J. Exp. Med. 144: 788–809.

Heinzen, R.A., Hayes, S.F., Peacock, M.G., and Hackstadt, T. 1993. Directional actin polymerization associated with spotted fever group *Rickettsia* infection of Vero cells. Infect Immun. 61: 1926–1935.

Helenius, A., Kartenbeck, J., Simons, K., and Fries, E. 1980. On the entry of Semliki forest virus into BHK-21 cells. J. Cell. Biol. 84: 404–420.

Hickey, T.E., Baqar, S., Bourgeois, A.L., Ewing, C.P., and Guerry, P. 1999. *Campylobacter jejuni*-stimulated secretion of interleukin-8 by INT407 cells. Infect. Immun. 67: 88–93.

Hu, L., and Kopecko, D.J. 1999. *Campylobacter jejuni* associates with microtubules and dynein during invasion into human intestinal cells. Infect. Immun. 67: 88–93

Hu, L. and Kopecko, D.J. 1999b. *Campylobacter jejuni* induces host protein phosphorylation events and a transient calcium flux during entry into human intestinal cells. In Proceedings of the US–Japan Conference on Cholera and Other Bacterial Enteric Infections (Vol. 35), pp. 158–163.

Hu, L. and Kopecko, D.J. 2000. Interactions of *Campylobacter* with eukaryotic cells: gut luminal colonization and mucosal invasion mechanisms. In *Campylobacter,* 2nd edn. Nachamkin, I. and Blaser, M.J., eds. ASM Press, Washington, DC, pp. 191–215.

Huang, X.Z., Tall, B., Schwan, W.R., and Kopecko, D.J. 1998. Physical limitations on *Salmonella typhi* entry into cultured human intestinal epithelial cells. Infect. Immun. 66: 2928–2937.

Isberg, R., and Falkow, S. 1985. A single genetic locus encoded by *Yersinia pseudotuberculosis* permits invasion of cultured animal cells by *Escherichia coli* K-12. Nature. 317: 262–264.

Isberg, R.R., and Van Nhieu, G.T. 1994. Two mammalian cell internalization strategies used by pathogenic bacteria. Annu. Rev. Genet. 28: 395–422.

Isberg, R.R., Voorhis, D.L., and Falkow, S. 1987. Identification of invasin: a protein that allows enteric bacteria to penetrate cultured mammalian cells. Cell 50: 769–778.

Ketley, J.M. 1995. The 16th C. L. Oakley Lecture. Virulence of *Campylobacter* species: a molecular genetic approach. J. Med. Microbiol. 42: 312–327.

Ketley, J.M. 1997. Pathogenesis of enteric infection by *Campylobacter*. Microbiology 143: 5–21.

Kihlstrom, E., and Nilsson, L. 1977. Endocytosis of *Salmonella typhimurium* 395 MS and MR10 by HeLa cells. Acta. Pathol. Microbiol. Scand. [B] 85B: 322–328.

Konkel, M.E., and Joens, L.A. 1989. Adhesion to and invasion of HEp-2 cells by *Campylobacter* spp. Infect. Immun. 57: 2984–2990.

Konkel, M.E., and Joens, L.A. 1990. Effect of enteroviruses on adherence to and invasion of HEp-2 cells by *Campylobacter* isolates. Infect. Immun. 58: 1101–1105.

Konkel, M.E., and Cieplak, W. Jr. 1992. Altered synthetic response of *Campylobacter jejuni* to cocultivation with human epithelial cells is associated with enhanced internalization. Infect. Immun. 60: 4945–4949.

Konkel, M.E., Corwin, M.D., Joens, L.A., and Cieplak, W. 1992a. Factors that influence the interaction of *Campylobacter jejuni* with cultured mammalian cells. J. Med. Microbiol. 37: 30–37.

Konkel, M.E., Hayes, S.F., Joens, L.A., and Cieplak, W.Jr. 1992b. Characteristics of the internalization and intracellular survival of *Campylobacter jejuni* in human epithelial cell cultures. Microb. Pathogen. 13: 357–370.

Konkel, M.E., Mead, D.J., Hayes, S.F., and Cieplak, W. Jr. 1992c. Translocation of *Campylobacter jejuni* across human polarized epithelial cell monolayer cultures. J. Infect. Dis. 166: 308–315.

Konkel, M.E., Mead, D.J., and Cieplak, W. Jr. 1993. Kinetic and antigenic characterization of altered protein synthesis by *Campylobacter jejuni* during cultivation with human epithelial cells. J. Infect. Dis. 168: 948–954.

Konkel, M.E., Garvis, S.G., Tipton, S.L., Anderson, D.E. Jr., Cieplak, W. .Jr. 1997. Identification and molecular cloning of a gene encoding a fibronectin-binding protein (CadF) from *Campylobacter jejuni*. Mol. Microbiol. 24: 953–963.

Konkel, M.E., Grey, S.A., Kim, B.J., Garvis, S.G., and Yoon, J. 1999a. Identification of the enteropathogens *Campylobacter jejuni* and *Campylobacter coli* based on the *cadF* virulence gene and its product. J. Clin. Microbiol. 37: 510–517.

Konkel, M.E., Kim, B.J., Rivera-Amill, V., and Garvis, S.G. 1999b. Bacterial secreted proteins are required for the internalization of *Campylobacter jejuni* into cultured mammalian cells. Mol. Microbiol. 32: 691–701.

Kopecko, D.J., Hu, L., and Zaal, K.J.M. 2001. *Campylobacter jejuni* – microtubule-dependent invasion. Trends Microbiol. 9: 389–396.

Lara-Tejero M, and Galan, J.E. 2000. A bacterial toxin that controls cell cycle progression as a deoxyribonuclease I-like protein. Science 290 (5490): 354–357.

Lee, A., O'Rourke, J., Barrington, P. J. and Trust, T.J. 1986. Mucus colonization as a determinant of pathogenicity in intestinal infection by *Campylobacter jejuni*: a mouse cecal model. Infect. Immun. 51: 536–546.

Majeed, M., and Kihlstrom, E. 1991. Mobilization of F-actin and clathrin during redistribution of *Chlamydia trachomatis* to an intracellular site in eucaryotic cells. Infect. Immun. 59: 4465–4472.

Majeed, M., Gustafsson, M., Kihlstrom, E., and Stendahl, O. 1993. Roles of Ca^{2+} and F-actin in intracellular aggregation of *Chlamydia trachomatis* in eucaryotic cells. Infect. Immun. 61: 1406–1414.

Makino, S., van Putten, J.P., and Meyer, T.F. 1991. Phase variation of the opacity outer membrane protein controls invasion by *Neisseria gonorrhoeae* into human epithelial cells. EMBO J. 10: 1307–1315.

Matlin, K.S., Reggio, H., Helenius, A., and Simons, K. 1981. Infectious entry pathway of influenza virus in a canine kidney cell line. J. Cell. Biol. 91: 601–613.

McSweegan, E., and Walker, R.I. 1986. Identification and characterization of two *Campylobacter jejuni* adhesins for cellular and mucous substrates. Infect. Immun. 53: 141–148.

McSweegan, E., Burr, D.H., and Walker, R.I. 1987. Intestinal mucus gel and secretory antibody are barriers to *Campylobacter jejuni* adherence to INT 407 cells. Infect. Immun. 55: 1431–1435.

Mengaud, J., Ohayon, H., Gounon, P., Mege, R.M., and Cossart, P. 1996. E-cadherin is the receptor for internalin, a surface protein required for entry of *L. monocytogenes* into epithelial cells. Cell 84: 923–932.

Miller, V.L., Finlay, B.B., and Falkow, S. 1988. Factors essential for the penetration of mammalian cells by *Yersinia*. Curr. Top. Microbiol. Immunol. 138: 15–39.

Monteville, M.R., Yoon, J.E., and Konkel, M.E. 2003. Maximal adherence and invasion on INT407 cells by *Campylobacter jejuni* requires the CadF outer membrane protein and microfilament reorganization. Microbiology 149: 153–165.

Newell, D.G., and Pearson, A. 1984. The invasion of epithelial cell lines and the intestinal epithelium of infant mice by *Campylobacter jejuni/coli*. J. Diarrhoeal. Dis. Res. 2: 19–26.

Oelschlaeger, T.A., Guerry, P., and Kopecko, D.J. 1993. Unusual microtubule-dependent endocytosis mechanisms triggered by *Campylobacter jejuni* and *Citrobacter freundii*. Proc. Natl. Acad. Sci. USA 90: 6884–6888.

Parkhill, J., Wren, B.W., Mungall, K., Ketley, J.M., Churcher, C., Basham, D., Chillingworth, T., Davies, R.M., Feltwell, T., Holroyd, S., Jagels, K., Karlyshev, A.V., Moule, S., Pallen, M.J., Penn, C.W., Quail, M.A., Rajandream, M.A., Rutherford, K.M., van Vliet, A.H.M., Whitehead, S., and Barrell, B.G. 2000. The genome sequence of the food-borne pathogen *Campylobacter jejuni* reveals hypervariable sequences. Nature 403: 665–668.

Pei, Z.H., Ellison, R.T.dD, and Blaser, M.J. 1991. Identification, purification, and characterization of major antigenic proteins of *Campylobacter jejuni*. J. Biol. Chem. 266: 16363–16369.

Pei, Z., and Blaser, M.J. 1993. PEB1, the major cell-binding factor of *Campylobacter jejuni*, is a homologue of the binding component in gram-negative nutrient transport systems. J. Biol. Chem. 268: 18717–18725.

Pei, Z., Burucoa, C., Grignon, B., Baqar, S., Huang, X.Z., Kopecko, D. J., Bourgeois, A.L., Fauchere, J-L., Blaser, M.J. 1998. Mutation in *peb1A* of *Campylobacter jejuni* reduces interactions with epithelial cells and intestinal colonization of mice. Infect. Immun. 66: 938–943.

Pickett, C.L. 2000. *Campylobacter* toxin and their role in pathogenesis. In: *Campylobacter*, 2nd edn. Nachamkin, I. and Blaser, M.J., eds. ASM Press, Washington DC, pp. 179–190,

Richardson, W.P., and Sadoff, J.C. 1988. Induced engulfment of *Neisseria gonorrhoeae* by tissue culture cells. Infect. Immun. 56: 2512–2514.

Rivera-Amill, V., Kim, B.J., Seshu, J., and Konkel, M.E. 2001. Secretion of the virulence-associated *Campylobacter* invasion antigens from *Campylobacter jejuni* requires a stimulatory signal. J. Infect. Dis. 183: 1607–1616.

Robins-Browne, R.M., Tzipori, S., Gonis, G., Hayes, J., Withers, M., and Prpic, J.K. 1985. The pathogenesis of *Yersinia enterocolitica* infection in gnotobiotic piglets. J. Med. Microbiol. 19: 297–308.

Ruiz-Palacios, G. 1992. Pathogenesis of *Campylobacter* infection: *in vitro* models, In: Nachamkin, I., Blaser, M.J., and Tompkins, L.S. eds. *Campylobacter jejuni*. Current Status and Future Trends. ASM Press, Washington DC, pp. 158–159.

Russell, R.G., and Blake, D.C., Jr. 1994. Cell association and invasion of Caco-2 cells by *Campylobacter jejuni*. Infect. Immun. 62: 3773–3779.

Russell, R.G., O'Donnoghue, M., Blake, D.C., Jr., Zulty, J., and DeTolla, L.J. 1993. Early colonic damage and invasion of *Campylobacter jejuni* in experimentally challenged infant *Macaca mulatta*. J. Infect. Dis. 168: 210–215.

Sansonetti, P.J. 1991. Genetic and molecular basis of epithelial cell invasion by *Shigella* species. Rev. Infect. Dis. 13 (Suppl 4): S285–292.

Sansonetti P.J, Tran Van Nhieu G, Egile C. 1999. Rupture of the intestinal epithelial barrier and mucosal invasion by *Shigella flexneri*. Clin. Infect. Dis. 28: 466–475.

Takeuchi, A. 1967. Electron microscope studies of experimental *Salmonella* infection. I. Penetration into the intestinal epithelium by *Salmonella typhimurium*. Am. J. Pathol. 50: 109–136.

Takeuchi, A., Formal, S.B., and Sprinz, H. 1968. Experimental acute colitis in the Rhesus monkey following peroral infection with *Shigella flexneri*. An electron microscope study. Am. J. Pathol. 52: 503–529.

van Spreeuwel, J.P., Duursma, G.C., Meijer, C.J., Bax, R., Rosekrans, P.C., and Lindeman, J. 1985. *Campylobacter* colitis: histological immunohistochemical and ultrastructural findings. Gut 26: 945–951.

Vogel J.P, Isberg R.R. 1999. Cell biology of *Legionella pneumophila*. Curr. Opin. Microbiol. 2: 30–34.

Walker R.I, Schmauder-Chock E.A, Parker J.L, Burr D. 1988. Selective association and transport of *Campylobacter jejuni* through M cells of rabbit Peyer's patches. Can. J. Microbiol. 34: 1142–1147.

Wassenaar, T.M., Bleumink-Pluym, N.M., and van der Zeijst, B.A. 1991. Inactivation of *Campylobacter jejuni* flagellin genes by homologous recombination demonstrates that *flaA* but not *flaB* is required for invasion. EMBO J. 10: 2055–2061.

Wooldridge, K.G., Williams, P.H., and Ketley, J.M. 1996. Host signal transduction and endocytosis of *Campylobacter jejuni*. Microb. Pathogen. 21: 299–305.

Yanagawa, Y., Takahashi, M., and Itoh, T. 1994. The role of flagella of *Campylobacter jejuni* in colonization in the intestinal tract in mice and the cultured-cell infectivity. Nippon Saikingaku Zasshi. 49: 395–403.

Yao, R., Burr, D.H., Doig, P., Trust, T.J., Niu, H., and Guerry, P. 1994. Isolation of motile and non-motile insertional mutants of *Campylobacter jejuni*: the role of motility in adherence and invasion of eukaryotic cells. Mol. Microbiol. 14: 883–893.

Yao, R., Burr, D.H., and Guerry, P. 1997. CheY-mediated modulation of *Campylobacter jejuni* virulence. Mol. Microbiol. 23: 1021–1031.

Young, G.M, Schmiel, D.H, and Miller, V.L. 1999. A new pathway for the secretion of virulence factors by bacteria: the flagellar export apparatus functions as a protein-secretion system. Proc. Natl. Acad. Sci. USA 96: 6456–6461.

Chapter 20

Cytolethal Distending Toxin

Carol L. Pickett and Robert B. Lee

Abstract

Many toxic activities have been reported to be produced by *Campylobacter jejuni*, but current work is centred on an exotoxin that is a member of the cytolethal distending toxin (CDT) family of toxins. CDT is encoded by three adjacent genes, termed *cdtA*, *cdtB*, and *cdtC*. The CDT holotoxin consists of one copy of each gene product. CDT is capable of causing eucaryotic cells of different lineages to become irreversibly blocked in the G_1 or G_2 phase of the cell cycle, and recent work has shown that the CdtB subunit carries a DNase activity that is responsible for bringing about cell cycle arrest. Current research includes studies of the role of the CdtA and CdtC subunits in receptor binding, as well as on CDT uptake, trafficking, secretion, and its role in disease production.

INTRODUCTION

In recent years, CDT has been the focus of most *C. jejuni* exotoxin research. There are several reasons for this lack of attention to other putative toxins, but one of the most compelling is the failure to identify other toxin encoding genes in the genome sequence of *C. jejuni* (Parkhill *et al.*, 2000). Wassenaar (1997) published a review of the *C. jejuni* toxin literature in which much of the early work is discussed. A subsequent review reported on additional toxic activities produced by CDT, but since that time, no new reports on these non-CDT activities have been published (Pickett, 2000). CDT appears to be toxic to a wide variety of human cultured cell lines, and it is possible that many of the early reports of *C. jejuni* toxicity were actually reporting CDT effects. Thus, this review is essentially a review of the CDT literature; it will principally cover the *C. jejuni* CDT, but also includes key work on other members of the CDT family of toxins.

CYTOLETHAL DISTENDING TOXIN

The earliest reports that identified CDT as a novel toxin were made by Johnson and Lior in 1987 and 1988 (Johnson and Lior, 1987; Johnson and Lior, 1988). They showed that certain *E. coli*, *Shigella* spp., and *Campylobacter* spp. produced a toxin that caused several cell lines, including HeLa, Chinese hamster ovary (CHO), and Vero cells to slowly distend and die. They further showed that the activity appeared to be distinct from other known toxins produced by these species. Since that time these authors' findings have been corroborated and extended by many researchers.

GENETICS OF CYTOLETHAL DISTENDING TOXIN

In 1994, it was discovered that CDT is encoded by three adjacent genes, *cdtA*, *cdtB*, and *cdtC* (Scott and Kaper, 1994; Pickett *et al.*, 1994). At that time, database searches revealed no similarities of the predicted CDT proteins to known proteins, thus indicating that CDT was indeed a novel toxin. These first *cdt* sequences were obtained from two different *E. coli* strains, and interestingly, their predicted amino acid sequences were significantly divergent, with the CdtA, CdtB, and CdtC proteins exhibiting 48%, 61%, and 42% identical and similar residues, respectively. The sequence of the *cdt* genes from *C. jejuni* strain 81-176 was reported in 1996 (Pickett *et al.*, 1996) and while the Cdt proteins encoded by these genes were obviously related to the *E. coli* proteins, their sequences ranged between 31% and 63% for identical and similar amino acids. The highest similarity was always between the different CdtB sequences, suggesting the need for greater conservation in CdtB, than in CdtA or CdtC.

More recently, *cdt* sequences have been reported from additional *E. coli* strains (Pérès *et al.*, 1997; Janka *et al.*, 2003; Tóth *et al.*, 2003), *C. jejuni* strains CH5 and NCTC 11168 (Hickey *et al.*, 2000; Parkhill *et al.*, 2000), *Haemophilus ducreyi* (Cope *et al.*, 1997), *Actinobacillus actinomycetemcomitans* (Sugai *et al.*, 1998; Mayer *et al.*, 1999), and various non-pylori *Helicobacter* sp. (Young *et al.*, 2000; Chien *et al.*, 2000; Kostia *et al.*, 2003). It is clear from this work that CDT activity is encoded by three closely linked genes, and that the predicted proteins have type II leader sequences for secretion.

The extent of *cdt* sequence relatedness between and within the different species varies depending upon which species is being examined. However, the nucleotide sequences of the *cdt* genes from the three *C. jejuni* strains are virtually identical, suggesting that in *C. jejuni* the *cdt* genes are highly conserved. Data from studies examining CDT production by *C. jejuni* strains from a variety of sources suggest that over 90% of all isolates produce active CDT (Pickett *et al.*, 1996; Eyigor *et al.*, 1999a; Bang *et al.*, 2001). The relatedness of the *cdt* genes amongst isolates has been studied using polymerase chain reaction (PCR) amplification of various portions of the *cdt* genes, restriction fragment length polymorphism (RFLP) analysis of PCR products amplified from the *cdt* genes, or by hybridizations (Eyigor *et al.*, 1999b; Bang *et al.*, 2001; Dorrell *et al.*, 2001). These studies reveal two consistent features. First, CDT-negative strains still appear to carry at least some portion of the *cdt* coding region, and secondly, the genes encoding fully active CDT are likely identical or nearly identical to that of the strains already sequenced. These observations suggest that the presence of the *cdt* genes in *C. jejuni* is not of recent origin, and support the notion that there may be some positive selective pressure for maintenance of the *cdt* genes. Unfortunately, sequence information on *C. jejuni* natural isolates with little or no CDT activity has not been published. It will be of interest to see if CDT-negative strains possess similar or identical mutational changes, or if most of these isolates contain unique mutations leading to inactivity.

Evidence for the presence of genes encoding cytolethal distending toxin have been found in *Campylobacter coli*, *C. fetus*, *C. upsaliensis*, *C. lari* and *C. hyointestinalis*. Studies using RFLP analysis of PCR products (Eyigor *et al.*, 1999b; Bang *et al.*, 2001) have shown that most, if not all, *C. coli* strains contain the *cdt* genes, despite evidence that most *C. coli* strains appear to produce little or no toxic activity. Of considerable interest in this regard is a recent Danish study (Bang *et al.*, 2003) showing that *C. coli* strains isolated from pigs produced substantially higher CDT titres than those isolated from cattle. If this observation is confirmed by analysis of additional isolates from diverse geographical locations, it poses some fascinating questions relating host animal to CDT activity.

C. fetus and *C. upsaliensis* have been the subject of a small number of studies investigating the prevalence of CDT production or the presence of *cdt* genes. In one study,

30 of 30 *C. upsaliensis* strains contained *cdt* sequences (Mooney *et al.*, 2001), while CDT activity was produced by 25 of 26 *C. fetus* subspecies *fetus* strains (Ohya *et al.*, 1993). Sequencing of a partial *cdtB* fragment demonstrated that the type strain of *C. fetus* (ATCC 27374) contains *cdt* sequences and that the sequence of *cdtB* in *C. fetus* is significantly diverged from that of *C. jejuni* (Pickett *et al.*, 1996). Finally, representative strains of *C. lari* and *C. hyointestinalis* produced PCR products with degenerative *cdtB* primers (Pickett *et al.*, 1996). Thus, it appears that at least some, if not all, isolates within these *Campylobacter* species possess *cdt* genes.

BIOLOGICAL ACTIVITY OF CDT

The initial observations that led to an understanding of the novel nature of CDT's activity occurred when it was shown that both *E. coli* and *C. jejuni* CDT could cause HeLa cells to become irreversibly blocked in the G_2 phase of the cell cycle (Pérèz *et al.*, 1997; Comayras *et al.*, 1997; Whitehouse *et al.*, 1998). Extracts from wild-type *C. jejuni* strain 81-176 were capable of causing 64% of treated HeLa cells to be blocked in G_2 within 24 hours, compared to only 10% of HeLa cells after treatment for a similar length of time with extracts from an isogenic *cdt* null mutant. These authors further showed that recombinant *C. jejuni* CDT produced in *E. coli* has a similar G_2-blocking effect on both HeLa cells and the Caco-2 human intestinal epithelial cell line (Whitehouse *et al.*, 1998). DNA content analysis that allows for the detection of cell cycle arrest is now routinely used as a test for CDT activity in sensitive cells; all CDTs have been reported to have this capability (Pérèz *et al.*, 1997; Whitehouse *et al.*, 1998; Sugai *et al.*, 1998; Cortes-Bratti *et al.*, 1999; Young *et al.*, 2000).

An examination of *C. jejuni* CDT-treated HeLa cells for possible evidence of entry into mitosis found an almost complete lack of chromatin condensation, no tubulin reorganization into normal spindle structures, and a predominantly phosphorylated (inactive) CDC2 (Whitehouse *et al.*, 1998), all indicative of a G_2 block. Activation of CDC2 is required for entry into mitosis, and this CDT-induced inactivation has also been noted for CDTs produced by *E. coli* (Comayras *et al.*, 1997), *H. ducreyi* (Cortes-Bratti *et al.*, 1999) and *A. actinomycetemcomitans* (Shenker *et al.*, 1999). More recently, it has become clear that CDT can produce a G_1 phase block; the determining factor appears to be whether the cell line being tested has a functional p53 (Cortes-Bratti *et al.*, 2001; Hassane *et al.*, 2003). That is, cell lines with abnormal or absent p53 function will arrest exclusively in G_2.

In 2000, a major breakthrough in our understanding of CDT action occurred when it was reported that the *E. coli* CdtB protein has homology to type I mammalian DNases (Elwell and Dreyfus, 2000). These investigators mutagenized CdtB at five amino acid positions predicted to be critical for DNase activity (Figure 20.1). Four of these mutations completely abolish the ability of CDT to cause G_2 arrest in HeLa cells; the fifth had reduced G_2 arrest activity. Similar analyses have been done with the *C. jejuni* CdtB protein. First, the His-152 and Asp-185 residues, amino acids predicted to be in the active site and involved in magnesium-binding, respectively, were specifically mutagenized (see Figure 20.1; Lara-Tajero and Galán, 2000). Both mutations led to inactive CDT as judged by cell distension, DNA content analysis, and chromatin appearance following transient transfection of COS-1 cells. Similarly, mutation of the *C. jejuni* CdtB Asp-222 residue, which is predicted to be involved in catalysis, led to a completely inactive CDT (Figure 20.1; Hassane *et al.*, 2001). We have made mutations in the *C. jejuni* Glu-60 and Asn-

Alignment of CdtB

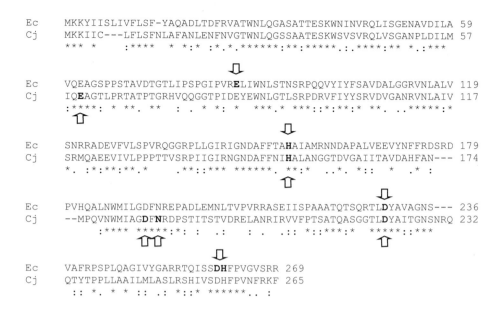

```
Ec    MKKYIISLIVFLSF-YAQADLTDFRVATWNLQGASATTESKWNINVRQLISGENAVDILA  59
Cj    MKKIIC---LFLSFNLAFANLENFNVGTWNLQGSSAATESKWSVSVRQLVSGANPLDILM  57
      *** *      :****  * *:* :*.*.******:**:*****.:.****:** *.:***

Ec    VQEAGSPPSTAVDTGTLIPSPGIPVRELIWNLSTNSRPQQVYIYFSAVDALGGRVNLALV  119
Cj    IQEAGTLPRTATPTGRHVQQGGTPIDEYEWNLGTLSRPDRVFIYYSRVDVGANRVNLAIV  117
      :****:  *  **.  **    : . * *:  *    ***.* ***::*:**:* **.  ..*****:*

Ec    SNRRADEVFVLSPVRQGGRPLLGIRIGNDAFFTAHAIAMRNNDAPALVEEVYNFFRDSRD  179
Cj    SRMQAEEVIVLPPPTTVSRPIIGIRNGNDAFFNIHALANGGTDVGAIITAVDAHFAN---  174
      *. .:*:**:**.*     .**::*** ******. **:*   ..*. *::  *   .* :

Ec    PVHQALNWMILGDFNREPADLEMNLTVPVRRASEIISPAAATQTSQRTLDYAVAGNS---  236
Cj    --MPQVNWMIAGDFNRDPSTITSTVDRELANRIRVVFPTSATQASGGTLDYAITGNSNRQ  232
      :**** *****:*: :   .:    :  .  .::  *::***:*    *****::***

Ec    VAFRPSPLQAGIVYGARRTQISSDHFPVGVSRR  269
Cj    QTYTPPLLAAILMLASLRSHIVSDHFPVNFRKF  265
      ::  *. * *  ::   .:  *::*  ******.. :
```

Figure 20.1 Alignment of CdtB predicted protein sequences from *C. jejuni* 81-176 (Cj) and *E. coli* 9142–88 (Ec) (Pickett *et al.*, 1996; Pickett *et al.*, 1994). The conserved amino acid residues in CdtB that have been mutagenized and identified as critical for DNase activity are shown in bold (see text for specific references). Arrows represent residues that were substituted in *C. jejuni* (upward) and *E. coli* (downward). Residues predicted to be critical for catalysis include Glu-86 (Ec), His-152 (Cj), Asp-222 (Cj), and His-261 (Ec). Residues predicted to be involved in metal ion binding include Glu-60 (Cj), Asp-185 (Cj), and Asp-260 (Ec). Asn-187 is predicted to bind DNA. Identical (*), high similarity (:), and low similarity (.) amino acids are indicated. The alignment was performed using the ClustalW program (Thompson *et al.*, 1994).

187 residues of the *C. jejuni* CdtB subunit; these residues are predicted to be involved in metal-binding and DNA contact, respectively (C.L. Pickett and R.B. Lee, unpublished). Both of these mutations produced inactive CDT when tested on HeLa cells. Figure 20.1 summarizes the residues in CdtB that have been mutagenized. These mutagenesis data clearly indicate that CdtB depends upon these individual residues for its activity, but by themselves do not prove that CdtB is a nuclease.

The initial studies examining the ability of CdtB to cause DNA damage all involved artificial systems of intoxication. For example, transient expression of the *C. jejuni* *cdtB* gene, but not *cdtA* or *cdtC*, in COS-1 cells caused chromatin condensation and fragmentation (Lara-Tejero and Galan, 2000). Similarly, the activity of the *C. jejuni* CdtB after microinjection into COS-1 cells, and the effects of electroporation of an *E. coli* CdtB into HeLa cells, were examined (Lara-Tejero and Galán, 2000; Elwell *et al.*, 2001). In both cases, chromatin abnormalities were observed only when wild-type CdtB was used; use of CdtA, CdtC, or CdtB carrying an active site mutation caused no chromatin alteration.

In a related type of analysis, but in a completely different system, the effects of expression of individual *C. jejuni* *cdt* genes in *Saccharomyces cerevisiae* were examined (Hassane *et al.*, 2001). Expression of *cdtB*, but not *cdtA* or *cdtC*, caused *S. cerevisiae* cells to arrest in G_2 with an accumulation of large, budded cells. Further work illustrated that this arrest was accompanied by upregulation of RNR2, a yeast gene known to be upregulated by DNA-damaging agents. And finally, expression of *cdtB*, but not *cdtB*[D222A], led to partial degradation of the yeast chromosomes. Taken together, these data indicate that CdtB alone is capable of recapitulating the phenotypic effects of CDT, and suggest that CdtB is likely a nuclease.

Nonetheless, the amount of CdtB used, or expressed, in these experiments is considerably greater than what is needed to bring about distension and cell cycle arrest following exogenous addition of the toxin. Evidence of DNA damage resulting from this kind of extracellular intoxication has been best demonstrated indirectly. Recently, two groups have shown that the exogenous addition of CDT holotoxin to sensitive cells led to the formation of DNA repair responses. CDT from *Haemophilus ducreyi* was used to intoxicate HeLa cells, as well as non-proliferating dendritic cells, and the authors were able to demonstrate that CDT causes the formation of repair complexes containing Mre11 as soon as 1 hour after treatment (Li *et al.*, 2002). Similarly, *C. jejuni* CDT treatment of normal fibroblasts causes the formation of Rad50 containing foci (Hassane *et al.*, 2003). Both of these results indicate that CDT likely causes double-stranded breaks (DSBs), since the Rad50/Mre11/Nbs1 complex forms as part of the nonhomologous end joining repair process in response to DSBs. In addition, CDT treatment was shown by both groups to cause phosphorylated H2AX (γ-H2AX) formation, an event triggered by, and located at, DSBs. The formation of Rad50/Mre11/Nbs1 complexes and γ-H2AX following treatment with *C. jejuni* CDT was completely dependent upon the enzymatic activity of CdtB, since CDT containing CdtB[D222A] was devoid of any ability to promote these processes that follow DSBs (Hassane *et al.*, 2003). These data indicate that while CDT has an apparently subtle ability to cause DNA damage, such damage is nonetheless rapidly detected and acted upon by the cell.

Demonstrations of CdtB or CDT nuclease activity *in vitro* have been reported for some CDTs. The use of as little as 0.025μg of purified His-tagged CdtB from *E. coli* stain 9142–88 was reported to be sufficient to observe at least some relaxation of supercoiled plasmid DNA (Elwell *et al.*, 2001). However, investigators working with CDT from *A. actinomycetemcomitans* and *H. ducreyi* reported that nearly two log greater amounts of CDT were required to observe similar levels of plasmid relaxation (Mao and DiRienzo, 2002; Li *et al.*, 2002). Since these experiments were not performed in exactly identical conditions with the same enzymes, it is difficult to reach any firm conclusions based on a comparison of the data presented by each group. In any case, it seems likely that CdtB carries nuclease activity, albeit at a level substantially less than that seen for mammalian type I DNases.

Experiments such as those described above, in which the ability of CDT to cause observable DNA damage often seems difficult, led to the recent proposal that CDT may exemplify a type of virulence factor that acts by causing only limited direct action, but that the action then leads to a cellular response which can ultimately have a profound effect on the normal function of the cell (Lara-Tejero and Galán, 2002). Alternatively, it may be that investigators have not yet determined the optimal conditions for demonstrating CDT activity *in vitro*. While the exogenous addition of CDT to sensitive cells appears to readily cause cell cycle disruption, there is essentially no information in the literature that would allow determination of the percentage of the exogenously applied toxin that enters the cell, nor what percentage of internalized CdtB makes it to the target site.

CDT HOLOTOXIN STRUCTURE

It has been clear for several years that expression of all three *cdt* genes is required in order to produce CDT activity (Pickett *et al.*, 1994). However, the actual composition of the CDT holotoxin was only recently determined. The *C. jejuni* CDT was shown to comprise CdtA, CdtB, and CdtC with each subunit present in just one copy per holotoxin molecule (Lara-Tejero and Galán, 2001). These investigators purified each subunit individually as either His$_6$-tagged or GST-tagged fusion proteins, and then successfully mixed the subunits together to reconstitute activity. They concluded that all three subunits are required to cause G$_2$ arrest in Henle-407 cells; individual subunits and any combination of two subunits failed to cause arrest. They then used combinations of coimmunoprecipitation assays, GST pull-down assays and gel filtration chromatography to determine that each subunit is capable of interacting with each other subunit, and that the CDT holotoxin has an apparent molecular mass of approximately 80 kDa. This would correspond to a holotoxin structure containing a single copy of each Cdt subunit.

More recently, it was shown that *C. jejuni* CDT activity can be reconstituted from just CdtB and CdtC, although these subunits were only about 25% as efficient as the CDT holotoxin on a per microgram basis (Lee *et al.*, 2003). This observation has also been reported for the *H. ducreyi* and the *A. actinomycetemcomitans* CDTs (Deng *et al.*, 2001; Akifusa *et al.*, 2001). Thus, CDT has maximal activity when all three subunits are present, and this form of the toxin is probably that which is found most often in nature.

FUNCTION OF CDTA AND CDTC

While a considerable amount of work by several different investigators has concentrated on elucidating the activity of CdtB, there have been only a few reports concerning the possible roles of CdtA and CdtC. Confocal microscopy of *A. actinomycetemcomitans* His$_6$-tagged Cdt subunits revealed that CdtA apparently bound to Chinese hamster ovary cells, but no binding of CdtB or CdtC or holotoxin could be demonstrated (Mao and DiRienzo, 2002).

However, somewhat different results were obtained in a recent study of the binding capabilities of the *C. jejuni* subunits (Lee *et al.*, 2003). Individually purified *C. jejuni* Cdt subunits were biotinylated and tested for binding to HeLa cells in an enzyme-linked immunosorbent assay (CELISA). This work indicated that both CdtA and CdtC bound in a specific manner to HeLa cells. CdtB gave no evidence of binding. In addition, binding of CDT holotoxin was observed. Competitive binding experiments suggested that CdtA, CdtC, and holotoxin all bound to the same surface receptor. These results indicate that both CdtA and CdtC likely mediate binding of the CDT holotoxin, and that both subunits together should be considered the B, or binding portion, of this A:B toxin.

A similar conclusion was reached in a study of *H. ducreyi* CDT using an entirely different method. In this study, the investigators isolated CdtA and CdtC as a complex and showed that this complex could bind to HeLa cells and block killing of the cells by holotoxin (Deng *et al.*, 2003). These authors could not detect binding of individual CdtA or CdtC subunits to HeLa cells, but both the methods and toxin used were different from those in the *C. jejuni* study. Although these discrepancies cannot yet be explained, it seems reasonable to conclude that both CdtA and CdtC mediate binding of CDT to sensitive cells.

IS THERE A ROLE FOR CDT IN HUMAN DISEASE?

C. jejuni

While much progress has been made in understanding the genetics and biology of CDT, only a small number of papers have appeared that address a possible role for CDT in human disease. The paucity of animal models that reliably mimic *C. jejuni* diarrhoeal disease has made advances is this field difficult. However, an interesting study in immunodeficient mice suggested CDT may assist spread of the organism in infected animals (Purdy *et al.*, 2000). In this study, the authors compared the abilities of wild-type or isogenic *cdtB* mutants to spread to blood, liver or spleen, following intragastric inoculation of C.B-17-SCID-Biege mice. Their results indicated that CDT may play a role in spread of the organism, particularly to the spleen. Unfortunately, this model does not mimic human disease, and thus it is difficult to apply these results directly to natural *C. jejuni* infection in humans.

CDT producers other than *C. jejuni*

There are a few reports of animal model studies with other members of the CDT family. A study testing *S. dysenteriae* CDT in a suckling mouse model indicated that CDT may contribute to diarrhoea, tissue necrosis and reparative hyperplasia in the descending colon of these animals (Okuda *et al.*, 1997). The *H. ducreyi* CDT has been tested in both rabbit models and in human volunteers. In a temperature-dependent rabbit model system, a *H. ducreyi cdtC* mutant was found to form dermal lesions similar to that of the wild-type strain (Stevens *et al.*, 1999). However, as the authors point out, this model is limited in its ability to mimic natural human infection. In a separate study, *H. ducreyi* CDT was injected intradermally into naive and immune rabbits (Wising *et al.*, 2002). While injection of the individual subunits produced no changes, the holotoxin caused dose-dependent skin reactions characterized by erythema and the presence of granulocytes and macrophages. The skin reactions of immune rabbits exhibited erythema and oedema, with increased presence of inflammatory cells. Finally, the ability of *H. ducreyi* wild-type and *cdtC* mutants to cause the formation of pustules in human volunteers was examined (Young *et al.*, 2001). This model only examines early stage disease, since the study is terminated once painful pustules have formed. The results of this study clearly indicated that CDT is not required for pustule formation. Given these results and the apparent interactions of CDT with immune cells (see below), the authors speculate that CDT plays no role in chancre formation, but may instead be involved in reducing the effectiveness of the immune response to *H. ducreyi*.

CDT INTERACTIONS WITH THE IMMUNE SYSTEM

CDT can suppress T cell function

In 1999, two very interesting reports appeared in which CDT was shown to be active against T cells (Gelfanova *et al.*, 1999; Shenker *et al.*, 1999). *H. ducreyi* CDT was shown to both suppress proliferation and induce apoptosis of Jurkat T cells (Gelfanova *et al.*, 1999) leading these authors to speculate that CDT may actively interfere with the T cell response to *H. ducreyi*. Similarly, the *A. actinomycetemcomitans* CDT was shown to cause G_2 arrest in both CD4$^+$ and CD8$^+$ T cells (Shenker *et al.*, 1999). Since a prominent feature of *A. actinomycetemcomitans* associated disease is impaired immune response to the infection, it seems appropriate for the authors to speculate that the principal role of CDT in these infections may be to participate in suppression of the immune response.

More recently, the CDT produced by *C. upsaliensis* was reported to be capable of arresting T cells in G_2 (Mooney *et al.*, 2001), and we have determined that the *C. jejuni* CDT likewise can cause G_2 arrest in human T cells (Pickett *et al.*, unpublished). These results, along with those described for *H. ducreyi* and *A. actinomycetemcomitans* suggest that all CDTs may be capable of affecting T cells. Thus, a common feature of CDT-associated disease may be the ability of the toxin to interfere with the host's immune response through impairment of T cell function.

CDT and release of proinflammatory cytokines

C. jejuni has been reported to induce release of certain cytokines. The first study indicated that live *C. jejuni* could stimulate secretion of IL-8 from the intestinal epithelial cell line, INT 407 (Hickey *et al.*, 1999). These authors further showed that the ability to induce IL-8 secretion appeared to be directly related to how well a particular *C. jejuni* strain could adhere and invade cultured cells. In a subsequent paper, the same group further investigated whether CDT might be involved in this induction process (Hickey *et al.*, 2000). Interestingly, this work indicated that there are likely multiple players in the stimulation of IL-8 release. That is, live CDT-negative mutants, capable of adherence and invasion were still able to stimulate IL-8 release, implying that CDT is not required for this process. However, membranes from *C. jejuni* strains containing CDT were able to stimulate IL-8 secretion, while membranes from CDT-negative strains were not.

A second group recently investigated the ability of *C. jejuni* to induce release of IL-1α, IL-1β, IL-6, IL-8, and TNF-α (Jones *et al.*, 2003). In this study, both live and killed *C. jejuni* were able to induce secretion of cytokines from the monocytic cell line, THP-1. CDT was not required for this induction.

These two reports are not necessarily contradictory in terms of the involvement of CDT, since the investigators used different cell lines and membrane fractions were not tested in the THP-1 study. Since the ability of CDT to stimulate cytokine production is of obvious importance in terms of the immune response to *C. jejuni*, it will be of interest to see if future work with additional CDT family members and other cell lines can confirm and extend these studies.

SUMMARY AND FUTURE DIRECTIONS

The last few years have provided a wealth of new data about the biology of CDT. The ability to purify the individual subunits as fusion proteins and subsequently reconstitute activity has allowed investigators to bypass what had been the apparently intractable problem of purifying sufficient quantities of holotoxin to perform studies requiring purified toxin. We now know that CDT comprises one molecule of each of the three Cdt subunits and that CdtB carries the enzymatic activity responsible for the principle phenotypic properties associated with CDT. Recent work has shown that CdtA and CdtC both participate in receptor recognition and binding of the toxin to sensitive cells. Two studies have addressed aspects of trafficking of CDT, although these did not specifically address the *C. jejuni* CDT. The *H. ducreyi* CDT was reported to be taken up by a receptor-mediated endocytic process, and subsequent retrograde transport of the toxin via the Golgi complex was indicated (Cortes-Bratti *et al.*, 2000). Recently, a report that an *A. actinomycetemcomitans* CdtB subunit was demonstrated to carry information for directing itself to the nucleus

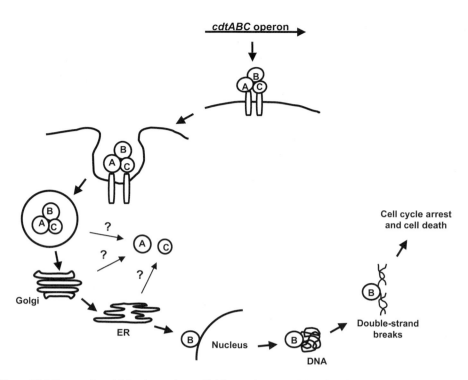

Figure 20.2 Proposed model for the uptake, trafficking, and target activity of CDT. The CDT operon consists of three adjacent genes, *cdtA*, *cdtB*, and *cdtC*. The three gene products interact to form the CDT holotoxin. CdtA and CdtC apparently mediate CDT binding to an unidentified host cell receptor, after which the toxin is internalized via receptor-mediated endocytosis (Cortes-Bratti *et al.*, 2001). The trafficking of CDT, and CdtB in particular, is poorly delineated. Available evidence suggests that the toxin undergoes retrograde transport through the Golgi complex, and that CdtB ultimately enters the nucleus. The question marks indicate our lack of knowledge about the fate of CdtA and CdtC. Once in the nucleus, CdtB apparently binds to DNA and causes double-stranded breaks, which lead to invocation of the DNA damage checkpoint pathway and ultimately cell cycle arrest and cell death.

furthers our understanding of CdtBs involvement in reaching its target (Nishikubo *et al.*, 2003). These two works are just the beginning of what is likely to be a growth area of CDT research. The receptor for CDT has not yet been identified, nor have the CDT residues involved in receptor recognition been identified. Many details of the uptake process and trafficking remain unknown. For example, are all three subunits taken-up into the cell? If so, where do CdtA and CdtC go? What is the precise route CdtB takes to reach the nucleus? What specific residues of CdtB are responsible for nuclear localization? Figure 20.2 is a model summarizing some of our current knowledge of CDT; it includes some speculative ideas based on current results from several investigators.

Just as we have much to learn about entry of CDT into sensitive cells, we also know little about secretion of CDT from the bacterial species that produce this toxin. It is not even clear to what extent this process may be similar in each CDT-producing species. In *C. jejuni*, the bulk of CDT activity appears to remain associated with the bacterial membrane (Pickett, 2000; Hickey *et al.*, 2000). While it is suspected that CDT may be principally in the outer membrane, this has not been formally established. Nor do we know exactly how CDT is presented to, or comes into contact with its eucaryotic cell receptor.

Finally, we still do not have a clear understanding of the role CDT plays in *C. jejuni* pathogenesis. The reports suggesting that CDT may be important for immune suppression is clearly of interest, but much remains to be tested before we can say how important this may be in *C. jejuni* diarrhoeal disease. *In vitro* studies clearly support the notion that *C. jejuni* CDT can kill intestinal epithelial cells. However, reliable animal model tests are needed to elucidate CDTs role in diarrhoeal pathogenesis.

References

Akifusa, S., Poole, S., Lewthwaite, J., Henderson, B., and Nair, S.P. 2001. Recombinant *Actinobacillus actinomycetemcomitans* cytolethal distending toxin proteins are required to interact to inhibit human cell cycle progression and to stimulate human leucocyte cytokine synthesis. Infect. Immun. 69: 5925–5930.

Bang, D.D., Møller Nielsen, E., Scheutz, F., Pedersen, K., Handberg, K., and Madsen, M. 2003. PCR detection of seven virulence and toxin genes of *Campylobacter jejuni* and *Campylobacter coli* isolates from Danish pigs and cattle and cytolethal distending toxin production of the isolates. J. Appl. Microbiol. 94: 1003–1014.

Bang, D.D., Scheutz, F., Ahrens, P., Pedersen, K., Blom, J., and Madsen, M. 2001. Prevalence of cytolethal distending toxin (*cdt*) genes and CDT production in *Campylobacter* spp. isolated from Danish broilers. J. Med. Microbiol. 50: 1087–1094.

Chien, C-C., Taylor, N.S., Ge, Z., Schauer, D.B., Young, V.B., and Fox, J.G. 2000. Identification of *cdtB* homologues and cytolethal distending toxin activity in enterohepatic *Helicobacter* spp. J. Med. Microbiol. 49: 525–534.

Comayras, C., Tasca, C., Pérès, S.Y., Ducommun, B., Oswark, E., and De Rycke, J. 1997. *Escherichia coli* cytolethal distending toxin blocks the HeLa cell cycle at the G$_2$/M transition by preventing cdc2 protein kinase dephosphorylation and activation. Infect. Immun. 65: 5088–5095.

Cope, L.D., Lumbley, S., Latimer, J.L., Klesney-Tait, J., Stevens, M.K., Johnson, L.S., Purven, M., Munson, Jr., R.S., Lagergard, T., Radolf, J.D., and Hansen, E.J. 1997. A diffusible cytotoxin of *Haemophilus ducreyi*. Proc. Natl. Acad. Sci. USA 94: 4056- 4061.

Cortes-Bratti, X., Chaves-Olarte, E., Lagergård, T., and Thelestam, M. 1999. The cytolethal distending toxin from the chancroid bacterium *Haemophilus ducreyi* induces cell-cycle arrest in the G2 phase. J. Clin. Invest. 103: 107–115.

Cortes-Bratti, X., Chaves-Olarte, E., L Lagergård, T., and Thelestam, M. 2000. Cellular internalization of cytolethal distending toxin from *Haemophilus ducreyi*. Infect Immun. 68: 6903–6911.

Cortes-Bratti, X., Karlsson, C., Lagergård, T., Thelestam, M., and Frisan, T. 2001. The *Haemophilus ducreyi* cytolethal distending toxin induces cell cycle arrest and apoptosis via the DNA damage checkpoint pathways. J. Biol. Chem. 276: 5296–5302.

Deng, L., Latimer, J.L., Lewis, D.A., and Hansen, E.J. 2001. Investigation of the interaction among the components of the cytolethal distending toxin of *Haemophilus ducreyi*. Biochem. Biophys. Res. Commun. 285: 609–615.

Dorrell, N., Mangan, J.A., Laing, K.G., Hinds, J., Linton, D., Al-Ghusein, H., Barrell, B.G., Parkhill, J., Stoker, N.G., Karlyshev, A.V., Butcher, P.D., and Wren, B.W. 2001. Whole genome comparison of *Campylobacter jejuni* human isolates using a low-cost microarray reveals extensive genetic diversity. Genome Res. 11: 1706–1715.

Elwell, C.A., and Dreyfus, L.A. 2000. DNase I homologous residues in CdtB are critical for cytolethal distending toxin-mediated cell cycle arrest. Mol. Microbiol. 37: 952–963.

Elwell, C., Chao, K., Patel, K., and Dreyfus, L. 2001. *Escherichia coli* CdtB mediates cytolethal distending toxin cell cycle arrest. Infect. Immun. 69: 3418–3422.

Eyigor, A., Dawson, K.A., Langlois, B.E., and Pickett, C.L. 1999a. Detection of cytolethal distending toxin activity and *cdt* genes in *Campylobacter* spp. isolated from chicken carcasses. Appl. Environ. Microbiol. 65: 1501–1505.

Eyigor, A., Dawson, K.A., Langlois, B.E., and Pickett, C.L. 1999b. Cytolethal distending toxin genes in *Campylobacter jejuni* and *Campylobacter coli* isolates: Detection and analysis by PCR. J. Clin. Microbiol. 37: 1646–1650.

Gelfanova, V., Hansen, E.J, and Spinola, S.M. 1999. Cytolethal distending toxin of *Haemophilus ducregi* induces apoptotic death of Jurkat T cells. Infect. Immun. 67: 6394–6402.

Hassane, D.C., Lee, R.B., Mendenhall, M.D., and Pickett, C.L. 2001. Cytolethal distending toxin demonstrates genotoxic activity in a yeast model. Infect. Immun. 69: 5752–5759.

Hassane, D.C., Lee, R.B., and Pickett, C.L. 2003. *Campylobacter jejuni* cytolethal distending toxin promotes DNA repair responses in normal human cells. Infect. Immun. 71: 541–545.

Hickey, T.E., Baqar, S., Bourgeois, A.L., Ewing, C.P., and Guerry, P. 1999. *Campylobacter jejuni*-stimulated secretion of interleukin-8 by INT407 cells. Infect. Immun. 67: 88–93.

Hickey, T.E., McVeigh, A.L., Scott, D.A., Michielutti, R.E., Bixby, A., Carroll, S.A., Bourgeois, A.L., and Guerry, P. 2000. *Campylobacter jejuni* cytolethal distending toxin mediates release of interleukin-8 from intestinal epithelial cells. Infect. Immun. 68: 6535–6541.

Janka, A., Bielaszewska, M., Dobrindt, U., Greune, L., Schmidt, M.A., and Karch, H. 2003. Cytolethal distending toxin gene cluster in enterohemorrhagic *Escherichia coli* O157: H⁻ and O157: H7: Characterization and evolutionary considerations. Infect. Immun. 71: 3634–3638.

Johnson, W.M., and Lior, H. 1987. Response of Chinese hamster ovary cells to a cytolethal distending toxin (CDT) of *Escherichia coli* and possible misinterpretation as heat-labile (LT) enterotoxin. FEMS Microbiol. Lett. 43: 19–23.

Johnson, W.M., and Lior, H. 1988. A new heat-labile cytolethal distending toxin (CLDT) produced by *Campylobacter* spp. Microbial Pathogen. 4: 115–126.

Jones, M.A., Totemeyer, S., Maskell, D.J., Bryant, C.E., and Barrow, P.A. 2003. Induction of proinflammatory responses in the human monocytic cell line THP-1 by *Campylobacter jejuni*. Infect. Immun. 71: 2626–2633.

Kostia, S., Veijalainen, P., Hirvi, U., and Hänninen, M.-L. 2003. Cytolethal distending toxin B gene (*cdtB*) homologues in taxa 2, 3 and 8 and in six canine isolates of *Helicobacter* sp. *flexispira*. J. Med. Microbiol. 52: 103–108.

Lara-Tejero, M., and Galán, J.E. 2000. A bacterial toxin that controls cell cycle progression as a deoxyribonuclease I-like protein. Science. 290: 354–357.

Lara-Tejero, M., and Galán, J.E. 2001. CdtA, CdtB, and CdtC form a tripartite complex that is required for cytolethal distending toxin activity. Infect. Immun. 69: 4358–4365.

Lara-Tejero, M., and Galán, J.E. 2002. Cytolethal distending toxin: limited damage as a strategy to modulate cellular functions. Trends Microbiol. 10: 147–152.

Lee, R.B., Hassane, D.C., Cottle, D.L., and Pickett, C.L. 2003. Interactions of *Campylobacter jejuni* cytolethal distending toxin subunits CdtA and CdtC with HeLa cells. Infect. Immun. 71: 4883–4890.

Li, L., Sharipo, A., Chaves-Olarte, E., Masucci, M.G., Levitsky, V., Thelestam, M., and Frisan, T. 2002. The *Haemophilus ducreyi* cytolethal distending toxin activates sensors of DNA damage and repair complexes in proliferating and non-proliferating cells. Cell. Microbiol. 4: 87–99.

Mao, X., and DiRienzo, J.M. 2002. Functional studies of the recombinant subunits of a cytolethal distending holotoxin. Cell. Microbiol. 4: 245–255.

Mayer, M.P.A., Bueno, L.C., Hansen, E.J., and DiRienzo, J.M. 1999. Identification of a cytolethal distending toxin gene locus and features of a virulence-associated region in *Actinobacillus actinomycetemcomitans*. Infect. Immun. 67: 1227–1237.

Mooney, A., Clyne, M., Curran, T., Doherty, D., Kilmartin, B., and Bourke, B. 2001. *Campylobacter upsaliensis* exerts a cytolethal distending toxin effect on HeLa cells and T lymphocytes. Microbiology 147: 735–743.

Nishikubo, S., Ohara, M., Ueno, Y., Ikura, M., Kurihara, H., Komatsuzawa, H., Oswald, E., and Sugai, M. 2003. An N-terminal segment of the active component of the bacterial genotoxin cytolethal distending toxin B (CDTB) directs CDTB into the nucleus. J. Bacteriol. Chem. 278: 50671–50681.

Ohya, T., Tominaga, K., and Nakazawa, M. 1993. Production of cytolethal distending toxin (CLDT) by *Campylobacter fetus* subsp. *fetus* isolated from calves. J. Vet. Med. Sci. 55: 507–509.

Okuda, J., Fukumoto, M., Takeda, Y., and Nishibuchi, M. 1997. Examination of diarrhoeagenicity of cytolethal distending toxin: Suckling mouse response to the products of the *cdtABC* genes of *Shigella dysenteriae*. Infect. Immun. 65: 428–479.

Parkhill, J., Wren, B.W., Mungall, K., Ketley, J.M., Churcher, C., Basham, D., Chillingworth, T., Davies, R.M., Feltwell, T., Holroyd, S., Jagels, K., Karlyshev, A.V., Moule, S., Pallen, M.J., Penn, C.W., Quail, M.A., Rajandream, M-A., Rutherford, K.M., van Vliet, A.H.M., Whitehead, S., and Barrell, B.G. 2000. The genome sequence of the food-borne pathogen *Campylobacter jejuni* reveals hypervariable sequences. Nature. 403: 665–668.

Pérès, S.Y., Marchès, O., Daigle, F., Nougayrède, J.-P., Hèrault, F., Tasca, C., De Rycke, J., and Oswald, E. 1997. A new cytolethal distending toxin (CDT) from *Escherichia coli* producing CNF2 blocks HeLa cell division in G2/M phase. Mol. Microbiol. 24: 1095–1107.

Pickett, C.L. 2000. *Campylobacter* toxins and their role in pathogenesis. *Campylobacter*, 2nd Ed. Nachamkin, I., and Blaser, M.J., editors. American Society for Microbiology, Washington, D.C.

Pickett, C.L., Cottle, D.C., Pesci, E.C., and Gikah, G. 1994. Cloning, sequencing, and expression of the *Escherichia coli* cytolethal distending toxin genes. Infect. Immun. 62: 1046–1051.

Pickett, C.L., Pesci, E.C., Cottle, D.C., Russell, G., Erdem, A.N., and Zeytin, H. 1996. Prevalence of cytolethal distending toxin production in *Campylobacter jejuni* and relatedness of *Campylobacter* sp. *cdtB* genes. Infect. Immun. 64: 2070–2078.

Purdy, D., Buswell, C.M., Hodgson, A.E., McAlpine, K., Henderson, I., and Leach, S.A. 2000. Characterization of cytolethal distending toxin (CDT) mutants of *Campylobacter jejuni*. J. Med. Microbiol. 49: 473–479.

Scott, D.A., and Kaper, J.B. 1994. Cloning and sequencing of the genes encoding *Escherichia coli* cytolethal distending toxin. Infect. Immun. 62: 244–251.

Shenker, B.J., McKay, T.,Datar, S., Miller, M., Chowhan, R., and Demuth, D. 1999. *Actinobacillus actinomycetemcomitans* immunosuppressive protein is a member of the family of cytolethal distending toxins capable of causing a G_2 arrest in human T cells. J. Immunol. 162: 4773–4780.

Stevens, M.K., Latimer, J.L., Lumbley, S.R., Ward, C.K., Cope, L.D., Lagergard, T., and Hansen, E.J. 1999. Characterization of a *Haemophilus ducreyi* mutant deficient in expression of cytolethal distending toxin. 67: 3900–3908.

Sugai, M., Kawamoto, T., Pérès, S.Y., Ueno, Y., Komatsuzawa, H., Fujiwara, T., Kurihara, H., Suginaka, H., and Oswald, E. 1998. The cell cycle-specific growth- inhibitory factor produced by *Actinobacillus actinomycetemcomitans* is a cytolethal distending toxin. Infect. Immun. 66: 5008–5019.

Thompson, J.D., Higgins, D.G., and Givson, T.J. 1994. CLUSTAL W: improving the sensitivity of progressive multiple sequence alignment through sequence weighting, position-specific gap penalties and weight matrix choice. Nucleic Acids Res. 22: 4673- 4680.

Tóth, I., Hérault, F., Beutin, L., and Oswald, E. 2003. Production of cytolethal distending toxins by pathogenic *Escherichia coli* strains isolated from human and animal sources: Establishment of the existence of a new *cdt* variant (Type IV). J. Clin. Invest. 41: 4285–4291.

Wassenaar, T.M. 1997. Toxin production by *Campylobacter* spp. Clin. Microbiol. Rev. 10: 466–476.

Whitehouse, C.A., Balbo, P.B., Pesci, E.C., Cottle, D.L., Mirabito, P.M., and Pickett, C.L. 1998. *Campylobacter jejuni* cytolethal distending toxin causes a G_2-phase cell cycle block. Infect. Immun. 66: 1934–1940.

Wising, C., Svensson, L.A., Ahmed, H.J., Sundaeus, V., Ahlman, K., Jonsson, I-M., Mölne, L., and Lagergård, T. 2002. Toxicity and immunogenicity of purified *Haemophilus ducreyi* cytolethal distending toxin in a rabbit model. Microb. Pathogen. 33: 49–62.

Young, R.S., Fortney, K.R., Gelfanova, V., Phillips, C.L., Katz, B.P., Hood, A.F., Latimer, J.L., Munson, Jr., Hansen, E.J., and Spinola, S.M. 2001. Expression of cytolethal distending toxin and hemolysin is not required for pustule formation by *Haemophilus ducreyi* in human volunteers. Infect. Immun. 69: 1938–1942.

Young, V.B., Knox, K.A., and Schauer, D.B. 2000. Cytolethal distending toxin sequence and activity in the enterohepatic pathogen *Helicobacter hepaticus*. Infect. Immun. 68: 184–191.

Chapter 21

Interactions of *Campylobacter jejuni* with Non-professional Phagocytic Cells

Brian H. Raphael, Marshall R. Monteville, John D. Klena, Lynn A. Joens and Michael E. Konkel

Abstract

Campylobacteriosis is a multifactorial process involving the organism's translocation of the intestinal epithelium, followed by adherence to host cells, secretion of virulence proteins, and epithelial cell invasion. *In vitro* evidence suggests that *C. jejuni* migrates across polarized epithelial cells, adheres to the basolateral surface of host cells, and invades preferentially at sites of extracellular matrix–host cell contact. Adherence of *C. jejuni* to host cells is mediated by a number of constitutively synthesized factors including PEB1A, JlpA, and the fibronectin binding protein, CadF. Adhesins play a major role in colonization *in vivo* as supported by the inability of a *cadF* mutant to colonize the ceca of experimentally inoculated newly hatched Leghorn chickens. Maximal invasion of epithelial cells requires the secretion of the Cia (*Campylobacter* invasion antigens) proteins. Mutations in *ciaB* reduce the organism's invasiveness and prevent the secretion of the other Cia proteins. Genetic evidence indicates that the Cia proteins are secreted via the flagellar export system and that the minimum secretion-competent structure requires a basal body, hook, and at least one of the filament proteins, FlaA or FlaB. Uptake of *C. jejuni* by epithelial cells requires both microfilaments and microtubules and induces the phosphorylation of paxillin, a focal adhesion molecule. Intracellular calcium also plays a role in *C. jejuni* uptake either by its involvement with microfilament rearrangement or its effects on cell signalling systems. Finally, *C. jejuni* are able to survive intracellularly for extended time periods (> 96 hours), however, the cellular compartment in which the organism resides has not yet been elucidated.

INTRODUCTION

Campylobacteriosis is characterized by fever, diarrhoea accompanied by blood and leucocytes, and abdominal pain. The disease is normally self-limiting and incidence of systemic infection is extremely low (0.4%) (Allos and Blaser, 1995). Invasion of epithelial cells has been documented in both *in vitro* and *in vivo* models. Colonic biopsies from individuals infected with *C. jejuni* (van Spreeuwel *et al.*, 1985) and data obtained from experimentally infected monkeys (Russell *et al.*, 1993) and piglets (Babakhani *et al.*, 1993) have documented the presence of bacteria within surface epithelial cells and in the lamina propria. Fauchere *et al.* (1986) observed that *C. jejuni* isolated from individuals with fever and diarrhoea adhered to cultured cells at a greater efficiency than those cultured from asymptomatic individuals. Invasion of cultured epithelial cells by isolates from individuals with colitis was significantly higher than those recovered from individuals with noninflammatory diarrhoea (Everest *et al.*, 1992). These observations suggest that host cell adherence and invasion are major factors contributing to the ability of *C. jejuni* to cause disease.

This review will focus on the various stages of *C. jejuni*–host cell interactions including the organism's translocation of the epithelial cell barrier, host cell binding with emphasis on the CadF adhesin, and the role of secreted proteins in host cell invasion. Also discussed is the involvement of the host cell's cytoskeletal components and signalling pathways in bacterial uptake and the intracellular survival of *C. jejuni*.

ADHERENCE

The ability of *C. jejuni* to colonize a host is influenced by an isolate's motility, chemotactic properties, surface charge, and the synthesis of multiple adhesins. Adhesins are defined as surface-exposed molecules that facilitate a microorganism's attachment to host cell receptors. *C. jejuni* adhesins are synthesized constitutively as indicated by experiments in which metabolically inactive (heat-killed and sodium azide-killed) *C. jejuni* were found to bind to cultured cells at levels equivalent to metabolically active organisms (Konkel and Cieplak, 1992). In addition, treatment of *C. jejuni* with chloramphenicol, a specific inhibitor of bacterial protein synthesis, had no effect on adherence (Konkel and Cieplak, 1992).

Pei and Blaser (1993) cloned a gene encoding a 28 kDa protein termed PEB1. This protein shares similarity with amino acid transport proteins of other Gram-negative bacteria. Null mutants in *peb1A* show reduced adherence to epithelial cells *in vitro* and colonized the intestine of experimentally infected mice for a shorter duration than the wild-type strain (Pei *et al.*, 1998). The host cell receptor for PEB1 is not known.

A 42.3 kDa surface-exposed lipoprotein is encoded by *jlpA* (Jin *et al.*, 2001). Mutants in *jlpA* were approximately 5-fold less adherent to HEp-2 cells than the wild-type isolate while the invasive ability of such mutants was unaffected. Both purified JlpA protein and JlpA-specific antibodies decreased adherence of *C. jejuni* to HEp-2 cells in a dose-dependent manner. Also, the addition of purified JlpA to HEp-2 cells induced the phosphorylation of IκBα and translocation of NF-κB into the nucleus (Jin *et al.*, 2003). However, NF-κB activation occurred in the *jlpA* mutant, indicating that *C. jejuni* induction of NF-κB activation can occur by alternative mechanisms. JlpA also induced phophorylation of p38 MAP kinase. While the activation of NF-κB and p38 MAP kinase are events leading to the inflammatory response of host cells, the precise chemokines or other cellular products produced in response to JlpA have not be determined. The epithelial cell receptor for JlpA is the heat shock protein, Hsp90α (Jin *et al.*, 2003).

We identified a 37 kDa outer membrane protein in *C. jejuni* termed CadF, for <u>C</u>ampylobacter <u>ad</u>hesion to <u>f</u>ibronectin, that mediates the organism's binding to the extracellular matrix component, fibronectin (Fn) (Konkel *et al.*, 1997). Fn is a 220 kDa glycoprotein that is present at regions of cell-to-cell contact in the gastrointestinal epithelium (Quaroni *et al.*, 1978). *In vitro* binding assays revealed a significant decrease (50%) in the binding and entry of the *C. jejuni* strain F38011 *cadF* mutant to INT 407 cells when compared to the *C. jejuni* F38011 wild-type isolate. The specificity of CadF for Fn was demonstrated by the reduction in binding of the *C. jejuni* F38011 and 81-176 clinical isolates to Fn in the presence of a CadF-specific antibody (Monteville *et al.*, 2003). In the presence of the CadF-specific antibody, the binding of *C. jejuni* F38011 was reduced by 54% and that of *C. jejuni* 81-176 was reduced by 56%. The binding of *C. jejuni* to epithelial cells via CadF also plays a role in host cell signalling pathways. Phosphorylation of paxillin was detected after inoculation of INT 407 cells with wild-type *C. jejuni* F38011 but not after inoculation with the *cadF* mutant indicating an interaction between bacterial adherence via CadF and host cell signalling (Monteville *et al.*, 2003) (see section below; '*C. jejuni* uptake involves host cell signalling events'). Ziprin *et al.* (1999) reported that a *C. jejuni* F38011 *cadF* mutant was unable to colonize the intestinal tract of newly hatched Leghorn chickens ($n=60$), thus providing evidence that CadF plays an *in vivo* role in mediating the organism's binding to the intestinal epithelium. The *cadF* gene is conserved amongst *C. jejuni* and *C. coli* isolates (Konkel *et al.*, 1999).

Other molecules proposed to act as adhesins include the flagellum (Pavlovskis *et al.*, 1991; Wassenaar *et al.*, 1991; Grant *et al.*, 1993; Wassenaar *et al.*, 1993), lipooligosaccharide (LOS) (McSweegan and Walker, 1986), the major outer membrane protein (MOMP, also called OmpE) (Moser *et al.*, 1997; Schroder and Moser, 1997), and a protein termed P95 (Kelle *et al.*, 1998).

Mutants in the *N*-linked protein glycosylation pathway (*pglB* and *pglE*) of *C. jejuni* 81-176 have reduced levels of adherence and invasion (Szymanski *et al.*, 2002). As host cell binding is a prerequisite for invasion and some bacterial surface structures are glycosylated (capsule, LOS, flagellin), reduced invasion by protein glycosylation mutants may be an effect of reduced adherence.

C. JEJUNI SECRETION AND HOST CELL INVASION

The ability of *C. jejuni* to enter, survive, and replicate in mammalian cells has been studied extensively using tissue culture models (Newell *et al.*, 1985; De Melo *et al.*, 1989; Konkel and Joens, 1989; Konkel *et al.*, 1990; Wassenaar *et al.*, 1991; Everest *et al.*, 1992; Konkel *et al.*, 1992a; Konkel *et al.*, 1992b; Grant *et al.*, 1993; Oelschlaeger *et al.*, 1993; Yao *et al.*, 1994; Pei *et al.*, 1998; Konkel *et al.*, 1999). Newell *et al.* (1985) found that *C. jejuni* environmental isolates were much less invasive than clinical isolates as determined by immunofluorescence and electron microscopic examination of infected HeLa cells. Unlike pathogens known to constitutively synthesize proteins to induce their own uptake, evidence indicates that *C. jejuni* must engage in a differential synthetic response to facilitate its uptake. More specifically, chloramphenicol, a specific inhibitor of bacterial protein synthesis, retards the internalization of *C. jejuni* by host cells (Konkel and Cieplak, 1992; Oelschlaeger *et al.*, 1993). In addition, metabolically inactive (sodium azide-killed) *C. jejuni* are not internalized (Konkel and Cieplak, 1992). In subsequent work, we demonstrated by both one- and two-dimensional electrophoretic analyses that *C. jejuni* cultured in the presence and absence of epithelial cells exhibited different protein profiles. More specifically, *C. jejuni* cultured in the presence of epithelial cells synthesized a number of proteins, either exclusively or preferentially, that were not synthesized by *C. jejuni* cultured in tissue culture medium alone (Konkel and Cieplak, 1992; Konkel *et al.*, 1993). The newly synthesized proteins were unique from those proteins induced by thermal stress (Konkel *et al.*, 1998). Concurrent with early work revealing the differential synthetic response of *C. jejuni*, Panigrahi *et al.* (1992) reported that *C. jejuni* grown in a rabbit ileal loop model synthesized a set of proteins that were not produced when the organism was cultured on standard laboratory medium. Two of the newly synthesized proteins, with apparent molecular masses of 84 and 47 kDa, were detectable using convalescent sera from *C. jejuni*-infected individuals. Combined, these results suggest that *C. jejuni* synthesize entry-promoting proteins upon exposure to host cells and that these proteins are part of a coordinated bacterial response to the epithelial cell environment.

Studies in our laboratory have demonstrated that a set of the proteins synthesized by *C. jejuni* in response to co-cultivation with eukaryotic cells are also secreted. By differentially screening *C. jejuni* F38011 genomic DNA-phage expression libraries with antisera generated in rabbits against *C. jejuni* F38011 cultured in the presence of INT 407 cells (Cj+INT) and *C. jejuni* F38011 cultured in the absence of INT 407 cells (Cj-INT), plaques were identified that reacted only with the Cj+INT antiserum. This screening resulted in the identification of a gene termed *ciaB* (Konkel *et al.*, 1999). *In vitro* assays revealed that a *C. jejuni ciaB* null mutant bound to INT 407 cells in numbers equal to or greater than the wild-type isolate, but exhibited a significant reduction (~100-fold) in internalization. Moreover, confocal microscopy studies using a CiaB-specific antiserum revealed an intense fluorescence signal in the cytoplasm of the *C. jejuni*-infected INT 407 cells but not in the cytoplasm of mock-infected INT 407 cells. Based on these observations, the secreted proteins were termed Cia (s), for <u>C</u>ampylobacter <u>i</u>nvasion <u>a</u>ntigens.

Metabolic labelling experiments showed that Cia protein synthesis and secretion are separable, and that secretion is the rate-limiting step of these processes (Rivera-Amill *et al.*, 2001). Secretion can be induced in the absence of host-cell contact by the addition of fetal bovine serum (FBS) (Figure 21.1). Additional work indicated that Cia protein synthesis is induced in response to the bile salt deoxycholate and various eukaryotic host cell components, whereas the latter is capable of also inducing Cia protein secretion.

We propose that the incubation of *C. jejuni* in the presence of 0.1% sodium deoxycholate, which is similar to the concentration of bile salts found in the lumen of the small intestine (Spiro, 1983; Elliott, 1985), provides an environment that partially mimics the *in vivo* environment encountered by *C. jejuni*. Culturing *C. jejuni* on plates supplemented with sodium deoxycholate retards the inhibitory effect of chloramphenicol on *C. jejuni* invasion as judged by the gentamicin-protection assay (Rivera-Amill *et al.*, 2001). These data suggest that the coordinate expression of the genes encoding the Cia proteins is subject to environmental regulation. In addition, *C. jejuni* pre-cultured with deoxycholate immediately invade epithelial cells at maximal levels while bacteria cultured in the absence of deoxycholate show a lag in host cell invasion (M.E. Konkel *et al.*, unpublished data).

Biochemical analysis of the Cia proteins has been problematic since induction of secretion requires the organism's co-cultivation with host cells or the addition of FBS to the culture medium. More specifically, FBS in the culture medium hinders purification of the supernatant proteins in a large enough quantity for N-terminal sequencing or antibody generation. We have determined that picogram quantities of the Cia proteins are secreted per bacterium during a three hour period by comparing the protein content of the labelling medium including bacteria versus medium without bacteria (M.E. Konkel *et al.*, unpublished data). To date, the function of the Cia proteins and their host cell substrates is not known.

Because the Cia proteins are induced and secreted in response to an environmental stimulus and are not processed, they conform to the criteria of type III proteins (Kubori *et al.*, 1998). However, the only type III export system encoded by *C. jejuni* NCTC 11168 is the flagellar apparatus (Parkhill *et al.*, 2000). Consequently, experiments were performed to test the hypothesis that the flagellum is used for Cia export. Secretion of virulence

(a)   (b)

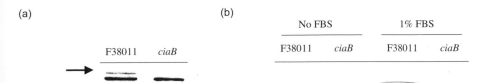

Figure 21.1 Immunoblot of *C. jejuni* whole cell lysates (a) and supernatant fluids (b) using a CiaB-specific antibody generated in a New Zealand white rabbit. Shown are *C. jejuni* strain F38011 and a *ciaB*::kan^R mutant. Arrow points to a band corresponding to CiaB (73 kDa).

proteins via the flagellar apparatus has been reported for a variety of bacteria including *Yersinia enterocolitica* (Young *et al.*, 1999) and *Bacillus thuringiensis* (Ghelardi *et al.*, 2002). To determine if the flagellar export apparatus could serve to export Cia proteins, a number of flagellar biosynthetic genes were mutated. A mutation was initially generated in the *flhB* flagellar export gene of both *C. jejuni* F38011 and 81-176. Additional mutations were generated in genes encoding the flagellar basal body (*flgB* and *flgC*), hook (*flgE2*), and filament cap (*fliD*) genes in *C. jejuni* strain F38011. Mutations that affected either the export of flagellar components or the non-filament structural components resulted in a secretion-negative (sec⁻) phenotype (Konkel *et al.*, 2004). Comparable results were obtained using a second *C. jejuni* strain, 81116, in which a double mutation affecting both of the flagellin genes (*flaA-flaB-*) was generated.

We have also performed assays to determine the contribution of motility and Cia protein secretion in *C. jejuni* uptake. Independent of the *C. jejuni* isolate used, a correlation existed between Cia protein export and invasion. More specifically, a significant (*t*-test: $P < 0.01$) decrease was noted in the internalization efficiency of the *C. jejuni* flagellar protein export apparatus (*flhB*) mutant and flagellar basal body (*flgB* and *flgC*), hook (*flgE2*) and filament (*flaA⁻ flaB⁻*) structural mutants; these mutants were all non-motile and did not secrete the Cia proteins (mot⁻, sec⁻). We found that assembly of a filament composed exclusively of either FlaA or FlaB was sufficient for Cia secretion from the *C. jejuni* flagellar apparatus, and that a *C. jejuni flaA⁻* (*flaB⁺*) mutant (mot⁻, sec⁺) was more invasive than a *C. jejuni flaA⁻ flaB⁻* mutant.

Several groups of investigators have demonstrated the importance of bacterial motility, and the product of the *flaA* gene, in *C. jejuni* colonization (Newell *et al.*, 1985; Pavlovskis *et al.*, 1991; Nachamkin *et al.*, 1993; Wassenaar *et al.*, 1993). Motility and/or the product of the *flaA* gene have also been found to be essential for both the invasion of eukaryotic cells and the translocation of polarized cell monolayers by *C. jejuni* (Wassenaar *et al.*, 1991; Grant *et al.*, 1993). Both Grant *et al.* (1993) and Wassenaar *et al.* (1991) reported differences in the invasive potential of the *C. jejuni* 81116 *flaA⁻* (*flaB⁺*) and *flaA⁻ flaB⁻* isolates, with the former strain being more invasive. Based on the difference in the invasive potential of the *C. jejuni* strains, it was concluded that the flagellar structure played a role in internalization that was independent of motility. We propose that the difference in the invasiveness of the *C. jejuni* 81116 *flaA⁻* (*flaB⁺*) and *C. jejuni* 81116 *flaA⁻ flaB⁻* strains is the result of Cia secretion.

Complementation of a *C. jejuni* 81116 *flaA⁻flaB⁻* mutant with a shuttle vector harbouring either the *flaA* or *flaB* gene resulted in transformants that displayed characteristics similar to that of the *C. jejuni* 81116 wild-type strain or the *C. jejuni* 81116 *flaA⁻* (*flaB⁺*) strain, respectively. However, transformation of the *C. jejuni* 81116 *flaA⁻ flaB⁻* mutant with a shuttle plasmid harbouring the *flaA* or *flaB* gene did not fully restore Cia secretion or the percent of bacteria internalized to the wild-type levels. A possible explanation for the diminished secretion by complemented strains, and their corresponding reduction in internalization efficiency, is that overexpression of FlaA or FlaB within a cell may interfere with Cia secretion.

Since binding is a prerequisite for host cell invasion, the invasive potential of *C. jejuni* isolates is variable. Therefore, we transformed the data from binding and internalization assays to an I/A ratio ([number of internalized bacteria]/[number of cell-associated bacteria] × 100) to minimize the influence of an organism's binding potential on entry. We observed that a *C. jejuni* mot⁻, sec+ strain (*flaA⁻* (*flaB⁺*)) had an I/A ratio that was consistently one-half log unit of that seen with the wild-type strain (mot⁺, sec⁺). In addition, we observed that the mot⁻, sec⁻ strain (F38011 *flgB⁻*, *flgC⁻* and *flgE2⁻*) and mot⁺, sec⁻ strain (F38011 *ciaB⁻*) had an I/A ratio 1.5 to 2 log units less than the mot⁺, sec⁺ wild-type strain.

To investigate whether there was biological relevance to the I/A ratio, piglets were infected with *C. jejuni* F38011 and its *ciaB⁻* derivative (Konkel *et al.*, 2001). The *C. jejuni ciaB* mutant harbouring the pMEK100 shuttle vector, which contains the entire *ciaB* gene from *C. jejuni* F38011, was used to ensure that the phenotype associated with the *C. jejuni ciaB* mutant was attributable to the mutation in the *ciaB* gene. *C. jejuni* were detected in the faeces of all pigs, except *E. coli* DH5α inoculated piglets, within 24 hours of inoculation and their excretion persisted throughout the experiment (Table 21.1). The piglets inoculated with the *C. jejuni* F38011 wild-type and complemented *C. jejuni ciaB* isolates developed diarrhoea within a 24 hour period, whereas diarrhoea did not develop in the piglets infected with the *C. jejuni ciaB* mutant until three days post-inoculation.

The small and large intestines of the *C. jejuni* and *E. coli* infected piglets were examined at the macroscopic and microscopic levels for the presence of lesions. Gross histopathological lesions were confined to the small intestine of each of the *C. jejuni* inoculated piglets that developed symptoms. The most severe gross lesions, characterized by oedema and haemorrhage, were observed in the intestines of piglets infected with either the wild-type isolate or complemented *ciaB* mutant (*ciaB*::kan^R (*ciaB⁺*)). The presence of microscopic lesions correlated with the observations made by gross examination of the tissues (Figure 21.2). The lesions consisted of congested blood vessels and damaged host cells as evidenced by villi blunting and the presence of a cellular exudate. While these data indicate that bacterial determinants other than the Cia proteins are sufficient for the development of campylobacteriosis, the Cia proteins contribute to the severity of *Campylobacter*-mediated enteritis.

Table 21.1 Experimental infection of piglets.

		C. jejuni			
	E. coli	M129	F38011	*ciaB*::kan^R [a]	*ciaB*::kan^R (*ciaB⁺*)[a]
Diarrhoea, onset	None	<24h	< 24 h	>72h	<24h
Deaths	0/2	3/5	2/5	0/6	3/6
Gross lesions[b]	0/2	3/4	2/4	1/6	2/4
Microscopic lesions[c]	0/2	2/4	1/4	1/6	2/4

[a]Mutant and complemented strain constructed in *C. jejuni* F38011.
[b]Gross lesions = presence of oedema and/or heemorrhage in the small or large intestine. Animals that died overnight were not necropsied due to concerns of postmortem histological changes.
[c]Microscopic lesions = presence of necrosis and/or haemorrhage.

TRANSLOCATION

The ability of pathogens to migrate across a cell barrier can be an important virulence attribute as it permits a pathogen to access the underlying tissues and to disseminate throughout a host (Finlay and Falkow, 1990; Cruz *et al.*, 1994; Kops *et al.*, 1996; Nataro *et al.*, 1996). The *in vivo* relevance of translocation by *C. jejuni* is unclear since it is unknown whether the organism must translocate the intestinal epithelial cells in order to gain access to deeper tissues. *C. jejuni* were reported to migrate across the intestinal barrier via M cells in a rabbit ileal loop model (Everest *et al.*, 1993) suggesting that translocation can occur *in vivo*. *In vitro*, *C. jejuni* translocate a Caco-2 polarized cell monolayer without a concomitant loss in transepithelial electrical resistance (TER) (Konkel *et al.*, 1992c; Grant *et al.*, 1993; Bras and Ketley, 1999). Stability of *C. jejuni*-inoculated cells indicates that the integrity of the cell monolayer remains intact. Everest *et al.* (1992) and Harvey *et al.* (1999) found no correlation between an isolate's invasive and translocation potential. This observation suggested that the genes encoding the products responsible for invasion in *C. jejuni* are distinct from those that confer translocation ability.

In vitro, the ability of *C. jejuni* to translocate epithelial cell monolayers is independent of M cells. To address the specific route of epithelial cell monolayer translocation, we initially performed a binding and internalization experiment with a *C. jejuni ciaB* (non-invasive) mutant and T84 polarized cells (Monteville and Konkel, 2002). Consistent with the work of Everest *et al.* (1992) and Harvey *et al.* (1999), we did not observed

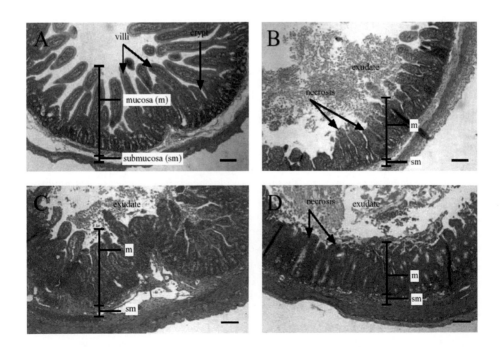

Figure 21.2 Hematoxylin and eosin stained sections of the small intestines of *Escherichia coli*- and *C. jejuni*-inoculated piglets. Tissue samples were acquired from *C. jejuni*-inoculated piglets exhibiting gross lesions. Note the presence of villi blunting and exudates (B–D) resulting from tissue necrosis. (A) *E. coli*-inoculated piglet; (B) *C. jejuni* F38011-inoculated piglet; (C) *C. jejuni ciaB* mutant-inoculated piglet; and (D) *C. jejuni ciaB* complemented-inoculated piglet. Bar = 100 µm.

a difference in the kinetics of translocation for the *C. jejuni* F38011 wild-type isolate, *ciaB* mutant, and a *trans*-complemented *ciaB* mutant. Since translocation ability was not dependent on the organism's invasive potential, it appears that *C. jejuni* translocates an intact cell monolayer via a paracellular (between cells) rather than a transcellular (through a cell) route. Supporting the possibility that *C. jejuni* could migrate across a cell barrier via a paracellular route with only a transient change in cellular integrity, others have reported that tight junctions temporally relax to allow regulated passage of both solutes and neutrophils (Madara, 1998) and can reseal following penetration by certain pathogens (Takeuchi, 1967).

Modifications were made in the standard translocation assay to assess whether *C. jejuni* that are initially bound to polarized cells migrate into the basolateral medium or remain cell-associated. Briefly, one set of T84 polarized cells was inoculated with the *C. jejuni* clinical isolates F38011 and 81-176 and incubated for four hours, after which time the monolayers were rinsed and cell-association quantified. A second set of *C. jejuni*-inoculated polarized cells was rinsed after the four hour incubation period and placed in a 24-well plate containing fresh media. *C. jejuni* present in the basolateral medium were then quantified hourly over the course of a three hour period. Following this incubation, cell-associated *C. jejuni* were determined. The results of this assay revealed that the majority of the cell-associated *C. jejuni* remained cell-associated rather than translocating to the basolateral chamber. More specifically, 96.9% of *C. jejuni* F38011 and 85.2% of *C. jejuni* 81-176 that were associated with the polarized cell monolayers after four hours remained cell-associated after the additional three hour incubation period. Thus, it appears that while certain strains of *C. jejuni* possess the ability to translocate across a cell monolayer, the migration of the organism from the apical (e.g. lumen) to the basolateral (e.g. lamina propria) chamber may not be, *per se*, reflective of the organism's preference.

Based on the finding that *C. jejuni* preferentially remain cell-associated, assays were conducted to compare the translocation kinetics of the *C. jejuni* F38011 wild-type isolate and an isogenic *cadF* mutant (Monteville and Konkel, 2002). Experiments were performed to determine host-cell association at different multiplicities of infection (MOI). At a MOI of 300, both *C. jejuni* isolates translocated across the T84 polarized cell monolayers, into the basolateral medium, with approximately the same efficiency. As the MOI decreased from 300 to 3, a significant decrease was noted in the ability of *C. jejuni* F38011 to translocate into the basolateral medium compared to the *cadF* mutant. We speculate that the *C. jejuni* wild-type strain did not translocate to the basolateral medium as efficiently as the *C. jejuni cadF* mutant at the lowest MOI because the former bound to Fn. It is possible that at a higher MOI, a difference was not apparent in the translocation of the two strains because the accessible host-cell receptors to which *C. jejuni* bind are occupied. To confirm that Fn serves as a host-cell receptor for *C. jejuni*, competitive cell-association assays were performed using T84 polarized cells. *C. jejuni* 81-176 was not significantly inhibited from associating with the polarized monolayer in the presence of 50 and 90-fold excess of the *cadF* mutant. In contrast, *C. jejuni* F38011 significantly reduced the association of *C. jejuni* 81-176 with the polarized cells in a dose-dependent manner. These data further support the notion that the binding of *C. jejuni* to Fn promotes bacterial-host cell association.

Given that the *C. jejuni* F38011 and 81-176 wild-type strains were found to compete with one another for cell-association (adherence and internalization) whereas the *C. jejuni cadF* isolate was incapable of competing with the *C. jejuni* 81-176 wild-type isolate for cell-association, additional experiments were performed to determine whether adherence and internalization could be retarded by blocking receptors with Fn-specific antibodies. Because Fn predominates at cell-substrate contacts, monolayers were pre-treated with EGTA prior to infection to expose the extracellular matrix (ECM). When antibodies against Fn were added to the non-EGTA treated polarized cells there was no effect on *C. jejuni* F38011 adherence and invasion, presumably due to the lack of antibody access to Fn localized at the basolateral surface. However, pretreatment of the T84 cell monolayers with EGTA resulted in a significant increase in the number of adherent (185%) and internalized (367%) bacteria compared to non-EGTA treated monolayers. Since a 1.8-fold increase was observed in binding compared to a 3.6-fold increase in internalization, it appears that a greater proportion of the bacteria bound to receptors involved in *C. jejuni* uptake. No difference in invasion was noted in *Salmonella typhimurium* invasion of EGTA-treated and untreated T84 host cells. Inclusion of Fn-specific antibodies with the infection inoculum following EGTA treatment of T84 polarized cells significantly decreased adherence and invasion from 185% to 35% and 367% to 67%, respectively. Noteworthy is that the addition of Fn-specific antibodies following EGTA treatment of the polarized cells resulted in a proportional decrease in adherence (5.3-fold decrease) and internalization (5.5-fold decrease).

PARTICIPATION OF THE HOST CELL IN *C. JEJUNI* UPTAKE

We have found that *C. jejuni* clinical isolates F38011 and 81-176 bind to Fn via the CadF adhesin. Because Fn is associated with microfilaments at focal adhesions, we examined the role of microfilaments in *C. jejuni* uptake. While the host cell cytoskeletal components involved in *C. jejuni* F38011 uptake had not been previously examined, *C. jejuni* 81-176 was reported to be internalized via a novel pathway exclusively involving microtubules (Oelschlaeger *et al.*, 1993; Hu and Kopecko, 1999; Bacon *et al.*, 2000; Kopecko *et al.*, 2001). To the best of our knowledge, *C. jejuni* 81-176 is the only isolate to date that has been reported to be internalized exclusively via a microtubule-dependent pathway (Oelschlaeger *et al.*, 1993; Kopecko *et al.*, 2001). In these studies, extracellular bacteria were killed using a 2-hour incubation with $100 \mu g/mL$ of gentamicin. In our hands, treatment of *C. jejuni* 81-176 with $100 \mu g/mL$ of gentamicin for 2 hours typically resulted in the recovery of 3 to 12% of the bacteria in the original suspension. Thus, it was found necessary to increase both the time of exposure and the concentration of the antibiotic to ensure bacterial death.

Using $250 \mu g/mL$ of gentamicin for 3 hours to kill the extracellular bacteria, we noted a decrease in the invasiveness of *C. jejuni* 81-176 with a microtubule inhibitor and microfilament inhibitors. The reduction of *C. jejuni* uptake with the microfilament inhibitor, cytochalasin D, appeared specific as treatment of the INT 407 cells with this drug had no effect on INT 407 cell viability as judged by staining with trypan blue. Moreover, the effect of cytochalasin D, which inhibits actin polymerization and transient integrin-stimulated FAK activation (Schlaepfer *et al.*, 1999), was reversible. Mycalolide B inhibits microfilament polymerization by severing microfilaments and sequestering G-actin. Treatment of INT 407 cells with mycalolide B also inhibited the uptake of both *C. jejuni* isolates. Biswas *et al.* (2000) also observed a reduction in *C. jejuni* entry with both microfilament and microtubule inhibitors with every clinical isolate tested (*n*=9). Moreover, the investigators observed that the most invasive isolates examined utilized microfilaments (Biswas *et al.*, 2000). Based on the data currently available, it appears that the uptake of most *C. jejuni* isolates involves cooperation of both microfilaments and microtubules.

The mechanism by which treatment of cells with microtubule inhibitors causes a reduction in *C. jejuni*-cell uptake is not known. Microtubules are known to regulate the turnover of focal adhesion contacts and modulate a cell's adhesive strength to the ECM (Ballestrem *et al.*, 2000). Also, treatment of cells with microtubule-inhibitors leads to an increase in a cell's adherence to the ECM (Ballestrem *et al.*, 2000; Sastry and Burridge, 2000). Thus, the effect of a microtubule inhibitor, such as nocodazole, on *C. jejuni*-cell uptake may be indirect, as the turnover of the focal adhesion sites is retarded. Alternatively, following the uptake of *C. jejuni* using microfilaments, microtubules could be required for the intracellular trafficking of the organism to the interior of the cell as has been proposed for *Citrobacter freundii* (Badger *et al.*, 1999). Regardless, there is clearly functional cooperation between host cytoskeletal elements (Ballestrem *et al.*, 2000; Goode *et al.*, 2000; Sastry and Burridge, 2000). We also found that the effects of cytochalasin D and nocodazole on *C. jejuni* uptake were not additive when used together, thus, it appears that the two *C. jejuni* isolates examined in our study utilized microfilaments and microtubules in concert.

We found that the uptake of *C. jejuni* 81-176 and F38011 by INT 407 cells was reduced by 56 and 66%, respectively, in the presence of nocodazole and cytochalasin D relative to the untreated control. These results are consistent with those reported elsewhere, even though the experimental protocols varied (Oelschlaeger *et al.*, 1993; Bacon *et al.*, 2000). What is unclear is why in the presence of both inhibitors a significant number of *C. jejuni* are still internalized. In comparison, *Salmonella* invasion was reduced by greater than 99% by cytochalasin D compared to untreated INT 407 cells. Comparable results for *Salmonella* invasion under these conditions have been reported by others (Oelschlaeger *et al.*, 1993; Hu and Kopecko, 1999; Bacon *et al.*, 2000; Biswas *et al.*, 2000).

C. JEJUNI UPTAKE INVOLVES HOST CELLS SIGNALLING EVENTS

Published work indicates a relationship between pericellular fibronectin and the cytoskeleton (Gumbiner, 1996; Miyamoto *et al.*, 1998; van der Flier and Sonnenberg, 2001). In mammalian cells, the actin cytoskeleton is necessary for a variety of cellular processes including control of cell-to-cell and cell-to-substrate interactions. Actin nucleation occurs at membrane-associated sites called focal adhesions. Focal adhesions are the sites at which the bundles of actin filaments (stress fibres) are cross-linked with membrane-associated adhesion molecules (e.g. integrins) and extracellular molecules (Gumbiner, 1996; Sarkar, 1999). The integrins are transmembrane glycoprotein receptors comprising heterodimeric $\alpha\beta$ subunits (Hynes, 1992; Vuori, 1998; Danen and Yamada, 2001; van der Flier and Sonnenberg, 2001). The two subunits are noncovalently associated with one another (1:1) in the membrane. There are eighteen known α subunits and eight known β subunits (van der Flier and Sonnenberg, 2001). Different α and β subunit combinations dictate the specificity of cell-to-cell and cell-to- ECM recognition. The $\alpha_5\beta_1$ integrin receptor specifically binds Fn (Hynes, 1992; Miyamoto *et al.*, 1998). The integrin molecules bind extracellularly to matrix components and intracellularly associate with protein complexes consisting of vinculin, talin, α-actinin, paxillin, tensin, zyxin, and focal adhesion kinase (FAK) (Tachibana *et al.*, 1995; Miyamoto *et al.*, 1998). Integrin occupancy and clustering is associated with tyrosine phosphorylation of cellular cytoplasmic proteins including FAK and paxillin, and is a means of regulating host signal transduction events leading to actin rearrangements (Tachibana *et al.*, 1995; Miyamoto *et al.*, 1998).

Wooldridge *et al.* (1996) reported that *C. jejuni* uptake is reduced in the presence of protein tyrosine phosphorylation inhibitors. We chose to examine whether paxillin was phosphorylated upon infection of INT 407 cells with *C. jejuni* because protein tyrosine phosphorylation is one of the earliest events upon integrin stimulation (Clark and Brugge, 1995). Consistent with the idea that the binding of *C. jejuni* leads to integrin stimulation, an increase in phosphorylated paxillin was observed between 30 and 45 minutes following *C. jejuni* F38011 infection. Noteworthy is that the increase in phosphorylated paxillin occurred just prior to and concomitant with an increase in *C. jejuni* internalization (Monteville *et al.*, 2003). In contrast to the *C. jejuni* F38011 wild-type strain, an increase was not observed in phosphorylated paxillin over the course of the assay with cells inoculated with the *C. jejuni* F38011 *cadF* mutant at a MOI of 100. However, upon infection of INT 407 cells with a *C. jejuni* F38011 *cadF* mutant at a MOI of greater than 2000, the pattern of phosphorylated paxillin in cells inoculated with the *C. jejuni* F38011 *cadF* mutant mirrored that obtained with the *C. jejuni* wild-type strain. A possible explanation for this finding was that *C. jejuni* adherence is multifactorial, and that several adhesins simultaneously function to promote the host cell binding, after which cell signalling events are stimulated. Thus, CadF does not appear to be required to induce host cell signalling events, but appears to promote signalling events by facilitating the organism's binding to appropriate host cell receptors. Watarai *et al.* (1996) observed an increase in the tyrosine phosphorylation of FAK and paxillin 20 and 30 minutes after infection of Chinese hamster ovary cells with *Shigella flexneri*.

Investigators have reported bacterial invasion into non-professional phagocytic cells is accompanied by a transient increase in cytoplasmic concentrations of calcium (Pace *et al.*, 1993; Tran Van Nhieu *et al.*, 2003). This increase in intracellular calcium, $[Ca^{2+}]_i$, may initially result via extracellular calcium entry through receptor-operated calcium channels, and secondarily from release of calcium from intracellular pools following synthesis and hydrolysis of phosphorylated phosphatidylinositol lipids (Janmey, 1994). Intracellular calcium subsequently acts as a global messenger within a cell triggering multiple cellular processes including cytoskeletal rearrangement. Calcium regulated actin-binding proteins (e.g. gelsolin) are involved in severing of filamentous actin, nucleation of globular actin to filamentous actin, as well as contribute to Ca^{2+}-dependent phosphorylation of myosin light chain by either myosin light chain kinase or protein kinase C. Thus, an increase in $[Ca^{2+}]_i$, may lead to the rearrangement of the host cytoskeleton, which is required for bacterial endocytosis.

Based on the indirect interaction of calcium and microfilaments, assays were performed to determine if invasion of *C. jejuni* into INT 407 cells required $[Ca^{2+}]_i$. Briefly, INT 407 cells were treated with the extracellular calcium chelating agent BAPTA ($15\,\mu M$) prior to infection with *C. jejuni* F38011. Treatment with BAPTA resulted in an increase in bacterial adherence to the host cells and a disproportional increase (*t* test: $P \le 0.01$) in *C. jejuni* uptake (Table 21.2). These data are in agreement with that previously observed upon treatment of T84 cultured cells with the chelating agent EGTA (Monteville and Konkel, 2002). We speculate that the observed increase in adherence and invasion in the presence of the extracellular calcium chelator BAPTA is a result of increased exposure of the extracellular matrix prior to infection. These data suggest that extracellular calcium is not critical for uptake of *C. jejuni* by cultured epithelial cells. In contrast, treatment of INT 407 cells with the intracellular calcium chelating agent BAPTA-AM ($15\,\mu M$) had no effect on binding of *C. jejuni* to host cells but did result in a significant decrease in bacterial uptake. Therefore, intracellular calcium appears essential in the uptake of *C. jejuni* by cultured non-professional phagocytic cells. The precise role of intracellular calcium on *C. jejuni* uptake is unknown. We speculate that rearrangement of the cytoskeleton, specifically microfilaments, in the absence of intracellular calcium would be impaired and thus result in a significant decrease in bacterial uptake. Nevertheless, calcium acts as global messenger within a cell and influences additional signalling processes.

To address whether other host cell signalling processes are involved in *C. jejuni* uptake, inhibitors of specific signalling molecules were tested (Table 21.2). In addition to BAPTA-AM, inhibitors of protein kinase A, protein kinase C, phosphatidylinositol 3-kinase, and mitogen-activated protein kinase resulted in a significant decrease in *C. jejuni* uptake while having no effect on bacterial adherence. These data further support the idea that the role of intracellular calcium on *C. jejuni* uptake may extend beyond strictly promoting cytoskeletal rearrangement and impact additional signalling processes required for bacterial endocytosis.

SURVIVAL OF *C. JEJUNI* WITHIN EPITHELIAL CELLS

To study the survival of *C. jejuni* within nonprofessional phagocytes, investigators have modified the basic gentamicin-protection assay. Briefly, *C. jejuni* infected epithelial cells are allowed to incubate for extended periods after the initial antibiotic treatment (250 µg/ mL). In one study, medium containing a low level of gentamicin (50 µg/mL) was added after the initial three hour antibiotic treatment to ensure killing of extracellular bacteria (De Melo *et al.*, 1989). With the four *C. jejuni* isolates tested, the investigators observed a steady decrease in the number of bacteria within the HEp-2 cells throughout a 36 hour time course experiment (De Melo *et al.*, 1989). Using a modified protocol where antibiotic-free medium was added to *C. jejuni* infected INT 407 cells after the initial period of antibiotic treatment (i.e. 250 µg/mL of gentamicin for 3 hours followed by 10 µg/mL gentamicin for 18 hours), three *C. jejuni* isolates were found to survive throughout a 96-hour period. In addition, there was an increase in the number of intracellular bacteria at 72 hours post-

Table 21.2 *C. jejuni* F38011 adherence and invasion in the presence of calcium chelating agents and inhibitors of signalling proteins

Treatment[a]	Adherence (%)[b]	Internalization (%)[c]
Untreated	100 ± 13	100 ± 12
H-89 (PKA inhibitor)	89 ± 2	39 ± 12^{d}
Calphostin C (PKC inhibitor)	102 ± 17	17 ± 5^{d}
PD 098,059 (MEKK inhibitor)	81 ± 9	62 ± 9^{d}
Wortmannin (PI-3K inhibitor)	123 ± 24	3 ± 0.4^{d}
BAPTA (extracellular Ca^{2+} chelator)	114 ± 26	196 ± 14^{d}
BAPTA-AM (intracellular Ca^{2+} chelator)	88 ± 20	29 ± 4^{d}

[a]INT 407 cells were pretreated with various reagents 45 min prior to infection with *C. jejuni* F38011.
[b]Determined after 30 min incubation period. Monolayers were rinsed three times with PBS and lysed with 0.25% (w/v) sodium deoxycholate. Viable bacteria were determined by direct plate counts. Values represent the percentages of adherent bacteria ± standard deviation relative to the untreated control.
[c] Determined after 3 h incubation period followed by an additional 3 hr incubation in the presence of a bactericidal concentration of gentamicin. Viable bacteria were enumerated as outlined above. Values represent the percentages of adherent bacteria ± standard deviation relative to the untreated control.
[d]The value was significantly different (P-value < 0.01) from that obtained using *C. jejuni* F38011 in the absence of treatment.

infection (Konkel *et al.*, 1992b). Such an increase in intracellular bacteria may not be solely due to intracellular replication since bacteria could be released allowing re-infection of epithelial cells. Consistent with this possibility, vacuoles containing single and multiple bacteria were observed by TEM in infected cell cultures at 72 hours. Survival declined after 72 hours, which was coincident with both a decrease in the number and viability of the epithelial cells. In both the De Melo *et al.* (1989) and Konkel *et al.* (1992b) studies, *C. jejuni* were observed within membrane-bound vacuoles of non-professional phagocytes as judged by transmission electron microscopy examination of the infected cells.

Several detoxification genes are encoded by the *C. jejuni* genome that may mediate intracellular survival of this organism. Catalase, encoded by *katA*, and alkyl hydroperoxide reductase, encoded by *aphC*, are involved in the oxidative stress response. Both genes are members of the PerR regulon (van Vliet *et al.*, 1999). PerR, a homologue of Fur, is an iron-responsive negative regulator of oxidative stress response genes. In the presence of iron, PerR inhibits transcription of *ahpC* and *katA*. In *C. jejuni*, a *katA* insertional mutant was more sensitive to hydrogen peroxide than the wild-type strain, M129 (Day *et al.*, 2000). Interestingly, the ability of a *katA* mutant to survive in epithelial cells was equal to that of the wild-type, however, the mutant was significantly inhibited for survival in J774A.1 murine macrophages. Intra-macrophage survival of the *katA* mutant was rescued when either the oxidative burst or nitric oxide synthesis was inhibited. Although an *ahpC* mutant was more sensitive to cumin hydroperoxide and was less aerotolerant than the wild-type strain, studies on its ability to survive intracellularly have not been performed (Baillon *et al.*, 1999).

C. jejuni also encodes a gene (*sodB*) for an iron containing superoxide dismutase. SodB converts the highly toxic superoxide radical to hydrogen peroxide and oxygen. A *sodB* null mutation was inhibited for survival within epithelial cells by approximately 10-fold (Pesci *et al.*, 1994). Unexpectedly, this mutant survived similarly to the wild-type strain in J774A.1 macrophages (L.A. Joens, unpublished data).

SUMMARY

We propose a model of *C. jejuni* pathogenesis in which bacteria migrate to the basolateral surface of the host cell and bind fibronectin via CadF. Uptake of *C. jejuni* at focal adhesion sites may be facilitated by type III effectors that induce cell signalling events leading to reorganization of the cytoskeleton. Uptake into an endocytic vesicle involves the accumulation of microfilaments at the focal adhesion site. Subsequent trafficking of the *Campylobacter* containing vesicle may involve both microfilaments and microtubles. The mechanism of survival of *C. jejuni* within the cell is an area of active investigation but may involve modification of the endosomal compartment or the expression of detoxification factors.

Progress in understanding *C. jejuni* pathogenesis has accelerated in part due to the availability of new genetic tools (transposons, reporter genes, etc.), genome sequences, and methods such as microarrays to study the differential expression of genes. The relative scarcity of known virulence factors in the annotated genome sequence implies that *C. jejuni* may have novel mechanisms of pathogenesis. Characterization of the repertoire of genes with unknown function will play a key role in defining the unique mechanisms of pathogenesis of this organism.

Acknowledgments

We thank Brent House for review of the manuscript. Work in the Konkel lab is supported by grants from the NIH (DK58911) and USDA National Research Initiative Competitive Grants Program (USDA/NRICGP 99–35201–8579).

References

Allos, B.M., and Blaser, M.J. 1995. *Campylobacter jejuni* and the expanding spectrum of related infections. Clin. Infect. Dis. 20: 1092–1099.

Babakhani, F.K., Bradley, G.A., and Joens, L.A. 1993. Newborn piglet model for campylobacteriosis. Infect. Immun. 61: 3466–3475.

Bacon, D.J., Alm, R.A., Burr, D.H., Hu, L., Kopecko, D.J., Ewing, C.P., Trust, T.J., and Guerry, P. 2000. Involvement of a plasmid in virulence of *Campylobacter jejuni* 81-176. Infect. Immun. 68: 4384–4390.

Badger, J.L., Stins, M.F., and Kim, K.S. 1999. *Citrobacter freundii* invades and replicates in human brain microvascular endothelial cells. Infect. Immun. 67: 4208–4215.

Baillon, M.L., van Vliet, A.H., Ketley, J.M., Constantinidou, C., and Penn, C.W. 1999. An iron-regulated alkyl hydroperoxide reductase (AhpC) confers aerotolerance and oxidative stress resistance to the microaerophilic pathogen *Campylobacter jejuni*. J. Bacteriol. 181: 4798–4804.

Ballestrem, C., Wehrle-Haller, B., Hinz, B., and Imhof, B.A. 2000. Actin-dependent lamellipodia formation and microtubule-dependent tail retraction control-directed cell migration. Mol. Biol. Cell. 11: 2999–3012.

Biswas, D., Itoh. K., and Sasakawa, C. 2000. Uptake pathways of clinical and healthy animal isolates of *Campylobacter jejuni* into INT-407 cells. FEMS Immunol. Med. Microbiol. 29: 203–211.

Bras, A. and Ketley, J. 1999. Transcellular translocation of *Campylobacter jejuni* across human polarised epithelial cell monolayers. FEMS Microbiol. Lett. 179: 209–215.

Clark, E.A. and Brugge, J.S. 1995. Integrins and signal transduction pathways: the road taken. Science 268: 233–239.

Cruz, N., Lu, Q., Alvarez, X., and Deitch, E.A. 1994. Bacterial translocation is bacterial species dependent: results using the human Caco-2 intestinal cell line. J. Trauma 36: 612–616.

Danen, E.H. and Yamada, K.M. 2001. Fibronectin, integrins, and growth control. J. Cell. Physiol. 189: 1–13.

Day, W.A., Sajecki, J.L., Pitts, T.M., and Joens, L.A. 2000. Role of catalase in *Campylobacter jejuni* intracellular survival. Infect. Immun. 68: 6337–6345.

De Melo, M., Gabbiani, G., and Pechere, J. 1989. Cellular events and intracellular survival of *Campylobacter jejuni* during infection of HEp-2 cells. Infect. Immun. 57: 2214–2222.

Elliott, W.H. 1985. Metabolism of bile acids in liver and extrahepatic tissues. In: Sterols and Bile Acids. eds. H. Danielsson and J. Sjovall. New York, Elsevier North-Holland Biomedical Press, pp. 303–329.

Everest, P.H., Butzler, J.P., Lloyd, D., Knutton, S., Ketley, M., and Williams, P.H. 1992. Differentiated Caco-2 cells as a model for enteric invasion of *Campylobacter jejuni* and *C. coli*. J. Med. Microbiol. 37: 319–325.

Everest, P.H., Goossens, H., Sibbons, P., Lloyd, D.R., Knutton, S., Leece, R., Ketley, J.M., Williams, P.H. 1993. Pathological changes in the rabbit ileal loop model caused by *Campylobacter jejuni* from human colitis. J. Med. Microbiol. 38: 316–321.

Fauchere, J.L., Veron, M., Moyen, E.N., Richard, S., and Pfister, A. 1986. Association with HeLa cells of *Campylobacter jejuni* and *Campylobacter coli* isolated from human feces. Infect. Immun. 54: 283–287.

Finlay, B.B. and Falkow, S. 1990. *Salmonella* interactions with polarized human intestinal Caco-2 epithelial cells. J. Infect. Dis. 162: 1096–1106.

Ghelardi, E.F., Celandroni, F., Salvetti, D.J., Beecher, M., Gominet, D., Lereclus, A.C., Wong, L., and Senesi, S. 2002. Requirement of *flhA* for swarming differentiation, flagellin export, and secretion of virulence-associated proteins in *Bacillus thuringiensis*. J. Bacteriol. 184: 6424–6433.

Goode, B.L., Drubin, D.G., and Barnes, G. 2000. Functional cooperation between the microtubule and actin cytoskeletons. Curr. Op. Cell. Biol. 12: 63–71.

Grant, C.C., Konkel, M.E., Cieplak, W., and Tompkins, L.S. 1993. Role of flagella in adherence, internalization, and translocation of *Campylobacter jejuni* in nonpolarized and polarized epithelial cell cultures. Infect. Immun. 61: 1764–1771.

Gumbiner, B.M. 1996. Cell adhesion: the molecular basis of tissue architecture and morphogenesis. Cell. 84: 345–357.

Harvey, P., Battle, T. and Leach, S. 1999. Different invasion phenotypes of *Campylobacter* isolates in Caco-2 cell monolayers. J. Med. Microbiol. 48: 461–469.

Hu, L. and Kopecko, D.J. 1999. *Campylobacter jejuni* 81-176 associates with microtubules and dynein during invasion of human intestinal cells. Infect. Immun. 67: 4171–4182.

Hynes, R.O. 1992. Integrins: versatility, modulation, and signaling in cell adhesion. Cell. 69: 11–25.

Janmey, P.A. 1994. Phosphoinositides and calcium as regulators of cellular actin assembly and disassembly. Annu. Rev. Physiol. 56: 169–191.

Jin, S., Joe, A., Lynett, J., Hani, E.K., Sherman, P., and Chan, V.L. 2001. JlpA, a novel surface-exposed lipoprotein specific to *Campylobacter jejuni*, mediates adherence to host epithelial cells. Mol. Microbiol. 39: 1225–1236.

Jin, S., Song, Y.C., Emili, A., Sherman, P.M., and Chan, V.L. 2003. JlpA of *Campylobacter jejuni* interacts with surface-exposed heat shock protein 90α and triggers signalling pathways leading to the activation of NF-κB and p38 MAP kinase in epithelial cells. Cell. Microbiol. 5: 165–174.

Kelle, K., Pages, J.M., and Bolla, J.M. 1998. A putative adhesin gene cloned from *Campylobacter jejuni*. Res. Microbiol. 149: 723–733.

Konkel, M.E., Babakhani, F., and Joens, L.A. 1990. Invasion-related antigens of *Campylobacter jejuni*. J. Infect. Dis. 162: 888–895.

Konkel, M.E. and Cieplak, W. 1992. Altered synthetic response of *Campylobacter jejuni* to cocultivation with human epithelial cells is associated with enhanced internalization. Infect. Immun. 60: 4945–4949.

Konkel, M.E., Corwin, M.D., Joens, L.A., and Cieplak, W. 1992a. Factors that influence the interaction of *Campylobacter jejuni* with cultured mammalian cells. J. Med. Microbiol. 37: 30–37.

Konkel, M.E., Garvis, S.G., Tipton, S.L., Anderson, D.E., and Cieplak, W. 1997. Identification and molecular cloning of a gene encoding a fibronectin-binding protein (CadF) from *Campylobacter jejuni*. Mol. Microbiol. 24: 953–963.

Konkel, M.E., Grey, S.A., Kim, B.J., Garvis, S.G., and Yoon, J. 1999. Identification of the enteropathogens *Campylobacter jejuni* and *Campylobacter coli* based on the *cadF* virulence gene and its product. J. Clin. Microbiol. 37: 510–517.

Konkel, M.E., Hayes, S.F., Joens, L.A., and Cieplak, W. 1992b. Characteristics of the internalization and intracellular survival of *Campylobacter jejuni* in human epithelial cell cultures. Microb. Pathogen. 13: 357–370.

Konkel, M.E. and Joens, L.A. 1989. Adhesion to and invasion of HEp-2 cells by *Campylobacter* spp. Infect. Immun. 57: 2984–2990.

Konkel, M.E., Kim, B.J., Klena, J.D., Young, C.R., and Ziprin, R. 1998. Characterization of the thermal stress response of *Campylobacter jejuni*. Infect. Immun. 66: 3666–3672.

Konkel, M.E., Kim, B.J., Rivera-Amill, V., and Garvis, S.G. 1999. Bacterial secreted proteins are required for the internaliztion of *Campylobacter jejuni* into cultured mammalian cells. Mol. Microbiol. 32: 691–701.

Konkel, M.E., Mead, D.J. and Cieplak, W. 1993. Kinetic and antigenic characterization of altered protein synthesis by *Campylobacter jejuni* during cultivation with human epithelial cells. J. Infect. Dis. 168: 948–954.

Konkel, M.E., Mead, D.J., Hayes, S.F., and Cieplak, W. 1992c. Translocation of *Campylobacter jejuni* across human polarized epithelial cell monolayer cultures. J. Infect. Dis. 166: 308–315.

Konkel, M.E., Monteville, M.R., Rivera-Amill, V., and Joens, L.A. 2001. The pathogenesis of *Campylobacter jejuni*-mediated enteritis. Curr. Issues Intest. Microbiol. 2: 55–71.

Konkel, M.E., Klena, J.D., Rivera-Amill, R., Monteville, M.R., Biswas, D., Raphael, B. and Mickelson, J. 2004. Secretion of virulence proteins from *Campylobacter jejuni* is dependent on a functional Flagellar export apparatus. J. Bacteriol. 186: 3296–3303.

Kopecko, D.J., Hu, L., and Zaal, K.J. 2001. *Campylobacter jejuni* – microtubule-dependent invasion. Trends Microbiol. 9: 389–396.

Kops, S.K., Lowe, D.K., Bement, W.M., and West, A.B. 1996. Migration of *Salmonella typhi* through intestinal epithelial monolayers: an *in vitro* study. Microbiol. Immunol. 40: 799–811.

Kubori, T., Matsushima, Y., Nakamura, D., Uralil, J., Lara-Tajero, M., Sukhan, A., Galan, J.E., and Aizawa, S.I. 1998. Supramolecular structure of the *Salmonella typhimurium* type III protein secretion systems. Science 280: 602–605.

Madara, J.L. 1998. Regulation of the movement of solutes across tight junctions. Annu. Rev. Physiol. 60: 143–159.

McSweegan, E. and Walker, R.I. 1986. Identification and characterization of two *Campylobacter jejuni* adhesins for cellular and mucous substrates. Infect. Immun. 53: 141–148.

Miyamoto, S., Katz, B.Z., LaFrenie, R.M., and Yamada, K.M. 1998. Fibronectin and integrins in cell adhesion, cell signaling, and morphogenesis. Ann. N. Y. Acad. Sci. 857: 119–129.

Monteville, M.R. and Konkel, M.E. 2002. Fibronectin-facilitated invasion of T84 eukaryotic cells by *Campylobacter jejuni* occurs preferentially at the basolateral cell surface. Infect. Immun. 70: 6665–6671.

Monteville, M.R., Yoon, J.E., and Konkel, M.E. 2003. Maximal adherence and invasion of INT 407 cells by *Campylobacter jejuni* requires the CadF outer-membrane protein and microfilament reorganization. Microbiol. 149: 153–165.

Moser, I., Schroeder, W. and Salnikow, J. 1997. *Campylobacter jejuni* major outer membrane protein and a 59-kDa protein are involved in binding to fibronectin and INT 407 cell membranes. FEMS Microbiol. Lett. 157: 233–238.

Nachamkin, I., Yang, X.H., and Stern, N.J. 1993. Role of *Campylobacter jejuni* flagella as colonization factors for three-day-old chicks: analysis with flagellar mutants. Appl. Environ. Microbiol. 59: 1269–1273.

Nataro, J.P., Hicks, S., Phillips, A.D., Vial, P.A. and Sears, C.L. 1996. T84 cells in culture as a model for enteroaggregative *Escherichia coli* pathogenesis. Infect. Immun. 64: 4761–4768.

Newell, D.G., McBride, H., Saunders, F., Dehele, Y., and Pearson, A.D. 1985. The virulence of clinical and environmental isolates of *Campylobacter jejuni*. J. Hyg. 94: 45–54.

Oelschlaeger, T.A., Guerry, P., and Kopecko, D.J. 1993. Unusual microtubule-dependent endocytosis mechanisms triggered by *Campylobacter jejuni* and *Citrobacter freundii*. Proc. Natl. Acad. Sci. USA 90: 6884–6888.

Pace, J., Hayman, M.J., and Galan, J.E. 1993. Signal transduction and invasion of epithelial cells by *S. typhimurium*. Cell. 72: 505–514.

Panigrahi, P., Losonsky, G., DeTolla, L.J., and Morris, J.G. 1992. Human immune response to *Campylobacter jejuni* proteins expressed *in vivo*. Infect. Immun. 60: 4938–4944.

Parkhill, J., Wren, B.W., Mungall, K., Ketley, J.M., Churcher, C., Basham, D., Chillingworth, T., Davies, R.M., Feltwell, T., Holroyd, S., Jagels, K., Karlyshev, A.V., Moule, S., Pallen, M.J., Penn, C.W., Quail, M.A., Rajandream, M.A., Rutherford, K.M., van Vliet, A.H., Whitehead, S., and Barrell, B.G. 2000. The genome sequence of the food-borne pathogen *Campylobacter jejuni* reveals hypervariable sequences. Nature. 403: 665–668.

Pavlovskis, O.R., Rollins, D.M., Haberberger, R.L., Green, A.E., Habash, L., Strocko, S., and Walker, R.I. 1991. Significance of flagella in colonization resistance of rabbits immunized with *Campylobacter* spp. Infect. Immun. 59: 2259–2264.

Pei, Z. and Blaser, M.J. 1993. PEB1, the major cell-binding factor of *Campylobacter jejuni*, is a homologue of the binding component in gram-negative nutrient transport systems. J. Biol. Chem. 268: 18717–18725.

Pei, Z., Burucoa, C., Grignon, B., Baqar, S., Huang, Z.X., Kopecko, D.J., Bourgeois, A.L., Fauchere, J.L., and Blaser, M.J. 1998. Mutation in the *peb1A* locus of *Campylobacter jejuni* reduces interactions with epithelial cells and intestinal colonization of mice. Infect. Immun. 66: 938–943.

Pesci, E., Cottle, D., and Pickett, C. 1994. Genetic, enzymatic, and pathogenic studies of the iron superoxide dismutase of *Campylobacter jejuni*. Infect. Immun. 62: 2687–2694.

Quaroni, A., Isselbacher, K.J., and Ruoslahti, E. 1978. Fibronectin synthesis by epithelial crypt cells of rat small intestine. Proc. Natl. Acad. Sci. USA 75: 5548–5552.

Rivera-Amill, V., Kim, B.J., Seshu, J., and Konkel, M.E. 2001. Secretion of the virulence-associated *Campylobacter* invasion antigens from *Campylobacter jejuni* requires a stimulatory signal. J. Infect. Dis. 183: 1607–1616.

Russell, R.G., O'Donnoghue, M., Blake, D.C., Zulty, J., and DeTolla, L.J. 1993. Early colonic damage and invasion of *Campylobacter jejuni* in experimentally challenged infant *Macaca mulatta*. J. Infect. Dis. 168: 210–215.

Sarkar, A. 1999. Focal adhesions. Curr. Biol. 9: R428.

Sastry, S.K. and Burridge, K. 2000. Focal adhesions: a nexus for intracellular signaling and cytoskeletal dynamics. Exp. Cell. Res. 261: 25–36.

Schlaepfer, D.D., Hauck, C.R., and Sieg, D.J. 1999. Signaling through focal adhesion kinase. Prog. Biophys. Mol. Biol. 71: 435–478.

Schroder, W. and Moser, I. 1997. Primary structure analysis and adhesion studies on the major outer membrane protein of *Campylobacter jejuni*. FEMS Microbiol. Lett. 150: 141–147.

Spiro, H.M. 1983. Clinical gastroenterology. New York, Macmillan Publishing Co., Inc.

Szymanski, C.M., Burr, D.H., and Guerry, P. 2002. *Campylobacter* protein glycosylation affects host cell interactions. Infect. Immun. 70: 2242–2244.

Tachibana, K., Sato, T., D'Avirro, N., and Morimoto, C. 1995. Direct association of pp125^{FAK} with paxillin, the focal adhesion-targeting mechanism of pp125^{FAK}. J. Exp. Med. 182: 1089–1100.

Takeuchi, A. 1967. Electron microscope studies of experimental *Salmonella* infection. I. Penetration into the intestinal epithelium by *Salmonella typhimurium*. Am. J. Pathol. 50: 109–136.

Tran Van Nhieu, G., Clair, C., Bruzzone, R., Mesnil, M., Sansonetti, P., and Combettes, L. 2003. Connexin-dependent inter-cellular communication increases invasion and dissemination of *Shigella* in epithelial cells. Nat. Cell. Biol. 5: 720–726.

van der Flier, A. and Sonnenberg, A. 2001. Function and interactions of integrins. Cell. Tissue. Res. 305: 285–298.

van Spreeuwel, J.P., Duursma, G.C., Meijer, C.J., Bax, R., Rosekrans, P.C., and Lindeman, J. 1985. *Campylobacter* colitis: histological immunohistochemical and ultrastructural findings. Gut. 26: 945–951.

van Vliet, A.H., Baillon, M.L., Penn, C.W., and Ketley, J.M. 1999. *Campylobacter jejuni* contains two fur homologues: characterization of iron-responsive regulation of peroxide stress defense genes by the PerR repressor. J. Bacteriol. 181: 6371–6376.

Vuori, K. 1998. Integrin signaling: tyrosine phosphorylation events in focal adhesions. J. Membr. Biol. 165: 191–199.

Wassenaar, T.M., Bleumink-Pluym, N.M., and van der Zeijst, B.A. 1991. Inactivation of *Campylobacter jejuni* flagellin genes by homologous recombination demonstrates that *flaA* but not *flaB* is required for invasion. EMBO J. 10: 2055–2061.

Wassenaar, T.M., van der Zeijst, B.A., Ayling, R., and Newell, D.G. 1993. Colonization of chicks by motility mutants of *Campylobacter jejuni* demonstrates the importance of flagellin A expression. J. Gen. Microbiol. 139: 1171–1175.

Watarai, M., Funato, S., and Sasakawa, C. 1996. Interaction of Ipa proteins of *Shigella flexneri* with $\alpha_5\beta_1$ integrin promotes entry of the bacteria into mammalian cell. J. Exp. Med. 183: 991–999.

Wooldridge, K.G., Williams, P.H., and Ketley, J.M. 1996. Host signal transduction and endocytosis of *Campylobacter jejuni*. Microb. Pathog. 21: 299–305.

Yao, R., Burr, D.H., Doig, P., Trust, T.J., Niu, H., and Guerry, P. 1994. Isolation of motile and non-motile insertional mutants of *Campylobacter jejuni*: the role of motility in adherence and invasion of eukaryotic cells. Mol. Microbiol. 14: 883–893.

Young, G.M., Schmiel, D.H., and Miller, V.L. 1999. A new pathway for the secretion of virulence factors by bacteria: the flagellar export apparatus functions as a protein-secretion system. Proc. Natl. Acad. Sci. USA 96: 6456–6461.

Ziprin, R.L., Young, C.R., Stanker, L.H., Hume, M.E., and Konkel, M.E. 1999. The absence of cecal colonization of chicks by a mutant of *Campylobacter jejuni* not expressing bacterial fibronectin-binding protein. Avian Dis. 43: 586–589.

Chapter 22

Campylobacter jejuni Interactions with Professional Phagocytes

Lynn Joens

Abstract

Campylobacter jejuni interacts with inflammatory cells during the course of infection. In experimental *Campylobacter* enteritis, large numbers of infiltrating leucocytes are observed in the lamina propria. *In vitro*, *C. jejuni* taken up by macrophages survive for extended periods of time (>72 h). At least one bacterial product, catalase (KatA), is required for optimal intra-macrophage survival of *C. jejuni*. However, variation in intracellular survival between environmental and clinical isolates of *C. jejuni* suggests the presence of additional bacterial survival enhancing factors. While the majority of *C. jejuni* phagocytosed by J774A.1 macrophages appear to co-localize with markers of the normal endocytic pathway which becomes acidified, a subpopulation may exist in a modified endosome. However, it is unclear which of these populations survive for extended periods. Finally, bacterial induced macrophage killing appears to occur by a process involving both apoptosis and necrosis, with the latter occurring in a bacterial dose-dependent manner.

INTRODUCTION

Campylobacter jejuni is an enteric pathogen which infects the small and large intestine of susceptible hosts. The bacterium invades the intestinal epithelium of the host and thereby gains access to deeper tissue. It is in the deeper tissue of the lamina propria and submucosa that *Campylobacter* comes into contact with inflammatory cells such as the macrophage. Electron microscopy and light microscopy studies have demonstrated the presence of the organism in granulocytes, parenchymal cells and mononuclear cells located within the lamina propria and submucosa of an infected host (Ruiz-Palacios *et al.*, 1981; Newell *et al.*, 1984; Kiehlbauch *et al.*, 1985; van Spreeuwel *et al.*,1985; Russel *et al.*, 1989; Babakhani *et al.*, 1993). While this might allow a pathogen to propagate and disseminate throughout the host, interactions between the macrophage and *C. jejuni* likely remain confined to the lamina propria since *Campylobacter* bacteraemia is notably infrequent. An intracellular niche is thought to shelter microorganisms from immune surveillance.

STAGES OF *CAMPYLOBACTER* COLITIS

van Spreeuwel *et al.* (1985) divided the histology of *Campylobacter* colitis in man into three stages and found abundant numbers of macrophage as well as other inflammatory cells infiltrating into colonic tissue. In the first stage, as represented by an onset of bloody diarrhoea, they reported the presence of crypt abscesses filled with granulocytes. The lamina propria appeared oedematous and infiltrated with granulocytes, histiocytes, and lymphocytes with histiocytes and granulocytes the most dominant. The second stage of infection was characterized by a reduction in inflammation, an oedematous lamina propria with an inflammatory infiltrate of mononuclear cells. In the third stage of the disease, inflammatory changes were minimal or absent, with an increase population of lymphocytes and plasma cells, a response indicating recovery of the patient. Non-human primates also develop colitis, which is similar to experimentally infected humans, with a diffuse infiltration of neutrophils, plasma cells and lymphocytes in the colonic lamina propria and a moderate to diffuse infiltration of monocytes in the submucosa (Russel *et al.*, 1989). Clearly, as noted by this work, the histiocyte/macrophage plays an important role in the inflammatory response of primates to infection with *C. jejuni,* but this cell type may also play a role in the overall survival of the bacterium and a source for continuous infection of the host.

IN VIVO EVIDENCE FOR *C. JEJUNI*–MACROPHAGE INTERACTION

In non-primates infected with *C. jejuni,* intestinal infiltrates consisting of macrophages and neutrophils have also been observed. Ruiz-Palacios *et al.* (1981) described a moderate oedema in the colonic lamina propria of chickens inoculated with *C. jejuni* along with an infiltration of mononuclear cells. The presence of *Campylobacter* inside the phagocytic cells was confirmed at 24h post-infection by immunofluorescence. The newborn piglet serves as a model which mimics human campylobacteriosis when orally infected with clinical isolates of *C. jejuni*. Electron microscopy studies of infected porcine intestinal tissue have identified macrophages and neutrophils in the lamina propria and submucosa within three days of infection, with numerous cells containing internalized *C. jejuni* (Babakhani *et al.*, 1993). These results correlate with those of the primate host and demonstrate an interaction of *C. jejuni* with the host macrophage. In addition, intraperitoneal injection of BALB/c mice with *C. jejuni* synthesizing Gfp were found to be associated with CD11b$^+$ Gr-1$^-$ lavage macrophages as early as four hours after infection and remained associated with this cell population for more than 30h (Mixter *et al.*, 2003).

C. JEJUNI UPTAKE BY MACROPHAGES

Numerous laboratory studies have been conducted on the uptake and survival of *C. jejuni* in macrophages. Uptake of *C. jejuni* by macrophages appears to be independent of complement opsonization, but is strain and host dependent. Avirulent strains of *C. jejuni* defined by their inability to invade the chorioallantoic membrane of chicken embryos were cleared *in vivo* and engulfed *in vitro* by peritoneal macrophages from intravenously infected Balb/c mice at a significantly higher rate than virulent strains (Myszewski and Stern, 1991). Resistance to phagocytosis was also demonstrated by *C. jejuni* strains exposed to guinea pig resident peritoneal macrophages, while avirulent *C. coli* strains were engulfed at significantly higher numbers (Banfi *et al.*, 1986). The authors suggested that the differences in uptake of the two strains were due to the presence of an antiphagocytic capsular-like material in *C. jejuni* that was absent in *C. coli* (Banfi *et al.*, 1986). In contrast, Wassenaar *et al.* (1997) reported a linear correlation between infective dose and internalized *C. jejuni* in monocytes from human blood donors. Supporting results were obtained by Myszewski and Stern (1991) when they examined the uptake of two different strains of *C. jejuni* by chicken peritoneal macrophages. *C. jejuni* was readily internalized by chicken macrophages within a 30 min incubation period with increased internalization when serum or macrophages from previously colonized hosts were added to the assay. Day *et al.* (2000) demonstrated a significant uptake of a clinical isolate of *C. jejuni* M129 in mouse and porcine peritoneal macrophages as well as a mouse macrophage cell line (J774A.1) independent of serum or complement. These results indicate that uptake is strain and host dependent and probably increases in the presence of immune serum.

Recently, Jones *et al.* (2003) reported the cytokine response of the THP-1 human macrophage cell line to *C. jejuni* infection. Proinflammatory cytokines interleukin 1 alpha, IL-1 beta, IL-6, IL-8, and tumour necrosis factor alpha were secreted in response to either live or killed bacteria.

C. JEJUNI SURVIVAL IN MACROPHAGES

Survival of *C. jejuni* in host macrophages for an extended period of time may contribute to the severity of the disease, longevity of clinical signs and the reoccurrence of campylobacteriosis. However, survival studies have been inconsistent and may reflect strain, assay, and host cell differences. Wassenaar *et al.* (1997) examined the ability of 16 isolates of *C. jejuni* to resist killing by activated human peripheral blood monocytes. Survival assays conducted for 72 h demonstrated the killing of *C. jejuni* by 90% of the donors' monocytes within 24–48 h, with *C. jejuni* surviving in 10% of the monocytes. Myszewski and Stern (1991) examined the ability of a high passage clinical isolate and a poultry isolate to resist killing by macrophages. They found that both isolates were killed by chicken peritoneal macrophages within a 6 h period. Similar findings were found by Joens (L. A. Joens, unpublished) in examining the survival of five poultry isolates of *C. jejuni* when internalized by a mouse macrophage cell line (J774A.1). Two of the six isolates were inactivated at 24 h and the remaining three isolates were killed within 72 h after engulfment (Pitts, 2001). Yet other studies have demonstrated the survival of *C. jejuni* in macrophages. Kiehlbauch *et al.* (1985) examined the survival of a clinical isolate of *C. jejuni* in a J774G8 murine macrophage cell line, resident BALB/c peritoneal macrophages, and human peritoneal blood monocytes. The researchers were able to recover *C. jejuni* from the three phagocytic cell types over a 6 day period. The microorganism was able to replicate within the phagosomes and survive for a longer period of time than *C. jejuni* grown in medium alone. Day *et al.* (2000) examined the survival of a *C. jejuni* clinical isolate in various macrophages. A clinical isolate, M129, was able to survive for 72 h in porcine and murine peritoneal macrophages and in the J774A.1 macrophage cell line. Nonetheless, there was a noticeable reduction in the number of *C. jejuni* recovered from the three phagocytic cell types at 72 h post-inoculation. Joens (L. A. Joens, unpublished) confirmed the above findings by demonstrating the survival of clinical isolates 81-176 and F38011 in a J774A.1 macrophage cell line over a 72 h period. These findings support the view that strain differences, rather than the host macrophage, is the critical factor involved in *C. jejuni* intracellular survival and that isolates obtained from clinical cases appear to survive at a higher rate than environmental or poultry isolates.

BACTERIAL PRODUCTS INFLUENCING ANIMAL INTRA-MACROPHAGE SURVIVAL

Many bacteria are able to circumvent antigen processing by altering the intracellular trafficking of the phagolysosome, preventing acidification, or inhibiting the effects of phagolysosome toxic radicals. This allows bacteria to occupy a modified phagosome and create a niche which provides bacterial nutrients for continued existence and shelter from the host immune system. Reactive oxygen species such as superoxide and hydrogen peroxide from the respiratory burst of the phagolysosome can be extremely toxic to bacteria. But bacteria express defensive mechanisms which can circumvent these effects. *Campylobacter* encodes superoxide dismutase (*sodB*) and catalase (*katA*) that may allow *C. jejuni* to survive the respiratory burst. Unlike *Salmonella*, the synthesis of superoxide dismutase by *C. jejuni* had no effect on intra-macrophage survival (Joens, unpublished data). The *sodB* mutant of strain 81-176 was able to survive for over 72 h in a J774A.1 macrophage cell line at a level equal to that of the parent strain (L.A. Joens, unpublished). In contrast, catalase production was shown to be an important factor in *C. jejuni* intra-macrophage survival. A *katA* mutant of *C. jejuni* was susceptible to killing by a J774A.1 macrophage cell line, whereas the parent strain *C. jejuni* M129 remained viable after 72 h of internalization (Day *et al.*, 2000). However, if the respiratory burst, or production of nitric oxide, is inhibited in J774A.1 cells, than the number of bacteria recovered from cells inoculated with the *C. jejuni katA* mutant was equal to that of the parent strain at 72 h (Day *et al.*, 2000).

C. JEJUNI TRAFFICKING WITHIN MACROPHAGES

Salmonella interferes with antigen processing and the endocytic pathway by preventing the maturation of the phagolysosome. Pitts (2001) examined the effect of phagolysosome development of the J774A.1 macrophage following engulfment of either a virulent clinical isolate or avirulent environmental isolate of *C. jejuni*. Macrophages were infected with the two *C. jejuni* strains at a multiplicity of infection of ~10 bacteria per macrophage and the phagosome examined by confocal microscopy for the presence of early and late proteins (Rab5 and Rab7, respectively) during a 72 h time period. These experiments demonstrated that approximately 80–85% of internalized virulent *C. jejuni* co-localized with early and late proteins of the endocytic pathway. The pH determination using Lysotracker (manufacturer) = (Invitrogen, Eugene, OR, USA) as an indicator demonstrated that 90% of the phagolysosomes containing *C. jejuni* underwent normal acidification. The majority of the avirulent environmental *C. jejuni* were processed through

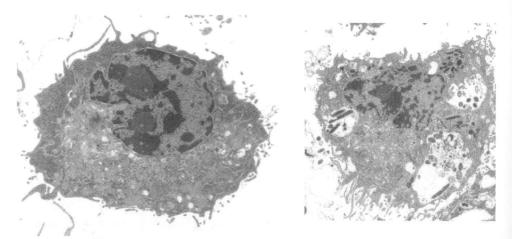

Figure 22.1 TEM photograph of control (left) and *C. jejuni* infected J774A.1 cells (right) at 24 h post-exposure. Changes in cells infected with *C. jejuni* were due to necrosis as identified by annexin V–FITC and propidium iodide staining.

the normal endocytic pathway. This finding indicates that the majority of phagocytosed *C. jejuni* undergo normal processing through the endocytic pathway, but a small fraction of isolates do not co-localize with endocytic proteins and perhaps live intracellularly in modified mononuclear phagocytes. It is still unclear, which of the two populations survive prolonged incubation.

C. JEJUNI INDUCE MACROPHAGE NECROSIS AND APOPTOSIS
Cell death occurs as a natural process through apoptosis or through necrosis from cytopathic effects. Invasive bacteria can induce death in the macrophage through either apoptosis or necrosis or through a combination of the two mechanisms (Figure 22.1). Konkel and Mixter (2000) examined cell death of professional phagocytes containing internalized *C. jejuni*. They demonstrated that THP-1 monocytes underwent apoptosis as a direct result of infection with *C. jejuni*. Joens (L.A. Joens, unpublished) demonstrated both apoptosis and necrosis in human U937 phagocytic cells infected with clinical isolates of *C. jejuni*. In contrast, murine macrophages infected with the same isolates were killed through necrosis as determined by measuring the release of the cytoplasmic enzyme, lactate dehydrogenase, in a dose-dependent fashion (Figure 22.2). These results indicate that the mechanism by which the macrophage dies following infection with *C. jejuni* may be more host dependent than bacterial strain dependent since human macrophages undergo apoptosis at higher levels and earlier time points than murine derived macrophages.

Summary
Although the data indicate a role for the macrophage in *C. jejuni* survival in the host, this assumption will need to be confirmed by molecular biology experiments. Identification of the gene (s) involved with survival and macrophage cell death will allow us to produce isogenic mutants to test this hypothesis in the newborn piglet model system.

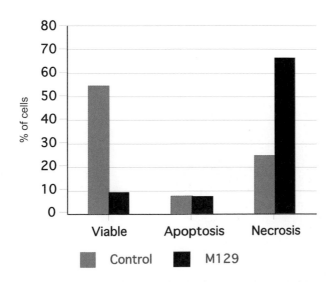

Figure 22.2 Flow cytometry measurement of cell death in J774A.1 cells uninfected or infected with *C. jejuni* strain M129. Cell death of Infected J774.A1 cells was attributed more so to necrosis rather then apoptosis.

References

Babakhani, F.K. and Joens, L.A. 1993. Primary intestinal cells as a model for studying *Campylobacter jejuni* invasiveness. Infect. Immun. 61: 2723–2726.

Banfi, E., Cinco, M., and Zabucchi, G. 1986. Phagocytosis of *Campylobacter jejuni* and *C. coli* by peritoneal macrophages. J. Gen. Microbiol. 132: 2409–2412.

Day, W.A., Jr., Sajecki, J.L., Pitts, T. M., and Joens, L.A. 2000. Role of catalase in *Campylobacter jejuni* intracellular survival. Infect. Immun. 68: 6337–6345.

Jones M.A., Totemeyer, S., Maskell, D.J., Bryant, C.E., and Barrow, P.A. 2003. Induction of proinflammatory responses in the human monocytic cell line THP-1 by *Campylobacter jejuni*. Infect Immun. 71: 2626–33.

Kiehlbauch, J.A., Albach, R.A., Baum, L.L., and Chang, K.P. 1985. Phagocytosis of *Campylobacter jejuni* and its intracellular survival in mononuclear phagocytes. Infect. Immun. 48: 446–451.

Konkel, M.E., and Mixter, P.F. 2000. Flow cytometric detection of host cell apoptosis induced by bacterial infection. Methods Cell Sci. 22: 209–215.

Mixter, P.F., Klena, J.D., Flom, G.A., Siegesmund, A.M., and Konkel, M.E. 2003. *In vivo* tracking of *Campylobacter jejuni* by using a novel recombinant expressing green fluorescent protein. Appl. Environ. Microbiol. 69: 2864–2874.

Myszewski, M.A., and Stern, N.J. 1991. Phagocytosis and intracellular killing of *Campylobacter jejuni* by elicited chicken peritoneal macrophages. Avian Dis. 35: 750–755.

Newell, D.G. and Pearson, A. 1984. The invasion of epithelial cell lines and the intestinal epithelium of infant mice by *Campylobacter jejuni*. J. Diarrhoeal. Dis. Res. 2: 19–26.

Pitts, T. 2001. MS. University of Arizona.

Ruiz-Palacios, G.M., Escamilla, E., and Torres, N. 1981. Experimental *Campylobacter* diarrhoea in chickens. Infect. Immun. 34: 250–255.

Russell, R.G., Blaser, M.J., Sarmiento, J.I., and Fox, J. 1989. Experimental *Campylobacter jejuni* infection in *Macaca nemestrina*. Infect. Immun. 57: 1438–1444.

van Spreeuwel, J.P., Duursma, G.C., Meijer, C.J., Bax, R., Rosekrans, P.C., and Lindeman, J. 1985. *Campylobacter* colitis: histological immunohistochemical and ultrastructural findings. Gut 26: 945–951.

Wassenaar, T., Engelskirchen, M., Park, S., and Lastovica, A. 1997. Differential uptake and killing potential of *Campylobacter jejuni* by human peripheral monocytes/macrophages. Med. Microbiol. Immunol. 186: 139–144.

Chapter 23

Campylobacter spp. and the Ability to Elicit Intestinal Inflammatory Responses

Paul Everest

Abstract

Campylobacters, particularly *C. jejuni* and *coli*, cause acute gastrointestinal infection which presents clinically as diarrhoea and intestinal epithelial damage. It has been proposed that primary infection for patients in both developed and developing countries is inflammation of the host intestinal tissues and these inflammatory events may be modified by subsequent exposure and maturing immunity. The organism can initiate these inflammatory signals by interaction (adhesion and invasion) with host cells and activate signalling pathways that lead to inflammatory cytokine production and recruitment of phagocytes, particularly neutrophils, into infected tissue. The molecular basis of this host/bacterial interaction is beginning to be understood which should facilitate the investigation of the role that inflammation plays in instigating the clinical symptoms of *Campylobacter*-induced diarrhoeal disease.

INTRODUCTION

The purpose of this short review is to summarize how *Campylobacter* spp. that cause enteric disease interact with the infected human host to elicit intestinal inflammation resulting in tissue damage and diarrhoea. Most material is drawn from current literature and combined with personal speculation that draws on known experimental studies to attempt to create disease models on which future experiments may be based. The clinical presentation of *Campylobacter* infection is one of diarrhoea and intestinal epithelial damage. It is not yet known whether the loss of fluid absorption and/or net secretion in the intestine is due to the action of toxin (s), bacterial invasion of intestinal cells, tissue inflammation or a combination of these factors. However, in the vast majority of animals colonized with *C. jejuni* no intestinal inflammation or disease occurs. This suggests that this pathogen may be recognized differently by immune effector or epithelial cells within the gastrointestinal tract of humans or young animals that develop disease, as opposed to colonized non-diseased farm animals or poultry. It is not yet understood why there are such differences in bacterial recognition or inflammatory responses between different host species infected or colonized with campylobacters.

ACUTE INFLAMMATION

Inflammation is a complex series of homeostatic reactions involving cells and molecular mechanisms coordinated so as to protect the host from pathogens. Pathological changes occurring early in disease become clinically recognizable when there is a marked acute response or when a chronic reaction leads to loss of function (Heumann *et al.*, 2001). Several phases can be distinguished in the inflammatory response to infection. The initial vascular reaction, controlled by vasoactive mediators, is followed by a cellular reaction with recruitment of inflammatory cells, which include immune competent leucocytes and neutrophils. This process requires cellular adhesion and migration and depends on the expression of host adhesion molecules. Leucocytes already present in the tissue (lymphocytes monocyte/macrophage cells, eosinophils, and mast cells) and newly recruited cells (neutrophils, eosinophils, monocyte/macrophage cells and lymphocytes) will release a variety of inflammatory mediators with both positive and negative (tissue destructive) effects. Epithelial cells, fibroblasts and smooth muscle cells are actively involved in the reaction. Normally inflammation ends by healing.

ACUTE DISEASE IN MAN – DEVELOPED COUNTRIES
A well-established spectrum of *Campylobacter*-mediated disease is seen in humans ranging from mild diarrhoea to severe dysentery-like inflammatory diarrhoea (Wassenaar and Blaser, 1999). There is good evidence that the status of the specific immune response is a critical factor in determining the clinical picture (Everest *et al.*, 1992). Although not supported by all studies (Everest *et al.*, 1992; Harvey *et al.*, 1999), there is also increasing evidence of differences in pathogenic potential between strains able to cause these types of disease. These differences in strain phenotype were observed using a variety of assays, including adhesion/invasion of cell lines, toxin production, animal models, that are believed to mimic virulence potential and hence disease causation. In developed countries the pattern of diarrhoeal disease is manifested by the presence of faecal leucocytes with a dysenteric type illness that reflects the process of inflammation occurring within the gastrointestinal tract. These findings were also reproduced in volunteer studies using clinical isolates (Black *et al.*, 1988). In humans the disease usually manifests as an unpleasant attack of acute diarrhoea lasting about 5 days (Skirrow and Blaser, 2000). Pathology of infected tissues reveals an acute inflammatory enterocolitis, with microscopy of stool from infected patients showing the presence of cellular exudates, predominantly polymorphonuclear leucocytes (PMNs) and red blood cells. Frank blood is visible in the faeces of about 25% of patients (Skirrow and Blaser, 2000) and fever and other signs of acute inflammation, particularly a characteristic severe abdominal pain, are common early manifestations.

PATHOLOGY OF ACUTE DISEASE
Mortality due to *Campylobacter* infection is thankfully uncommon, but in patients who have died of *Campylobacter* enteritis and septicaemia the macroscopic pathology consists of haemorrhagic inflammation and congestion of the jejunum and first half of the ileum (Skirrow, 1986). Oedema and inflammation of the ileum have been described at laparotomy for acute appendicitis. The fact that in a significant proportion of patients stools contain fresh blood, pus or mucous, suggests that colitis and colorectal inflammation occur. The disease is really an acute enterocolitis and spread into the colon and rectum is usual (Skirrow, 1986; van Spreeuwel *et al.*, 1985; Price *et al.*, 1984; Price *et al.*, 1979; Loss *et al.*, 1980; Lambert *et al.*, 1979). Laparotomy, sigmoidoscopy, colonoscopy and post mortem examination performed on patients who have succumbed to disease have shown that mucosal damage occurs during the course of infection. Children with acute *C. jejuni*

colitis examined endoscopically (study cited in (Everest *et al.*, 1992); Dr Herman Goossens, personal communication) have tissue damage that correlated with the degree of colitis observed clinically. Thus, mild disease correlated with moderate mucosal hyperaemia, moderate disease with micronodular ulceration, moderate oedema and hyperaemia, and severe disease with large nodular ulceration, severe oedema and hyperaemia with exudates of mucous and pus.

Microscopic pathological features of *Campylobacter* infection may vary from small changes in the mucosa to inflammation and oedema of the full thickness of the intestinal wall, haemorrhagic lesions or frank necrosis, gangrene and perforation (Skirrow and Blaser, 2000). The microscopic histology of mucosal biopsies from patients with *Campylobacter*-mediated enterocolitis is indistinguishable from that of salmonellosis or shigellosis (Skirrow, 1986). Tissue injury is observed in the jejunum, ileum and colon with similar microscopic pathology. Van Spreeuwel *et al.* (1985) concluded that campylobacter infection produces a histological picture of acute infectious colitis and were able to observe invasive organisms in the colonic mucosa. There is an acute inflammatory reaction consisting of focal collections of polymorphonuclear leucocytes in the lamina propria and lumen of mucosal capillaries, often with crypt abscess formation. Mucosal oedema is seen as a separation of the crypts of Lieberkuhn and a widening of the gap between crypts and muscularis mucosae. The epithelium may be flattened and sometimes eroded and ulcerated. Biopsies taken from children with acute *C. jejuni* colitis examined endoscopically (see above) revealed that, similar to gross tissue damage, the degree of microscopic histological damage correlated with severity of disease. Therefore, mild disease demonstrated moderate infiltration of PMNs and lymphocytes, moderate disease demonstrated marked infiltration of PMNs and lymphocytes and severe disease showed marked infiltration of PMNs and lymphocytes with oedema and cryptitis. Late in disease the presence of chronic inflammatory cells may give rise to a histological picture indistinguishable from inflammatory bowel disease (Loss *et al.*, 1980), although this has been disputed.

ACUTE DISEASE IN MAN – DEVELOPING COUNTRIES

In developing countries, *Campylobacter* enteritis is a disease of young children (Oberhelman and Taylor, 2000) with watery diarrhoea being the most common manifestation of infection. Watery diarrhoea, however, is not exclusively observed in developing countries. In our study (Everest *et al.*, 1992) looking at the cellular invasive ability of *C. jejuni* strains isolated from patients with different clinical presentations (colitis versus non-inflammatory diarrhoea), all the non-inflammatory isolates were from children presenting with watery diarrhoea who had normal recto-colonoscopy and no faecal leucocytes or erythrocytes detected in stained stool preparations.

Watery diarrhoea implies a lack of an acute inflammatory response within the mucosa (although this may not always be so as the inflammation may be milder and our ability to detect it less well developed) suggesting a change in either the pathophysiology of the disease, differences in the strains causing disease or modification of the effect of the pathogen on the host by prior exposure. This last scenario envisages development of a protective host response to bacterial factors that initiate inflammation within the intestinal mucosa, thus preventing inflammation within the tissue. This 'anti-inflammatory immunity' against *C. jejuni* cell components, if it exists, has several consequences. Firstly, we should be able to identify those bacterial components that are responsible for the initiation of intestinal inflammation (see below) and secondly diarrhoea in these patients is independent of bacterial components inducing inflammation. The two processes of diarrhoea and inflammation, which are inextricably linked to disease presentation in developed countries, may be initiated by separate bacterial virulence determinants. Thus, the host can be immune to the bacterial molecules that induce acute intestinal inflammation, but not immune to those bacterial components responsible for inducing secretory diarrhoea. As yet we do not know which bacterial virulence factors are responsible for inducing diarrhoea and we can only guess (with some exceptions) those factors that contribute to acute inflammation.

A study undertaken in Thailand suggested that *Campylobacter* infection in the first year of life was most often associated with bloody diarrhoea and that this was a manifestation of primary infection (Oberhelman and Taylor, 2000). Subsequently, as children are constantly exposed and re-infected with campylobacters, these observations suggest that increasing immunity may prevent bloody diarrhoea from developing. Consequently, immunity modifies the clinical consequences of infection by different strains to a watery diarrhoea, and eventually prevents any disease symptoms from occurring. This hypothesis was expanded by Ketley (Ketley *et al.*, 1997) in the following manner. He proposed that in the immunologically naïve host, *C. jejuni* colonizes the epithelial cell surface of the gut and the action of toxin (s) and host cell invasion result in epithelial cell damage. Host cell damage and the presence of bacteria in the lamina propria induce an inflammatory response with the recruitment of predominantly neutrophils, but also macrophages and lymphocytes to the site of infection. Tissue damage and inflammatory responses may result in net loss of fluid containing inflammatory cells, erythrocytes and serosal protein into the intestinal lumen. A partial immune response leads to the presence of antibodies against bacterial components responsible for acute inflammation due to presence of IgA and IgG. On infection, host cell damage and tissue penetration is limited because of the presence of these antibodies. Restriction of host cell damage and tissue invasion reduces the inflammatory response resulting in non-inflammatory diarrhoea. A completely protective immune response prevents tissue invasion and limits the organism to asymptomatic colonization of the mucous layer or prevents colonization completely. This hypothesis considered that diarrhoea was due to loss of absorptive surface from the mucosa or leakage of fluid from a damaged intestine. A secretory component to the diarrhoea was not considered.

Other factors which must be considered when looking at clinical presentation of *Campylobacter*-mediated diarrhoeal disease in developing countries is co-infection with other bacterial and viral pathogens and the presence of co-existent helminth infection. In developing countries *Campylobacter* spp. are isolated relatively frequently with another enteric pathogen in patients with diarrhoea (Albert *et al.*, 1999; Coker *et al.*, 2002). Of particular interest is *C. jejuni* co-infection with *Vibrio cholerae* or enterotoxigenic *Escherichia coli* (ETEC) producing heat-labile toxin. In one study in Bangladesh (Albert *et al.*, 1999), *C. jejuni* was present with *V. cholerae* or ETEC in 7% and 19% of patients, respectively. The enterotoxin of *V. cholerae* and heat-labile toxin of ETEC not only induce watery diarrhoea, but can down regulate host inflammatory responses (Hirst, 1999), therefore potentially modifying any inflammatory manifestation of *Campylobacter* infection in these patients. The majority of helminth infections are associated with host responses that reflect the release of cytokines from Th2 cells and prominent among these responses is the increased production of IgE, mast cells and eosinophils. Th2 cytokines can modulate or inhibit Th1 responses (Cox and Liew, 1992). In a single study (Baquar *et al.*, 2001) where cytokines from an acutely infected patient afflicted with *C. jejuni* enteritis (who had a previous infection with a different serotype) were investigated, a Th1 type cytokine profile was obtained during and after recovery from infection, although it was not stated if Th2 type cytokines were investigated in this patient. Thus it may be speculated that co-infection by helminths in a population in developing nations may alter the acute inflammatory response to *C. jejuni* infection.

Persistent Disease

C. jejuni infection can induce chronic inflammation after acute infection has resolved. Spiller *et al* (Spiller *et al.*, 2000) demonstrated that post dysenteric irritable bowel syndrome (PD-IBS) develops in up to 25% of patients following acute *Campylobacter* infection. This was characterized by abnormal bowel habits characterized by frequent loose stools often associated with urgency. This syndrome was characterized by increased T lymphocyte counts in the lamina propria (CD3, CD4 and CD8) and increased intraepithelial lymphocytes (IEL), increased enteroendocrine cells (EC) particularly associated with 5-hydroxytryptamine (5-HT) and increased gut permeability. This syndrome continued for over a year post acute infection in some patients, demonstrating that this organism can cause persisting chronic inflammation characterized by T-cell pathology.

Guillain–Barré syndrome (GBS), an autoimmune disorder of the peripheral nervous system follows *Campylobacter* infection with an annual incidence of 1.3 cases per 100 000 (Mishu and Blaser, 1993; Rees *et al.*, 1993; Koski, 1997; Nachamkin *et al.*, 1998). GBS can be divided into several types but the most common type observed in Europe and North America is one of acute inflammatory demyelinating polyneuropathy, characterized by an immune mediated attack on myelin and varying degrees of lymphocyte infiltration. This disorder is thus characterized by cross-reacting antibodies to bacterial structures and nerve proteins and recognition by reactive immune cells which is primarily T-cell mediated.

ACUTE INFLAMMATION IN ANIMAL MODELS OF *C. JEJUNI* DISEASE

Inflammation has been demonstrated in several animal models of *C. jejuni* mediated disease with histological features similar to those observed in inflammatory disease in man (Newell, 2001). Descriptions of the inflammatory changes have mainly arisen from experimental models infection using non-human primates, ferrets, rabbits and piglets. Of particular relevance to disease in man was the development of a non-human primate model of *C. jejuni* diarrhoeal disease.

Tissue damage and inflammation is a feature of the non-human primate model. Using *Macaca mulatta* primates, Russell *et al* (Russell *et al.*, 1993) observed bacteria invading the colonic mucosa before the development of an inflammatory response and determined that mucosal cell invasion by *C. jejuni* was the triggering event for the host inflammatory response. Colitis was characterized by intense diffuse infiltration of neutrophils into the infected mucosa. This model generated a number of other interesting observations regarding potential parallels with human disease. Intracellular *C. jejuni* were located both in membrane-bound vacuoles and free in the cell cytoplasm and extracellular *C. jejuni* were present in the mucosa and submucosa. Damaged epithelial cells exhibited morphological features associated with apoptosis. Similar results were reported following infection of *Macaca nemestrina* with *C. jejuni* (Russell *et al.*, 1989).

Ferrets have been used to study disease induced by *C. jejuni* (Fox *et al.*, 1987) and these animals also show the acute intestinal inflammation seen in human disease. The rabbit model used with various surgical procedures, for example the removable intestinal tie adult rabbit diarrhoeal model (RITARD (Walker *et al.*, 1988)) or rabbit ileal loop test (RILT (Everest *et al.*, 1993a)), results in acute inflammatory lesions of intestinal tissue when infected by *C. jejuni*. In contrast to the modified rabbit models, pig models are notable because of the close physiology between the human and pig gastrointestinal tract. Colostrum-deprived newborn piglets develop diarrhoea with blood after challenge (Vitovec *et al.*, 1989; Konkel *et al.*, 2001) with a histopathology similar to that observed in man.

HOST FACTORS INVOLVED IN ACUTE INFLAMMATION INDUCED BY *C. JEJUNI* INFECTION

In the absence of a host response against bacteria, the recognizable responses of acute inflammation within an infected host would not occur. Thus, for *C. jejuni*, interaction with mucosal tissue initiates the clinical symptoms that reflect gastrointestinal inflammation. This interaction induces host responses that presumably consist of bacteria coming into close contact to the cell surface, host cell invasion, translocation across the epithelium with multiplication within the lamina propria and/or elaboration of cytopathic toxins. There are varying degrees of evidence for these interactions from *in vitro* tissue culture models and *in vivo* models of infection. *In vitro*, Caco-2 cells (a human colonic immortal enterocyte-like cell line) infected with clinical isolates of *C. jejuni* exhibit a stress response with upregulation of molecules involved in glycolysis, anti-apoptotic mechanisms, oxidative stress and inflammatory kinase signalling pathway activity (P. Everest, unpublished).

In epithelial cells (Mellits *et al.*, 2002) *in vitro C. jejuni* can activate the transcription factor NF-kB (Mellits *et al.*, 2002) which is responsible for activation of numerous inflammation-associated genes, including production of interleukin 8 (IL-8). The innate immune response, via the activation of the transcription factor NF-kB, involves secretion of cytokines and other mediators leading to an inflammatory response. NF-kB, which is rapidly activated by chemically diverse agents and cellular stress conditions including bacterial lipopolysaccharides (LPS), cytokines and growth factors, is a critical transcriptional activator of cytokines including TNF-α and IL-1 and serves to protect against apoptosis and supports cell cycle progression.

Host cell receptor molecules respond to patterns or motifs present on the bacterial cell surface (pattern recognition) or molecules secreted by the infecting organism (toxins?), as well as soluble host molecules induced by these early events in infection, to coordinate the induction of the cellular infiltrate characteristic of acute inflammation. Not limited to specialised cell types, infected epithelial cells can also generate pro-inflammatory cytokines resulting in recruitment of inflammatory cells to the sites of bacterial multiplication. Thus, incubation of INT-407 cells with clinical isolates of *C. jejuni* resulted in secretion of IL-8 from these cells (Hickey *et al.*, 1999). Secretion of IL-8 required live bacteria, was dependent on *de novo* protein synthesis and mutants defective in adhesion and invasion induced less IL-8 than wild-type or hyperadherent or hyperinvasive mutants. A non-invasive mutant of *C. jejuni* 81-176 showed marked reduction in IL-8 secretion from INT-407 cells. The authors suggested cell invasion is required for high levels of IL-8 secretion. IL-8 is a potent neutrophil chemotactic factor, and accumulating evidence has established it as a crucial mediator in neutrophil-dependent acute inflammation. Neutrophils produce large amounts of leukotriene B4 (LTB4), which, in turn, is also a potent chemoattractant for neutrophils. Infection of rabbit ileal loops with strains of *C. jejuni* from inflammatory diarrhoea in man caused elevation of cyclic AMP, prostaglandin E2, and LTB4 levels in tissue and fluids (Everest *et al.*, 1993b). Consistent with LTB4 being a neutrophil chemoattractant, in this study a correlation was observed between the numbers of neutrophils in loop fluids from infected animals and the amount of LTB4 present in the fluid. In addition, neutrophil infiltration of the mucosa from infected loops correlated with fluid LTB4 levels. By contrast, loops inoculated with strains from non-inflammatory human disease or a non-colonizing mutant strain showed no neutrophil infiltrate in the mucosa or fluid accumulation. It was concluded that in this rabbit model *C. jejuni* strains from cases of inflammatory human diarrhoea induce histological inflammation, white cell infiltrate and large amounts of LTB4 compared to uninfected controls or non-inflammatory strains.

The infiltration of an organ or tissue by neutrophils is the hallmark of acute inflammation. The microscopic pathology of *Campylobacter* enteritis reveals the predominant cell type interacting with the organism is the neutrophil. Recruitment of neutrophils helps to limit and control infection within a confined affected area. Production of pro-inflammatory cytokines and chemokines by infected epithelial cells leads to attraction and transepithelial cell migration of neutrophils, thereby disrupting the permeability of the epithelium. It is therefore perhaps surprising that few studies have looked at the *C. jejuni*–neutrophil interaction. In one such study, Walan *et al* (1992) showed that four different clinical isolates varied in their interaction with human neutrophils. The ability of the intracellular organisms to induce neutrophil production of oxidative metabolites as measured by chemiluminescence (CL) varied between the strains. All strains evoked an intracellular CL response and three strains evoked an extracellular response, which might contribute to local tissue damage by toxic oxygen species release from phagocytosing neutrophils. The inflammatory response may be especially tissue damaging secondary to excessive neutrophil accumulation, superoxide and protease release. Phagocytosis was increased by opsonization with normal human serum with an increase in the intracellular CL response whereas the extracellular response disappeared or decreased considerably upon opsonization. These studies demonstrated that complement opsonized bacteria are readily attacked by the oxidative defence system of the neutrophil but neutrophil product release may be involved in tissue damage in natural disease. Injury to the vascular endothelium is a critical event in acute inflammatory disease processes and in acute inflammation, endothelial cell injury is frequently mediated by activated neutrophils (Heumann *et al.*, 2001). However, with infection by *C. jejuni*, Ullmann and Krause (1987) found no correlation between the neutrophil CL response and bacterial killing.

C. jejuni is phagocytosed by monocytes and macrophages *in vitro* (Wassenaar *et al.*, 1997; Myszewski and Stern, 1991). Some studies have shown killing of the organism in these cells while others have demonstrated prolonged intracellular survival. A study using a human clinical isolate and a strain from a chicken found that these organisms were killed by chicken peritoneal macrophages within 6 hours. Wassenaar *et al.* (1997) examined the survival of strains within human peripheral monocytes and demonstrated killing of the organisms within 48 hours, however, some monocytes failed to kill phagocytosed *C. jejuni*. Other investigators demonstrated survival of *C. jejuni* for 72 hours or more after uptake by mononuclear phagocytic cells (Kielbauch *et al.*, 1985) with some surviving over a 6-day period. The potential relevance of these findings to ongoing inflammation is that if *C. jejuni* survive in monocytes and macrophages, subsequent cytokine release will prolong clearance or enhance ensuing immune responses for clearance and may provoke further immunopathological damage.

C. jejuni can induce apoptosis in macrophages although not to the same extent as that observed by *Salmonella* or *Shigella* (Kielbauch *et al.*, 1985). In these studies features of apoptosis and necrotic cell death were observed in human and murine macrophage-like cell lines (Kielbauch *et al.*, 1985) and also in a human T-cell line (Zhu *et al.*, 1999). Apoptotic and necrotic events within cells has been associated with the development of acute inflammatory episodes but as yet the significance of these observations to acute disease by *C. jejuni* is not known.

Walker *et al* using the RITARD model demonstrated that *C. jejuni* traversed the submucosa via M cells (Walker *et al.*, 1988). Presumably *C. jejuni* may also traverse similar cells overlying colonic lymphoid follicles although this has never been directly demonstrated. In the rabbit ligated loop model for Shigellosis, organisms traverse the M cells overlying lymphoid follicles and can also invade villus epithelial cells (reviewed in Sansonetti, 2002), but the latter requires bacterial inocula that are much higher than those known to cause natural disease in man. This invasion occurs later (8 hours) than that known to occur via the follicle associated epithelium (FAE) (2 hours). Experimental *C. jejuni* infection in humans (Black *et al.*, 1988) has revealed that low doses of bacteria (500–800 viable cells) causes diarrhoeal disease and that increasing the dose will give a simultaneous rise in volunteers acquiring disease. If we apply these observations to *C. jejuni* infection we could argue that the low numbers of organisms as used in the volunteer experiments may traverse the epithelium via the M-cell route in colonic epithelia, cause inflammation in the submucosa and set up the subsequent histological picture of acute inflammation. Subsequent growth of the organisms in the lamina propria, neutrophil migration leading to increased epithelial permeability and loss of tight junction integrity would facilitate invasion of the intestinal villus epithelial cells by *C. jejuni*. Therefore, in volunteers receiving higher numbers of bacteria, infecting *C. jejuni* would penetrate the epithelium both via the FAE and via apical enterocytes. Such a hypothesis would help explain why such low numbers of *C. jejuni* can cause clinical inflammation and diarrhoeal disease. It might also be one explanation for why some *C. jejuni* isolates invade to low levels in epithelial cell invasion assays, whereas for other clinical isolates this is not always the case (Hu and Kopecko, 1999; Kopecko *et al.*, 2001).

The outcome of the major inflammatory response to infection is resolution of inflammation with eradication of bacteria from the tissues. The fact that in the majority of subjects *Campylobacter* enteritis is a self limiting disease suggests that the inflammatory response is effective in controlling and initiating a lasting immune response to that particular serotype of *C. jejuni*. We do not know why some patients do not evoke an inflammatory response to some strains yet still manifest symptoms of diarrhoea, but it may be that the organism evades or does not provoke the inflammatory response within the mucosa, or does not invade to such numbers as to provoke an inflammatory cell infiltrate. Alternatively, the site of bacterial replication within the tissue could decide the magnitude of the response, with those strains able to replicate quickly within the lamina propria provoking the severest inflammatory responses, most severe microscopic and endoscopic pathology and hence more tissue damage resulting in more severe symptoms. Those strains that replicate elsewhere within the tissue or outside the tissues (extracellularly, or within cellular spaces between cells?) would not provoke such severe cellular responses and hence less inflammation. This hypothesis has yet to be tested by experiment.

The ability of *C. jejuni* to be phagocytosed by macrophages in animal models with subsequent liberation of inflammatory cytokines has been investigated in the mouse. Pancorbi *et al* (1999) injected various toxin producing and non-producing strains into murine peritoneal cavities. These organisms provoked an inflammatory response with an increase in peritoneal phagocyte oxidative activity. One non-toxin producing strain enhanced Tumour necrosis factor alpha (TNF-α) production in this model. Murine plasma cytokines have also been investigated using an intraperitoneal infection model (Pancorbo *et al.*, 1999). Interferon gamma production progressively rose and TNF-α peaked at days 2–3 post infection. Surprisingly, IL-6 was not detected in this mouse model, but has been detected upon infection of pig intestinal cells (Parthasarathy and Mansfield, 2001).

Baquar *et al* (2001) described the cellular immune response, and hence the inflammatory cytokine profile, of a patient with pre-existing humoral and cellular recognition of *C. jejuni* antigens. This patient clinically manifested with watery diarrhoea, fever, faecal leucocytes and gross blood in the faeces. The patient had rapid responses to *C. jejuni* antigens by lymphocyte proliferation assays and cytokine and C-reactive protein (CRP) responses were also rapid. Plasma inflammatory cytokines and CRP levels were increased with IL-8 levels peaking at day 3 and IL-1 and CRP on day 10 post infection. IL-6 and interferon gamma (IFN-γ) were not present on day 3 but were increased at day 10. Plasma TNF-α levels were never elevated.

C. JEJUNI MOLECULES ACTIVATING HOST ACUTE INFLAMMATORY RESPONSES

Bacterial cell components are the primary stimulus for an acute inflammatory response to develop in an infected host. The bacterial molecules responsible for these events are as yet not completely understood. Live cells of *C. jejuni* and *C. coli* can induce the release of IL-8 from INT407 cells (Hickey *et al.*, 1999). Mutants shown to be defective in adhesion and invasion resulted in reduced levels of IL-8 and *cheY* mutants, which are hyperadherent and hyperinvasive resulted in higher levels of IL-8 secretion from INT 407 cells (Yao *et al.*, 1997). A non-invasive mutant of *C. jejuni* 81-176 that adhered to 43% of wild-type levels also showed reduction in IL-8 secretion. From these data the authors suggested that invasion is necessary for the induction of high levels of IL-8 secretion by *C. jejuni*. This is perhaps the first evidence, albeit *in vitro*, that the ability of *C. jejuni* to interact with enterocytes can lead to differences in inflammatory cell recruitment. Thus the scenario emerging from these studies leads us to propose greater inflammation may occur from those strains that adhere, invade and transcytose than from those that just adhere without invasion due to the amount of IL-8 secreted in response to the bacterial inflammatory stimulus. These responses are amplified *in vivo* because on encountering *C. jejuni* neutrophils secrete LTB4 (Hickey *et al.*, 1999), which in turn recruits more neutrophils, and thus the cycle of inflammation in terms of neutrophil recruitment increases. For those strains that simply adhere without invading epithelial cells the amount of IL-8 generated is less, less neutrophil recruitment occurs and so the cycle of amplifying events in terms of LTB4 generation is also less. Such a hypothesis may explain the differences between inflammatory and non-inflammatory diarrhoea in *C. jejuni* enteritis.

Production of cytotoxins during infection is one obvious source of tissue damaging molecules that may contribute to host inflammation within the gastrointestinal mucosa.

The genome sequence of *C. jejuni* contains four potentially cytotoxic proteins, namely cytolethal distending toxin (CDT, *cdtA*, *cdtB*, *cdtC*), a putative integral membrane protein with a haemolysin domain, a putative haemolysin (*tlyA*), and phospholipase A (*pldA*). A role for CDT in IL-8 induction has been investigated. Membrane fractions of *C. jejuni* 81-176, but not membrane fractions of *C. coli* strains, induced release of IL-8. Membrane preparations from 81-176 mutants defective in any of the three membrane-associated protein subunits of CDT were unable to induce IL-8. The presence of the three *cdt* genes on a shuttle plasmid *in trans* restored both CDT activity and the ability to release IL-8 to membrane fractions. However, CDT mutations did not affect the ability of 81-176 to induce IL-8 during adherence to or invasion of INT407 cells (Hickey *et al.*, 2000). These data suggest that other components of the cell are also involved in mediating IL-8 release from infected epithelial cells and hence recruitment of neutrophils to sites of bacterial multiplication within tissue. *Serpulina hyodysenteriae*, the cause of swine dysentery, has a similar *tlyA* gene which has haemolytic and cytotoxic functions, and is thus implicated in tissue damage (Hyatt *et al.*, 1994). The role of the *Serpulina hyodysenteriae* haemolysin encoded by the *tlyA* gene in the pathogenesis of swine dysentery (SD) has been studied. *tlyA* mutants of two *S. hyodysenteriae* strains (B204 and C5) were tested for virulence in pigs. None of the animals developed SD. However, after infection with wild-type strain B204 or C5, the incidence of SD was 100% or 60%, respectively. Thus, the *tlyA*-encoded haemolysin of *S. hyodysenteriae* is an important virulence factor in SD. It is unknown if *tlyA* performs a similar role in virulence in *C. jejuni* mediated diarrhoeal disease. *C. jejuni* has a phospholipase A2 (*pldA*) whose role in disease is unknown, but a similar molecule in *Yersinia* contributes to inflammation in a mouse model of infection (Schmiel *et al.*, 1998).

    C. jejuni secretes proteins named *Campylobacter* invasion antigens (Cia proteins) when in the presence of eukaryotic cells or in serum supplemented medium (Konkel *et al.*, 1999). Infection of newborn piglets with wild-type and Cia secretion-deficient mutants has shown that these proteins contribute to the observed pathology of *C. jejuni* enteritis (Konkel *et al.*, 2001). More severe histological lesions are observed when piglets are infected with the *ciaB* complemented isolate than when infected with the *ciaB* mutant. It is not yet clear what is the function of individual Cia proteins in terms of them being secretory apparatus or cell effectors. *C. jejuni* strain 81178 has a putative type IV secretion system encoded upon a plasmid which has been demonstrated to have a role in invasion of epithelial cells and mutants in this system have reduced virulence in the ferret diarrhoeal model, producing less diarrhoea and presumably less tissue inflammation (Bacon *et al.*, 2000). The secretion system may secrete target proteins which have these effects *in vitro* and *in vivo*. The bacterial cell surface lipoprotein JlpA is another identified adhesin in *C. jejuni*. This protein interacts with a cell surface heat shock protein Hsp-90 and initiates inflammatory pathways in cells, namely nuclear translocation of NK-κB and activation of p38 MAP kinase (Jin *et al.*, 2003).

    Recently the genome sequence of *C. jejuni* NCTC 11168 (Parkhill *et al.*, 2000) has shown that this organism has a greater capacity for polysaccharide synthesis than previously realized and subsequent analysis has shown that *C. jejuni* produces a capsule (Karlyshev *et al.*, 2001). *C. jejuni* directly visualized in loop fluids from rabbits exhibiting inflammatory diarrhoea have large halos around the bacteria demonstrating that the bacterium expresses capsule in the face of an acute inflammatory response (P. Everest, unpublished). The loop fluids contained macroscopic blood with many neutrophils present, however, it is currently unknown whether this capsule is involved in evading phagocytosis by neutrophils with or without complement present. One obvious bacterial molecule that may provoke inflammation within tissue is lipopolysaccharide (LPS) or lipooligosacharide (LOS) in *C. jejuni*. Moran (1995) tested lipopolysaccharides from *C. jejuni* for their ability to induce lethality in sensitized mice, pyrogenicity in rabbits and TNF-α secretion from murine peritoneal macrophages. Compared with *Salmonella* LPS, lethal toxicity was 50% lower, pyrogenicity was 30–50 fold lower and the ability to induce TNF-α was 100-fold lower. Thus *C. jejuni* LPS may not be as potent as LPS from Enterobacteriaceae but can exhibit biological activity in these assays. Perhaps the capsule or other polysaccharides that could mimic host molecules may delay recognition of *C. jejuni* by immune cells thus facilitating colonization and growth itself within intestinal tissues. Sialic acid present on surface structures would also aid this delay in recognition and the LOS of some serotypes mimics host ganglioside structures (Linton *et al.*, 2000). Bacon *et al* (2001) showed that the capsule (*kpsM*) of *C. jejuni* is involved in virulence as the mutant is also significantly reduced in invasion of INT407 cells and reduced in virulence in a ferret diarrhoeal disease model. The animals fed mutant bacteria developed less severe disease without blood and at lower doses no disease compared to wild-type organisms.

CONCLUSIONS

The host response to acute *C. jejuni* infection contributes to the clinical and laboratory features of the disease. It is assumed that primary infection by this organism may be inflammatory but this is not certain and especially in children clinical features may be modified or inflammation may not be clinically obvious or detected by laboratory testing. The acute inflammatory response seems to localize the infection to the gastrointestinal tract in most affected persons but in immunocompromised patients such as those with HIV infection progressing to AIDS, infection will spread and persist, with bacteraemia and wide dissemination, indicating that for elimination of bacteria from the tissues intact cell mediated immunity and inflammatory cytokines are also required.

In this chapter we have highlighted the ability of *C. jejuni* to cause acute inflammation in the intestine. It does this by activating host cytokines and recruiting phagocytic cells to sites of bacterial invasion. The bacterial molecules that elicit this response are only partially known and it is hypothesized that future research should be able to discriminate bacterial molecules involved in acute inflammation from those involved in causing acute diarrhoea in the affected host.

Acknowledgements

I wish to acknowledge the helpful collaborations and discussions of all the people with whom I have worked over the years particularly Julian Ketley, Brendan Wren, Andrey Karlyshev, Dennis Linton, Simon Hardy, Peter Williams, Herman Goossens, Karl Wooldridge, Mark Roberts, Andy Stevenson, Amanda MacCallum and the financial support of the BBSRC and Royal Society.

References

Albert, M.J., Faruque, A.S.G., Faruque, S.M., Sack, R.B., and Mahalanabis, D. 1999. Case control study of enteropathogens associated with childhood diarrhoea in Dhaka, Bangladesh. J. Clin. Microbiol. 37: 3458–3464.

Bacon, D.J., Alm, R.A., Burr, D., Hu, L., Kopecko, D., Ewing, C., Trust, T.J., and Guerry, P. 2000. Involvement of a plasmid in virulence of *Campylobacter jejuni* 81-176. Infect. Immun. 68: 4384–4390.

Bacon, D.J., Szymanski, C., Burr, D.H., Silver, R., Alm, R.A., and Guerry, P. 2001. A phase variable capsule is involved in virulence of *Campylobacter jejuni* 81-176. Mol. Microbiol. 40: 769–777.

Baquar, S., Rice, B., Lee, L., Bourgeois, Noor El Din, A., Heresi, G., Mourad, A., and Murphy, J. 2001. *Campylobacter jejuni* enteritis. Clin. Infect. Dis. 33: 901–905.

Black, R.E., Levine, M.M., Clements, M.L., Hughes, T.P., and Blaser, M.J. 1988. Experimental *Campylobacter jejuni* infection in humans, J. Infect. Dis. 157: 472–9.

Coker, A.O., Isokpehi, R.D., Thomas, B.N., Amisu, K.O., and Obi, L.C. 2002. Human campylobacteriosis in developing countries. Emerging Infect. Dis. 8: 237–243.

Cox, F.E.G., and Liew, F.Y. 1992. T-cell subsets and cytokines in parasitic infections, Immunol. Today 13: 445–8.

Everest, P.H., Goossens, H., Sibbons, P., Lloyd, D.R., Knutton, S., Leece, R., Ketley, J.M. and Williams, P.H. 1993a. Pathological changes in the rabbit ileal loop model caused by *Campylobacter jejuni* from human colitis. J. Med. Microbiol. 38: 316–321.

Everest, P.H., Cole, A.T., Hawkey, C.J., Goossens, H., Knutton, S., Butzler, J.P, Ketley, J.M. and Williams, P.H. 1993b. Roles of leukotriene B4, prostaglandin E2 and cyclic AMP in *Campylobacter jejuni*-induced intestinal fluid secretion. Infect. Immun. 61: 4885–4887.

Everest, P.H., Goossens, H., Butzler, J.P., Lloyd, D., Knutton, S., Ketley, J.M., and Williams, P.H. 1992. Differentiated Caco-2 cells as a model for enteric invasion by *Campylobacter jejuni* and *Campylobacter coli*. J. Med. Microbiol. 37: 319–325.

Fox, J., Ackermann, J.I., Taylor, N., Claps, M., and Murphy, J.C. 1987. *Campylobacter jejuni* infection in the ferret: an animal model of human campylobacteriosis. Am. J. Vet. Res. 48 (1): 85–90.

Harvey, P., Battle, T., and Leach, S. 1999. Different invasion phenotypes of *Campylobacter* isolates in Caco-2 cell monolayers. J. Med. Microbiol. 48: 461–469.

Heumann, D., Glauser, M.P., and Calandra, T. 2001. The generation of inflammatory responses. In: Molecular Medical Microbiology. M. Sussman, ed. Academic Press, pp. 687–727.

Hickey, T.E. Baqar, S., Bourgeois, A.L., and Guerry, P. 1999. *Campylobacter jejuni* stimulated secretion of interleukin-8 by INT-407 cells. Infect. Immun. 67: 88–93.

Hickey, T.E., McVeigh, A.L., Scott, D.A., Michielutti, R.E., Bixby, A., Carroll, S.A., Bourgeois, A.L., and Guerry, P. 2000. *Campylobacter jejuni* cytolethal distending toxin mediates release of interleukin-8 from intestinal epithelial cells. Infect. Immun. 68: 635–6541.

Hirst, T. 1999. Cholera toxin and heat labile enterotoxin. P104–12. In Bacterial Protein Toxins. Alouf, J.E and Freer, J., eds. Academic Press, pp. 104–112.

Hu, L. and Kopecko, D.J. 1999. *Campylobacter jejuni* induces host phosphorylation events and a transient calcium flux during entry into human intestinal cells. In Proceedings of the US–Japan conference on cholera and other bacterial enteric infections (vol 35), pp. 158–163.

Hyatt, D.R., ter Huurne, AA., van der Zeijst, B.A., and Joens, L.A. 1994. Reduced virulence of *Serpulina hyodysenteriae* haemolysin negative mutants in pigs and their potential to protect pigs against challenge with a virulent strain. Infect. Immun. 62: 2244–2248.

Jin, S., Song, Y., Emilli, A., Sherman, P., and Chan, V.L. 2003. JlpA of *Campylobacter jejuni* interacts with surface-exposed heat shock protein 90alpha and triggers signalling pathways leading to the activation of NF-kB and p38 MAP kinase in epithelial cells. Cell. Microbiol. 5: 165–174.

Karlyshev, A.V., McCrossan, M.V., and Wren, B.W. 2001. Demonstration of polysaccharide capsule in *Campylobacter jejuni* using electron microscopy. Infect. Immun. 69 (9): 5921–4.

Ketley, J.M. 1997. Pathogenesis of enteric infection by *Campylobacter*. Microbiology 143: 5–21.

Kiehlbauch, J.A., Albach, R.A., Baum, L.L., and Chang, K.P. 1985. Phagocytosis of *Campylobacter jejuni* and its intracellular survival in mononuclear phagocytes. Infect Immun. 48: 446–451.

Kielbauch, J.A., Albach, R.A., Baum, L.L., and Chang, K.P. 1985. Phagocytosis of *Campylobacter jejuni* and its intracellular survival in mononuclear phagocytes. Infect. Immun. 48: 446–451.

Konkel, M.E., Kim, B.J., Rivera-Amill, V., and Garvis, S.G. 1999. Bacterial secreted proteins are required for the internalization of *Campylobacter jejuni* into cultured mammalian cells. Mol. Microbiol. 32: 691–701.

Konkel, M.E., Monteville, M.R., Rivera-Amill, V., and Joens, L. 2001. The Pathogenesis of *Campylobacter jejuni*-mediated enteritis. Curr. Issues Int. Microbiol. 2 (2): 55–71.

Kopecko, D.J., Hu, L., and Zaal, K.J.M. 2001. *Campylobacter jejuni* – microtubule-dependent invasion. Trends Microbiol. 9 (8): 389–396.

Koski, C.L. 1997. Mechanisms of Schwann cell damage in inflammatory neuropathy. J. Infect. Dis. 176: (Supplement 2): S169–172.

Lambert, M.E., Schofields, P.F., Ironside, A.G., and Mandal, B.K. 1979. *Campylobacter colitis*. Br. Med. J. 1: 857–859.

Linton, D., Karlyshev, A.V., Hitchen, P.G., Morris, H.R., Dell, A., Gregson, N.A., and Wren, B.W. 2000. Multiple N-acetyl neuraminic acid synthetase (*neuB*) genes in *Campylobacter jejuni*: identification and characterization of the gene involved in sialylation of lipo-oligosaccharide. Mol. Microbiol. 35: 1120–34.

Loss, R.W., Mangla, J.C., and Pereira, M. 1980. *Campylobacter colitis* presenting as inflammatory bowel disease with segmental colonic ulcerations. Gastroenterol. 79: 138–140.

Mellits, K., Mullen, J., Wand, M., Armbruster, G., Patel, A., Connerton, P., Skelly, M. and Connerton, I. 2002. Activation of the transcription factor NF-kB by *Campylobacter jejuni*. Microbiology 148: 2753–2763.

Mishu, B., and Blaser, M.J., 1993 Role of infection due to *Campylobacter jejuni* in the initiation of Guillain–Barré syndrome, Clin. Infect. Dis. 17: 104–8.

Moran, A.P. 1995. Biological and serological characterization of *Campylobacter jejuni* lipopolysaccharides with deviating core and lipid A structures. FEMS Immunol. Med. Microbiol. 11 (2): 121–130.

Myszewski, M.A. and Stern, N.J. 1991. Phagocytosis and intracellular killing of *Campylobacter jejuni* by elicited chicken peritoneal macrophages. Avian Dis. 35: 750–755.

Nachamkin, I., Mishu Allos, B., and Ho, T. 1998. *Campylobacter* species and Guillain–Barré syndrome. Clin. Micro. Rev. 11: 555–567.

Newell, D.G. 2001. Animal models of *Campylobacter jejuni* colonization and disease and the lessons learned from similar *Helicobacter pylori* models. J. App. Microbiol. 90, 57S–67S.

Oberhelman, R.A., and Taylor, D.N. 2000. *Campylobacter* infections in developing countries. In: Nachamkin, I., Blaser, M.J. (eds) *Campylobacter*. Washington DC: ASM Press, pp. 27–44.

Pancorbo, P.L., de Pablo, M.A., Ortega, E., Gallego, A.M., Alvarez, C., and Alvarez de Cienfuegos, G. 1999. Evaluation of cytokine production and phagocytic activity in mice infected with *Campylobacter jejuni*. Curr. Microbiol. 39 (3): 129–133.

Parkhill, J., Wren, B.W., Mungall, K., Ketley, J.M., Churcher, C., Basham, D., Chillingworth, T.,Davies, R.M., Feltwell, T., Holroyd, S., Jagels, K., Karlyshev, A.V., Moule, S., Pallen, M.J., Penn, C.W., Quail, M.A., Rajandream, M.A., Rutherford, K.M., van Vliet, A.H., Whitehead, S., and Barrell, B.G. 2000. The genome sequence of the food borne pathogen *Campylobacter jejuni* reveals hypervariable sequences. Nature 403: 665–668.

Parthasarathy, G., and Mansfield, L.S. 2001. Upregulation of IL-6 from intestinal pig epithelial cells (IPEC-1) by *Campylobacter jejuni*. In '11th International Workshop on *Campylobacter, Helicobacter* and related organisms, Abstracts of Scientific Presentations. Int. J. Med. Microbiol. 291: 24. (Supplement no. 31).

Price, A.B., Dolby, J.M., Dunscombe, P.R., and Stirling, J. 1984. Detection of *Campylobacter* by immunofluorescence in stools and rectal biopsies of patients with diarrhoea. J. Clin. Pathol. 37: 1007–1013.

Price, A.B., Jewkes, J., and Sanderson, P.J. 1979. Acute diarrhoea: *Campylobacter colitis* and the role of rectal biopsy. J. Clin. Pathol. 32: 990–996

Rees, J.H., Gregson, N.A., Griffiths, P.L., and Hughes, R.A. 1993. *Campylobacter jejuni* and Guillain–Barré syndrome. Q. J. Med. 86: 623–34.

Russell, R.G., Blaser, M., Sarmiento, J.I., and Fox, J. 1989. Experimental *Campylobacter jejuni* infection in *Macaca nemestrina*. Infect. Immun. 57: 1438–1444.

Russell, R.G., O'Donnoghue, M., Blake Jr, D.C., Zulty, J., and DeTolla, L.J. 1993. Early colonic damage and invasion of *Campylobacter jejuni* in experimentally challenged infant *Macaca mulatta*. J. Infect. Dis. 168: 210–215.

Sansonetti, P. 2002. Host–pathogen interactions: the seduction of molecular cross talk. Gut 50 (Supplement 111: iii2–iii18).

Schmiel, D.H., Wagar, E., Karamanou, L., Weeks, D., and Miller, V.L. 1998. Phospholipase A of *Yersinia enterocolitica* contributes to pathogenesis in a mouse model. Infect. Immun. 66: 3941–3951.

Skirrow, M. 1986. *Campylobacter* enteritis. In Medical Microbiology Vol 4. C.S.F. Easmon, ed. Academic Press.

Skirrow, M. and Blaser, M. 2000. Clinical aspects of *Campylobacter* infection. In: Nachamkin,I., Blaser, M.J., eds. *Campylobacter*. Washington DC: ASM Press, pp. 68–88.

Spiller R.C., Jenkins, D., Thornley, J.P., Hebden, J.M., Wright, T., Skinner, M., and Neal, K.R. 2000. Increased rectal mucosal enteroendocrine cells, T lymphocytes and increased gut permeability following acute *Campylobacter enteritis* and in post dysenteric irritable bowel syndrome. Gut 47: 804–811.

Ullmann U., and Krause, R. 1987. The reaction of *Campylobacter* species on the chemiluminescence, chemotaxis and haemagglutination. Zentralbl. Bakteriol. Mikrobiol. Hyg. 266 (1–2): 178–190.

Van Spreeuwel, P., Duursma,G.C., Meijer, C.J., Bax, R., Rosekrans, P.C.M. and Lindeman, J. 1985. *Campylobacter* colitis: histological immunohistochemical and ultrastructural findings. Gut 26: 945–951.

Vitovec, J., Koudela, B., Sterba, J., Tomancova, I., Matyas, Z., and Vladik, P. 1989. The gnotobiotic piglet as a model for the pathogenesis of *Campylobacter jejuni* infection. Zentralbl. Bakteriol. 271 (1): 91–103.

Walan, A., Dahlgren, C., Kihlstrom, E., Stendahl, O., and Lock, R. 1992. Phagocyte killing of *Campylobacter jejuni* in relation to oxidative activation. APMIS 100: 424–430.

Walker, R.I., Schmauder-Chock, E.A., Parker, J.L., and Burr, D. 1988. Selective association and transport of *Campylobacter jejuni* through M cells of rabbit Peyer's patches. Can. J. Microbiol. 34: 1142–1147.

Wassenaar, T.M., and Blaser M.J. 1999. Pathophysiology of *Campylobacter jejuni* infections of humans. Microbes and Infection 1: 1023–1033.

Wassenaar, T.M., Engelskirchen, M., Park, S., and Lastovica, A. 1997. Differential uptake and killing potential of *Campylobacter jejuni* by human peripheral monocytes/macrophages. Med. Microbiol. Immunol. 186 (2–3): 139–144.

Yao, R., Burr, D.H., and Guerry, P. 1997. Che-Y mediated modulation of *Campylobacter jejuni* virulence. Mol. Microbiol. 23: 1021–1031.

Zhu, J. Meinersmann, R.J., Hiett, K.L., and Evans, D.L. 1999. Apoptotic effect of outer membrane proteins from *Campylobacter jejuni* on chicken lymphocytes. Curr. Microbiol. 38: 244–249.

Chapter 24

Campylobacter jejuni Interaction with Enterocytes – Using Host Gene Expression Analysis to Unravel Potential Disease Mechanisms

Amanda MacCallum, Pawel Herzyk, Catriona Young, Julian Ketley and Paul Everest

Abstract

Campylobacter jejuni is a common food-borne pathogen causing diarrhoeal disease worldwide. Little is known of the mechanisms responsible for the clinical symptoms associated with infection. Focusing on data derived using microarray technology and comparisons with published literature, possible events involved in the interaction of *C. jejuni* with host cells are highlighted. The identification of host cell responses to *C. jejuni* can be used to develop testable models and highlight novel areas for future research.

DERIVING CLUES TO THE BIOLOGY HOST–PATHOGEN INTERACTIONS

Microarrays have revolutionized our ability to investigate both pathogen and host cell gene expression. They enable the identification of transcriptional changes within an infected host cell and while some changes reflect direct responses to the infecting organism, others relate to cell stress or metabolic changes induced by an infectious insult. Although genome-wide analysis of host cell transcriptional responses facilitates the understanding of host–pathogen interactions (Xia *et al.*, 2003) it is of particular value where focused studies of individual changes have not yielded great progress. Models based on the overview obtained from microarray analysis clearly need to be interpreted in the light of published data, but microarray studies can highlight previously unexpected changes in infected host cells. It is the provision of a comprehensive description of host cell responses and possible identification of novel events that makes the technology so appealing for use with enigmatic pathogens like *Campylobacter jejuni*.

In contrast to the asymptomatic infection of animals, the symptoms associated with *C. jejuni*-mediated human infection ranges from mild watery diarrhoea to severe dysentery-like disease (Wassenaar and Blaser 1999). Little is known of the molecular mechanisms that cause diarrhoea, however, toxin production and host cell invasion are candidate factors (Ketley 1997, Konkel *et al.*, 2001, Wassenaar and Blaser 1999); such factors could induce net fluid loss by direct stimulation of ion secretion pathways, inhibition of absorptive mechanisms and/or disruption of the intestinal epithelium. Given the lack of knowledge of host mechanisms, microarray analysis of *C. jejuni*-infected mammalian cells similar to those encountered by the bacterium during infection may highlight the molecular mechanisms involved in *C. jejuni*-mediated diarrhoea. To this end, human gene arrays were probed with RNA from Caco-2 cells infected with one of four *C. jejuni* strains of differing clinical backgrounds (NCTC 11168, sequenced strain (Parkhill *et al.*, 2000); 81-176, human volunteer study strain (Black *et al.*, 1988); L115, inflammatory diarrhoea isolate (Everest *et al.*, 1992); N82, watery diarrhoea isolate (Everest *et al.*, 1992, Everest, unpublished data). RNA was isolated at 4 and 24 hours post infection and genes were identified whose expression was increased more than two-fold versus that of uninfected cells using dChip, SAM and Genespring software. In this chapter we model (Figure 24.1) potential molecular interactions based both on our microarray analysis of *C. jejuni* infected Caco-2 cells and previous published work. Development of such models will focus future investigations on the molecules involved in the disease process.

C. JEJUNI INTERACTIONS WITH HOST CELLS: RECEPTORS

A number of host cell receptor like-molecules are increased in terms of gene expression in *C. jejuni* infected Caco-2 cells (Table 24.1). Previously identified host cell receptors

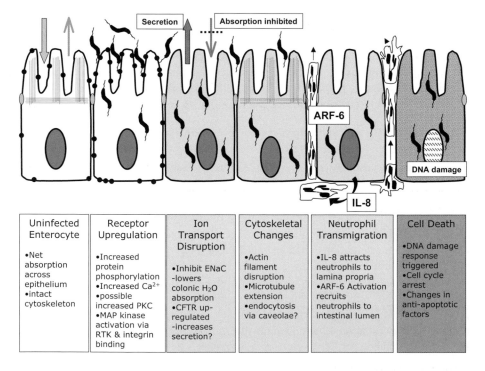

Uninfected Enterocyte	Receptor Upregulation	Ion Transport Disruption	Cytoskeletal Changes	Neutrophil Transmigration	Cell Death
•Net absorption across epithelium •intact cytoskeleton	•Increased protein phosphorylation •Increased Ca²⁺ •possible increased PKC •MAP kinase activation via RTK & integrin binding	•Inhibit ENaC -lowers colonic H₂O absorption •CFTR up-regulated -increases secretion?	•Actin filament disruption •Microtubule extension •endocytosis via caveolae?	•IL-8 attracts neutrophils to lamina propria •ARF-6 Activation recruits neutrophils to intestinal lumen	•DNA damage response triggered •Cell cycle arrest •Changes in anti-apoptotic factors

Figure 24.1 Proposed interactions between *C. jejuni* and host enterocytes. The highlighted host events are based on data from microarray analysis of transcriptional response of Caco-2 cells to *C. jejuni* infection. For clarity each stage is shown separately but we envisage the stages to be dynamic and interrelated. *C. jejuni* upregulates host receptors on infected cells leading to tyrosine phosphorylation, increases in intracellular calcium and possibly activation of protein kinase C. Cell mediators activated are NF-κB and MAP kinase signalling pathways. *C. jejuni* upregulates molecules that interfere with electrogenic sodium transport and hence disrupts water absorption in the colon. Infection also results in upregulation of CFTR expression to possibly increase levels of this chloride ion channel in the apical membrane. Activation of ARF-6 recruits neutrophils to the intestinal lumen. DNA damage is induced upon infection with resulting cell cycle arrest.

include the extracellular matrix protein fibronectin and heat shock protein 90 (Hsp90), which bind to the bacterial CadF outer membrane protein and JlpA, respectively (Jin *et al.*, 2003, Konkel *et al.*, 1997, Monteville *et al.*, 2003). In addition, several host proteins are tyrosine phosphorylated when *C. jejuni* invades experimental cell cultures (for example, Caco-2 and INT407) (Wooldridge *et al.*, 1996) and inhibition of protein tyrosine kinases markedly reduces invasion of *C. jejuni* (Wooldridge *et al.*, 1996). The cell surface receptors identified may explain this general tyrosine phosphorylation event upon infection. Whether *C. jejuni* binds to any of the receptors observed by microarray analysis is currently unknown but some have been demonstrated to bind other bacterial pathogens. For example, carcinoembryonic antigen 1 (CD66d) and biliary glycoprotein (CD66a) mediate *Neisseria gonorrhoeae* transcellular passage across T84 colonic epithelial cells (Wang *et al.*, 1998). In addition, diffuse adhering *Escherichia coli* recruit CD66e (carcinoembryonic antigen) and decay accelerating factor (DAF) for complement at attachment sites on Caco-2 cells (Guignot *et al.*, 2000). *C. jejuni* is also capable of transcytosis across polarized monolayers (Everest *et al.*, 1992, Konkel *et al.*, 1992), but it is unclear if any of the identified molecules (Table 24.1), which include CD66d, CD66a and DAF, mediate this process.

C. JEJUNI INVASION OF HOST CELLS: CYTOSKELETON/ SIGNALLING

Internalization of *C. jejuni* into host cells was shown to require actin filaments alone, microtubules alone, both or neither (reviewed in Kopecko *et al.*, 2001). One suggested explanation for the involvement of both microfilaments and microtubules is that signalling cascades from invading bacteria trigger localized disruption of cortical actin filaments

Table 24.1 Changes in host cell gene expression following infection with *C. jejuni*. Below are highlighted examples of genes significantly up-regulated or down-regulated in Caco-2 cells infected by *C. jejuni*. Potential models of molecular interactions occurring during *C. jejuni* infection of host cells are discussed in the text. Models are based on microarray analysis of *C. jejuni*-infected Caco-2 cells and previous published work.

Regulators of ENaC	
Isoprenylcysteine carboxyl methyltransferase	Casein kinase 2
WW domain binding protein 2	Serum/glucocorticoid regulated kinase
Mitogen-activated protein kinase kinase 1	Nedd4 WW-binding protein
Endothelin 1	Syntaxin 3A
Cell receptors	
Carcinoembryonic antigen CD66 a and d	PAR-1 (thrombin receptor)
Decay acceleration factor for complement (DAF)	Hsp-90
Plasminogen activator urokinase receptor	CXCR4
Epithelial growth factor receptor (EGFR)	Fibronectin receptor
Epithelial membrane proteins 1 and 3	Low-density lipoprotein receptor
Interleukin 6 receptor	Hepatitis A virus receptor
Transporters/ion channel proteins	
CFTR	Sulphate
Chloride/bicarbonate exchanger	Calcium
Sodium/pottasium/2 chloride	Monocarboxylic acid
Sodium hydrogen exchanger	Organic ion
Creatine	Cholesterol
Glucose	Transferrin
Copper	NRAMP 2 (iron transport)
Glutamate	

and extension of microtubules in contact with cell surface bound bacteria (Kopecko *et al.*, 2001). Interaction with putative receptors in membrane caveolae triggers endocytosis in which the organism enters the cell. Wooldridge *et al.* (1996) showed that disruption of caveolae and inhibition of host cell tyrosine kinases, heterotrimeric G proteins or phosphatidylinositol 3-kinase (PI3K) inhibits *C. jejuni* invasion. The activation of receptor tyrosine phosphorylation and signalling through integrin receptors via fibronectin both have common signalling molecules and may converge on the same cellular pathways (Plopper *et al.*, 1995) by activation of ERK (Kioka *et al.*, 2002). Caveolins 1 and 2 were upregulated suggesting a role for caveolae, further supporting a role for the interaction of *C. jejuni* with the host cell. Additional support of a role for caveolae comes from the upregulation of the expression of receptor molecules present in caveolae including EGFR, mitogen-activated protein kinase signalling molecules, G-protein subunits and the insulin receptor (reviewed in van Deurs *et al.*, 2003).

EPITHELIAL CELL DAMAGE: CELL CYCLE

C. jejuni has a cytolethal distending toxin (CDT) which has the ability to disrupt cell cycle progression in eukaryotic cells (Lara-Tejero and Galan, 2002). CDT arrests cells in G_2/M phase of cell cycle progression. However, in other cell types growth arrest has been documented in G_1 and G_2 cell cycle phase and, in B cells, apoptosis was observed. Thus, CDT may cause host cell death or modulate host cell function by triggering a DNA damaging response, leading to cell cycle arrest. CdtB has also been shown to have DNase-1 like activity (Lara-Tejero and Galan 2000). A number of genes associated with host cell cycle progression are transcriptionally upregulated in Caco-2 cells infected with *C. jejuni*. GADD45 (growth arrest and DNA damage-inducible protein 45) (el-Deiry, 1998; Sheikh *et al.*, 2000) is involved in inflammation, DNA damage, hypoxia, apoptosis and cell cycle arrest. Also induced upon *C. jejuni* infection is kruppel like factor 4 (KLF4). Under the control of p53, KLF4 is activated by DNA damage and arrests G1/S phase of cell cycle progression specifically in colonic mucosa (el-Deiry, 1998; Yoon *et al.*, 2003). Up-regulated in *C. jejuni* infected cells is p57/kip2 that belongs to the Cip/Kip family and is known as one of the universal negative regulators of the cell cycle. P57/kip2 directly regulates cyclin-dependent kinase 2 (CDK2) activity (Hashimoto *et al.*, 1998). Transcription of cyclin G2 is also upregulated and this protein interacts with phosphatase 2A and is increased as cells undergo cell cycle arrest or apoptosis in response to p53-independent stimuli (Bennin *et al.*, 2002). Finally, cyclin-dependent kinase 4 (CDK4) expression is downregulated, affecting the early G1 phase of the cell cycle. Based on these observations *C. jejuni* may disturb the cell cycle by affecting the expression of host cell molecules that control the cell cycle. Whether CDT induces these host molecules via DNA damage is yet to be determined and other *C. jejuni* factors may have a role in disrupting the cell cycle. The consequence of cell cycle arrest of infected immature colonic crypt epithelium may be fluid loss due to cell death or induction of inflammation.

EPITHELIAL CELL DAMAGE: APOPTOSIS/CELL STRESS

Apoptosis

Stimulated cells of the macrophage cell line THP-1 were shown in a number of assays to undergo apoptosis after *C. jejuni* infection whereas induction of apoptosis was not observed with non-invasive *ciaB* mutants (Konkel and Mixter, 2000). Induction of apoptosis has not been shown in epithelial cells, but it may be the relative abundance of different pro or anti apoptotic mediators in a cell that determines its fate. If invasion of enterocytes by *C. jejuni* does lead to apoptosis and exfoliation, then such processes will compromise the integrity of the mucosa and result in a reduction of the available absorptive surface; this pattern is consistent with the pathology seen in the macaque model (Russell *et al.*, 1993).

The responses seen in infected Caco-2 cells by microarray analysis do not show a clear pattern with respect to apoptosis as transcription of both host genes encoding factors with anti-apoptotic effects as well as some potential inducers of apoptosis were changed. NF-κB, seen to be upregulated by our microarray analysis, is activated in response to *C. jejuni* infection (Mellits *et al.*, 2002) and serves to protect against apoptosis and support cell cycle progression (Hatada *et al.*, 2000). Transcription of the anti-apoptotic Bcl-x and Bcl-2 like proteins is also upregulated; these proteins are blocked by pro-apoptotic members of the Bcl family, which according to the arrays, were not increased. Proliferating cell nuclear antigen protein (PCNA) is one of the central molecules responsible for the control of cell fate (Paunesku *et al.*, 2001). The PCNA gene is induced by p53, which interacts with other proteins (including GADD45). It is the interactions of p53 with these other proteins that determine the fate of the cell. Consequently, when PCNA protein levels are high in the absence of p53, replication occurs, if PCNA protein levels are high in the presence of p53 DNA repair occurs and cell apoptosis results when PCNA is non-functional, or present in low quantities. Although the array analysis showed PCNA to be downregulated 24 hours post-infection, infected cells can survive for longer than this time point (up to 8 days), suggesting that PCNA levels in infected cells do not result in cell death at 24 hours. In our microarray analysis, pro-apoptotic molecules were generally found to be down regulated, including the effector caspase 6 and death effector domain 8 (DED). Some other anti-apoptotic molecules were increased, including TRAF-2, a tumour necrosis factor receptor-associated protein that mediates activation of NF-.B (Hilger-Eversheim *et al.*, 2000) and the transcription factor AP-2 that negatively regulates c-Myc and is involved in apoptosis, cell cycle control and differentiation (Nathan, 2003).

On balance, the transcriptional response to *C. jejuni* infection of Caco-2 cells in culture appears not to favour apoptosis of infected epithelial cells. However, on intestinal villi enterocytes are at differing stages of differentiation. Thus invasion of villous tip cells by *C. jejuni* may result in apoptosis and exfoliation, whereas with invasion of less differentiated crypt cells, anti-apoptotic factors may prevail.

Cell stress

Low oxygen levels were recently shown to stimulate inflammatory responses in immune cells (Nathan 2003) and it is tempting to suggest that this may be important for inducing inflammation in the infected epithelium. The induction of genes encoding factors normally associated with hypoxia suggests that infection by *C. jejuni* renders cells hypoxic. HIF-2 alpha is a heterodimeric transcription factor, which is hypoxia inducible and essential for the maintenance of cellular oxygen homeostasis (Freeburg *et al.*, 2003). In response to hypoxia, genes are expressed that alleviate hypoxic stress including those promoting vascularization. HIF-2 (and HIF-1), which stimulates transcription of vascular endothelial growth factor (VEGF), is also increased in the array.

Also increased in infected cells was the transcript encoding p300/CREB binding protein, which interacts with HIF-1, under the control of mitogen activated protein kinase signalling (Sang *et al.*, 2003). Uncontrolled hypoxia may induce apoptosis (or cell death by necrosis) and therefore this stress may contribute to the apoptotic effects noted in the primate model of *C. jejuni* infection (Russell *et al.,* 1993).

DISRUPTION OF NET FLUID UPTAKE: ION CHANNELS/ TRANSPORT

The intestinal epithelium functions largely for fluid and electrolyte absorption and, when disrupted, diarrhoeal disease results. *C. jejuni* causes diarrhoea by as yet unknown mechanisms, but direct disruption of host ion channels is a candidate mechanism for net fluid loss. The genes encoding major transport proteins upregulated in infected cells are detailed in Table 24.1. The cystic fibrosis transmembrane conductance regulator (CFTR) is involved in chloride secretion providing the osmotic gradient down which water moves into the lumen. In diarrhoeal disease due to toxin producing enteropathogens, CFTR becomes activated for excessive secretion via second messengers resulting in phosphorylation of the channel (Greger, 2000). In Caco-2 cells CFTR is downregulated by *C. jejuni* infection in terms of RNA (Table 24.1) and protein expression at 24 hours post infection (Everest, unpublished). However, CFTR protein expression was increased in infected cells at earlier time points post infection. The functional relevance of this observation is not clear as a general upregulation of receptors is observed in infected cells (Table 24.1) and this may include CFTR. In addition, although CFTR is a receptor for several pathogens (Lyczak and Pier, 2002; Schroeder *et al.,* 2002), this is not known to be the case for *C. jejuni* and it has not been determined if CFTR is phosphorylated in *C. jejuni* infected cells. If CFTR is both upregulated and phosphorylated on infection, this would suggest a secretory component in diarrhoeal disease caused by *C. jejuni*.

In the distal colon, electrogenic Na^+ absorption takes place through an aldosterone- and amiloride-sensitive electrogenic sodium channel (ENaC) with paracellular water and ion movement (Rossier *et al.,* 2002, Sellin 1998). The channel transports Na^+ electrogenically through tight epithelia and water absorption follows, resulting in very low faecal water content. ENaC can be regulated by a number of molecules including CFTR. Indeed, upregulation of CFTR inhibits ENaC function via mechanisms that are as yet unclear (Gormley *et al.,* 2003). Any disruption of ENaC expression would result in decreased electrogenic sodium transport and thus failure of water absorption in the distal colon, clinically manifested as diarrhoea. Several genes affecting ENaC function or expression were either positively or negatively regulated (Table 24.1). From these observations we hypothesise that *C. jejuni* cause diarrhoea by directly affecting pathways that lead to ion channel regulation and induce or inhibit molecules that regulate their functions. This may occur in several ways including via: (1) CFTR activation, which leads to water secretion and inhibition of ENaC function; (2) ENaC downregulation directly, which results in disruption of the process of normal water absorption in the distal colon; or (3) the general downregulation of host ion transport proteins in response to *C. jejuni* infection. The latter would not only abolish secretory events, but would inhibit water absorption by preventing the flow of ions into cells thereby also inhibiting the flow of water paracellularly. The bacterial molecules that may mediate these effects are yet to be identified.

DISRUPTION OF THE EPITHELIUM: INNATE RESPONSES/ INFLAMMATION

The transcription of a number of host molecules are upregulated that have effects on inflammation, cytokine production and innate responses. It is established that *C. jejuni* upregulates expression of the early response transcription factor NF-κB, which triggers expression of genes associated with cellular immune and inflammatory responses (Hickey *et al.*, 1999, Hickey *et al.*, 2000). Host cells infected with viable *C. jejuni* cells or treated with CDT secrete interleukin 8 (IL-8) basolaterally from the cell thus recruiting neutrophils into the lamina propria of infected tissue (Hickey *et al.*, 2000). As seen in animal models (reviewed in Newell, 2001; Russell *et al.*, 1993), neutrophils can also transit through epithelium to the luminal surface. In *Salmonella* infection of cultured cells, the apical (not basolateral) application of bacteria to polarized monolayers, induces neutrophil transmigration controlled by ADP ribosylation factor 6 (ARF-6) (Criss *et al.*, 2001). ARF-6 is a small GTPase and appears to be a signal transducer downstream of G protein coupled receptors and receptor tyrosine kinases (Criss *et al.*, 2001). ARF-6 regulates a phospholipase D-dependent lipid signalling system (also upregulated, Table 24.1) that is necessary for recruitment of protein kinase C to the apical cell surface and subsequently PMN transmigration. Microarray analysis of strain L115 infection of Caco-2 cells showed upregulation of ARF-6; L115 was isolated from a case of clinically and histologically verified inflammatory diarrhoea (Everest *et al.*, 1992). N82 was isolated from a case of non-inflammatory diarrhoea (Everest *et al.*, 1992) and array analysis of this strain did not show significant upregulation of ARF-6. In addition, array analysis of N82 indicated that a monocyte/neutrophil elastase inhibitor (MNEI) is upregulated along with the RelA inhibitor, an inhibitor of NF-κB. MNEI is a specific inhibitor of neutrophil granule proteases, enzymes which when released into infected tissue sites cause tissue damage. Some of the transcriptional responses of genes involved in inflammatory or innate responses by host cells to *C. jejuni* infection may be indirect and therefore greater bacterial numbers or more efficient tissue invasion may affect such host molecules to alter disease presentation. As a reaction to *C. jejuni*-mediated changes in gene expression, other host molecules are upregulated that tend to decrease tissue inflammation or limit tissue damage; these include tissue factor pathway inhibitor, inhibitors of metalloproteases and TGF-β.

The combination of previous data and those from the microarray of host transcriptional responses to infection indicate that bacterial and host molecules causing the diarrhoeal process can be separated from those involved in tissue inflammation. Therefore, disease presentation in terms of inflammation can be explained by induction of host molecules on infection by *C. jejuni*.

CONCLUSIONS

The use of microarray-based studies to investigate genome-wide transcriptional responses is a powerful approach to derive an overview of the host response to microbial insult. It is a particularly valuable tool for pathogens like *C. jejuni* where the molecular basis of host–pathogen interactions has not been elucidated. Such an overview has revealed a number of potential cell surface receptors for *C. jejuni* and with respect to the molecular mechanisms of disease, there is evidence of changes to the host cell cycle, an upregulation of anti-apoptotic factors, an induction of inflammatory mediators and most interestingly, the regulation of host ion channels.

Microarray analysis generates an enormous amount of information, much of which may reflect host cell changes that have little to do with the disease mechanism under investigation and is only a snapshot of the host cell–pathogen interaction. Nevertheless, the identification of novel host responses to infection does generate hypotheses that can be tested using conventional experimentation. Moreover, further microarray analysis of the kinetics of the host–pathogen interaction and investigation of complementary post-transcriptional responses such as protein levels and modification are essential.

Acknowledgements
Thanks to Dr Simon Hardy, University of Brighton, for valuable discussion. P.E. and J.M.K. are grateful to the Royal Society, BBSRC and Wellcome Trust for support.

References
Bennin, D. A., A.S. Don, T. Brake, J.L. McKenzie, H. Rosenbaum, L. Ortiz, A. A. DePaoli-Roach, and M.C. Horne. 2002. Cyclin G2 associates with protein phosphatase 2A catalytic and regulatory B' subunits in active complexes and induces nuclear aberrations and a G1/S phase cell cycle arrest. J. Biol. Chem. 277: 27449–27467.

Black, R.E., M.M. Levine, M.L. Clements, T.P. Hughes, and M. . Blaser. 1988. Experimental *Campylobacter jejuni* infection in humans. J. Infect. Dis. 157: 472–479.

Criss, A.K., M. Silva, J.E. Casanova, and B.A. McCormick. 2001. Regulation of *Salmonella*-induced neutrophil transmigration by epithelial ADP-ribosylation factor 6. J. Biol. Chem. 276: 48431–48439.

el-Deiry, W.S. 1998. Regulation of p53 downstream genes. Semin. Cancer Biol. 8: 345–357.

Everest, P.H., H. Goossens, J.P. Butzler, D. Lloyd, S. Knutton, J.M. Ketley, and P.H. Williams. 1992. Differentiated caco-2 cells as a model for enteric invasion by *Campylobacter jejuni* and *Campylobacter coli*. J. Med. Microbiol. 37: 319–325.

Freeburg, P.B., B. Robert, P.L. St John, and D.R. Abrahamson. 2003. Podocyte expression of hypoxia-inducible factor (HIF)-1 and HIF-2 during glomerular development. J. Am. Soc. Nephrol. 14: 927–938.

Gormley, K., Y. Dong, and G.A. Sagnella. 2003. Regulation of the epithelial sodium channel by accessory proteins. Biochem. J. 371: 1–14.

Greger, R. 2000. Role of CFTR in the colon. Annu. Rev. Physiol. 62: 467–491.

Guignot, J., I. Peiffer, M. F. Bernet-Camard, D.M. Lublin, C. Carnoy, S.L. Moseley, and A.L. Servin. 2000. Recruitment of CD55 and CD66e brush border-associated glycosylphosphatidylinositol-anchored proteins by members of the Afa/Dr diffusely adhering family of *Escherichia coli* that infect the human polarized intestinal Caco-2/TC7 cells. Infect. Immun. 68: 3554–3563.

Hashimoto, Y., K. Kohri, Y. Kaneko, H. Morisaki, T. Kato, K. Ikeda, and M. Nakanishi. 1998. Critical role for the 310 helix region of p57 (Kip2) in cyclin-dependent kinase 2 inhibition and growth suppression. J. Biol. Chem. 273: 16544–16550.

Hatada, E.N., D. Krappmann, and C. Scheidereit. 2000. NF-kappaB and the innate immune response. Curr. Opin. Immunol. 12: 52–58.

Hickey, T.E., S. Baqar, A.L. Bourgeois, C.P. Ewing, and P. Guerry. 1999. *Campylobacter jejuni*-stimulated secretion of interleukin-8 by INT407 cells. Infect. Immun. 67: 88–93.

Hickey, T.E., A.L. McVeigh, D.A. Scott, R.E. Michielutti, A. Bixby, S.A. Carroll, A.L. Bourgeois, and P. Guerry. 2000. *Campylobacter jejuni* cytolethal distending toxin mediates release of interleukin-8 from intestinal epithelial cells. Infect. Immun. 68: 6535–6541.

Hilger-Eversheim, K., M. Moser, H. Schorle, and R. Buettner. 2000. Regulatory roles of AP-2 transcription factors in vertebrate development, apoptosis and cell-cycle control. Gene 260: 1–12.

Jin, S., Y.C. Song, A. Emili, P. M. Sherman, and V.L. Chan. 2003. JlpA of *Campylobacter jejuni* interacts with surface-exposed heat shock protein 90alpha and triggers signalling pathways leading to the activation of NF-kappaB and p38 MAP kinase in epithelial cells. Cell. Microbiol. 5: 165–174.

Ketley, J.M. 1997. Pathogenesis of enteric infection by *Campylobacter*. Microbiology 143: 5–21.

Kioka, N., K. Ueda, and T. Amachi. 2002. Vinexin, CAP/ponsin, ArgBP2: a novel adaptor protein family regulating cytoskeletal organization and signal transduction. Cell Struct. Funct. 27: 1–7.

Konkel, M.E., S.G. Garvis, S. L. Tipton, D.E. Anderson, and W. Cieplak. 1997. Identification and molecular cloning of a gene encoding a fibronectin-binding protein (CadF) from *Campylobacter jejuni*. Mol. Microbiol. 24: 953–963.

Konkel, M. E., D.J. Mead, S. F. Hayes, and W.J. Cieplak. 1992. Translocation of *Campylobacter jejuni* across human polarized epithelial cell monolayer cultures. J. Infect. Dis. 166: 308–315.

Konkel, M.E., and P.F. Mixter. 2000. Flow cytometric detection of host cell apoptosis induced by bacterial infection. Methods Cell Sci. 22: 209–215.

Konkel, M.E., M.R. Monteville, V. Rivera-Amill, and L.A. Joens. 2001. The pathogenesis of *Campylobacter jejuni*-mediated enteritis. Curr. Issues Intest. Microbiol. 2: 55–71.

Kopecko, D.J., L. Hu, and K.J. Zaal. 2001. *Campylobacter jejuni* – microtubule-dependent invasion. Trends Microbiol. 9: 389–96.

Lara-Tejero, M., and J.E. Galan. 2000. A bacterial toxin that controls cell cycle progression as a deoxyribonuclease I-like protein. Science 290: 354–357.

Lara-Tejero, M., and J.E. Galan. 2002. Cytolethal distending toxin: limited damage as a strategy to modulate cellular functions. Trends Microbiol. 10: 147–152.

Lyczak, J.B., and G.B. Pier. 2002. *Salmonella enterica* serovar typhi modulates cell surface expression of its receptor, the cystic fibrosis transmembrane conductance regulator, on the intestinal epithelium. Infect. Immun. 70: 6416–6423.

Mellits, K.H., J. Mullen, M. Wand, G. Armbruster, A. Patel, P.L. Connerton, M. Skelly, and I.F. Connerton. 2002. Activation of the transcription factor NF-kappaB by *Campylobacter jejuni*. Microbiology 148: 2753–2763.

Monteville, M.R., J.E. Yoon, and M.E. Konkel. 2003. Maximal adherence and invasion of INT 407 cells by *Campylobacter jejuni* requires the CadF outer-membrane protein and microfilament reorganization. Microbiology 149: 153–165.

Nathan, C. 2003. Immunology: Oxygen and the inflammatory cell. Nature 422: 675–676.

Newell, D.G. 2001. Animal models of *Campylobacter jejuni* colonization and disease and the lessons to be learned from similar *Helicobacter pylori* models. J. Appl. Microbiol. 90: 57S–67S.

Parkhill, J., B.W. Wren, K. Mungall, J.M. Ketley, C. Churcher, D. Basham, T. Chillingworth, R.M. Davies, T. Feltwell, S. Holroyd, K. Jagels, A. V. Karlyshev, S. Moule, M.J. Pallen, C.W. Penn, M.A. Quail, M. A. Rajandream, K.M. Rutherford, A.H.M. van Vliet, S. Whitehead, and B.G. Barrell. 2000. The genome sequence of the food-borne pathogen *Campylobacter jejuni* reveals hypervariable sequences. Nature 403: 665–668.

Paunesku, T., S. Mittal, M. Protic, J. Oryhon, S. V. Korolev, A. Joachimiak, and G.E. Woloschak. 2001. Proliferating cell nuclear antigen (PCNA): ringmaster of the genome. Int. J. Radiat. Biol. 77: 1007–1021.

Plopper, G.E., H.P. McNamee, L.E. Dike, K. Bojanowski, and D.E. Ingber. 1995. Convergence of integrin and growth factor receptor signaling pathways within the focal adhesion complex. Mol. Biol. Cell 6: 1349–1365.

Rossier, B.C., S. Pradervand, L. Schild, and E. Hummler. 2002. Epithelial sodium channel and the control of sodium balance: interaction between genetic and environmental factors. Annu. Rev. Physiol. 64: 877–897.

Russell, R.G., Blake, D.C. Jr., J. Zulty, and L.J. DeTolla. 1993. Early colonic damage and invasion of *Campylobacter jejuni* in experimentally challenged infant *Macaca mulatta*. J. Infect. Dis. 168: 210–215.

Sang, N., D.P. Stiehl, J. Bohensky, I. Leshchinsky, V. Srinivas, and J. Caro. 2003. MAPK signaling up-regulates the activity of hypoxia-inducible factors by its effects on p300. J. Biol. Chem. 278: 14013–14019.

Schroeder, T.H., M.M. Lee, P.W. Yacono, C.L. Cannon, A. A. Gerceker, D.E. Golan, and G.B. Pier. 2002. CFTR is a pattern recognition molecule that extracts *Pseudomonas aeruginosa* LPS from the outer membrane into epithelial cells and activates NF-kappa B translocation. Proc. Natl. Acad. Sci. USA 99: 6907–12.

Sellin, J.H. 1998. Intestinal electrolyte absorption and secretion. In: Sleisenger and Fordtran's Gastrointestinal and Liver Disease, Sleisenger M.H., Fordtran J.M., Feldman, M., and Scharschmidt, B., eds. W.B. Saunders, Philadelphia, p. 86.

Sheikh, M.S., M.C. Hollander, and A.J. Fornance, Jr. 2000. Role of Gadd45 in apoptosis. Biochem. Pharmacol. 59: 43–45.

van Deurs, B., K. Roepstorff, A.M. Hommelgaard, and K. Sandvig. 2003. Caveolae: anchored, multifunctional platforms in the lipid ocean. Trends Cell. Biol. 13: 92–100.

Wang, J., .D. Grey-Owen, A. Knorre, T. F. Meyer, and C. Dehio. 1998. Opa binding to cellular CD66 receptors mediates the transcellular traversal of *Neisseria gonorrhoeae* across polarized T84 epithelial cell monolayers. Mol. Microbiol. 30: 657–671.

Wassenaar, T.M., and M.J. Blaser. 1999. Pathophysiology of *Campylobacter jejuni* infections of humans. Microbes and Infection 1: 1023–1033.

Wooldridge, K.G., P.H. Williams, and J.M. Ketley. 1996. Host Signal-Transduction and Endocytosis of *Campylobacter jejuni*. Microbial Pathogen. 21: 299–305.

Xia, M., R.E. Bumgarner, M.F. Lampe, and W.E. Stamm. 2003. *Chlamydia trachomatis* infection alters host cell transcription in diverse cellular pathways. J. Infect. Dis. 187: 424–434.

Yoon, H.S., X. Chen, and V.W. Yang. 2003. Kruppel-like factor 4 mediates p53-dependent G1/S cell cycle arrest in response to DNA damage. J. Biol. Chem. 278: 2101–2105.

Index

Index

Index